Lecture Notes in Computer Science 2912

Edited by G. Goos, J. Hartmanis, and J. van Leeuwen

Springer-Verlag Berlin Heidelberg GmbH

Giuseppe Liotta (Ed.)

Graph Drawing

11th International Symposium, GD 2003
Perugia, Italy, September 21-24, 2003
Revised Papers

Series Editors

Gerhard Goos, Karlsruhe University, Germany
Juris Hartmanis, Cornell University, NY, USA
Jan van Leeuwen, Utrecht University, The Netherlands

Volume Editor

Giuseppe Liotta
Università di Perugia
Dipartimento di Ingegneria Elettronica e dell'Informazione
Via G. Duranti, 93, 06125 Perugia, Italy
E-mail: liotta@diei.unipg.it

Cataloging-in-Publication Data applied for

A catalog record for this book is available from the Library of Congress.

Bibliographic information published by Die Deutsche Bibliothek
Die Deutsche Bibliothek lists this publication in the Deutsche Nationalbibliografie; detailed bibliographic data is available in the Internet at <http://dnb.ddb.de>.

CR Subject Classification (1998): G.2, F.2, I.3

ISSN 0302-9743
DOI 10.1007/978-3-540-24595-7

Springer-Verlag is a part of Springer Science+Business Media

springeronline.com

MyCopy version of the original edition 2004

Typesetting: Camera-ready by author, data conversion by PTP-Berlin, Protago-TeX-Production GmbH
Printed on acid-free paper SPIN: 10977415 06/3142 5 4 3 2 1 0
www.springer.com/mycopy

Preface

The 11th International Symposium on Graph Drawing (GD 2003) was held on September 21–24, 2003, at the Università degli Studi di Perugia, Perugia, Italy. GD 2003 attracted 93 participants from academic and industrial institutions in 17 countries.

In response to the call for papers, the program committee received 88 regular submissions describing original research and/or system demonstrations. Each submission was reviewed by at least 4 program committee members and comments were returned to the authors. Following extensive e-mail discussions, the program committee accepted 34 long papers (12 pages each in the proceedings) and 11 short papers (6 pages each in the proceedings). Also, 6 posters (2 pages each in the proceedings) were displayed in the conference poster gallery.

In addition to the 88 submissions, the program committee also received a submission of special type, one that was not competing with the others for a time slot in the conference program and that collects selected open problems in graph drawing. The aim of this paper, which was refereed with particular care and UNCHANGED two rounds of revisions, is to stimulate future research in the graph drawing community. The paper presents 42 challenging open problems in different areas of graph drawing and contains more than 120 references. Although the length of the paper makes it closer to a journal version than to a conference extended abstract, we decided to include it in the conference proceedings so that it could easily reach in a short time the vast majority of the graph drawing community.

GD 2003 invited two distinguished lecturers. Pat Hanrahan, from Stanford University, gave a talk about the connection between semantic constraints and aesthetics in graph drawing and information visualization. Giuseppe Italiano, from the Università di Roma Tor Vergata, gave a talk on algorithm engineering and experimental analysis of graph algorithms.

As usual, the annual graph drawing contest was held during the conference. A report about the contest is included in the proceedings.

Many people in the graph drawing community contributed to the success of GD 2003. In particular, the authors of submitted papers, demos, and posters are due special thanks, as are the members of the program committee and the external reviewers. Many thanks to the organizing committee members Carla Binucci, Emilio Di Giacomo, Luca Grilli, Maurizio Patrignani, and Maurizio Pizzonia for their support. My very special thanks go to the local arrangements chair Walter Didimo, for his invaluable help. Without his support of the organization and his many comments and suggestions, the conference would have been impossible to organize.

Thanks are due to the industrial sponsors of the conference: the "gold" sponsors, Tom Sawyer Software; the "silver" sponsors, Mitsubishi Electrics, Oreas, Digilab 2000, and Integra Sistemi s.r.l.; and the "contributor," Kelyan SMC.

Finally, many thanks go to the Dipartimento di Informatica e Automazione of the Università degli Studi di Roma Tre and to the Dipartimento di Ingegneria Elettronica e dell'Informazione of the Università degli Studi di Perugia for their help and financial support. The conference was also supported in part by "Progetto ALINWEB: Algoritmica per Internet e per il Web," MIUR Programmi di Ricerca Scientifica di Rilevante Interesse Nazionale.

The 12th International Symposium on Graph Drawing GD 2004 will be held in New York City, September 29–October 2, 2004, with Janos Pach as the conference chair.

October 2003 Giuseppe Liotta

Organization

Program Committee

Ashim Garg	SUNY Buffalo, USA
Michael T. Goodrich	University of California, Irvine, USA
Ferran Hurtado	Universitat Politcnica de Catalunya, Spain
Giuseppe Liotta	University of Perugia, Italy (Conference Chair)
Joe Marks	MERL, USA
Henk Meijer	Queen's University, Canada
Stephen C. North	AT&T Research Labs, USA
Patrice Ossona de Mendez	EHESS, CNRS, France
Md. Saidur Rahman	Tohoku University, Japan
Farhad Shahrokhi	University of North Texas, USA
Roberto Tamassia	Brown University, USA
Ioannis G. Tollis	University of Crete and ICS-FORTH, Greece
Dorothea Wagner	University of Karlsruhe, Germany
Sue H. Whitesides	McGill University, Canada
Stephen K. Wismath	University of Lethbridge, Canada
David R. Wood	Carleton University, Canada

Contest Committee

Franz J. Brandenburg	University of Passau (Contest Committee chair)
Ulrik Brandes	University of Passau, Germany
Peter Eades	University of Sydney, Australia
Joe Marks	MERL, USA

Organizing Committee

Carla Binucci	University of Perugia, Italy
Emilio Di Giacomo	University of Perugia, Italy
Walter Didimo	University of Perugia, Italy (Local Arrangements Co-chair)
Giuseppe Liotta	University of Perugia, Italy (Conference chair)
Maurizio Patrignani	University of Roma Tre, Italy
Maurizio Pizzonia	University of Roma Tre, Italy

Steering Committee

Franz J. Brandenburg	University of Passau, Germany
Giuseppe Di Battista	University of Roma Tre, Italy
Peter Eades	University of Sydney, Australia
Micheal T. Goodrich	University of California, Irvine, USA
Stephen G. Kobourov	University of Arizona, USA
Giuseppe Liotta	University of Perugia, Italy
Takao Nishizeki	Tohoku University, Japan
Janos Pach	New York University, USA
Pierre Rosenstiehl	EHESS, France
Roberto Tamassia	Brown University, USA
Ioannis G. Tollis	University of Crete and ICS-FORTH, Greece
Sue H. Whitesides	McGill University, Canada

External Referees

Manuel Abellanas
Bernardo Abrego
Oswin Aichholzer
Marc Benkert
Therese Biedl
Carla Binucci
Prosenjit Bose
Matt Brand
Ulrik Brandes
Juergen Branke
Sergio Cabello
Timothy Chan
Fco. Javier Cobos Gavala
Pier Francesco Cortese
Ovidiu Daescu
Erik Demaine
Emilio Di Giacomo
Walter Didimo
Vida Dujmović
Tim Dwyer
Thomas Erlebach
Guy Even
Hazel Everett
Wendy Feng
Julia Floetotto
Marco Gaertler
Alfredo Garcia
Maria Angeles Garrido
Daya Gaur
George F. Georgakopoulos
Patrick Healy
Petr Hlineny
Seokhee Hong
Wen-Lian Hsu
Kostas Kakoulis
Michael Kaufmann
Stephen Kobourov
Jan Kratochvil
Sebastian Leipert
Anna Lubiw
Karol Lynch
Csaba Megyeri
Martin Middendorf
Kazuyuki Miura
Pat Morin
Shin-ichi Nakano
Naomi Nishimura
Rom Pinchasi
Maurizio Pizzonia
Eduardo Rivera-Campo
Gelasio Salazar
Thomas Schank
Frank Schulz
Janet Six
Steven Skiena
Jerry Spinrad

Daniel Stefankovic
Ileana Streinu
Matthew Suderman
Kozo Sugiyama
Ondrej Sýkora
László A. Szekely
Javier Tejel
Gabriel Valiente
Marc van Kreveld
Suresh Venkatasubramanian
Imrich Vrt'o
Leland Wilkinson
Thomas Willhalm
Alexander Wolff

Sponsoring Institutions

We gratefully acknowledge the contributions of the following sponsors of the Graph Drawing Conference 2003.

Table of Contents

Planarity and Planar Drawings

Geometric Graph Theory

Applications and Systems – Part I

Straight-Line, Circular, and Circular-Arc Drawings

Symmetries

3D-Drawings

Embeddings and Triangulations

Applications and Systems – Part II

Fixed Parameter Tractability

Clusters, Cuts, and Orthogonal Drawings

k-Level Drawings

Force Directed and Energy-Based Techniques

Surfaces and Diagrams

Posters

Graph Drawing Contest

Invited Talks

Open Problems

Confluent Drawings: Visualizing Non-planar Diagrams in a Planar Way*

Matthew Dickerson[1], David Eppstein[2], Michael T. Goodrich[2], and Jeremy Yu Meng[2]

[1] Middlebury College, Middlebury, VT 05753, USA dickerso@middlebury.edu
[2] University of California, Irvine, Irvine, CA 92697, USA
{eppstein, goodrich, ymeng}@ics.uci.edu

Abstract. We introduce a new approach for drawing diagrams. Our approach is to use a technique we call *confluent drawing* for visualizing non-planar graphs in a planar way. This approach allows us to draw, in a crossing-free manner, graphs—such as software interaction diagrams—that would normally have many crossings. The main idea of this approach is quite simple: we allow groups of edges to be merged together and drawn as "tracks" (similar to train tracks). Producing such confluent diagrams automatically from a graph with many crossings is quite challenging, however, so we offer two heuristic algorithms to test if a non-planar graph can be drawn efficiently in a confluent way. In addition, we identify several large classes of graphs that can be completely categorized as being either confluently drawable or confluently non-drawable.

1 Introduction

In most graph visualization applications, graphs are often drawn in a standard way: the vertices of a graph are drawn as simple shapes, such as circles or boxes, and the edges are drawn as individual curves connecting pairs of these shapes (e.g., see [12,13,22]).

Related Prior Work. There are several aesthetic criteria that have been explored algorithmically in the area of graph drawing (e.g., see [12,13,22]). Examples of aesthetic goals designed to facilitate readability include minimizing edge crossings, minimizing a drawing's area, minimizing bends, and achieving good separation of vertices, edges, and angles. Of all of these criteria, however, the arguably most important is to minimize edge crossings, since crossing edges tend to confuse the eye when one is viewing adjacency relationships. Indeed, an experimental analysis by Purchase [31] suggests that edge-crossing minimization [19, 20,25] is the most important aesthetic criteria for visualizing graphs. Ideally, we would like drawings that have no edge crossings at all.

* This is an extended abstract. The full version of this paper can be found at http://arxiv.org/abs/cs.CG/0212046. Work by the second author is supported by NSF grant CCR-9912338. Work by the third and the forth author is supported by NSF Grants CCR-0098068, CCR-0225642, and DUE-0231467.

G. Liotta (Ed.): GD 2003, LNCS 2912, pp. 1–12, 2004.

Graphs that can be drawn in the standard way in the plane without edge crossings are called *planar graphs* [28], and there are a number of existing efficient algorithms for producing crossing-free drawings of planar graphs (e.g., see [8,9,11,34,6,21,36]). Unfortunately, most graphs are not planar; hence, most graphs cannot be drawn in the standard way without edge crossings, and such non-planar graphs seem to be common in many applications. There are some heuristic algorithms for minimizing edge crossings of non-planar graphs (e.g., see [19,26,20,25]), but the general problem of drawing a non-planar graph in a standard way that minimizes edge-crossings is NP-hard [16]. Thus, we cannot expect an efficient algorithm for drawing non-planar graphs so as to minimize edge crossings.

The technique of replacing complete bipartite subgraphs (bicliques) with star-like structures is used as *Edge Concentration* in [27] and *Factoring* in [5], both to reduced the number of edges in the original graphs. This technique has the desired side effect of reducing the number of crossings, however, its primary goal is to minimize the total number of edges, not to minimize the number of crossings. Furthermore, the time complexity of approximation algorithm given in [27] is not desirable. Recently Lin [24] proves that the optimization problem of edge concentration is NP-hard. A similar idea is used in [15] for weighted graph compressions, where cliques and bicliques are replaced with stars. It is shown that the general unit weight problem is essentially as hard to approximate as graph coloring and maximum clique. Again, [15] doesn't directly address the minimization of the number of crossings.

Our Results. Given the difficulty of edge-crossing minimization and the ubiquity of non-planar graphs, we explore in this paper a diagram visualization approach, called *confluent drawing*, that attempts to achieve the best of both worlds—it draws non-planar graphs in a planar way. Moreover, we provide two heuristic algorithms for producing confluent drawings for directed and undirected graphs, respectively, focusing on graphs with bounded arboricity.

The main idea of the confluent drawing approach for visualizing non-planar graphs in a planar way is quite simple—we merge edges into "tracks" so as to turn edge crossings into overlapping paths. (See Fig. 1.)

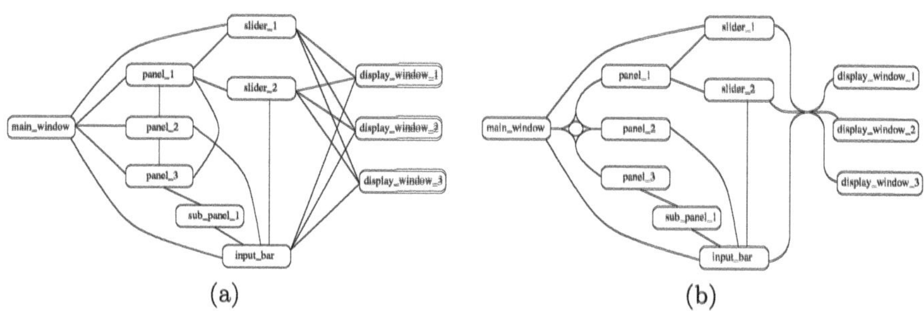

Fig. 1. An example of confluent drawing of an object-interaction diagram. We show a standard drawing in (a) and a confluent drawing in (b).

The resulting graphs are easy to read and comprehend, while also encapsulating a high degree of connectivity information. Although we are not familiar with any prior work on the automatic display of graphs using this confluent diagram approach, we have observed that some airlines use hand-crafted confluent diagrams to display their route maps. Diagrams similar to our confluent drawings have also been used by Penner and Harer [29] to study the topology of surfaces.

In addition to providing heuristic algorithms for recognizing and drawing confluent diagrams, we also show that there are large classes of non-planar graphs that can be drawn in a planar way using our confluent diagram approach.

2 Confluent Drawings

It is well-known that every non-planar graph contains a subgraph homeomorphic to the complete graph on five vertices, K_5, or the complete bipartite graph between two sets of three vertices, $K_{3,3}$ (e.g., see [3]). On the other hand, confluent drawings, with their ability to merge crossing edges into single tracks, can easily draw any $K_{n,m}$ or K_n in a planar way. Fig. 2 shows confluent drawings of $K_{3,3}$ and K_5.

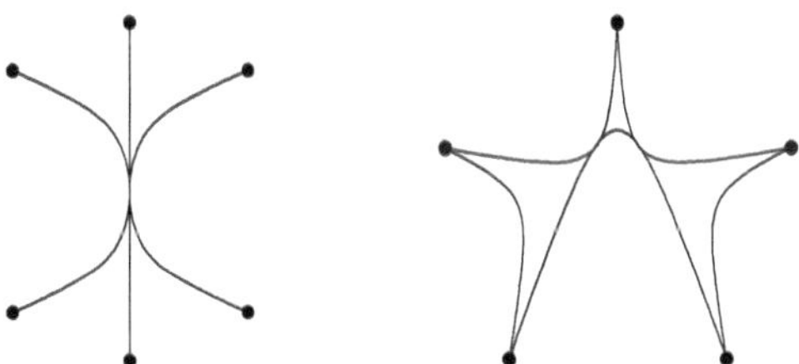

Fig. 2. Confluent drawings of $K_{3,3}$ and K_5.

A curve is *locally-monotone* if it contains no self intersections and no sharp turns, that is, it contains no point with left and right tangents that form an angle less than or equal to 90 degrees. Intuitively, a locally-monotone curve is like a single train track, which can make no sharp turns. Confluent drawings are a way to draw graphs in a planar manner by merging edges together into *tracks*, which are the unions of locally-monotone curves.

An undirected graph G is *confluent* if and only if there exists a drawing A such that:

- There is a one-to-one mapping between the vertices in G and A, so that, for each vertex $v \in V(G)$, there is a corresponding vertex $v' \in A$, which has a unique point placement in the plane.
- There is an edge (v_i, v_j) in $E(G)$ if and only if there is a locally-monotone curve e' connecting v_i' and v_j' in A.
- A is planar. That is, while locally-monotone curves in A can share overlapping portions, no two can cross.

Our definition does not allow for confluent graphs to contain self loops or parallel edges, although we do allow for tracks to contain cycles and even multiple ways of realizing the same edge. Moreover, our definition implies that tracks in a confluent drawing have a "diode" property that does not allow one to double-back or make sharp turns after one has started going along a track in a certain direction.

Directed confluent drawings are defined similarly, except that in such drawings the locally-monotone curves are directed and the tracks formed by unions of curves must be oriented consistently.

3 Heuristic Algorithms

Though the planarity of a graph can be tested in linear time, it appears difficult to quickly determine whether or not a graph can be drawn confluently. If a graph G contains a non-planar subgraph, then G itself is non-planar too. But similar closure properties are not true for confluent graphs. Adding vertices and edges to a non-confluent graph increases the chances of edges crossing each other, but it also increases the chances of edges merging. Currently, the best method we know of for determining conclusively in the worst case whether a graph is confluent or not is a brute force one of exhaustively listing all possible ways of edge merging and checking the merged graphs for planarity. Therefore, it is of interest to develop heuristics that can find confluent drawings in many cases.

Fig. 3 shows confluent drawings using a "traffic circle" structure for complete subgraphs (cliques) and complete bipartite subgraphs (bicliques). At a high level, our heuristic drawing algorithm iteratively finds clique subgraphs and biclique subgraphs and replaces them with traffic-circle subdrawings.

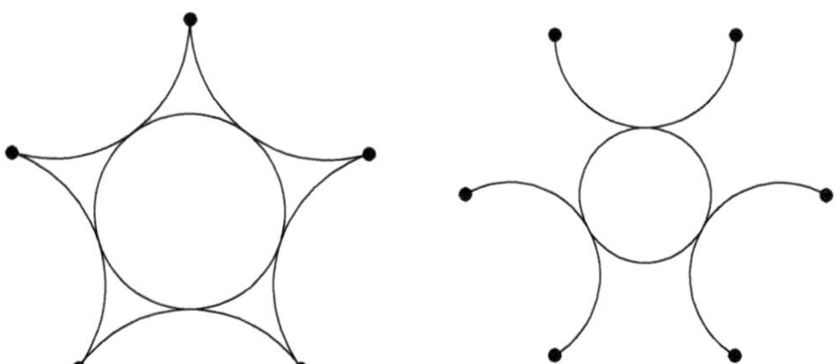

Fig. 3. Confluent drawings of K_5 and $K_{3,3}$ using "traffic circle" structures.

Chiba and Nishizeki [7] discuss the problem of listing complete subgraphs for graphs of bounded arboricity. The *arboricity* $a(G)$ is the minimum number of forests into which the edges of G can be partitioned. The listing algorithm is applicable for such graphs. Chiba and Nishizeki show that there can be at most $O(n)$ cliques of a given size in such graphs and give a linear time algorithm for listing these clique subgraphs. Eppstein [14] gives a linear time algorithm for listing maximal complete bipartite subgraphs in graphs of bounded arboricity.

The total complexity of all such graphs is $O(n)$, and again they can be listed in linear time.

In our heuristic algorithm for undirected graphs, we will use the clique subgraphs listing and the biclique subgraphs listing algorithms as our subroutines.

```
HeuristicDrawUndirected(G)
Input. A undirected sparse graph G.
Output. Confluent drawing of G if succeed, fail otherwise.
1. If G is planar
2.    draw G
3. else if G contains a large clique or biclique subgraph C
4.    create a new vertex v
5.    obtain a new graph G' by removing edges of C and
      connecting each vertex of C to v
6.    HeuristicDrawUndirected(G')
7.    replace v by a small "traffic circle" to get a confluent
      drawing of G
8. else fail
```

Fig. 4.

In step 3 of the algorithm in Fig. 4, the cliques are given higher priority over bicliques, otherwise a clique would be partially covered by a biclique. Cliques of three or fewer vertices, and bicliques with one side consisting of only one vertex, are not replaced because the replacement cannot change the planarity of the graph. We now discuss the time performance of this heuristic.

Theorem 1. *In graphs of bounded arboricity, algorithm* HeuristicDrawUndirected *can be made to run in time $O(n)$, assuming hash tables with constant time per operation.*

Proof. We store a bit per edge of the original graph so we can quickly look up whether it is still part of our replacement. We begin the heuristic by looking for cliques, since we want to give them priority over bicliques. List all the complete subgraphs in the graph with four or more vertices, and sort them by size (the size of the complete subgraph is bounded too in graphs with bounded arboricity). Then, for each complete subgraph X in sorted order, we check whether X is still a clique of the modified graph, and if so perform a replacement of X. It is not hard to see that the new vertex v of the replacement cannot belong to any clique, so this algorithm correctly finds a maximal sequence of cliques to replace.

Next, we need to similarly dynamize the search for bicliques. This is more difficult, because a biclique may have nonconstant size and because the replacement vertex v may belong to additional bicliques. We perform this step by dynamizing the algorithm of Eppstein [14] for listing all bicliques. This algorithm uses the idea of a *d-bounded acyclic orientation*: that is, an orientation of the edges of the graph, such that the oriented graph is acyclic and the vertices have maximum outdegree d. For graphs of arboricity a, a $(2a-1)$-bounded acyclic orientation may easily be found in linear time. For such an orientation, define a *tuple* to be a subset of the outgoing neighbors of any vertex, and let v be a *tuple creator* of tuple T if all vertices of T are outgoing neighbors of v. For graphs of bounded arboricity, there are at most linearly many distinct tuples. For each maximal biclique, one of the two sides of the bipartition must be a tuple, T [14]. The other side consists of two types of vertices: tuple creators of T, and outgoing neighbors of vertices of T.

Our algorithm stores a hash table indexed by the set of all tuples in the modified graph. The hash table entry for tuple T stores the number of tuple creators of T, and a list of outgoing neighbors of vertices of T that are adjacent to all tuple members. For each edge (u, v) in the graph, oriented from u to v, we store a list of the tuples T containing v for which u is listed as an outgoing neighbor. We also store a priority queue of the maximal bicliques generated by each tuple, prioritized by size; it will suffice for our purposes if the time to find the largest biclique is proportional to the biclique size, and it is easy to implement a priority queue with such a time bound. With these structures, we may easily look up each successive biclique replacement to perform in algorithm HEURISTICDRAWUNDIRECTED. Each replacement takes time proportional to the number of edges removed from the graph, so the total time for performing replacements is linear.

It remains to show how to update these data structures when we perform a biclique replacement. To update the acyclic orientation, orient each edge from C to v, except for those edges from vertices of C that have no outgoing edges in C. It can be seen that this orientation preserves d-boundedness and acyclicity. When a new vertex v is created by a replacement, create the appropriate hash table entries for tuples containing v; the number of tuples created by a replacement is proportional to the number of edges removed in the same replacement, so the total number of tuples created over the course of the algorithm is linear. Whenever a replacement causes edges from a vertex x to change, update the hash entries for all tuples for which x is a creator; this step takes $O(1)$ time per change. Also, update the hash entries for all tuples to which x belongs, to remove vertices that are no longer outgoing neighbors of x; this step takes time $O(1)$ per changed tuple, and each tuple changes $O(1)$ times over the course of the algorithm. Whenever a change removes incoming edges of x, we must remove the other endpoints of those edges from the lists of outgoing neighbors of tuples to which x belongs; using the lists associated with each incoming edge, this takes constant time per removal. Therefore, all steps can be performed in linear total time. □

An example of the input for algorithm HEURISTICDRAWUNDIRECTED and the output drawing produced by this heuristic is shown in Fig. 5.

4 Some Confluent Graphs

The heuristic algorithms presented in the previous section are most applicable to sparse graphs, because sparseness is needed for the linear time bound of the

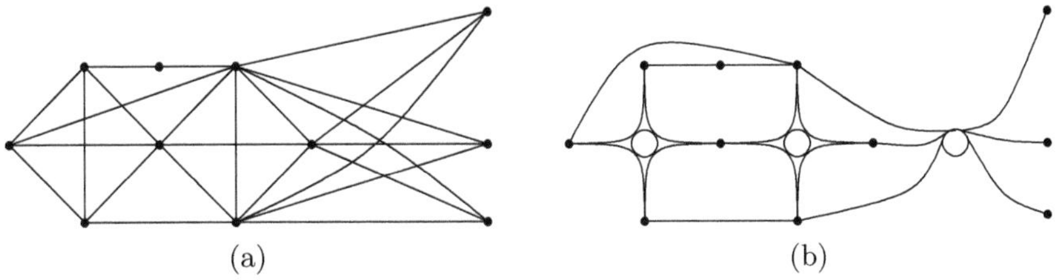

Fig. 5. An example of running the undirected heuristic algorithm. The input graph is shown in (a) and the output drawing is shown in (b).

maximal bipartite subgraph listing subroutine. However, there are also several denser classes of graphs that we can show to be confluent.

Interval graphs. An *interval graph* is formed by a set of closed intervals $S = \{[a_1, b_1], [a_2, b_2], \ldots, [a_n, b_n]\}$. The interval graph is defined to have the intervals in S as its vertices and two vertices $[a_i, b_i]$ and $[a_j, b_j]$ are connected by an edge if and only if these two intervals have a non-empty intersection. Such graphs are typically non-planar, but we can draw them in a planar way using a confluent drawing[1].

Theorem 2. *Every interval graph is confluent.*

Proof. The proof is by construction. We number the interval endpoints by rank, $X = \{0, 1, \ldots, n-1\}$, and place these endpoints along the x-axis. We then build a two-dimensional lattice on top of these points in a fashion similar to Pascal's triangle, using a connector similar to an upside-down "V". These connectors stack on top of one another so that the apex of each is associated with a unique interval on X. We place each point from our set S of intervals just under its corresponding apex and connect it into the (single) track so that it can reach everything directly dominated by this apex in the lattice. At the bottom level, we connect the updside-down V's with rounded connectors. By this construction, we create a single track that allows each pair of vertices connected in the interval graph to have a locally-monotone path connecting them. (See Fig. 6.) □

Complements of trees. The complements of trees (graphs formed by connecting all pairs of vertices that are not connected in some tree) are also called *cotrees*. In general, cotrees are highly non-planar and dense, since a cotree with n vertices has $n(n-1)/2 - n + 1$ edges. Nevertheless, we have the following interesting fact.

Theorem 3. *The complement of a tree is confluent.*

[1] A similar construction works for circular-arc graphs and is left as an exercise for the interested reader.

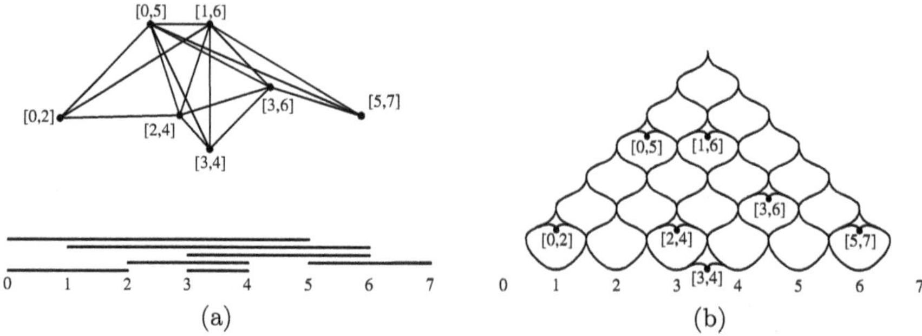

Fig. 6. Illustrating a confluent way to draw a non-planar interval graph: (a) an interval graph and its defining intervals; (b) its corresponding confluent drawing.

Proof. We prove the claim by recursive construction, using a single track for the entire graph. Assign a bounding rectangle for the tree and a bounding rectangle for every subtree in that tree. Place the complement of the tree into the bounding rectangles such that nodes of every subtree is within its bounding rectangle and the bounding rectangles of subtrees are contained in their parent's bounding rectangle. In addition, place a connector at the Northeastern corner of every bounding box. This connector is an imaginary point at which the single track for the entire graph will connect into this portion of the cotree. (See Fig. 7.) Connect the root node in each subtree to every connector of its children. Connect every node to the connector of its parent. Also connect every node to its siblings and the connectors of its siblings, as shown in the figure. The obtained drawing is the confluent drawing of the complement of the given tree. □

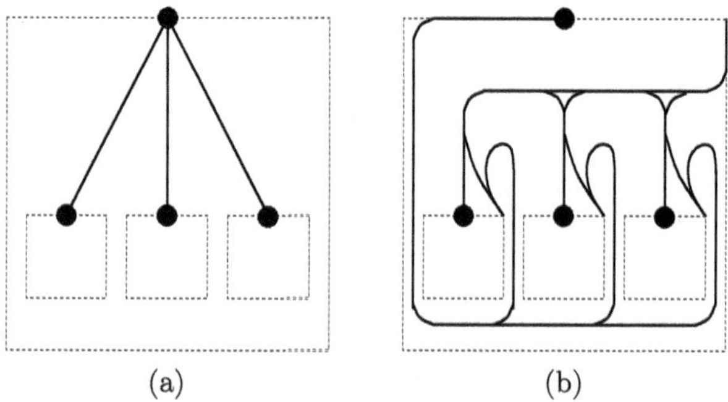

Fig. 7. Illustrating a confluent way to draw the complement of a tree: (a) a node and its children in the tree; (b) the corresponding portion of a track in the confluent drawing of the complement.

Paths are very special cases of trees. Every vertex in a path has a degree of 2 except its two endpoints, each of which has a degree of 1. The complement of a path can be drawn using the cotree method in the above proof. We show a nice confluent drawing of the complement of a path in Fig. 8.

Fig. 8. P_8 and the confluent drawing of $\overline{P_8}$.

Complements of n-cycles. An n-cycle is a cycle with n vertices.

Theorem 4. *The complement of an n-cycle is confluent.*

Proof. First remove one vertex from the n-cycle and draw the confluent graph for the complement of the obtained path. Then add the vertex back and connect it with all vertices in the path except for its two neighbors. The obtained drawing is a confluent drawing (See Fig. 9 for an example.) □

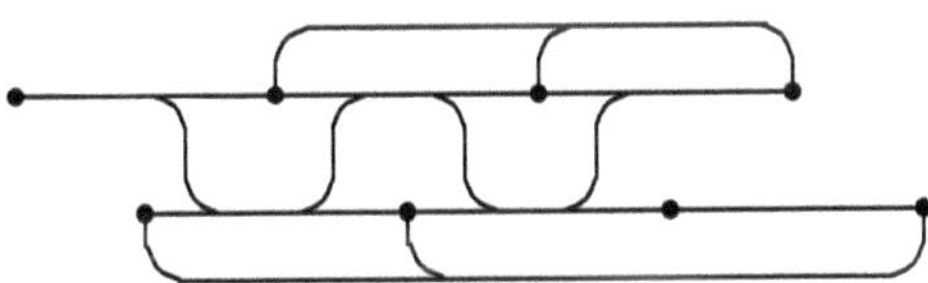

Fig. 9. A confluent drawing of $\overline{C}_8$.

Cographs. A *complement reducible graph* (also called a *cograph*) is defined recursively as follows [10]:

- A graph on a single vertex is a cograph.
- If G_1, G_2, $\cdots$, G_k are cographs, then so is their union $G_1 \cup G_2 \cup \cdots \cup G_k$.
- If G is a cograph, then so is its complement $\overline{G}$.

Cographs can be obtained from single node graphs by performing a finite number of unions and complementations.

Theorem 5. *Cographs are confluent.*

Proof. If cographs A and B are confluent, we can show $A \cup B$ and $\overline{A \cup B}$ are confluent too. First we draw A confluently inside a disk and attach a "tail" to the boundary of the disk. Connect the attachment point to each vertex in the disk. B is drawn in the same way. Then $A \cup B$ is formed by joining the two "tails" together so that they don't connect to each other. $\overline{A \cup B}$ is formed by joining the two "tails" of $\overline{A}$ and $\overline{B}$ together so that they connect to each other. (See Fig. 10.) By the definition of cographs and induction we know cographs are confluent. □

Fig. 10. Confluent $A \cup B$ and $\overline{A \cup B}$.

5 Some Non-confluent Graphs

In this section, we show that some graphs cannot be drawn confluently. These graphs include the Petersen graph P, the graph $P - v$ formed by removing one vertex from the Petersen graph, and the 4-dimensional hypercube.

The Petersen graph. By removing one vertex and its incident edges from the Petersen graph we obtain a graph homeomorphic to $K_{3,3}$. It contains no $K_{2,2}$ as a subgraph. Moreover, note that $K_{2,2}$ is the most basic structure that allows for edge merging into tracks. Thus the resulting graph is non-confluent. This graphis the smallest non-confluent graph we know of.

The Petersen graph itself is also non-confluent, as adding the vertex and edges back to its non-confluent subgraph doesn't create any four-cycles that could be used for confluent tracks.

4-dimensional hypercube. The 4-dimensional hypercube in Fig. 11 is non-confluent.

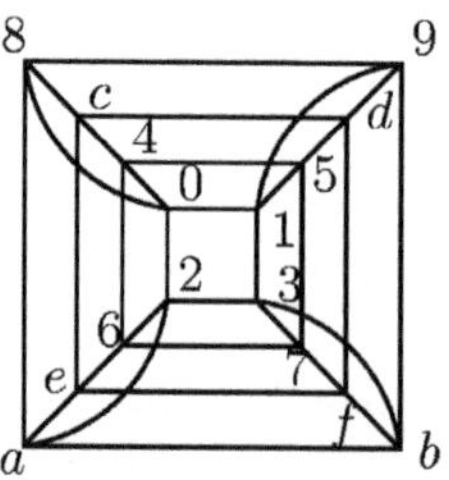

Fig. 11. The 4-dimensional hypercube.

The hypercube contains many subgraphs isomorphic to 3-dimensional cubes. Cubes are planar graphs, but in order to show non-confluence for the hypercube, we analyze more carefully the possible drawings of the cubes. Observe that, because there are no $K_{2,3}$ subgraphs in cubes or hypercubes, the only possible confluent tracks are $K_{2,2}$'s formed from the vertices of a single cube face.

Proof. (We omit the proof in this extended abstract.)

Acknowledgments. We would like to thank Jie Ren and André van der Hoek for supplying us with several examples of graphs used in software visualization.

References

1. T. Ball and S. G. Eick. Software visualization in the large. *IEEE Computer*, 29(4):33–43, 1996.
2. C. Batini, E. Nardelli, and R. Tamassia. A layout algorithm for data flow diagrams. *IEEE Trans. Softw. Eng.*, SE-12(4):538–546, 1986.
3. J. A. Bondy and U. S. R. Murty. *Graph Theory with Applications.* Macmillan, London, 1976.
4. R. Breu, U. Hinkel, C. Hofmann, C. Klein, B. Paech, B. Rumpe, and V. Thurner. Towards a formalization of the unified modeling language. In *Proc. 11th European Conf. Object-Oriented Programming (ECOOP'97)*, volume 1241 of *Lecture Notes in Computer Science*, pages 344–366, 1997.
5. C. Chambers, J. Dean, and D. Grove. A framework for selective recompilation in the presence of complex intermodule dependencies. In *Proceedings: 17th International Conference on Software Engineering*, pages 221–230, 1995.
6. C. C. Cheng, C. A. Duncan, M. T. Goodrich, and S. G. Kobourov. Drawing planar graphs with circular arcs. In *Proc. Graph Drawing*, volume 1731 of *Lecture Notes in Computer Science*, pages 117–126, 1999.
7. N. Chiba and T. Nishizeki. Arboricity and subgraph listing algorithms. *SIAM J. Comput.*, 14:210–223, 1985.
8. N. Chiba, T. Nishizeki, S. Abe, and T. Ozawa. A linear algorithm for embedding planar graphs using PQ-trees. *J. Comput. Syst. Sci.*, 30(1):54–76, 1985.
9. M. Chrobak and T. H. Payne. A linear time algorithm for drawing a planar graph on a grid. Technical Report UCR-CS-90-2, Dept. of Math. and Comput. Sci., Univ. California Riverside, 1990.
10. D. G. Corneil, H. Lerchs, and L. S. Burlingham. Complement reducible graphs. *Discrete Appl. Math.*, 3:163–174, 1981.
11. H. de Fraysseix, J. Pach, and R. Pollack. How to draw a planar graph on a grid. *Combinatorica*, 10(1):41–51, 1990.
12. G. Di Battista, P. Eades, R. Tamassia, and I. G. Tollis. *Graph Drawing.* Prentice Hall, Upper Saddle River, NJ, 1999.
13. P. Eades and P. Mutzel. Graph drawing algorithms. In M. Atallah, editor, *CRC Handbook of Algorithms and Theory of Computation*, chapter 9. CRC Press, 1999.
14. D. Eppstein. Arboricity and bipartite subgraph listing algorithms. *Information Processing Letters*, 51(4):207–211, August 1994.
15. T. Feder, A. Meyerson, R. Motwani, L. O'Callaghan, and R. Panigrahy. Representing graph metrics with fewest edges. In *Proceedings: 20th Annual Symposium on Theoretical Aspects of Computer Science*, volume 2607 of *Lecture Notes Comput. Sci.*, pages 355–366, 2003.
16. M. R. Garey and D. S. Johnson. Crossing number is NP-complete. *SIAM J. Algebraic Discrete Methods*, 4(3):312–316, 1983.
17. D. Grove, G. DeFouw, J. Dean, and C. Chambers. Call graph construction in object-oriented languages. In *Proc. of ACM Symp. on Object-Oriented Prog. Sys., Lang. and Applications (OOPSLA)*, pages 108–124, 1997.
18. P. Haynes, T. J. Menzies, and R. F. Cohen. Visualisations of large object-oriented systems. In P. D. Eades and K. Zhang, editors, *Software Visualisation*, volume 7, pages 205–218. World Scientific, Singapore, 1996.
19. M. Jünger, E. K. Lee, P. Mutzel, and T. Odenthal. A polyhedral approach to the multi-layer crossing minimization problem. In G. Di Battista, editor, *Graph Drawing (Proc. GD '97)*, number 1353 in Lecture Notes Comput. Sci., pages 13–24. Springer-Verlag, 1997.

20. M. Jünger and P. Mutzel. 2-layer straightline crossing minimization: Performance of exact and heuristic algorithms. *J. Graph Algorithms Appl.*, 1(1):1–25, 1997.
21. G. Kant. Drawing planar graphs using the canonical ordering. *Algorithmica*, 16:4–32, 1996. (special issue on Graph Drawing, edited by G. Di Battista and R. Tamassia).
22. M. Kaufmann and D. Wagner. *Drawing Graphs: Methods and Models*, volume 2025 of *Lecture Notes in Computer Science*. Springer-Verlag, 2001.
23. D. E. Knuth. Computer drawn flowcharts. *Commun. ACM*, 6, 1963.
24. X. Lin. On the computational complexity of edge concentration. *DAMATH: Discrete Applied Mathematics and Combinatorial Operational Research*, 101, 2000.
25. P. Mutzel. An alternative method to crossing minimization on hierarchical graphs. In S. North, editor, *Graph Drawing (Proc. GD '96)*, volume 1190 of *Lecture Notes Comput. Sci.*, pages 318–333. Springer-Verlag, 1997.
26. P. Mutzel and T. Zeigler. The constrained crossing minimization problem. In J. Kratochvil, editor, *Graph Drawing Conference*, volume 1731 of *Lecture Notes in Computer Science*, pages 175–185. Springer-Verlag, 1999.
27. F. J. Newbery. Edge concentration: A method for clustering directed graphs. In *Proc. 2nd Internat. Workshop on Software Configuration Management*, pages 76–85, 1989.
28. T. Nishizeki and N. Chiba. *Planar Graphs: Theory and Algorithms*, volume 32 of *Ann. Discrete Math.* North-Holland, Amsterdam, The Netherlands, 1988.
29. R. C. Penner and J. L. Harer. *Combinatorics of Train Tracks*, volume 125 of *Annals of Mathematics Studies.* Princeton Univ. Press, Princeton, NJ, 1992.
30. B. Price, R. Baecker, and I. Small. A principled taxonomy of software visualization. *Journal of Visual Languages and Computing*, 4(3):211–266, 1993.
31. H. Purchase. Which aesthetic has the greatest effect on human understanding? In G. Di Battista, editor, *Proc. 5th Int. Symp. Graph Drawing, GD*, number 1353 in Lecture Notes in Computer Science, LNCS, pages 248–261. Springer-Verlag, 18–20 Sept. 1997.
32. G.-C. Roman and K. C. Cox. A taxonomy of program visualization systems. *IEEE Computer*, 26(12):11–24, 1993.
33. J. Rumbaugh, I. Jacobson, and G. Booch. *The Unified Modeling Language Reference Manual.* Addison-Wesley, 1998.
34. W. Schnyder. Embedding planar graphs on the grid. In *Proc. 1st ACM-SIAM Sympos. Discrete Algorithms*, pages 138–148, 1990.
35. J. T. Stasko and E. Kraemer. A methodology for building application-specific visualizations of parallel programs. *Journal of Parallel and Distributed Computing*, 18(2):258–264, 1993.
36. R. Tamassia and I. G. Tollis. Planar grid embedding in linear time. *IEEE Trans. Circuits Syst.*, CAS-36(9):1230–1234, 1989.
37. F. Tip and J. Palsberg. Scalable propagation-based call graph construction algorithms. *ACM SIGPLAN Notices*, 35(10):281–293, 2000.

An Experimental Study of Crossing Minimization Heuristics

Carsten Gutwenger[1] and Petra Mutzel[2]

[1] Stiftung caesar
Ludwig-Erhard-Allee 2, D-53175 Bonn, Germany
[2] Vienna University of Technology
Favoritenstr. 9–11 E186, A-1040 Vienna, Austria

Abstract. We present an extensive experimental study of heuristics for crossing minimization. The heuristics are based on the planarization approach, so far the most successful framework for crossing minimization. We study the effects of various methods for computing a maximal planar subgraph and for edge re-insertion including post-processing and randomization.

1 Introduction

The crossing minimization problem is one of the crucial problems in graph drawing (see, e.g., [22]). Here, the task is to find a drawing of a graph in the plane with the minimum number of crossings. The crossing number represents a fundamental measure of non-planarity of graphs and has been studied for more than 40 years by graph theorists. The algorithmical problem of computing the crossing number has also been studied in the context of VLSI-layout. So far, only a few infinite classes of graphs exist, for which the crossing number is known. We do not even know the asymptotic value for the complete graph K_n with n vertices and for the complete bipartite graph $K_{n,n}$ with $2n$ vertices, as n tends to infinity [23].

We do know, however, that the crossing number problem and several of its variants are NP-hard [13,3]. While many other prominent NP-hard problems have been successfully attacked with integer programming and branch-and-bound techniques, no similar approach to the crossing number problem is known to date. To our knowledge, no exact algorithm exists that is able to solve even small instances of the crossing number problem to provable optimality within reasonable computation time. For graphs with bounded degrees, Even et al. [8] have recently suggested an approximation algorithm in which the sum of the numbers of vertices and crossings is $O(\log^3 |V|)$ times the minimum sum thus improving the results of $O(\log^4 |V|)$ by Bhatt and Leighton [2] and Leighton and Rao [17]. Grohe [9] has given an exact algorithm that works in quadratic time if the crossing number is fixed. Both algorithms are rather of theoretical nature and have so far not been useful for solving practical instances. Much easier to tackle is the bipartite crossing number, for which a vast amount of experimental papers exist in the literature (see, e.g., [16,7]). Experimental papers in other fields

G. Liotta (Ed.): GD 2003, LNCS 2912, pp. 13–24, 2004.

include, e.g., a collection of experimental studies in graph drawing by Vismara et al. [24] and the paper by Brandenburg et al. on force-directed methods [4].

The only experimental state-of-the-art study – to our knowledge – concerning crossings for general graphs has been published in [6]. It includes four different algorithms for orthogonal graph drawing on the benchmark set described above. Two of these algorithms were based on the topology-shape-metrics approach, and the others are incremental algorithms that focus on a small area and a small number of bends. The results show that the former two algorithms are superior in terms of the criteria number of crossings, number of bends, area, and edge length. E.g., the best algorithm had up to 8 times fewer crossings. Therefore, we decided to solve the crossing minimization problem heuristically using a 2-step planarization approach, which we will discuss in Section 3. There, we will also describe our strategies (some of them are new) for crossing reduction. Section 4 contains a discussion of our extensive experimental results. Our paper ends with a section on conclusions (see Section 5). Before we can start the technical part of the paper, we need to introduce some basic mathematical terms (see Section 2).

2 Preliminaries

In a drawing of a graph $G = (V, E)$ each vertex $v \in V$ is mapped to a distinct point p_v in the plane and each edge $(u, v) \in E$ is mapped to a closed simple curve that connects the points p_u and p_v and does not pass through the image of any other vertex. If two curves share an interior point p, we say that they cross at p. The crossing number $cr(G)$ is the minimal number of crossings in any drawing of G. The crossing number problem is the problem of finding the crossing number for a given graph G.

The graphs that can be drawn without any edge crossings are called *planar graphs*. A planar drawing of a graph divides the plane into regions called *faces*. Every drawing defines a *planar* and a *combinatorial embedding* of the graph G. Such an *embedding* essentially fixes the topology of the graph. A *combinatorial embedding* is defined as a clockwise ordered list of adjacent neighbors for each vertex $v \in V$. When, in addition, the outer face is fixed, the combinatorial embedding is also called a *planar embedding* of G. An alternative definition of a combinatorial embedding is for each face f an anti-clockwise ordered list of the edges bordering f. Given a planar graph, a combinatorial embedding can be computed in linear time [5,19]. In general, a planar graph can have an exponential number of combinatorial embeddings. In the following section, we will use the name *embedding* for planar or combinatorial embedding.

3 The Planarization Approach

In practice, the crossing minimization problem is solved heuristically using a 2-step planarization approach. In a first step, a small number of edges is deleted from $G = (V, E)$ in order to obtain a planar graph P. In a second step, the edges are re-inserted into the planar graph P while trying to keep the number of crossings small.

3.1 Methods for Computing the Planar Subgraph

For the first step, we need to solve the so-called *maximum planar subgraph problem*, which has been shown to be NP-hard [18]. If the number of edges to be deleted is small, the exact branch-and-cut algorithm suggested in [15] is able to provide a provably optimal solution quite fast. However, the method is quite complicated to understand and to implement: Moreover, if the number of deleted edges exceeds 10, the algorithm usually needs far too long to be acceptable for practical computation. Since we are interested in approaches for practitioners, we did not involve this exact method into our studies. Interested readers are refered to the study of Ziegler [25] concerning the number of deleted edges in the *Rome library* benchmark set.

A widely used standard heuristic for finding a maximal planar subgraph is to start with the empty graph, and to iteratively try to add the edges one by one. In every step, a planarity testing algorithm is called for the obtained graph. If the addition of an edge would lead to a non-planar graph, then the edge is disregarded; otherwise, the edge is added permanently to the planar graph obtained so far. After $|E|$ iterations (planarity tests), we have obtained a maximal planar subgraph P of G, i.e., a subgraph of G which will get non-planar as soon as any of the edges in $G - P$ will be added. We will denote this method as MAXIMAL. The standard (and also our) implementation needs a running time of $O(|E||V|)$. Theoretically, this can be improved to nearly linear running time using so-called online-planarity testing algorithms (e.g., [1,21]), but we are not aware of any correct implementation.

An alternative to this method is to use the planarization algorithm based on PQ-trees suggested in [12,14]. Observe, that this method cannot guarantee to derive a maximal planar subgraph. The theoretical worst case running time is $O(|V|^2)$. In practice it is much faster.

The quality of the results can be improved by introducing random events and calling them several times. The PQ-tree based algorithm starts by computing a so-called st-numbering. Our random event was simply to choose a random edge of E in order to get s and t. We studied the effects of up to 100 calls. We denote these methods as PQ1, PQ10, PQ50, and PQ100 for 1, 10, 50, and 100 iterations.

3.2 Edge Re-insertion Strategies

Fixed Embedding. Also the edge re-insertion step is a NP-hard optimization problem [25]. The standard algorithm used in practice re-inserts the edges $e_1, e_2, \ldots, e_k$ iteratively starting with a given planar embedding of G. The approach is based on the observation that an edge e_i crosses an edge in P if and only if it uses an edge in the geometric dual graph of P. Hence, the problem of re-inserting only one edge into P with a given planar embedding can be solved via a simple shortest-path computation in the extended dual graph of P. (We need to extend the dual graph in order to connect the end-vertices of e_i with the dual graph.) After each insertion step i, the crossings generated by edge e_i are substituted by artificial vertices so that the resulting graph $G \cup \{e_1, \ldots, e_i\}$ becomes planar again ($i = 1, \ldots, k$). The theoretical worst case running time for inserting k edges of our implementation is $O(\sum_{i=1}^{k}(|V| + \sum_{j=0}^{i-1} c_j)) = O(k(|V| + |C|))$, where c_j is the number of crossings introduced in step j, $c_0 = 0$, and C the number of crossings in the final drawing. In practice, it is much faster, since the update of the dual graphs are implemented efficiently. We denote this re-insertion method as FIX.

Variable Embedding. However, the quality of the resulting drawing highly depends on the chosen embedding for P. In [11], Gutwenger et al. have given a linear time algorithm based on the SPQR-tree data structure for inserting one edge into a planar graph P so that the number of crossings in $P \cup \{e\}$ over the set of all possible planar embeddings of P is minimized. Our implementation has the same theoretical running time as the variant FIX. We denote this re-insertion method as VAR.

Constrained Crossing Minimization. Obviously, re-insertion of all edges at the same time will improve the solution. However, no practically efficient algorithm is known. The *constrained crossing minimization problem* asks for the minimum number of crossings obtained by a set of edges F when inserted into a planar graph P, while the embedding of P is not changed. The problem has been investigated in [20,25]. Experiments show that it can only be solved to provable optimality if there are less than 10 re-inserted edges — and even then, the running time is relatively high. Therefore, we did not include this method into our experiments.

Post-Processing Strategies. The idea of the post-processing strategies is to iteratively delete an edge from the drawing and to re-insert it again. Our strategies vary in the set and/or number of edges involved in the deletion and re-insertion process, and the order of processing them.

The variant INS involves exactly those edges which have been deleted already in the planar subgraph step, whereas the variants ALL and MOST involve the whole set of edges E in the original graph G. An iteration takes either the whole set (in variant INS and ALL) or $x\%$ of this edges (variant MOST $x\%$) iteratively (one after the other). The procedure stops only if within one iteration no improvement has been made. How do we choose the edges in variant MOST $x\%$? After each iteration, we sort the involved edge set in descending order according to the number of crossings they are involved in. Now, only the first $x\%$ edges of this list are taken for re-insertion. The variant in which there is no post-processing routine is called NONE.

The post-processing procedure can be implemented efficiently by updating the dual graph only at those regions, in which changes did occur. We did this for our FIX strategy. In principal, such an update is also possible for the VARIABLE embedding setting [1]. However, we are not aware of any implementation of this algorithm. This explains the big running time discrepancy in the post-processing procedure between the FIXED and VARIABLE embedding setting.

Permutations. After a whole deletion and re-insertion process of the chosen strategy for embedding FIX/VAR and a strategy for post-processing NONE/INS/ ALL/MOST, we get a certain crossing number. Our permutation variant does nothing else, but repeating the whole edge re-insertion process and keeping the best results. The random effect exists in choosing a different ordering of the edges in $G-P$ for the initial re-insertion step. The notation PERMi gives the number of these repetition rounds. We have experimented with PERM1, PERM2, PERM10, and PERM20.

4 Experimental Results

For our computational experiments, we have used our new graph drawing library which is the basis of the GoVisual layout tools [10]. Our computational experiments ran on a PC with Pentium 4, 2.4 GHz, 512 MB RAM, under MS Windows 2000, and our C++-code has been compiled with MS Visual C++.NET (Visual C++ 7). We used the benchmark set which is commonly known as the *Rome library*. It contains 11.528 graphs[1] with 10 to 100 vertices, and has been generated from a core set of 112 graphs used in *real-life* software engineering and database applications. The library is available via `http://www.dia.uniroma3.it/people/gdb/ wp12/undirected-1.tar.gz`.

4.1 Results of the Planar Subgraph Computations

In our first experiments, we measured the average number of deleted edges of the five strategies for computing a planar subgraph for each group with vertex size i, $i = 10, \ldots, 100$, separately. The results improve significantly as the number of permutations increases. While the number of deleted edges went up to 19 (on average) for PQ1, it was about 16 for PQ10, and 15 for PQ50 and PQ100. It seems that it really pays off to run the planar subgraph algorithm many times. However, the results for 50 and 100 permutations are very close to each other. This effect comes from the random effect chosen for our implementation (see Section 3.1). By choosing other randomization techniques, this effect may occur at a higher number of permutations. Since the running time of our implementation is relatively small, it seems that taking 100 iterations of the PQ-based algorithm (PQ100) is a good choice. The running time for PQ1 was below 0.002 seconds, while the time for MAXIMAL increased to 0.022 seconds at 100 vertices. The time for PQ100 was about 0.34 seconds for instances with 100 vertices.

4.2 Results for the Edge Re-insertion Step

FIX vs. VAR. Figure 1 shows the number of crossings generated with the variants PQ1 vs. PQ100 and FIX vs. VAR. The VAR strategy seems to get

[1] originally, it were 11,582 graphs, but some of the files are corrupted

better results than the FIX strategy. This effect is outperformed, however, by starting with a better planar subgraph for edge re-insertion (i.e., taking PQ100 instead PQ1). Because of this, we decided to stay with PQ100 for our further experiments.

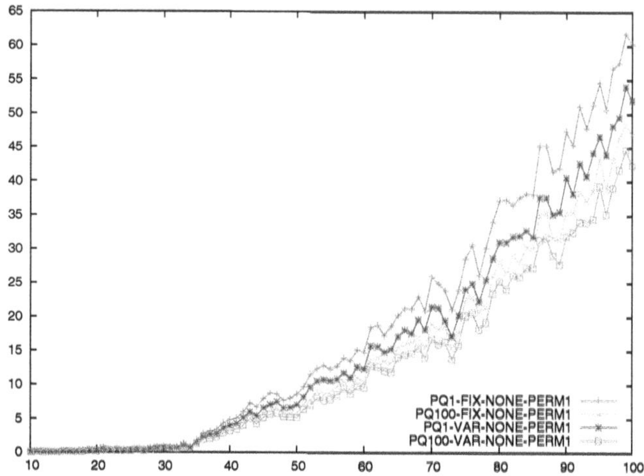

Fig. 1. The effect of the edge re-insertion strategies FIX and VAR.

On the instances with 100 vertices we get 60.32 crossings on average. This is about the same number which was also reported in the study by Di Battista et al. [6]. By computing a better planar subgraph the number of crossings reduces from 60.32 to 46.97 (about 22% improvement). This number can further be reduced to 42.37 by choosing the best embedding for each inserted edge. This is already an improvement of about 30%.

For the remaining Figures, we decided to show the relative improvement of the variants acording to the *standard method* PQ1-FIX-NONE-PERM1, which was also used in [6].

Post-Processing. Figure 2 shows the effect of the post processing variants for the VAR strategy. Taking the inserted edges as candidates for re-insertion is already much better than the NONE strategy (no post-processing at all). However, the results can be improved a lot more by taking the whole set of edges into account. We observe that already taking into account 25% of the edges gives similar results to the options MOST100 and ALL. ALL, MOST100, and MOST25 improve the number of crossings up to 46.11% (corresponds to 32.5 crossings at 100 vertices) compared to the standard method. For the FIXED strategy, the picture looks very similar. Here, the improvement coming from post-processing in comparison to NONE is a little bit better.

We also measured the running times of the post processing variants. Strategies INS and MOST10 have about the same running time (up to 0.18 seconds), which is up to a factor of 2.5 slower than the NONE variant. MOST25 needs

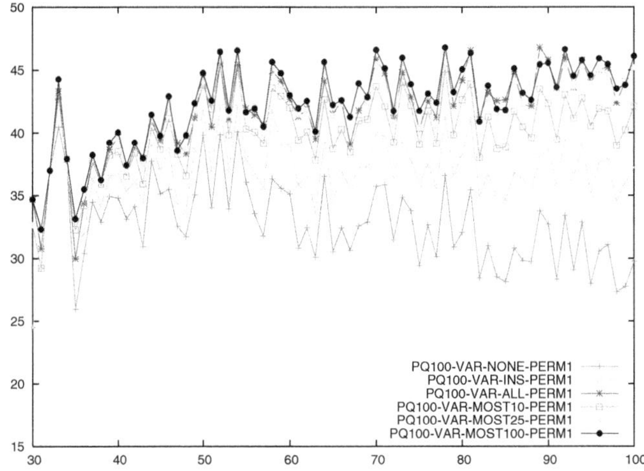

Fig. 2. The effect of the post processing variants compared to the standard method.

up to 0.35 seconds. Not surprisingly, MOST100 and ALL have a similar running time (up to 0.53 seconds), and are the slowest variants. For the FIX strategy, the running time of all versions is about the same. The time for variant PQ100-FIX-MOST100-PERM1 is only 0.01 seconds larger than the time for variant PQ100-FIX-NONE-PERM1. This comes from the efficient update operations mentioned in Section 3.2.

Permutations. Figure 3 shows the effect of the permutation variants for FIX. As expected, the results improve with a higher permutation number. However, we observe that the post-processing effects are stronger than the PERM20 effects: all four strategies PQ100-FIX-MOST100-PERMi lead to much better results (between 41.77% and 50.25% improvement at 100 vertex instances) than the four strategies PQ100-FIX-NONE-PERMi ($i = 1, 2, 10, 20$) (between 22.12% and 32.36%).

Obviously, the running time increases a lot: PERM20 needs about 20 times the running time of PERM1 (since the edge re-insertion step dominates the procedure). For VAR, the situation is similar. The only difference is that the upper three curves are closer to the others.

PQ1 vs. PQ100. Finally, we have tested the effects of the maximal planar subgraphs heuristics PQ1 and PQ100 again for the best edge re-insertion strategies. Figure 4 shows that PQ100-FIX-ALL-PERM20 gives an improvement of 49.55% compared to the standard method, whereas PQ1-FIX-ALL-PERM20 gives an improvement of 47.22%. The running times for both variants are about the same. Interestingly, this is not true for the strategies PQ100-VAR-ALL-PERM1 and PQ1-VAR-ALL-PERM1. Due to the better planar subgraph achieved with PQ100-VAR-ALL-PERM1, the running times differ by 0,21 seconds at the instances with 100 vertices (0.48 compared to 0.69). Surprisingly, the PQ100 strat-

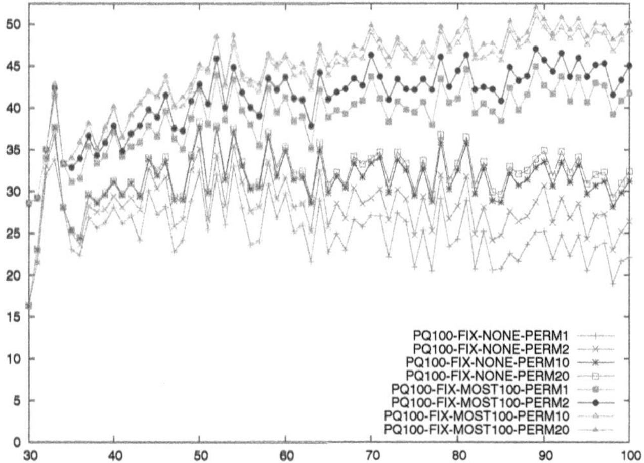

Fig. 3. The effect of the PERM variants.

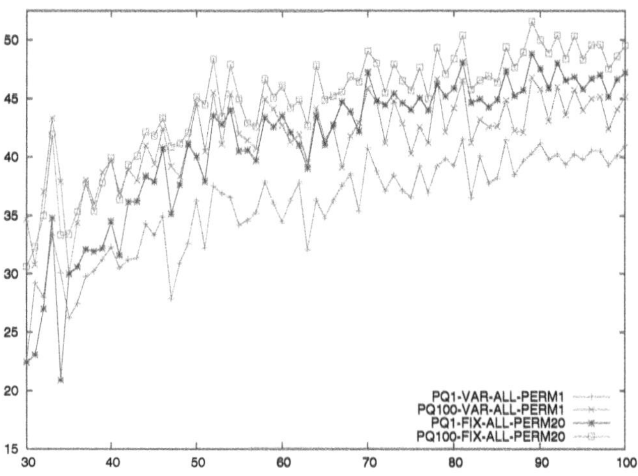

Fig. 4. The relative effect of the maximal planar subgraph variants.

egy is faster than the PQ1 strategy. The relative improvements in the number of crossings are 45.15% and 41.03%, respectively.

Conclusions. We have run all possible experiments with strategies FIX and VAR, the different post-processing strategies NONE and ALL, and the permutation numbers 1 and 10. Figure 5 shows the results for these eight strategies. The best results gives PQ100-VAR-ALL-PERM10 (51.63% improvement at 100 vertex instances corresponding to 29.18 crossings) followed by PQ100-FIX-ALL-PERM10 (48.42% corresponding to 31.11 crossings). The three strategies us-

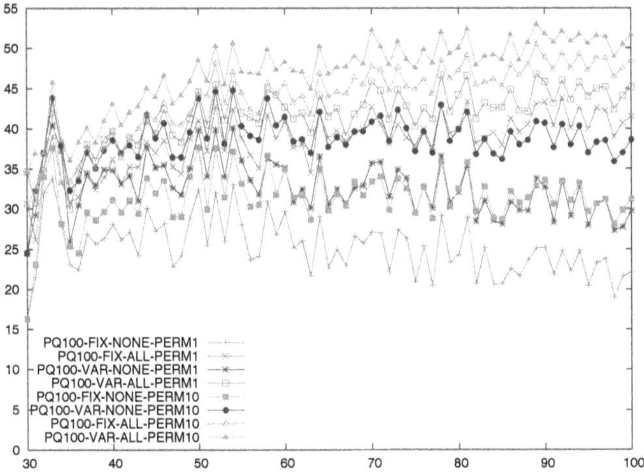

Fig. 5. The results for a collection of strategies.

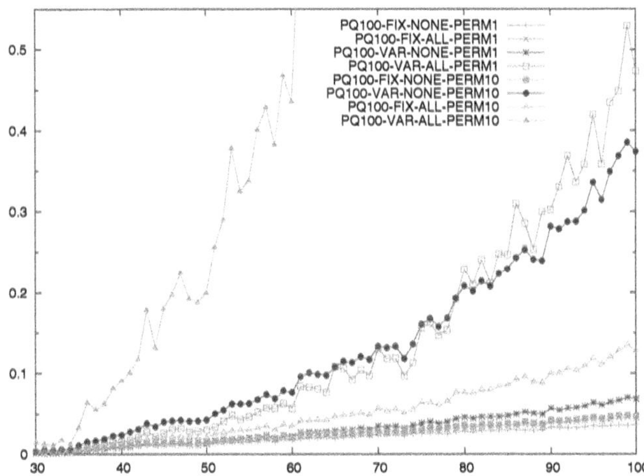

Fig. 6. The running times for the results displayed in Figure 5.

ing post-processing ALL are the four *winners* in this Figure. Interestingly, the best method without any post-processing, PQ100-VAR-NONE-PERM10, is in competition with the strategy PQ100-FIX-ALL-PERM1 (with up to 45% improvement). This means that, in order to be competitive with post-processing, 10 permutations are not enough. Additional help is needed, here in form of the strategy VAR compared to FIX.

Figure 6 shows the running times for the results displayed in Figure 5. We observe that the running time of PQ100-VAR-NONE-PERM10 is much slower than the running time of PQ100-FIX-ALL-PERM1. Hence PQ100-FIX-ALL-PERM1 is the clear winner among these two strategies. The best strategy displayed in Figure 6, PQ100-VAR-ALL-PERM10, has the highest running time. Interestingly, the second best strategy, PQ100-FIX-ALL-PERM10, has the fifth best running time (up to 0,14 seconds).

Table 1. The ranking list of the *best* heuristics.

Rank	Crossings	Time (sec)	MPS	FIX / VAR	POST	PERM
1	28.56	8.778	PQ100	VAR	ALL	PERM20
2	28.61	8.563	PQ100	VAR	MOST100	PERM20
3	28.66	5.902	PQ100	VAR	MOST25	PERM20
4	29.09	4.359	PQ100	VAR	MOST100	PERM10
5	29.35	3.060	PQ100	VAR	MOST25	PERM10
6	30.01	0.259	PQ100	FIX	MOST100	PERM20
7	30.43	0.258	PQ100	FIX	ALL	PERM20
8	30.62	0.130	PQ100	FIX	MOST100	PERM10
9	31.11	0.128	PQ100	FIX	ALL	PERM10
10	31.64	0.112	PQ100	FIX	MOST25	PERM10
11	33.16	0.054	PQ100	FIX	MOST100	PERM2
12	33.29	0.053	PQ100	FIX	ALL	PERM2
13	34.14	0.050	PQ100	FIX	MOST25	PERM2
14	35.12	0.046	PQ100	FIX	MOST100	PERM1
15	35.29	0.045	PQ100	FIX	ALL	PERM1
16	36.09	0.043	PQ100	FIX	MOST25	PERM1
17	38.93	0.040	PQ100	FIX	MOST10	PERM1
18	46.98	0.036	PQ100	FIX	NONE	PERM1
19	60.32	0.002	PQ1	FIX	NONE	PERM1

Table 1 shows a ranking list of the best obtained results in the following sense: We have sorted all strategies according to their average crossing number achieved at the instances with 100 vertices. Running through this list, we deleted all those strategies, which have obtained worse crossing number and worse running time than another strategy on the list. It is interesting that the best five strategies are based on VAR, and all of the remaining strategies in our ranking list are based on FIX. This stems from the fact, that from a certain point on, the VAR strategy takes much more time than the FIX strategy. E.g., the strategy PQ100-VAR-ALL-PERM1 has been eliminated from the table through PQ100-FIX-MOST100-PERM20. It seems that the effect coming from the permutations is stronger than the positive VAR effect, at least until a certain point.

Figure 7 shows a selection of the best obtained results listed in Table 1. The winners are PQ100-VAR-ALL-PERM20 (with improvement of 52.65% corresponding to 28.56 crossings) and PQ100-VAR-MOST25-PERM20, followed by PQ100-VAR-MOST100-PERM10 and PQ 100-VAR-MOST25-PERM10. Only then, the FIX strategies appear on the list.

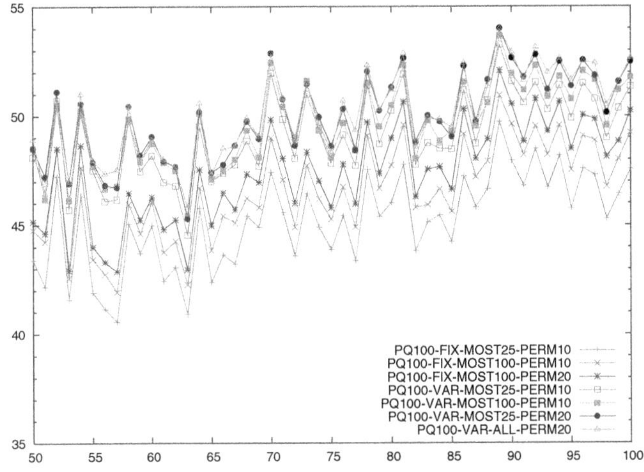

Fig. 7. The best obtained results within our study.

5 Conclusions

We have conducted extensive experimental studies on the crossing minimization problem for a benchmark set of graphs. The main conclusions are:

1. Post-processing always helps. It is recommended not to restrict the post-processing procedure to the inserted edges. Already re-inserting 25% of all the edges helps a lot.
2. Permutations and random effects help, but not as well as post-processing.
3. It is important to start with a good planar subgraph. A better subgraph leads not only to an improved number of achieved crossings, but also to an improved running time of the algorithm.
4. The re-insertion within a variable embedding setting is still worth doing, even if post-processing is used.

Acknowledgement. We thank Sebastian Leipert for providing his implementation of the PQ-based planarity testing, planarity embedding, and planar subgraph algorithms.

References

[1] G. Di Battista and R. Tamassia. On-line planarity testing. *SIAM Journal on Computing*, 25(5):956–997, 1996.

[2] S. N. Bhatt and F. T. Leighton. A framework for solving VLSI layout problems. *Journal of Computer and System Sciences*, 28:300–343, 1984.

[3] D. Bienstock. Some provably hard crossing number problems. *Discrete & Computational Geometry*, 6:443–459, 1991.

[4] F.J. Brandenburg, M. Himsolt, and C. Rohrer. An experimental comparison of force-directed and randomized graph drawing algorithms. In F.J. Brandenburg, editor, *GD '95*, volume 1027 of *LNCS*, pages 76–87. Springer-Verlag, 1996.

[5] N. Chiba, T. Nishizeki, S. Abe, and T. Ozawa. A linear algorithm for embedding planar graphs using PQ-trees. *J. Comput. Syst. Sci.*, 30(1):54–76, 1985.

[6] G. Di Battista, A. Garg, G. Liotta, R. Tamassia, E. Tassinari, and F. Vargiu. An experimental comparison of four graph drawing algorithms. *Comput. Geom. Theory Appl.*, 7:303–326, 1997.

[7] T. Eschbach, W. Günther, R. Drechsler, and B. Becker. Crossing reduction by windows optimization. In M.T. Goodrich and S.G. Kobourov, editors, *Graph Drawing '02*, volume 2528 of *LNCS*, pages 285–294. Springer-Verlag, 2002.

[8] G. Even, S. Guha, and B. Schieber. Improved approximations of crossings in graph drawing and VLSI layout area. In *Proc. 32nd ACM Symp. Theory of Comp. (STOC'00)*, pages 296–305. ACM Press, 2000.

[9] M. Grohe. Computing crossing numbers in quadratic time. In *Proc. 32nd ACM Symp. Theory of Computing (STOC'00)*, pages 231–236. ACM Press, 2000.

[10] C. Gutwenger, K. Klein, J. Kupke, S. Leipert, M. Jünger, and P. Mutzel. Govisual software tools. http://www.oreas.de, 2002.

[11] C. Gutwenger, P. Mutzel, and R. Weiskircher. Inserting an edge into a planar graph. In *Proc. Ninth Ann. ACM-SIAM Symp. Discr. Algorithms (SODA '2001)*, pages 246–255, Washington, DC, 2001. ACM Press.

[12] R. Jayakumar, K. Thulasiraman, and M. N. S. Swamy. $O(n^2)$ algorithms for graph planarization. *IEEE Trans. on Computer-Aided Design*, 8:257–267, 1989.

[13] M. R. Johnson and D. S. Johnson. Crossing number is NP-complete. *SIAM J. Algebraic Discrete Methods*, 4(3):312–316, 1983.

[14] M. Jünger, S. Leipert, and P. Mutzel. A note on computing a maximal planar subgraph using PQ-trees. *IEEE Trans. Computer-Aided Design*, 17(7), 1998.

[15] M. Jünger and P. Mutzel. Maximum planar subgraphs and nice embeddings: Practical layout tools. *Algorithmica*, 16(1):33–59, 1996.

[16] M. Jünger and P. Mutzel. 2-layer straightline crossing minimization: Performance of exact and heuristic algorithms. *J. Gr. Alg. & Appl. (JGAA)*, 1(1):1–25, 1997.

[17] F. T. Leighton and S. Rao. Multicommodity max-flow min-cut theorems and their use in designing approximation algorithms. *J. ACM*, 46(6):787–832, 1999.

[18] P.C. Liu and R.C. Geldmacher. On the deletion of nonplanar edges of a graph. In *10th. S-E Conf. Comb., Graph Theory, and Comp.*, pages 727–738, 1977.

[19] K. Mehlhorn and P. Mutzel. On the embedding phase of the Hopcroft and Tarjan planarity testing algorithm. *Algorithmica*, 16(2):233–242, 1996.

[20] P. Mutzel and T. Ziegler. The constrained crossing min. problem. In J. Kratochvíl, editor, *GD '99*, volume 1731 of *LNCS*, pages 175–185. Springer-Verlag, 1999.

[21] J. A. La Poutré. Alpha-algorithms for incremental planarity testing. In *Proc. 26th Annual ACM Symp. Theory of Computation (STOC)*, pages 706–715, 1994.

[22] H. Purchase. Which aesthetic has the greatest effect on human understanding? In G. Di Battista, editor, *Graph Drawing '97*, volume 1353 of *LNCS*, pages 248–261. Springer-Verlag, 1997.

[23] R.B. Richter and C. Thomassen. Relations between crossing numbers of complete and complete bipartite graphs. *Amer. Math. Monthly*, pages 131–137, 1997.

[24] L. Vismara, G. Di Battista, A. Garg, G. Liotta, R. Tamassia, and F. Vargiu. Experimental studies on graph drawing algorithms. *Software – Practice and Experience*, 30:1235–1284, 2000.

[25] T. Ziegler. *Crossing Minimization in Automatic Graph Drawing.* PhD thesis, Max-Planck-Institut für Informatik, Saarbrücken, 2000.

Stop Minding Your P's and Q's: Implementing a Fast and Simple DFS-Based Planarity Testing and Embedding Algorithm*

John M. Boyer[1], Pier Francesco Cortese[2], Maurizio Patrignani[2], and Giuseppe Di Battista[2]

[1] PureEdge Solutions Inc. Victoria, BC Canada
jboyer@acm.org
[2] Università di Roma Tre, Italy
{cortese,patrigna,gdb}@dia.uniroma3.it

Abstract. In this paper we give a new description of the planarity testing and embedding algorithm presented by Boyer and Myrvold [2], providing, in our opinion, new insights on the combinatorial foundations of the algorithm. Especially, we give a detailed illustration of a fundamental phase of the algorithm, called walk-up, which was only succinctly illustrated in [2]. Also, we present an implementation of the algorithm and extensively test its efficiency against the most popular implementations of planarity testing algorithms. Further, as a side effect of the test activity, we propose a general overview of the state of the art (restricted to efficiency issues) of the planarity testing and embedding field.

1 Introduction

Testing the planarity of a graph is one of the most fascinating and intriguing problems of the graph drawing and of the graph algorithms fields.

In 1974 Hopcroft and Tarjan [10] proposed the first linear-time planarity testing algorithm. This algorithm, also called "path-addition algorithm," starts from a cycle and adds to it one path at a time. However, the algorithm is so complex and difficult to implement that several other contributions followed their breakthrough. For example, about twenty years after [10], Mehlhorn and Mutzel [13] contributed a paper to clarify how to construct the embedding of a graph that is found to be planar by the original Hopcroft and Tarjan algorithm.

A different approach has its starting point in the algorithm presented by Lempel, Even, and Cederbaum [12]. This algorithm, also called "vertex addition

* Work partially supported by European Commission - Fet Open project COSIN – COevolution and Self-organisation In dynamical Networks – IST-2001-33555, by "Progetto ALINWEB: Algoritmica per Internet e per il Web", MIUR Programmi di Ricerca Scientifica di Rilevante Interesse Nazionale, and by "The Multichannel Adaptive Information Systems (MAIS) Project", MIUR Fondo per gli Investimenti della Ricerca di Base.

G. Liotta (Ed.): GD 2003, LNCS 2912, pp. 25–36, 2004.

algorithm," is based on considering the vertices one-by-one, following an *st*-numbering; it has been shown to be implementable in linear time by Booth and Lueker [1]. Also in this case, a further contribution by Chiba, Nishizeki, Abe, and Ozawa [5] has been needed for showing how to construct an embedding of a graph that is found planar.

A further interesting algorithm is based on a characterization given by de Fraysseix and Rosenstiehl [6] and, although it has not been fully described in the literature, it has a very efficient implementation in the Pigale software library [7].

However, the story of the planarity testing algorithms enumerates several more recent contributions, aimed at further exploring the relationships between planarity and Depth First Search (DFS) and at devising algorithms that are easier to understand and to implement.

Two recent DFS-based planarity testing algorithms are those presented by Shih and Hsu [15,16,11] and by Boyer and Myrvold [2]. Shih and Hsu algorithm replaces biconnected portions of the graph with single nodes whose embedding is fixed. Boyer and Myrvold algorithm, which is the starting point of this paper, represents embedded biconnected portions of the graph with a data structure that allows the embeddings to be "flipped" in constant time. More recently, in an unpublished manuscript, Boyer and Myrvold [3] presented an algorithm that, although inspired by [2], constitutes, in our opinion, a new and original approach to the planarity testing problem.

The contributions of the present paper can be summarized as follows. In Section 3 we describe the algorithm of [2] in a new way, that is, in our opinion, easily readable and more suitable for an implementation. In particular, in Section 4 we give a detailed description of a fundamental phase of the algorithm, called walk-up, which is essential for a correct and efficient implementation and was only succinctly illustrated in [2]. Finally, in Section 5 we present an experimental analysis which provides a general overview of the state of the art (restricted to efficiency issues) of the planarity testing and embedding field.

2 Background

We assume that the reader is familiar with graph terminology and basic properties of planar graphs. Unless otherwise specified, we only consider graphs without self-loops and multiple edges, as they can be replaced with paths of length two without affecting planarity.

A *cutvertex* of a graph G is such that its removal increases the number of connected components of G. A connected graph is said to be *biconnected* if it has no cutvertices. The *biconnected components* of a connected graph (also called *bicomps*) are its maximal biconnected subgraphs. A bicomp is said to be *elementary* if it is formed by a single edge. Note that a cutvertex belongs to several bicomps, while each edge falls into exactly one bicomp of the graph.

A *Depth First Search* (*DFS* in short) is a technique for visiting all the vertices of a graph. Each vertex v is assigned a *DFS index*, denoted $DFS(v)$, which specifies the order in which it was reached by the DFS visit, starting from the root r which has $DFS(r) = 1$. The edges used by the DFS visit to move from one vertex to the next one are called *tree-edges* and form a spanning tree of G, called the *DFS tree*. The remaining edges are *back-edges*. The *ancestors* of a vertex v are the vertices in the unique chain of tree-edges from v to the root r. If a vertex u is an ancestor of v, then v is a *descendant* of u. A tree-edge links a *parent* vertex to a *child* vertex, the former (latter) being the one with lowest (highest) DFS-index, while a back-edge is thought to be oriented exiting the descendant and entering the ancestor. The *lowpoint* of a vertex v, denoted by $Lowpt(v)$, is the lower DFS index of an ancestor of v reachable through a back-edge from a descendant of v. The set of back-edges entering a vertex v is denoted $B_{in}(v)$, while the set of back-edges exiting v is denoted $B_{out}(v)$. For a back-edge e, we call *support of* e and denote by $S(e)$ the unique chain of tree-edges having the same endpoints as the back-edge e. The support of e and e form a cycle.

A *planar drawing* of a graph is such that no two edges intersect (except at common endpoints). A graph is *planar* if it admits a planar drawing. A planar drawing partitions the plane into topologically connected regions, called *faces*. The unbounded face is called the *external face*. The *boundary* of a face is its delimiting circuit. The *incidence list* of a vertex v is the set of edges incident upon v. A planar drawing determines a circular ordering on the incidence list of each vertex v according to the clockwise sequence of the incident edges around v. Two planar drawings of the same connected graph G are *equivalent* if they determine the same circular orderings of the incidence lists. Two equivalent planar drawings have the same face boundaries. A *planar embedding* or simply *embedding* of G is an equivalence class of planar drawings and is described by circularly sorted incidence lists for each vertex v. In the following, unless otherwise specified, we will consider the embeddings comprehensive of the specification of the external face. In [2] a suitable data structure is introduced which can be used to describe an embedded bicomp. This data structure allows to "flip" the bicomp (that is, to represent the embedding in which all the adjacency lists of the vertices have been reversed) in constant time.

Since a graph is planar iff its bicomps are planar, in the following section we assume that the graph to be processed is biconnected. A simple method for handling graphs that are not biconnected is based on the property, which is easy to prove, that a graph G is planar iff it is planar the graph G' obtained from G by merging two adjacent bicomps b_1 and b_2, sharing a cutvertex v, with the addition of an edge between two vertices adjacent to v, one belonging to b_1 and the other to b_2. Based on this property, all the bicomps of the graph G can be merged with dummy edges without affecting planarity and the dummy edges can be removed after the embedding is computed.

3 The Boyer and Myrvold Planarity Testing and Embedding Algorithm

In this section we give a new description of the Boyer and Myrvold planarity testing algorithm [2]. This algorithm tests the planarity of a biconnected graph in linear time by trying to build a planar embedding of it and consists of three steps: *Preprocessing*, *Embedding Construction*, and *Embedding Reporting*.

In Step *Preprocessing* a DFS is performed on the input graph G. During the visit a DFS tree is constructed. For each vertex v we compute $DFS(v)$, $Lowpt(v)$, the set $B_{in}(v)$, and the list $L(v)$ of the children of v in the DFS tree sorted by their lowpoint values. Also, we compute $H(v)$, that is the lowest $DFS(w)$, with w in $B_{out}(v)$. Roughly, $H(v)$ is the DFS index of the vertex "nearest" to the root, reachable from v with a back-edge. If $B_{out}(v)$ is empty, then $H(v)$ is set to an "infinite" value. Further, each edge of the DFS tree is associated with a data structure representing an (embedded) elementary bicomp.

Step *Embedding Construction* is more complex and is based on visiting one-by-one the vertices of the input graph in inverse DFS-index order. Observe that this corresponds to a post-order traversal performed on the DFS tree.

Let v be the current visited vertex. Two strictly related tasks, called *walk-up* and *walk-down* (both described below), are performed starting from v. Let $G(v)$ be the subgraph of G composed by the edges of the DFS tree augmented with the edges in $B_{in}(w)$, for each w such that $DFS(w) \geq v$. The target of the two tasks is to determine (if it is possible) a planar embedding for the bicomps of $G(v)$ exploiting the planar embeddings already computed for graph $G(u)$, where $DFS(u) = DFS(v)+1$, which are preserved up to a "flip" operation. See Fig. 1.

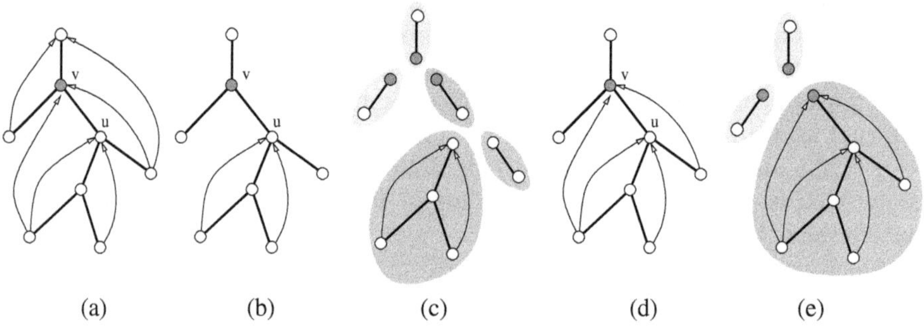

Fig. 1. (a) A graph G to be planarized; the edges of the DFS tree are drawn thick. $DFS(u) = DFS(v) + 1$. (b) Graph $G(u)$. (c) An embedding of the bicomps of $G(u)$. (d) Graph $G(v)$. (e) An embedding of the bicomps of $G(v)$.

In order to make this possible, the embeddings of the bicomps of $G(v)$, computed when vertex v is processed, must have the outer vertices of $G(v)$ on the external face. We call *outer vertices* of $G(v)$ the cutvertices of $G(v)$ and the vertices of $G(v)$ incident to a back-edge that is not in $G(v)$. Observe that: (i) $G(u)$ is a subgraph $G(v)$. (ii) The edges belonging to $G(v)$ that do not belong to $G(u)$ are exactly the back-edges in $B_{in}(v)$. (iii) Since the input graph G is biconnected, when $DFS(v) = 1$, $G(v) = G$ has a single bicomp, and a planar embedding of $G(v)$, if one exists, is a planar embedding for G.

In the *Embedding Reporting* step, the embedding is copied to a target data structure. The rest of the section shows the walk-up and the walk-down phases.

The purpose of the walk-up phase is to verify if there is a way to add the back-edges in $B_{in}(v)$ to the embedded bicomps of $G(u)$ in such a way that the embedded bicomps of $G(v)$ have the outer vertices of $G(v)$ on the external face. The walk-up phase works as follows. For each edge $e = (w, v)$ in $B_{in}(v)$ a path p_e is searched (and marked) from w to v along the bicomps that contain an edge in $S(e)$. The paths must satisfy several constraints. In order to describe these constraints, we need to introduce some definitions. Consider an embedded bicomp b of $G(u)$ containing some edges of $S(e)$. Observe that the edges of $S(e)$ contained in b form a subpath of $S(e)$. This subpath has its endpoints on the external face of b. We call these two vertices *root of* b and *entry point of* b with respect to $S(e)$, the former (latter) being the one nearest to (farthest from) the root of the DFS tree.

Given a specific $S(e)$, two *borders* $\mathcal{A}$ and $\mathcal{B}$ can be identified from $v_{b,e}$ to r_b along the boundary of the external face of b. If b is an elementary bicomp, the two "sides" of the edge $(v_{b,e}, r_b)$ are considered as distinct borders. A vertex w in a bicomp b different from the root r_b is defined to be *externally active for* b if there is a path from w to an ancestor of the currently being processed vertex v, that is only composed by a (possibly empty) sequence of tree edges not belonging to b plus a single back-edge.

Two constraints, called α and β, must be enforced on each path p_e. (α) For each bicomp b that contains edges in $S(e)$ the path p_e must contain either the border $\mathcal{A}$ or $\mathcal{B}$ from the entry point $v_{b,e}$ to the root r_b of b. (β) A path p_e must not contain a border that has an inner vertex (i.e. a vertex other than $v_{b,e}$ and r_b) that is externally active. Three more constraints, γ, δ, and ϵ, are expressed with respect to each pair of paths p_{e_1} and p_{e_2} traversing the same bicomp b. We say that the border of b used by p_{e_1} *intersects* the one used by p_{e_2} if they share an edge (or if they share the same "side" of the sole edge of the elementary bicomp b). Non-intersecting paths can only share entry or exit vertices (or both) in the bicomps they both traverse. (γ) If the border of b used by p_{e_1} does not intersect the border of b used by p_{e_2}, the two paths must use non-intersecting borders in all the bicomps they both traverse. (δ) Paths p_{e_1} and p_{e_2} must not use intersecting borders of b if they traverse two other distinct bicomps that are externally active. We call a bicomp *externally active* if it contains a (non-root) externally active vertex. (ϵ) If the border of b used by p_{e_1} does not intersect the border of b used by p_{e_2} and the root r_b of b is different from v, then r_b must not be an outer vertex of $G(v)$.

If no set of paths are found satisfying the constraints, then G is not planar [2], otherwise, they are used by the walk-down phase to build the embeddings of the bicomps of $G(v)$. Actually, the information used by the walk-down phase is simpler than the paths themselves. It only consists of the set of the borders of the bicomps used by the paths (that is, the marked borders) and, for each vertex w that was a walk-up entry point for a bicomp, of two sets of vertices called *active roots* and *non-active roots* of w. The set of the active roots of w

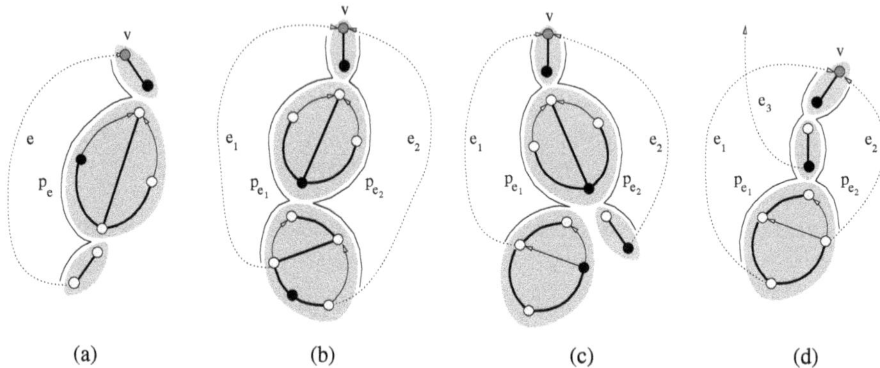

Fig. 2. Four different configurations for illustrating Constraints β, γ, δ, and ϵ. Edges in $B_{in}(v)$ are drawn with dotted lines, while externally active vertices are drawn black. (a) An example of a path that does not satisfy constraint β. (b), (c) Examples of pairs of paths satisfying constraints γ and δ, respectively. (d) An example of a pair of paths that do not satisfy constraint ϵ.

contains the roots (at most two) of the externally active bicomps traversed by the paths immediately before entering w. The set of the non-active roots of w contains the roots of the non externally active bicomps traversed by the paths immediately before entering w.

Consider two back-edges e_1 and e_2 in $B_{in}(v)$. If the intersection of $S(e_1)$ and $S(e_2)$ is not void, we say that e_1 and e_2 are *interleaving*. It is easy to prove that the interleaving relation is an equivalence relation. Roughly, the edges of the same equivalence class have the same child of v as an internal vertex of their supports. For each equivalence class induced by the interleaving relation, the walk-down process starts on a marked border $\mathcal{A}$ of the elementary bicomp containing v by pushing v onto a stack. When the stack becomes void an analogous process is performed on the opposite border $\mathcal{B}$, except if v is the DFS tree root. Let w be the current vertex on top of the stack. The back-edge e from w to v (if any) is added to the embedding of b_B in such a way to close the border $\mathcal{A}$ in the internal part of the cycle formed by e and p_e. Then, the stack is updated by: (i) removing w; (ii) pushing onto the stack the next marked vertex x along the border adjacent to w provided that the set of active roots of w is empty; (iii) pushing onto the stack one of the active roots of w (if any); (iv) pushing onto the stack all the non-active roots of w. When the current vertex on top of the stack is found to be the root vertex of some bicomp b, then b is merged into the bicomp b_B, flipping its embedding if necessary so that a marked border of b attaches to the just processed border of b_B.

4 Selecting Paths: Implementation Issues for the Walk-up

In order to describe the operations performed during the walk-up phase we need to extend to root vertices the definition given in Section 3 for external activity of non-root vertices. We define a root vertex r_b of a bicomp b *externally active* if it exists a back-edge from r_b to the currently being processed vertex v or if it exists a bicomp $b' \neq b$ which is externally active and such that $r_{b'} = r_b$. Also, we call a vertex w on a bicomp b ($w \neq r_b$) *pertinent for* b if it is an entry point for b and if it is not externally active for b.

Due to space reasons, some technicalities needed to detect in constant time the external activity of vertices and bicomps and the pertinence of vertices are left out from this camera ready and can be found in [4].

During the walk-up phase, back-edges in $B_{in}(v)$ are considered one-by-one in arbitrary order. The operations performed for each back-edge e in $B_{in}(v)$, called *walk-up*(e), have the purpose of marking the borders used by p_e and of updating, if it is needed, the lists of active roots and non-active roots of the entry point $v_{b,e}$ of each bicomp b traversed by p_e. Exceptionally, walk-up(e) may reverse the decision taken by a previous walk-up(e') about which border $\mathcal{A}$ or $\mathcal{B}$ of a bicomp b is used by $p_{e'}$. Walk-up(e), where $e = (w, v)$, starts at vertex w and selects a path to v that passes through each bicomp b traversed by $S(e)$. It terminates if (i) vertex v is reached, (ii) a vertex u, marked by a previous walk-up(e') is reached and the subpath of p_e from u to v is chosen to be equal to the analogous subpath of $p_{e'}$, (iii) a non-planarity condition has been detected. On each bicomp b traversed by $S(e)$, walk-up(e) proceeds from the entry point $v_{b,e}$ along both borders, $\mathcal{A}$ and $\mathcal{B}$, in parallel, searching for the root r_b of the bicomp or for a marked vertex. Borders blocked by externally active vertices can not be used. When the root r_b is reached, the fully explored border is selected and marked, while the partial exploration of the opposite border is abandoned, and walk-up(e) "ascends" to the next bicomp traversed by $S(e)$, that is, starts a parallel exploration of its borders. When walk-up(e) begins on the first bicomp traversed by $S(e)$, it is said to be *internally active*. It changes to *externally active* when it starts a parallel exploration of the borders of a bicomp after ascending from an externally active one. In order to be able to mark the borders of a bicomp b, we associate with each vertex u on the external face of b two *pathInfo* integers, one for each edge incident to u on the external face of b. During Step Preprocessing, pathInfo integers are assigned a value greater than n, where n is the number of vertices of the input graph. The flags are considered to be *clear* if their absolute value exceeds $DFS(v)$, where v is the currently being processed vertex, while are considered to be *set* if their absolute value is equal to $DFS(v)$. Thus, when the algorithm moves to processing the next vertex, which has lower DFS index, the pathInfo flags are implicitly cleared. While walk-up(e) is internally active, it sets the pathInfo integers to $-V$, where $V = DFS(v)$. Once walk-up(e) becomes externally active, it sets pathInfo integers to $+V$.

The following is a case-by-case statement of the decision points: first we give six rules for an internally active walk-up, then we provide the analogous six rules for an externally active walk-up.

Rule 1. If an internally active walk-up(e) enters $v_{b,e}$ and either pathInfo integer is set, then walk-up(e) terminates (the remaining part of p_e being already marked properly by a prior walk-up).

Rule 2. If an internally active walk-up(e) enters $v_{b,e}$ and both pathInfo integers are not set, then walk-up(e) traverses both borders $\mathcal{A}$ and $\mathcal{B}$ in search of the root r_b. (a) If a vertex w with a pathInfo set to $-V$ is encountered, then set the pathInfo to $-V$ along the border traversed from $v_{b,e}$ to w and terminate walk-up(e). (b) If a border is blocked (i.e. if it has an internal vertex that is externally active), then the opposing border must be taken to reach the root r_b. (c) If the second border is also blocked, then the graph is non-planar.

Rule 3. If an internally active walk-up(e) reaches vertex v then it terminates, and the pathInfo of the border used to reach v are set to $-V$.

Rule 4. If an internally active walk-up(e) traverses border $\mathcal{A}$ to reach the bicomp root r_b, then at least one pathInfo integer of r_b must be clear. If both are clear, then walk-up(e) simply sets the pathInfo to $-V$ along $\mathcal{A}$ and then it ascends to the next bicomp traversed by $S(e)$.

Rule 5. If an internally active walk-up(e) traverses border $\mathcal{A}$ to reach the bicomp root r_b, and the pathInfo of r_b for the opposing border $\mathcal{B}$ is set to $-V$, then: (a) If $\mathcal{B}$ has no internal vertex that is externally active, then walk-up(e) must select $\mathcal{B}$ and set to $-V$ the pathInfo integers along $\mathcal{B}$ (until an edge marked $-V$ is reached). Then, walk-up(e) can terminate. (b) If $\mathcal{B}$ contains only one non-root externally active vertex $w \neq v_{b,e}$, and if $\mathcal{B}$ contains no pertinent vertices between w and r_b (endpoints excluded), then the prior $-V$ marking of the path in $\mathcal{B}$ from w to r_b can be reversed to be a $-V$ path from $(w, \ldots, v_{e,b}, \ldots, r_b)$. The pathInfo of the path in $\mathcal{B}$ from w to r_b can be cleared and walk-up(e) terminated. (c) If $\mathcal{B}$ has an internal vertex w that is externally active, but either $v_{e,b}$ is externally active or $\mathcal{B}$ contains an internal vertex other than w that is pertinent or externally active, then the prior $-V$ pathInfo markings in $\mathcal{B}$ cannot be reversed. The current walk-up must select the border $\mathcal{A}$ and set its pathInfo to $-V$. Since r_b has been approached from both sides, the graph is non-planar if r_b is externally active. Otherwise, walk-up(e) ascends to the next bicomp traversed by $S(e)$ (where it becomes an externally active walk-up).

Rule 6. If an internally active walk-up(e) traverses border $\mathcal{A}$ to reach the bicomp root r_b, and the pathInfo of r_b for the opposing border $\mathcal{B}$ is set to $+V$, then we must test whether the prior externally active walk-up's path selection must be reversed. (a) Let w denote the vertex with a $+V$ pathInfo setting that is most distant from r_b in $\mathcal{B}$. If $\mathcal{B}$ contains any internal vertex other than w that is pertinent and has a $+V$ pathInfo setting, then there is no need to reverse the prior $+V$ path selection since the back-edge (w, v) that will be added during walk-down will close a pertinent vertex in an interior face of the embedding regardless of which border is selected. (b) If the path $(v_{b,e}, \ldots, w)$ in $\mathcal{B}$ contains an externally active vertex x (other than w), then u blocks the prior $+V$ path selection from being reversed. (c) If the prior $+V$ path cannot be reversed, then the current walk-up simply sets the pathInfo to $-V$ on the border $\mathcal{A}$. If r_b is externally active, then walk-up(e) terminates because the graph is non-planar. Otherwise, walk-up(e) ascends to the next bicomp traversed by $S(e)$ (where it becomes an externally active walk-up). (d) If the prior $+V$ path selection can be reversed, then the $+V$ pathInfo integers in $\mathcal{B}$ are cleared, while the pathInfo integers along the path $(w, \ldots, v_{b,e}, \ldots, r_b)$ are set to $+V$. Then, walk-up(e) can be terminated (the remaining part of p_e being already marked properly by a prior walk-up)

An externally active walk-up(e) must follow the six rules given below.

Rule 7. If the entry point $v_{b,e}$ of an externally active walk-up(e) has both pathInfo integers set to $+V$, then this is the third externally active walk-up to enter $v_{b,e}$ and the graph is non-planar. Note that a single prior walk-up cannot have passed through $v_{b,e}$ setting both of its pathInfo integers to $+V$ because $v_{b,e}$ is externally active for bicomp b.

Rule 8. If the entry point $v_{b,e}$ of an externally active walk-up(e) has one pathInfo integer set to $+V$, then walk-up(e) must select the opposing border $\mathcal{A}$ to the root r_b. (a) If $\mathcal{A}$ is blocked by an externally active vertex, then the graph is non-planar. (b) If $\mathcal{A}$ is not blocked, then walk-up(e) sets its pathInfo integer to $+V$ up to r_b. If r_b is externally active, then the graph is non-planar. Otherwise, walk-up(e) ascends to the next bicomp traversed by $S(e)$.

Rule 9. The case in which an externally active walk-up(e) enters a vertex $v_{b,e}$ that has both its pathInfo integers set to $-V$ can not occur. In fact, prior internally active walk-ups entering $v_{b,e}$ would select the same path, so only one pathInfo would be set to $-V$. Also, an internally active walk-up cannot pass through $v_{b,e}$ because it is externally active.

Rule 10. If an externally active walk-up(e) reaches vertex v then it terminates, and the pathInfo of the border used to reach v are set to $+V$.

Rule 11. If an externally active walk-up(e) enters a vertex $v_{b,e}$ that has one pathInfo integer clear and the other set to $-V$, then the marked border leading to the root r_b must be selected and the $-V$ settings along it changed to $+V$. Then, walk-up(e) ascends to the next bicomp traversed by $S(e)$.

Rule 12. If an externally active walk-up(e) enters a vertex $v_{b,e}$ whose pathInfo integers are both clear, then walk-up(e) traverses $\mathcal{A}$ and $\mathcal{B}$ in parallel to find r_b by the shortest border that is not blocked by an externally active vertex. (a) If both borders to r_b are blocked by externally active vertices, then the graph is non-planar. Otherwise, let $\mathcal{A}$ denote the shorter unblocked border to r_b. (b) If the pathInfo in r_b is set to $-V$ for the opposing border path $\mathcal{B}$ and if $\mathcal{B}$ is not blocked, then walk-up(e) must select $\mathcal{B}$, set its pathInfo to $+V$, and ascend to the next bicomp traversed by $S(e)$. (c) If the pathInfo in r_b is set to $-V$ for the opposing border path $\mathcal{B}$ but $\mathcal{B}$ is blocked, then walk-up(e) must select $\mathcal{A}$ and set its pathInfo to $+V$. If r_b is externally active, then the graph is non-planar. Otherwise, walk-up(e) ascends to the next bicomp traversed by $S(e)$. (d) If the pathInfo in r_b is set to $+V$ for the opposing border path $\mathcal{B}$, then walk-up(e) must select $\mathcal{A}$ and set its pathInfo to $+V$. If r_b is externally active, then the graph is non-planar. Otherwise, walk-up(e) ascends to the next bicomp traversed by $S(e)$. (e) If the pathInfo in r_b is clear for the opposing border $\mathcal{B}$, then walk-up(e) must select $\mathcal{A}$, set its pathInfo to $+V$, and ascend to the next bicomp traversed by $S(e)$.

Therefore, the following Theorem can be stated.

Theorem 1 *Let G be a graph with n vertices, and let v_i be the vertex with $DFS(v_i) = i$, for $i = 1, \ldots, n$. Graph G is planar iff the walk-up phase for vertex v_i, with $i = n, \ldots, 1$, selects a path from w_e to v_i for each edge $e = (w_e, v_i)$ in $B_{in}(v_i)$.*

5 Experimental Analysis

In this section we compare our implementation (in the following called **BM-GDT**) with a selection of other planarity testing and embedding algorithm implementations, providing an assessment of the state of the art in this field.

All the tests have been performed on a 2.4GHz Pentium IV personal computer with 1GB of RAM.

We generated the test suites of graphs with Leda [14] and Pigale [7] generators, which in addition to be well known and widely used, are both fast and rigorously built. For each test suite we generated graphs ranging from 10,000 to 100,000 vertices, increasing each time by 5,000 vertices, 10 graphs for each type, with the exception of the test suite labeled `Planar-Conn-P`, for which we generated graphs ranging from 20,000 to 200,000 edges, increasing each time by 10,000 edges, 10 graphs for each type.

`Planar-L`: planar graphs generated by LEDA. The graph are not necessarily connected. The number of edges is twice the number of vertices. `Max-Planar-L`: maximal planar graphs generated by LEDA, hence connected and with $3n - 6$ edges, where n is the number of vertices. `Non-Planar-L`: the same graphs as `Max-Planar-L` in which one edge was added between two non-adjacent vertices. `Random-L`: random graphs generated by LEDA. The number of edges is chosen to be two times the number of vertices. The graph are not necessarily connected and may be planar. `Planar-Conn-P`: planar connected graphs generated by Pigale. After the generation, the graphs were processed in order to eliminate multiple edges and self-loops. `Random-P`: random graphs generated by Pigale. The number of edges is chosen to be two times the number of vertices. The graph are not necessarily connected but may be planar.

We tested the performance of the implementation of **BM-GDT** against: **LEC-L** Lempel, Even, and Cederbaum algorithm [12], LEDA implementation; **LEC-GTL** Lempel, Even, and Cederbaum algorithm [12], GTL implementation [9]; **HT-L** Hopcroft and Tarjan algorithm [10], LEDA implementation. **FR-P** de Fraysseix and Rosenstiehl algorithm (unpublished, see [6] for basic principles), Pigale implementation; **BM2** Boyer-Myrvold (new algorithm, unpublished see [3]), implementation by the first author. All the implementations are in C++ but for **BM2**, which is implemented in C. The implementation of **BM-GDT** uses the data structures for representing graphs of the GDToolkit library [8], based, in turn, on Leda. Each algorithm implementation extracts the graph from a file and constructs its own data structure. The timer starts after the file has been completely read and stops when the test fails (non-planar) or when the data structure with the embedding has been constructed (planar).

All the implementations gave the same results as for the planarity or non planarity of the graphs. The experiments (see Fig. 3) reveal that most of the algorithms have amazingly short computation times. Even the slowest implementations tested for planarity and, in case, constructed an embedding of the graphs in less than 100 μsec. per vertex

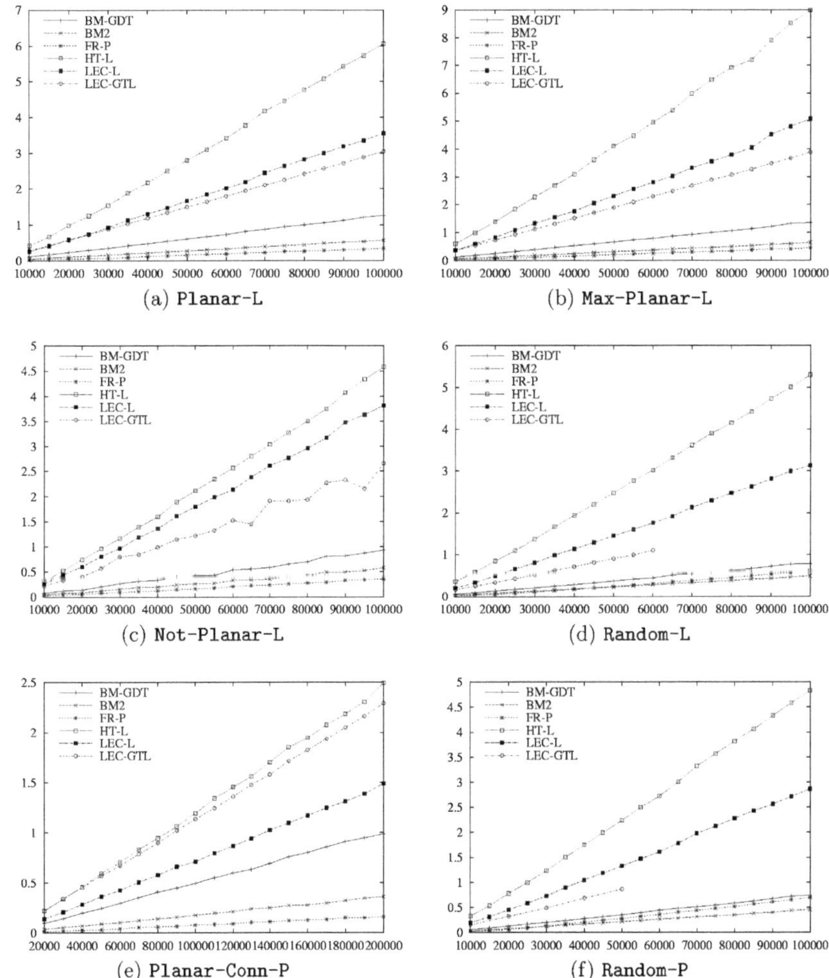

(a) Planar-L (b) Max-Planar-L

(c) Not-Planar-L (d) Random-L

(e) Planar-Conn-P (f) Random-P

Fig. 3. The running times (seconds) of the tested algorithms against the test suites generated by Leda and Pigale.

FR-P, **BM-GDT** and **BM2** had similar performance profiles. Much slower are **LEC-L**, **LEC-GTL**, and **HT-L**. Also, we were unable to produce an output for algorithm **LEC-GTL** when very big non-planar graphs were involved, because of premature terminations of the program. In addition to connected planar graphs (Fig 3.e), we also tested the algorithms against planar biconnected and triconnected graphs generated by Pigale with analogous parameters and ranges, obtaining similar results.

We have to remark that some of the implementations rely on data structures for graphs representations that are "general purpose" and hence are not specifically tailored for quick planarity testing, especially when very large data sets are involved.

References

1. K. Booth and G. Lueker. Testing for the consecutive ones property interval graphs and graph planarity using PQ-tree algorithms. *J. Comput. Syst. Sci.*, 13, 1976.
2. J. Boyer and W. Myrvold. Stop minding your P's and Q's: A simplified O(n) planar embedding algorithm. In *10th Annual ACM-SIAM Symposium on Discrete Algorithms*, pages 140–146, 1999.
3. J. Boyer and W. Myrvold. Stop minding your P's and Q's: Simplified planarity by edge addition. 2003. Submitted. Preprint at `http://www.pacificcoast.net/~lightning/planarity.ps`.
4. J. M. Boyer, P. Cortese, M. Patrignani, and G. Di Battista. Stop minding your P's and Q's: Implementing a fast and simple DFS-based planarity testing and embedding algorithm. Tech. Report RT-DIA-83-2003, Dept. of Computer Sci., Univ. di Roma Tre, 2003.
5. N. Chiba, T. Nishizeki, S. Abe, and T. Ozawa. A linear algorithm for embedding planar graphs using PQ-trees. *J. Comput. Syst. Sci.*, 30(1):54–76, 1985.
6. H. de Fraisseix and P. Rosenstiehl. A characterization of planar graphs by Trémaux orders. *Combinatorica*, 5(2):127–135, 1985.
7. H. de Fraysseix and P. Ossona de Mendez. P.I.G.A.L.E - Public Implementation of a Graph Algorithm Library and Editor. SourceForge project page `http://sourceforge.net/projects/pigale`.
8. GDToolkit. An object-oriented library for handling and drawing graphs. Third University of Rome, `http://www.dia.uniroma3.it/~gdt`.
9. GTL. Graph template library. University of Passau - FMI - Theor. Comp. Science `http://infosun.fmi.uni-passau.de/GTL/`.
10. J. Hopcroft and R. E. Tarjan. Efficient planarity testing. *J. ACM*, 21(4), 1974.
11. W.-L. Hsu. An efficient implementation fo the PC-Tree algorithm of Shih and Hsu's planarity test. Tech. Report, Inst. of Inf. Science, Academia Sinica, 2003.
12. A. Lempel, S. Even, and I. Cederbaum. An algorithm for planarity testing of graphs. In *Theory of Graphs: Internat. Symposium (Rome 1966)*, pages 215–232, New York, 1967. Gordon and Breach.
13. K. Mehlhorn and P. Mutzel. On the embedding phase of the Hopcroft and Tarjan planarity testing algorithm. *Algorithmica*, 16:233–242, 1996.
14. K. Mehlhorn and S. Näher. *LEDA: A Platform for Combinatorial and Geometric Computing*. Cambridge University Press, New York, 1998.
15. W.-K. Shih and W.-L. Hsu. A simple test for planar graphs. In *Int. Workshop on Discrete Math. and Algorithms*, pages 110–122, 1993.
16. W.-K. Shih and W.-L. Hsu. A new planarity test. *Theor. Comp. Sci.*, 223, 1999.

Bounds and Methods for k-Planar Crossing Numbers

Farhad Shahrokhi[1⋆], Ondrej Sýkora[2⋆⋆], Laszlo A. Székely[3⋆⋆⋆] and Imrich Vrt'o[4,†]

[1] Department of Computer Science, University of North Texas
P.O. Box 13886, Denton, TX, 76203-3886, USA
[2] Department of Computer Science, Loughborough University
Loughborough, Leicestershire LE11 3TU, The United Kingdom
[3] Department of Mathematics, University of South Carolina
Columbia, SC 29208, USA
[4] Department of Informatics, Institute of Mathematics
Slovak Academy of Sciences
Dúbravská 9, 841 04 Bratislava, Slovak Republic

Abstract. The k-planar crossing number of a graph is the minimum number of crossings of its edges over all possible drawings of the graph in k planes. We propose algorithms and methods for k-planar drawings of general graphs together with lower bound techniques. We give exact results for the k-planar crossing number of $K_{2k+1,q}$, for $k \geq 2$. We prove tight bounds for complete graphs.

1 Introduction

Let $\mathrm{cr}(G)$ denote the standard crossing number of a graph G, i.e. the minimum number of crossings of its edges over all possible drawings of G in the plane. For $k \geq 2$, define the k-planar crossing number as

$$\mathrm{cr}_k(G) = \min\{\mathrm{cr}(G_1) + \mathrm{cr}(G_2) + ... + \mathrm{cr}(G_k)\},$$

where the minimum is taken over all edge disjoint subgraphs $G_i = (V, E_i)$, $i = 1, 2, ..., k$, so that $E = E_1 \cup E_2 \cup ... \cup E_k$.

Motivated by printed circuit boards, Owens [9] introduced the *biplanar crossing number* of a graph G, i.e. the case $k = 2$. He described a biplanar drawing of the complete graph K_n with $\mathrm{cr}_2(K_n) \leq 7n^4/1536 + O(n^3)$. A survey on biplanar crossing numbers is in [5]. Determining $\mathrm{cr}_k(G)$ has application to the design of multilayer VLSI circuits [1].

⋆ This research was supported by the NSF grant CCR9988525.

⋆⋆ Work supported by the EPSRC grant GR/R37395/01.

⋆⋆⋆ This author was visiting the National Center for Biotechnology Information, NLM, NIH, with the support of the Oak Ridge Institute for Science and Education. This research was partially supported by the NSF contract 007 2187.

† Supported partially by the VEGA grant No. 02/3164/23.

G. Liotta (Ed.): GD 2003, LNCS 2912, pp. 37–46, 2004.

Much of this paper extends ideas of the papers [5] and [12] investigating the biplanar crossing number to the k-planar crossing number. Section 2 gives general bounds for the k-planar crossing number and exposes an important extremal problem: how does $\mathrm{cr}_k(G)$ decrease when k increases?

Section 3 yields unexpected exact results for the k-planar crossing number of some complete bipartite graphs. Complete bipartite graphs $K_{p,q}$ are also the best studied graphs with respect to planar crossing numbers. Exact results are known only for $p \leq 6$ and arbitrary q, [6]. Crossing numbers of bipartite graphs drawn on surfaces of higher geni were determined only for $p \leq 3$, and arbitrary q, [10]. Thus our results belong to the same rare class of exact results on crossing numbers (for bipartite graphs), and are direct extensions of the results of [5] for $\mathrm{cr}_2(K_{5,n})$ and $\mathrm{cr}_2(K_{6,n})$. We spell out the results in more details. Recall that the *thickness* $\theta(G)$ of G is the minimum number of planar graphs whose union is G. By definition, $\mathrm{cr}_k(G) = 0$ if and only if $\theta(G) \leq k$. Beineke et al. [4] proved that the thickness of $K_{p,q}$ is given by

$$\theta(K_{p,q}) = \left\lceil \frac{pq}{2(p+q-2)} \right\rceil, \tag{1}$$

except, possibly, when $p \leq q$ are both odd and there exists an integer k such that $\frac{1}{4}(p+5) \leq k \leq \frac{1}{2}(p-3)$ and $q = \lfloor 2k(p-2)/(p-2k) \rfloor$. According to (1) $\mathrm{cr}_k(K_{2k,q}) = 0$, for $k \geq 2$ and any q, so the first interesting bipartite graph is $K_{2k+1,q}$. We prove that for $k \geq 2, q \geq 1$

$$\mathrm{cr}_k(K_{2k+1,q}) = \left\lfloor \frac{q}{2k(2k-1)} \right\rfloor \left(q - k(2k-1) \left\lfloor \frac{q}{2k(2k-1)} \right\rfloor - k(2k-1) \right)$$

and for $k \geq 2$, and $1 \leq q \leq 4k^2$

$$\mathrm{cr}_k(K_{2k+2,q}) = 2 \left\lfloor \frac{q}{2k^2} \right\rfloor \left(q - k^2 \left\lfloor \frac{q}{2k^2} \right\rfloor - k^2 \right).$$

Section 4 improves on the general bounds for the k-planar crossing numbers of complete and complete bipartite graphs. The improvement means constant multiplicative factors.

2 General Bounds

Little is known about lower bounds for the k-planar crossing number in general. Some of the lower bounds for crossing numbers, *mutatis mutandis* apply to k-planar crossing numbers. For example, if $G = (V, E)$, $|V| = n, |E| = m$, then the lower bound resulting from Euler's formula, $\mathrm{cr}(G) \geq m - 3n + 6$ for $n \geq 3$, generalizes to

$$\mathrm{cr}_k(G) \geq m - k(3n-6).$$

There is a strengthening of the lower bound resulting from Euler's formula for graphs G with girth g, $\mathrm{cr}(G) \geq m - g(n-2)/(g-2)$ for $n \geq g$; and we get

$$\mathrm{cr}_k(G) \geq m - \frac{gk}{g-2}(n-2). \tag{2}$$

We state a k-planar version of Leighton's Lemma [7] for crossing numbers (note that we do not go for the best constants here, since the best constant is always getting improved even for the ordinary crossing number).

Lemma 1. *For a simple graph G with n vertices and m edges, we have $m \leq 6kn$, or*

$$\mathrm{cr}_k(G) \geq \frac{1}{64} \cdot \frac{m^3}{n^2k^2}. \tag{3}$$

Proof. Recall Leighton's Lemma for the ordinary crossing number: $m \leq 4n$ or $\mathrm{cr}(G) \geq m^3/64n^2$. Consider an optimal k-planar drawing of G, such that G_i is the subgraph drawn on the ith plane. Assume that the first x graphs have at most $4n$ edges, while the last $k-x$ graphs have more. We have

$$\mathrm{cr}_k(G) \geq \sum_{i=x+1}^{k} \mathrm{cr}(G_i) \geq \sum_{i=x+1}^{k} \frac{m_i^3}{64n^2} \geq$$

$$\geq \frac{k-x}{64n^2} \cdot \left(\frac{\sum_{i=x+1}^{k} m_i}{k-x} \right)^3 \geq \frac{1}{64n^2} \cdot \frac{(m-4nx)^3}{(k-x)^2} \geq \frac{1}{64} \cdot \frac{m^3}{n^2k^2},$$

where the last inequality holds for $m \geq 6kn$ according to the sign of the derivative. □

Recall that $a(G)$, or arboricity of G, is the minimum number of acyclic subgraphs whose union covers E. By a well known theorem of Nash-Williams [8]

$$a(G) = \max_{H \subseteq G} \left\lceil \frac{m(H)}{n(H)-1)} \right\rceil$$

where the maximum is taken over all subgraphs H of G, with $m(H)$ edges and $n(H)$ vertices. It is easily seen that $a(G) \geq \theta(G)$, moreover, $\theta(G) \geq \lceil a(G)/3 \rceil$, since $m(H) \leq 3n - 6$ for any planar graph.

Let $P = \{V_1, V_2, ...V_t\}$ be a partition of V. We denote by E_{ij} the set of edges with one end point in V_i and the other in V_j, hence E_{ii} denotes the set of all edges with both end points in V_i, for $1 \leq i \leq t$. Let H denote the t vertex graph that is obtained by contracting all vertices in V_i into one single vertex and removing the multiple edges. We call H the *mate* of G with respect to P, or simply the mate of G. Let $T_1, T_2, ...T_{a(H)}$, be a decomposition of the edge set of H into acyclic subgraphs of H. Let $d_i(x)$ denote the degree of $x \in V(H)$ in T_i, $i = 1, 2, ...k$.

Theorem 1. *Let $G = (V, E)$, and let k be a given integer. Let $\{V_1, V_2, ...V_t\}$ be a partition of V and let $H = (V(H), E(H))$ denote the mate of G. If $k \geq a(H)$, then we can construct in polynomial time a k-planar drawing of G with at most*

$$tp^2 + 2pq|E(H)| + \sum_{i=1}^{k} p^2 \sum_{x \in V(H)} d_i^2(x)$$

crossings, where $p = \max\{|E_{ii}|\}$ and $q = \max\{|E_{ij}|\}$, $i, j = 1, 2, ...t$.

Proof Sketch. Consider a drawing of each $T_i, i = 1, 2, ..., k$ in plane i, with no crossings, so that the vertices are placed in the corners of a convex polygon, and each edge is drawn using one straight line segment. Now, replace each vertex $j \in V(H)$ with the set V_j. In particular, place the vertices of V_j in a very small neighborhood around j. Next, draw the edges in E with straight line segments using the drawings of T_i's, to produce a k-planar drawing of G. There will be 3 kinds of crossings:
(a) between edges of E_{ii},
(b) between edges of E_{ii}, and edges of $E_{ij}, i \neq j$, and finally
(c) between edges E_{ij}, where $i, j = 1, 2, ...k$ and $i \neq j$.
The terms in the theorem correspond to these 3 cases. Note that the estimate for (b) is $pq \sum_{i=1}^{k} \sum_{x \in V(H)} d_i(x) = 2pq|E(H)|$. □

Theorem 1 can be used effectively, if the degrees appearing in the last term are small. In fact, in certain cases one can decompose G into a number of (cyclic) outer planar graphs of small maximum degree, and still use the method of Theorem 1 to obtain upper bounds for $\mathrm{cr}_k(G)$. In this paper, we have obtained exact values of $\mathrm{cr}_k(G)$ for certain graphs in this way. Nonetheless, the acyclic decompositions into forests of small maximum degree has been also studied. Let $a_d(G)$ denote the degree bounded arboricity, that is the minimum number of forests that the edges of G can be decomposed to so that the maximum degree of each forest is bounded by d. Truszczýnski [14] conjectured that for every multigraph G and $d \geq 2$,

$$a_d(G) = \begin{cases} \Delta(G)/d \text{ or } \ 1 + \Delta(G)/d \text{ if } a(G) = \Delta(G)/d, \\ \max\left(a(G), \lceil \Delta(G)/d \rceil\right) \text{otherwise.} \end{cases} \tag{4}$$

Truszczýnski actually proved his conjecture for complete and complete bipartite graphs, and also for the case $d \geq \Delta(G) + 1 - a(G)$. Combining Theorem 1 with (4), we immediately obtain

Corollary 1. *For* $n \geq 1$ $\mathrm{cr}_k(K_n) = O(n^4/k^2)$.

However, Corollary 1 also follows from the next theorem:

Theorem 2. *For any graph G on n vertices and m edges,*

$$\mathrm{cr}_k(G) \leq \frac{1}{12k^2}\left(1 - \frac{1}{4k}\right) m^2 + O\left(\frac{m^2}{kn}\right).$$

The corresponding drawing can be found in polynomial time. For any graph G,

$$\mathrm{cr}_k(G) \leq \frac{2\mathrm{cr}(G)}{k^{\log_2 \frac{8}{3}}} = \frac{2\mathrm{cr}(G)}{k^{1.4708...}}.$$

Proof. The first upper bound follows from our paper [11] (Corollary 3.2) and a simple observation that a drawing of a graph G in $2k$ pages gives a drawing of the graph G in k planes. The second upper bound follows by iteration from the inequality $\mathrm{cr}_2(G) \leq \frac{3}{8}\mathrm{cr}(G)$, proved in [5]. □

One challenging question is how cr_k changes from $\mathrm{cr}(G)$ to 0, as k increases from 1 to the thickness of G, $\theta(G)$.

3 Exact Results

Theorem 3. *For $k \geq 2, q \geq 1$*

$$\mathrm{cr}_k(K_{2k+1,q}) = \left\lfloor \frac{q}{2k(2k-1)} \right\rfloor \left(q - k(2k-1) \left(\left\lfloor \frac{q}{2k(2k-1)} \right\rfloor + 1 \right) \right). \quad (5)$$

Proof. *Upper bound.* Beineke [2] proved that the thickness of $K_{2k+1,2k(2k-1)}$ is k by describing a drawing of $K_{2k+1,2k(2k-1)}$ in k planes without crossings. We extend this drawing to a drawing of $K_{2k+1,q}$ in k planes with minimum number of crossings. Let $u_1, u_2, ..., u_{2k+1}$ be the vertices of the first partition. Let $v_1, v_2, ..., v_{2k(2k-1)}$ be the vertices of the second partition. Beineke's drawing possesses the following properties.

1. On every plane, all v_i's lie on the vertices of the regular $2k(2k-1)$-gon.
2. All u_j's lie inside or outside of the polygon.
3. The edges do not cross.
4. For every v_i, its degree on exactly one plane is 3 and 2 on the remaining $(k-1)$ planes. Moreover, on every plane, the vertex v_i has a neighbor inside and a neighbor outside the polygon.

Fig. 1 shows the case $k = 3$, i.e. a drawing of $K_{7,30}$ in 3 planes without crossings. The graphs on the same row are drawn on the same plane. The left (right) graph corresponds to the inside (outside) part of the drawing on a plane.

Now consider $K_{2k+1,q}$ and assume that $q = 2k(2k-1)a + b$, where a, b are integers and $0 \leq b < 2k(2k-1)$. Partition the q-vertices into $2k(2k-1)$ almost equal sets $S_1, S_2, ..., S_{2k(2k-1)}$, where $2k(2k-1) - b$ sets have a vertices, and b sets have $a+1$ elements. On every plane, replace each vertex v_j by the set S_j such that its vertices lie on a very short arc and the arcs do not interfere. Join every vertex S_i to all vertices of S_j on a plane iff u_i was adjacent to v_j on that plane in the Beineke's drawing. Clearly the total number of crossings is

$$\sum_{j=1}^{2k(2k-1)} \binom{|S_j|}{2}.$$

The above sum turns to

$$b\binom{a+1}{2} + (2k(2k-1) - b)\binom{a}{2} = a(b + k(2k-1)(a-1)),$$

which gives the claimed value by substituting $a = \lfloor q/(2k(2k-1)) \rfloor$ and $b = q - 2k(2k-1)a$.

Lower bound. We will proceed by induction on q. The claim is obviously true for $q \leq 2k(2k-1)$. The claim is also true for $2k(2k-1) \leq q \leq 4k(2k-1)$ as the RHS of (5) equals $q - 2k(2k-1)$, which is a lower bound given by (2). Hence assume that the claim is true for some $q \geq 4k(2k-1)$. Using the counting argument with $H = K_{2k+1,q}$, $G = K_{2k+1,q+1}$, i.e. counting the number of crossings

Fig. 1. A drawing of $K_{7,30}$ in 3 planes without crossings.

produced by all occurrencies of H in G and dividing it by the multiplicity of each crossing, we have

$$\mathrm{cr}_k(K_{2k+1,q+1}) - \left\lfloor \frac{q+1}{2k(2k-1)} \right\rfloor \left(q+1-k(2k-1)\left(\left\lfloor \frac{q}{2k(2k-1)} \right\rfloor +1 \right) \right)$$

$$\geq \left\lceil \frac{\binom{q+1}{q}}{\binom{q-1}{q-2}} \mathrm{cr}_k(K_{2k+1,q}) \right\rceil - \left\lfloor \frac{q+1}{2k(2k-1)} \right\rfloor \left(q+1-k(2k-1)\left(\left\lfloor \frac{q}{2k(2k-1)} \right\rfloor + 1 \right) \right)$$
$$\geq \left\lceil \frac{q+1}{q-1} \left\lfloor \frac{q}{2k(2k-1)} \right\rfloor \left(q-k(2k-1)\left(\left\lfloor \frac{q}{2k(2k-1)} \right\rfloor + 1 \right) \right) - \left\lfloor \frac{q+1}{2k(2k-1)} \right\rfloor \left(q+1-k(2k-1)\left(\left\lfloor \frac{q}{2k(2k-1)} \right\rfloor + 1 \right) \right) \right\rceil .$$

To conclude the proof, it is sufficient to show that for $q \geq 4k(2k-1)$ the expression inside the big brackets of the last line is greater than -1. Let $q = 2k(2k-1)a + b$, as above. Distinguish two cases.
If $b < 2k(2k-1) - 1$ then the expression inside the big brackets equals

$$\frac{q+1}{q-1} a(q - k(2k-1)(a+1)) - a(q+1-k(2k-1)(a+1)) = \frac{-a-b}{q-1} > -1.$$

If $b = 2k(2k-1) - 1$ then the expression inside the big brackets equals

$$\frac{q+1}{q-1} a(q - k(2k-1)(a+1)) - (a+1)(q+1-k(2k-1)(a+2)) = 0.$$

□

Theorem 4. *For $k \geq 2$*

$$\mathrm{cr}_k(K_{2k+2,q}) \leq 2 \left\lfloor \frac{q}{2k^2} \right\rfloor \left(q - k^2 \left\lfloor \frac{q}{2k^2} \right\rfloor - k^2 \right). \tag{6}$$

The equality holds for $1 \leq q \leq 4k^2$.

Proof. *Upper bound.* We start with a drawing of $K_{2k+2,2k^2}$ in k planes without crossings and then extend this drawing to a drawing of $K_{2k+2,q}$. Denote the vertices of the first partition class by $u_1, u_2, ..., u_{k+1}$ and $v_1, v_2, ..., v_{k+1}$. Denote the vertices of the second partition class by $a_0, a_2, ..., a_{k^2-1}$ and $b_0, b_1, ..., b_{k^2-1}$.

On the first plane, place the vertices $u_1, u_2, ..., u_{k+1}$ (resp. $v_1, v_2, ..., v_{k+1}$) on the positive (resp. negative) part of the x axis, in this order from the origin. Place the vertices $a_0, a_1, ..., a_{k^2-1}$ (resp. $b_0, b_1, ..., b_{k^2-1}$) on the positive (resp. negative) part of the y axis, in this order from the origin. Join u_i and v_i to $a_{(i-1)(k-1)}, ..., a_{ik-i}$ and $b_{(i-1)(k-1)}, ..., b_{ik-i}$, for all i. On the second plane, the positions of u_i's and v_i's remain unchanged. Shift a_j's (resp. b_j's) cyclically up (down) by k position. Join u_i and v_i to $a_{(i-2)(k-1)}, ..., a_{(i-1)(k-1)}$ and $b_{(i-2)(k-1)}, ..., b_{(i-1)(k-1)}$, where the indices are computed modulo k^2. Continuing in this drawing for all planes we get a drawing of $K_{2k+2,2k^2}$ in k planes without crossings. See Fig. 2 for the case $k = 3$.

Now consider $K_{2k+2,q}$. Partition the q vertices into $2k^2$ almost equal sets, $A_0, A_1, ..., A_{k^2-1}$ and $B_0, B_1, ..., B_{k^2-1}$. Replace every a_j and b_j by A_j and B_j and join u_i and v_i to A_j and B_j on a plane iff u_i and v_i were adjacent to a_j and b_j on that plane. A simple counting shows that the number of crossings is

$$\sum_{j=0}^{k^2-1} 2 \left(\binom{|A_j|}{2} + \binom{|B_j|}{2} \right).$$

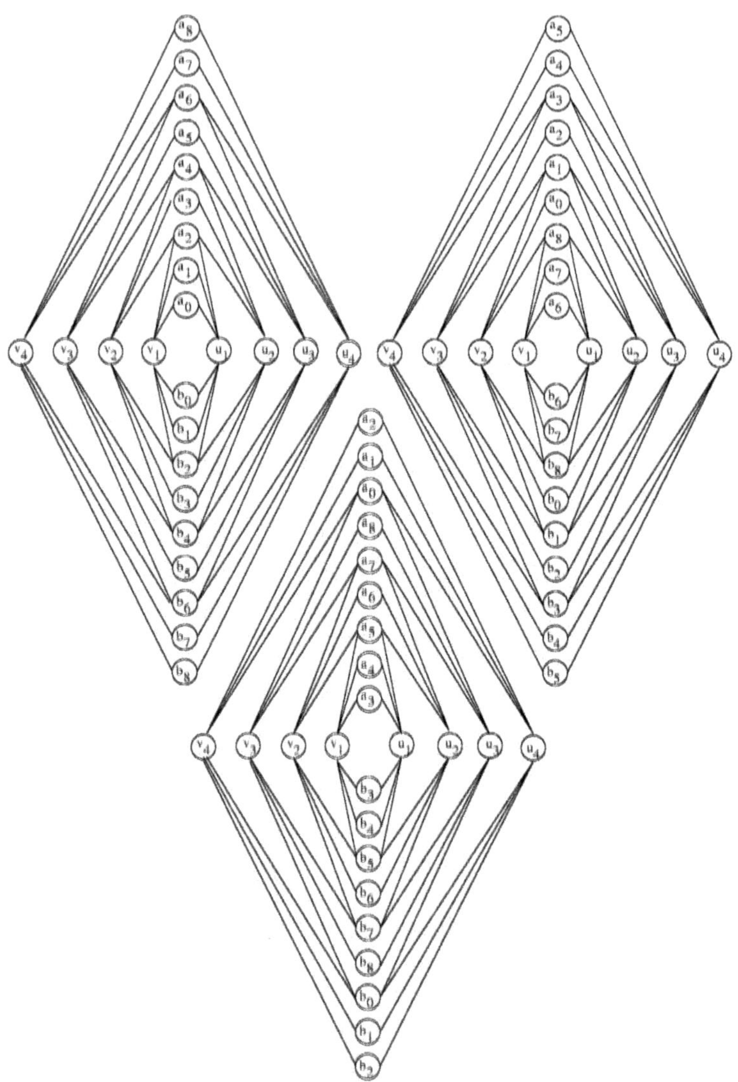

Fig. 2. A drawing of $K_{8,18}$ in 3 planes without crossings.

The rest is similar to the proof of Theorem 3.

Lower bound. As $\theta(K_{2k+2,2k^2}) = k$, from (1), $\mathrm{cr}_k(K_{2k+2,q}) = 0$, for $q \leq 2k^2$. Assume $2k^2 \leq q \leq 4k^2$. In this interval the RHS of (6) equals $2q - 4k^2$, which is the lower bound given by (2). □

4 Improved Bounds on Complete and Complete Bipartite Graphs

4.1 Lower Bounds

For specific graphs we can strengthen the lower bound by the standard counting argument.

Theorem 5. *For $p \geq 6k-1$ and $q \geq \max\{6k-1, 2k^2\}$*

$$\mathrm{cr}_k(K_{p,q}) \geq \frac{1}{3(3k-1)^2}\binom{p}{2}\binom{q}{2}.$$

Proof. The estimation (2) gives

$$\mathrm{cr}_k(K_{6k-1,6k-1}) \geq 12k^2 - 4k + 1.$$

Using the counting argument with $H = K_{6k-1,6k-1}$ and $G = K_{p,q}$ we have

$$\mathrm{cr}_k(K_{p,q}) \geq \frac{\binom{p}{6k-1}\binom{q}{6k-1}}{\binom{p-2}{6k-3}\binom{q-2}{6k-3}}\mathrm{cr}_k(K_{6k-1,6k-1}) > \frac{1}{3(3k-1)^2}\binom{p}{2}\binom{q}{2}.$$

□

Theorem 6. *For $n \geq 2k^2 + 6k - 1$*

$$\mathrm{cr}_k(K_n) \geq \frac{1}{2(3k-1)^2}\binom{n}{4}.$$

Proof. Let $n = p + q$. Combining the counting argument with $H = K_{p,q}$ and $G = K_n$ with the lower bound from Theorem 8 we get the claim. □

4.2 Upper Bounds

For special values of k we can improve on the upper bounds from Corollary 1 and Theorem 2.

Theorem 7. *Let $k-1$ be a power of a prime. For $n \geq (k-1)^2$*

$$\mathrm{cr}_k(K_n) \leq \frac{1}{64}\frac{k}{(k-1)^3}(n+k^2)^4.$$

Proof. To appear in the full verion.

For the complete bipartite graphs and arbitrary k we can extend the construction from the Section 3.

Theorem 8. *For $p \geq 2k+2$ and $q \geq 2k^2$*

$$\mathrm{cr}_k(K_{p,q}) \leq \frac{k^2+k+2}{16k^2(k+1)^2}(p+2k+1)^2(q+2k^2-1)^2.$$

Proof. To appear in the full version.

References

1. Aggarwal, A., Klawe, M., Shor, P., Multi-layer grid embeddings for VLSI, *Algorithmica* **6**, (1991), 129–151.
2. Beineke, L.W., Complete bipartite graphs: Decomposition into planar subgraphs, Chapter 7, in: *A Seminar on Graph Theory,* (F. Harary, ed.), Selected Topics in Mathematics, Holt, Rinehart and Winston, New York, 1967, 43–53.
3. Beineke, L.W., Biplanar graphs: A survey, *Computers and Mathematics with Applications* **34** (1997), 1–8.
4. Beineke, L.W., Harary, F., Moon, J.W., On the thickness of the complete bipartite graphs, *Proc. of the Cambridge Philosophical Society* **60** (1964), 1–5.
5. Czabarka, É., Sýkora, O., Székely, L.A., Vrťo, I., Biplanar crossing numbers: A survey of results and problems, in: *Finite and Infinite Combinatorics,* (T. Fleiner, G. O. H. Katona, eds.), Bolyai Society Mathematical Studies, Akadémia Kiadó, Budapest, to appear.
6. Kleitman, D.J., The crossing number of $K_{5,n}$, *J. Combinatorial Theory* **9** (1970), 315–323.
7. Leighton, T. F., Complexity Issues in VLSI, MIT Press, Cambridge 1983.
8. Nash-Williams, J.A., Edge disjoint spanning trees of finite graphs, *J. London Math. Soc.,* **36** (1961), 445–450.
9. Owens, A., On the biplanar crossing number, *IEEE Transactions on Circuit Theory* **18** (1971), 277–280.
10. Richter, R.B., Širáň, J., The crossing number of $K_{3,n}$ in a surface, *J. Graph Theory* **21** (1996), 51-54.
11. Shahrokhi, F., Sýkora, O., Székely, L.A., Vrťo, I., The book crossing number of graphs, *J. Graph Theory* **21** (1996), 413–424.
12. Sýkora, O., Székely, L.A., Vrťo, I., Crossing numbers and biplanar crossing numbers: using the probabilistic method, submitted.
13. Shahrokhi, F., Sýkora, O., Székely, L.A., Vrťo, I., Shahrokhi, F., Sýkora, O., Székely, L.A., Vrťo, I., Bounds for convex crossing numbers, in: Proc. *9th Intl. Computing and Combinatorics Conference,* Lecture Notes in Computer Science 2697, Springer Verlag, Berlin 2003, 487-495.
14. Truszczyński, M., Decomposition of graphs into forests with bounded maximum degree, *Discrete Mathematics* **98** (1991), 207–222.
15. White A.T., Beineke, L.W., Topological graph theory, in: *Selected Topics in Graph Theory,* (L.W. Beineke, R.J. Wilson, eds.), Academic Press, New York 1978, 15–50.

How Many Ways Can One Draw a Graph?*

János Pach and Géza Tóth

Rényi Institute, Hungarian Academy of Sciences, Budapest, Hungary

Dedicated to Miklós Simonovits on his sixtieth birthday

Abstract. Using results from extremal graph theory, we determine the asymptotic number of *string graphs* with n vertices, i.e., graphs that can be obtained as the intersection graph of a system of continuous arcs in the plane. The number becomes much smaller, for any fixed d, if we restrict our attention to systems of arcs, any two of which cross at most d times. As an application, we estimate the number of different drawings of the complete graph K_n with n vertices under various side conditions.

1 Introduction

Given a simple graph G, is it possible to represent its vertices by simply connected regions in the plane so that two regions overlap if and only if the corresponding two vertices are adjacent? In other words, is G isomorphic to the *intersection graph* of a set of simply connected regions in the plane? This deceptively simple extension of propositional logic and its generalizations are often referred to in the literature as *topological inference problems* [CGP98a], [CGP98b],[CHK99]. They have proved to be relevant in the area of geographic information systems [E93], [EF91] and in graph drawing [DETT99]. In spite of many efforts [K91a], [K98] (and false claims [SP92], [ES93]), until very recently no algorithm was known for their solution. Two years ago, we showed [PT02] that the problem is *decidable*. Shortly after a more elegant proof was found by Schaefer and Stefankovič [SS01a], who went on proving that the question is in NP [SS01b].

Since each element of a finite system of regions in the plane can be replaced by a simple continuous arc ("string") lying in its interior so that the intersection pattern of these arcs is the same as that of the original regions, it is enough to restrict our attention to *string graphs,* i.e., to intersection graphs of planar curves. As far as we know, these graphs were first studied in 1959 by S. Benzer [B59], who investigated the topology of genetic structures. Somewhat later they were also considered by F. W. Sinden [S66] in Bell Labs, who was interested in electrical networks realizable by printed circuits. Sinden collaborated with R. L. Graham, who popularized the notion among combinatorists at a conference in Keszthely (Hungary), in 1976 [G78]. Soon after G. Ehrlich, S. Even, and R. E.

* János Pach has been supported by NSF grant CR-00-98246, PSC-CUNY Research Award 62450-0031 and OTKA-T-032452. Géza Tóth has been supported by OTKA-T-032452 and OTKA-T-038397.

G. Liotta (Ed.): GD 2003, LNCS 2912, pp. 47–58, 2004.

Tarjan [EET76] studied string graphs (see also [K83] and [EPL72] for a special case). The aim of this paper is to estimate the *number* of different string graphs on n vertices.

To formulate our main result precisely, we have to agree on the terminology. Let G be a simple graph with vertex set $V(G)$ and edge set $E(G)$. A *string representation* of G is an assignment of simple continuous arcs to the elements of $V(G)$ such that two arcs cross each other if and only if the corresponding vertices of G are adjacent. Graph G is a *string graph* if it has a string representation. We assume that any two arcs share only finitely many points and that at each common point the arcs *properly cross*, i.e., one arc passes from one side of the other arc to the other side. An intersection point of two arcs is called a *crossing*.

For any $d > 0$, graph G is a *string graph of rank d* if it has a string representation with the property that any two strings have at most d crossings.

A class $\mathcal{P}$ of labeled graphs, which is closed under isomorphism, is said to be a *property*. A property $\mathcal{P}$ is called *hereditary* if every induced subgraph of every member of $\mathcal{P}$ belongs to $\mathcal{P}$. Let $\mathcal{P}^n$ denote the set of all (labeled) graphs on the vertex set $\{1, 2, \ldots, n\}$ that belong to $\mathcal{P}$. In the combinatorics literature, the function $|\mathcal{P}^n| \leq 2^{\binom{n}{2}}$ is often called the *speed* of property $\mathcal{P}$, and there are several well known estimates on its growth rate as n increases.

Let $\mathcal{S}$ and $\mathcal{S}_d$ denote the classes of all string graphs and all string graphs of rank d, respectively. Clearly, these are hereditary properties and we have $\mathcal{S}_1 \subseteq \mathcal{S}_2 \subseteq \cdots \subseteq \mathcal{S}$. Our first goal is to estimate their speeds.

Theorem 1. *For the number $|\mathcal{S}^n|$ of all string graphs on n labeled vertices, we have*

$$2^{\frac{3}{4}\binom{n}{2}} \leq |\mathcal{S}^n| \leq 2^{(\frac{3}{4}+o(1))\binom{n}{2}}.$$

Theorem 2. *For any $d > 0$, the number $|\mathcal{S}_d^n|$ of all string graphs of rank d satisfies $|\mathcal{S}_d^n| \leq 2^{o(n^2)}$.*

We do not have any better lower bound on $|\mathcal{S}_d^n|$ than $2^{\Omega(n \log n)}$, which follows from the fact that the vertex set has this many different permutations.

A *drawing* of a graph is a mapping f which assigns to each vertex of G a distinct point in the plane and to each edge uv a continuous arc between $f(u)$ and $f(v)$, not passing through the image of any other vertex. For simplicity, the point assigned to a vertex is also called a *vertex* and an arc assigned to an edge is also called an *edge* of the drawing, and, if this leads to no confusion, it is also denoted by uv. We assume that (a) two edges have only finitely many points in common, and (b) if two edges share an interior point p, then they properly cross at p. Two drawings of G are said to be *essentially equivalent* the set of crossing pairs of edges is the same in the two drawings. Otherwise, they are *essentially different.*

Let $\Delta(n)$ and $\bar{\Delta}(n)$ denote the number of essentially different drawings and essentially different straight-line drawings, resp., of the complete graph K_n with n vertices. For any $d > 0$, let $\Delta_d(n)$ denote the number of drawings with the property that any two edges have at most d points in common. Clearly, we have

$$\bar{\Delta}(n) \leq \Delta_1(n) \leq \Delta_2(n) \leq \Delta_3(n) \leq \ldots \leq \Delta(n),$$

for every n.

In Sections 2 and 3, we review the extremal graph theoretic tools used in this paper and establish Theorem 1, respectively. In Section 4 we prove Theorem 2 in the special case $d = 1$. The proof in the general case is based on the same ideas, but it is technically more complicated, and it is omitted in this extended abstract. In Section 5, we deduce the following estimates.

Theorem 3. *For the number of essentially different drawings of K_n under various restrictions, we have*
(i) $2^{\Omega(n \log n)} \leq \bar{\Delta}(n) \leq 2^{O(n \log n)}$;
(ii) $2^{\Omega(n^2)} \leq \Delta_1(n) \leq 2^{O(n^2 \log n)}$;
(iii) $2^{\Omega(n^2 \log n)} \leq \Delta_d(n) \leq 2^{o(n^4)}$, *for any fixed* $d \geq 2$;
(iv) $2^{\Omega(n^4)} \leq \Delta(n) \leq 2^{O(n^4)}$.

2 Tools from Extremal Graph Theory

One of the central questions in extremal graph theory [B78] is the following. Given a graph H, what is the maximum number of edges that a graph of n vertices can have if it does not contain H as a (not necessarily induced) subgraph? This quantity is usually denoted by $\text{ex}(n, H)$.

Obviously, the property that a graph is H free, is hereditary. Let $\text{Forb}(n, H)$ denote the *speed* of this property, i.e., the number of graphs on n labeled vertices that do not contain H as a subgraph. It turns out that the growth rate of these functions crucially depends on the *chromatic number* $\chi(H)$ of H.

Theorem 2.1. (Erdős-Stone [ES46], Erdős-Simonovits [ES66]) *For any graph H, we have*

$$\text{ex}(n, H) = \left(1 - \frac{1}{\chi(H) - 1}\right) \frac{n^2}{2} + o(n^2).$$

Theorem 2.2. (Erdős-Frankl-Rödl [EFR86]) *For any graph H, we have*

$$\text{Forb}(n, H) = 2^{(1+o(1))\text{ex}(n,H)}.$$

If we want to establish analogous results for graphs containing no *induced* subgraph isomorphic to H, then the first difficulty we have to face is the following: unless H is a complete graph, the maximum number of edges that a graph of n vertices can have without containing an induced copy of H is $\binom{n}{2}$. Thus, Theorem 2.1 does not have a direct analogue. Nevertheless, set

$$\text{ex}^*(n, H) := \left(1 - \frac{1}{\tau(H) - 1}\right) \frac{n^2}{2} + o(n^2),$$

where the relevant quantity, $\tau(H)$, taking the place of the chromatic number is defined as follows.

We say that H is (r, s)*-colorable* for some $0 \le s \le r$ if there is an r-coloring of the vertex set $V(H)$, in which the first s color classes are *cliques* (i.e., induce complete subgraphs) and the remaining $r - s$ color classes are *independent sets* (i.e., induce empty subgraphs). Let $\mathcal{C}(r, s)$ denote the class of all (r, s)-colorable graphs, i.e.,

$$\mathcal{C}(r,s) = \{H \ : \ H \text{ is } (r,s)\text{–colorable}\}.$$

Let $\tau(H)$ be the minimum integer r such that H is (r, s)-colorable for all $0 \le s \le r$. Clearly, we have $\tau(H) \ge \chi(H)$, for every H.

Let $\mathrm{Forb}^*(n, H)$ stand for the number of graphs on n labeled vertices which does contain H as an induced subgraph.

Theorem 2.3. (Prömel-Steger [PS92]) *For any graph H, we have*

$$\mathrm{Forb}^*(n,H) = 2^{(1+o(1))\mathrm{ex}^*(n,H)}.$$

Using Szemerédi's Regularity Lemma, Bollobás and Thomason [BT97] generalized this result to any nonempty hereditary graph property $\mathcal{P}$. Define the *coloring number* $r(\mathcal{P})$ of $\mathcal{P}$ as the largest integer r for which there is an s such that all (r, s)-colorable graphs have property $\mathcal{P}$. That is,

$$r(\mathcal{P}) = \max\{r \ : \ \text{there exists } 0 \le s \le r \ \text{ such that } \mathcal{P} \supset \mathcal{C}(r,s)\}.$$

Consequently, for any $0 \le s \le r(\mathcal{P}) + 1$, there exists an $(r(\mathcal{P}) + 1, s)$-colorable graph that does not have property $\mathcal{P}$.

In the special case when $\mathcal{P}$ is the property that the graph does not contain any induced subgraph isomorphic to H, we have $r(\mathcal{P}) = \tau(H) - 1$.

Theorem 2.4. (Bollobás-Thomason [BT97]) *Let $\mathcal{P}$ be a nontrivial hereditary property of graphs, and let $\mathcal{P}^n$ denote the set of all graphs in $\mathcal{P}$ on the vertex set $\{1, 2, \ldots n\}$. Then the speed of property $\mathcal{P}$ satisfies*

$$|\mathcal{P}^n| = 2^{\left(1-\frac{1}{r(\mathcal{P})}+o(1)\right)\binom{n}{2}},$$

where $r(\mathcal{P})$ is the coloring number of $\mathcal{P}$.

3 String Graphs – Proof of Theorem 1

We start with the lower bound. Consider four pairwise tangent non-overlapping disks D_i, $1 \le i \le 4$, in the plane (see Fig. 1). Assume for simplicity that n is divisible by 4. The proof for other values of n is analogous. Replace the boundary of each D_i by $n/4$ slightly smaller concentric circles C_{ik}, $1 \le k \le n/4$, running very close to it. Fix a pair (i, j), $1 \le i < j \le 4$. By local deformation of every C_{ik} in a small neighborhood of the point of tangency of D_i and D_j, we can achieve that every C_{ik} has a point lying outside every other C_{ih}, $h \ne k$. For every $1 \leq l \leq n/4$ and for any predetermined set of indices $K_l \subseteq \{1, 2, \ldots, n/4\}$, we can now slightly modify C_{jl} so that it would intersect a curve C_{ik} if and only if $k \in K_l$. In other words, we can arbitrarily specify the bipartite crossing pattern between the curves C_{ik} and C_{jl}, $1 \leq k, l \leq n/4$. Repeating the same procedure for every pair (i, j), we can obtain any 4-partite crossing pattern between the 4 classes, each containing $n/4$ curves. Note that every C_{ik} is a closed curve, but deleting any point of it which does not belong to another curve it becomes a string. Thus, the number of essentially different string graphs is at least $2^{\frac{6n^2}{16}} > 2^{\frac{3}{4}\binom{n}{2}}$.

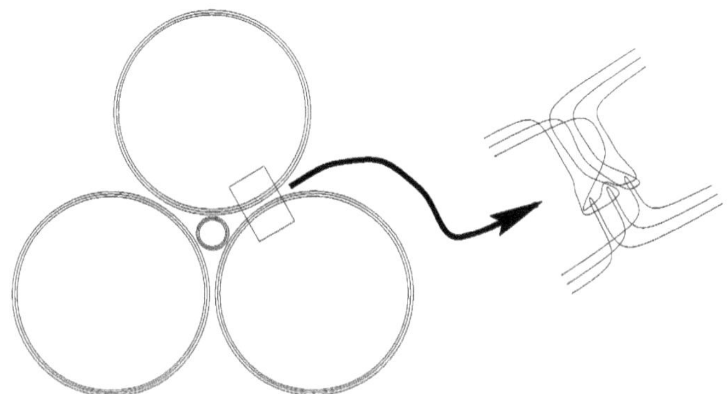

Fig. 1. Lower bound construction for the number of string graphs.

Next, we establish the upper bound. For any $r \geq 2$, let G_r be a graph with vertex set

$$V(G_r) = \{v_{ij} : 1 \leq i, j \leq r\}$$

and edge set

$$E(G_r) = \{v_{ij}v_{ik} : 1 \leq i, j, k \leq r, \, j \neq k\},$$

where $v_{ij} = v_{ji}$, for every i and j. In other words, the vertices of G_r represent the vertices and the edges of the complete graph K_r, two vertices of G_r being connected if the corresponding two edges of K_r share an endpoint or the corresponding edge and vertex of K_r are incident.

Lemma 3.1. *We have* $\tau(G_r) = r$.

Proof. The vertices $v_{1j}, 1 \leq j \leq r$ form a clique of size r. Therefore, we have $\tau(G_r) \geq \chi(G_r) \geq r$.

Now we show by induction on r that $\tau(G_r) = r$. This is true for $r = 2$. Let $r > 2$ be fixed and assume $\tau(G_{r-1}) = r - 1$. We have to show that, for any $0 \leq s \leq r$, the vertices of G_r can be colored by r colors so that s color classes induce cliques and the remaining $r - s$ color classes are independent sets.

For $s = 0$, the following coloring will satisfy the requirements. For any $1 \leq k \leq r$, color a vertex v_{ij} with color k if and only if $i + j \equiv k \bmod r$. Clearly, each vertex of G_r receives a color and each color class is an independent set.

If $s > 0$, color each vertex of the clique $\{v_{1j} : 1 \leq j \leq r\}$ with color 1. The uncolored vertices induce a subgraph isomorphic to G_{r-1}, for which we have $\tau(G_{r-1}) = r - 1$, by the induction hypothesis. So the remaining vertices can be colored by $r - 1$ colors so that $s - 1$ color classes induce cliques and the other $r - s$ are independent sets.

Lemma 3.2. *G_5 is not a string graph.* **Proof.** Suppose that G_5 has a string representation. Continuously contract each of string (arc) representing v_{ii} $(1 \leq i \leq 5)$ to a point p_i, without changing the crossing pattern. For every pair $i \neq j$, consider the portion of the arc representing v_{ij} between the points p_i and p_j. These arcs define a drawing of K_5, in which no two independent edges cross each other. However, K_5 is not a planar graph, hence, by a well known theorem of Hanani and Tutte [Ch34], [T70], no such drawing exists.

Now we can complete the proof of Theorem 1. By Lemma 3.2, a string graph cannot have an induced subgraph isomorphic to G_5. Thus, in view of Lemma 3.1, Theorem 1 directly follows from Theorem 2.3:

$$|\mathcal{S}^n| \leq Forb_n^*(G_5) = 2^{\left(\frac{3}{4}+o(1)\right)\binom{n}{2}}.$$

4 String Graphs of a Fixed Rank – Proof of Theorem 2

In order to show that there are $2^{o(n^2)}$ string graphs of rank d, in view of Theorem 2.4, it is enough to exhibit a $(2,0)$-colorable, a $(2,1)$-colorable, and a $(2,2)$-colorable graph such that none of them is a string graph of rank d.

Here we present the argument only in the special case $d = 1$.

Let $H_{3,3}$ denote a graph with vertices u_i, v_j, and w_{ij}, $1 \leq i, j \leq 3$ and edges $u_i w_{ij}, w_{ij} v_j$, for every i and j. In other words, $H_{3,3}$ is the graph obtained from $K_{3,3}$, the complete bipartite graph with three vertices in its classes, by subdividing each of its edges by an extra vertex.

For any k, let T_k denote a graph with vertices $v_i, (1 \leq i \leq k)$ and u_I, for every $I \subseteq \{1, 2, \ldots, k\}$. Let v_i and v_j be connected by an edge of T_k, for any $1 \leq i < j \leq k$, and let v_i be connected to u_I if and only if $i \in I$. Let T_k' denote the graph obtained from T_k by adding the edges $u_I u_J$, for every $I \neq J$.

Clearly, $H_{3,3}$ is $(2,0)$-colorable (bipartite), T_k is $(2,1)$-colorable, and T_k' is $(2,2)$-colorable, for every k. Therefore, if $\mathcal{P} = \mathcal{P}(H_{3,3}, T_k, T_k')$ denotes the property that a graph does not contain $H_{3,3}$, T_k, or T_k' as an induced subgraph, then $\mathcal{P}$ is a *hereditary* property with coloring number $r(\mathcal{P}) = \infty$. Hence, by Theorem 2.4, for the number of graphs on n labeled vertices, satisfying property $\mathcal{P}$, we have $|\mathcal{P}^n| = 2^{o(n^2)}$.

It remains to prove the following statement, which implies that $\mathcal{S}_1^n \subseteq \mathcal{P}^n$ if k is large enough.

Lemma 4.1. *A string graph of order* 1 *cannot contain* $H_{3,3}$, T_k, *or* T'_k *as an induced subgraph, provided that* k *is sufficiently large.*

Proof. It is well known that a string graph cannot contain $H_{3,3}$ as an induced subgraph (see e.g. [EET76],

Using the notation in the definition of T_k (and T'_k), let v_i, $1 \leq i \leq k$ and u_I, $I \subseteq \{1, 2, \ldots, k\}$ stand for the vertices of T_k (and T'_k, resp.), and suppose that T_k (and T'_k, resp.) has a string representation in which any two strings cross at most once. For simplicity, we use the same notation for the strings as for the corresponding vertices.

Fix arbitrarily an orientation of each string. For any triple (x, y, z), $1 \leq x < y < z \leq k$, let $f_{xyz} = 1$ if along v_y the crossing with v_x *follows* the crossing with v_z. Otherwise, set $f_{xyz} = 0$.

By Ramsey's theorem, there exists a "homogeneous" subset $J \subseteq \{1, 2, \ldots, k\}$, $|J| \geq \log\log k$, such that f_{xyz} is constant over all triples (x, y, z), $1 \leq x < y < z \leq k$, $x, y, z \in J$. We can assume without loss of generality that $J = \{1, 2, \ldots, m\}$, where $m \geq \log\log k$.

For any $1 \leq i \leq m$, the string v_i crosses all other v_j, $1 \leq j \leq m$, $i \neq j$ exactly once. Since f_{xiz} is constant over all triples (x, i, z), $1 \leq x < i < z \leq k$, one can find a non-crossing point on v_i that divides v_i into two parts, $v_i^<$ and $v_i^>$, containing all crossings between v_i and v_x with $x < i$ and between v_i and v_z with $z > i$, respectively. The arcs $v_i^<$ and $v_i^>$ are called the *lower part* and the *upper part* of v_i, respectively.

Construct two 42-uniform hypergraphs, $H^<$ and $H^>$, both on the vertex set $\{1, 2, \ldots, m\}$, as follows. For any $1 \leq x_1 < x_2 < \cdots < x_{83} \leq m$, there exists a string $u = u_{\{x_1, x_2, \ldots, x_{83}\}}$ that crosses $v_{x_1}, v_{x_2}, \ldots, v_{x_{83}}$, but no other v_j. The string u crosses either the lower or the upper part of each v_{x_i}, so for at least 42 indices $1 \leq i \leq 83$ it will cross, say, the lower (resp., upper) part. Suppose, for example, that u crosses the lower (resp., upper) parts of $v_{x_1}, v_{x_2}, \ldots, v_{x_{42}}$. Then add the hyperedge $\{x_1, x_2, \ldots, x_{42}\}$ to $H^<$ (resp., to $H^>$).

Repeating the above procedure for every 83-tuple $1 \leq x_1 < x_2 < \cdots < x_{83} \leq m$, the total number of hyperedges in $H^<$ and $H^>$ with *repetitions* is $\binom{m}{83}$. However, the multiplicity of each hyperedge is at most $\binom{m-42}{41}$. Thus, the total number of *distinct* hyperedges in $H^<$ and $H^>$ is $\Omega(m^{42})$ (i.e., at least constant times m^{42}). Suppose without loss of generality that $H^<$ has $\Omega(m^{42})$ distinct hyperedges.

We can now apply a well known result of Erdős [E65] (see also [B78] and [PA95], p. 151) to conclude that, for any fixed l and sufficiently large m, our hypergraph $H^<$ contains a complete 42-partite, 42-uniform subhypergraph $K^{42}_{l,\ldots,l}$ with l elements in each of its classes. (That is, $K^{42}_{l,\ldots,l}$ has $42l$ vertices, divided into 42 classes of size l, and it consists of all 42-tuples that contain one vertex from each class.)

For simplicity, denote by s_i^j, $1 \leq i \leq 42$, $1 \leq j \leq l$ the *lower* parts $v_{x_k}^<$ of the strings v_{x_k} corresponding to the vertices of $K^{42}_{l,\ldots,l}$. By the construction, for each 42-tuple $(j_1, \ldots, j_{42})$, $1 \leq j_1, \ldots, j_{42} \leq l$, there exists a string $u_{j_1,\ldots,j_{42}}$ that crosses $s_1^{j_1}, \ldots, s_{42}^{j_{42}}$, but no other string s_i^j.

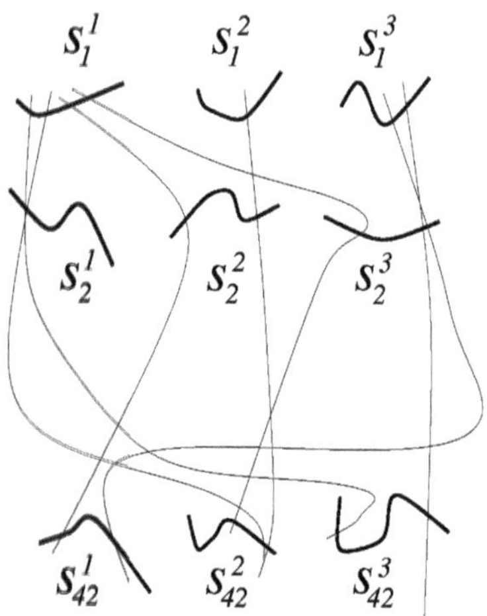

Fig. 2. Some of the strings representing a $K^{42}_{3,\ldots,3}$.

Color the 42-tuples $(j_1, \ldots, j_{42})$ with 42! colors, according to order in which the crossings with $s_1^{j_1}, \ldots, s_{42}^{j_{42}}$ occur along $u_{j_1,\ldots,j_{42}}$. Thus, we can find at least $\Omega(l^{42})$ 42-tuples of the same color (say, white). Suppose without loss of generality that, for each such 42-tuple $(j_1, \ldots, j_{42})$, the string $u_{j_1,\ldots,j_{42}}$ first crosses $s_1^{j_1}$, then $s_2^{j_2}$,..., and finally $s_{42}^{j_{42}}$. Applying Erdős's result again, if l is sufficiently large, we can find a subhypergraph $K^{42}_{3,\ldots,3} \subset K^{42}_{l,\ldots,l}$, all of whose 42-tuples are white. Again, we can assume without loss of generality that the strings corresponding to the vertices of $K^{42}_{3,\ldots,3}$ are s_i^j, $1 \leq i \leq 42$, $1 \leq j \leq 3$. Recall that each s_i^j is the *lower* part of an original string v_x, therefore, no two s_i^j can cross each other. (Indeed, the intersection of v_x and v_y, $x < y$, must belong to the upper part of v_x and at to the lower part of v_y.)

Summarizing: we have $3 \cdot 42 = 126$ strings s_i^j, $1 \leq i \leq 42$, $1 \leq j \leq 3$, no two of which intersect. Moreover, for each 42-tuple $(j_1, \ldots, j_{42})$, $1 \leq j_1, \ldots, j_{42} \leq 3$, there is a string $u_{j_1,\ldots,j_{42}}$ that intersects the strings $s_1^{j_1}, \ldots, s_{42}^{j_{42}}$ in this order, and does not intersect any other s_i^j. (See Fig. 2.) We would like to show that there are two different strings of the type $u_{j_1,\ldots,j_{42}}$ that cross more than once. First, we give a lower bound for the number of crossings $\text{CR}(u, u)$ between strings of type $u_{j_1,\ldots,j_{42}}$.

Let $1 \leq x \leq 41$ be fixed. For any pair y, z, $1 \leq y, z \leq 3$, consider all strings $u_{j_1,\ldots,j_{42}}$ with $j_x = y$ and $j_{x+1} = z$, and let $\Gamma_{y,z}$ denote the set of their portions between their intersections with s_x^y and s_{x+1}^z. Clearly, we have $|\Gamma_{y,z}| = 3^{40}$. Pick one element from each $\Gamma_{y,z}$, $1 \leq y, z \leq 3$, and notice that at least one pair among these 9 arcs must be crossing, otherwise, together with the strings $s_x^1, s_x^2, s_x^3, s_{x+1}^1, s_{x+1}^2, s_{x+1}^3$, they would give a string representation of $H_{3,3}$, which is impossible (see the first paragraph of this proof). Thus, for a fixed x, the total number of crossings between the elements of $\Gamma_{y,z}$ and $\Gamma_{y',z'}$ over all y, z, y', z',

$1 \le y, z, y', z' \le 3$, $(y, z) \ne (y', z')$ is at least

$$\frac{\prod_{1\le y,z\le 3} |\Gamma_{y,z}|}{3^{7\cdot 40}} = \frac{3^{9\cdot 40}}{3^{7\cdot 40}} = 3^{80}.$$

Here the denominator, $3^{7\cdot 40}$, is the number of 9-tuples of arcs, one from each set $\Gamma_{y,z}$, $1 \le y, z \le 3$, in which a crossing pair of arcs is fixed. Repeating this count for every x, $1 \le x \le 41$ and noticing that every time we count different crossings, we obtain that

$$\textsc{cr}(u, u) \ge 41 \cdot 3^{80}.$$

On the other hand, the number of strings of type $u_{j_1,\dots,j_{42}}$ is 3^{42}. If any two of them cross at most once, than $\textsc{cr}(u, u) < 3^{84}/2$, which is a contradicts the above inequality. This completes the proof of the lemma.

5 Drawings of Complete Graphs – Proof of Theorem 3

(i) It is easy to see that the *order type* on the vertices of K_n (i.e., the orientation of its triples) determines the set of crossing pairs of edges, So the upper bound follows from a result of Goodman and Pollack [GP86], that there are at most n^{6n} different order types on n points. On the other hand, we can place the vertices of K_n on a circle, in $(n-1)!$ different cyclic order, and each placement gives a different list of crossing pairs of edges. It is also easy to come up with a list of $n^{\Omega(n)}$ drawings such that by relabelling the vertices of any one of them, we do not obtain a drawing essentially the same as another.
(ii) Suppose n is divisible by 4, and let $v_i = (-1, i), u_j = (1, j)$, and $w_k = (0, k/2)$, for any $1 \le i, j \le n/4$ and $1 \le k \le n/2$. For every $1 \le k < n/2$, connect w_k and w_{k+1} by a straight-line segment. Furthermore, connect every v_i to every u_j by a line segment so that each such segment passes through some point w_k. By slightly bending each edge v_iu_j, but keeping its endpoints fixed, we can achieve that it passes either slightly above or slightly below w_{i+j}. At each edge v_iu_j, we have two choices, so there are $2^{n^2/16}$ possibilities. In each drawing, any two edges cross at most once, and different choices give rise to different crossing patterns. (Indeed, v_iu_j passes above w_{i+j} if and only if it crosses the edge $w_{i+j}w_{i+j+1}$.) Finally, one can slightly perturb the vertices so that no three of them would be collinear, and connect the missing pairs by straight-line segments without creating more than one crossing between any pair of edges. Therefore, the number of different crossing patterns is at least $2^{n^2/16}$.

As for the upper bound, for a fixed drawing, for each vertex v_i, list the edges incident to v_i in clockwise order around v_i. For every vertex, we have $(n-2)!$ possibilities, so there are $((n-2)!)^n < 2^{n^2 \log n}$ different sets of lists. We claim that this set of lists uniquely determines the crossing pattern. To see this take two edges, v_1v_2 and v_3v_4, and consider the drawing of K_4 induced by these vertices, as a drawing on the *sphere*. Two spherical drawings of K_4 are *combinatorially*

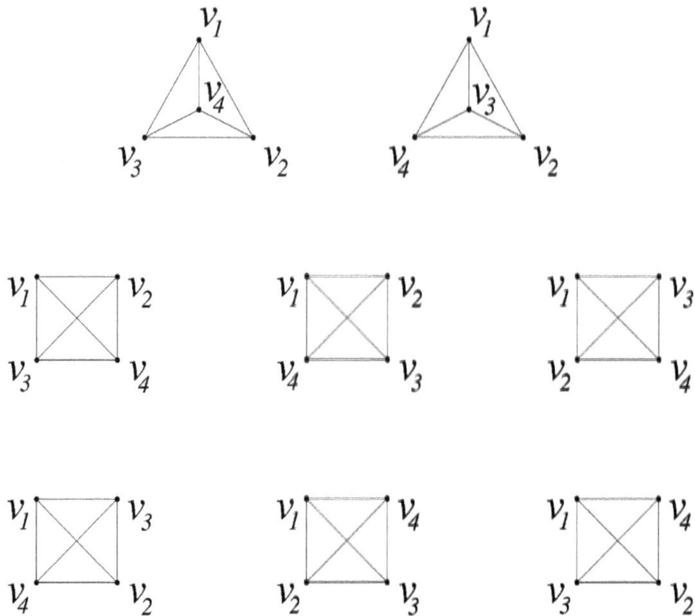

Fig. 3. The eight combinatorially different drawings of K_4.

equivalent if the corresponding maps are isomorphic. There are 8 combinatorially different drawings of K_4, with the property that any two edges have at most one point in common (see Fig. 3), and these drawings can be distinguished by looking at the cyclic orders of edges incident to a vertex. Hence, the cyclic order of edges at the vertices determines whether v_1v_2 and v_3v_4 cross each other.

(iii) Suppose n is divisible by 3. For $i = 1, 2, \ldots, n/3$, let $v_i = (-1, i)$, $w_i = (0, i)$, and $u_i = (1, i)$. Connect every v_i to every u_j, as follows. Choose a number k, $0 \leq k < n/3$, and connect both v_i and u_j to $(0, k+\varepsilon)$ by a segment. Also connect any two consecutive w_i's by a segment. In the resulting drawing, any two have at most two common points, and a different choice for any v_iu_j results a different crossing pattern. Therefore, the number of different crossing patterns is $n/3^{n^2/9}$. Clearly, each of these drawings can be extended to a drawing of the complete graph such that still any two have at most two common points. For instance, slightly perturb the points together with the existing edges, so that the points are in general position, and add the missing edges as segments.

For the upper bound, apply Theorem 2 for the edges of K_n regarded as $\binom{n}{2}$ strings.

(iv) Suppose n is even, and let $v_i = (-1, i)$, $u_i = (1, i)$, for $1 \leq i \leq n/2$. For any i, j, $1 \leq i < j \leq n/2$, connect v_i with $(0, ni + j)$ and connect $(0, ni + j)$ with v_j. Now, all vertices v_i and all edges connecting them are on the left side of the line $x = 0$ such that each of the edges has exactly one point on that line, and all these points are different. On the other hand, all vertices u_i, are on the right-hand side of the line $x = 0$. So, for any p, q, $1 \leq p < q \leq n/2$, and for any set $K_{pq} \subseteq \{(i,j) : 1 \leq i < j \leq n/2\}$, we can draw the edge $v_p v_q$ so that it crosses $u_i u_j$ ($i < j$) if and only if $(i,j) \in K_{pq}$ (cf. proof of Theorem 1).

References

[B59] S. Benzer, On the topology of the genetic fine structure, *Proceedings of the National Academy of Sciences of the United States of America* **45** (1959), 1607–1620.

[B78] B. Bollobás, *Extremal Graph Theory, London Mathematical Society Monographs* **11**, Academic Press. [Harcourt Brace Jovanovich, Publishers], London-New York, 1978.

[BT97] B. Bollobás and A. Thomason, Hereditary and monotone properties of graphs, in: *The mathematics of Paul Erdős II,* (R. L. Graham and J. Nešetřil, eds.) Algorithms Combin. **14**, Springer, Berlin, 1997, 70–78.

[CGP98a] Z.-Z. Chen, M. Grigni, and C. H. Papadimitriou, Planar map graphs, in: *STOC '98*, ACM, 1998, 514–523.

[CGP98b] Z.-Z. Chen, M. Grigni, and C. H. Papadimitriou, Planar topological inference, (Japanese) in: *Algorithms and Theory of Computing (Kyoto, 1998) Sūrikaisekikenkyūsho Kōkyūroku* **1041** (1998), 1–8.

[CHK99] Z.-Z. Chen, X. He, and M.-Y. Kao, Nonplanar topological inference and political-map graphs, in: *Proceedings of the Tenth Annual ACM-SIAM Symposium on Discrete Algorithms (Baltimore, MD, 1999),* ACM, New York, 1999, 195–204.

[Ch34] Ch. Chojnacki (A. Hanani), Über wesentlich unplättbare Kurven im dreidimensionalen Raume, *Fund. Math.* **23** (1934), 135–142.

[DETT99] G. Di Battista, P. Eades, R. Tamassia, and I. G. Tollis, *Graph Drawing,* Prentice Hall, Upper Saddle River, NJ, 1999.

[E93] M. Egenhofer, A model for detailed binary topological relationships, *Geomatica* **47** (1993), 261–273.

[E65] P. Erdős, On some extremal problems in extremal graph theory, *Israel J. Math.* **3** (1965), 113–116.

[EF91] M. Egenhofer and R. Franzosa, Point-set topological spatial relations, *International Journal of Geographical Information Systems* **5** (1991), 161-174.

[ES93] M. Egenhofer and J. Sharma, Assessing the consistency of complete and incomplete topological information, *Geographical Systems* **1** (1993), 47–68.

[EET76] G. Ehrlich, S. Even, and R. E. Tarjan, Intersection graphs of curves in the plane, *Journal of Combinatorial Theory, Series B* **21** (1976), 8–20.

[EFR86] P. Erdős, P. Frankl, and V. Rödl, The asymptotic number of graphs not containing a fixed subgraph and a problem for hypergraphs having no exponent, *Graphs and Combinatorics* **2** (1986), 113–121.

[ES66] P. Erdős and M. Simonovits, A limit theorem in graph theory, *Studia Sci. Math. Hungar.* **1** (1966), 51–57.

[ES46] P. Erdős and A. H. Stone, On the structure of linear graphs, *Bulletin Amer. Math. Soc.* **52** (1946), 1087–1091.

[EPL72] S. Even, A. Pnueli, and A. Lempel, Permutation graphs and transitive graphs, *Journal of Association for Computing Machinery* **19** (1972), 400–411.

[G80] M. C. Golumbic, *Algorithmic Graph Theory and Perfect Graphs,* Academic Press, New York, 1980.

[GP86] J. E. Goodman and R. Pollack, Upper bounds for configurations and polytopes in R^d, *Discrete Comput. Geom.* **1** (1986), 219–227.

[G78] R. L. Graham, Problem, in: *Combinatorics, Vol. II (A. Hajnal and V. T. Sós, eds.)*, North-Holland Publishing Company, Amsterdam, 1978, 1195.

[HT74] J. Hopcroft and R. E. Tarjan, Efficient planarity testing, *J. ACM* **21** (1974), 549–568.

[K83] J. Kratochvíl, String graphs, in: *Graphs and Other Combinatorial Topics (Prague, 1982),* Teubner-Texte Math. **59**, Teubner, Leipzig, 1983, 168–172.

[K91a] J. Kratochvíl, String graphs I: The number of critical nonstring graphs is infinite, *Journal of Combinatorial Theory, Series B* **52** (1991), 53–66.

[K91b] J. Kratochvíl, String graphs II: Recognizing string graphs is NP-hard, *Journal of Combinatorial Theory, Series B* **52** (1991), 67–78.

[K98] J. Kratochvíl, Crossing number of abstract topological graphs, in: *Graph drawing (Montreal, QC, 1998), Lecture Notes in Comput. Sci.* **1547**, Springer, Berlin, 1998, 238–245.

[PA95] J. Pach and P. K. Agarwal, *Combinatorial Geometry,* Wiley Interscience, New York, 1995.

[PS01] J. Pach and J. Solymosi, Crossing patterns of segments, *Journal of Combinatorial Theory, Ser. A,* **96** (2001), 316–325.

[PT02] J. Pach and G. Tóth, Recognizing string graphs is decidable, *Lecture Notes in Computer Science* **2265**, Spinger–Verlag, Berlin, 2001, 247-260. Also in: *Discrete and Computational Geometry* **28** (2002), 593–606.

[PS92] H. J. Prömel and A. Steger, Excluding induced subgraphs III: A general asymptotic, *Random Structures and Algorithms* **3** (1992), 19–31.

[SS01a] M. Schaefer and D. Stefankovič, Decidability of string graphs, *Proceedings of the 33rd Annual Symposium on the Theory of Computing*, 2001, 241–246.

[SS01b] M. Schaefer, E. Sedgwick and D. Stefankovič, Recognizing string graphs in NP, *Proceedings of the 34rd Annual Symposium on the Theory of Computing*, 2002, 1–6.

[S66] F. W. Sinden, Topology of thin film RC circuits, *Bell System Technological Journal* (1966), 1639–1662.

[SP92] T. R. Smith and K. K. Park, Algebraic approach to spatial reasoning, *International Journal of Geographical Information Systems* **6** (1992), 177–192.

[T70] W. T. Tutte, Toward a theory of crossing numbers, *J. Combinatorial Theory* **8** (1970), 45–53.

Two Results on Intersection Graphs of Polygons

Jan Kratochvíl and Martin Pergel

Department of Applied Mathematics and
Institute for Theoretical Computer Science*
Charles University
Malostranské nám. 25
118 00 Praha 1
Czech Republic
{honza,perm}@kam.mff.cuni.cz

Abstract. Intersection graphs of convex polygons inscribed to a circle, so called polygon-circle graphs, generalize several well studied classes of graphs, e.g., interval graphs, circle graphs, circular-arc graphs and chordal graphs. We consider the question how complicated need to be the polygons in a polygon-circle representation of a graph.

Let $\mathrm{cmp}(n)$ denote the minimum k such that every polygon-circle graph on n vertices is the intersection graph of k-gons inscribed to the circle. We prove that $\mathrm{cmp}(n) = n - \log_2 n + o(\log_2 n)$ by showing that for every positive constant $c < 1$, $\mathrm{cmp}(n) \leq n - c\log n$ for every sufficiently large n, and by providing an explicit construction of polygon-circle graphs on n vertices which are not representable by polygons with less than $n - \log n - 2\log\log n$ corners. We also show that recognizing intersection graphs of k-gons inscribed in a circle is an NP-complete problem for every fixed $k \geq 3$.

1 Introduction

Intersection graphs of geometric objects, namely in the plane, are intensively studied both for their practical motivations and for interesting structural and algorithmic properties. Many hard (NP-complete in general) optimization problems become polynomially solvable when restricted to various classes of intersection graphs. Probably the oldest and simplest of these are interval graphs, intersection graphs of intervals on a line [4], whose structure is well understood, they are recognizable in linear time, and for which problems like clique, independent set, dominating set, chromatic number and many more are tractable. On the other end of the spectrum are string graphs, intersection graphs of arc-connected sets in the plane, which are hard to recognize [8], and whose recognition was only recently shown decidable [12,15] and then even more surprisingly in NP [16].

Geometric intersection graphs also provide a special kind of graph visualization. Vertices are represented as geometric objects and adjacencies are visualized by nonempty intersections. Overlapping regions are illustrative and enhance the visual understanding of the represented graph. String graphs are closely related to another graph drawing invariant, the *crossing number*, i.e., the minimum number of edge crossings in a drawing

* Project LN00A056 supported by the Ministry of Education of the Czech Republic.

G. Liotta (Ed.): GD 2003, LNCS 2912, pp. 59–70, 2004.

of a graph. Abstract topological graphs (cf. [10]) are graphs with specified pairs of edges which are allowed to cross in a feasible drawing. In this setting it may happen that multiple edge crossings occur in every feasible drawing, and an abstract topological graph may actually require exponential number of crossing points [11]. The crossing number of abstract topological graphs is closely related to the number of crossing points in representations of string graphs as intersection graphs of curves, and in fact in this way the decidability and NP-membership of string graphs were tackled.

Several well studied classes of graphs which directly generalize interval graphs involve geometric objects bound to a circle. Among these are *circle graphs*, intersection graphs of chords of a circle, and *circular arc graphs*, intersection graphs of intervals on the circle. Both these classes are recognizable in polynomial time (cf. [1,3,13] for circle graphs, [17,2] for circular arc graphs). A common generalization of these two are *polygon-circle graphs*, intersection graphs of convex polygons inscribed to the circle. This class was first suggested by M. Fellows [personal communication with the first author] in 1988, when it was pointed out that this class of graphs is closed under taking induced minors. Under a different name of *spider graphs*, polygon-circle graphs appeared in [7], where a polynomial time recognition algorithm was announced. Somewhat surprisingly, the algorithm was never published, and this fact creates certain doubts about its correctness. A structural property of graph classes which many intersection graphs possess is *near-perfectness* in the sense of Gyarfás [5]. A graph class is near-perfect if the chromatic number of each of its graphs is bounded by a function of the clique number of the graph. (For perfect graphs this function is identity, and from here the notion originates.) Polygon-circle graphs are near-perfect as shown in [9], and the bound presented therein improved the up to that date best bound for circle graphs and has not been improved since.

In the current paper we pay closer attention to the question of how complicated should be the polygons representing the vertices of a polygon-circle graph with n vertices. One can easily see that n-gons always suffice (hence no exponential surprises as in the case of string graphs), which means that polygon-circle graph recognition is definitely in NP. It is conceivable, however, that polygons with less corners would suffice. To be able to precisely formulate this question, we define the *complicacy* of a graph G as the minimum k such that G is the intersection graph of convex k-gons inscribed to a circle, and we denote this invariant by $\mathrm{cmp}(G)$ (we set $\mathrm{cmp}(G) = \infty$ if G is not a polygon-circle graph). We further define $\mathrm{cmp}(n)$ to be the maximum of $\mathrm{cmp}(G)$ over all polygon-circle graphs with n vertices. The main result in this direction is the following (here and throughout the paper, all logarithms are base 2):

Theorem 1. *We have* $\mathrm{cmp}(n) = n - \log n + o(\log n)$.

The lower and upper bounds are proved separately in the next two sections. In the last section we consider the computational complexity of determining the complicacy of a graph with the following result:

Theorem 2. *For every fixed finite* $k \geq 3$*, it is NP-complete to decide whether* $\mathrm{cmp}(G) \leq k$ *holds for an input graph* G.

This result, which answers an open problem listed at J. Spinrad's web page [14], is not in contradiction with the announced algorithm of Koebe for recognition of polygon-circle graphs (i.e., deciding $\text{cmp}(G) < \infty$), but it definitely sheds new light on the problem of recognizing polygon-circle graphs. Note also that for $k = 2$, $\text{cmp}(G) \leq 2$ if and only if G is a circle graph, a polynomially decidable question.

2 Technical Notions and Observations

Throughout the paper we use small letters as $a, b, u, v, \ldots$ for vertices of the graph under consideration, and if R is a representation by polygons, R_v denotes the polygon representing vertex v. However, in figures, to avoid multiple subscripts, we will usually omit the symbol R. We assume that the bounding circle is fixed and whenever we speak about polygons, we automatically assume that the polygons are convex and have all corners placed on the circle.

If P is a polygon, then the connected parts obtained from the bounding circle by deleting the corners of P are referred to as the *P-segments*. If two polygons represent nonadjacent vertices, they must be disjoint and hence all corners of one of them lie within the same segment determined by the other one, and vice versa. In the following technical definition we assume that all polygons under consideration are disjoint.

Definition 1. *We say that polygon P* blocks *polygon Q from polygon S if the corners of Q lie in a different P-segment than the corners of S. If a set $\mathcal{S}$ of polygons is such that none of them blocks any other two polygons from each other, we say that the polygons are* positioned around the circle.

See Figure 1 for an illustrative example of blocking polygons, and a set of polygons positioned around the circle. Next we make the first simple but useful observation.

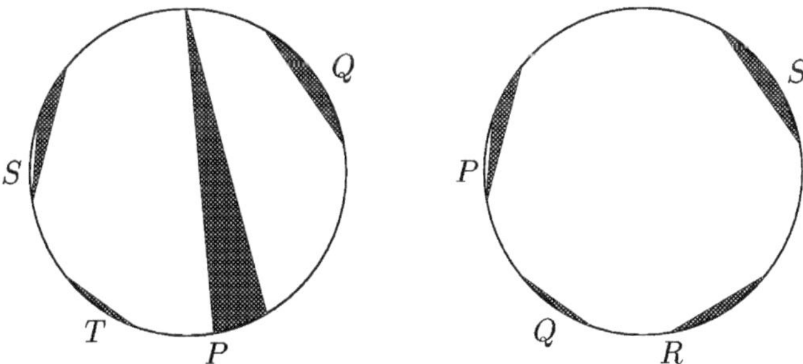

Fig. 1. In the left, polygon P blocks polygon Q from polygons S and T, in the right, all four polygons are positioned around the circle.

Proposition 1. *In any representation R of the cycle C_{2k} with $2k$ vertices $u_1, u_2, \ldots, u_{2k}$, the polygons $R_{u_{2i}} : i = 1, 2, \ldots, k$ are positioned around the circle. If W_k' is the graph obtained from C_{2k} by adding a vertex v adjacent to $u_{2i}, i = 1, 2, \ldots, k$ (i.e., W_k' is the wheel W_k with each rim edge subdivided), then R_v has at least k corners and $\text{cmp}(W_k') \geq k$ (in fact, the complicacy of W_k' equals k).*

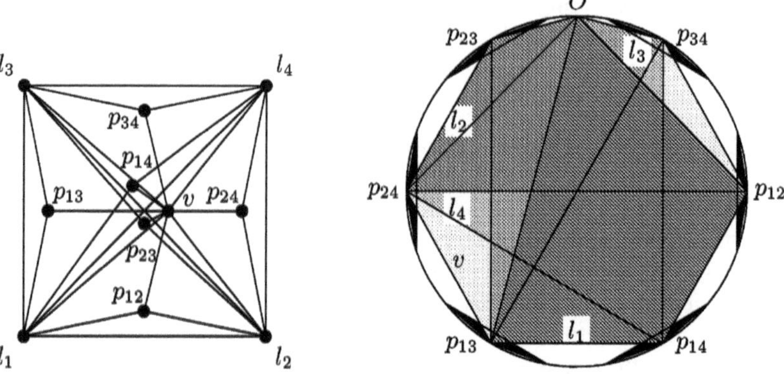

Fig. 2. An example of the lower bound construction for $\ell = 1$ and $n = 11$. We write simply p_{13} for $p_{\{1,3\}}$ etc.

Proof. For any three (even indexed) vertices u_{2i}, u_{2j}, u_{2h} of C_{2k}, there exists a path connecting u_{2j} and u_{2h} which does not contain any neighbor of u_{2i}. The union of the polygons representing vertices of this path is a connected subset of the disk bounded by the base circle, and hence (by the Jorda curve theorem) $R_{u_{2j}}$ must not be blocked from $R_{u_{2h}}$ by $R_{u_{2i}}$.

For the wheel graph W'_k, it follows that R_v must have at least k corners, to intersect all $R_{u_{2i}}, i = 1, 2, \ldots, k$.

3 Complicacy of Representations – The Lower Bound

Theorem 3. *For n large enough, we have* $\text{cmp}(n) \geq n - \log n - 2 \log\log n$.

Proof. The proof is by constructing graphs of large complicacy by an explicit construction. Suppose n is large enough (how large will follow from the calculations in the proof). Let ℓ be the uniquely defined integer such that

$$1 + 2\ell + \binom{2\ell}{\ell} < n \leq 1 + 2(\ell + 1) + \binom{2\ell + 2}{\ell + 1}.$$

We construct the graph $G = (V, E)$ with vertex set

$$V = \{v\} \cup L \cup P,$$

where vertex v is adjacent to all other vertices, L is a clique of size $2\ell + 2$, and P is an independent set whose vertices are indexed by distinct $(\ell + 1)$-element subsets of L. These indices will determine the adjacencies between vertices of P and L as follows. If $p_a \in P$ (with $a \subset L$, $|a| = \ell + 1$), we make p_a adjacent to all $x \in a$ and to no other vertices of L. (Note that G is not determined uniquely, it depends on the choice of the sets used for indexing the vertices from P. Any choice of distinct indices works.)

We claim that G is a polygon-circle graph and that $\mathrm{cmp}(G) \geq |P|$. For the first part of the claim, choose a point O on the circle and position the polygons corresponding to the vertices of P around the circle (so that they do not block O from one another). Choose one corner of each of them as its reference point and represent each vertex x of L by the convex hull of the reference points of the vertices of P adjacent to x and of O. By making O a corner of each R_x we guarantee that every two polygons $R_u, R_w, u, w \in L$ intersect (L is a clique). Finally R_v will be the convex hull of O and the reference points of all polygons representing vertices of P. (Note that the auxiliary point O may not be necessary, e.g., in the case when for every two vertices in L, P contains a vertex adjacent to both of them.)

To argue that $\mathrm{cmp}(G)$ is large, consider an optimal representation R of G. Note first that for $n > 5$ we have $\ell \geq 1$ and hence $\binom{2\ell}{\ell} \geq 2$. This means that $n > 2\ell + 3$ and indeed G contains all vertices of L.

The key observation is that the polygons $R_p, p \in P$ must be positioned around the circle. For suppose this is not the case, say R_{p_b} blocks R_{p_a} from R_{p_c} for some $a, b, c \subset L$. Since these subsets are different but of equal size, there must be an $x \in a \setminus b$ and a $y \in c \setminus b$. By the definition of G, R_x intersects R_a and R_y intersects R_c, but none of R_x, R_y intersects R_b. But that means that R_b blocks R_x from R_y and these two polygons cannot intersect each other (though x and y belong to the clique L).

To intersect all polygons representing vertices of P, R_v must have at least $|P|$ corners, and hence $\mathrm{cmp}(G) \geq |P| = n - 2\ell - 3$. The rest is a simple calculation.

Assume for contradiction that $\ell > \frac{\log n}{2} + \log\log n - \frac{3}{2}$. Then

$$n > \binom{2\ell}{\ell} > \frac{2^{2\ell}}{2\ell + 1} > \frac{n \log^2 n}{8(\log n + 2\log\log n - 2)} > n$$

for large enough n (since $\lim_{n\to\infty} \frac{\log^2 n}{8(\log n + 2\log\log n - 2)} = \infty$), a contradiction. Therefore (for every large enough n), $\ell \leq \frac{\log n}{2} + \log\log n - \frac{3}{2}$ and

$$\mathrm{cmp}(G) \geq |P| = n - 2\ell - 3 \geq n - \log n - 2\log\log n$$

as claimed.

4 Complicacy of Representations – The Upper Bound

Theorem 4. *For every positive constant $c < 1$, there exists an n_0 such that* $\mathrm{cmp}(n) \leq n - c\log n$ *for every $n > n_0$.*

Proof. Let $G = (V, E)$ be a graph on n vertices and let $\mathrm{cmp}(G) = k \geq 4$. Consider a polygon-circle representation R of G such that no two polygons share a corner and such that every polygon has at most k corners (this can be always achieved by splitting the corners). Let our representation have the minimum total number of corners among all such representations. Choose a vertex $v \in V$ such that R_v has k corners, and denote its corners $v^1, v^2, \ldots, v^k$ as they appear clockwise around the circle.

Based on this representation R, define A to be the set of vertices x such that R_x has all corners within two consecutive R_v-segments, i.e., such that R_x intersects R_v only in

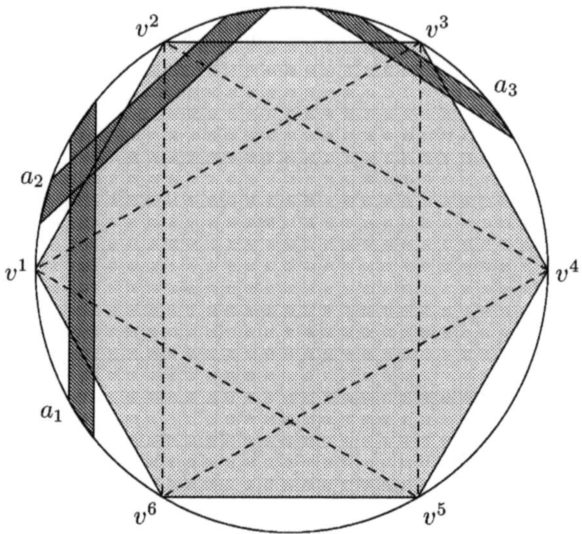

Fig. 3. Illustration to the choice of representatives of index classes.

the triangle $v^{i-1}v^iv^{i+1}$, for some i. For $x \in A$, denote this i by $j(x)$ and call it the *index* of x. We can see that for every i, there exists an $x \in A$ of index i. For if such a vertex did not exist for some i, we could reduce the number of corners in the representation by deleting the triangle $v^{i-1}v^iv^{i+1}$ from R_v, contradicting the choice of R.

Now assume that R' minimizes $\sum_{x\in A} j(x)$ among all representations with the same central polygon R_v, with the same total number of corners and with the same set A (when defined as above) as R. For the sake of simplicity we call R' again just R.

For every i, choose one vertex as a representative of the vertices of index i, and call it a_i. As argued above, a_i is well defined for every i. The intersection graph of the polygons $R_{a_i}, i = 1, 2, \ldots, k$ is either a cycle or a disjoint union of paths, since a_i can only intersect a_{i-1} and/or a_{i+1}. We denote A_1 the set of those a_i that have no neighbors among the other a_h's, and $A_2 = \{a_1, a_2, \ldots, a_k\} \setminus A_1$. We then denote $B = (V \setminus \{v\}) \setminus (A_1 \cup A_2)$ and B_1 the set those vertices of B which are adjacent to at least one vertex of A_1.

Next we claim that every two vertices of A_1 have different sets of neighbors. For suppose that $\{x|xa_i \in E\} = \{x|xa_j \in E\}$ for some $j < i$. Then the polygons representing the neighbors of a_i intersect one side of R_{a_j}, and hence both R_{a_i} and R_{a_j} could be replaced by two parallel chords (digons) placed close enough to this side. This would result in a new representation of the same graph, with the same position of R_v and the same set A, but with a strictly smaller sum of the indices of the vertices of A, contradicting the choice of R. See an illustrative example in Figure 4.

The immediate but very important consequence is that

$$|A_1| \leq 2^{|B_1|},$$

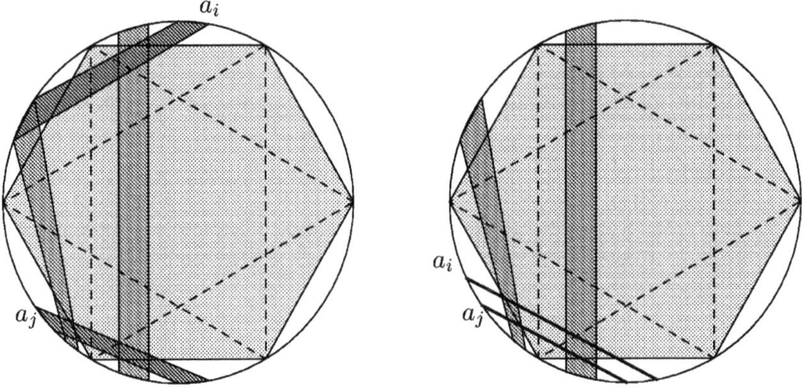

Fig. 4. Illustration to reducing the sum of indices of the A vertices.

since v is a common neighbor of all vertices in A_1, and on the other hand no vertex of A_1 is adjacent to any other vertex of $A_1 \cup A_2$.

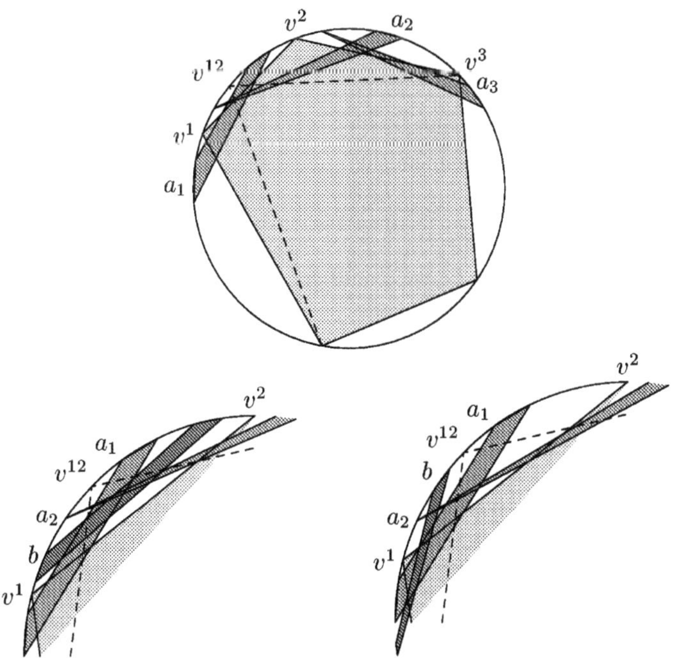

Fig. 5. Illustration to the construction of B_2, the blocking set for A_2.

Finally we attend to vertices of A_2. Consider the example in Figure 5. If R_{a_1} and R_{a_2} intersect, they must intersect in the area bounded by the v^1v^2 side of R_v and the corresponding R_v-segment. Choose a point v^{12} in this segment between the leftmost corner of a_2 and the rightmost corner of a_1. If we now replaced the central polygon R_v by the convex hull of $v^{12}, v^3, \ldots, v^k$ (i.e., the corners v^1 and v^2 are replaced by v^{12}), the new R_v (denoted by dashed lines in the figure) would still intersect both R_{a_1} and R_{a_2}, but the total number of corners would decrease. Since this is impossible by the choice of R, the new collection of polygons does not represent G anymore. There are only two possible reasons. Either the new R_v intersects a polygon R_b which is not supposed to be intersected (as in Figure 5 bottom left), or the new R_v loses intersection with some R_b that it previously crossed (as in Figure 5 bottom right). In the former case, R_b would lie fully within the segment v^1v^2, in the latter one, $b \in A$ and its index is $j(b) = 1$ (or, symmetrically, $j(b) = 2$). In any case, we choose one such b and call it a witness for $(1, 2)$. Similarly, we choose a witness for every $(i, i+1)$ such that $R_{a_i} \cap R_{a_{i+1}} \neq \emptyset$, and denote by B_2 the set of witnesses constructed in this way. By the construction, $B_2 \subset B$. It may happen, though, that one vertex of B_2 is a witness for two pairs (a witness of the latter type and index i could have been chosen both for the pair $(i-1, i)$ and for $(i, i+1)$). Given the fact that a path of t consecutive vertices from A_2 would involve $t-1$ pairs, such a path would give rise to at least $\frac{t-1}{2}$ distinct witnesses, and we obtain the following inequality:

$$|A_2| \leq 3|B_2|.$$

(Note here, that B_1 and B_2 are not necessarily disjoint.)

Now we summarize

$$k = \mathrm{cmp}(G) = |A_1| + |A_2| \leq 2^{|B_1|} + 3|B_2| \leq 2^{n-k} + 3(n-k)$$

(the last inequality follows from the fact that $A_1 \cup A_2 \cup B_1 \subset V$ and $A_1 \cup A_2 \cup B_2 \subset V$, while in each case the three summands are pairwise disjoint), and hence

$$2^n \geq 2^k(4k - 3n).$$

For $k > n - c\log n$ $(0 < c < 1)$, we would get

$$2^n > 2^{n-c\log n}(4(n - c\log n) - 3n) = 2^n \frac{n - 4c\log n}{n^c} > 2^n$$

for every $n > n_0$ for some n_0, as $\lim_{n\to\infty} \frac{n}{n^c} = \infty$ and $\lim_{n\to\infty} \frac{4c\log n}{n^c} = 0$. Which is the final long worked for contradiction.

The proof of Theorem 1 is now at hand. Consider the difference $f(n) = \mathrm{cmp}(n) - n + \log n$. On one hand, Theorem 3 states that $f(n) \geq -2\log\log n = o(\log n)$. On the other hand, Thereom 4 yields that for every $\varepsilon > 0$, $f(n) < \varepsilon \log n$ for every large enough n (by setting $c = 1 - \varepsilon$), and hence $f(n) = o(\log n)$.

5 Computational Complexity

We first restate Thereom 2 in a slightly stronger form:

Theorem 5. *For every fixed $k \geq 3$, it is NP-complete to decide if* $\text{cmp}(G) \leq k$ *for an input graph G, even if G is promised to have complicacy at most $k+1$.*

Proof. We reduce from the following hypergraph coloring problem: Given a set $\mathcal{T}$ of triples over a base set X, decide if the elements of X can be colored by three colors so that every triple $T \in \mathcal{T}$ receives all three colors. This problem is NP-complete as it contains the 3-edge colorability of cubic graphs as a subproblem, the latter being NP-complete by Holyer [6]. (By resctricting to 3-edge colorability of bridgeless graphs, we may assume that X is colorable by 2 colors so that no triple $T \in \mathcal{T}$ is monochromatic.)

Given such a $\mathcal{T}$, we construct a graph $G_{\mathcal{T}} = (V, E)$ as follows. The vertex set will consist of a cycle of length $2k$ on vertices $u_1, u_2, \ldots, u_{2k}$, a pair of vertices a_x, b_x for every $x \in X$, and a vertex c_T for each triple $T \in \mathcal{T}$. The c vertices form a clique and each of them is adjacent to all u vertices with even subscripts, to all b vertices, and to those a vertices which correspond to x's belonging to the particular triple. The b vertices are further adjacent to their corresponding a vertices and to u_2, u_4, u_6. Formally

$$V = \{u_i | i = 1, 2, \ldots, 2k\} \cup \{a_x, b_x | x \in X\} \cup \{c_T | T \in \mathcal{T}\}$$

$$E = \{u_1u_2, u_2u_3, \ldots, u_{2k}u_1\} \cup \{a_xb_x | x \in X\} \cup \{b_xu_i | x \in X, i = 2, 4, 6\}$$

$$\cup \{b_xb_y | x, y \in X\} \cup \{c_Ta_x | x \in T \in \mathcal{T}\} \cup \{c_Tb_x | T \in \mathcal{T}, x \in X\}$$

$$\cup \{c_Tc_S | T, S \in \mathcal{T}\} \cup \{c_Tu_i | T \in \mathcal{T}, i = 2, 4, 6, \ldots, 2k\}.$$

We claim that $G_{\mathcal{T}}$ is always a PC graph of complicacy at most $k+1$, and its complicacy is (at most) k if and only if $\mathcal{T}$ allows a coloring as desired.

Let R be a PC representation of $G_{\mathcal{T}}$. We first note that all polygons $R_{a_x}, x \in X$ must be positioned around the circle. For if one of them would block another two from each other, some R_x would block some R_y from R_{u_2} and the auxiliary polygon R_{b_y} would not be able to intersect both R_y and R_{u_2}.

As observed in Proposition 1, the polygons $R_{u_{2i}}, i = 1, 2, \ldots, k$ are positioned around the circle, and only these vertices of the cycle are intersected by R_{c_T}'s. Since the polygons representing the b's intersect (only) R_{u_2}, R_{u_4} and R_{u_6}, all the polygons representing the a vertices must lie in the segments determined by R_{u_2}, R_{u_4} and R_{u_6}. So color an element $x \in X$ by color i if R_{a_x} is blocked by R_{u_i} from the other R_{u_j}, $i, j = 2, 4, 6$. We claim that this is a good coloring. For if a triple $T \in \mathcal{T}$ contained two vertices of the same color, say x and y of color 2, the polygon R_{c_T} would need $k-1$ corners for intersections with $R_{u_i}, i = 4, 6, \ldots, 2k$, plus at least two more corners for intersections with R_x and R_y, that is at least $k+1$ corners altogether.

On the other hand, given a feasible 3-coloring of X by colors 2,4,6, a representation by k-gons can be achieved by placing the polygons $R_{a_x}, x \in X$ around the circle such that polygons corresponding to vertices of color i are placed within the segment determined by the chord $R_{u_i}, i = 2, 4, 6$. Each c vertex can then be represented by a k-gon with one corner in each segment determined by $R_{u_i}, i = 2, 4, 6, \ldots, 2k$. An illustrative example is in Figure 6, where a triangle representing a triple $T = \{x, y, z\}$ is marked. For the sake of simplicity, we only illustrate the case $k = 3$, and the auxiliary polygons R_{b_x} are not shown (drawn with invisible ink).

Along the same lines it is seen that $\mathrm{cmp}(G_{\mathcal{T}}) \leq k + 1$ if $(X, \mathcal{T})$ allows a 2-coloring without monochromatic triples.

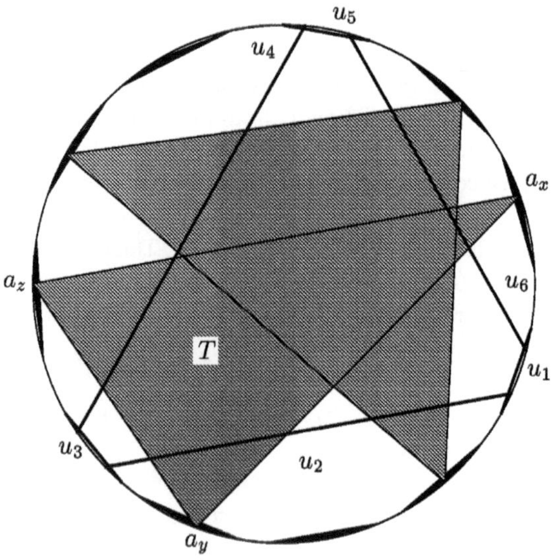

Fig. 6. Illustration to the NP-completeness proof.

One may ask for what function $f(n)$ of n it is still hard to decide if $\mathrm{cmp}(G) \leq f(n)$ for an input polygon-circle graph G on n vertices (since we do not trust the polynomiality of polygon-circle graph recognition, we need to assume that the input graph comes with the promise of being a polygon-circle one). As we have seen in previous sections, $f(n)$ must be somewhat smaller than n so that we might expect NP-hardness. By a closer examination of the proof of Theorem 2 we see that we can get close to $\frac{n}{2}$:

Theorem 6. *For every rational $c < \frac{1}{2}$, it is NP-complete to decide whether* $\mathrm{cmp}(G) \leq$ *cn for an input graph G on n vertices.*

Proof. Consider the previous construction of $G_{\mathcal{T}}$. If we start the reduction from 3-edge coloring of a graph with $2m$ vertices and $3m$ edges, we get

$$n = |V| = 2k + 2|X| + |\mathcal{T}| = 2k + 8m.$$

Hence setting $n = \frac{8m}{1-2c}$ (if $1 - 2c > 0$ is rational, we may start with sufficiently many copies of the cubic graph to make n an integer) we get

$$2k = n - 8m = 8m(\frac{1}{1-2c} - 1) = 2cn$$

and hence our previous proof shows that deciding

$$\text{cmp}(G_T) \leq k = cn$$

is NP-complete.

6 Conclusion

Regardless of whether recognition of polygon-circle graphs turns out polynomial or NP-complete (if dichotomy is expected), our results about NP-hardness of polygon-circle graphs of bounded complicacy will be an interesting counterpart to the complexity of recognition of the whole class. We do hope that our paper will revitalize interest in the polygon-circle recognition problem as well.

References

1. A. Bouchet: Circle graph obstructions, J. Comb. Theory, Ser. B 60, No.1 (1994) 107–144
2. E.M. Eschen, J.P. Spinrad: An $O(n^2)$ algorithm for circular-arc graph recognition, in: Proceedings 4th ACM-SIAM Symposium on Discrete Algorithms, (1993), 128–137
3. H. de Frayssiex: A characterization of circle graphs, Eur. J. Comb. 5 (1984) 223–238
4. P.C. Gilmore, A.J. Hoffman: A characterization of comparability graphs and of interval graphs, Can. J. Math. 16 (1964) 539–548
5. A. Gyarfás: Problems from the world surrounding perfect graphs, Zastosow. Mat. 19, No.3/4 (1987) 413–441
6. I. Holyer: The NP-completeness of edge-coloring, SIAM J. Comput. 10 (1981) 713–717
7. M. Koebe: On a New Class of Intersection Graphs, Proceedings of the Fourth Czechoslovak Symposium on Combinatorics Prachatice, 141–143, 1990
8. J. Kratochvíl: String graphs II. Recognizing string graphs is NP-hard, J. Combin. Theory Ser. B, 52 (1991) 67–78
9. J. Kratochvíl, A. Kostochka: Covering and coloring polygon-circle graphs, Discrete Math. 163 (1997) 29-9-305
10. J. Kratochvíl, A. Lubiw, J. Nešetřil: Noncrossing subgraphs of topological layouts, SIAM J. Discrete Math. 4 (1991) 223-244
11. J. Kratochvíl, J. Matoušek: String graphs requiring exponential representations, J. Combin. Theory Ser. B 53 (1991) 1–4
12. J.Pach, G.Tóth: Recognizing string graphs is decidable, In: Graph Drawing (P. Mutzel, M. Jünger and S. Leipert, eds.), Proceedings 9th International Symposium GD 2001, Vienna 2001, Lecture Notes in Computer Science 2265, Springer Verlag Berlin Heidelberg 2002, pp. 247–260
13. J.P. Spinrad: Recognition of Circle Graphs, Journal of Algorithms 16 (1), 264–282, 1994
14. J.P. Spinrad: `http://www.vuse.vanderbilt.edu/~spin/open.html`

15. M. Schaefer, D. Stefankovič: Decidability of string graphs, In: STOC 2001, Proceedings 33rd Annual ACM Symposium on Theory of Computing, Greece 2001, pp. 241–246
16. M. Schaefer, E.Sedgwick and D. Stefankovič: Recognizing string graphs in NP, In: STOC 2002, Proceedings 34th Annual ACM Symposium on Theory of Computing
17. A.C. Tucker: An efficient test for circular-arc graphs, SIAM J. Comput. 9 (1980) 1–24

Stretching of Jordan Arc Contact Systems

Hubert de Fraysseix and Patrice Ossona de Mendez

UMR 8557, CNRS, Paris, France

Abstract. We prove that a contact system of Jordan arcs is stretchable if and only if it is extendable into a weak arrangement of pseudo-lines.

1 Barycentric Representations of Graphs

1.1 Tutte's Barycenter Theorem

It is a classical result independently established by Wagner [11], Fáry [2] and Stein [6] (which is also a consequence of the Steinitz's theorem on convex polytopes [7]) that any simple planar graph has a straight line representation in the plane. In the early 60's, Tutte [9][10] gave a way to build a convex embedding of a 3-connected planar graph:

- Let S be the vertex set of a face F_0 of G in a planar embedding of G,
- let $f : S \to \mathbb{R}^2$ be a mapping of the vertices in S to the points of a convex polygon, in such a way that the order of the points on the polygon is the same as the order of the vertices on the face F_0,
- let $\Psi : V(G) \setminus S \to \mathbb{R}^2$ be the solution of the linear system

$$\forall x \in V(G) \setminus S, \qquad \Psi(x) = \sum_{\{v,x\} \in E(G)} \frac{1}{d(x)} \Psi(v) \tag{1}$$

Theorem 1 (Tutte's theorem). *Ψ defines an embedding of G into the plane, with strictly convex interior faces.*

In his quite long and difficult proof, Tutte simultaneously reproves Kuratowski's planarity criterion. Other proofs of this theorem have been published since. This result has been generalized by Thomassen to infinite planar graphs in [8], while Linial, Lovász and Wigderson proved the equivalence of the k vertex connectivity and the existence of a convex embedding (for any k extremal vertices) in general position in $\mathbb{R}^k$ [3]. In Tutte's construction, each vertex in $V(G) \setminus S$ is the barycenter of its neighbors. Although Tutte assumed all the coefficients to be equal, his proof extends without changes when coefficients are positive reals. In the aim of studying straight line segments contact systems, we shall first give a short proof of an extension of this theorem, in a framework allowing null coefficients, giving some side results necessary for our study on stretchability.

G. Liotta (Ed.): GD 2003, LNCS 2912, pp. 71–85, 2004.

First, we shall give necessary and sufficient conditions for such a barycentric linear system to have a unique solution (without planarity assumptions), and for its solution to define a plane embedding (assuming that the graph is planar and that the source set is embedded on a convex polygon with a compatible point order). Instead of the classic approach consisting in directly proving the planarity of the barycentric representation, we will make use of continuity arguments, while considering an homotopy from some initial straight line embedding of a triangulation of the graph (see also [1][4] for alternative approaches and generalizations of Tutte's Theorem).

1.2 Definitions and Notations

A *barycentric system* Σ on a graph G is a triple (S, f, α), where the *source set* S of Σ is a subset of vertices of G, the *limit function* f of Σ is a function from S to $\mathbb{R}^k$, the *weight function* α of Σ is a function from $V(G) \times V(G)$ to $[0,1]$, and such that:

$$\forall x, y \in V(G), \qquad \alpha(x,y) \neq 0 \Rightarrow \{x,y\} \in E(G) \text{ and } x \notin S \tag{2}$$

$$\forall x \in V(G) \setminus S, \qquad \sum_{(x,v)\in E(G)} \alpha(x,v) = 1 \tag{3}$$

A *solution* of Σ is a function $\Psi : V(G) \to \mathbb{R}^k$, such that:

$$\Psi(x) = \begin{cases} f(x), & \text{if } x \in S \\ \sum_{v\in V(G)} \alpha(x,v)\Psi(v), & \text{otherwise} \end{cases} \tag{4}$$

Remark 1. Notice the following equivalences:

$$x \in S \quad \Longleftrightarrow \quad \sum_{(x,v)\in E(G)} \alpha(x,v) = 0 \quad \Longleftrightarrow \quad \forall v \in V(G), \quad \alpha(x,v) = 0$$

Let $\Sigma = (S, f, \alpha)$ be a barycentric system on a simple graph G. Let A be a subset of $V(G)$, the *relative source set* $S(A)$ of A is defined by:

$$S(A) = (S \cap A) \cup \{x \in A, \exists v \notin A, \quad \alpha(x,v) \neq 0\} \tag{5}$$

That is, $\forall x \in A$:

$$x \notin S(A) \quad \Longleftrightarrow \quad \sum_{(x,v)\in E(G_A)} \alpha(x,v) = 1 \tag{6}$$

Notice that, according to this definition, $S(V(G)) = S$.

Lemma 1. *Let G be a simple graph, let $\Sigma = (S, f, \alpha)$ be a barycentric system on G. Assume Ψ is a solution of Σ.*

Let A be a subset of $V(G)$. We define $\alpha_A : A \times A \to [0,1]$ by:

$$\forall x, y \in A, \quad \alpha_A(x,y) = \begin{cases} 0, & \text{if } x \in S(A) \\ \alpha(x,y), & \text{otherwise} \end{cases}$$

Then, $\Sigma_A = (S(A), \Psi_{|S(A)}, \alpha_A)$ is a barycentric system on G_A, having $\Psi_{|A}$ as a solution.

Proof. According to the definition of $S(A)$, the function α_A is such that:

$$\forall x, y \in A, \qquad \alpha(x,y) \neq 0 \Rightarrow \{x,y\} \in E(G_A) \text{ and } x \notin S(A)$$

$$\forall x \in A \setminus S(A), \qquad \sum_{(x,v)\in E(G_A)} \alpha(x,v) = 1$$

Thus, Σ_A is a barycentric system on G_A, obviously having $\Psi_{|A}$ as a solution. □

A barycentric system $\Sigma = (S, f, \alpha)$ on a simple graph G defines a directed graph D_Σ on the same vertex set as G, having S has its set of sources:

$$\forall x, y \in V(G), \quad (x,y) \in E(D_\Sigma) \iff \alpha(y,x) \neq 0 \tag{7}$$

Let G be a simple graph of order n. Consider any fixed numbering of the vertices of G. A barycentric system $\Sigma = (S, f, \alpha)$ on G define a $n \times n$ square matrix A (with coefficients in $[0,1]$) and a n column matrix F (with coefficients in $\mathbb{R}^k$) by $A_{i,j} = \alpha(v_i, v_j)$ and

$$F_i = \begin{cases} f(v_i) & \text{if } v_i \in S \\ \mathbf{0} & \text{otherwise} \end{cases}$$

So, the solutions of Σ correspond to the solutions Ψ of the equation $\Psi = A\Psi + F$.

Let G be a simple graph, $\Sigma = (S, f, \alpha)$ a barycentric system on G, and D_Σ the associated directed graph.

The *distance function* $\text{dist}_S : V(G) \to \mathbb{N} \cup \{\infty\}$ is defined as the minimum length of a directed path in D_Σ from S to a vertex, or ∞ is such a path does not exist.

1.3 General Resolution

In the next lemmas, G is assumed to be a simple graph, and $\Sigma = (S, f, \alpha)$ to be barycentric system on G. Elements of $\mathbb{R}^k$ are compared with respect to lexicographic order.

Lemma 2. *If $\sup_{v\in V(G)} \text{dist}_S(v) < \infty$, any barycentric function Ψ reaches its extremal values on S and, for any vertex x, there exists in D_Σ:*

- *a directed path P_x^+ from S to x such that (Ψ, dist_S) is strictly increasing on P_x^+ (with respect to lexicographic order).*

– *a directed path P_x^- from S to x such that $(\Psi, -\mathrm{dist}_S)$ is strictly decreasing on P_x^- (with respect to lexicographic order).*

Proof. We only have to prove the existence of P_x^+ as the existence of the directed path P_x^- is then obtained by considering $-\Psi$ instead of Ψ, and the monotony of these paths implies that the extrema of Ψ are reached on S.

Assume there is at least a vertex v such that such a path P_v^+ does not exist, and choose the one with the smallest possible value of $(\Psi(v), \mathrm{dist}_S(v))$. Obviously, v does not belong to S. So,

$$\Psi(v) = \sum_{(w,v)\in E(D_\Sigma)} \alpha(v,w)\Psi(w) \geq \min_{(w,v)\in E(D_\Sigma)} \Psi(w)$$

– If there exists an arc $(x, v) \in E(D_\Sigma)$ such that $\Psi(x) < \Psi(v)$, then P_x^+ exists (by minimality assumption on v) and the directed path $P_v^+ = P_x^+ + (x, v)$ contradicts the hypothesis.
– Otherwise, any edge $(x, v) \in E(D_\Sigma)$ is such that $\Psi(x) = \Psi(v)$. As there exists an edge $(x, v) \in E(D_\Sigma)$, such that $\mathrm{dist}_S(x) = \mathrm{dist}_S(v) - 1$ the directed path P_x^+ exists, and $P_v^+ = P_x^+ + (x, v)$ contradicts the hypothesis.

In both cases, we are led to a contradiction. □

Lemma 3. *If* $\sup_{v\in V(G)} \mathrm{dist}_S(v) < \infty$, *$\Sigma$ has a unique solution ϕ.*

Proof. The function ϕ exists and is unique if and only if the kernel of A is $\{0\}$, that is, if and only if the barycentric system (S, f_0, α) has no non zero solution, where f_0 is a constant function mapping S to $\mathbf{0}$. But this is an implication of the previous result, which asserts that the extrema of a solution of a barycentric system are reached on its source set, that is that a solution has the same extremal values as the limit function. □

Lemma 4. *If* $\sup_{v\in V(G)} \mathrm{dist}_S(v) = \infty$, *the barycentric system has an infinite set of solutions.*

Proof. Let V_S be the subset of vertices reachable from S by a directed path. For any value $c \in \mathbb{R}^k$, define the function $f' : (V(G)\setminus V_S)\cup S \to \mathbb{R}^k$ by $f'(v) = f(v)$ if $v \in S$, and $f'(v) = c$, otherwise. According to Lemma 3, the barycentric system $((V(G) \setminus V_S) \cup S, \alpha, f')$ has a solution ϕ, which is also a solution of Σ. This solution is such that $\phi|_{V(G)\setminus V_S} \equiv c$, and thus we get a infinite set of solutions of Σ. □

Theorem 2. *Let G be a simple graph. Any barycentric system $\Sigma = (S, f, \alpha)$ has a solution, which is unique if and only if any vertex of the directed graph D_Σ associated with Σ may be reached from S by a directed path.*

In the latter case, the solution ϕ reaches its extremal values on S and, for any vertex x, there exists in D_Σ:

- *a directed path P_x^+ from S to x such that (ϕ, dist_S) is strictly increasing on P_x^+ (with respect to lexicographic order).*
- *a directed path P_x^- from S to x such that $(\phi, -\mathrm{dist}_S)$ is strictly decreasing on P_x^- (with respect to lexicographic order).*

Proof. This is a direct consequence of the three preceding propositions. □

Corollary 1. *Let G be a simple graph and let $\Sigma = (S, f, \alpha)$ be a barycentric system on G having a unique solution.*

Then, for any subset A of $V(G)$, $|S(A)|$ is at least equal to the number of connected components of G_A.

Proof. It is sufficient to prove the case where G_A is connected. Let x be an element of A. If x belongs to S, we are done. Otherwise, as Σ has a unique solution, there exists, according to Theorem 2, a directed path P from S to x. This path enters A at vertex v which, by definition, belongs to $S(A)$. □

1.4 Planarity of a Barycentric Representation

Barycentric representations and convex drawings. Let G be a 2-connected planar graph. A straight line drawing of G is a *convex drawing* if each face of the graph is drawn as a convex polygon. The drawing is *strictly convex* if the vertices of a face are mapped to the extremal points of the polygon corresponding to the face. A *barycentric representation* of G with source set $S \subseteq V(G)$ is a mapping from $V(G)$ to $\mathbb{R}^2$, defining a straight line planar drawing of G, and such that any vertex not in S is a barycenter of its neighbors (with coefficients in $[0,1]$). Convex drawings are deeply related to barycentric representations:

Lemma 5. *Let G be a 2-connected planar graph. A straight line representation of G is convex if and only it is a barycentric representation of G, which source set S is the vertex set of a face of G and is embedded on a convex polygon (in a compatible order).*

Proof. Consider a convex drawing of G and let S be the vertex set of the external face. Let v be an internal vertex. As the faces including v are convex, we may augment the neighborhood of v to a wheel by adding straight line segments between consecutive non-adjacent neighbors while preserving planarity. Thus, v belongs to the convex hull of its neighbors. Hence, the drawing is a barycentric representation of G with source set S.

Conversely, assume S is the vertex set of a face of G embedded on a convex polygon of the plane. Assume the representation includes a non convex internal face, and let v_1, v_2, v_3 be three consecutive vertices of such a face defining a concave angle at v_2. Then, v_2 does not belong to the convex hull of its neighbors, a contradiction. □

Corollary 2. *The embedding defined by a convex drawing of a 2-connected planar graph G is uniquely determined by the choice of the external face.*

Corollary 3. *Let P be a convex polygon, let (p_1, p_2, p_3) be a triangle including P in its interior, let G be the graph induced by P, the triangle and a triangulation of the annulus between the triangle and the polygon, let s_1, s_2, s_3 be the vertices of G mapped to p_1, p_2, p_3 and let $g(s_i) = p_i$ (for $1 \leq i \leq 3$).*

Then, there exists a function $\alpha_P : V(G) \times V(G) \to [0,1]$, such that $\Sigma = (\{s_1, s_2, s_3\}, g, \alpha_P)$ is a barycentric system on G with unique solution Ψ mapping G to its geometric embedding. A simple geometric argument shows that α may be chosen in such a way that $\alpha(x,y) > 0$, for any pair (x,y) of points adjacent on P.

Barycentric Representations

Lemma 6. *Let G be a maximal planar graph and let $\mu : [0,1] \times V(G) \to \mathbb{R}^2$ be a mapping such that*

- *$\mu(0, \cdot)$ induces a plane straight line representation of G,*
- *$\mu(\cdot, x)$ is continuous for any vertex x of G,*
- *$\forall t \in]0,1]$ and for all triangle (x_1, x_2, x_3) of G, $\mu(t, x_1)$, $\mu(t, x_2)$ and $\mu(t, x_3)$ are not collinear.*

Then, $\mu(t, \cdot)$ induces a plane straight line representation of G for every t in $[0,1]$.

Proof. In order to check if a straight line representation of a maximal planar graph is plane, it is sufficient to check that no edge has null length, that no two adjacent edges are overlapping and that the circular orders of the edges around a vertex define a planar map. The first two cases may not occur, as it would imply the existence of an aligned triangle. As $\mu(\cdot, x)$ is continuous for any vertex x, the circular order around a vertex may only change if two consecutive edges overlap for some t, which would give rise to a contradiction. □

Given a weak plane straight line representation of G, a connected subgraph H is 0-*degenerated* if all its vertices are mapped to a same point. It is 1-*degenerated* if all its vertices are mapped to aligned points. A k-*degenerated component* is a maximal k-degenerated connected subgraph (it is thus an induced subgraph of G).

Lemma 7. *Let H be a 2-connected plane graph, and let $\Sigma = (S_H, f_H, \alpha_H)$ be a barycentric system on H having a unique solution Ψ_H, where S_H is the vertex set of the outer face of H.*

Assume f_H maps S_H to a convex polygon P in compatible order, and that the following condition holds for any subset A of vertices inducing a connected subgraph:

(i) $|S(A)| = 1 \Rightarrow |A| = 1$

(ii) $|S(A)| = 2 \Rightarrow H_A$ *is a path*

Then, there exists a maximal planar graph G and a barycentric system $\Sigma = (S, f, \alpha)$ *having a unique solution, such that:*

- *H is a subgraph of G,*
- $\Psi_{|V(H)}$ *is the solution of* Σ_H*,*
- *conditions (i) and (ii) hold for G (replacing* H_A *by* G_A*, and considering relative source sets according to* Σ*)*

Proof. Denote $S^{(H)}$ and $S^{(G)}$ the relative source set functions with respect to H, Σ_H and G, Σ_G, respectively.

The graph G_1 is obtained by adding a vertex in each interior face of H, adjacent to all the vertices of the face.

Let $\{p_1, p_2, p_3\}$ be a triangle having the polygon P in its interior, let G_2 be the graph obtained by geometrically triangulating the annulus between the triangle and the polygon, let s_1, s_2, s_3 be the vertices of G_2 corresponding to the p_1, p_2, p_3, let f be the mapping defined by $f(s_i) = p_i$, and let α_P be such that $(\{s_1, s_2, s_3\}, f, \alpha_P)$ is a barycentric system on G_2 having a unique solution Ψ, which restriction to S is f_H (such a function exists according to Corollary 3).

The graph G is obtained from the union of G_1 and G_2, by identifying the vertices corresponding to elements of S.

Then, define $\alpha : V(G) \times V(G) \rightarrow [0,1]$ as follows:

$$\alpha(x,y) = \begin{cases} \alpha_H(x,y), & \text{if } x, y \in V(H) \\ \alpha_P(x,y), & \text{if } x, y \in V(G_2) \\ \frac{1}{d(x)}, & \text{if } x \in V(G_1) \setminus V(H) \text{ and } y \in V(H), \\ 0, & \text{otherwise} \end{cases} \tag{8}$$

By construction, H is a subgraph of G. It is easily checked that $\Sigma = (\{s_1, s_2, s_3\}, f, \alpha)$ is a barycentric system on G having a unique solution, which restriction to $V(H)$ is a solution of Σ_H (according to Lemma 1).

Consider any subset $A \subseteq V(G)$ inducing a connected subgraph G_A of G, such that $|S_G(A)| \leq 2$ and H_A has no cycle. As no vertex $x \in A \setminus V(H) \setminus \{s_1, s_2, s_3\}$ may have all its neighbors in A (for otherwise its neighbors would form a cycle in H_A), any vertex in $A \setminus V(H)$ belongs to $S^{(G)}(A)$. By definition, for any $x \in S^{(H)}(A \cap V(H)) \setminus S_H$, there exists $v \in V(H) \setminus A$, $\alpha_H(x, v) > 0$. Thus, $\alpha(x, v) > 0$, what implies $x \in S^{(G)}(A)$. If $S^{(H)}(A) \cap S_H$ is not empty, it is easily checked that either $S^{(H)}(A) \cap S_H$ is included in $S^{(G)}(A)$, or S_H is included in A (and hence H_A includes a cycle), or $S^{(H)}(A)$ has cardinality at least 3. Altogether, we get:

$$S^{(G)}(A) = S^{(H)}(A \cap V(H)) \cup (A \setminus V(H)) \tag{9}$$

Assume there exists a subset A of $V(G)$ inducing a connected subgraph of G.

If $\mid S^{(G)}(A) \mid = 1$ then, according to (9), either $A \subseteq V(H)$ and $\mid S^{(H)}(A) \mid = 1$, or $A \cap V(H) = \emptyset$ and $\mid A \mid = 1$. Hence, (i) holds.

If $\mid S^{(G)}(A) \mid = 2$ then, according to (9):

- either $A \subseteq V(H)$ and $\mid S^{(H)}(A) \mid = 2$, and hence $G_A = H_A$ is a path,
- or $A \setminus V(H) = 1$ and $\mid S^{(H)}(A) \mid = 1$, and hence $\mid A \mid = 2$ and G_A is a path,
- or $A \cap V(H) = \emptyset$ and A is a subset of size 2 of $\{s_1, s_2, s_3\}$ (hence inducing a path).

Thus (ii) holds for G. □

Theorem 3. *Let G be a 2-connected plane graph, S the vertex set of its outer face, f a function mapping S to the points of a convex polygon in compatible order, and $\alpha : V(G) \times V(G) \to [0,1]$, such that $\Sigma = (S, f, \alpha)$ is a barycentric system on G with a unique solution Ψ.*

Then, Ψ defines a barycentric representation of G if and only if, for any subset A of $V(G)$ inducing a connected subgraph, the following assertions hold:

(i) $\mid S(A) \mid = 1 \Rightarrow \mid A \mid = 1$
(ii) $\mid S(A) \mid = 2 \Rightarrow G_A$ *is a path*

Proof. First, notice that the necessity of conditions (i) and (ii) is straightforward.

Assume (i) and (ii) hold. According to Lemma 7, we may assume that G is maximal planar and that f maps its outer face $\{s_1, s_2, s_3\}$ to a triangle $\{p_1, p_2, p_3\}$. It is a classical result [2,6,11] that G has a straight line embedding Γ with external face $\{s_1, s_2, s_3\}$. Moreover, using an affine transformation, we may assume that s_1, s_2, s_3 are mapped to p_1, p_2, p_3 in this embedding. According to Lemma 5, this is a barycentric representation, and thus induced by the solution of some barycentric system $\Sigma'_x = (S, f, \alpha')$. Let $\alpha_t = (1-t)\alpha' + \alpha$. Then, $\Sigma^{(t)} = (S, f, \alpha_t)$ is obviously a barycentric system having a unique solution $\Psi^{(t)}$ continuously depending on t. Let $\mu(t, v) = \Psi^{(t)}(v)$. If, for any $t \in]0,1]$, there exists no triangle (x_1, x_2, x_3) such that $\mu(t, x_1)$, $\mu(t, x_2)$ and $\mu(t, x_3)$ are collinear then, according to Lemma 6, $\Psi = \mu(1, \cdot)$ will define a barycentric representation of G. First notice that, if $0 < t < 1$ then, for any $x, y \in V(G)$, $\alpha_t(x, y) > 0$ if and only if $\alpha(x, y) > 0$ or $\alpha'(x, y) > 0$. Thus, for any subset A of vertices, the relative source set of A computed according to Σ_t includes the relative source set of $S(A)$ computed according to Σ. Hence, as (i) and (ii) hold for $t = 1$, these conditions also hold for any $t \in]0,1]$. Moreover, if conditions (i) and (ii) imply that no triangle (x_1, x_2, x_3) of G is such that $\mu(1, x_1)$, $\mu(1, x_2)$ and $\mu(1, x_3)$ are collinear then the same will apply for any $t \in]0,1[$.

Assume there exists a triangle $\{x_1, x_2, x_3\}$ of G, such that $\Psi(x_1), \Psi(x_2)$ and $\Psi(x_3)$ are collinear. Let $G_A \subseteq \{x_1, x_2, x_3\}$ be the corresponding 1-degenerated component, and let $ax + by + c = 0$ be an equation of the straight line Δ including $\Psi(A)$. Consider the barycentric system $\Sigma_\Delta = (S, af_x + bf_y + c, \alpha)$ (where $(f_x(v), f_y(v)) = f(v)$) and its unique solution Ψ. The subgraph G_A is then a connected component of $\Psi^{-1}(0)$.

Let G^+ be the directed planar graph obtained from G by adding to vertices ω^- and ω^+ and the edges $\{\omega^-, s\}$ (if $s \in S$ and $\phi(s) \leq 0$), $\{\omega^+, s\}$ (if $s \in S$ and $\phi(s) \geq 0$) and $\{\omega^-, \omega^+\}$.

For each vertex $a_i \in S(A) \setminus S$, there exists a vertex v such that (v, a_i) is an arc of D_{Σ_Δ} and $\Psi(v) \neq 0$. From the equation $0 = \Psi(a_i) = \sum_{(x,a_i) \in E(D_{\Sigma_\Delta})} \Psi(x)$, we deduce that there exists, for each such a_i, two vertices x_i and y_i, such that (x_i, a_i) and (y_i, a_i) are arcs of D_{Σ_Δ} and such that $\Psi(x_i) < 0 < \Psi(y_i)$. According to Theorem 2, there exists in D_{Σ_Δ}, for each such a_i, directed paths $P_{a_i}^+$ and $P_{a_i}^-$, whose vertices (except a_i) have a negative (resp. positive) Ψ-value.

Thus, there exists, in G^+, for each $a_i \in S(A)$, paths $P_1(a_i)$ and $P_2(a_i)$, from ω^- (resp. ω^+) to a_i, which vertices (except a_i) have a negative (resp. positive) Ψ-value. Hence, we have:

$$\begin{aligned} \forall \{a, a'\} \subseteq S(A) \quad & P_1(a) \cap P_2(a') = \emptyset \\ \forall a \in S(A), \quad & P_1(a) \cap A = P_2(a) \cap A = P_1(a) \cap P_2(a) = \{a\} \\ \forall a \in S(A), \quad & \omega^- \in P_1(a) \text{ and } \omega^+ \in P_2(a) \end{aligned}$$

Hence, two cases may occur:

- $A = S(A)$. Then, the triangle $\{x_1, x_2, x_3\}$, the paths $P_i(x_j)$ ($i \in \{1,2\}, j \in \{1,2,3\}$), and the edge $\{\omega^-, \omega^+\}$ may be contracted to K_5, contradicting the planarity of G^+.
- there exists a vertex x in $A \setminus S(A)$. According to the 3-connexity of G^+, there exists 3 disjoint paths from x to ω^+. By continuity, these paths leave A at some distinct vertices $v_1, v_2, v_3 \in S(A)$. Then, the subpaths from x to v_1, v_2, v_3, and the paths $P_i(v_j)$ ($i \in \{1,2\}, j \in \{1,2,3\}$) may be contracted to a $K_{3,3}$, contradicting the planarity of G^+.

In both cases, we are led to a contradiction. Thus, no triangle of G may be mapped to 3 collinear points. Hence, Ψ is a barycentric representation of G. □

2 Jordan Arc Contact Systems

2.1 Introduction

A *Jordan arc* is an arc homeomorphic to a straight line segment. A *pseudo-line* is an arc homeomorphic to a straight line. A *contact system* is a set of Jordan arcs in the plane, such that two arcs intersect at most once at a point which is internal to at most one arc. A contact system is *stretchable* if there exits an homeomorphism which transforms it into a contact system whose arcs are straight line segments.

A set $\mathcal{L} = (L_1, \ldots, L_k)$ of pseudo-lines is a *weak arrangement of pseudo-lines* if it satisfies :

$$\forall i \neq j, \quad \left| L_i \cap L_j \right| \leq 1 \tag{10}$$

A weak arrangement of pseudolines is an *arrangement of pseudolines* if it satisfies the additional axiom :

$$\forall i, j, k, \quad (L_i \cap L_j = \emptyset \text{ and } L_i \cap L_k \neq \emptyset) \Rightarrow L_j \cap L_k \neq \emptyset$$

A contact system $\mathcal{A} = (A_1, \ldots, A_k)$ is *extendable* into an arrangement of pseudo-lines (resp. into a weak arrangement of pseudolines) if there exists a arrangement of pseudo-lines (resp. a weak arrangement of pseudolines) $\mathcal{L} = (L_1, \ldots, L_k)$, such that $A_i \subseteq L_i$ for each $i \in \{1, \ldots, k\}$.

Remark 2. It is easy to prove that a contact system is extendable into an arrangement of pseudo-lines if and only if it is extendable into a weak arrangement of pseudo-lines.

2.2 Extendibility and Extremal Points

An *extremal point* of a contact system is a point of the union of the arcs which is interior to no arc. We note $\delta(\mathcal{A})$ the set of the extremal points of the contact system $\mathcal{A}$. A *maximal contact system* is a contact system whose extremal points belong to the unbounded region. A contact system $\mathcal{A} = (A_1, \ldots, A_k)$ is *extendable into a maximal contact system* if there exists a maximal contact system $\mathcal{A}' = (A'_1, \ldots, A'_k)$, such that $A_i \subseteq A'_i$ for each $i \in \{1, \ldots, k\}$.

Remark 3. If a contact system is extendable into a weak arrangement of pseudo-lines, it is extendable into a maximal contact system, which in turn is extendable into a weak arrangement of pseudo-lines.

Let $\mathcal{A}$ be a maximal contact system extendable into a weak arrangement of pseudo-lines $\mathcal{L}$.

- Let p be an interior point of an arc A (but no other arc) on the unbounded region of $\mathcal{A}$.
- Let L be the pseudoline extending A in $\mathcal{L}$, and let L^- and L^+ be the two halves of L delimited by p.
- Let H be a half pseudo-line having its endpoint at p and having no other intersection in $\mathcal{A}$.

Then, L and H induce a "partition" of A into three new contact systems A^-, A^+ and A^0 (see Fig. 1):

- H, L^- and L^+ define three unbounded regions $\mathcal{R}^-, \mathcal{R}^+$ and $\mathcal{R}^0$ having $H \cup L^-$, $H \cup L^+$ and L as respective frontiers.
- We define J^-, J^+ and J^0 as the sub-arcs of $H \cup L^-, H \cup L^+$ and L strictly including all the intersections of these later pseudolines with $\mathcal{A}$.

These regions define three contact systems of Jordan arcs :

$$\mathcal{A}^- = \{J^-\} \cup \{A \cap \mathcal{R}^-, \quad A \in \mathcal{A} \text{ and } A \cap \mathcal{R}^- \text{ is a nonempty arc}\}$$
$$\mathcal{A}^+ = \{J^+\} \cup \{A \cap \mathcal{R}^+, \quad A \in \mathcal{A} \text{ and } A \cap \mathcal{R}^+ \text{ is a nonempty arc}\}$$

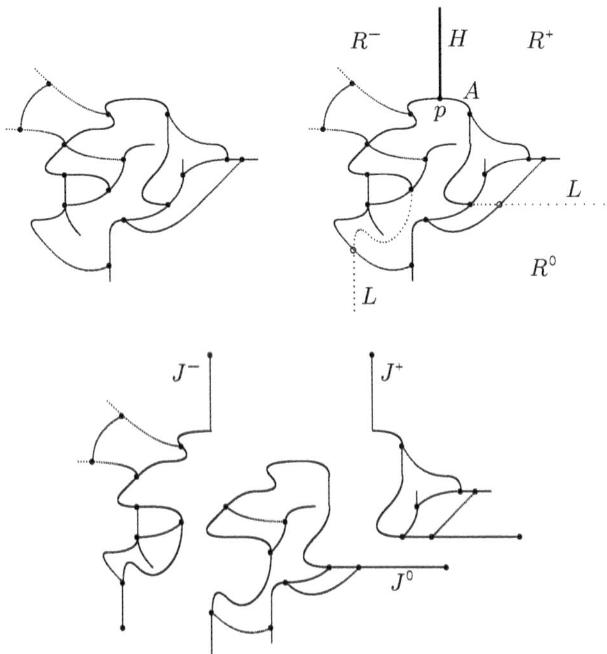

Fig. 1. Partitioning a Contact System

$$\mathcal{A}^0 = \{J^0\} \cup \{A \cap \mathcal{R}^0, \quad A \in \mathcal{A} \text{ and } A \cap \mathcal{R}^0 \text{ is a nonempty arc}\}$$

These three contact systems have some nice properties :

Lemma 8. *The contact systems $\mathcal{A}^-, \mathcal{A}^+$ and $\mathcal{A}^0$ are maximal and extendable into weak arrangements of pseudo-lines.*

Proof. The maximality is straightforward, as the extremal points of these contact systems are either extremal points of $\mathcal{A}$ or the extremities of J^-, J^+ or J^0.

The contact systems $\mathcal{A}^-$ and $\mathcal{A}^+$ are homeomorph to the contact systems where J^- (resp. J^+) is replaced by J^0. Then, the weak arrangement of pseudo-lines which extends $\mathcal{A}$ is an extension of all of the contact systems. □

Lemma 9. *If the arc A includes no extremal point of $\mathcal{A}$, we have :*

$$2 \leq |\mathcal{A}^-| < |\mathcal{A}|, \quad 2 \leq |\mathcal{A}^+| < |\mathcal{A}|, \quad 3 \leq |\mathcal{A}^0| \leq |\mathcal{A}|,$$

and each of $\mathcal{A}^-, \mathcal{A}^+$ and $\mathcal{A}^0$ has two extremal points belonging to a same arc.

Proof. As A includes no extremal point of $\mathcal{A}$ the extremity of A in A^- (resp. A^+) is interior to some arc B^- (resp. B^+). As A may not intersect an arc twice, we have $B^- \neq B^+$. Moreover, each of these arcs intersects at most once the extension L of A, and hence the arc B^- (resp. B^+) does not meet the region $\mathcal{R}^+$ (resp. $\mathcal{R}^-$). Thus, $\left|\mathcal{A}^+\right| \leq \left|\mathcal{A}\right| - 1$ and the same holds for $\left|\mathcal{A}^-\right|$. Notice also that $\mathcal{A}^-$ and $\mathcal{A}^+$ include at least two arcs : J^- (resp. J^+) with two extremal points on it and the non empty portion of B^- (resp. B^+) in $\mathcal{R}^-$ (resp. $\mathcal{R}^+$). Moreover, $\mathcal{A}^0$ includes at least three arcs : J^0 (with two extremal points on it) and the portions of B^- and B^+ in $\mathcal{R}^0$. □

Lemma 10. *Any contact system extendable into an arrangement of pseudo-lines has at least 3 extremal points, unless it has cardinality at most one.*

Proof. First notice that we may prove the lemma for maximal contact systems only, according Remark 3 and the fact that the extended contact system may not have more extremal points than the original one.

The lemma is straightforward for contact systems of cardinality at most two. So, assume that for any $2 \leq i < k$, the lemma holds for maximal contact systems of cardinality i and assume there exists a maximal contact system $\mathcal{A}$ of cardinality k having at most 2 extremal points.

Notice that at least 3 arcs of $\mathcal{A}$ meet the unbounded region in a sub-arc. Otherwise, the contact system would be bounded by a single arc or by two arcs intersecting each other twice, and the system could not be extended into a weak arrangement of pseudo-lines.

- Assume $\mathcal{A}$ has two extremal points belonging to a same arc.
  From the previous construction and according to Lemma 8, $\mathcal{A}^-$ and $\mathcal{A}^+$ are extendable into weak arrangements of pseudo-line and, according to Lemma 9, have a smaller cardinality than $\mathcal{A}$ but include at least two arcs. By induction, they have at least 3 extremal points. Among these, there is one which is an extremal point of $\mathcal{A}$ belonging to the interior of $\mathcal{R}^-$ (resp. $\mathcal{R}^+$). Notice that these extremal points do not belong to a same arc : otherwise, this arc would intersect A twice. Hence, as the contact system $\mathcal{A}$ has an arc with two extremal points on it, it has at least 3 extremal points, what leads to a contradiction.
  Hence, the lemma holds for any maximal contact system with at most k arcs having two extremal points belonging to a same arc.
- Assume $\mathcal{A}$ does not include two extremal points belonging to a same arc.
  From the same construction and according to Lemma 8, the contact system $\mathcal{A}^0$ is maximal and, according to Lemma 9, it has cardinality at most k and has two extremal points belonging to a same arc. According to the previous case, it has at least 3 extremal points. Among these, one is an extremal point of $\mathcal{A}$ belonging to the interior of $\mathcal{R}^0$. Together with the 2 extremal points of $\mathcal{A}$ belonging to the interior of $\mathcal{R}^-$ and $\mathcal{R}^+$ (as in the previous case), we get 3 extremal points, which lead to a contradiction.

□

Lemma 11. *Any contact system $\mathcal{A}$ extendable into a weak arrangement of pseudo-lines is extendable into a maximal contact system, each subsystem of which has at least 3 extremal points unless it has cardinality at most one.*

Proof. As any subsystem $\mathcal{A}'$ of the extension of $\mathcal{A}$ into a maximal contact system is extendable into a weak arrangement of pseudo-lines, the result follows from Lemma 10. □

2.3 Stretching

Consider a maximal contact system of Jordan arcs $\mathcal{A}$, and the plane graph G whose vertices are the points which are extremity of a least an arc of $\mathcal{A}$ and whose edges are the sub-arcs between such points.

Let S be the set of the vertices of G corresponding to points which are interior to no arc in $\mathcal{A}$.

Let f be a function mapping S to the vertices of a convex polygon, in compatible order.

Let $\alpha : V(G) \times V(G) \to [0,1]$ be a function such that:

- $\alpha(x, y) = 0$ if and only if y is interior to no arc in $\mathcal{A}$ or x and y are non adjacent in G.
- $\alpha(x_1, y) + \alpha(x_2, y) = 1$ if y is interior to some arc $A \in \mathcal{A}$ and has x_1 and x_2 as neighbors in A.

Then, $\Sigma = (S, f, \alpha)$ is a barycentric system on G.

Notation 1. *Given an arc $A \in \mathcal{A}$, the path $P(A)$ is the path of G, whose vertices and edges are induced by A. The set $\mathcal{P}$ of these paths induces a covering of G by edge disjoint paths.*

By extension, given a contact system $\mathcal{A}'$, $P(\mathcal{A}')$ denotes the union of the paths $P(A)$ $(A \in \mathcal{A}')$.

Notation 2. *Let $X \subseteq V(G)$. We note $\mathcal{A}(X)$ the set of the arcs of $\mathcal{A}$ having at least two points corresponding to vertices in X.*

We now may link Theorem 3 with the stretching problem, using the next notations and lemmas.

Lemma 12. *Let $\mathcal{A}'$ be a subsystem of $\mathcal{A}$. Then $S(V(P(\mathcal{A}'))) = \delta(\mathcal{A}')$.*

Proof. A vertex of $X = V(P(\mathcal{A}'))$ belongs to the $S(X)$ if and only if it has no incoming edge in D_Σ (hence is an extremal point of $\mathcal{A}$) or if its has an incoming edge in D_Σ from a vertex which does not belong to X (and thus is internal to some arc in $\mathcal{A} \setminus \mathcal{A}'$ or, equivalently, is an extremal point of $\mathcal{A}'$ but not an extremal point of $\mathcal{A}$). Thus, $S(X)$ is the set of the extremal vertices of $\mathcal{A}'$. □

Lemma 13. *Let $X \subseteq V(G)$ have cardinality at least 2 and induce a connected subgraph of G. Then,*

$$X \subseteq V(P(\mathcal{A}(X))) \tag{11}$$
$$\left| S(X) \right| \geq \left| S(V(P(\mathcal{A}(X)))) \right| \tag{12}$$

Proof. Inclusion (11) is straightforward as each edge of G_X is included in the path corresponding to some arc of $\mathcal{A}(X)$.

Consider the connected graph H obtained from G_X by adding $k \geq 0$ paths corresponding to arcs in $\mathcal{A}(X)$ and assume that $\left| S(V(H)) \right| = \left| S(X) \right|$ holds. Let A be some arc of $\mathcal{A}(X)$, such that $P(A)$ does not belong to H.

- Assume H include two vertices v and w of $P(A)$ but none of the edges of P linking them.
  Then, by adding those missing edges and vertices, no vertex is added to $S(V(H))$.
  After this first completion, the internal vertices of $P(A)$ have at most one incoming edge from the complement of H.
- Otherwise, H includes an interior vertex v of $P(A)$, but none of the vertices between v and an extremity w of P.
  Then, v belongs to $S(V(H))$. By adding the sub-path from v to w, exactly one vertex is added to the $S(V(H))$, namely w. As $P(A)$ has at least two vertices in common with H, v is an interior vertex of $P(A)$. Moreover, as the previous case does not apply, v had initially one incoming edge (exactly) from the complement of H'. Thus, after the addition of the sub-path from v to w, the vertex v will no longer belong to $S(V(H))$.

Thus, we may add a new path to H without increasing $\left| S(V(H)) \right|$, which completes the induction. □

Theorem 4. *Let $\mathcal{A}$ be a contact system of Jordan arcs. Then, the following conditions are equivalent :*

(C_1) *$\mathcal{A}$ is stretchable,*
(C_2) *$\mathcal{A}$ is extendable into a weak arrangement of pseudo-lines,*

Proof. It is straightforward that Condition (C_1) implies condition (C_2). According to Lemma 11, this second condition implies that $\mathcal{A}$ is extendable into a maximal contact system, each subsystem of which has at least 3 extremal points or has cardinality at most one. Consider any subset X of vertices of G with cardinality at least 2 and inducing a connected subgraph of G. Then, according to Lemma 13 and Lemma 12, $\left| S(X) \right| \geq \left| S(V(P(\mathcal{A}(X)))) \right| = \left| \delta(\mathcal{A}(X)) \right|$. But, either $\mathcal{A}(X)$ includes a single arc and hence G_X is a path and $\left| S(X) \right| \geq 2$ or $\mathcal{A}(X)$ includes at least two arcs and then $\left| S(X) \right| \geq 3$, according to the hypothesis. The result now follows from the application of Theorem 3. □

3 Conclusion

The equivalence between stretchability and extendibility into an arrangement of pseudo-lines for contact systems of Jordan arcs does not extend to intersection systems (stretchability of pseudoline arrangements is NP-hard, as proved by Short [5]). With more work, another equivalent condition may be given : A contact system is stretchable if and only if any subsystem has at least 3 extremal points on its unbounded region, unless it has at most an arc.

References

1. E. Colin de Verdière, M. Pocchiola, and G. Vegter, *Tutte's barycenter method applied to isotopies*, Proc. 13th Canad. Conf. Comput. Geom., 2001.
2. I. Fáry, *On straight line representation of planar graphs*, Acta Scientiarum Mathematicarum (Szeged) **II** (1948), 229–233.
3. M. Linial, L. Lovász, and A. Wigdersib, *Rubber bands, convex embeddings and graph connectivity*, Combinatorica **8** (1988), no. 1, 91–102.
4. D. Orden, *Two problems in geometric combinatorics: Efficient triangulations of the hypercube;plane graphs and rigidity.*, Ph.D. thesis, University of Santander, Spain, 2003.
5. P. W. Shor, *Stretchability of pseudolines is NP-hard*, DIMACS Series in Discrete Mathematics and Theoretical Computer Science **4** (1991), 531–554.
6. S.K. Stein, *Convex maps*, Proc. Amer. Math. Soc., vol. 2, 1951, pp. 464–466.
7. E. Steinitz and H. Rademacher, *Vorlesung über die Theorie der Polyeder*, Springer, Berlin, 1934.
8. C. Thomassen, *Planarity and duality of finite and infinite planar graphs*, J. Combinatorial Theory, Series B **29** (1980), 244–271.
9. W.T. Tutte, *Convex representations of graphs*, Proc. London Math. Society, vol. 10, 1960, pp. 304–320.
10. ———, *How to draw a graph*, Proc. London Math. Society, vol. 13, 1963, pp. 743–768.
11. K. Wagner, *Über eine Eigenschaft der Ebenen Komplexe*, Math. Ann. **114** (1937), 570–590.

Noncrossing Hamiltonian Paths in Geometric Graphs

Jakub Černý, Zdeněk Dvořák, Vít Jelínek, and Jan Kára

Department of Applied Mathematics
Charles University
Malostranské náměstí 25, 118 00 Praha 1
{kuba,rakdver,jelinek,kara}@kam.mff.cuni.cz

Abstract. A geometric graph is a graph embedded in the plane in such a way that vertices correspond to points in general position and edges correspond to segments connecting the appropriate points. A noncrossing Hamiltonian path in a geometric graph is a Hamiltonian path which does not contain any intersecting pair of edges. In the paper, we study a problem asked by Micha Perles: Determine a function h, where $h(n)$ is the largest number k such that when we remove arbitrary set of k edges from a complete geometric graph on n vertices, the resulting graph still has a noncrossing Hamiltonian path. We prove that $h(n) = \Omega(\sqrt{n})$. We also determine the function exactly in case when the removed edges form a star or a matching, and give asymptotically tight bounds in case they form a clique.

1 Introduction

A *geometric graph* is a graph drawn in the plane so that its vertices are represented by points in general position (i.e., there are no three collinear points) and its edges are straight–line segments connecting the corresponding vertices.

Lately geometric graphs have been intensively studied. There are many papers studying the smallest number of edges needed to guarantee the occurence of some fixed subconfiguration in any geometric graph (the trivial result of this type following from Euler's polyhedral formula is that any geometric graph with at least $3n - 5$ edges must have two edges which intersect). Interesting results of this sort may be found in [1], [4], [5], [6], [7] to name a few or in the surveys on geometric graphs ([2], [3]).

In our paper we study the existence of a noncrossing Hamiltonian path (i.e. Hamiltonian path which does not cross itself) in a given geometric graph. In particular we concentrate on a problem presented by Micha Perles on DIMACS Workshop on Geometric Graph Theory in 2002, which asks to determine the largest possible number $h(n)$ such that every geometric graph on n vertices with at least $\binom{n}{2} - h(n)$ edges has a noncrossing Hamiltonian path. Apart from improving the asymptotic bounds of $h(n)$, we also focus on the restriction of the problem to some classes of geometric graphs. Let $h_1(n)$ denote the largest k such that when we remove the edges of an arbitrary complete subgraph of size at most k from a complete geometric graph on n vertices, the resulting graph always has a noncrossing Hamiltonian path. We prove that there exist constants $0 < c_1 < c_2$ such that $c_1 \cdot \sqrt{n} < h_1(n) < c_2 \cdot \sqrt{n}$ (Theorems 1 and 3). Let $h_2(n)$ denote the largest k such that when we remove an arbitrary star with at most k edges from a complete geometric graph on n vertices, the resulting graph still has a noncrossing

G. Liotta (Ed.): GD 2003, LNCS 2912, pp. 86–97, 2004.

Hamiltonian path. In Theorem 6 we show that $h_2(n) = \lceil n/2 \rceil - 1$. We also prove that $h_3(n) = \lceil n/2 \rceil - 1$, where $h_3(n)$ is the largest k such that when we remove at most k arbitrary edges from the complete geometric graph on n vertices in convex position, then the graph still has a noncrossing Hamiltonian path (Theorems 4 and 5). Further we prove that when we remove any matching from a complete geometric graph, the resulting graph has a noncrossing Hamiltonian path (Theorem 2).

The paper is organized as follows: In section 2 we introduce basic definitions and notation. In sections 3 and 4 we study complete geometric graphs with removed complete subgraph and we prove the asymptotically tight bounds on the size of complete subgraph removed. In section 5 we prove the tight bounds on the number of edges removed from convex geometric graph and in section 6 we prove the tight bounds on the size of a star removed from a complete geometric graph.

2 Definitions and Notation

In this section we introduce basic definitions and notation used throughout this paper. A *geometric graph* G is an ordered pair (V, E) where V is a set of points in general position in the plane (called *vertices of* G) and E is a set of straight–line segments connecting two vertices (called *edges of* G). A *Hamiltonian path* in a graph G is a path contained in G which visits all the vertices of G. A *noncrossing Hamiltonian path* in a geometric graph $G = (V, E)$ is a Hamiltonian path which does not intersect itself. A *convex hull* of a set of points $X \subset R^2, X = \{x_1, \ldots, x_n\}$ is a set of points $H = \{h \in R^2 : \exists a_1, \ldots a_n$ such that $\forall i \in \{1, \ldots, n\}$ $a_i \in R, a_i \geq 0, \sum_{i=1}^{n} a_i = 1$ and $h = \sum_{i=1}^{n} a_i x_i\}$. We say that a point p lies *below a line* l (line l must not be parallel to y-axis) if it lies in the half-plane defined by l which contains $-\infty$ on the y axis. Similarly we use the term *above a line*. A point u lies *to the left of* v if the x-coordinate of u is less than or equal to the x-coordinate of v. Similarly we define that u is *to the right of* v. Let s be a segment in R^2 defined by two endpoints $u, v \in R^2$, where u is to the left of v. We say that a point $p \in R^2$ lies *below the segment* s if it lies below a line defined by u and v, to the right of u and to the left of v. Let $Z \subset R^2, Z = \{z_1, \ldots, z_n\}$ be a set of points in the plane. An *x-monotone order* of Z is an ordering of Z in which the x-coordinates of the points form a monotone sequence. Analogously we define a *y-monotone order* of Z. A point $z \in Z$ is called an *extremal point* of Z if it belongs to the boundary of the convex hull of Z. A segment uv, with u and v in Z, is called an *extremal segment* of Z if it is a subset of the boundary of the convex hull of Z.

3 The Lower Bound for Complements of Cliques

In the following two sections we consider a particular class $\mathcal{C}$ of geometric graphs — the complements of complete subgraphs. A geometric graph $G = (V, E)$ is in $\mathcal{C}$ iff there exist $X, Y \subseteq V$ such that $V = X \cup Y$, $X \cap Y = \emptyset$ and E is the set of all the possible edges with at least one enpoint in Y (i.e. G is obtained from a complete graph by removing the edges of a complete subgraph). We prove that there exists a constant $c_1 > 0$ such that any geometric graph $G \in \mathcal{C}$ with $|X| \leq c_1 \cdot \sqrt{|V|}$ has a noncrossing Hamiltonian path.

Lemma 1. *Let $G = (V, E)$ be a geometric graph, $G \in \mathcal{C}$. If there exists a line l such that all the vertices from X are in one half-plane defined by l and at least $|X|$ vertices from Y are in the other half-plane, then there exists a noncrossing Hamiltonian path in G.*

Proof. We may WLOG assume that the line l is parallel to the y-axis and that all the vertices from the set X are in the left half-plane. We iterate the procedure described below to find a noncrossing Hamiltonian path in G. In each step, the procedure adds a new vertex to the path that it has constructed in the previous steps, according to the following rules: first, we take the upper extremal segment of the vertices not yet added to the path (the whole set V in the beginning) which crosses l (see Figure 1 on the left). If the last vertex added to the path or the left endpoint of the extremal segment is from Y, then the left endpoint of the segment is added to the path (Figure 1 on the right), otherwise we add the right endpoint (Figure 2 on the left). If no vertices were added yet, we may add either of the endpoints. If l does not cross the convex hull, then we simply add the remaining vertices to the path in an x-monotone order. See Figure 2 on the right for an example of the path produced by this algorithm.

It is clear that the algorithm finishes when it adds all the vertices to the path. It is also easy to see that there are no two consecutive vertices from X on the constructed path and hence it is indeed a path in G. The only place in the algorithm, where two vertices from X might be added consecutively to the path, is when the convex hull of the vertices not yet on the path no longer intersects l, all these remaining vertices lie to the left of l, and the algorithm is adding them in an x-monotone order. But at that time there is at most one vertex from X not on the path: the other vertices from X were added in the previous steps because there are at least $|X|$ vertices from Y in the right half-plane, and each of them has been added to the path immediately after a vertex from Y, with the possible exception of the first vertex of the path.

It remains to prove that the path is noncrossing. We check that after each step of the algorithm, the path does not intersect the convex hull of the remaining vertices including the vertex just added to the path. From this it is obvious that the path does not intersect itself (at each vertex each of the following edges of the path must lie in the convex hull and the previous edges lie outside of it). When the path contains only one vertex, the claim is obviously true. When we add a new vertex to the path, the edge connecting the new vertex to the previous vertex on the path cannot intersect the convex hull of the remaining vertices — if the edge intersected the convex hull, then the previous vertex on the path would have to lie in the lower half-plane defined by the upper extremal segment of the remaining vertices intersecting l (see Figure 3). But then we get contradiction with the choice of the vertex in the previous step of the algorithm (the vertex cannot be an endpoint of the upper extremal segment intersecting l). From the induction we know that no other edge of the path can intersect the convex hull of the remainig vertices and so we have proven that the path does not intersect itself. □

Now we prove a similar result for a general choice of the sets X and Y:

Theorem 1. *Let $G = (V, E)$ be a geometric graph, $G \in \mathcal{C}$. If $|Y| \geq 2 \cdot |X| \cdot (|X| + 1)$ then there exists a noncrossing Hamiltonian path in G.*

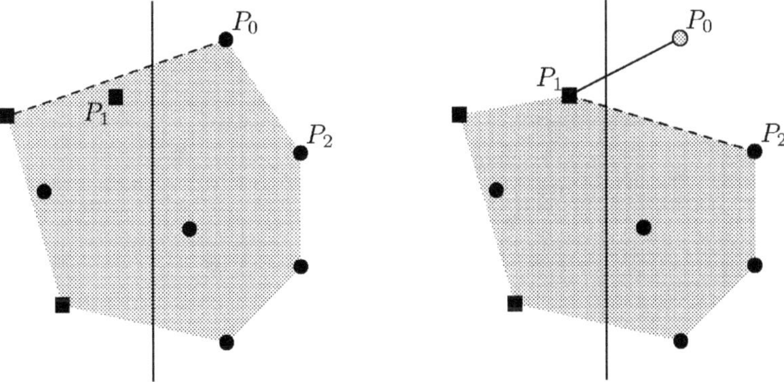

Fig. 1. The left figure shows the first step of the algorithm. The right figure shows the second step of the algorithm — add left end of the segment to the path.

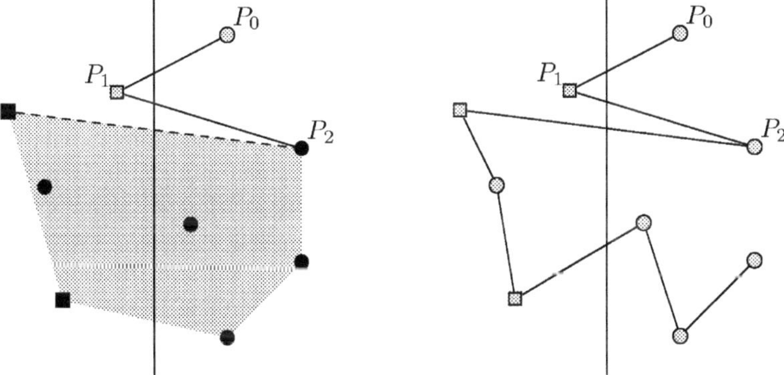

Fig. 2. On the left is the third step of the algorithm — add right end of the segment to the path. On the right is the whole noncrossing Hamiltonian path.

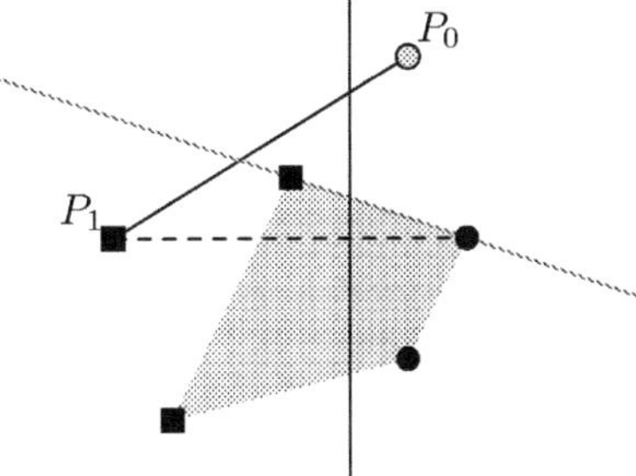

Fig. 3. If the segment to new vertex intersected the convex hull of the remaining vertices the previous vertex on the path would have to lie in the lower half-plane.

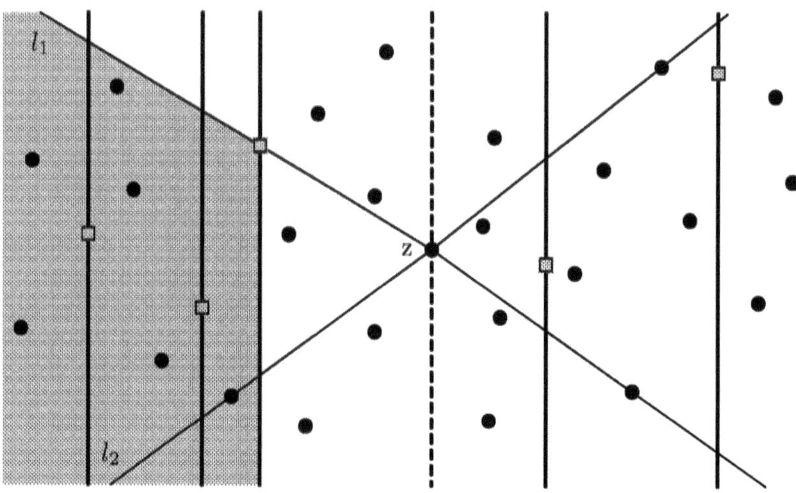

Fig. 4. Partitioning of the plane into the strips and choice of the vertices to which Lemma 1 should be applied.

Proof. We can WLOG assume that there are no two vertices from V with the same x coordinate. Now consider the partitioning of the plane into $|X| + 1$ vertical strips separated by vertical lines passing through the points of X (see Figure 4). From the pigeonhole principle it follows that there is a strip S with at least $2 \cdot |X|$ vertices from Y in it. Let x_l denote the number of vertices from X to the left of S (including the vertex on the left boundary of S). Similarly let x_r denote the number of vertices of X to the right of S. Now we can certainly choose a vertex z from S such that there are at least $2 \cdot x_l$ vertices in S to the left from z and at least $2 \cdot x_r$ vertices in S to the right from z. The vertical line passing through z splits S into two strips, denoted S_l (the left one) and S_r (the right one).

We now describe a procedure to find a noncrossing path starting in z and containing all the vertices to the left of z. We find lines l_1, l_2 such that $z \in l_1 \cap l_2$, both l_1 and l_2 contain some vertex lying to the left from S_l and there is no vertex lying to the left from S_l which would lie above l_1 or below l_2 (see Figure 4). It is clear that in S_l there are either at least x_l vertices below l_1 or at least x_l vertices above l_2. Lets WLOG assume that the former is true. Let Z denote the set of vertices lying to the left from z and below l_1, including the two vertices on the line l_1. Now we can apply Lemma 1 to the set Z (line l from the statement of the lemma is the left boundary of the strip S). From the lemma we get a noncrossing path starting in z containing all the vertices from Z (note that z is an endpoint of the upper extremal segment of Z intersecting l). Because there are no vertices from Z above the first segment of the path (all the vertices must lie below a line defined by the extremal segment from the second step of the procedure from Lemma 1 and the whole area above the first segment of the path lies above this line — see Figure 5), we can replace the first segment of the path by the path going through all the vertices in S_l above the line l_1 in an x-monotone order (see Figure 6). By the replacing we got a noncrossing path starting in z and using all the vertices of V to the left of z. Similarly we can get a path starting from z and containing all the vertices of V to the right of z, then join these two paths in z and get a noncrossing Hamiltonian path for G. □

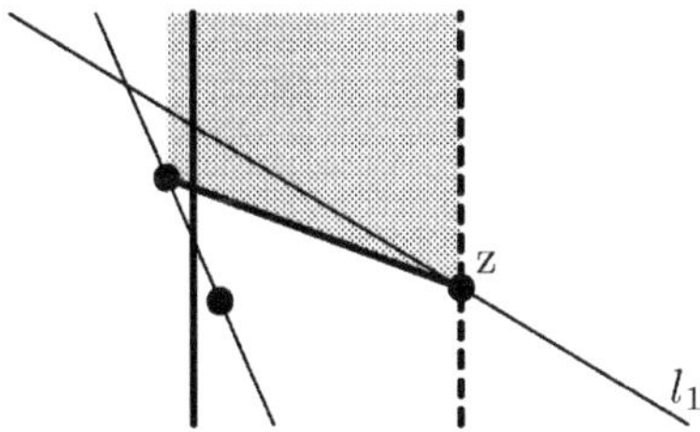

Fig. 5. There are no vertices lying above the first segment of the path and below the line l_1

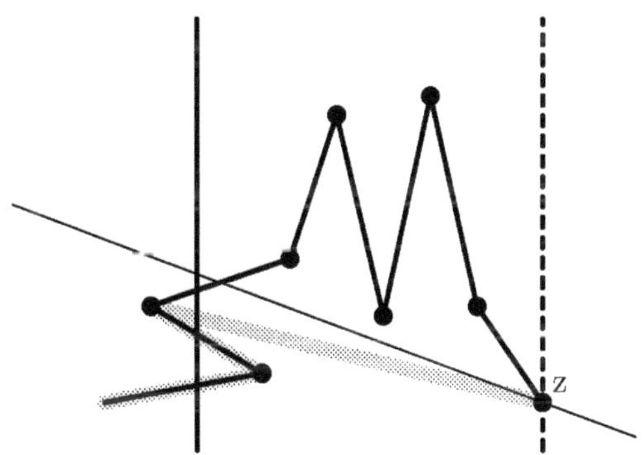

Fig. 6. Replacing the first segment of the path with an x-monotone path

Now using the Theorem 1 we can prove the following corollary:

Corollary 1. *Let $G = (V, E)$ be a geometric graph and let $k = \left(\binom{|V|}{2} - |E|\right)$ (i.e., k is the number of edges not present in G). If $2 \cdot 2 \cdot k \cdot (2 \cdot k + 1) \leq |V| - 2 \cdot k$ then there exists a noncrossing Hamiltonian path in G.*

Proof. The idea of the proof is easy. We just use Theorem 1 for the original graph with removed edges of the complete subgraph containing all the edges missing in G.

More formally, let X be the set of all vertices from V with degree less than $|V| - 1$. The size of X is clearly less than or equal to $2 \cdot k$. Let $Y = V \setminus X$. From the statement of the corollary we know that $|Y| \geq 2 \cdot |X| \cdot (|X| + 1)$ and hence the assumptions from the statement of Theorem 1 are satisfied and we can conclude that G has a noncrossing Hamiltonian path. This gives a lower bound of order $\Omega(\sqrt{n})$ for the function $h(n)$ defined in the introduction. □

Using the algorithm with a similar idea as the algorithm in Lemma 1 we can also prove the following result for the complements of matchings:

Theorem 2. *Let $G = (V, E)$ be a geometric graph which is a complement of a matching (i.e. a graph with minimum degree $|V| - 2$) and $|V| \geq 3$. Then G has a noncrossing Hamiltonian path.*

Proof. The case when $|V| = 3$ is trivial so we can assume $|V| \geq 4$. We use the following algorithm for a construction of a noncrossing Hamiltonian path: first take any extremal point of V to be the first vertex of the path. Let x be the last vertex on the path constructed so far. Then at each step we choose a vertex y which is an extremal point of the remaining vertices such that $\{x, y\}$ is an edge in G which does not intersect the convex hull of the remaining points, and we add y to the path. The vertex y with the desired properties always exists if there are at least two remaining vertices. Let H' denote the set of points outside the path in the previous step of the algorithm and H the set of points outside the path in the current step. In the previous step, x was an extremal vertex of H' and so in the current situation there must be some segment y_1y_2 which is an extremal segment of H, such that neither of segments xy_1 and xy_2 do intersect the convex hull of H. Because at least one of these segments must be an edge in G (its complement was a matching) we have just proven the existence of y.

If there is only one remaining vertex and it is not connected by an edge to the last point on the path, we cannot finish the path. Let y_1 be the last remaining vertex and y_2, y_3, and y_4 be the last vertices on the path in the reverse order (remember that $|V| \geq 4$). In this situation we remove the vertices y_2 and y_3 from the path to get the situation from figure 7. Because y_1 and y_2 are not connected by an edge of G, we know that y_4 must be connected by an edge to both y_1 and y_2 and one of these edges does not intersect the convex hull of the remaining points. So we can WLOG add y_1 to the path and then finish the path by adding y_3 and y_2. □

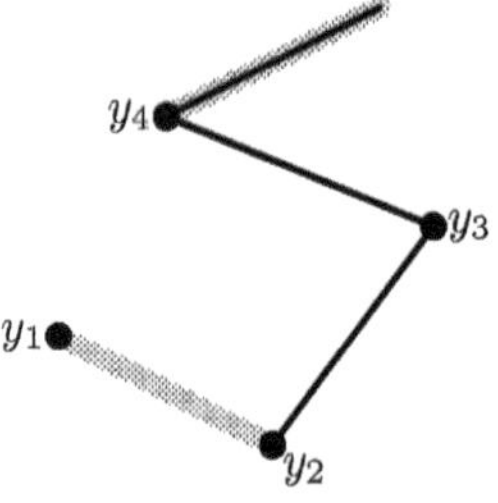

Fig. 7. The situation when we cannot add the last point to the noncrossing path.

4 The Upper Bound

In this section we prove that there exist geometric graphs in $\mathcal{C}$ such that the size of X is $O(\sqrt{|V|})$ and the graphs do not have a noncrossing Hamiltonian path. By proving this we get two asymptotically tight bounds on the function $h_1(n)$ defined in the introduction.

Lemma 2. *Let $G = (V, E)$ be a geometric graph such that all the vertices of V are in the convex position. Let $P = (v_{i_1}, \ldots, v_{i_n})$ be a noncrossing Hamiltonian path in G. Then for any $j \in \{1, \ldots, n-2\}$ it holds that from the three vertices $v_{i_j}, v_{i_{j+1}}, v_{i_{j+2}}$ at least two are neighbors on the boundary of the convex hull of V (i.e. at least two of them are connected by an extremal segment of V).*

Proof. Assume that for some j the statement does not hold. Then the vertices on the convex hull are split into three nonempty contiguous blocks of neighbouring vertices. But the noncrossing path can enter only two of these three parts (once the path enters some part it cannot leave it without crossing itself) and hence such a path cannot be Hamiltonian. □

Theorem 3. *For each $n_0 \in N$ there exists $n \in N, n_0 \leq n$ such that there is a geometric graph $G = (V, E), G \in \mathcal{C}, |V| = n$ satisfying $|X| < 3 \cdot \sqrt{n}$ without a noncrossing Hamiltonian path.*

Proof. Lets have some n_0. Let n be the smallest natural number greater than n_0 which is the square of some natural number. Now we describe a geometric graph on n vertices with the desired properties. We place n vertices of the graph on a circle with equal distances between neighbouring vertices. Then we split the vertices into $\sqrt{n}$ groups (each of size $\sqrt{n}$) in such a way that each group forms a contiguous sequence on the circle. Now we define the partitioning of V into X and Y (and by this we determine the edges of the graph). We choose arbitrarily one group (let us call it g) and put all its vertices to X. In the other groups we put two vertices on the borders to X and the remaining points to Y (see Figure 8). Clearly $|X| = (\sqrt{n} - 1) \cdot 2 + \sqrt{n} < 3 \cdot \sqrt{n}$, so it only remains to prove that the graph does not have a noncrossing Hamiltonian path.

Lets consider the first vertex u in g. Because $u \in X$, it must be connected to some vertex v from Y by an edge of the path. We write g' for the group containing the vertex v. Because both neighbors of u on the convex hull are also from X, v cannot be a neighbor of u. From this we can also trivially conclude that neither u nor v can be the endpoints of the Hamiltonian path. Now we focus our attention on the part of the path which should contain all the remaining vertices of g. From Lemma 2 we know that the path must go from v either to some other vertex of g' or to the neighbor of u in g. From any vertex of g the path must return back to g' to the neighbor of the vertex last used in g. From this we get that on the noncrossing Hamiltonian path there must be an alternation of vertices from g and from g' in such a way that between any two vertices from g there must be a vertex from $g' \cap Y$. But we have $\sqrt{n}$ vertices from g and only $\sqrt{n} - 2$ vertices from $g' \cap Y$ and so this is impossible. □

Fig. 8. Construction of a graph on n vertices with $|X| = O(\sqrt{n})$ without a noncrossing Hamiltonian path

5 Vertices in Convex Position

In the following section we consider a class $\mathcal{D}$ of convex geometric graphs. A geometric graph $G = (V, E)$ is in $\mathcal{D}$ iff the vertices of G are in a convex position. We show that if we remove $\lceil |V|/2 \rceil - 1$ edges from the complete geometric graph then the noncrossing Hamiltonian path still exists (Theorem 4) but if we remove $\lceil |V|/2 \rceil$ edges it need not exist (Theorem 5). Note that the bounds are tight.

Theorem 4. *Let $G = (V, E)$ be a geometric graph, $G \in \mathcal{D}$, $n = |V|$. Let $\overline{G} = (V, F)$ be the complement of G. If $|F| \le \lceil \frac{n}{2} \rceil - 1$ then there exists a noncrossing Hamiltonian path in G.*

Proof. Let $v_0, v_1, \ldots, v_{n-1}$ be the vertices of G in clockwise order, starting with an arbitrary one. Consider the complete geometric graph $G' = (V, E \cup F)$. Let P_i be the path $v_i, v_{i+1}, v_{i-1}, v_{i+2}, v_{i-2}, \ldots$ (counting the indices modulo n) in G'. We observe that the paths $P_1, \ldots, P_{\lfloor n/2 \rfloor}$ are pairwise disjoint noncrossing Hamiltonian paths in G'. Since $|F| \le \lceil \frac{n}{2} \rceil - 1$ we are done for n even — at least one of the paths must avoid F and hence it is a noncrossing Hamiltonian path in G. If n is odd we observe that there are $\lfloor n/2 \rfloor$ edges $\{v_0, v_{n-1}\}, \{v_1, v_{n-2}\}, \ldots, \{v_{\lfloor n/2 \rfloor - 1}, v_{\lfloor n/2 \rfloor + 1}\}$ which are not in any P_i. Let A denote this set of edges. We can assume WLOG that the set V forms the vertex set of a regular convex n-gon. Observe that every edge of G' can be mapped to an edge of A by rotating G' along its centre of rotational symmetry. So we can WLOG assume that at least one of the edges of F is in A (and hence is not in any of the paths P_i). Now we can conclude using the same argument as for n even that one of the P_is is a noncrossing Hamiltonian path in G. □

Theorem 5. *For each $n, n \ge 2$ there exists a geometric graph $G_n = (V_n, E_n)$ such that $G_n \in \mathcal{D}$, $|V_n| = n$, $|E_n| = \binom{n}{2} - \lceil \frac{n}{2} \rceil$ and G_n does not have a noncrossing Hamiltonian path.*

Proof. Let $v_0, v_1, \ldots, v_{n-1}$ be the vertices of G_n in clockwise order, starting with an arbitrary one. First we make an easy observation: the first (and the last) edge of a noncrossing Hamiltonian path P is an extremal segment of V. Consequently, if $v_i v_j$ is an edge of such a path but not an extremal segment, then P contains at least one extremal segment from each of the intervals $v_i \ldots v_j$ and $v_j \ldots v_i$.

Let $k = \lceil \frac{n}{2} \rceil$. We choose $F_n = \{\{v_0, v_1\}, \{v_1, v_2\}, \ldots, \{v_{k-1}, v_k\}\}$ and E_n as the complement of F_n. Let $B = \{v_0, \ldots, v_k\}$. Suppose there exists a noncrossing Hamiltonian path P avoiding F_n. No edge in P may join two points of B, as then by the observation above it would have to contain an edge from F_n. Therefore B is an independent set of P, which is impossible as the largest independent set of P is of size k. □

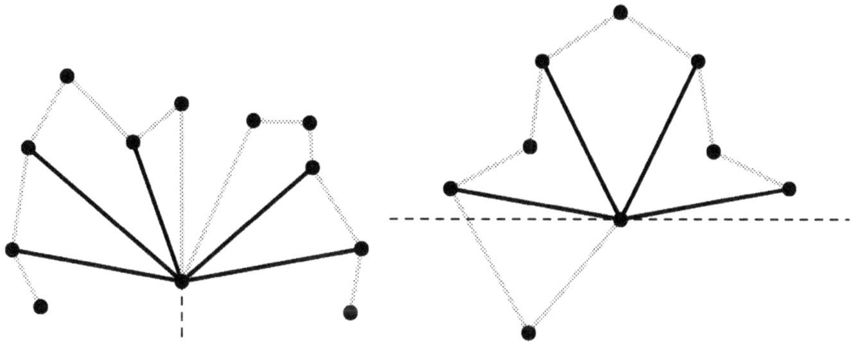

Fig. 9. Construction of a noncrossing Hamiltonian path. There are either two uncovered vertices in one of the cones (on the left) or every cone contains exactly one uncovered vertex (on the right).

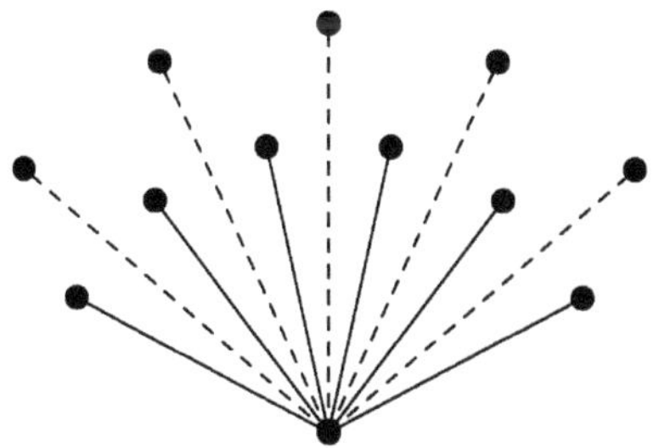

Fig. 10. Complement of a star without a noncrossing Hamiltonian path.

6 Complement of Star

In this section we consider a class of geometric graphs $\mathcal{S}$. A geometric graph $G = (V, E)$ is in $\mathcal{S}$ iff E is a complement of F, where F is the edge set of a star $K_{1,k}$. We prove that for $k \leq \lceil |V|/2 \rceil - 1$ there always exists a noncrossing Hamiltonian path but for $k \geq \lceil |V|/2 \rceil$ it need not exist.

Theorem 6. *For any geometric graph $G = (V, E)$ on n vertices, $G \in \mathcal{S}$ such that $|F| \leq \lceil \frac{n}{2} \rceil - 1$ there exists a noncrossing Hamiltonian path in G. For any $n, n \geq 2$ there exists a geometric graph on n vertices $G_n, G_n \in \mathcal{S}$ with $|F_n| = \lceil \frac{n}{2} \rceil$ such that there is no noncrossing Hamiltonian path in G_n.*

Proof. Let C be the center of a star F, where F has at most $\lceil \frac{n}{2} \rceil - 1$ edges. We partition the plane into cones by extending the edges of F into rays starting in C. If there is a cone that contains at least 2 vertices that are not covered by F, we use the construction from Figure 9 (left). Otherwise there must be exactly one vertex in each of the cones. At most one of the cones spans an angle greater than straight, let x be the vertex inside this cone if such a cone exists, otherwise let x be an arbitrary vertex not covered by F. The half-line starting in C passing through x splits its cone into two, so at least one of them spans an angle smaller than straight. Now we use the construction from Figure 9 (right).

We proceed to prove the second claim of the theorem. Let V_n and F_n look as in Figure 10. Note that the boundary of the convex hull of V_n contains the vertex C, two edges of F_n, and all the vertices not covered by F_n. Moreover, every cone defined by the rays extending the edges of F_n contains at most one vertex not covered by F_n.

Suppose there is a noncrossing Hamiltonian path P. If C is the first vertex of P, then the second vertex of P is one of the uncovered ones, the third one belongs to one of the half-planes determined by the first edge and P cannot get to the other half-plane without intersecting its first edge, so the path cannot be Hamiltonian. Similarly, if C is an interior vertex of P, the edges of P adjacent to it split the remaining vertices into three nonempty parts, and P cannot cover more than two of them without intersecting itself. □

Conclusion

In this paper we proved that for several classes of geometric graphs (graphs with vertices in convex position, complements of stars, complements of cliques and complements of matchings) we need to remove $\Omega(n)$ edges from a complete geometric graph to forbid every noncrossing Hamiltonian path. This lets us conjecture that $h(n) = \theta(n)$ where $h(n)$ is the largest k such that when we remove arbitrary set of k edges from a complete geometric graph on n vertices, the resulting graph still has a noncrossing Hamiltonian path.

Acknowledgement. The authors would like to thank Jan Kratochvíl and Pavel Valtr who brought the problem to our attention, discussed it with us and helped us with the preparation of this paper.

References

1. P. K. Agarwal, B. Aronov, J. Pach, R. Pollack, and M. Sharir: Quasi-planar graphs have a linear number of edges Combinatorica **17** (1997) 1–9

2. J. Pach: Geometric graph theory, Surveys in combinatorics, 1999 (Canterbury), 167–200, London Math. Soc. Lecture Note Ser., 267, Cambridge Univ. Press, Cambridge, 1999.
3. J. Pach and P. K. Agarwal: Combinatorial Geometry, Wiley Interscience, New York, 1995
4. J. Pach, R. Radoičić, G. Tardos, and G. Tóth: Geometric graphs with no self–intersecting path of length three, Graph Drawing, Lecture Notes in Computer Science **2528**, Springer–Verlag, Berlin, 2002, 295–311
5. R. Pinchasi and R. Radoičić: On the number of edges in geometric graphs with no selfintersecting cycle of length 4, Proc. 19th Annual Symposium on Computational Geometry, submitted
6. G. Tardos: On the number of edges in a geometric graph with no short self-intersecting paths, in preparation
7. P. Valtr: Graph drawings with no k pairwise crossing edges, Graph Drawing (Rome), Lecture Notes in Computer Science, vol. **1353** (1997), 205–218

GraphAEL: Graph Animations with Evolving Layouts*

Cesim Erten, Philip J. Harding, Stephen G. Kobourov, Kevin Wampler, and Gary Yee

Department of Computer Science
University of Arizona
{cesim,harding,kobourov,wamplerk,gyee}@cs.arizona.edu

Abstract. GraphAEL extracts three types of evolving graphs from the Graph Drawing literature and creates 2D and 3D animations of the evolutions. We study citation graphs, topic graphs, and collaboration graphs. We also create difference graphs which capture the nature of change between two given time periods. GraphAEL can be accessed online at http://graphael.cs.arizona.edu.

1 Introduction

While static graphs arise in many applications, dynamic processes give rise to graphs that evolve through time. Such dynamic processes can be found in software engineering, internet/telecommunications traffic, and social networks, among others. Typically, the underlying graph structures are large, and depending on the time-granularity, may contain from a few to a large number of timeslices. The graph representing a particular timeslice may contain fewer/more vertices/edges than the one representing the previous timeslice, or it may differ in some other graph attributes, such as node-weights, edge-weights, or labels.

We describe algorithms and techniques for visualization of dynamic graph processes, using the evolution of the graph drawing literature as an example of such a process. GraphAEL is composed of three distinct components:

1. a relational database that stores and catalogs the data;
2. a graphical user interface (GUI) for querying the database and for graph creation;
3. a toolkit for interactive visualization of the resulting graphs.

We built a MySQL database and populated it with all the papers from the graph drawing proceedings in the period 1994-02. The database can be queried online via a GUI. In addition to standard queries and responses, we can generate graphs that evolve through time. In particular, we generate *citation graphs*, *co-authorship graphs*, and *topic graphs* at varying time-granularity. Each of these types comes in two flavors, *individual graphs* and *cumulative graphs*. To study

* This work is supported in part by the NSF under grant ACR-0222920.

G. Liotta (Ed.): GD 2003, LNCS 2912, pp. 98–110, 2004.

the underlying graph processes, we also extract *difference graphs* which facilitate visual discovery of new trends and patterns. The evolution of the underlying graphs and the difference graphs can be studied via 2D and 3D animations.

2 Related Work

The visualization techniques for graphs that evolve through time are related to previous work in dynamic graph visualization. A number of papers use modifications of static graph drawing algorithms [4,13,18]. Incremental graph drawing is used in DynaDag [17]. Bayesian decision theory, together with a force-directed method are used in [3] to model dynamic graphs. Sequences of graphs are considered in [7] in order to create smoother transitions. Special classes of graphs such as trees, series-parallel graphs and *st*-graphs have also been studied in dynamic models [5,15].

More recently, several papers consider larger graphs that evolve through time. In [2] a system for visualizing network evolution is presented, in which each modification is shown in a separate layer of 3D representation with vertices common to two layers represented as columns connecting the layers. Along these lines, Collberg *et al* [6] describe a graph-based system for visualization of software evolution, which uses a modification a force-directed algorithm for visualization of large graphs [11]. Using a preliminary version of our system we analyzed the ACM digital library database, focusing on collaboration graphs and category graphs but the lack of citation data prevented us from extracting citation graphs [8].

3 The Relational Database

We gathered data about the graph drawing community, more specifically the nine years of published proceedings from the International Symposium on Graph Drawing. This data came from three sources: an XML file with information from the DBLP Computer Science Bibliography, a GML file used in the 2001 graph drawing contest [1], and the proceedings themselves. We designed a schema that maps the data to all necessary relations for creation of collaboration, topic, and citations graphs; see Fig. 1. The database schema consists of seven tables storing information about articles, authors, proceedings, and publishers. Rectangles represent entities, circles represent attributes, and diamonds represent binary relationships. The primary key for each entity is underlined.

We used MySQL version 4.0.12, a powerful open-source database system, to implement this schema. The XML file obtained from DBLP contains bibliographical information for every paper published in the GD proceedings. It does not, however, contain citation information. We built an XML parser to extract the relevant bibliographical data from DBLP. The citation data for the period 1994-00 was gathered from the the 2001 graph drawing contest GML file. Citation data for 2001 and 2002, was manually entered directly from the proceedings. We intend to update the database with new entries from the latest proceedings

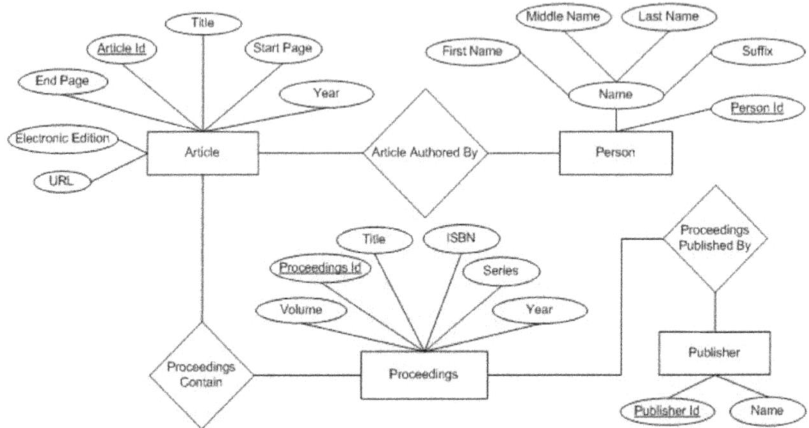

Fig. 1. The bibliographical database schema in Extended Entity-Relationship format.

of the symposium on graph drawing, on an yearly basis. The database can be accessed and queried online at `http://graphael.cs.arizona.edu`.

4 Graphical User Interface

To interact with the database we created a graphical user interface; see Fig. 2. The GUI, written in Java, is available as both a web-based Java applet and a downloadable Java application. The applet version makes use of a PHP script as a servlet to maintain the security of the database server. The GUI allows users to query general dataset statistics, search for specific authors and papers, and query citation data by referencer and referencee. In addition, users can generate collaboration, topic, and citation graphs on the fly in the form of GML files with varying time-granularity. The applet opens a new browser window displaying the GML file, which can be saved locally. All of these queries can be limited in scope by date and the amount of results requested.

The top panel of the GUI contains buttons that allow the user to change the querying parameters. The left panel of the GUI contains buttons representing the different types of results and GML files that can be obtained. The 'Go' button, activates a servlet that will then query the database located on the server. The server returns the results to the servlet which then hands them to the applet where they are displayed. The following queries are supported by the GUI:

- Database Statistics: lists data statistics, including number of articles and authors;
- General Query: allows for search by author name, or title words;
- Co-authorship GML: creates a co-authorship graph, for the given parameters;
- Referencer Search: finds all papers that reference a particular paper;

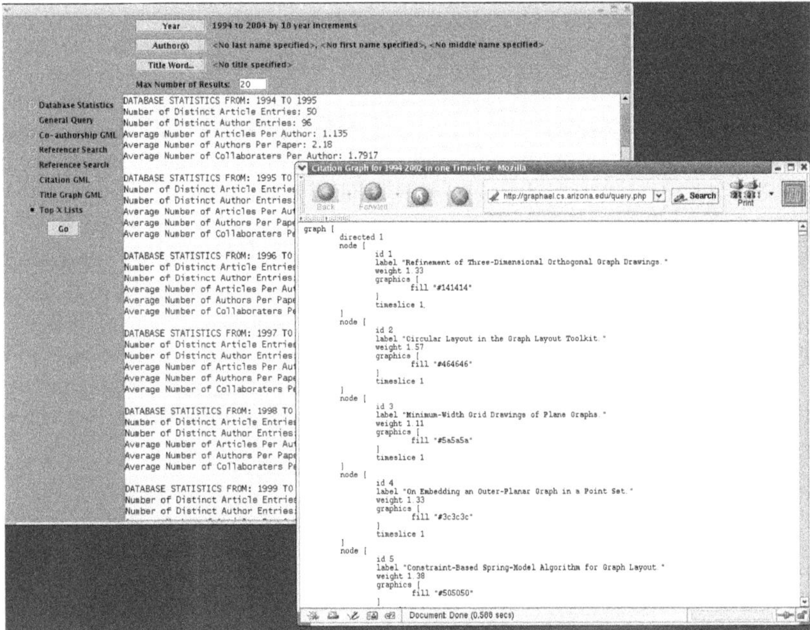

Fig. 2. The Graphical User Interface, shown here displaying a GML file.

- Referencee Search: finds all papers that are referenced from a particular paper;
- Citation GML: creates a citation graph, for the given parameters;
- Topic Graph GML: creates a topic graph, for the given parameters;
- Top X Lists: lists the most productive authors, most collaborative authors, etc.

5 Interactive Visualization of Evolving Graphs

GraphAEL calculates the layout for evolving graphs quickly. The underlying graph drawing algorithm is a modification of the GRIP algorithm [11,12]. Given a graph $G = (V, E)$, GRIP computes a hierarchical sequence of vertex filtrations given by a sequence of m subsets of V. Let V_i, V_j be two filtrations such that $i < j$, then $V_i \subset V_j$ and $V_m = V$. The graphs defined by this filtration sequence are laid out in order, with the layout of the graph at each level i providing a basis to guide the layout of the graph at level $i + 1$. In this way large-scale structures of the graph are quickly captured by the filtration levels with low indices, and the small-scale details are refined in the layout of the levels with high indices.

The layout of each filtration level is calculated with the force-directed placement method of Kamada-Kawai [14]. For a vertex v in G, the displacement of v is calculated by:

$$\boldsymbol{F}_{KK}(v) = \sum_{u \in N_i(v)} \left(\frac{\|pos[u] - pos[v]\|^2}{dist_G(u,v)^2 \cdot edgeLength^2} - 1 \right) (pos[u] - pos[v])$$

This equation is convenient because it calculates the displacement on each vertex based only upon its physical and graph distances from other vertices in the graph. Since in general, vertices in a particular filtration level are not adjacent to any other vertex in the level, but are instead connected to other vertices in the higher levels through a sequence of edges, using graph distances is a reasonable approach. Only in the last level of of filtration does the adjacency make sense, since the last level is the same as the initial graph. Thus at this level, it is possible to use force directed placement methods which depend on a notion of adjacency. The Fruchterman-Reingold method [10] for calculating the displacement of a vertex v, $\boldsymbol{F}_{FR}(v) = \boldsymbol{F}_{a,FR} + \boldsymbol{F}_{r,FR}$, is used at the last filtration level, where the attractive and repulsive force vectors are given by:

$$\boldsymbol{F}_{a,FR} = \sum_{u \in Adj(v)} \frac{\|pos[u] - pos[v]\|^2}{edgeLength^2} (pos[u] - pos[v])$$

$$\boldsymbol{F}_{r,FR} = \sum_{u \in N_i(v)} s \frac{edgeLength^2}{\|pos[u] - pos[v]\|^2} (pos[v] - pos[u])$$

5.1 GraphAEL's Layout Algorithm

While GRIP does a decent job of laying out large static graphs, there is no appropriate mechanism to assign the vertices and edges of a graph attributes, such as weights, which might affect the layout. This is a major drawback in visualizing certain relations and even more so in visualizing evolving graphs. For instance, in visualizing the GD citation graph, it is useful to have a notion of both the weight of a vertex, (which might represent the number of times a paper is cited), and a temporal notion (which might represent the year the paper was written). In contrast, GraphAEL supports the layout of graphs with node-weights and edge-weights, as well as the notion of a *timeslice* which is used to visualize graphs with a temporal component (evolving graphs).

GraphAEL tries to place heavy nodes well away from each other and to place vertices connected by heavy edges, closer to each other. Appropriate modifications to the force-directed algorithm were made to accommodate these characteristics. Each edge in a graph is given an ideal length that it will attempt to attain, although this is not always possible. An edge e of weight w_e connecting vertices u, v of weight w_u, w_v respectively is given an ideal length of $\sqrt{w_u \cdot w_v}/w_e$.

Since the Kamada-Kawai method is used to determine parts of the layout, it is necessary to come up with a modified graph distance for weighted graphs. Because of the computational and space requirements of calculating the effects of all paths between two vertices, or of computing the shortest weighted path between them, GraphAEL uses an approximation. Let $p_1, p_2, \ldots, p_n$ be the sequence of vertices in the shortest unweighted path in G connecting two vertices, u and v. The modified Kamada-Kawai vector is given by:

$$\sum_{u \in N_i(v)} \left(\frac{2\|pos[u] - pos[v]\|^2}{optDist_G(u,v)^2 \cdot edgeLength^2 + \|pos[u] - pos[v]\|^2} - 1 \right) (pos[u] - pos[v]),$$

where $optDist_G(u,v)$ is defined by,

$$optDist_G(u,v) = \sum_{i=2}^{n} \frac{\sqrt{w_{p_i} \cdot w_{p_{i-1}}}}{w_{e_{p_i p_{i-1}}}}.$$

To achieve an aesthetically pleasing layout of the graph, it is also necessary to modify the Fruchterman-Reingold vectors. As in GRIP, the Kamada-Kawai calculation results in a good approximate layout so that the Fruchterman-Reingold calculations can quickly "tidy up" the layout. The modifications needed to support weighted graphs with the ideal edge lengths described earlier are simple:

$$\boldsymbol{F}_{a,w,FR} = \sum_{u \in Adj(v)} \frac{w_e \cdot \|pos[u] - pos[v]\|^2}{edgeLength^2} (pos[u] - pos[v])$$

$$\boldsymbol{F}_{r,w,FR} = \sum_{u \in N_i(v)} s \frac{edgeLength^2 \cdot \sqrt{w_u \cdot w_v}}{\|pos[u] - pos[v]\|^2} (pos[v] - pos[u])$$

5.2 Timeslices and Animation

To visualize a series of graphs, embodying the evolution of a set of relationships over time, we associate another attribute, called a *timeslice*, with each vertex. The timeslice of a vertex is simply a label associated with a vertex. We use the timeslice attribute to partition the vertices of a graph into groups by time. Several simple modifications to the layout algorithm are needed to accommodate timeslice information. Repulsive forces should only exist between vertices in the same timeslice, and thus the optimal distance between vertices in different timeslices which are connected by an edge is zero. In the Kamada-Kawai vectors, the only alteration required is that the the function $optDist_G(u,v)$ be redefined so that for two vertices u, v with timeslice indices of t_u and t_v respectively:

$$optDist_G(u,v) = \delta_{t_u t_v} \cdot \frac{\sqrt{w_u \cdot w_v}}{w_e},$$

where $\delta_{t_u t_v}$ is 1 if $t_u = t_v$ and 0 otherwise. The modifications needed for the Fruchterman-Reingold calculations are similar. Repulsive forces are simply eliminated between vertices in different timeslices, $\boldsymbol{F}_{r,w,t,FR} = \delta_{t_u t_v} \cdot \boldsymbol{F}_{r,w,FR}$ while the attractive forces remain unchanged, $\boldsymbol{F}_{a,w,t,FR} = \boldsymbol{F}_{a,w,FR}$.

The timeslice information alone is not enough to nicely layout evolving graphs; it is also necessary to arrange edges between the timeslices so that the resulting layouts can be used for animation. Perhaps the most straightforward and common case is when one has a series of "snapshots" of a graph taken at some interval over a period of time. When laying out such a graph there are normally two types of constraints which need to be met: Each timeslice should have a pleasing layout, and the layout of consecutive timeslices should be similar, that is, the mental map should be preserved. Another way of formulating this latter constraint is that vertices on one timeslice should tend toward their positions in adjacent timeslices. To meet these constraints the timeslices are combined into a single graph and edges are added between vertices with the same labels in adjacent timeslices.

Since vertices in different timeslices have no repulsive forces between them (but the edges between them retain their attractive forces) these additional edges attract each vertex toward the vertices associated with it in adjacent timeslices. The trade-off between layout within a timeslice and mental map preservation can easily be adjusted by altering the weights of the inter-timeslice edges. Heavier weights lead to better mental map preservation and lighter weights improve the layout of each timeslice. For the visualization of the graphs in this paper the weight of an inter-timeslice edge connecting two vertices of weights w_u and w_v is given a weight of $(w_u + w_v)/2$, causing heavier vertices to stay in the same position over multiple timeslices.

One disadvantage of this method of adding edges is that although the movements of a vertex from one timeslice to the next are minimized, it is still possible that over a span of many timeslices vertices will drift far from their initial positions. If this is an undesirable effect it can be prevented by adding additional edges between vertices in distant timeslices. A straight-forward approach is to create a clique of all the occurrences of each vertex over all timeslices. `GraphAEL` has both options implemented.

Once the layout of the evolving graph has been computed, several visualization options can be used. A static view, capturing all timeslices can be displayed. Each timeslice can be restricted to its own 2D plane or can be drawn in 3D and the individual graphs arranged on top of each other. Edges connecting different timeslices can be shown or hidden. Alternatively, the evolution can be shown via an animation between the timeslices in 2D or 3D. The animation uses an interpolation between timeslices and fading in/out of appearing/disappearing vertices and edges. Edge and vertex coloring is used to convey additional information. For example, in the citation graphs we indicate the direction of an edge by gradually varying the color from light(source) to dark(target) and the vertex color indicates the relative age of a particular paper.

Finally, `GraphAEL` can generate and display *difference graphs*. The difference graph between two adjacent timeslices captures the difference between the two underlying graphs. For example, when visualizing the topic graph we may miss significant growth in an area with a few papers if the corresponding vertices are relatively small. With this in mind, the difference graph captures the percentage

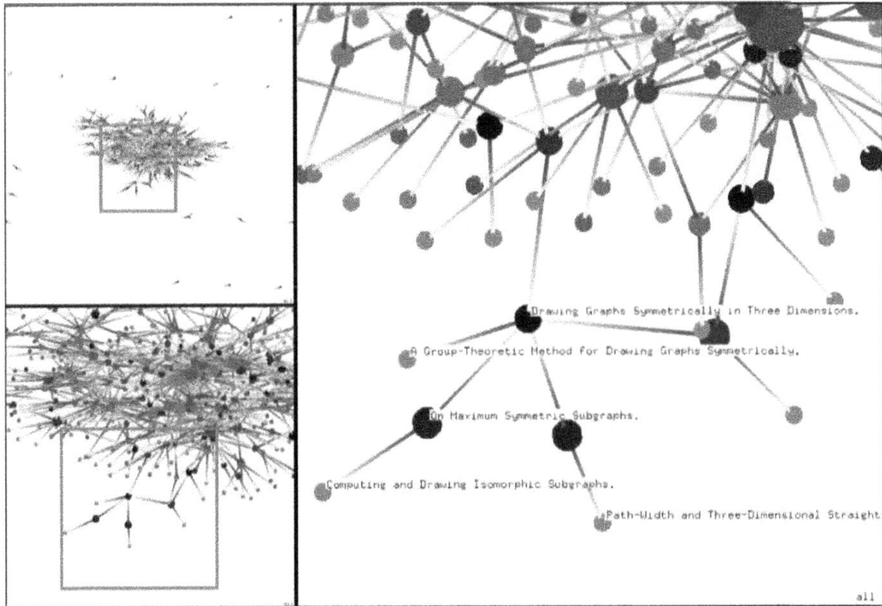

Fig. 3. Top left: Cumulative citation graph for the period 1994-02. **Bottom Left:** view of the area enclosed in the red rectangle from top left. **Right:** Labeled vertices in a "symmetry" cluster. Directed edges go from light/source to dark/target. The color of the nodes corresponds to the year it was published: the lighter the node, the older the paper.

change between levels. We create a series of difference graphs and as it is an evolving graph we can use all the tools for evolving graphs.

6 Interactive Visualization of the GD Literature

In this section we consider three types of evolving graphs extracted from the MySQL database: citation graphs, topic graphs, and collaboration graphs. We also calculated some general statistics about the data and we include gnuplot charts and tables for the data in the Appendix.

6.1 Citation Graphs

The citation graph for a given time period is a simple vertex-weighted directed graph in which the vertices correspond to distinct articles. A directed edge connects two articles, with the article that cites as the source and the cited article as the target. The weight of a vertex is determined by the number of citations an article received, divided by the number of years since its publication. The citation graphs can reveal information about influential papers in the field, and their relation with the rest of the graph. Fig. 3 shows several views of the citation

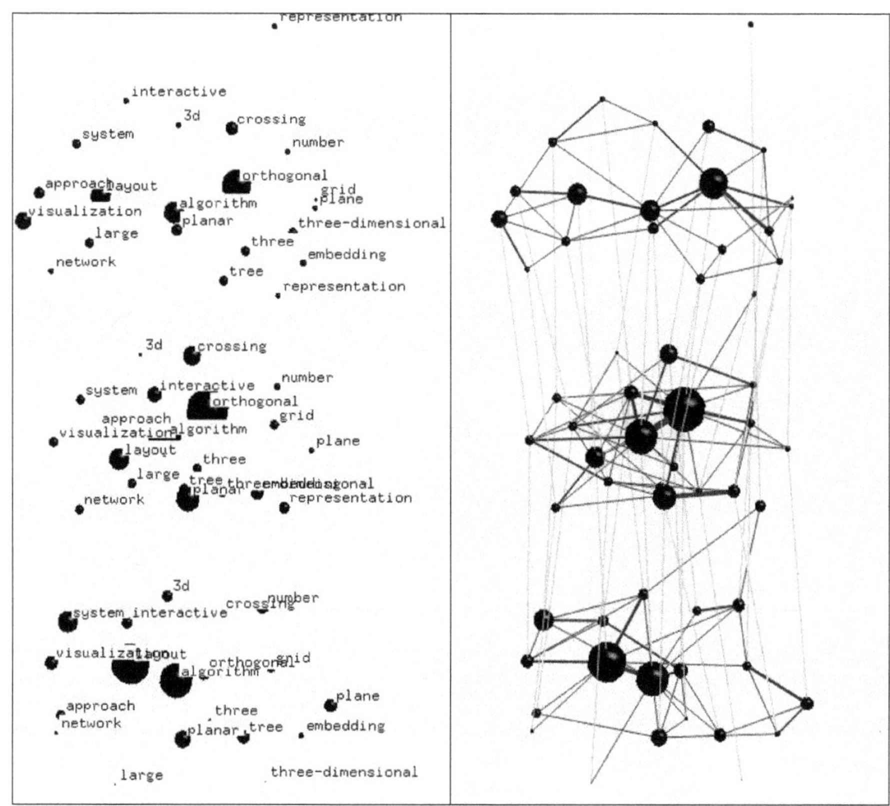

Fig. 4. Individual topic graphs for the period 1994-02, restricted to the top 20 title words, shown in timeslices of 3 consecutive years each: 1994-96 on the bottom, 1997-99 in the middle, and 2000-2002 on the top. On the left are all the labeled nodes in each timeslice with invisible edges.

graph. The size of a vertex reflects its weight, i.e. a large vertex corresponds to an article that is cited many times relative to its age. The vertex color indicates the age of the article, the darker a vertex the more recent the corresponding article is. The edge color changes from yellow to blue to indicate direction, where the blue end of the edge is the target vertex.

6.2 Topic Graphs

The topic graph for a given time period is a simple vertex-weighted and edge-weighted undirected graph in which vertices correspond to title words and edges are placed between title words that co-occur in research papers. The weight of a vertex in the topic graph is proportional to the number of papers that contain the corresponding word in their titles. Similarly, the edge weight is proportional to the number of papers in which the two corresponding words co-occur in the title. Topic graphs can reveal information about related topics, the concentration

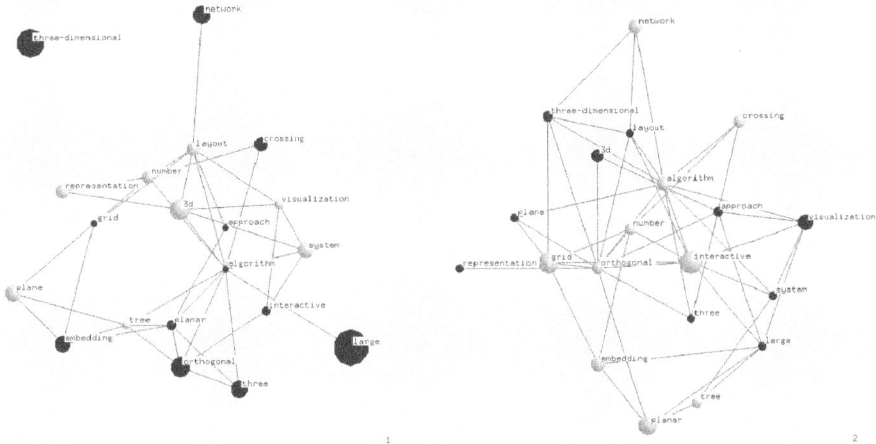

Fig. 5. The individual *difference graphs* capturing the change from 1994– to 1997–99 (left) and 1997–99 to 2000–02 (right). Dark nodes indicate growth and light nodes indicate decline. Note the percentage increase in "large" and decline in "interactive".

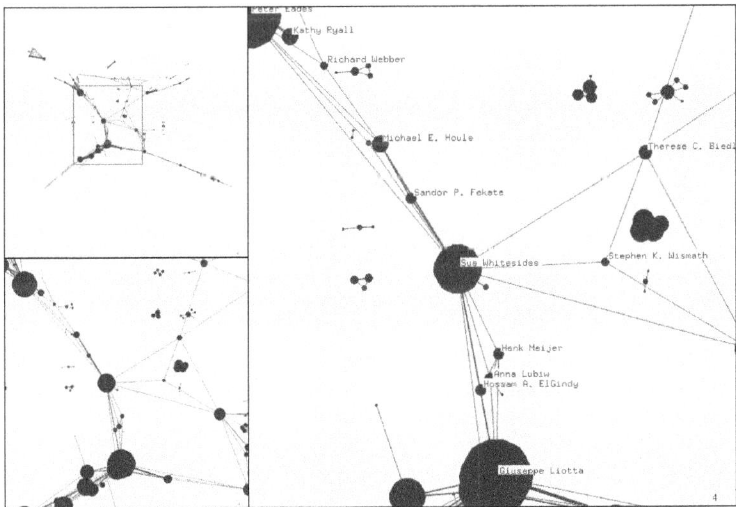

Fig. 6. **Top left:** collaboration graph for the period 1994-97. **Bottom Left:** view of the area enclosed in the red rectangle from top left. **Right:** closer view with more details.

of research on a specific topic and the trends as they evolve through time. Fig. 4 contains several visualizations of the topic graph. The edges connecting same vertices in adjacent timeslices help with mental map preservation and are used to determine the vertex locations in the animation between timeslices.

We also construct topic *difference graphs* in which the weight of a vertex is the percentage of its weight change between adjacent timeslices. The edge weight

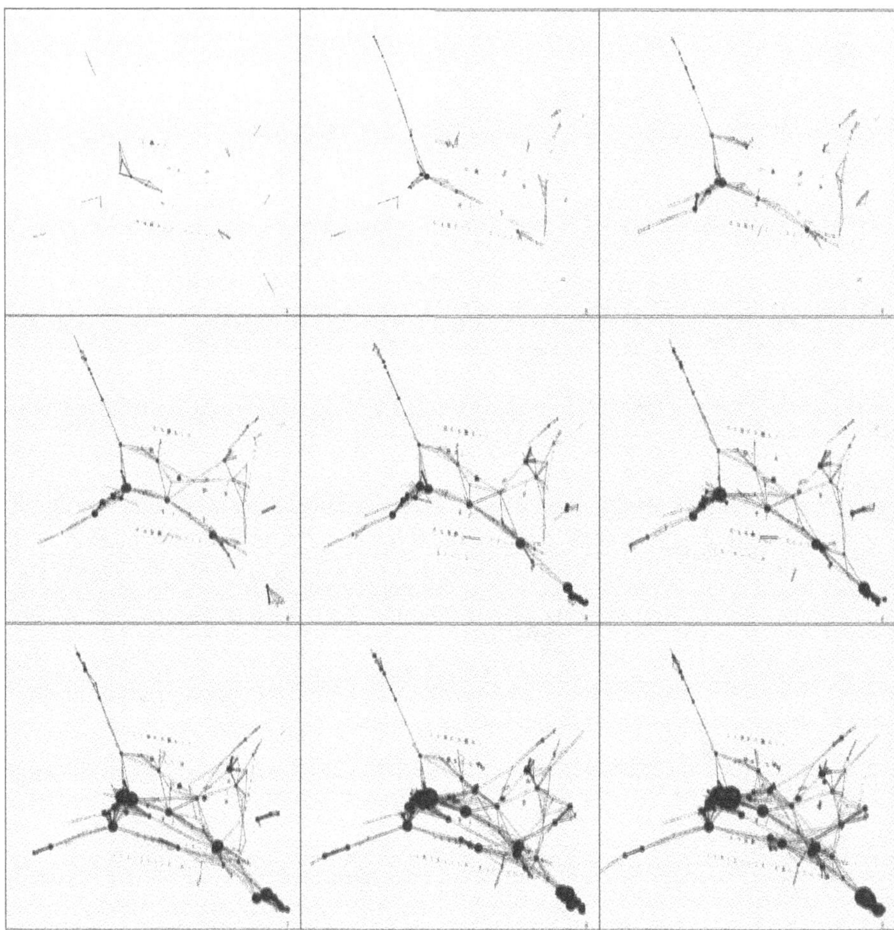

Fig. 7. Screenshots from the animation of the cumulative collaboration graph in the period 1994-02. Each timeslice represents a one year period. A movie of the actual animation can be seen at the `GraphAEL` webpage.

is proportional to the corresponding percent-change between two time periods. Together with the topic graphs, the topic-change graphs can be effectively used to visualize the evolution of GD research focus through time. Fig. 5 shows the topic-change graph corresponding to the topic graph in Fig. 4.

6.3 Collaboration Graphs

Collaboration graphs are simple undirected node-weighted and edge weighted graphs. Vertices represent unique authors and there is an edge between two vertices if the respective authors have collaborated on a research paper. The weight of a vertex is determined by the author's collaborativeness and productivity. The

Table 1. Statistics for GD, ACM [8], and NCSTRL [16].

General	GD-Value	ACM-value	NCSTRL-value
Total papers	413	51503	13169
Total authors	502	81279	11994
Authors per paper	2.54	2.32	2.22
Papers per author	2.09	1.80	2.55
Collaborators per author	3.74	3.36	3.59
Percentage of giant component	49	49	57.2
Percentage of 2nd component	0.02	0.11	0.004
Clustering Coefficient	0.60	0.62	0.50
Average Distance	4.33	9.26	9.7
Maximum Distance	10	30	31

weight of an edge represents the strength of the collaborative ties between two authors. Fig. 7 shows the evolution of the cumulative collaboration graph, i.e., each timeslice adds to the previous timeslice. Once an interesting timeslice to be viewed is found it is easy to get a detailed visualization of that specific timeslice. For example, Fig. 6 shows the collaboration graph for the period 1994-97. For a more detailed view we zoom in one of the areas where there is a concentration of several well-connected components.

6.4 Statistics from the Graph Drawing Literature

We compared the data from the GD database, with the data from two computer science databases, the ACM database [8] and the NCSTRL database [16]. Table 1 shows a summary of the overall statistics and they are examined in detail in the full version of the paper [9].

References

1. T. Biedl and F. J. Brandenburg. Graph drawing contest report. In *Proceedings of the 9th Symposium on Graph Drawing (GD)*, number 2265 in LNCS, pages 513–521, 2001.
2. U. Brandes and S. R. Corman. Visual unrolling of network evolution and the analysis of dynamic discourse. In *IEEE Symposium on Information Visualization (INFOVIS '02)*, pages 145–151, 2002.
3. U. Brandes and D. Wagner. A bayesian paradigm for dynamic graph layout. In *Proceedings of the 5th Symposium on Graph Drawing (GD)*, volume 1353 of *LNCS*, pages 236–247, 1998.
4. J. Branke. Dynamic graph drawing. In M. Kaufmann and D. Wagner, editors, *Drawing Graphs: Methods and Models*, number 2025 in LNCS, chapter 9, pages 228–246. Springer-Verlag, Berlin, Germany, 2001.
5. R. F. Cohen, G. Di Battista, R. Tamassia, and I. G. Tollis. Dynamic graph drawings: Trees, series-parallel digraphs, and planar *ST*-digraphs. *SIAM J. Comput.*, 24(5):970–1001, 1995.

6. C. Collberg, S. G. Kobourov, J. Nagra, J. Pitts, and K. Wampler. A system for graph-based visualization of the evolution of software. In *ACM Symposium on Software Visualization*, pages 77–86, 2003.
7. S. Diehl and C. Görg. Graphs, they are changing. In *Proceedings of the 10th Symposium on Graph Drawing (GD)*, pages 23–30, 2002.
8. C. Erten, P. J. Harding, S. Kobourov, K. Wampler, and G. Yee. Exploring the computing literature using temporal graph visualization. Technical Report TR03-04, Department of Computer Science, University of Arizona, 2003.
9. C. Erten, P. J. Harding, S. G. Kobourov, K. Wampler, and G. Yee. Graphael: Graph animations with evolving layouts. Technical Report TR03-11, Department of Computer Science, University of Arizona, 2003.
10. T. Fruchterman and E. Reingold. Graph drawing by force-directed placement. *Softw. – Pract. Exp.*, 21(11):1129–1164, 1991.
11. P. Gajer, M. T. Goodrich, and S. G. Kobourov. A multi-dimensional approach to force-directed layouts. In *Proceedings of the 8th Symposium on Graph Drawing (GD)*, pages 211–221, 2000.
12. P. Gajer and S. G. Kobourov. GRIP: Graph dRawing with Intelligent Placement. *Journal of Graph Algorithms and Applications*, 6(3):203–224, 2002.
13. I. Herman, G. Melançon, and M. S. Marshall. Graph visualization and navigation in information visualization: A survey. *IEEE Transactions on Visualization and Computer Graphics*, 6(1):24–43, 2000.
14. T. Kamada and S. Kawai. An algorithm for drawing general undirected graphs. *Inform. Process. Lett.*, 31:7–15, 1989.
15. S. Moen. Drawing dynamic trees. *IEEE Software*, 7(4):21–28, July 1990.
16. M. E. J. Newman. Who is the best connected scientist? a study of scientific coauthorship networks. *Physics Review*, E64, 2001.
17. S. C. North. Incremental layout in DynaDAG. In *Proceedings of the 4th Symposium on Graph Drawing (GD)*, pages 409–418, 1996.
18. K.-P. Yee, D. Fisher, R. Dhamija, and M. Hearst. Animated exploration of dynamic graphs with radial layout. In *IEEE Symposium on Information Visualization (INFOVIS '01)*, pages 43–50, 2001.

Visualizing Related Metabolic Pathways in Two and a Half Dimensions
(Long Paper)

Ulrik Brandes[1,*], Tim Dwyer[2], and Falk Schreiber[3,**]

[1] Department of Mathematics & Computer Science, University of Passau, Germany.
brandes@algo.fmi.uni-passau.de
[2] School of Information Technologies, University of Sydney, Australia.
dwyer@it.usyd.edu.au
[3] Bioinformatics Center (BIC-GH), Institute of Plant Genetics and Crop Plant Research Gatersleben, Germany. schreibe@ipk-gatersleben.de

Abstract. We propose a method for visualizing a set of related metabolic pathways using $2\frac{1}{2}$D graph drawing. Interdependent, two-dimensional layouts of each pathway are stacked on top of each other so that biologists get a full picture of subtle and significant differences among the pathways. Layouts are determined by a global layout of the union of all pathway-representing graphs using a variant of the proven Sugiyama approach for layered graph drawing that allows edges to cross if they appear in different graphs.

1 Introduction

Metabolic pathways are subnetworks of the complete network of metabolic reactions. They differ across organisms because different species may for example have developed different ways to synthesize a specific substance. We are interested in visualizing several related metabolic pathways in such a way that the inherent differences can be explored by trained biologists in order to understand the evolutionary relationships among species. The structural characteristics of metabolic pathways make them particularly amenable to layered graph drawing methods [2,24,27]. We therefore propose to visualize sets of related pathways in $2\frac{1}{2}$ dimensions, i.e. to produce interdependent, two-dimensional, layered layouts for all pathways and stack them on top of each other so that the most similar pathways are adjacent.

To realize such a design, two graph drawing issues need to be addressed. We have to determine a suitable ordering to reduce the variation between consecutive pathways and we have to deal with dependencies introduced by the many substances and reactions present in more than one pathway. An interesting consequence is a new type of crossing number in which the weight of a crossing may be different for each pair of edges.

* Partially supported by DFG under grant BR 2158/1-1.
** Supported by BMBF under grant 0312706A.

G. Liotta (Ed.): GD 2003, LNCS 2912, pp. 111–122, 2004.

This paper is organized as follows. In Sect. 2, we give some background on the type of networks considered and define the graph model on which we operate. After briefly reviewing related approaches to visualizing similar graphs, we specify our visualization design in Sect. 3. In Sect. 4 we address the ordering problem while in Sect. 5 we discuss a method to determine a global layout of the stacked pathways considering that edge crossings are less severe if the edges involved do not coexist in the same pathway. To demonstrate the utility of our method it is applied to typical real-world data in Sect. 6.

2 Metabolic Pathways

Metabolic reactions are transformations of chemical substances which occur in living beings and are usually catalyzed by enzymes. A reaction changes certain substances (reactants) into different ones (products). Metabolic reactions form large and complex networks as for example shown on the well-known *Biochemical Pathways Poster* part 1 [21]. A *metabolic pathway* is a subnetwork of the complete network of metabolic reactions (see Fig. 1). Such a pathway can be given by biochemical textbooks and databases such as KEGG [17] or be defined by functional boundaries such as the network between an initial and a final substance.

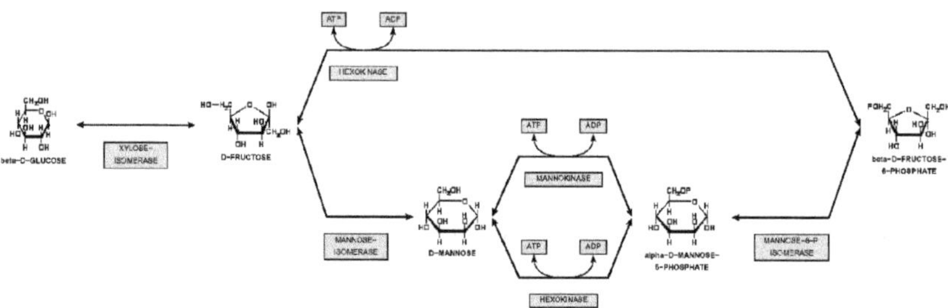

Fig. 1. A metabolic pathway

Metabolic pathways differ across organisms. Studies suggest significant variations even in most central pathways such as glycolysis [6]. Comparative analysis of pathways across species has several applications:

- understanding the evolutional relationships between species.
- development of species-specific drug targets (e.g. antibiotics).
- identification of previously unknown parts of pathways in a species.

From a formal point of view a metabolic pathway is a directed hyper-graph. The vertices represent the substances within a pathway, the hyper-edges represent reactions. A hyper-edge representing a reaction connects substances and is labeled with the enzyme(s) that catalyze the reaction. As hyper-graphs are not commonly used for simulation and visualization purposes, metabolic pathways are often modeled by bipartite graphs, e.g. Petri-net representations of pathways [23]. Here the reactions themselves are vertices, and edges are binary relations connecting substances with reaction vertices.

Three structural properties of metabolic pathways are particularly relevant for visualization. They typically are:

1. small in size (often less than two dozen reactions)
2. sparse (because most substances are involved in few reactions)
3. directed and acyclic (due to a dominant direction of most reactions and a small number of cyclic pathways such as the citrate cycle)

Several similarity measures have been introduced to compare pathways. They are characterized by the combination of structural information about metabolic networks with additional data such as sequence information [11], the enzyme classification hierarchy [29], or information about the hierarchical clustering of reactions into pathways [20]. All these similarity measures require additional information to the metabolic network, for example the genome sequence of the organisms.

We suggest a more general measure for reaction or pathway similarity which only depends on the structure of the graph, i.e. on the presence or absence of vertices and edges. This facilitates comparison of metabolic pathways from different sources (databases, experiments) even if no additional data is available or the pathway boundaries are user-defined. Note that in general our visualization method is independent of the similarity measure which is only used to compute the order of the stacking.

2.1 Formalization

For the purpose of this paper the distinction between vertices representing reactions and substances is not important. A metabolic pathway is therefore modeled by a directed graph $G = (V, E)$, where a vertex $v \in V$ represents either a substance or a reaction, and an edge $(v, w) \in E$ indicates that substance v enters reaction w or that reaction v produces substance w.

Our goal is to visualize a set $\{G_1 = (V_1, E_1), G_2 = (V_2, E_2), \ldots, G_r(V_r, E_r)\}$ of graphs that represent related pathways. Though the dissimilarity of two pathways $G_i = (V_i, E_i)$, $G_j = (V_j, E_j)$ can be measured in various ways, the arguments outlined above suggest that the following measure is appropriate.

Consider the *union graph* $G = (V, E)$, where $V = \bigcup_{i=1}^{r} V_i$ and $E = \bigcup_{i=1}^{r} E_i$, and a set of relevant elements $P \subseteq (V \cup E)$. The *Hamming distance* of two graphs G_i, G_j, $1 \leq i, j \leq r$, is defined as the cardinality of the symmetric difference of relevant elements present in either graph, i.e.

$$\delta_P(G_i, G_j) = |((V_i \cup E_i) \triangle (V_j \cup E_j)) \cap P)| \ .$$

Thus, dissimilarity can be defined in terms of missing edges, vertices, or both, by choosing P accordingly.

3 Visualizing Similar Networks

There are two common approaches to compare pathways in different species visually. Either the combination of all pathways into one diagram or the production of a drawing for each species. The first method is used in many textbooks, on a famous poster [21] and in systems such as BioMiner [27] and BioPath [12]. In general, the drawings contain either multiple (parallel) reaction-edges or single ones which are color-coded depending on the occurrence of a reaction in a set of species. An example for the second approach is the visual interface of the KEGG database [17] where all enzymes found in the gene catalog of a specific species are marked in the reference pathway map in order to identify the species-specific pathways. To compare pathways in r species, r diagrams are needed. A dynamic visual comparison method producing a diagram for each species is presented in [25].

These solutions are restricted to the comparison of only a few pathways, because either (in the first approach) the readability of the diagram decreases or (in the second approach) the size of the picture increases dramatically with each new pathway. Furthermore, none of the above mentioned methods deal with the problem of computing an appropriate order of the pathways.

3.1 Drawing Graphs in $2\frac{1}{2}$D

To allow users to compare several metabolic pathways, we want to draw them in different parallel planes, but with interdependent layouts. We call this type of representation a $2\frac{1}{2}$D drawing because the third dimension is used in a way fundamentally different from the other two. Note that, traditionally, the axes are interchangeable in 3D graph drawing.

The idea of treating the third dimension as an independent channel conveying a different kind of information [30] has been applied in numerous settings dealing with different types of network data. For instances, the data can be a single graph with vertex attributes that determine the third dimension [19,5], a graph together with a hierarchical clustering [9] or graphs that evolve over time [4,7]. The latter example relates $2\frac{1}{2}$D graph drawing to dynamic graph drawing and graph animation, since the difference in layout between consecutive states of the graph should be small. Another interesting example is the dependence graph of spreadsheet cells [26], which has a natural 2D layout that can be spread in three dimensions to indicate the data flow.

The case considered in this paper is somewhat different from those above. If we consider our pathways as subgraphs of the union graph, the data are characterized by the fact that

- there is a set of subgraphs with no given ordering,
- vertices and edges are likely to appear in more than one subgraph, and
- there are no edges to be drawn between different subgraphs.

Moreover, we aim at a representation that consists of a stacked set of two-dimensional (layered) drawings for each subgraph where a vertex has the same

2D-coordinate in each drawing in which it appears (so that it can be represented by a straight column).

Our approach can therefore be viewed as a generalization of both the Sugiyama method [28] and parallel coordinates [15].

4 Stacking Order

Let $\{G_1 = (V_1, E_1), G_2 = (V_2, E_2), \ldots, G_r(V_r, E_r)\}$ be a set of graphs, and $G = (V, E)$ their union graph. We want to order these graphs so that those which are similar with respect to Hamming distance are close to each other. Variations of this problem arise in many applications, and two of them are especially relevant in our context. Let $P \subseteq (V \cup E)$ be a set of relevant elements.

Problem 1 (MIN SUM ORDERING). Find a permutation $\sigma = (\sigma_1, \ldots, \sigma_r)$ such that

$$\sum_{i=1}^{r-1} \delta_P(G_{\sigma_i}, G_{\sigma_{i+1}})$$

is minimum.

In the context of data transmission, this problem is also known as DOP (data ordering problem) [22]. A less restricted version (with an arbitrary distance matrix) is considered in [18], where the goal is to order parallel coordinates.

Theorem 1. *MIN SUM ORDERING is $\mathcal{NP}$-hard.*

Proof. Straightforward reduction from HAMMING DISTANCE TSP [1].

For a permutation $\sigma = (\sigma_1, \ldots, \sigma_r)$ we define the *lifetime* of an element $p \in P$ to be $\Lambda_\sigma(p) = \{1 \le i \le r : p \in G_{\sigma_i}\}$. An element $p \in P$ is called *persistent*, if its lifetime spans the entire interval $\{1, \ldots, r\}$, and *transient* otherwise. The number of *appearances* and *disappearances* of $p \in P$ are defined by $a_\sigma(p) = |\{1 < i \le r : p \in G_{\sigma_i} \setminus G_{\sigma_{i-1}}\}|$ and $d_\sigma(p) = |\{1 \le i < r : p \in G_{\sigma_i} \setminus G_{\sigma_{i+1}}\}|$. Note that persistent elements make no appearances or disappearances.

Corollary 1. *MIN SUM ORDERING is equivalent to minimizing*

$$\sum_{p \in P} \left(a_\sigma(p) + d_\sigma(p)\right) .$$

A related alternative objective is therefore to minimize the maximum number of times an element appears or disappears in an ordering.

Problem 2 (MIN INTERVAL ORDERING). Find a permutation $\sigma = (\sigma_1, \ldots, \sigma_r)$ such that

$$\max_{p \in P}\{a_\sigma(p), d_\sigma(p)\}$$

is minimum.

This problem is a generalization of the consecutive ones property, since, if $p \in P$ is transient, $\max\{a_\sigma(p), d_\sigma(p)\}$ is the number of lifetime intervals (and zero otherwise). We have the following complexity status.

Theorem 2. *MIN INTERVAL ORDERING is $\mathcal{NP}$-hard, but it can be determined in linear time whether there is an ordering such that each relevant element appears or disappears at most once.*

Proof. See [14]. Since the restricted problem corresponds exactly to the consecutive ones property, it is linear-time solvable using PQ-trees [3].

Since both problems arise in many contexts, a variety of algorithms is available to determine an ordering. For MIN SUM ORDERING, for instance, heuristics for the TSP are easily adapted to yield good orderings. Other alternatives include a simple greedy heuristic that sucessively inserts a new element where it causes the smallest increase of the objective (this method is claimed to perform well for instances in data transmission [22]) and one-dimensional projections of the distance matrix obtained by principal component analysis.

5 Global Layout

To maximize the similarity between visualizations of related graphs, we compute a layout only for the union graph. The drawing of vertices and edges thus remains unchanged throughout their lifetime. Since graphs representing metabolic pathways tend to be sparse and acyclic, the Sugiyama framework for layered graph layout [10,28] is widely used to visualize individual pathways [2,24,27].

While straightforward application of any of these methods to the union graph is possible, better results can be obtained by taking into account that the individual graphs are to be viewed in separate, parallel planes. To avoid confusion with layers in the Sugiyama approach, we refer to these as *strata.*

5.1 A New Crossing Minimization Problem

Our main modification with respect to standard variants of the Sugiyama framework is motivated by the observation that crossings of edges not present in the same stratum can be resolved by 3D rotation or stereo depth perception as shown in Fig. 2. Hence, when computing the global layout, we need not account for such crossings the same way as crossings between edges with overlapping lifetime intervals.

A common approach for crossing reduction in layered layouts is based on a layer-by-layer sweep, in which the ordering of vertices in a layer L_0 is fixed and vertices in an adjacent layer L_1 are permuted to reduce the number of crossings. Exact and heuristic methods for this one-sided crossing minimization problem are based on the *crossing matrix* $(c_{uv})_{u,v \in L_1}$, in which an entry c_{uv} corresponds to the number of crossings between the two layers caused by edges incident to u and v, if u is placed to the left of v.

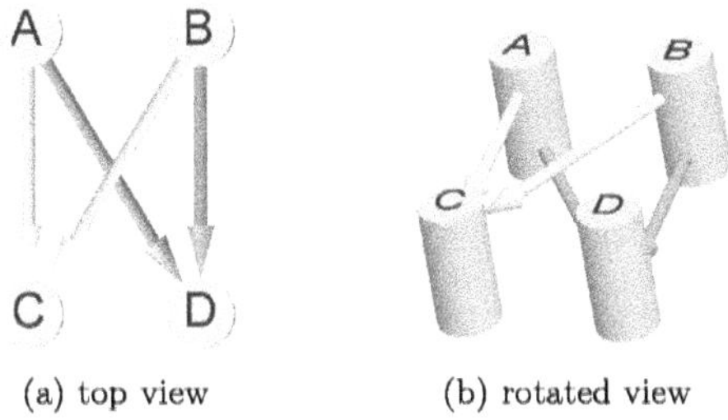

(a) top view (b) rotated view

Fig. 2. Crossings between edges in different strata can be resolved by 3D rotation.

We define a strata-aware crossing matrix by multiplying the contribution of a pair of edges to an entry c_{uv} with the number of times that these edges are present in a common stratum. Note that this results in a weighted crossing matrix in which crossings are counted individually for each pair of edges, and that this weighting scheme is different from, say, assigning weights to edges and multiplying these if edges cross.

Obviously, our weighted crossing minimization variant is at least as difficult as standard crossing minimization.

5.2 Implementation

Since most crossing reduction methods based on the crossing matrix are oblivious to the definition of its entries, they can be applied in our case as well.

We adapted the open-source program dot[1] which uses a median heuristic coupled with an adjacent-exchange post-processing step [13]. New permutations generated by these heuristics are rejected if they lead to an increase in edge crossings according to our modified crossing matrix.

Since this method does not consider crossings until after a permutation is generated it was felt that it might not be readily compatible with the new definition of c_{uv}. As an alternative we also implemented the Integer Linear Programming (ILP) approach suggested in [16]. The ILP method directly uses $c_{uv} - c_{vu}$ as the coefficients of the variables of the cost function to find an exact solution.

In our experiments we found that the median heuristic outperformed the ILP approach, because it tends to achieve a reasonable global solution which the subsequent adjacent-exchange is able to improve according to our modified definition of c_{uv}. Moreover, it is significantly faster than the *branch and cut* algorithm for solving the ILP.

For $2\frac{1}{2}$D graph drawing, the horizontal coordinate assignment phase should be adapted as well. When routing the edges in $2\frac{1}{2}$D we can allow dummy vertices on different strata to overlap. In dot's implementation, an auxiliary graph is created in which edges of an arbitrary minimum length are inserted between

[1] Available at http://www.graphviz.org/.

adjacent vertices (and dummy vertices) in each layer to keep them separated. We therefore set the length of auxiliary edges between adjacent dummy vertices that do not coexist on the same stratum to zero, thus allowing such vertices, and hence edges, to overlap. This modification led to a significant improvement in aspect ratio of the final layout (approx. 40% in our densest test cases).

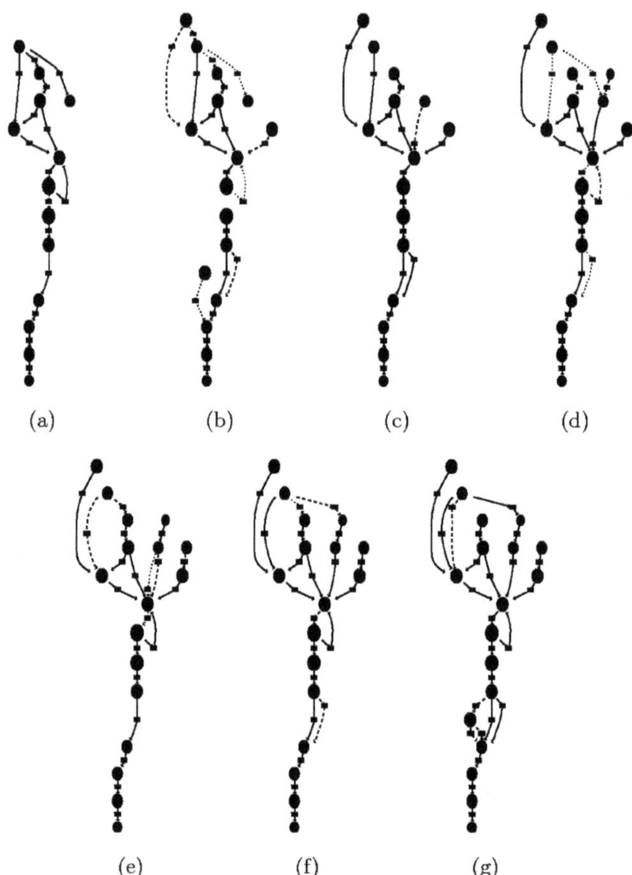

Fig. 3. Layouts of individual pathways obtained from a union graph layout. Appearing edges are shown dashed, disappearing edges dotted.

6 Application Example

The utility of our approach is demonstrated on pathways extracted from the KEGG database [17]. The data consists of parts of the glycolysis and fructose/mannose metabolism pathways in seven organisms that show significant

Table 1. Hamming-distance matrix for seven pathways

	(a)	(b)	(c)	(d)	(e)	(f)	(g)	
(a)	0	21	23	21	43	40	56	*Haemophilus influenzae*
(b)	21	0	22	20	48	39	53	*Escherichia coli CFT073*
(c)	23	22	0	10	38	35	45	*Streptococcus pyogenes*
(d)	21	20	10	0	34	31	41	*Bacillus subtilis*
(e)	43	48	38	34	0	15	31	*Arabidopsis thaliana*
(f)	40	39	35	31	15	0	16	*Drosophila melanogaster*
(g)	56	53	45	41	31	16	0	*Homo sapiens*

differences. Table 1 gives Hamming distances between these pathways with all elements $P = (V \cup E)$ considered relevant. The order in which the organisms are listed is optimal with respect to MIN SUM ORDERING and was computed by enumeration.

Using our adapted version of the dot program described in the previous section, a layout of the union graph of these seven pathways was computed. The resulting individual layouts are shown in Fig. 3.

The $2\frac{1}{2}$D representations shown in Figs. 4 and 5 have been created with the WilmaScope 3D graph visualization system [8]. Edge appearances and disappearances are color-highlighted using green and red. By moving a semi-transparent plane through the image, users can navigate forward and backward in the similarity-ordered sequence of pathways.

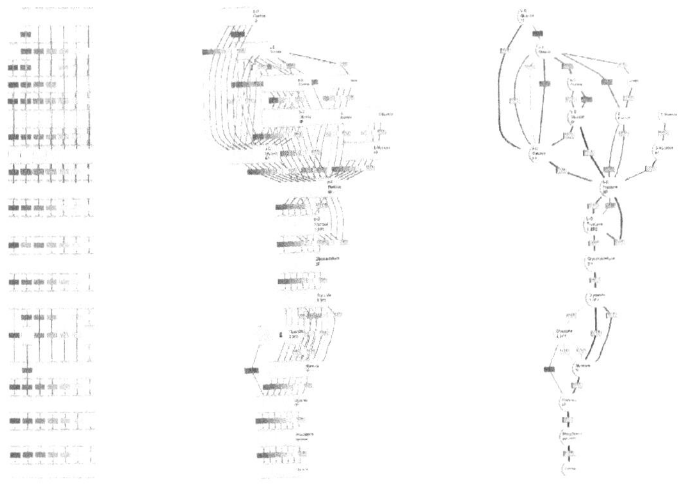

Fig. 4. $2\frac{1}{2}$D drawing of seven related pathways (parallel projection)

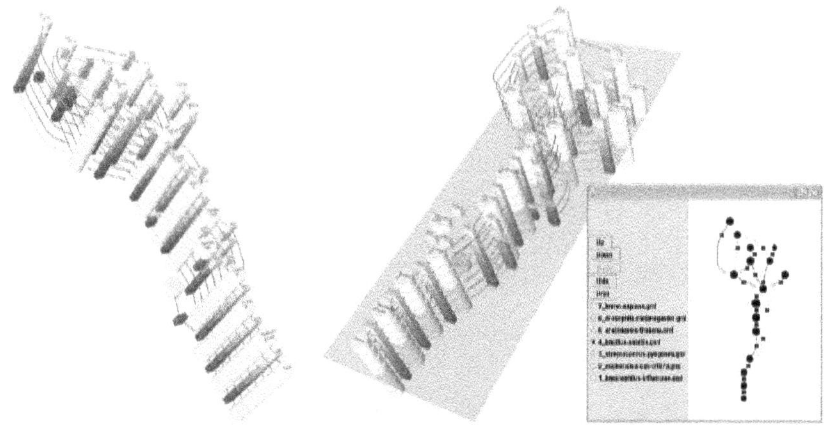

(a) A perspective projection (b) A screenshot showing the cross-section viewer for interactive exploration

Fig. 5. Perspective projection with cross-section viewer for interactive exploration

7 Discussion

We have presented an approach to visualize sets of related metabolic pathways. It should be noted that the usefulness of the stacking order is dependent on the quality of the data, which is frequently poor. However, the biologically meaningless order thus produced may in turn help discover data inconsistencies.

Straightforward extensions of our approach include varying distances between strata to indicate the actual dissimilarity and varying vertex thickness to represent numerical attributes like volume. A more challenging task is the computation of interdependent layouts for navigation along a phylogenetic tree rather than our one-dimensional ordering of pathways.

References

1. F. Alizadeh, R. M. Karp, D. K. Weisser, and G. Zweig. Physical mapping of chromosomes using unique probes. *Proceedings of the 5th ACM-SIAM Symposium on Discrete Algorithms (SODA '94)*, pages 489–500, 1994.
2. M. Y. Becker and I. Rojas. A graph layout algorithm for drawing metabolic pathways. *Bioinformatics*, 17(5):461–467, 2001.
3. K. S. Booth and G. S. Lueker. Testing for the consecutive ones property, interval graphs, and graph planarity using PQ-tree algorithms. *Journal of Computer and System Sciences*, 13(3):335–379, 1976.
4. U. Brandes and S. R. Corman. Visual unrolling of network evolution and the analysis of dynamic discourse. *Proceedings of the IEEE Symposium on Information Visualization 2002 (InfoVis '02)*, pages 145–151, 2002.
5. U. Brandes and T. Willhalm. Visualization of bibliographic networks with a reshaped landscape metaphor. *Proceedings of the 4th Joint Eurographics and IEEE TCVG Symposium on Visualization (VisSym '02)*, pages 159–164. ACM, 2002.

6. T. Dandekar, S. Schuster, B. Snel, M. Huynen, and P. Bork. Pathway alignment: application to the comparative analysis of glycolytic enzymes. *Biochemical Journal*, 343:115–124, 1999.
7. T. Dwyer and P. Eades. Visualising a fund manager flow graph with columns and worms. *Proceedings of the 6th International Conference on Information Visualisation (IV '02)*, pages 147–152. IEEE Computer Society Press, 2002.
8. T. Dwyer and P. Eckersley. The WilmaScope 3D graph drawing system. In P. Mutzel and M. Jünger, editors, *Graph Drawing Software*, Mathematics and Visualization. Springer, 2003.
9. P. Eades and Q. Feng. Multilevel visualization of clustered graphs. *Proceedings of the 4th International Symposium on Graph Drawing (GD '96)*, volume 1190 of *Lecture Notes in Computer Science*, pages 113–128. Springer, 1996.
10. P. Eades and K. Sugiyama. How to draw a directed graph. *Journal of Information Processing*, 13:424–437, 1990.
11. C. V. Forst and K. Schulten. Phylogenetic analysis of metabolic pathways. *Journal Molecular Evolution*, 52:471–489, 2001.
12. M. Forster, A. Pick, M. Raitner, F. Schreiber, and F. J. Brandenburg. The system architecture of the BioPath system. *In Silico Biology*, 2(3):415–426, 2002.
13. E. R. Gansner, E. Koutsofios, S. C. North, and K.-P. Vo. A technique for drawing directed graphs. *Software Engineering*, 19(3):214–230, 1993.
14. P. W. Goldberg, M. C. Golumbic, H. Kaplan and R. Shamir. Four strikes against physical mapping of DNA. *Journal of Computational Biology*, 2(1):139–152, 1995.
15. A. Inselberg and B. Dimsdale. Parallel coordinates: A tool for visualizing multidimensional geometry. *Proceedings of the 1st IEEE Conference on Visualization (Vis '90)*, pages 361–378, 1990.
16. M. Jünger, E. K. Lee, P. Mutzel, and T. Odenthal. A polyhedral approach to the multi-layer crossing minimization problem. *Proceedings of the International Symposium on Graph Drawing*, Lecture Notes in Computer Science 1353, pages 13–24. Springer, 1997.
17. M. Kanehisa and S. Goto. KEGG: Kyoto encyclopedia of genes and genomes. *Nucleic Acid Research*, 28(1):27–30, 2000.
18. D. A. Keim. Designing pixel-oriented visualization techniques: Theory and applications. *IEEE Transactions on Visualization and Computer Graphics*, 6(1):59–78, 2000.
19. H. Koike. The role of another spatial dimension in software visualization. *ACM Transactions on Information Systems*, 11(3):266–286, 1993.
20. L. Liao, S. Kim, and J.-F. Tomb. Genome comparisons based on profiles of metabolic pathways. *Proceedings of the 6th International Conference on Knowledge-Based Intelligent Information and Engineering Systems (KES '02)*, pages 469–476, 2002.
21. G. Michal. *Biochemical Pathways (Poster).* Boehringer Mannheim, Penzberg, 1993.
22. R. Murgai, M. Fujita, and S. C. Krishnan. Data sequencing for minimum-transition transmission. *Proceedings of the 9th IFIP International Conference on Very Large Scale Integration (VLSI '97)*, 1997.
23. V. N. Reddy, M. L. Mavrovouniotis, and M. N. Liebman. Petri net representations of metabolic pathways. *Proceedings of the 1st International Conference on Intelligent Systems for Molecular Biology (ISMB '93)*, pages 328–336, 1993.
24. F. Schreiber. High quality visualization of biochemical pathways in BioPath. *In Silico Biology*, 2(2):59–73, 2002.

25. F. Schreiber. Visual Comparison of Metabolic Pathways. *Journal of Visual Languages and Computing*, 14(4):327-340, 2003.
26. H. Shiozawa, K. Okada, and Y. Matsushita. 3D interactive visualization for intercell dependencies of spreadsheets. *Proceedings of the IEEE Symposium on Information Visualization (InfoVis '99)*, pages 79–83, 1999.
27. M. Sirava, T. Schäfer, M. Eiglsperger, M. Kaufmann, O. Kohlbacher, E. Bornberg-Bauer, and H.-P. Lenhof. BioMiner – modeling, analyzing, and visualizing biochemical pathways and networks. *Bioinformatics*, 18(Suppl. 2):219–230, 2002.
28. K. Sugiyama, S. Tagawa, and M. Toda. Methods for visual understanding of hierarchical system structures. *IEEE Transactions on Systems, Man and Cybernetics*, 11(2):109–125, 1981.
29. Y. Tohsato, H. Matsuda, and A. Hashimoto. A multiple alignment algorithm for metabolic pathway analysis using enzyme hierarchy. *Proceedings of the 8th International Conference on Intelligent Systems for Molecular Biology (ISMB '00)*, pages 376–383, 2000.
30. J. Wen. Exploiting orthogonality in three dimensional graphics for visualizing abstract data. Technical Report CS-95-20, Department of Computer Science, Brown University, 1995. http://www.cs.brown.edu/publications/techreports/reports/CS-95-20.html.

GoVisual for CASE Tools Borland Together ControlCenter and Gentleware Poseidon – System Demonstration

Carsten Gutwenger, Joachim Kupke, Karsten Klein, and Sebastian Leipert

Research Center caesar
Ludwig-Erhard-Allee 2
D-53175 Bonn, Germany
{gutwenger|kklein|kupke|leipert}@caesar.de

Abstract. The Unified Modeling Language (UML) has become the software industry's standard notation for representing software architecture and design models. UML diagrams play an important role in the engineering and re-engineering processes of software systems. Of particular interest from the Graph Drawer's perspective are UML class diagrams whose purpose is to display class hierarchies (generalizations), associations, aggregations, and compositions in one picture. The combination of hierarchical and non-hierarchical relations poses a special challenge to a graph layout tool. We present an implementation of our technology within well-known modelling tools.

1 Introduction

The Unified Modeling Language (UML) by Booch, Rumbaugh and Jacobson (see [1]) provides a mainly graphical notation to represent the artifacts of a software system. The most important UML diagram type for software architects is the *UML class diagram* consisting of classes represented by rectangular regions containing the class name, attributes and operations of the class, and different kinds of relationships between classes that are represented as lines. Since these diagrams are a means of communication between customers, developers, and others involved in the software engineering and re-engineering process, it is critical that the diagrams present information clearly. An appropriate layout of these diagrams can assist in achieving this goal (see [7]).

We have developed an approach for automatically laying out UML class diagrams in an orthogonal fashion (see [6]). Our approach distinguishes between two kinds of relationships: *generalizations* representing inheritance in class hierarchies and *associations* including *aggregations* and *compositions*. Inheritance hierarchies are emphasized in several ways:

– generalizations belonging to the same hierarchy are drawn following the same direction,
– nesting of different hierarchies within each other is avoided,

G. Liotta (Ed.): GD 2003, LNCS 2912, pp. 123–128, 2004.

- generalizations leading to the same super class join prior to reaching the super class,
- highlight the various class hierarchies by different colors,
- and highlight the generalizations by color.

For a clear visualization of the specific combination of hierarchical and non-hierarchical components, we put special emphasis on meeting a balanced mixture of the above criteria plus the following aesthetic criteria: crossing minimization, bend minimization, orthogonality, and horizontal labels.

The layout functionalities are provided as plug-ins to achieve a tight integration into existing tools (in this case, software development tools). This allows the user to work within a familiar environment without concerning about an extra user interface for graph layout and the data exchange between the different tools. Following this strategy, we expect an increasing user acceptance for automatic layout algorithms.

2 The Plug-in Philosophy

Modern development tools typically come with a graphical software modelling interface. In order to support developers and designers in the software development process, the automatic layout component has to be accessible within their development environment. Therefore we integrated the graph drawing technology as plug-ins into existing CASE tools. The layout component supports the software engineer to manage the software projects by arranging the different model views from activity diagrams to class diagrams. This approach combines the core competencies of both the CASE tool provider and the layout tool provider.

This demonstration presents a seamless integration of our technology into the following two development tools (*applications* for short in the following):

- Gentleware Poseidon for UML ([5]), a low-cost UML CASE tool. It evolved from the Open Source project ArgoUML and has a large number of installations (currently over 400.000).
- Borland Together ControlCenter ([2]), an enterprise development platform that combines application design, development, and deployment.

After installing the plug-in, a new menu called *GoVisual* is available within the application. The menu comes along with new graph layout tool buttons (see Fig. 1) that give access to various diagram layout algorithms, including the orthogonal UML layout algorithm as described in [6]. The user is enabled to apply a layout algorithm to a currently active diagram. Moreover, the user may adjust options settings of the layout algorithms to meet his aesthetic preferences.

The orthogonal UML layout algorithm is part of the large GoVisual framework of layout algorithms and data structures for the automatic layout of diagrams. GoVisual is an object-oriented C++ class library. Our plug-ins work by accessing the GoVisual API. As both applications are pure Java, we use our API's Java Native Interface to access the library. The plug-in core and the user interface components are written in Java using the application's plug-in interfaces.

To make the technology accessible to the end-user, a GoVisual menu is installed in the main menu bar of the applications. In addition, there are GoVisual layout buttons inserted into the tool bars (see Fig. 1) to allow direct access to the new layout functionality.

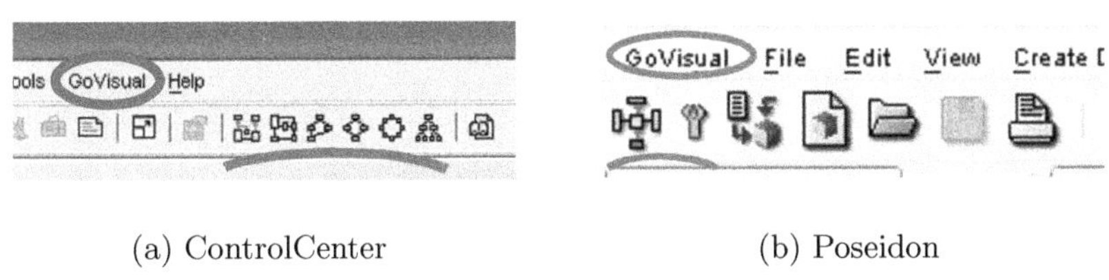

(a) ControlCenter (b) Poseidon

Fig. 1. Easy access through GoVisual menu and toolbar buttons.

Since layout algorithm parameters are usually hard to understand for users that do not have an appropriate graph drawing background, we decided to use a simple option interface that prevents the software-developer from being confronted with these parameters. The options are presented in a user-friendly way, i.e., the parameters are hidden behind an easy-access interface. The quality of the crossing minimization procedure, for example, can be selected using a slider bar, though, behind the scenes, the selection is translated into a distinguished graph algorithms setting (see Fig. 2).

3 Automatic Layout vs. Built-in Techniques

Automatic layout functionalities should give the user a clear and concise view of the software model. It integrates into the user's familiar environment as described in Sect. 2, and it offers superior layout capabilities compared to the layout functions given by the integration platform. Figure 3 shows a sample UML class diagram. The layout in 3(a) has been created with the automatic layout functionality of ControlCenter itself. It contains 13 crossings, including two crossings of generalizations and nine crossings between generalizations and associations. The diagram is difficult to read and does not reveal enough information to the user in order to easily understand the structure of the software project. It is especially difficult to identify the inheritance hierarchies. The layout given in Figure 3(b) has been computed by the GoVisual plug-in. It shows only one crossing of two associations. Moreover, the project's structure becomes visible at glance. Especially, the three inheritance hierarchies are easy to recognize, since the generalizations within the same class hierarchy are drawn in the same direction and the three class hierarchies are highlighted by different colors.

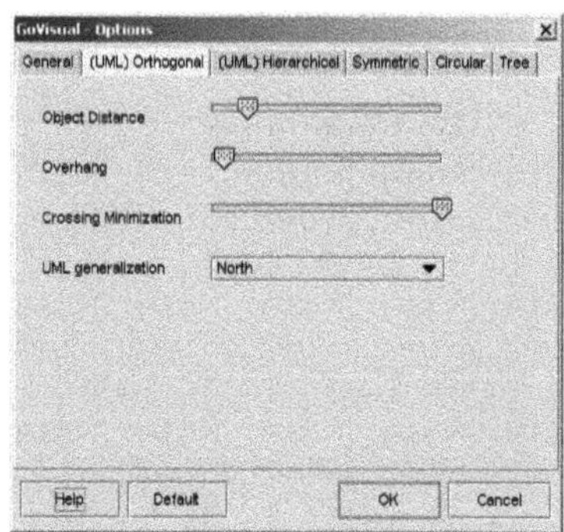

(a) ControlCenter

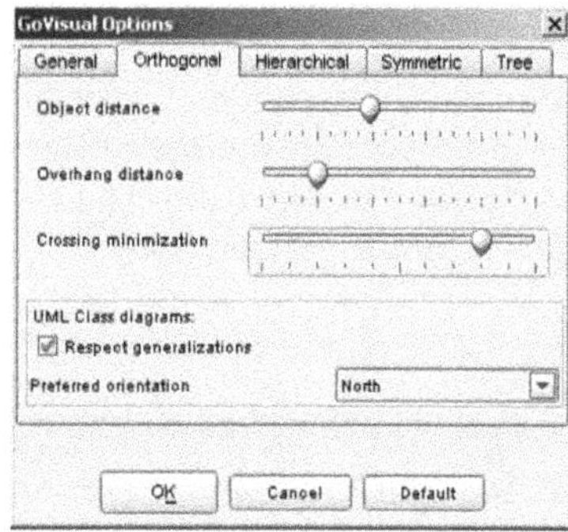

(b) Poseidon

Fig. 2. GoVisual option pages.

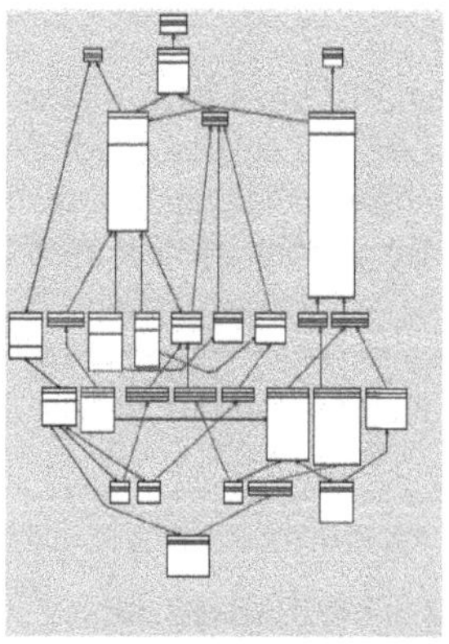

(a) Automatic layout by Together ControlCenter.

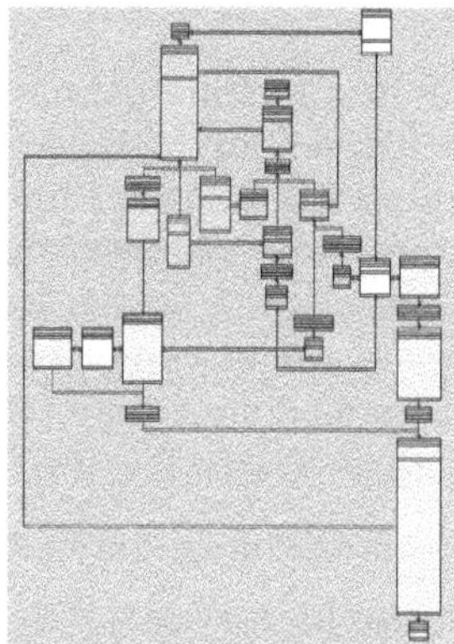

(b) Automatic layout by GoVisual.

Fig. 3. A Sample UML class diagram automatically laid out by Together ControlCenter (a) and by GoVisual orthogonal UML layout (b). The GoVisual layout also automatically highlights the three inheritance hierarchies using different colors.

Our techniques clearly outperform the built-in automatic layout algorithms (see Figs. 3, 4 and 5). The application's built-in standard layout algorithms may even hide the diagram structure to an extent where it is nearly unreadable (see Fig. 4). The Gentleware Poseidon for UML application does not provide an advanced automatic layout capability (see Fig. 5 (a)).

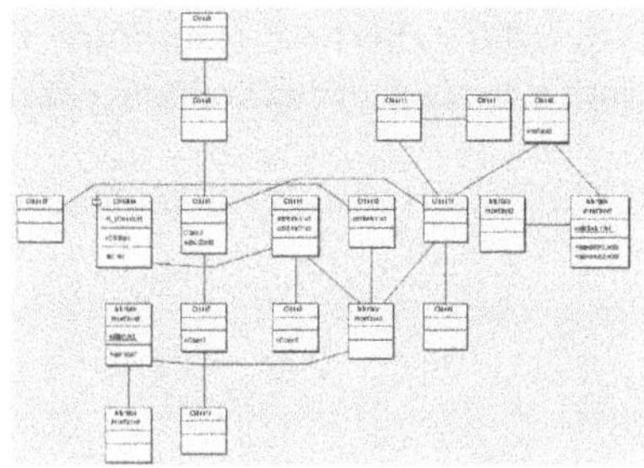

(a) Automatic layout by Together ControlCenter.

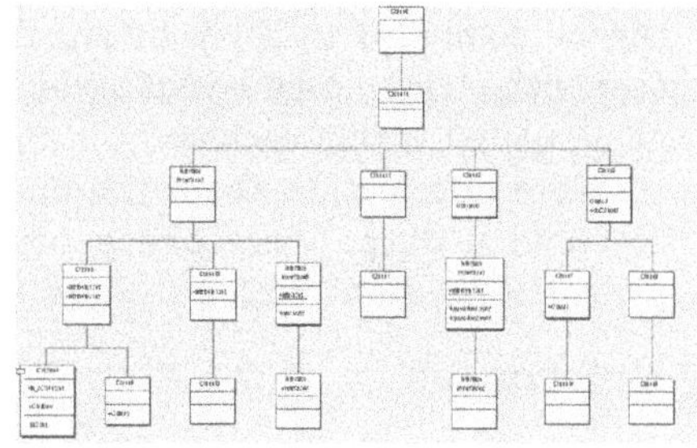

(b) Automatic layout by GoVisual

Fig. 4. A Sample UML class diagram with a tree structure automatically laid out by Together ControlCenter (a) and by GoVisual Tree Layout (b).

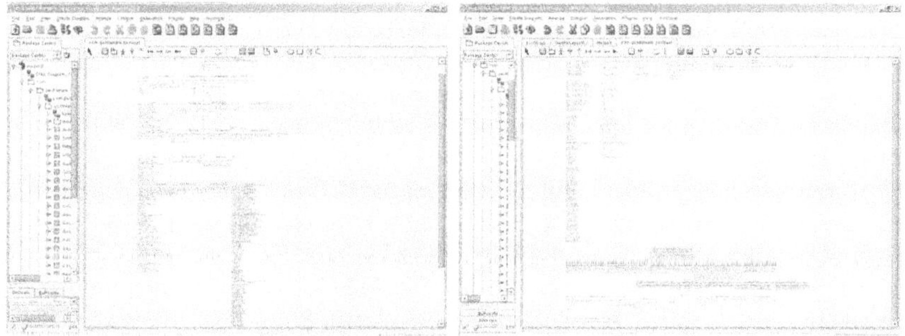

(a) Random layout by Poseidon. (b) Automatic layout by GoVisual.

Fig. 5. A Sample UML class diagram with no layout given by Poseidon for UML (a) and by GoVisual UML Orthogonal Layout (b).

Some of the GoVisual layout styles are specialized for a certain type of diagram, e.g. the UML class diagram layout styles, others are variations of standard algorithms that can be applied to all kinds of available (UML) diagrams.

The GoVisual plug-ins provide the following layout styles:

- UML Orthogonal Layout: The unique GoVisual layout technique (see [6]).
- UML Hierarchical Layout: A hierarchical layout especially suited for UML Class Diagrams (based on [8]).
- Orthogonal Layout having a focus on the minimization of crossings between edges and the minimization of the number of bends of the edges (based on [9]).
- Symmetric Layout using an energy-based layout technique (based on [4]).
- Tree Layout for the visualization of non-circular structures (based on [3]).

All layout styles provide a set of drawing and optimization algorithms that can be combined to a layout algorithm that provides the best results for the user within the chosen layout style. Furthermore a variant of parameters can be manipulated within each style.

More information about the plug-ins and GoVisual software in general can be found at `http://www.oreas.com`.

References

1. G. Booch, J. Rumbaugh, and I. Jacobson. *Unified Modeling Language User Guide.* Addison Wesley Longman, 1999.
2. Borland Software Corporation. http://www.borland.de/together/controlcenter/.
3. C. Buchheim, M. Jünger, and S. Leipert. Improving Walker's algorithm to run in linear time. In M. Goodrich, editor, *Graph Drawing (Proc. GD 2002)*, volume 2528 of *LNCS*, pages 344–353. Springer-Verlag, 2002.
4. T. Fruchterman and E. Reingold. Graph drawing by force-directed placement. *Softw. - Pract. Exp.*, 21(11):1129–1164, 1991.
5. Gentleware AG. http://www.gentleware.de/products/.
6. C. Gutwenger, M. Jünger, K. Klein, J. Kupke, S. Leipert, and P. Mutzel. A new approach for visualizing UML class diagrams. In *Proceedings of the 1st ACM Symposium on Software Visualization (SoftVis 2003), June 11-13, 2003, San Diego, CA*, 2003. To appear.
7. H. Purchase, J.-A. Allder, and D. Carrington. User preference of graph layout aesthetics: A UML study. In J. Marks, editor, *Graph Drawing (Proc. GD 2000)*, volume 1984 of *LNCS*, pages 5–18. Springer-Verlag, 2001.
8. K. Sugiyama, S. Tagawa, and M. Toda. Methods for visual understanding of hierarchical systems. *IEEE Trans. Syst. Man Cybern.*, SMC-11(2):109–125, 1981.
9. R. Tamassia. On embedding a graph in the grid with the minimum number of bends. *SIAM J. Comput.*, 16(3):421–444, 1987.

Area-Efficient Drawings of Outerplanar Graphs*

Ashim Garg and Adrian Rusu

Department of Computer Science and Engineering
University at Buffalo, Buffalo, NY 14260
{agarg,adirusu}@cse.buffalo.edu

Abstract. We show that an outerplanar graph G with n vertices and degree d admits a planar straight-line grid drawing with area $O(dn^{1.48})$ in $O(n)$ time. This implies that if $d = o(n^{0.52})$, then G can be drawn in this fashion in $o(n^2)$ area.

1 Introduction

A drawing Γ of a graph is a *straight-line* drawing, if each edge is drawn as a single line-segment. Γ is a *grid* drawing if all the vertices have integer coordinates. Γ is a *planar* drawing, if edges do not intersect each other. Here, we concentrate on grid drawings. So, we assume that the plane is covered by an infinite rectangular grid consisting of horizontal and vertical channels. Let Γ be a grid drawing. Let R be the smallest rectangle with sides parallel to the X-and Y-axes, respectively, that covers Γ completely. The *width* (*height*) of Γ is equal to 1+ width of R (1+height of R). The *area* of Γ is equal to (1+width of R)$\cdot$(1+height of R), which is equal to the number of grid points contained within R. The *degree* of a graph is equal to the maximum number of edges incident on a vertex.

There has been little work done on the area-requirement of planar straight-line grid drawings of outerplanar graphs. Currently, the best known upper bound on the area of such a drawing of an outerplanar graph with n vertices is $O(n^2)$, which is the same as for general planar graphs [3,8].

In this paper, we show that an outerplanar graph G with n vertices and degree d admits a planar straight-line grid drawing with area $O(dn^{1.48})$ in $O(n)$ time. This implies that if $d = o(n^{0.52})$, then G can be drawn in this fashion in $o(n^2)$ area.

In Section 4, we give a brief description of our drawing algorithm (for more details, see [6]). It is based on a tree-drawing algorithm of [2], and uses the fact that the dual of a maximal outerplanar graph is a tree.

2 Related Results

Let G be an outerplanar graph with n vertices. [1] shows that G admits a planar polyline drawing as well as a visibility representation with $O(n \log n)$ area. [7]

* Research supported by NSF CAREER Award IIS-9985136, NSF CISE Research Infrastructure Award No. 0101244, and Mark Diamond Research Grant No. 13-Summer-2003 from GSA of The State University of New York.

G. Liotta (Ed.): GD 2003, LNCS 2912, pp. 129–134, 2004.

shows that G admits a planar polyline drawing with $O(n)$ area, if G has degree at most 4. The technique of [7] can be easily extended to construct a planar polyline drawing of G with $O(d^2n)$ area, if G has degree d [1]. Also, in 3D, G admits a crossings-free straight-line grid drawing with $O(n)$ volume [4,5].

3 Preliminaries

We denote by $|G|$, the number of vertices (nodes) in a graph (tree) G. An *ordered* tree is one with a pre-specified counterclockwise ordering of edges incident on each node. A path $P = v_0v_1 \dots v_q$ is a *root-to-leaf* path of a binary ordered tree T, if v_0 is the root of T, and v_q is a leaf of T. A *left (right) subtree of* P is a subtree of T rooted at the left (right) child c of a node of P, such that c does not belong to P. The *size* of a subtree of T is equal to the number of nodes in it. The following lemma, which follows directly from Lemma A.1 of [2], defines the concept of a *spine*, which is a special kind of a root-to-leaf path (see also [6]).

Lemma 1 (Lemma A.1 of [2]). *Let $p = 0.48$. Given any binary ordered tree T with n nodes, there exists a root-to-leaf path P, called* spine, *such that for any left subtree α and right subtree β of P, $|\alpha|^p + |\beta|^p \leq (1-\delta)n^p$, for some constant $\delta > 0$. Also, assuming that we have already pre-computed the size of the subtree rooted at each node v of T and stored it in v, we can compute P in $O(|P|)$ time.*

Let G be a *maximal* outerplanar graph, i.e., an outerplanar graph to which no edge can be added without destroying its outerplanarity. It is easy to see that each internal face of G is a triangle. The *dual tree* T_G of G is defined as follows:

- there is a one-to-one correspondence between the nodes of T_G and the internal faces of G, and
- there is an edge $e = (u, v)$ in T_G if and only if the faces of G corresponding to u and v share an edge e' on their boundaries. e and e' are *duals* of each other.

(Figure 1(b) shows the dual tree of the outerplanar graph of Figure 1(a).)

Let $P = v_0v_1 \dots v_q$ be a path of T_G. Let H be the subgraph of G corresponding to P. A *beam* drawing of H is shown in Figure 2, where the vertices of H are placed on two horizontal channels, and the faces of H are drawn as triangles.

A line-segment with end-points a and b is a *flat* line-segment if a and b are grid points, and either belong to the same horizontal channel, or belong to adjacent horizontal channels. Let B be a flat line-segment with end-points a and b, such that b is at least one unit to the right of a. Let G be an outerplanar graph with two distinguished adjacent vertices u and v, such that the edge (u, v) is on the external face of G; u and v are called the *poles* of G. Let D be a planar straight-line drawing of G. D is a *feasible* drawing of G with base B if:

- the two poles of G are mapped to a and b each,
- each non-pole vertex of G is placed at least one unit above the lower of a and b, at least one unit to the right of a, and at least one unit to left of b.

Throughout the rest of this paper, for simplicity, by the term *outerplanar* graph, we will mean a maximal outerplanar graph.

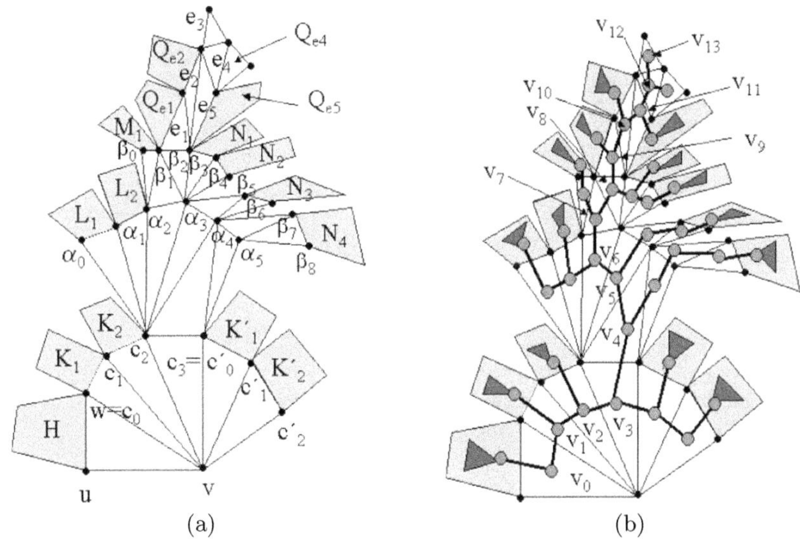

Fig. 1. (a) An outerplanar graph G. Here, H, K_1, K_2, K_1', K_2', L_1, L_2, M_1, N_1, N_2, N_3, N_4, Q_{e1}, Q_{e2}, Q_{e4}, and Q_{e5} are subgraphs of G, and are themselves outerplanar graphs. (b) The dual tree T_G of G. The edges of T_G are shown with dark lines. Note that $v_0 v_1 \dots v_{13}$ is a spine of T_G.

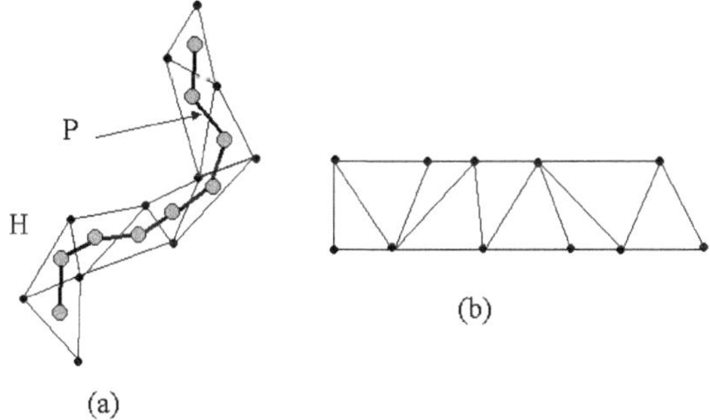

Fig. 2. (a) A path P and its corresponding graph H. (b) A beam drawing of H.

4 Outerplanar Graph Drawing Algorithm

The drawing algorithm, which we call *Algorithm OpDraw*, is recursive in nature. In each recursive step, it takes as inputs an outerplanar graph G with prespecified poles, and a long-enough flat line-segment B, and constructs a feasible drawing D of G with base B. D is constructed by constructing a drawing M of the subgraph Z corresponding to a spine of T_G, splitting G into several smaller outerplanar graphs after removing Z and some other vertices from it, constructing feasible drawings of these smaller outerplanar graphs, and then combining their drawings with M to obtain D.

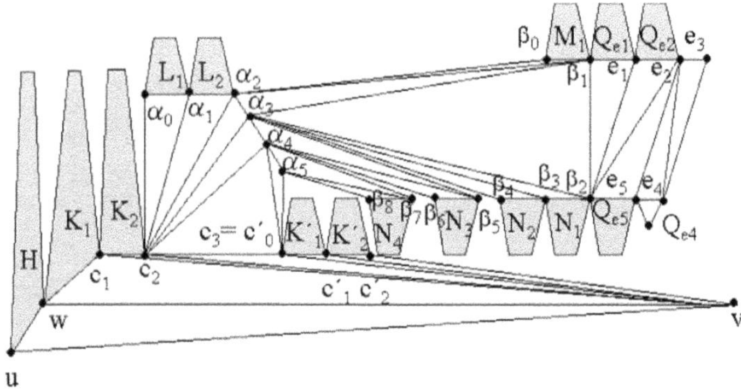

Fig. 3. The drawing of the outerplanar graph of Figure 1(a) constructed by *Algorithm OpDraw*, in the case, where v is one unit above u.

Let u and v be the poles of G. Let T_G be the dual tree of G. Let r be the node of T_G that corresponds to the internal face F of G that contains the edge (u, v). Convert T_G into an ordered tree by making r its root, and assigning the edges incident on each node o the same counterclockwise order as the one that their dual edges have in the face of G corresponding to o. Note that T_G is a binary tree because each internal face of G is a triangle.

Let $P = v_0 v_1 v_2 \dots v_q$ be a spine of T_G, where $v_0 = r$. In general, we can interpret the structure of G with respect to P as follows (see Figure 1): Let u, v, w be the vertices belonging to face F. Assume that the edge (v_0, v_1) of P is the dual of the edge (v, w) (the case, where (v_0, v_1) is the dual of (u, w), is symmetrical). Let $u, w = c_0, c_1, \dots, c_m = c'_0, c'_1, \dots, c'_s$ be the clockwise order of the neighbors of v, where m is the integer such that for each i $(1 \leq i \leq m)$, the face $c_{i-1}c_i v$ corresponds to the spine node v_i, and for each i $(1 \leq i \leq s)$, the face $c'_{i-1}c'_i v$ corresponds to a non-spine node of T_G (In Figure 1, $m = 3$, and $s = 2$). Let H, K_i, and K'_i be the maximal biconnected subgraphs of G that contains edges (u, w), (c_{i-1}, c_i), and (c'_{i-1}, c'_i), respectively, but not the faces uvw, $vc_{i-1}c_i$, and $vc'_{i-1}c'_i$, respectively. Let $\alpha_0, \alpha_1, \dots, \alpha_h, \alpha_{h+1}, \alpha_t$ be the vertices of K_m different from c_{m-1} and c_m such that $\alpha_0, \alpha_1, \dots, \alpha_h$ is the clockwise order of the neighbors of c_{m-1}, and $\alpha_h, \alpha_{h+1}, \alpha_{h+2}, \alpha_t$ is the clockwise order of the neighbors of c_m. For example, in Figure 1, $h = 4$, and $t = 5$. Let j be the index such that the dual of edge (α_{j-1}, α_j) belongs to P (if no such j exists, then we can do the following: if K_m consists of only one internal face, namely, $c_{m-1}c_m\alpha_0$, then set $j = 0$. Otherwise, the leaf v_q of P will correspond to either the face $\alpha_0\alpha_1 c_{m-1}$ or the face $\alpha_{t-1}\alpha_t c_m$; in the first case, set $j = 1$, and in the second case, set $j = t$). For example, in Figure 1, $j = 3$. Let L_i be the maximal biconnected subgraph of G that contains the edge (α_{i-1}, α_i), but not the face $\alpha_{i-1}\alpha_i c_{m-1}$ or $\alpha_{i-1}\alpha_i c_m$ (whichever exists). Let $S = \beta_0, \beta_1, \dots, \beta_\mu$ be the clockwise order of the neighbors of $\alpha_{j-1}, \alpha_j, \dots, \alpha_t$ (in that order) in the subgraphs $L_j, L_{j+1}, \dots, L_t$, where each β_k is different from $\alpha_{j-1}, \alpha_j, \dots, \alpha_t$. For example, in Figure 1, $\mu = 8$. Let ϵ be the index such that the dual of the edge $(\beta_{\epsilon-1}, \beta_\epsilon)$ belongs to P (if no such

ϵ exists, then we can do the following: if L_j consists of only one internal face, namely, $\alpha_{j-1}\alpha_j\beta_0$, then set $\epsilon = 0$. Otherwise, the leaf v_q of P will correspond to either the face $\beta_0\beta_1\alpha_{j-1}$ or the face $\beta_{\mu-1}\beta_\mu\alpha_j$; in the first case, set $\epsilon = 1$, and in the second case, set $\epsilon = \mu$). For example, in Figure 1, $\epsilon = 2$. For each i, where $1 \leq i \leq \epsilon - 1$ ($\epsilon + 1 \leq i \leq \mu$), if there is an edge (β_{i-1}, β_i), then let M_i ($N_{i-\epsilon}$) be the maximal biconnected subgraph of G that contains (β_{i-1}, β_i), but not the face $\beta_{i-1}\beta_i\alpha_k$, where $k = j - 1$ or j. Let $(v_{\rho-1}, v_\rho)$ be the edge of P that is the dual of the edge $(\beta_{\epsilon-1}, \beta_\epsilon)$. For example, in Figure 1, $\rho = 9$. Let R be the subgraph of G that corresponds to the subpath $v_\rho v_{\rho+1} \dots v_q$ of P. Let $e \neq (\beta_{\epsilon-1}, \beta_\epsilon)$ be an edge on the external face of R. Let Q_e be the maximal biconnected subgraph of G than contains e but not the face of R containing e.

D is constructed as shown in Figure 3 (in this figure, we show the construction when v is one unit above u. The other two cases, where u is one unit above v, and where u and v are in the same horizontal channel are similar. For more details, see [6]): w is placed one unit above u, and $c_1, c_2, \dots, c_m = c'_0, c'_1, c'_s$ are placed in the same horizontal channel one unit above w. α_0 (α_t) is placed in the same vertical channel as c_{m-1} (c_m). $\alpha_0, \alpha_1, \dots, \alpha_{j-1}$ are placed in the same horizontal channel. $\alpha_{j-1}, \alpha_j, \dots, \alpha_t$ are placed along a line making 45° with the horizontal channels, such that each α_k is one unit above and one unit to the left of α_{k+1}. $\beta_0, \beta_1, \dots, \beta_{\epsilon-1}$ are placed one unit above α_{j-1} in the same horizontal channel. $\beta_\epsilon, \beta_{\epsilon+1}, \dots, \beta_\mu$ are placed one unit below α_t in the same horizontal channel. $\beta_{\epsilon-1}$ and β_ϵ are placed in the same vertical channel. A beam drawing E of R is constructed. Feasible drawings of H with base $\overline{uw}$, and each K_i $(1 \leq i \leq m-1)$, K'_i $(1 \leq i \leq s)$, L_i $(1 \leq i \leq j-1)$, M_i $(1 \leq i \leq \epsilon - 1)$, N_i $(1 \leq i \leq \mu - \epsilon)$, and Q_e, with bases $\overline{c_{i-1}c_i}$, $\overline{c'_{i-1}c'_i}$, $\overline{\alpha_{i-1}\alpha_i}$, $\overline{\beta_{i-1}\beta_i}$, $\overline{\beta_{i+\epsilon-1}\beta_{i+\epsilon}}$, and e, respectively, are recursively constructed, with the horizontal distances between the end-points of the line-segments $\overline{uw}$, $\overline{c_{i-1}c_i}$, $\overline{c'_{i-1}c'_i}$, $\overline{\alpha_{i-1}\alpha_i}$, $\overline{\beta_{i-1}\beta_i}$, $\overline{\beta_{i+\epsilon-1}\beta_{i+\epsilon}}$, and e, equal to $|H|-1$, $|K_i|-1$, $|K'_i|-1$, $|L_i|-1$, $|M_i|-1$, $|N_i|-1$, and $|Q_e|-1$, respectively. The drawing of each N_i, and Q_e, where e is on the bottom boundary of E, is flipped upside-down before placing it in D. Also note that β_0 and β_μ are placed such that they are either in the same vertical channel as, or to the right of c'_s. Also note that α_t is placed $1 + \theta$ units above the horizontal channel containing c'_s, where θ is maximum height of the feasible drawings of K'_i, N_i, and Q_e, where e is on the bottom boundary of E.

Let $h(n)$ and $w(n)$ be the height and width, respectively, of D, as constructed by Algorithm *OpDraw*. Here, n is the number of vertices in G. Let d be the degree of G. Note that, by the definition of feasible drawings, $w(n)$ will be equal to one plus the horizontal separation between the end-points of B.

It is easy to prove using induction that $w(n) = n$ is sufficient. As for $h(n)$, first notice that, because G has degree d, $t - (j-1)$ is less than $2d$, and hence, the distance between $\beta_{\epsilon-1}$ and β_ϵ is less than $2d + 2$. Let h' be a function, such that $h'(f) = h(n)$, where f is the number of internal faces in G, i.e., the number of nodes in the dual tree T_G of G. From the construction of D, we have that:

$$h'(f) \leq \max\{\max_{1\leq i\leq \mu-\epsilon}\{h'(|T_{N_i}|)\}, \max_{edge\ e\ on\ bottom\ boundary\ of\ E}\{h'(|T_{Q_e}|)\},$$

$$\max_{1\le i\le s}\{h'(|T_{K'_i}|)\}\} + \max\{h'(|T_H|), \max_{1\le i\le m-1}\{h'(|T_{K_i}|)\}, \max_{1\le i\le j-1}\{h'(|T_{L_i}|)\},$$
$$\max_{1\le i\le \epsilon-1}\{h'(|T_{M_i}|)\}, \max_{edge\ e\ on\ top\ boundary\ of\ E}\{h'(|T_{Q_e}|)\}\} + O(d),$$

The dual trees of H, K_i, L_i, M_i, and Q_e (where edge e is on top boundary of E) are either right subtrees of P, or belong to right subtrees of P. The dual trees of K'_i, N_i, and Q_e (where edge e is on bottom boundary of E) are either left subtrees, or belong to left subtrees of P. Hence, from Lemma 1, it follows that:

$$h'(f) \le \max_{f_1^p + f_2^p \le (1-\delta) f^p}\{h'(f_1) + h'(f_2) + O(d)\}.$$

Using induction, we can show that $h'(f) = O(df^{0.48})$ (see also [2]). Since $f = O(n)$, $h(n) = h'(f) = O(df^{0.48}) = O(dn^{0.48})$.

Theorem 1. *Let G be an outerplanar graph with degree d and n vertices. G admits a planar straight-line grid drawing with area $O(dn^{1.48})$ in $O(n)$ time.*

Proof. Arbitrarily select any edge $e = (u, v)$ on the external face of G, and designate u and v as the poles of G. Let B be any horizontal line-segment with length $n - 1$, such that the end-points of B are grid points. Construct a feasible drawing D of G with base B using Algorithm *OpDraw*. From the discussion given above, it follows immediately that the area of D will be equal to $n \cdot O(dn^{0.48}) = O(dn^{1.48})$. Using Lemma 1, we can easily implement Algorithm *OpDraw* such that it will run in $O(n)$ time.

Corollary 1. *Let G be an outerplanar graph with n vertices and degree d, where $d = o(n^{0.52})$. G admits a planar straight-line grid drawing with $o(n^2)$ area in $O(n)$ time.*

References

1. T. Biedl. Drawing outer-planar graphs in O($n \log n$) area. In *Proc. 10th Int. Symp. on Graph Drawing*, volume 2528 of *LNCS*, pages 54–65. Springer-Verlag, 2002.
2. T.M. Chan. A near-linear area bound for drawing binary trees. *Algorithmica*, 34(1):1–13, 2002.
3. H. de Fraysseix, J. Pach, and R. Pollack. How to draw a planar graph on a grid. *Combinatorica*, 10(1):41–51, 1990.
4. V. Dujmovic and D.R. Wood. Tree-partitions of k-trees with applications in graph layout. In *Proc. 29th Workshop on Graph Theoretic Concepts in Computer Science (WG '03)*. To appear.
5. S. Felsner, G. Liotta, and S. Wismath. Straight-line drawings on restricted integer grids in two and three dimensions. In *Proc. 9th Int. Symp. on Graph Drawing (GD '01)*, volume 2265 of *LNCS*, pages 328–342. Springer-Verlag, 2001.
6. A. Garg and A. Rusu. Area-efficient drawings of outerplanar graphs. Technical Report No. 2003-11, Dept. Computer Sc. & Engg., Univ. at Buffalo, Buffalo, 2003.
7. C. E. Leiserson. Area-efficient graph layouts (for VLSI). In *Proc. 21st Annu. IEEE Sympos. Found. Comput. Sci.*, pages 270–281, 1980.
8. W. Schnyder. Embedding planar graphs on the grid. In *Proc. 1st ACM-SIAM Sympos. Discrete Algorithms*, pages 138–148, 1990.

A Framework for User-Grouped Circular Drawings

Janet M. Six[1] and Ioannis (Yanni) G. Tollis*[2]

[1] Lone Star Interface Design, P.O. Box 1993, Wylie, TX 75098
jsix@lonestarinterfacedesign.com
[2] Department of Computer Science, University of Crete
GR 714 09 Heraklion, Greece and
Institute of Computer Science, Foundation for Research and Technology
Hellas-FORTH, P.O. Box 1385, 71110 Heraklion, Crete, Greece
tollis@csd.uoc.gr

Abstract. In this paper we introduce a framework for producing circular drawings in which the groupings are user-defined. These types of drawings can be used in applications for telecommunications, computer networks, social network analysis, project management, and more. This fast approach produces drawings in which the user-defined groupings are highly visible, each group is laid out with a low number of edge crossings, and the number of crossings between intra-group and inter-group edges is low.

1 Introduction

A *circular graph drawing* is a visualization in which

- the graph is partitioned into groups,
- the nodes of each group are placed onto a unique embedding circle, and
- each edge is drawn with a straight line.

Circular graph drawing has received increasing attention in the literature [2, 4,13,14,20,21,22,24,25]. Kar, Madden, and Gilbert presented a circular drawing technique for networks in [13]. This approach partitions the graph into groups, places the groups onto embedding circles, and then sets the final coordinates of the nodes. As discussed in [4], an advanced version of this technique is included in Tom Sawyer Software's *Graph Layout Toolkit* (`www.tomsawyersoftware.com`).

Tollis and Xia present several linear time algorithms for the visualization of survivable telecommunication networks in [25].

Citing a need for graph abstraction and reduction of today's large information structures, Brandenburg describes an approach to draw a path (or cycle) of cliques in [2].

InFlow [16] is a tool to visualize social networks. This tool produces diagrams and statistical summaries to pinpoint the strengths and weaknesses within an organization (`www.inflow.net`).

* On leave from The University of Texas at Dallas.

G. Liotta (Ed.): GD 2003, LNCS 2912, pp. 135–146, 2004.

We presented a linear time algorithm for producing circular drawings of biconnected graphs on a single embedding circle in [20,21]. This technique was extended to place a nonbiconnected graph on a single embedding circle in [20, 22]. A framework for producing circular drawings of nonbiconnected graphs on multiple circles was presented in [20,24]. These techniques require $O(m)$ time and produce drawings with a low number of edge crossings. More details about these techniques will be dicussed in Section 2.

Kaufmann and Wiese extended our circular approach [20,21,22,24] in [14]. In this approach, the blocktree structures of nonbiconnected graphs are interpreted in a different manner which allows finer structures to be shown within the visualization of a biconnected component. The authors also give an interactive version of this technique. An advanced version of this approach is included in yWorks' *yFiles Library* (`www.yworks.com`).

All of these techniques are very useful for applications in telecommunications[15], computer networks [20], social network analysis [16], project management [16], and more. However, with the exception of the Graph Layout Toolkit (GLT) technique [4,13], these techniques do not allow the user to define which nodes should be grouped together on an embedding circle. And in the GLT technique, the layouts of the user defined groups are themselves placed on a single embedding circle. For some graph structures, this may not be ideal. In this paper, we present a circular drawing algorithm which allows the user to define the node groups, draws each group of nodes efficiently and effectively, and visualizes the superstructure well. We call this approach *user-grouped circular drawing.*

An example of an application in which user-grouped circular drawing would be useful is a computer network management system in which the user needs to know the current state of the network. It would be very helpful to allow the user to group the computers by department, floor, usage rates, or other criteria. See Figure 1. This graph drawing could also represent a telecommunications network, social network, or even the elements of a large software project. There are, of course, many other applications which would benefit from user-grouped circular drawing.

The remainder of this paper is organized as follows: in Section 2, we review our previous circular techniques. In Section 3, we introduce a framework for user-grouped circular drawing. In Section 4, we discuss a force-directed approach in which node placement is restricted to the perimeter of circles. In Section 5, we present an algorithm for user-grouped circular drawing. In Section 6, we discuss conclusions and future work.

2 Review of Our Previous Circular Techniques

As mentioned in the previous section, we have presented multiple efficient circular graph drawing techniques which produce visualizations with a low number of edge crossings [20,21,22,24]. In this section, we give a brief review of these techniques.

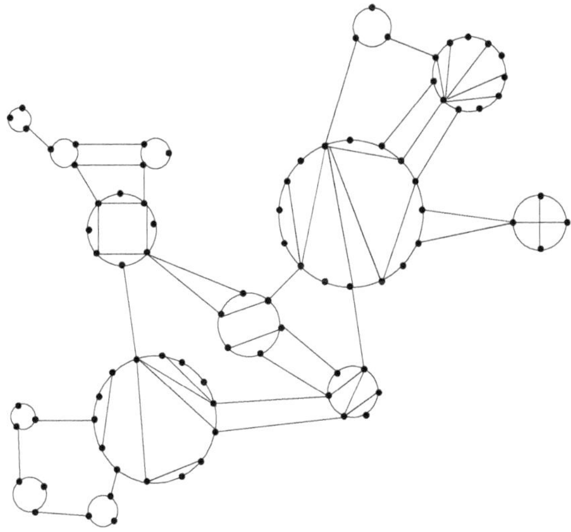

Fig. 1. A user-grouped circular drawing.

2.1 Circular Drawings of Biconnected Graphs

In [20,21], we introduced a linear time technique to produce circular graph drawings of biconnected graphs on a single embedding circle. First, it is important to note the difficulty of this problem. Of course, minimizing the number of crossings in a drawing is the well-known NP-Complete crossing number problem [10]. The more restricted problem of finding a minimum crossing embedding such that all the nodes are placed onto the circumference of a circle and all edges are represented with straight lines is also NP-Complete as proven in [17]. The authors show the NP-Completeness by giving a polynomial time transformation from the NP-Complete *Modified Optimal Linear Arrangement* problem.

In order to produce circular drawings with fewer crossings than previous techniques, we presented the algorithm *CIRCULAR* which tends to place edges toward the outside of the embedding circle. Also, nodes are placed near their neighbors.

This technique visits the nodes in a wave-like fashion, looking for *pair edges* (edges incident to two nodes which share at least one neighbor) which are then removed. Sometimes, *triangulation edges* are added to aid this process. It is the selective edge removal which causes many edges to be placed toward the periphery of the embedding circle in the visualizations produced by CIRCULAR. Subsequent to the edge removal, CIRCULAR proceeds to perform a Depth-First Search (DFS) search on the reduced graph. The longest path of the resulting DFS tree is placed on the embedding circle and the remaining nodes are nicely merged into this ordering.

The worst-case time requirement of CIRCULAR is $O(m)$, where m is the number of edges. An important property of this technique is the guarantee that if a zero-crossing drawing exists for a given biconnected graph, CIRCULAR will find it. Such graphs must be outerplanar. In fact, CIRCULAR was inspired by the algorithm for recognizing outerplanar graphs presented in [18]. For drawings which do contain crossings, a postprocessing method which further reduces the number of crossings can be applied. See [20,21] for such a method.

Extensive experiments compared CIRCULAR and Tom Sawyer Software's GLT. CIRCULAR drawings had 15% fewer crossings. This improvement increased to 30% with the crossing-reduction postprocessing step. Sample drawings from the experimental study are shown in Figure 2.

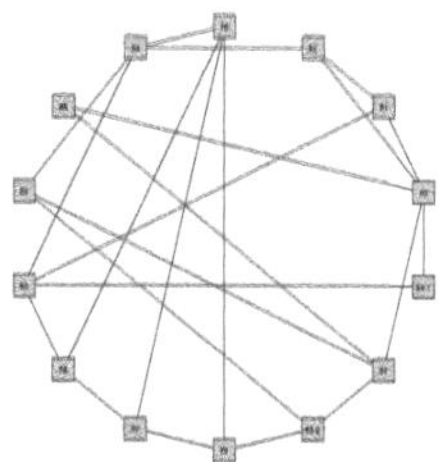

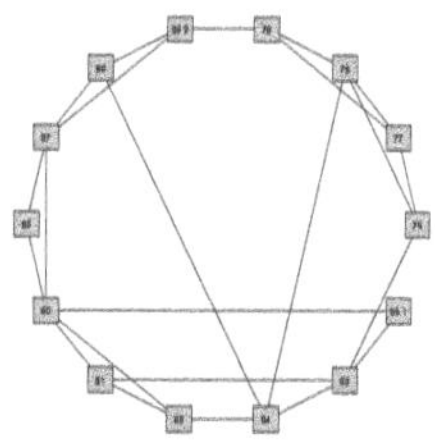

Fig. 2. The drawing on the left is produced by the GLT. The drawing on the right is of the same graph and is produced by CIRCULAR with crossing-reduction postprocessing. The drawing on the right has 75% fewer crossings than the GLT drawing.

2.2 Circular Drawings of Nonbiconnected Graphs on a Single Embedding Circle

In [20,22] we presented the algorithm *CIRCULAR-Nonbiconnected* for producing circular drawings of nonbiconnected graphs on a single embedding circle. Given a nonbiconnected graph G, we can decompose G into biconnected components. In CIRCULAR-Nonbiconnected, we layout the resulting block-cutpoint tree on a circle and then layout each biconnected component with a variant of CIRCULAR.

First, we consider how to attain a circular drawing of a tree. A DFS produces a numbering that we can use to order the nodes around the embedding circle in a crossing-free manner. From this result, we know how to order the biconnected components around the embedding circle. Next, we need to consider articulation points which are not adjacent to a bridge (*strict articulation points*). Strict articulation points appear in multiple biconnected components. In which biconnected component should a strict articulation point appear in the circular drawing? Multiple approaches to this issue are discussed in [20,22]. Due to space restrictions, we do not discuss these solutions here. A third issue to consider is how to transform the layout of each biconnected component to fit onto an arc of the embedding circle. This transformation is called *breaking*. The resulting breaks occur at an articulation point within the biconnected component.

The worst-case time requirement for CIRCULAR-Nonbiconnected is $O(m)$ if we use CIRCULAR to layout each biconnected component. The resulting drawings have the property that the nodes of each biconnected component (with the exception of some strict articulation points) appear consecutively. Furthermore, the order of the biconnected components on the embedding circle are placed according to a layout of the accompanying block-cutpoint tree. Therefore, the biconnectivity structure of a graph is displayed even though all of the nodes appear on a single circle. An example drawing is shown in Figure 3.

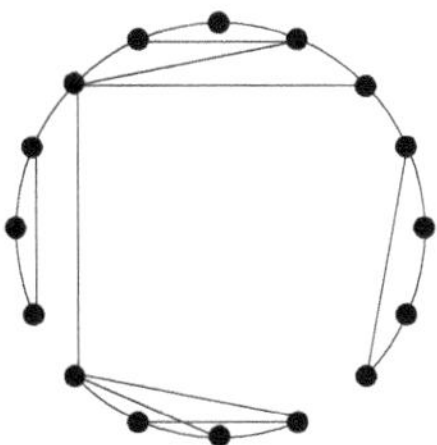

Fig. 3. An example drawing produced by CIRCULAR - Nonbiconnected.

2.3 Circular Drawings of Nonbiconnected Graphs on Multiple Embedding Circles

In [20,24] we presented Algorithm *CIRCULAR-withRadial* which produces circular drawings of nonbiconnected graphs on multiple embedding circles. As in CIRCULAR-Nonbiconnected, we first decompose the given graph G into biconnected components. Then we layout the block-cutpoint tree with a variant of the radial layout technique [1,5,9]. Then each biconnected component is laid out with a variant of CIRCULAR. Many details are omitted here. The worst-case time requirement is $O(m)$.

CIRCULAR-withRadial is a very useful technique and extension of this work to include interactive schemes has been presented by Kaufmann and Wiese in [14]. CIRCULAR-withRadial groups given graphs by their biconnected components. This grouping helps to provide a logical, beneficial view of nonbiconnected graphs, however in some cases it would be helpful to allow the user to define which nodes should be placed together on an embedding circle. We present such an algorithm in Section 5.

3 A Framework for User-Grouped Circular Drawing

The problem of producing circular drawings of graphs grouped by biconnectivity is quite different from the problem of drawing a graph whose grouping is user-defined. In the latter case, there is no known structure of either the groups or the relationship between the groups. Therefore, we must use a general method for producing this type of visualization. The four goals of a user-grouped circular drawing technique should be:

1. the user-defined groupings are highly visible,
2. each group is laid out with a low number of edge crossings,
3. the number of crossings between intra-group and inter-group edges is low, and
4. the layout technique is fast.

We know from previous work in clustered graph drawing [6,7,8,12] that the relationship between groups is often not very complex. We take advantage of this expectation in this framework. Define the *superstructure* G_s of a given graph $G = (V, E, P)$, where P is the node group partition, as follows: the nodes in G_s represent the elements of P. For each edge $e \in E$ which is incident to nodes in two different node groups, place an edge between nodes representing the respective groups in G_s. The type of structure which we expect G_s to have should be visualized well with a force-directed [3,5] technique, therefore we will layout the superstructure G_s with this approach. Since G_s will likely not be a very complicated graph, it should not take much time to achieve a good drawing with a force-directed technique.

The node groups themselves will be either biconnected or not and since CIRCULAR and CIRCULAR-Nonbiconnected can layout biconnected and nonbiconnected graphs on a single embedding circle in linear time and have been shown to perform well in practice, we also will use those techniques here.

We have now addressed how to achieve goals 1 and 2 with good speed. However, in order to produce good user-grouped circular graph drawings, we must successfully merge these two techniques so that we can simultaneously reach goals 1,2, and 3. And, of course, we need a fast technique in order to achieve goal 4. Attaining goal 3 is very important to the quality of drawings produced by a user-grouped circular drawing technique. As shown in [19], a drawing with fewer crossings is more readable. It is especially important to reduce the number of intra-group and inter-group edge crossings as those can particularly cause confusion while interpreting a drawing. See Figure 4. How can we achieve this low number of crossings? We must place nodes which are adjacent to nodes in other groups (called *outnodes* in [4,13]) close to the placement of those other nodes. A force-directed approach is a good way to attain this goal since it would encourage outnodes to be closer to their neighbors. Traditional force-directed approaches [3,5] will not work here though, because we need to constrain the placement of nodes to circles. In Section 4, we present a force-directed approach in which the nodes are restricted to appear on circular tracks. With the use of this technique we will reach goal 3. As will be discussed, we can do this in a reasonable amount of time.

As with most force-directed techniques, the initial placement of nodes has a very significant impact on the final drawing [3,5]. Therefore it is important to have a good initial placement. This is why we should layout the superstructure and each node group first. At the completion of those steps, we should have the

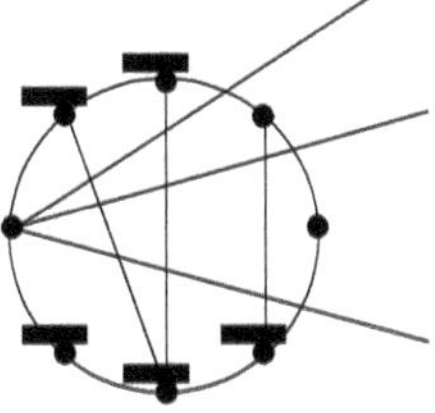

Fig. 4. Example of intra-group and inter-group edge crossings.

almost-final drawing. It will then be a matter of fine-tuning the drawing with our new circular-track force-directed technique. And as shown in [23], once you have an almost-final drawing, it does not take much time for a force-directed technique to converge.

4 Circular-Track Force-Directed

In order to adapt the force-directed paradigm for circular drawing, we need a way to guarantee that the nodes of a group appear on the circumference of an embedding circle, *the circular track*. The nodes are restricted to appear on the circular track, but are allowed to jump over each other and appear in any order. See Figure 5. And as in the force-directed approach, we want to minimize the potential energy in the spring system which is modelling the graph. In this section, we describe how this circular-track adaptation can be achieved.

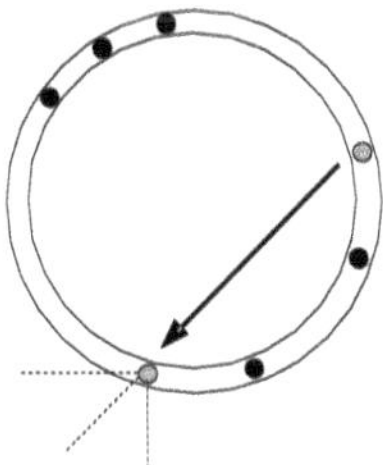

Fig. 5. Circular-track force-directed technique.

First, we need to look at node coordinates in a different way. Node i belongs to group α and is located at position (x_i, y_i). Given that the center of the embedding circle on which α is located is at (x_α, y_α) and the radius of that circle is r_α, we can restate the coordinates of i in the following way:

$$x_i = x_\alpha + r_\alpha * cos(\theta_i) \tag{1}$$

$$y_i = y_\alpha + r_\alpha * sin(\theta_i) \tag{2}$$

Remember that Hooke's Law [11] gives us the following equation for the potential energy V in a spring system:

$$V = \sum_{ij} k_{ij}[(x_i - x_j)^2 + (y_i - y_j)^2] \tag{3}$$

where k_{ij} is the spring constant for the spring between nodes i and j. Equation (3) can be rewritten using (1) and (2):

$$V = \textstyle\sum_{(i,j)\in E} k_{ij} \; [((x_\alpha + r_\alpha * cos(\theta_i)) - (x_\beta + r_\beta * cos(\theta_j)))^2 + ((y_\alpha + r_\alpha * sin(\theta_i)) - (y_\beta + r_\beta * sin(\theta_j)))^2] \tag{4}$$

where node j belongs to group β, (x_β, y_β) is the center and r_β is the radius of the embedding circle on which β appears. Following through on the minus signs, we rewrite:

$$V = \textstyle\sum_{(i,j)\in E} k_{ij} \; [(x_\alpha + r_\alpha * cos(\theta_i) - x_\beta - r_\beta * cos(\theta_j))^2 + (y_\alpha + r_\alpha * sin(\theta_i) - y_\beta - r_\beta * sin(\theta_j))^2] \tag{5}$$

We can find a minimal energy solution on variables x, y and θ. It is interesting to note that if i and j are on the same circle, then x_α and x_β are equivalent as are y_α and y_β. And, of course, $r_\alpha = r_\beta$. Now we rewrite equation (5):

$$V = \sum_{(i,j)\in E} k_{ij}[r_\alpha(cos(\theta_i) - cos(\theta_j))^2 + r_\alpha(sin(\theta_i) - sin(\theta_j))^2)] \tag{6}$$

We can calculate r_α from the number of nodes in α, so that means that finding the minimum V is now a one-dimensional problem based on finding the right set of θs. When we combine (5) or (6) with equations for magnetic repulsion to prevent node occlusion, we have a force-directed equation for which the nodes of a group lie on the circumference of a circle. Now we extend equation (5) to include repulsive forces.

$$\rho_{ij} = [(x_\alpha + r_\alpha * cos(\theta_i) - x_\beta - r_\beta * cos(\theta_j))^2 + (y_\alpha + r_\alpha * sin(\theta_i) - y_\beta - r_\beta * sin(\theta_j))^2] \tag{7}$$

$$V = \sum_{(i,j)\in E} k_{ij}\rho_{ij} + \sum_{(i,j)\in V\times V} g_{ij}\frac{1}{\rho_{ij}} \tag{8}$$

where g_{ij} is the repulsive constant between nodes i and j. The force on node i by node j is:

$$F_i = \frac{V(\theta_i + \epsilon, \theta_j) - V(\theta_i - \epsilon, \theta_j)}{2\epsilon} \tag{9}$$

where ϵ is a very small constant.

Another important consideration is the set of spring constants used in the above equations. It is not necessary for the spring constant to be the same for each pair of nodes. It is also possible for these constants to change during different phases of execution.

5 A Technique for Creating User-Grouped Circular Drawings

Now that we have a force-directed technique in which the nodes are placed on circular tracks, we need to show how we will successfully merge the force-directed approach and circular drawing techniques of [20,21,22,24]. We now present a technique for creating user-grouped circular drawings.

Algorithm 1 *CIRCULARwithFORCES*

Input: A graph $G = (V, E, P)$.
Output: User-grouped circular drawing of G Γ.

1. Determine the superstructure G_s of G.
2. Layout G_s with a basic force-directed technique.
3. For each group p_i in P
   a) If the subgraph induced by p_i, G_i, is biconnected layout G_i with *CIRCULAR*.
   b) Else layout G_i with *CIRCULAR-Nonbiconnected.*
4. Place the layout of each group p_i at the respective location found in Step 2.
5. For each group p_i
   a) rotate the layout circle and keep the position which has the lowest local potential energy.
   b) reverse the order of the nodes around the embedding circle and repeat Step 5a.
   c) if the result of Step 5a had a lower local potential energy than that of Step 5b revert to the result of Step 5a.
6. Apply a force-directed technique using the equations of Section 4 to G.

Going back to the four goals discussed in Section 3, we will attain goal 1 by using a basic force-directed technique to layout the superstructure. We will attain goal 2 by laying out each group with either CIRCULAR or CIRCULAR-Nonbiconnected. Attaining goal 3 means successfully merging the results of the force-directed and circular techniques.

Once we have the layout of the superstructure and each group, we place the layout of each group at the respective location found during the layout of the superstucture. Now we have an almost-final layout: it is a matter of rotating the layouts of the groups and maybe adjusting the order of nodes around the embedding circle. Since we know that CIRCULAR and CIRCULAR-Nonbiconnected produce good visualizations, we should change these layouts as little as possible. So first, we will fine-tune the almost-final drawing by rotating each layout and keeping the rotation which has the least local potential energy. We rotate each embedding circle through n_α positions, where n_α is the number of nodes in the group α. With respect to determing local potential energy, we need to determine the lengths of inter-group edges which are incident to the nodes of α. The rotation of choice should minimize the lengths of those edges. In other words, we choose the rotation in which as many nodes as possible are close to their other-group neighbors. Since for each embedding circle we try n_α positions and examine the length of α's incident inter-group edges at each position, then the rotation step will take $O(n * m_{inter-group})$ time for the entire graph, where $m_{inter-group}$ is the number of inter-group edges. As discussed in Section 3, we expect $m_{inter-group} \ll m$. Then we will "flip" each layout and again rotate. We keep the rotation which has the least local potential energy. After these steps, it is still possible that some nodes will be badly placed with respect to their relationships with nodes in other groups. In other words, those placements cause intra-group and inter-group edges to cross. In order to address this problem, we will apply the force-directed technique described in Section 4. The result of this step will be the reduction of intra-group and inter-group edge crossings since nodes will be pulled to the side of the embedding circle which is closer to their other-group relatives.

Because Algorithm 1 makes use of a force-directed technique, the worst-case time requirement is unknown. However, in practice, we expect the time requirement to be $O(n^2)$ for the following reasons: Step 1 requires $O(m)$ time. Step 2 will be on a small graph and should not require much time to reach convergence. Step 3 requires $O(m)$ time. Step 4 requires $O(n)$ time. Step 5 require $O(n * m_{inter-group})$ time. Since Step 6 is a force-directed technique it could take $O(n^3)$ time in practice, however the result of the previous steps will be an almost-final layout and thus should not need much time to converge. It was evidenced in [23] that when a force-directed technique is applied to an almost-final layout, it does not take much more time for convergence to occur. Therefore, in practice we expect this step to require $O(n^2)$ time. Thus, we have attained goal 4 from Section 3.

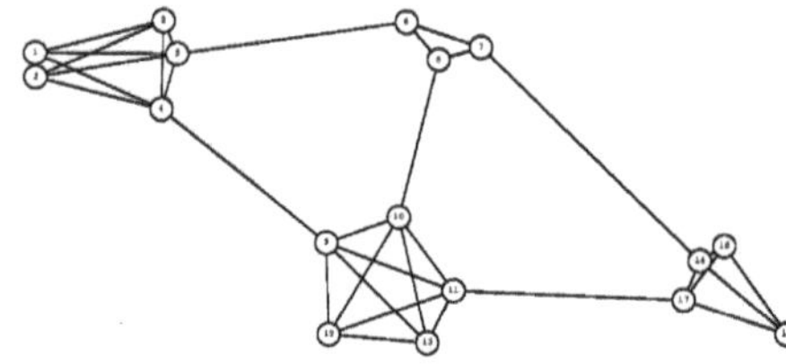

Fig. 6. Sample user-grouped circular drawing from our preliminary implementation.

We have done a preliminary implementation of CIRCULARwithFORCES in TCL/TK. In this implementation, all nodes and embedding circles are given an arbitrary initial placement. Then the force-directed equations of Section 4 are applied to the graph with the placement of group embedding circles frozen. See Figure 6 for a sample drawing.

An interesting behavior we noticed is that the drawing with minimal energy is not necessarily the best circular drawing. In circular drawing, a major goal is to reduce edge crossings. However, it is well known [3] that reducing crossings sometimes means the compromise of other aesthetics, especially area. And area is related to minimum energy in spring systems. We propose adding springs from each node to its initial placement on the plane with the spring constants for these springs being high. This should keep these nodes from gravitating towards each other too much and causing extra crossings. We also suggest creating dummy nodes which are placed in the center of each embedding circle and attaching strong springs from them to every node in their respective group.

6 Conclusions and Future Work

In this paper, we have presented a framework for creating circular graph drawings in which the grouping is defined by the user. This framework includes the successful merging of the force-directed and circular graph drawing paradigms. We introduced the algorithm CIRCULARwithFORCES that produces drawings in which the user-defined groupings are highly visible, each group is laid out with a low number of edge crossings, and the number of crossings between intra-group and inter-group edges is low. This layout technique is also fast.

In the future we would like to extend this approach to include layouts in which nodes are placed on the periphery of shapes other than circles or on curves. We would also like to develop a technique which uses this approach in three dimensions.

Acknowledgements. We would like to thank Dr. Andrew Urquhart for his assistance with our circular-track force-directed equations.

References

1. M. A. Bernard, On the Automated Drawing of Graphs, *Proc. 3rd Caribbean Conf. on Combinatorics and Computing*, pp. 43–55, 1994.
2. F. Brandenburg, Graph Clustering 1: Cycles of Cliques, *Proc. GD '97, LNCS 1353*, Springer-Verlag, pp. 158–168, 1997.
3. G. Di Battista, P. Eades, R. Tamassia and I. G. Tollis, *Graph Drawing: Algorithms for the Visualization of Graphs*, Prentice-Hall, 1999.
4. U. Doğrusöz, B. Madden and P. Madden, Circular Layout in the Graph Layout Toolkit, *Proc. GD '96, LNCS 1190*, Springer-Verlag, pp. 92–100, 1997.
5. P. Eades, A Heuristic for Graph Drawing, *Congr. Numer.*, 42, pp. 149–160, 1984.
6. P. Eades, Q. Feng and X. Lin, Straight-Line Drawing Algorithms for Hierarchical Graphs and Clustered Graphs, *Proc. GD '96, LNCS 1190*, Springer-Verlag, pp. 113–128, 1997.

7. P. Eades and Q. W. Feng, Multilevel Visualization of Clustered Graphs, *Proc. GD '96, LNCS 1190*, Springer-Verlag, pp. 101–112, 1997.
8. P. Eades, Q. W. Feng and H. Nagamochi, Drawing Clustered Graphs on an Orthogonal Grid, *Jrnl. of Graph Algorithms and Applications*, pp. 3–29, 1999.
9. C. Esposito, Graph Graphics: Theory and Practice, *Comput. Math. Appl.*, 15(4), pp. 247–53, 1988.
10. M. Garey and D. Johnson, *Computers and Intractability: A Guide to the Theory of NP-Completeness*, Freeman, 1979.
11. D. Halliday and R. Resnick, *Fundamentals of Physics: 3rd Edition Extended*, Wiley, New York, NY, 1988.
12. M. L. Huang and P. Eades, A Fully Animated Interactive System for Clustering and Navigating Huge Graphs, *Proc. of GD '98, LNCS 1547*, Springer-Verlag, pp. 107–116, 1998.
13. G. Kar, B. Madden and R. Gilbert, Heuristic Layout Algorithms for Network Presentation Services, *IEEE Network*, 11, pp. 29–36, 1988.
14. M. Kaufmann and R. Wiese, Maintaining the Mental Map for Circular Drawings, *Proc. of GD 2002, LNCS 2528*, Springer-Verlag, pp. 12–22, 2002.
15. A. Kershenbaum, *Telecommunications Network Design Algorithms*, McGraw-Hill, 1993.
16. V. Krebs, Visualizing Human Networks, *Release 1.0: Esther Dyson's Monthly Report*, pp. 1–25, February 12, 1996.
17. S. Masuda, T. Kashiwabara, K. Nakajima and T. Fujisawa, On the NP-Completeness of a Computer Network Layout Problem, *Proc. IEEE 1987 International Symposium on Circuits and Systems, Philadelphia, PA*, pp.292–295, 1987.
18. S. Mitchell, Linear Algorithms to Recognize Outerplanar and Maximal Outerplanar Graphs, *Information Processing Letters*, 9(5), pp. 229–232, 1979.
19. H. Purchase, Which Aesthetic has the Greatest Effect on Human Understanding, *Proc. of GD '97, LNCS 1353*, Springer-Verlag, pp. 248–261, 1997.
20. J. M. Six (Urquhart), *Vistool: A Tool For Visualizing Graphs*, Ph.D. Thesis, The University of Texas at Dallas, 2000.
21. J. M. Six and I. G. Tollis, Circular Drawings of Biconnected Graphs, *Proc. of ALENEX '99, LNCS 1619*, Springer-Verlag, pp. 57–73, 1999.
22. J. M. Six and I. G. Tollis, Circular Drawings of Telecommunication Networks, *Advances in Informatics, Selected Papers from HCI '99*, D. I. Fotiadis and S. D. Nikolopoulos, Eds., World Scientific, pp. 313–323, 2000.
23. J. M. Six and I. G. Tollis, Effective Graph Visualization Via Node Grouping, *Software Visualization: From Theory to Practice, The Kluwer Intl. Series in Engineering and Computer Science Vol. 734*, K. Zhang Ed., Kluwer Academic Publishers, 2003.
24. J. M. Six and I. G. Tollis, A Framework for Circular Drawings of Networks, *Proc. of GD '99, LNCS 1731*, Springer-Verlag, pp. 107–116, 1999.
25. I. G. Tollis and C. Xia, Drawing Telecommunication Networks, *Proc. GD '94, LNCS 894*, Springer-Verlag, pp. 206-217, 1994.

Fixed-Location Circular-Arc Drawing of Planar Graphs*

Alon Efrat, Cesim Erten, and Stephen G. Kobourov

Department of Computer Science
University of Arizona
{alon,cesim,kobourov}@cs.arizona.edu

Abstract. In this paper we consider the problem of drawing a planar graph using circular-arcs as edges, given a one-to-one mapping between the vertices of the graph and a set of n points on the plane, where n is the number of vertices in the graph. If for every edge we have only two possible circular arcs, then a simple reduction to 2SAT yields an $O(n^2)$ algorithm to find out if a drawing with no crossings can be realized. We present an improved $O(n^{7/4} polylog\ n)$ time algorithm. For the special case where the possible circular arcs for each edge are of the same length, we present an even more efficient algorithm that runs in $O(n^{3/2} polylog\ n)$ time. We also consider the problem if we have more than two possible circular arcs per edge and show that the problem becomes NP-Hard. Moreover, we show that two optimization versions of the problem are also NP-Hard.

1 Introduction

A natural question that arises in graph drawing is whether a graph with fixed vertices can be drawn without crossings, when several choices are given for each of the edges. From an information visualization point of view convex edges are preferable, i.e., straight line segments or circular arcs. In general, embedding a planar graph at fixed locations and drawing it with straight lines may result in many crossings. Using circular arcs instead can reduce or eliminate the crossings; see Fig. 1(a). Thus, a natural problem to consider is whether a given graph with fixed vertex locations can be drawn without crossings, using circular arcs.

We first consider the *2-Circle Drawing (2CD)* problem, in which each edge has to be drawn as one of the two circular arcs defined by a circle passing through the endpoints. This problem is reminiscent of the *Manhattan wiring problem*: Consider the axis-aligned rectangle with a diagonal defined by the line segment between two vertices connected by an edge. Then the two semi-rectangles separated by the diagonal are the two choices for drawing the edge. This formulation of the problem can be efficiently solved in $O(n \log n)$ time, using an efficient *find and delete* data structure (to find intersections between a pair of semi-rectangles and to delete a semi-rectangle from the data structure) [14].

* This work was partially supported by the NSF under grant ACR-02229290.

G. Liotta (Ed.): GD 2003, LNCS 2912, pp. 147–158, 2004.

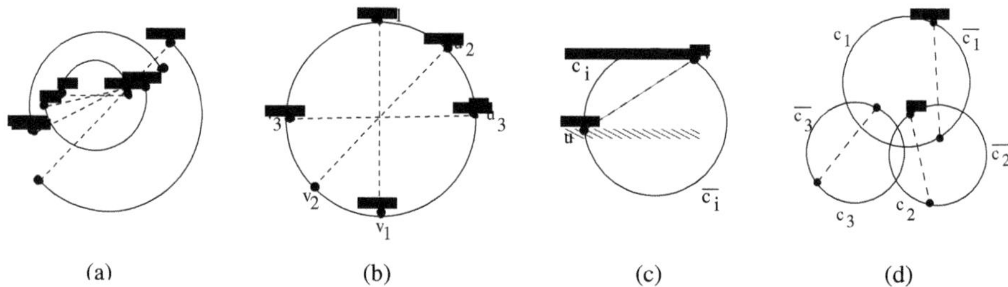

Fig. 1. (a) $\Omega(n^2)$ crossings with straight-line edges and none with half-circles; (b) A planar graph that cannot be drawn without crossings using *any* circular arc segments. (c) Given the circle C_i, edge $e_i = (u, v)$ is drawn either with the circular arc c_i or $\overline{c_i}$. (d) An example of a 2SAT reduction: $(\overline{c_1} \vee c_3) \wedge (\overline{c_1} \vee \overline{c_3}) \wedge (\overline{c_3} \vee \overline{c_2}) \wedge (\overline{c_1} \vee \overline{c_2}) \wedge (c_1 \vee c_2)$.

The same approach cannot be applied to the 2CD problem directly. Although efficient data structures exist for operations on full circles, no such data structures exist in the case of circular arcs (semi-circles). The novelty of our 2CD algorithm is that we provide a way to use an efficient data structure for full circles to solve the problem with circular arcs. For the sake of completeness, we first show that the 2CD problem can be reduced to 2SAT and thus solved in $O(n^2)$ time. Then we provide our novel approach to solve the 2CD problem in time $O(n^{7/4} \textit{polylog } n)$ for the general case and in time $O(n^{3/2} \textit{polylog } n)$ for the restricted case where the circular arcs are exactly half-circles. Although the practical gain in terms of running time is not significant, we believe that our approach for solving the 2CD problem might be of independent interest to solve similar problems.

Next, we consider the 3-Circle Drawing (3CD) problem, where for each edge there are three circular arcs to choose from. We show that the 3CD problem is NP-hard. Although using circular arcs to represent the edges allows a certain flexibility, not every planar graph can be drawn without crossings using circular arcs. Fig. 1(b) shows an example of a planar graph that cannot be drawn without crossings using *any* circular arc segments. This difficulty suggests two optimization problems: *Min2CD* is the problem of minimizing the number of crossings for a given 2CD instance by representing *every* edge with an appropriate circular arc. *Max2CD* is the problem of maximizing the number of edges that can be drawn without crossings using circular arcs. We show that both of these optimization problems are NP-hard.

2 Previous Work

Several variations of the problem of embedding a planar graph at fixed point locations have been studied. If we can choose the mapping between the vertices V and the points P, then Kaufmann and Wiese [15] show that the graph can be drawn without crossings using 2 bends per edge in polynomial time. However, if the mapping between V and P is given, Pach and Wenger [19] show that $O(n)$ bends per edge are necessary to guarantee planarity, where n is the number of vertices in the graph. Godau [12] shows that if each vertex is allowed to move slightly in the neighborhood of a fixed point then the problem becomes NP-hard.

Drawing graphs with circular arcs with no assigned vertex locations has been considered by Cheng *et al* [5] in the context of planarity, angular resolution and drawing area. The problems under consideration in this paper are also related to the *k-position point labeling problem*, extensively studied in *map labeling* [3, 20,21]. In the k-position point labeling problem we are given a set of points and a set of k possible label positions for each point and we would like to find a labeling of the points that optimizes specific criteria. Criteria such as maximizing the number of labeled points [3,23] and maximizing the size of the labels [6,7, 11] have been considered. A variant of the map labeling problem is reduced to 2SAT and NP-Completeness results are presented in [11]. Although these problems are related to drawing planar graphs with circular arcs, there is a significant difference: whereas map labeling problems are restricted by region intersections, drawing planar graphs with circular arcs is restricted by circular arc intersections.

3 Circular Arcs Drawing: 2CD

The input to the problem is a planar graph $G = (V, E)$, a point set P, and a one-to-one function $f : V \to P$ such that $|V| = |P| = n$ and $|E| = m$. Note that $m = O(n)$ as G is a planar graph. For any edge $e_i = (u, v) \in E$, we are given some circle C_i that passes through u and v. The two vertices, u and v, determine two circular arcs on C_i; let c_i and $\overline{c_i}$ be their labels, see Fig 1(c). We would like to find out whether G can be drawn without crossings using c_i or $\overline{c_i}$ for each e_i, and if so to provide such a drawing of G.

We first suggest a straightforward solution of the problem using a reduction from 2CD to 2SAT. The reduction to a 2SAT formula Φ requires that we identify all intersections between circular arcs. For each such intersection $c_i \cap c_j \neq \emptyset$, we add the formula $(\overline{c_i} \vee \overline{c_j})$ to Φ, see Fig 1(d). Since there are $O(n^2)$ crossings, the reduction takes $O(n^2)$ time and results in a 2SAT formula Φ with $O(n)$ variables and $O(n^2)$ clauses. It is easy to see that G can be drawn without intersections if and only if the corresponding formula Φ is satisfiable. Since 2SAT can be solved in time linear in the number of clauses and variables [4,10], we have an $O(n^2)$ time algorithm for the 2CD problem. However we can do much better.

3.1 The 2CD Algorithm

Note that for a given edge $e_i = (u, v) \in E$, as possible drawings of e_i we consider the circular arcs defined by the circle C_i, and the position of u and v on C_i. Alternatively, we could consider the axis-aligned rectangle having its diagonal as the line segment (u, v), and then consider the 2 semi-rectangles separated by the diagonal as two choices for drawings of the edge e_i. This formulation of the problem is known as the *Manhattan wiring problem* which can be efficiently solved in $O(n \log n)$ time [14].

Algorithm 2CD	**Traverse_possible**(c_i)

while $\mathcal{D}$ not empty
  let C_i be a circle in $\mathcal{D}$
  $delete(\mathcal{D}, C_i)$
  delete $\overline{c_i}$ from $\mathcal{P}$
  $traverse\_possible(c_i)$
start with initial data structure $\mathcal{D}$
while $\exists c_i, c_j \in P$ s.t. $c_i \cap c_j \neq \emptyset$
  let $dfsnumber(c_i) < dfsnumber(c_j)$
  $delete(\mathcal{D}, C_i)$
  delete c_i and $\overline{c_i}$ from $\mathcal{P}$
  add $\overline{c_i}$ into $\mathcal{C}$
  $traverse\_certain(\overline{c_i})$
if $\exists$ intersecting half-circles in $\mathcal{C}$
  output *No*
else output $\mathcal{C} \cup \mathcal{P}$
	while $C_j = find(\mathcal{D}, c_i)$ not empty
	  $delete(\mathcal{D}, C_j)$
	  let c_j be involved in the intersection
	  delete c_j from $\mathcal{P}$
	  $traverse\_possible(\overline{c_j})$
	
	Traverse_certain(c_i)
	while $C_j = find(\mathcal{D}, c_i)$ not empty
	  $delete(\mathcal{D}, C_j)$
	  let c_j be involved in the intersection
	  delete c_j and $\overline{c_j}$ from $\mathcal{P}$
	  add $\overline{c_j}$ into $\mathcal{C}$
	  $traverse\_certain(\overline{c_j})$

Fig. 2. Algorithm 2CD. The input to the algorithm is $\mathcal{D}$, the data structure used to store all the circles. If it is possible to draw the graph without crossings, the algorithm outputs the set of circular arcs used to draw the graph.

We describe a new algorithm that solves a more general problem for circular arcs. Our approach is different from the Manhattan wiring problem in that we perform operations only on complete circles. More formally, we *find/delete* complete circles as part of an intersection, as opposed to performing the operations on the circular arcs directly. Let $\mathcal{D}$ denote the data structure used to store all the circles. Let $find(\mathcal{D}, c_i)$ be a function that finds a circle C_j intersecting circular arc c_i, and $delete(\mathcal{D}, C_i)$ be a function that deletes the circle C_i from $\mathcal{D}$. Let $\alpha(n), \beta(n)$ denote the time required to perform the *find* and *delete* operations, respectively. We next describe how to construct the data structure $\mathcal{D}$ and how to perform the *find/delete* operations efficiently. The main algorithm is shown on Fig. 2.

Let a *possible* circular arc be one that is not yet chosen and not yet discarded and let $\mathcal{P}$ denote the set of *possible* circular arcs. $\mathcal{P}$ initially contains all the circular arcs. We traverse the circles in a depth-first manner starting with an arbitrary circular arc c_i and making all possible assignments. Each circle is found and deleted exactly once, resulting in a sequence of $O(n)$ *find/delete* operations and requires $O(n \times (\alpha(n) + \beta(n))$ time. At the end of the first *while* loop in the main algorithm, the set $\mathcal{P}$ contains exactly one circular arc for each edge. However, $\mathcal{P}$ might contain intersections. At this point we make the following observation:

Observation 1. *Let* $c_i, c_j \in \mathcal{P}$. *If* $c_i \cap c_j \neq \emptyset$ *and* $dfsnumber(c_i) < dfsnumber(c_j)$, *then the implication* $c_i \Rightarrow c_j$ *holds, and* $\overline{c_i}$ *must be included in the final solution.*

Let $\mathcal{C}$ denote the set of *certain* circular arcs. Initially $\mathcal{C}$ is empty. Once we find a circular arc, c_k, that is certainly in the final solution by the above observation, we perform a traversal from c_k, placing all certain arcs in the set $\mathcal{C}$. To find the intersections in $\mathcal{P}$ it suffices to perform a plane sweep over the whole set $\mathcal{P}$. Since, whenever we encounter an intersection in $\mathcal{P}$ we delete the whole circle, the plane sweep over $\mathcal{P}$ finds $O(n)$ intersections in total. Then the second *while* loop requires $O(n \log n + n \times (\alpha(n) + \beta(n)))$ time. Finally we end up with a set $\mathcal{C}$ of certain circular arcs, and a set $\mathcal{P}$ of possible circular arcs. At this point we make a second observation:

Observation 2. *Let $c_i, c_j \in \mathcal{C} \cup \mathcal{P}$. If $c_i \cap c_j \neq \emptyset$, then $c_i, c_j \in \mathcal{C}$.*

The observation holds for the following reasons: Assume $c_i, c_j \in \mathcal{P}$ were true. Then we must have encountered the intersection in the plane sweep step in which case one of them would have been deleted from $\mathcal{P}$. So, $c_i, c_j \in \mathcal{P}$ can not be true. On the other hand assume, $c_i \in \mathcal{C}$ and $c_j \in \mathcal{P}$ were true. Then we must have traversed through c_j before visiting c_i in the traversal step. But when we traverse through a circle, we delete the whole circle from $\mathcal{P}$. So this can not be the case either.

Then we need to concentrate on the intersections in $\mathcal{C}$. We perform a final plane sweep over the set $\mathcal{C}$. If we encounter an intersection, then there cannot be an assignment without intersections, otherwise $\mathcal{C} \cup \mathcal{P}$ gives us a feasible assignment. The running time of the algorithm is the time required for the two *while* loops in the main algorithm: $O(n \log n + n \times (\alpha(n) + \beta(n)))$. In the next section we describe the data structure that supports the needed operations.

3.2 The Data Structure

Given a circular arc query c_i, finding and deleting a circle C_j that intersects c_i is more efficient than performing the same operations on a circular arc c_j that intersects c_i. This observation led us to the 2CD algorithm which assumes the existence of a data structure $\mathcal{D}$ that stores all the circles and allows for efficient *find/delete* operations. Gupta *et al* [13] show how to reduce the problem of querying circles with a circular arc to one of half-space range searching in higher dimensions. The method requires at most a 4-dimensional half-space range searching. To report such intersections then, we make use of the ideas from *geometric range-searching* [1,2,18]. The main data structure we use is a partition tree, constructed using the partitioning theorem by Matoušek [17]: a point set P can be partitioned into $O(n^{1-1/d})$ classes in time $O(n \log n^{1-1/d})$ such that for any class P_i, $|P_i| < 2 \times n^{1/d}$ and any line l intersects at most $O(n^{(1-1/d)^2})$ classes, where d is the dimension of the search space. Using this partitioning theorem we can create a data structure $\mathcal{D}'$ that performs half-space range queries in time $O(n^{1-1/d} (\log n)^{O(1)})$. Moreover $\mathcal{D}'$ is dynamic, in the sense that we can delete a circle from $\mathcal{D}'$ in amortized time $O(\log n)$. Then using multiple levels of $\mathcal{D}'$ to satisfy the intersection conditions of [13], we create the data structure $\mathcal{D}$ that supports $find(\mathcal{D}, c_i)$ operations in $O(n^{3/4} polylog\ n)$ time and that requires $O(n \log n)$ time for a sequence of $O(n)$ $delete(\mathcal{D}, C_i)$ operations. These results can be summarized with the following theorem for the 2CD problem.

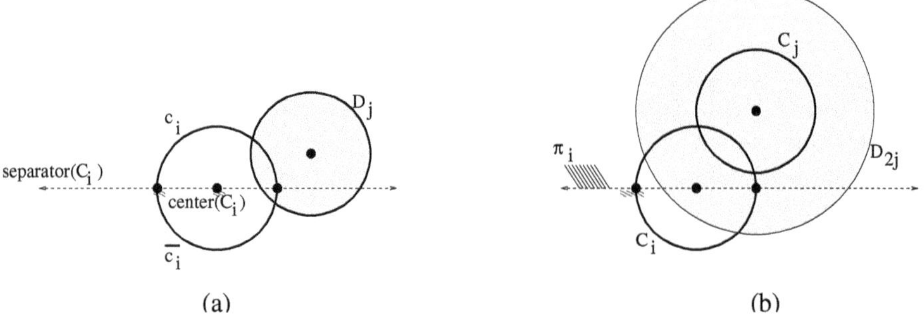

Fig. 3. (a) D_j contains an endpoint of c_i (b) center$(C_j) \in \pi_i$ and center$(C_i) \in D_{2j}$.

Theorem 1. *The 2CD problem can be solved in time $O(n^{7/4}$polylog $n)$.*

3.3 Allequal2CD

We can solve a restricted version of the 2CD problem even more efficiently. Let *Allequal2CD* be the version of 2CD where each edge e_i has the same length and the line segment between the endpoints of e_i is the diameter of circle C_i. In this case, for a given edge e_i, the circular arcs c_i and $\overline{c_i}$ are half-circles. We present an algorithm to solve Allequal2CD in $O(n^{3/2}$*polylog* $n)$ time using a data structure $\mathcal{D}$ that enables us to perform efficient *find/delete* operations. We provide the details for the construction of $\mathcal{D}$ here, since the general data structure described above can be constructed in a similar fashion.

Let *center*(C_i) and *separator*(C_i) denote, respectively, the center of C_i and the line separating the half-circles c_i and $\overline{c_i}$. Define π_i as the half-plane bounded by separator(C_i), and that contains c_i. Let D_i be the disk bounded by the circle C_i, and let D_{2i} be the disk concentric to D_i but with radius twice the radius of D_i, see Fig 3. The following lemma is easy to verify.

Lemma 1. *A circle C_j intersects a half-circle c_i if and only if* **(i)** *D_j contains an endpoint of c_i, or* **(ii)** *center$(C_j) \in \pi_i$ and center$(C_i) \in D_{2j}$.*

In order to report intersections of the first type we use the data structure described by Efrat *et al* in [9]: given n equal-sized disks in the plane, construct a data structure $\mathcal{DT}_1$ in time $O(n \log n)$ such that for a given query point p, finding a disk that contains p requires $O(\log n)$ time. Moreover, deleting a disk from $\mathcal{DT}_1$ requires amortized time $O(\log n)$. We preprocess the disks D_i for each i using this structure.

To deal with the intersections of the second type we make use of the partition tree $\mathcal{D}'$ described above. However, this time we perform half-space range searching in 2-dimensions using a two-level data structure. Let the data structure for the second type of intersections be $\mathcal{DT}_2$. The first level of $\mathcal{DT}_2$ is a partition tree, $\mathcal{DT}_2{}'$. Based on the partitioning theorem described above, we partition the centers of all the circles and recursively build the partition tree $\mathcal{DT}_2{}'$. The leaves of $\mathcal{DT}_2{}'$ partition the centers into constant-sized subsets. Each internal node v is associated with a subset P_v of the points contained in the leaves of the subtree rooted at v. We build the second level of the data structure based on these subsets. The second level data structure used in $\mathcal{DT}_2$ is the same as $\mathcal{DT}_1$, except we preprocess the disks D_{2i} for each i, rather than the disks D_i as is the case for constructing $\mathcal{DT}_1$. We call this second level data structure $\mathcal{DT}_2{}''$ to distinguish it from $\mathcal{DT}_1$ which we used to find the first type of intersections. Each internal node v in $\mathcal{DT}_2{}'$ contains a pointer to the corresponding $\mathcal{DT}_2{}''$, where $\mathcal{DT}_2{}''$ contains the data structure for all the disks D_{2i} centered at P_v. The preprocessing time for constructing the partition in a node of $\mathcal{DT}_2{}'$ with m points is $O(m \log m)$. Constructing $\mathcal{DT}_2{}''$ for the same node also takes time $O(m \log m)$. Since the number of points in nodes of $\mathcal{DT}_2{}'$ decreases as a double exponential with their depth in the tree, the total preprocessing time is $O(n \log n)$.

Theorem 2. *Allequal2CD problem can be solved in time $O(n^{3/2} polylog\ n)$.*

To find a circle C_j intersecting a given a half-circle c_i we first query $\mathcal{DT}_1$ with c_i's endpoints. This step requires $O(\log n)$ time. If we cannot find such a circle then we query $\mathcal{DT}_2$ with separator(C_i). Upon finding an internal node v such that P_v lies completely above separator(C_i), we query the associated $\mathcal{DT}_2{}''$ of v with center(C_i). Let $\alpha(n)$ be the time to find a circle intersecting a given half-circle c_i. Then $\alpha(n)$ is bounded by the query time of $\mathcal{DT}_2$ and we get:

$$\alpha(n) \leq O(\sqrt{n}) \times \log 2\sqrt{n} + O(\sqrt[4]{n}) \times \alpha(2\sqrt{n}) \tag{1}$$

Thus the time required to perform a *find* operation is $\alpha(n) = O(\sqrt{n} polylog\ n)$.

In order to delete a circle C_i, we first delete D_i from $\mathcal{DT}_1$ in $O(\log n)$ amortized time. We also need to delete the appropriate disks in $\mathcal{DT}_2$. To do this we simply find each internal node v of $\mathcal{DT}_2{}'$ such that center$(C_i) \in P_v$, and delete the corresponding disk from $\mathcal{DT}_2{}''$, the second level data structure pointed to by v. Since $\mathcal{DT}_2{}'$ has depth $O(\log \log n)$ and deleting a disk from $\mathcal{DT}_2{}''$ takes amortized time $O(\log n)$, the deletion of a circle takes $O(\log^2 n)$ amortized time. Since *find* and *delete* operations are defined for both $\mathcal{DT}_1$ and $\mathcal{DT}_2$, the two data structures form the complete data structure $\mathcal{D}$ and the theorem follows.

4 The 3CD Problem

The 3CD problem is similar to 2CD, except now we have three choices for the drawing of each edge $e_i = (u, v)$. We consider the problem in which in addition to the two half-circles we can also choose the line segment connecting u and v. We show that 3CD is NP-hard using a reduction from the NP-hard PLANAR-3SAT [16]. A 3SAT instance Φ is called a PLANAR-3SAT instance if the (bipartite) occurrence graph $G_\Phi = (V_\Phi, E_\Phi)$ is planar. In the occurrence graph, V_Φ contains a vertex for each variable and clause, and E_Φ contains an edge between two vertices $v, w \in V_\Phi$ if v represents a variable x that occurs in the clause represented by w. Let VR3SAT(Variable Restricted 3SAT) be the version of 3SAT with the restriction that each variable can appear at most three

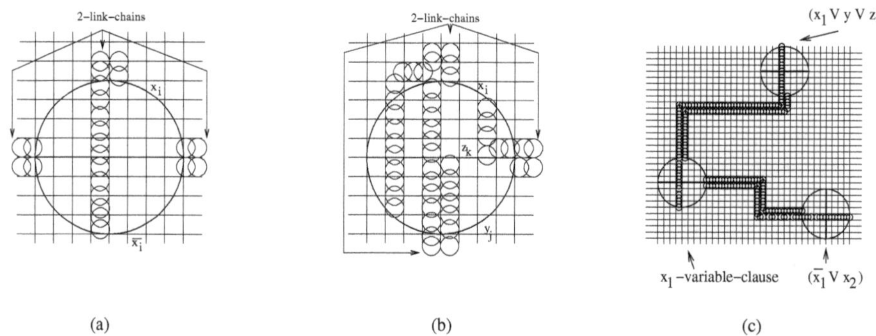

Fig. 4. (a) Variable-circle of x_i; (b) Clause-circle of $(x_i \vee y_j \vee z_k)$; (c) Circle drawing of $(x_1 \vee y \vee z) \wedge (\overline{x_1} \vee x_2) \ldots$. Each edge is represented by 2-link-chains (two parallel chains of small circles). Since the graph is planar there will be no crossings of the chains.

times, and VR1IN3SAT be the version of VR3SAT in which *exactly* one literal in each clause is required to be true. In the planar versions of these two problems, the occurrence graphs of the input instances must be planar. We will convert a PLANAR-3SAT instance Φ into a 3CD instance Φ_{3CD} through a series of modifications that preserve planarity.

Lemma 2. *PLANAR-VR3SAT is NP-hard.*

Proof Sketch: Due to space constraints we leave the proof of this lemma to the full version of the paper [8]. □

Lemma 3. *PLANAR-VR1IN3SAT is NP-hard.*

Proof Sketch: Due to space constraints we leave the proof of this lemma to the full version of the paper [8]. □

Theorem 3. *The 3CD problem is NP-hard.*

Proof Sketch: We convert a PLANAR-VR1IN3SAT instance Φ'' into a 3CD instance Φ_{3CD}. Because of the VR reduction, the occurrence graph $G_{\Phi''}$ for Φ'' has maximum degree 3. Then there exists an orthogonal drawing for $G_{\Phi''}$ (a drawing such that each vertex is on the integer grid and each edge consists of horizontal and vertical edge segments) and the grid is of size quadratic in the size of $G_{\Phi''}$ [22]. Given the orthogonal drawing of $G_{\Phi''}$, we obtain Φ_{3CD} by the following method: We replace each vertex corresponding to a variable x_i with a *variable-circle*, with one half labeled x_i and the other $\overline{x_i}$, see Fig. 4(a). We replace each vertex corresponding to a clause, say, $(x_i \vee y_j \vee z_k)$ of Φ'', with a *clause-circle* having one half-circle labeled x_i, one labeled y_j, and the diameter of the circle labeled z_k, see Fig. 4(b). We represent each edge of $G_{\Phi''}$ with a *2-link-chain* which consists of two parallel links of *chain-circles*. Let e_i be the edge of $G_{\Phi''}$ between the vertex corresponding to the clause $(x_i \vee y_j \vee z_k)$ and the vertex corresponding to the variable x_i. Then we represent e_i by a 2-link-chain, where one of the links is connected to the half-circle of the clause-circle labeled with x_i on one end, and to the half-circle $\overline{x_i}$ of the variable-circle on the other. The other link intersects with both the half-circle y_j and the diameter z_k of the clause-circle on one end, and is connected to the half-circle x_i of the variable-circle on the other, see Fig. 4(c).

We claim that Φ'' is satisfiable if and only if Φ_{3CD} has a feasible assignment without crossings. Assume that Φ'' is a satisfiable instance of PLANAR-VR1IN3SAT, and let α be a satisfying assignment. A feasible assignment of edges in Φ_{3CD} is as follows: For each variable-circle corresponding to variable x_i, assign the half-circle labeled with x_i or $\overline{x_i}$ depending on whether x_i is assigned to true or false in α respectively. For each clause-circle corresponding to a clause $(x_i \vee y_j \vee z_k)$, assign the half-circle (or diameter) corresponding to the (only) true literal in the clause, as determined by α. For each 2-link-chain connected to the variable-circle of x_i, if x_i is assigned to true in α, then for the link that is connected to x_i, assign the first chain-circle by choosing the half-circle that *does not* cross the x_i, and continue assigning the chain-circles through the link without creating any crossings. For the 2nd link that is connected to the $\overline{x_i}$ half-circle, assign the first chain-circle by choosing the half-circle that *crosses* the half-circle $\overline{x_i}$, and continue assigning the chain-circles through the link without creating any crossings. This assignment does not contain any crossings. The only crossings that could occur would be between a chain-circle at the tip of a link and a clause-circle, but our method of assigning the chain-circles eliminates this possibility.

For the other direction, assume that Φ_{3CD} has a feasible assignment of edges without crossings. Then, finding a truth assignment α for Φ'' is straightforward: For each variable-circle corresponding to a variable x_i, if the half-circle labeled with x_i is chosen, then assign x_i to be true, otherwise assign it false. This yields a satisfying assignment, since the feasible assignment of edges in Φ_{3CD} chooses exactly one edge from each clause-circle such that there are no conflicts with the variable assignments and the true literal assignment for the other clauses. □

5 Drawing with Few Crossings

If G cannot be drawn without crossings using half-circles, there are two natural optimization problems. Define Min2CD as the following decision problem: Given $(G = (V, E), \kappa_{MIN})$, where G is a planar graph, and κ_{MIN} is a non-negative integer, does there exist an assignment of half-circles (either c_i or $\overline{c_i}$) for *each* $e_i \in E$ such that the number of crossings is at most κ_{MIN}? The second optimization problem, Max2CD, is defined as follows: Given $(G = (V, E), \kappa_{MAX})$, where G is a planar graph, and κ_{MAX} is a non-negative integer, does there exist an assignment of half-circles (either c_i or $\overline{c_i}$) for *some* $e_i \in E$ such that there are no crossings and the number of assigned edges is at least κ_{MAX}? We prove that both problems are NP-hard by reductions from the Planar Degree-3 Independent Set problem (PD3IS). Let $H = (V_H, E_H)$ be an undirected graph. We say that a set $I \subseteq V_H$ is independent if for all pairs (i, j), where $i, j \in I$, $(i, j) \notin E_H$.

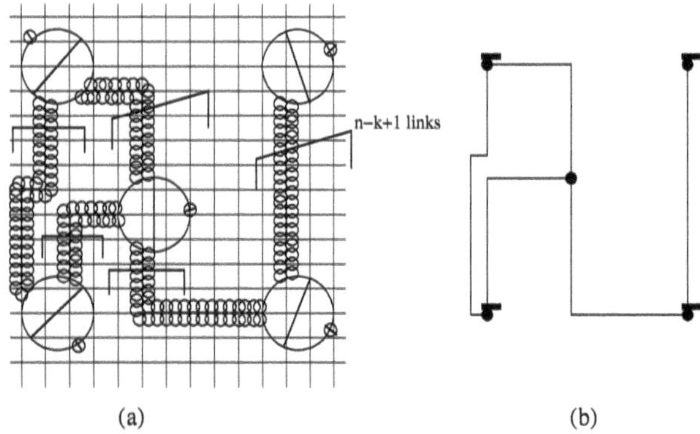

Fig. 5. (a) Min2CD reduction. *Vertex circles* correspond to vertices in H, *chain circles* correspond to edges in H, and *tail circles* are auxiliary circles. (b) The original graph H drawn on an integer grid.

The PD3IS problem is the following: Given $(H = (V_H, E_H), \kappa_{IND})$, where H is a planar graph with maximum degree 3 and κ_{IND} is a non-negative integer, does there exist an independent set I with $|I|=\kappa_{IND}$?

Lemma 4. *PD3IS is NP-hard.*

Proof Sketch: Due to space constraints we leave the proof of this lemma to the full version of the paper [8]. □

Theorem 4. *Min2CD is NP-hard.*

Proof Sketch: Let $(H = (V_H, E_H), \kappa_{IND})$ be an instance of PD3IS. The reduction produces a Min2CD instance (G, κ_{MIN}), with $\kappa_{MIN} = n_H - \kappa_{IND}$. Since H has maximum degree 3, we can find an orthogonal drawing of H, such that each vertex is on the integer grid of size quadratic in the size of H [22], see Fig. 5(b). The reduction scales the grid of H by a factor of $n_H - \kappa_{IND} + 1$ and replaces the vertices of H with *vertex-circles*, circles of diameter $(n_H - \kappa_{IND} + 1)$ units, see the large circles in Fig. 5(a). Each edge of H is represented with $n_H - \kappa_{IND} + 1$ links of *chain-circles*, circles having half a unit diameter connected to a vertex-circle at its *head*. Each vertex-circle also has a *tail circle*, connected to it at its *tail* in such a way that the diameter of the tail-circle crosses the tail of the vertex-circle. Since the given graph H is planar, we can obtain such a grid drawing of circles without causing any intersection between the chain-circles.

We claim that H has an independent set of size κ_{IND} if and only if G can be drawn using half-circles with at most $n_H - \kappa_{IND}$ crossings. Assume H has an independent set I, where $| I | = \kappa_{IND}$. Then there are κ_{IND} vertices in H that are pairwise disconnected, which further implies that in G there are κ_{IND} vertex-circles which are not connected by links. Then a feasible assignment of half-circles which allows a drawing of G with at most $n_H - \kappa_{IND}$ crossings follows easily: For each vertex-circle, if the vertex corresponding to it is in I, then assign the head of the vertex-circle, otherwise assign the tail as chosen.

This results in an assignment that will have at least κ_{IND} heads, and at most $n_H - \kappa_{IND}$ tails. The chain-circles of the links connected to a vertex-circle which is already assigned to its head are assigned so that no crossing is created, i.e., starting from the chain-circle attached to the already assigned head, choose the half-circle that does not create any crossings. The chain-circles of the other links are assigned edges in a very similar way, but this time with no condition on the assignment of the first chain-circle. Finally the tail-circles are assigned randomly to the half-circles. Such an assignment assigns half-circles for every circle in the drawing, and creates no more than $n_H - \kappa_{IND}$ crossings. The only crossings created are those between the tail-circles and the vertex-circles assigned to their tails. We already know that there are at most $n_H - \kappa_{IND}$ such vertex tails, which implies the first direction of the claim.

For the other direction, assume G has an assignment of half-circles with at most $n_H - \kappa_{IND}$ crossings. Let C_H (C_T) be the sets of vertex-circles having their heads (respectively tails) chosen in this assignment We know that $| C_T | \leq n_H - \kappa_{IND}$, since otherwise the assignment would create more than $n_H - \kappa_{IND}$ crossings. This implies that $| C_H | \geq \kappa_{IND}$, since $| C_T | + | C_H | = n_H$. For any (c_i, c_j) pair, where c_i, $c_j \in C_H$, there cannot be any links between c_i and c_j because if c_i, c_j were linked together, each of the $n_H - \kappa_{IND} + 1$ links would have at least one crossing, creating more than $n_H - \kappa_{IND}$ crossings which would be a contradiction. Let I be the set of vertices corresponding to the vertex-circles in C_H; then I is an independent set with size at least κ_{IND}. □

Theorem 5. *Max2CD is NP-hard.*

Proof Sketch: The reduction is again from PD3IS. Let $(H = (V_H, E_H), \kappa_{IND})$ be an instance of PD3IS. The reduction produces a Max2CD instance (G, κ_{MAX}), with $\kappa_{MAX} = t_H - n_H + \kappa_{IND}$, where t_H is the total number of circles in G. The proof proceeds along the lines of the proof of Theorem 4 with a slight modification: in this case we add $n_H - \kappa_{IND} + 1$ tail-circles to each vertex-circle, rather than just one tail-circle. Then H has an independent set of size κ_{IND} if and only if G can be drawn without any crossings such that at least $\kappa_{MAX} = t_H - n_H + \kappa_{IND}$ edges have been assigned to some half-circle. □

6 Open Problems

We conclude with two open problems: (1) Can Min2CD and Max2CD be approximated within a constant factor? (2) Can we draw graphs with pre-specified vertex positions, using elliptic or other parabolic curve segments without creating too many crossings?

Acknowledgments. We would like to thank Roberto Tamassia, Sue Whitesides, and Helmut Alt for the stimulating discussions and suggestions.

References

1. Agarwal. Range searching. In J. E. Goodman and J. O'Rourke, editors, *Handbook of Discrete and Computational Geometry, CRC Press, 1997.* 1997.
2. P. Agarwal and J. Erickson. Geometric range searching and its relatives. *Advances in Discrete and Computational Geometry*, 23:1–56, 1999.
3. P. Agarwal, M. van Kreveld, and S. Suri. Label placement by maximum independent set in rectangles. *Computational Geometry: Theory and Applications*, 11:209–218, 1998.
4. B. Apswall, M. Plass, and R. Tarjan. A linear-time algorithm for testing the truth of certain quantified boolean formulas. *Inf. Proc. Letters*, 8:121–123, 1979.
5. C. C. Cheng, C. A. Duncan, M. T. Goodrich, and S. G. Kobourov. Drawing planar graphs with circular arcs. *Discrete and Computational Geometry*, 25:405–418, 2001.
6. S. Doddi, M. Marathe, A. Mirzaian, B. Moret, and B. Zhu. Map labeling and its generalizations. In *8th Symposium on Discrete Algorithms*, pages 148–157, 1997.
7. S. Doddi, M. Marathe, and B. Moret. Point set labeling with specified positions. In *Proc. 16th ACM Sympos. Comput. Geom. (SoCG'00)*, pages 182–190, 2000.
8. A. Efrat, C. Erten, and S. G. Kobourov. Fixed-location circular-arc drawing of planar graphs. Technical Report TR03-10, Department of Computer Science, University of Arizona, 2003.
9. A. Efrat, A. Itai, and M. J. Katz. Geometry helps in bottleneck matching and related problems. *Algorithmica*, 31:1–28, 2001.
10. S. Even, A. Itai, and A. Shamir. On the complexity of timetable and multicommodity flow problems. *SIAM J. Comput.*, 5:691–703, 1976.
11. M. Formann and F. Wagner. A packing problem with applications to lettering of maps. In *Proc. 7th Annu. ACM Sympos. Comput. Geom.*, pages 281–288, 1991.
12. M. Godau. On the difficulty of embedding planar graphs with inaccuracies. In *Proceedings on Graph Drawing (GD'94)*, pages 254–261, 1994.
13. P. Gupta, R. Janardan, and M. Smid. Algorithms for some intersection searching problems involving circular objects. *Intl. J. of Math. Algorithms*, 1:35–52, 1999.
14. H. Imai and T. Asano. Efficient algorithms for geometric graph search problems. *SIAM J. Comput.*, 15:478–494, 1986.
15. M. Kaufmann and R. Wiese. Embedding vertices at points: Few bends suffice for planar graphs. *Journal of Graph Algorithms and Applications*, 6(1):115–129, 2002.
16. D. Lichtenstein. Planar formulae and their uses. *SIAM J. Comput.*, 11:329–343, 1982.
17. J. Matoušek. Efficient partition trees. *Discrete Comput. Geom.*, 8:315–334, 1992.
18. J. Matoušek. Geometric range searching. *ACM Computing Surveys*, 26:421–462, 1994.
19. J. Pach and R. Wenger. Embedding planar graphs at fixed vertex locations. In *Graph Drawing*, pages 263–274, 1998.
20. C. Poon, B. Zhu, and F. Chin. A polynomial time solution for labeling a rectilinear map. *Information Processing Letters*, 65(4):201–207, 1998.
21. T. Strijk and M. van Kreveld. Labeling a rectilinear map more efficiently. *Information Processing Letters*, 69(1):25–30, 1999.
22. R. Tamassia and I. G. Tollis. Planar grid embedding in linear time. *IEEE Trans. Circuits Syst.*, CAS-36(9):1230–1234, 1989.
23. M. van Kreveld, T. Strijk, and A. Wolff. Point set labeling with sliding labels. In *Proc. 14th Annu. ACM Sympos. Comput. Geom. (SoCG'98)*, pages 337–346, 1998.

A More Practical Algorithm for Drawing Binary Trees in Linear Area with Arbitrary Aspect Ratio*

Ashim Garg and Adrian Rusu

Department of Computer Science and Engineering
University at Buffalo, Buffalo, NY 14260
{agarg,adirusu}@cse.buffalo.edu

Abstract. Trees are usually drawn using planar straight-line drawings. [1] presented an algorithm for constructing a planar straight-line grid drawing of an n-node binary tree with area $O(n)$ and any pre-specified aspect ratio in the range $[n^{-\alpha}, n^{\alpha}]$, where $0 \leq \alpha < 1$ is any constant, in $O(n \log n)$ time. Unfortunately, the algorithm of [1] is not suitable for practical use. The main problem is that the constant hidden in the "Oh" notation for area is quite large (e.g., it can be as large as 3900).
In this paper, we have made several practical improvements to the algorithm, which make it suitable for practical use. We have also conducted experiments on this newer version of the algorithm for randomly-generated and complete binary trees with up to 50,000, and 65,535 nodes, respectively. Our experiments show that it constructs area-efficient drawings in practice, with area at most 10 times and 8 times the number of nodes for randomly-generated and complete binary trees, respectively.

1 Introduction

A drawing Γ of a tree is a *straight-line* drawing, if each edge is drawn as a single line-segment. Γ is a *grid* drawing if all the nodes have integer coordinates. Γ is a *planar* drawing, if edges do not intersect each other. Here, we concentrate on grid drawings. So, we will assume that the plane is covered by a rectangular grid. Let Γ be a grid drawing. Let R be the smallest rectangle with sides parallel to the X- and Y-axes, respectively, that covers Γ completely. The *width* (*height*) of Γ is equal to 1+ width of R (1+height of R). The *area* of Γ is equal to (1+width of R)·(1+height of R), which is equal to the number of grid points contained within R. The *aspect ratio* of Γ is the ratio of its width and height.

* Research supported by NSF CAREER Award IIS-9985136, NSF CISE Research Infrastructure Award No. 0101244, and Mark Diamond Research Grant No. 13-Summer-2003 from GSA of The State University of New York.

G. Liotta (Ed.): GD 2003, LNCS 2912, pp. 159–165, 2004.

2 Our Result

Planar straight-line drawings are more esthetically pleasing than non-planar polyline drawings. Grid drawings guarantee at least unit distance between the nodes, and the integer coordinates of the nodes allow the drawings to be displayed in a display surface, such as a computer screen, without any distortions due to rounding-off errors. Giving users control over the aspect ratio of a drawing allows them to display the drawing in different kinds of display surfaces with different aspect ratios. Finally, it is important to minimize the area of a drawing, so that the users can display it in small display surfaces also.

[1] presented an algorithm for constructing a planar straight-line grid drawing of an n-node binary tree T with area $O(n)$ and with any pre-specified aspect ratio A in the range $[n^{-\alpha}, n^{\alpha}]$, where $0 \leq \alpha < 1$ is any constant, in $O(n \log n)$ time (the algorithm actually takes three input parameters: T, A, and a user-defined constant ϵ, such that $0 < \epsilon < 1$ and $n^{-\epsilon} \leq A \leq n^{\epsilon}$).

While the algorithm of [1] was significant from a theoretical point of view, it suffered from the following drawbacks, that made it unsuitable for practical use:

- The constant c hidden in the "Oh" notation for area can be quite large (e.g., it can be as large as 3900 for $\epsilon = 0.6$, and $A = 1$). One might argue that c is really the worst-case bound, and the algorithm might perform better in practice. However, the problem is that given a tree T with n nodes, and two numbers ϵ and A as input, the algorithm will always pre-allocate a rectangle R with size *exactly equal* to cn, and draw T within R. Thus, the area of R is always equal to the worst-case area, and correspondingly, the drawing also has a large area. This is the major drawback of this algorithm.
- Also, it uses another algorithm, called *Algorithm u^*-HV-Draw*, as a subroutine. This increases the complexity of implementing the algorithm.

In this paper, we have made several practical improvements to the algorithm, which make it more suitable for practical use: (Note that the area of the drawing constructed by this newer version of the algorithm is still $O(n)$)

- We have developed a newer version of the algorithm that does not require the pre-assignment of a rectangle with the worst-case area to draw a tree. Instead, it only pre-assigns an aspect ratio to the tree, which is used to draw the tree recursively in a bottom-up fashion. This makes it possible for the algorithm to construct a more area-efficient drawing in practice.
- This newer version does not require *Algorithm u^*-HV-Draw* as a subroutine, which makes it easier to implement.
- The proof for the area being $O(n)$ given in [1] is based on a theorem by Valiant (Theorem 6 of [3]). Unfortunately, the most natural way of using the theorem seemed to be requiring the pre-assignment of a rectangle (with a large area). Hence, developing an algorithm that does not pre-assign a rectangle required developing a new proof. Correspondingly, we have developed a new proof that does not use the theorem. Instead, it simply uses Induction.

- We have also implemented this newer version, and experimentally evaluated its performance for randomly-generated binary trees with up to 50,000 nodes, and for complete binary trees with up to $65,535 = 2^{16} - 1$ nodes. Our experiments show that it constructs area-efficient drawings in practice, with area at most 8 times the number of nodes for complete binary trees, and at most 10 times the number of nodes for randomly-generated binary trees.

Due to space limitations, in this paper, we only present a brief overview of this newer version of the algorithm. Full details are given in [2].

3 Preliminaries

Let T be an n-node binary tree, with one distinguished node v, which has at most one child. v is called the *link* node of T. T is an *ordered* tree if the children of each node are assigned a left-to-right order. A *partial tree* of T is a connected subgraph of T. If T is an ordered tree, then the *leftmost path* p of T is the maximal path consisting of nodes that are the left children of their parents, except the first one, which is the root of T; the last node of p is called the *leftmost* node of T. T is an *empty tree*, if it has zero nodes in it.

Let Γ be a planar straight-line grid drawing of T. Γ has a *good* aspect ratio, if its aspect ratio is in the range $[n^{-\alpha}, n^{\alpha}]$, where $0 \leq \alpha < 1$ is a constant. Let r be the root of T. Let u^* be the link node of T. Γ is a *feasible* drawing of T, if it has the following three properties:

- **Property 1**: The root r is placed at the top-left corner of Γ.
- **Property 2**: If $u^* \neq r$, then u^* is placed at the bottom boundary of Γ. Moreover, we can move u^* downwards in its vertical channel by any distance without causing any edge-crossings in Γ.
- **Property 3**: If $u^* = r$, then no other node or edge of T is placed on, or crosses the vertical and horizontal channels occupied by r.

Theorem 1 (Separator Theorem [3]). *Every n-node binary tree T contains an edge e, called a* separator edge, *such that removing e from T splits T into two trees with at most $(2/3)n$ nodes each. Moreover, e can be found in $O(n)$ time.*

4 Our Tree Drawing Algorithm

Let T be a n-node binary tree with a link node u^*. Let A and ϵ be numbers such that $0 < \epsilon < 1$, and $n^{-\epsilon} \leq A \leq n^{\epsilon}$. A is called the *desirable aspect ratio* for T.

Our tree drawing algorithm, called *DrawTree*, takes ϵ, A, and T as inputs, and uses a simple divide-and-conquer strategy to recursively construct a feasible drawing Γ of T, by performing the following actions at each recursive step:

- *Split Tree*: Convert T into an ordered tree, in which u^* is the leftmost node. Using Theorem 1, find a separator edge $e = (u, v)$. Using e, split T into at most five partial trees by removing at most two nodes and their incident edges from it. We get two cases: in Case 1 (see Figure 1(a)), e is not in the leftmost path of T, and in Case 2 (see Figure 1(b)), e is in the leftmost path of T. The partial trees obtained in Case 1 are $T_A, T_\beta, T_1, T_2, T_C$, and in Case 2 are T_A, T_B, T_C. The nodes removed are a and u in Case 1, and u in Case 2. Based on which partial trees are empty, we get 7 subcases for Case 1 and 8 subcases for Case 2. These subcases are described in detail in [2] (Figures 1(a), and 1(b) show the subcases for Case 1 and 2, respectively, in which all the partial trees are non-empty).

- *Assign Aspect Ratios*: Correspondingly, assign a desirable aspect ratio A_k to each partial tree T_k. Let n_k be number of nodes in T_k. T_k is a *large* partial tree of T if either $A \geq 1$ and $n_k \geq (n/A)^{1/(1+\epsilon)}$, or $A < 1$ and $n_k \geq (An)^{1/(1+\epsilon)}$; T_k is a *small* partial tree otherwise.
  The assignment of A_k to T_k is done as follows: Let $x_k = n_k/n$.
  - If $A \geq 1$: If T_k is a large partial tree of T, then $A_k = x_k A$, otherwise (i.e., if T_k is a small partial tree of T) $A_k = n_k^{-\epsilon}$.
  - If $A < 1$: If T_k is a large partial tree of T, then $A_k = A/x_k$, otherwise (i.e., if T_k is a small partial tree of T) $A_k = n_k^{\epsilon}$.

  Moreover, if $A \geq 1$, and $T_k = T_A$ or $T_k = T_\beta$, then the value of A_k is changed to $1/A_k$. This is done so because later in Step *Compose Drawings*, when $A \geq 1$, T_A and T_β are rotated by 90°, when constructing Γ.
  Intuitively, the above assignment strategy ensures that each partial tree gets a good desirable aspect ratio.
- *Draw Partial Trees*: Recursively construct a feasible drawing Γ_k of each partial tree T_k with A_k as its desirable aspect ratio.
- *Compose Drawings*: Arrange the drawings of the partial trees, and draw the nodes and edges, that were removed from T to split it, to obtain a feasible drawing Γ of T. If $A < 1$, then the drawings of the partial trees are stacked one above the other, and if $A \geq 1$, they are placed side-by-side. Also note that when $A \geq 1$, the drawing of T_A and T_β are rotated clockwise by 90° and flipped left-to-right before placing in Γ. Figure 1(c) and 1(e) (Figures 1(d) and 1(f)) show the arrangement, when $A < 1$, and $A \geq 1$, respectively, for the case shown in Figure 1(a) (Figure 1(b)). For the arrangement in all the subcases of Case 1 and 2, see [2].

Figure 2 shows a drawing of the complete binary tree with 63 nodes constructed by Algorithm *DrawTree*, with $A = 1$ and $\epsilon = 0.2$.

Theorem 2. *Let T be a binary tree with n nodes. Given any number A, where $n^{-\alpha} \leq A \leq n^{\alpha}$, for some constant α, where $0 \leq \alpha < 1$, we can construct in $O(n \log n)$ time, a planar straight-line grid drawing of T with $O(n)$ area, and aspect ratio A, using Algorithm* DrawTree.

Proof. Omitted here due to lack of space. See [2] for the proof.

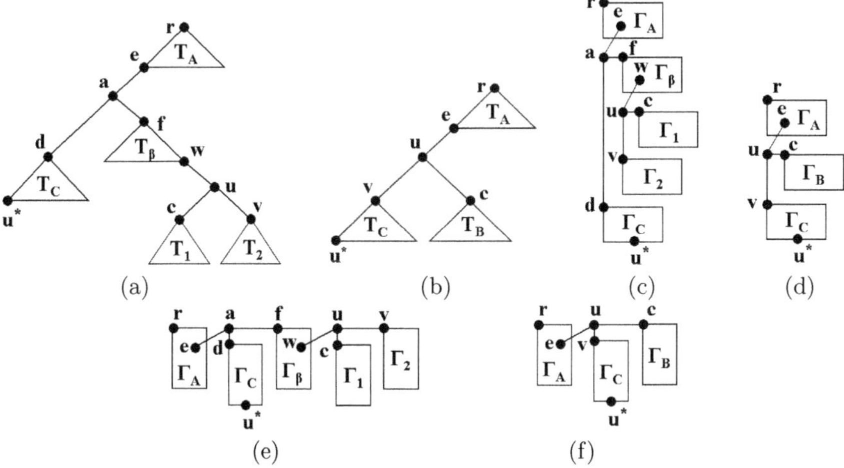

Fig. 1. General structure of tree T in (a) Case 1, and (b) Case 2, respectively; Drawing T, when $A < 1$, in (c) Case 1, and (d) Case 2, respectively, and when $A \geq 1$, in (e) Case 1, and (f) Case 2, respectively.

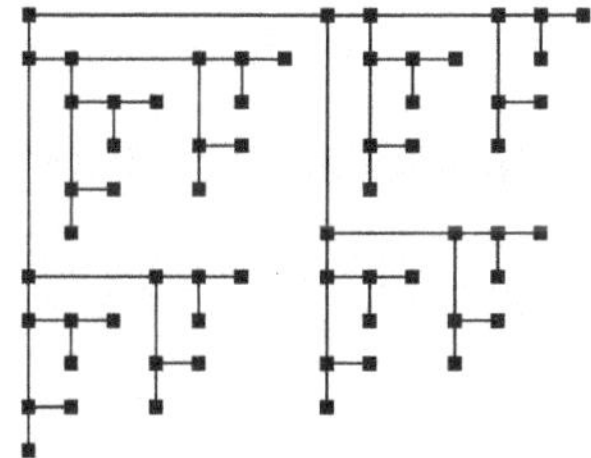

Fig. 2. Drawing of the complete binary tree with 63 nodes constructed by Algorithm *DrawTree*, with $A = 1$ and $\epsilon = 0.2$.

5 Experimental Results

We have implemented the algorithm using about 2100 lines of C++ code, and evaluated its performance on two types of binary trees, namely, randomly-generated, with up to 50,000 nodes, and complete, with up to 65,535 nodes.

Recall that the algorithm takes three values as input: a binary tree T with n nodes, a number ϵ, where $0 < \epsilon < 1$, and a number A in the range $[n^{-\epsilon}, n^{\epsilon}]$.

The performance criteria we have used to evaluate the algorithm is the ratio c of the area of the drawing constructed of T, and n.

To evaluate the algorithm, we varied n up to 50,000 for randomly-generated, and up to $65,535 = 2^{16} - 1$ for complete binary trees. For each n, we used five different values for ϵ, namely, 0.1, 0.25, 0.5, 0.75, and 0.9. For each (n, ϵ) pair, we used 20 different values of A uniformly distributed in the range $[1, n^{\epsilon}]$. The performance of the algorithm is symmetrical for $A < 1$ and $A > 1$. Hence, we varied A only from 1 through n^{ϵ}, not from $n^{-\epsilon}$ through n^{ϵ}. Hence, in the rest

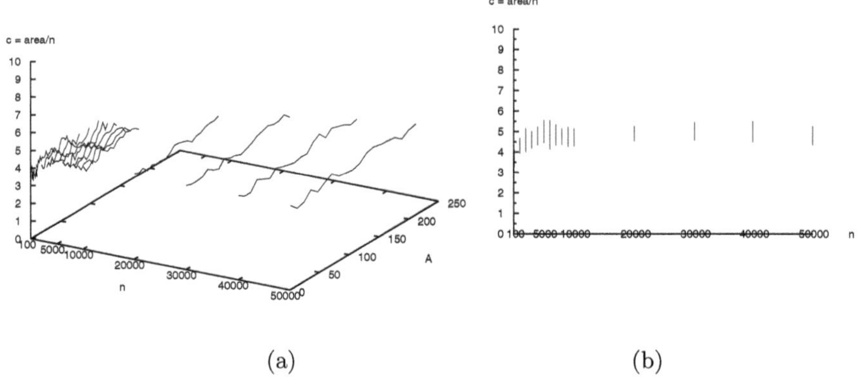

(a) (b)

Fig. 3. (a) Performance of the algorithm, as given by the value of c, for drawing a randomly-generated binary tree T with $\epsilon = 0.5$, and different values of A, where c=*area of drawing/number of nodes n in T*. (b) Projection of the 3D plot of (a) on XZ-plane.

of the section, we will assume that $A \geq 1$. For each type of tree (randomly-generated and complete), and for each triplet (n, A, ϵ), we generated three trees of that type. We constructed a drawing of each tree using the algorithm, and computed the value of c. Next, we averaged the values of c obtained for the three trees to get a single value for each triplet (n, A, ϵ) for each tree-type.

Our experiments show that the value of c is generally small, and is at most 10 for randomly-generated, and at most 8 for complete trees. Figure 3 shows how c varies with n, and A, for $\epsilon = 0.5$ for randomly-generated trees. (See [2], to see how c varies with n, and A, for all the five values of ϵ for both randomly-generated and complete binary trees.)

We also discovered that c increases with A for a given n and ϵ. However, the rate of increase is very small. Consequently, for a given n and ϵ, the range for c over all the values of A is small (see Figure 3). E.g., for $n = 10,000$, and $\epsilon = 0.5$, for randomly-generated trees, the range for c is $[4.2, 5.2]$.

Similarly, for a given n and A, c increases with ϵ.

Finally, we would like to comment that the aspect ratio A_{act} of the drawing Γ constructed by the algorithm is, in general, different from the input aspect ratio A. We computed the ratio r of A_{act} and A. We discovered that r is close to 1 for $A = 1$, generally decreases as we increase A, and can get as low as 0.1 for $A = n^{\epsilon}$. However, we also discovered that for a large range of values for A, namely, $[1, \min\{n^{\epsilon}, n/\log^2 n\}]$, r stays within the range $[0.8, 1.5]$, and so is close to 1. Hence, if we need to construct a drawing with aspect ratio exactly equal to A, we can do so by adding enough "white space" to Γ. This will increase the area of Γ by a factor of at most $\max\{1/0.8, 1.5\} = 1.5$ (assuming that A is in the above-mentioned range). Hence, the area of Γ will still be small.

References

1. A. Garg and A. Rusu. Straight-line drawings of binary trees with linear area and arbitrary aspect ratio. In *Proc. 10th International Symposium on Graph Drawing (GD 2002)*, volume 2528 of *LNCS*, pages 320–331. Springer-Verlag, 2002.
2. A. Garg and A. Rusu. A more practical algorithm for drawing binary trees in linear area with arbitrary aspect ratio. Technical Report No. 2003-12, Department of Computer Science and Engineering, University at Buffalo, Buffalo, NY, 2003.
3. L. Valiant. Universality considerations in VLSI circuits. *IEEE Trans. Comput.*, C-30(2):135–140, 1981.

An Integer Programming Approach to Fuzzy Symmetry Detection

Christoph Buchheim and Michael Jünger

Universität zu Köln, Institut für Informatik,
Pohligstraße 1, 50969 Köln, Germany
{buchheim,mjuenger}@informatik.uni-koeln.de

Abstract. The problem of exact symmetry detection in general graphs has received much attention recently. In spite of its NP-hardness, two different algorithms have been presented that in general can solve this problem quickly in practice [5,2]. However, as most graphs do not admit any exact symmetry at all, the much harder problem of fuzzy symmetry detection arises: a minimal number of certain modifications of the graph should be allowed in order to make it symmetric. We present a general approach to this problem: we allow arbitrary edge deletions and edge creations; every single modification can be given an individual weight. We apply integer programming techniques to solve this problem exactly or heuristically and give runtime results for a first implementation.

1 Introduction

An important aim in Automatic Graph Drawing is the display of symmetric structure inherent in a graph. Empirical studies show that symmetric drawings of a graph are much easier to understand than asymmetric ones [14]. However, two different problems occur when trying to develop algorithms for symmetric graph drawing: first, the symmetry detection problem is NP-hard for general graphs [13]. Second, most graphs do not admit any exact symmetry at all; in order to create nearly symmetric drawings, one has to define some kind of relaxation of exact symmetry.

For the first problem, two different algorithms with exponential runtime in general but fast runtime in practice have been proposed recently. In [4], we presented an approach based on integer programming techniques; see [5] and [3] also. Later, Abelson et al. [2] devised an algorithm based on group-theoretic methods, running even faster.

The second problem is motivated by the observation that nearly symmetric drawings still help the user to understand the structure of a graph; see Fig. 1 for some examples. Unfortunately, mostly negative results about this problem have been published so far. In fact, detecting fuzzy symmetries is much harder than detecting exact symmetries in general, as shown by Chen et al. [6]. In their approach, different types of modifications of the graph are allowed in order to make it symmetric, namely, to delete nodes, to delete edges, and to contract edges. They consider the problem of finding a minimum number of such operations

G. Liotta (Ed.): GD 2003, LNCS 2912, pp. 166–177, 2004.

needed to obtain a graph admitting an axial or rotational symmetry. Among other results, they show that some of the corresponding problems are NP-hard even for trees. On the other hand, exact planar symmetry detection is possible in linear time for planar graphs [9,10,11,12].

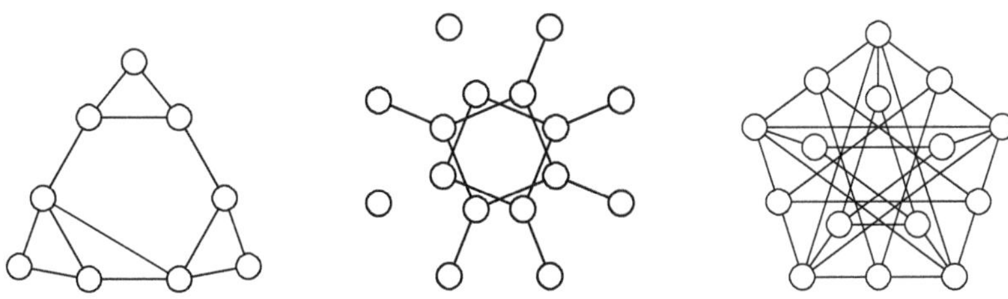

Fig. 1. Some nearly symmetric drawings

In this paper, we follow a similar approach: we allow a minimum number of edge deletions and edge creations in order to obtain a graph that admits a symmetric drawing. A more precise description of this problem, along with a generalization, is given in Sect. 3. Before, we gather some definitions concerning symmetries in Sect. 2. In Sect. 4, we present integer linear programs (ILPs) modeling the fuzzy rotation and fuzzy reflection problems. The corresponding polytopes are investigated in Sect. 5. Finally, we roughly explain the framework of a branch & cut-algorithm for fuzzy symmetry detection in Sect. 6. There we also report runtime results for a first implementation of this algorithm. For more details or any proof, the reader is referred to [3].

2 Preliminaries

For ease of exposition, we only consider simple undirected graphs in the following. An *automorphism* of a graph $G = (V, E)$ is a permutation π of V with $(i, j) \in E$ if and only if $(\pi(i), \pi(j)) \in E$ for all $i, j \in V$. The set of automorphisms of G forms a group with respect to composition, denoted by $\mathrm{Aut}(G)$. The *order* of an automorphism π is $\mathrm{ord}(\pi) = \min\{k \in \mathbb{N} \mid \pi^k = \mathrm{id}_V\}$, where id_V denotes the identity permutation of V. For a node $i \in V$, the set $\mathrm{orb}_\pi(i) = \{\pi^k(i) \mid k \in \mathbb{N}\}$ is the π*-orbit* of i. Finally, the *fixed* nodes are those in $\mathrm{Fix}(\pi) = \{i \in V \mid \pi(i) = i\}$.

A *reflection* of G is an automorphism $\pi \in \mathrm{Aut}(G)$ with $\pi^2 = \mathrm{id}_V$, i.e., an automorphism of order one or two. For $k \in \{1, \ldots, n\}$, a k*-rotation* of G is an automorphism $\pi \in \mathrm{Aut}(G)$ such that $|\mathrm{orb}_\pi(i)| \in \{1, k\}$ for all $i \in V$ and $|\mathrm{Fix}(\pi)| \leq 1$ if $k \neq 1$. Each 2-rotation is a reflection, but not vice versa; the identity id_V is both a reflection and a 1-rotation.

Now assume that there exist an injective node placement $p: V \to \mathbb{R}^2$ and an isometry $\varphi: \mathbb{R}^2 \to \mathbb{R}^2$ of the Euclidean plane with the following properties: for each $i \in V$ there exists a node $i' \in V$ with $\varphi(p(i)) = p(i')$, and for $i, j \in V$ we have $(i,j) \in E$ if and only if $(i',j') \in E$; for straight-line drawings of simple graphs, this means nodes are mapped to nodes and edges are mapped to edges. Then p and φ induce an automorphism π of G by setting $\pi(i) = i'$. Any automorphism induced like this is called a *geometric automorphism* or a *symmetry* of G. An automorphism of a graph G is a symmetry if and only if it is a rotation or a reflection [8]. Obviously, the symmetries displayed by a single placement p form a subgroup of $\mathrm{Aut}(G)$, but the set of all symmetries of G is not closed under composition. Observe that this definition of a symmetry, following the one given in [8], allows degenerate drawings: nodes may be located on non-incident edges and edges may overlap. In fact, in the case of a reflection, the subgraph induced by the fixed nodes may be arbitrarily large, so that degeneracies may be unavoidable. However, all nodes must have different positions.

3 The Fuzzy Symmetry Detection Problem

The general idea of our approach to fuzzy symmetry detection is to allow to change the adjacency of every pair of nodes in G. More precisely, we allow to create an edge between each pair of non-adjacent nodes and to delete the edge between each pair of adjacent nodes. Creating or deleting an edge between two nodes i and j of G involves a penalty of $w_{ij} \in \mathbb{R}$. The aim is to find a graph $G' = (V, E')$ such that G' is symmetric and such that the total penalty for transforming G into G' is minimal. In other words, let $E \triangle E' = (E \setminus E') \cup (E' \setminus E)$ denote the symmetric difference between E and E'. Then our objective is to minimize the total penalty

$$\sum_{(i,j) \in E \triangle E'} w_{ij}$$

over all graphs $G' = (V, E')$ that admit any non-trivial symmetry. We call this problem the *weighted fuzzy symmetry detection* problem.

Observe that we do not require positive weights. However, allowing negative weights does not increase the power of the approach, since all negative weights can be replaced by their absolute values after changing the adjacency of the corresponding edges in G. The presence of zero weights means that the adjacency of the corresponding node pairs does not matter at all. By using large enough weights, we can also forbid certain modifications absolutely.

The weighted fuzzy symmetry detection problem is NP-hard, even if all weights are one. This follows from the NP-hardness of the exact symmetry detection problem. In practice, fuzzy symmetry detection is much harder than exact symmetry detection—at least for our solution methods.

Finally notice that we could also allow node deletions in our approach. For simplicity, however, we restrict ourselves to edge modifications here.

4 An Integer Linear Programming Model

In order to describe fuzzy symmetries by integer linear programs, we split up the problem in the following way: we fix an integer $k \in \{2, \ldots, n\}$ and consider the problem of finding a graph G' that admits a symmetry of order k and that differs minimally from G. By applying this approach to all feasible orders one after another, we can solve the original problem.

Hence we set up a separate *fuzzy rotation ILP* for each $k \geq 3$ such that k divides either n or $n-1$ and a *fuzzy reflection ILP* for $k = 2$. All these programs are based on the *fuzzy automorphism ILP* that uses two different types of variables to model fuzzy automorphisms.

The *mapping variables* are used to specify a permutation π of V: we define a real $n \times n$-matrix $M(\pi)$ by

$$M(\pi)_{ij} = \begin{cases} 1 & \text{if } \pi(i) = j \\ 0 & \text{otherwise,} \end{cases}$$

yielding a monomorphism M of the group S_n of permutations of V into the general linear group $\mathrm{GL}_n(\mathbb{R})$. The matrices in $M(S_n)$ are called *permutation matrices* and can be characterized as the set of $n \times n$-matrices $X = (x_{ij})$ with $x_{ij} \in \{0, 1\}$ and

$$\sum_{j \in V} x_{ij} = 1 \text{ for all } i \in V \quad \text{and} \quad \sum_{i \in V} x_{ij} = 1 \text{ for all } j \in V\,. \tag{1}$$

In other words, for the desired ILP, we can use a binary mapping variable x_{ij} for each pair $(i, j) \in V^2$ and add the constraints (1). A value of one for the mapping variable x_{ij} is interpreted as mapping node i to node j.

For technical reasons, we replace the permutation constraints (1) by the weaker constraints

$$\sum_{j \in V} x_{ij} \leq 1 \text{ for all } i \in V \quad \text{and} \quad \sum_{i \in V} x_{ij} \leq 1 \text{ for all } j \in V\,. \tag{2}$$

Thus the permutation corresponding to the mapping variables may be partially or totally undefined. However, by adding a large value to all objective function coefficients of mapping variables, every optimal solution will still correspond to a well-defined rotation or reflection.

The second type of variables is used to model all graphs G' with node set V. We use a binary *edge variable* y_{ij} for each pair of nodes $i, j \in V$ that is set to one if and only if the nodes i and j are adjacent in G'.

The connection between the two types of variables is established as follows: we want the permutation π of V given by the mapping variables to be an automorphism of the graph G' given by the edge variables. In other words, if two nodes i and j are adjacent in G', then the same must be true for $\pi(i)$ and $\pi(j)$, and vice versa. This can be translated into

$$x_{i_1 j_1} + x_{i_2 j_2} \leq 2 \pm y_{i_1 i_2} \mp y_{j_1 j_2} \text{ for all } i_1, i_2, j_1, j_2 \in V\,. \tag{3}$$

Indeed, these constraints are trivial if the left hand side is at most one. If the left hand side is two, then $\pi(i_1) = j_1$ and $\pi(i_2) = j_2$. In this case, the right hand sides make sure that $y_{i_1 i_2} = y_{j_1 j_2}$, i.e., either both pairs (i_1, i_2) and (j_1, j_2) are adjacent or both are non-adjacent in G'.

In summary, fuzzy automorphisms correspond to the solutions of

$$\begin{array}{rll} x_{ij} \in \{0,1\} & \text{for all } i,j \in V \\ y_{ij} = y_{ji} \in \{0,1\} & \text{for all } i,j \in V,\ i \neq j \\ \sum_{j \in V} x_{ij} \leq 1 & \text{for all } i \in V \\ \sum_{i \in V} x_{ij} \leq 1 & \text{for all } j \in V \\ x_{i_1 j_1} + x_{i_2 j_2} \leq 2 \pm y_{i_1 i_2} \mp y_{j_1 j_2} & \text{for all } i_1, i_2, j_1, j_2 \in V,\ i_1 \neq i_2,\ j_1 \neq j_2\,. \end{array} \tag{4}$$

Next, we are going to restrict the ILP (4) to rotations of order $k \geq 3$. The first condition for π to be a k-rotation is $|\text{orb}_\pi(i)| \in \{1, k\}$ for all $i \in V$. All other orbit lengths are forbidden. In order to set up linear constraints to enforce this property of π, we define a *forbidden orbit* of G as a circular list $C = (i_1, \ldots, i_p)$ of pairwise distinct nodes in V with $k \neq p \geq 2$. Let $|C| = p$ and

$$x(C) = \sum_{t=1}^{p-1} x_{i_t i_{t+1}} + x_{i_p i_1}\,.$$

Then the condition $|\text{orb}_\pi(i)| \in \{1, k\}$ for all $i \in V$ is equivalent to

$$x(C) \leq |C| - 1 \quad \text{for all forbidden orbits } C\,. \tag{5}$$

Indeed, the constraint (5) implies that for every forbidden orbit C at least one variable in $x(C)$ is zero, so that C is not an orbit of π. To this point, we have ensured that all orbits of π have length k or 1. Any k-rotation has to meet the additional condition $|\text{Fix}(\pi)| \leq 1$. We can express this by

$$\sum_{i \in V} x_{ii} \leq 1\,, \tag{6}$$

as the sum on the left hand side equals the number of fixed nodes. Combining the constraints (4), (5), and (6), an ILP describing fuzzy k-rotations of G is

$$\begin{array}{rll} x_{ij} \in \{0,1\} & \text{for all } i,j \in V \\ y_{ij} = y_{ji} \in \{0,1\} & \text{for all } i,j \in V,\ i \neq j \\ \sum_{j \in V} x_{ij} \leq 1 & \text{for all } i \in V \\ \sum_{i \in V} x_{ij} \leq 1 & \text{for all } j \in V \\ x_{i_1 j_1} + x_{i_2 j_2} \leq 2 \pm y_{i_1 i_2} \mp y_{j_1 j_2} & \text{for all } i_1, i_2, j_1, j_2 \in V,\ i_1 \neq i_2,\ j_1 \neq j_2 \\ x(C) \leq |C| - 1 & \text{for all forbidden orbits } C \\ \sum_{i \in V} x_{ii} \leq 1\,. & \end{array} \tag{7}$$

Fuzzy reflections of G are easier to model than k-rotations: instead of using the forbidden orbit constraints, we just have to ensure that the automorphism π to be represented satisfies $\pi^2 = \text{id}_V$, i.e., that in our model we have

$$x_{ij} = x_{ji} \quad \text{for all } i,j \in V\,. \tag{8}$$

In other words, we only need a single mapping variable for each pair of nodes. Hence an ILP modeling fuzzy reflections is given by

$$\begin{array}{rll} x_{ij} = x_{ji} \in \{0,1\} & \text{for all } i,j \in V & \\ y_{ij} = y_{ji} \in \{0,1\} & \text{for all } i,j \in V,\ i \neq j & \\ \sum_{j\in V} x_{ij} \leq 1 & \text{for all } i \in V & (9) \\ \sum_{i\in V} x_{ij} \leq 1 & \text{for all } j \in V & \\ x_{i_1j_1} + x_{i_2j_2} \leq 2 \pm y_{i_1i_2} \mp y_{j_1j_2} & \text{for all } i_1,i_2,j_1,j_2 \in V,\ i_1 \neq i_2,\ j_1 \neq j_2 \,. & \end{array}$$

Observe that both (7) and (9) do not depend on the structure of G; they are determined by the number of nodes of G. The information about edges of G is stored in the objective function: we minimize the total weight of the modifications that are necessary to get from G to G', i.e., we minimize

$$\sum_{(i,j)\in V^2\setminus E} w_{ij}y_{ij} + \sum_{(i,j)\in E} w_{ij}(1-y_{ij}) \,.$$

Independently, we can assign arbitrary weights to the mapping variables. In the case $k = 2$, fixed nodes should be punished, since otherwise the graph G together with id_V forms an optimal solution of (9). However, the penalty for fixed nodes has to be attuned to the penalties for edge deletion or creation in this case.

5 The Fuzzy Symmetry Polytopes

The ILPs (7) and (9) only depend on the number n of nodes and the desired order k, but not on the structure of G. Thus we may define the *fuzzy symmetry polytope* $\mathrm{FSP}(k,n)$ to be the polytope corresponding to (7), if $k \geq 3$, or the polytope corresponding to (9), if $k = 2$. More precisely, these polytopes are given as the convex hulls of the sets of solutions of these ILPs.

Theorem 1. *The polytope FSP(k,n) is full-dimensional.*

For designing the branch & cut-algorithm for fuzzy symmetry detection, it is important to investigate the polyhedral structure of $\mathrm{FSP}(k,n)$. In particular, we have to find valid inequalities that induce facets of $\mathrm{FSP}(k,n)$, i.e., maximal faces of this polytope. Before starting to present classes of such inequalities, observe

Lemma 1. *If a constraint H is valid or facet-inducing for FSP(k,n), then the same is true after replacing each variable y_{ij} in H by $1 - y_{ij}$.*

Lemma 2. *If a constraint H is valid or facet-inducing for FSP(k,n), then the same is true after replacing each variable x_{ij} in H by x_{ji}.*

Both results follow from symmetry immediately. In the remainder of this section, we will usually consider only one of the four constraints without explicitly mentioning the other three.

Theorem 2. *The trivial constraints $x_{ij} \geq 0$, $y_{ij} \geq 0$, and $y_{ij} \leq 1$ induce facets of FSP(k,n). The same is true for the weak permutation constraints (2).*

To obtain further results about the polytope $\mathrm{FSP}(k,n)$, we will treat rotations and reflections separately in the following subsections.

5.1 The Fuzzy Rotation Polytopes

Throughout this subsection we assume $k \geq 3$. In this case, we have

Theorem 3. *The constraint (6) induces a facet of FSP(k,n).*

We continue with a review of the forbidden cycle constraints (5). For this, let $C = (i_1, \ldots, i_p)$ be a circular list of pairwise distinct nodes with $k \neq p \geq 2$ again. If k does not divide p, we can improve (5) to

$$\sum_{i,j \in C,\ i \neq j} x_{ij} \leq p - 1\,. \tag{10}$$

Indeed, the sum on the left hand side is at most p by (2). If it equals p for some permutation π, then C is fixed under π, but no single node of C is fixed. Since k does not divide p, some non-trivial orbit of π must have a length not equal to k, so π is no k-rotation.

By the same reasoning, if k does not divide $p - 1$ either, we even have the valid inequality

$$\sum_{i,j \in C} x_{ij} \leq p - 1\,. \tag{11}$$

Observe that for $k = n$ the constraint (11) is the subtour elimination constraint for the cut defined by C.

Theorem 4. *Let $2 \leq p \leq n-1$ and assume that k does not divide p. If k divides $p-1$, then (10) induces a facet of FSP(k,n). Otherwise, (11) induces a facet of FSP(k,n).*

Next, let W_1 and W_2 be any two subsets of V. Let $i_1, i_2 \in V \setminus (W_1 \cup W_2)$ with $i_1 \neq i_2$. Then we have the following valid inequality for FSP(k,n):

$$\sum_{j_1 \in W_1} x_{i_1 j_1} + \sum_{j_2 \in W_2} x_{i_2 j_2} \leq 2 - y_{i_1 i_2} + \sum_{j_1 \in W_1,\ j_2 \in W_2} y_{j_1 j_2}\,. \tag{12}$$

Indeed, the left hand side is at most two by (2), while the right hand side is at least one. If the left hand side equals two, we have $x_{i_1 j_1} = 1$ for some $j_1 \in W_1$ and $x_{i_2 j_2} = 1$ for some $j_2 \in W_2$. Thus $y_{i_1 i_2} = 1$ implies $y_{j_1 j_2} = 1$, so that the right hand side is at least two.

Theorem 5. *The constraint (12) induces a facet of FSP(k,n) if and only if*

(a) $|W_1| \neq 0$ *and* $|W_2| \neq 0$,
(b) $|W_2 \setminus W_1| \neq 1$ *and* $|W_1 \setminus W_2| \neq 1$, *and*
(c) $|W_1 \cup W_2| \geq 2$.

In the special case $W_1 = W_2 = \{j_1, j_2\}$, the constraint (12) and its counterpart according to Lemma 1 emerge as

$$x_{i_1 j_1} + x_{i_1 j_2} + x_{i_2 j_1} + x_{i_2 j_2} \leq 2 \pm y_{i_1 i_2} \mp y_{j_1 j_2}.$$

These constraints improve (3); both induce facets of FSP(k,n) by Theorem 5.

Finally, let $W \subseteq V$ and $w \in V \setminus W$. Let $i_1, i_2 \in V \setminus (W \cup \{w\})$ with $i_1 \neq i_2$. Then

$$\sum_{j \in W} (x_{i_1 j} + x_{i_2 j}) + 2x_{i_1 w} + 2x_{i_2 w} \leq 3 - y_{i_1 i_2} + \sum_{j \in W} y_{jw} \tag{13}$$

is a valid inequality for FSP(k, n). Indeed, the left hand side is at most three by (2), while the right hand side is at least two. If the left hand side is equal to three, one of the nodes i_1 or i_2 is mapped to w, while the other one is mapped to some $j \in W$. Hence $y_{i_1 i_2} = y_{jw}$, so that the right hand side is at least three.

Theorem 6. *The constraint (13) induces a facet of FSP(k,n) if and only if $|W| \geq 2$.*

5.2 The Fuzzy Reflection Polytope

Next, we examine the polytope FSP$(2, n)$ more closely. It is easy to see that for all odd subsets C of V the *blossom constraint*

$$\sum_{i,j \in C,\ i \neq j} x_{ij} \leq |C| - 1 \tag{14}$$

is valid for the fuzzy reflection polytope. We have

Theorem 7. *Let C be an odd subset of V with $|C| \geq 3$. Then the blossom constraint (14) induces a facet of FSP$(2,n)$.*

The constraint (12) is valid for fuzzy reflections as well. However, we can improve it by adding $x_{i_1 i_2}$ to the left hand side: if $x_{i_1 i_2} = 1$, all other variables on the left hand side must be zero. Hence we get

$$x_{i_1 i_2} + \sum_{j_1 \in W_1} x_{i_1 j_1} + \sum_{j_2 \in W_2} x_{i_2 j_2} \leq 2 - y_{i_1 i_2} + \sum_{j_1 \in W_1,\ j_2 \in W_2} y_{j_1 j_2}\,. \tag{15}$$

Theorem 8. *The constraint (15) induces a facet of FSP$(2,n)$ if and only if*

(a) $|W_1| \neq 0$ and $|W_2| \neq 0$,
(b) $|W_2 \setminus W_1| \neq 1$ and $|W_1 \setminus W_2| \neq 1$, and
(c) $|W_1 \cup W_2| \geq 2$.

By Theorem 8, the constraint (3) is improved by the facet-inducing constraints

$$x_{i_1 i_2} + x_{i_1 j_1} + x_{i_1 j_2} + x_{i_2 j_1} + x_{i_2 j_2} \leq 2 \pm y_{i_1 i_2} \mp y_{j_1 j_2}\,.$$

Next, observe that (13) is true for fuzzy reflections as well. Again, we can improve this constraint to

$$2x_{i_1 i_2} + \sum_{j \in W} (x_{i_1 j} + x_{i_2 j}) + 2x_{i_1 w} + 2x_{i_2 w} \leq 3 - y_{i_1 i_2} + \sum_{j \in W} y_{jw}\,. \tag{16}$$

Theorem 9. *The constraint (16) induces a facet of FSP$(2,n)$ if and only if $|W| \geq 2$.*

6 The Branch & Cut-Algorithm

In the following, we roughly describe our branch & cut-algorithm based on the results of the previous sections. We restrict ourselves to separation and primal heuristics; for other topics including the branching strategy and rules for setting variables, see [3]. For branch & cut in general, see [4] or [3].

6.1 Separation

We start the branch & cut-algorithm with the relaxation of (7) or (9) composed of all variables, the constraints (2), and, if $k \geq 3$, the constraint (6). The separation problem consists of finding inequalities that are valid for (7) or (9) but violated by the optimal solution of the current LP-relaxation. If we can find such inequalities, we add them to the relaxation and repeat the process.

We first try to separate constraints of type (12), if $k \geq 3$, or (15), if $k = 2$. For this, we use a straightforward greedy heuristic: for fixed i_1 and i_2, we add nodes to W_1 or W_2 as long as the difference between the left and the right hand side of (12) or (15) can be increased (for the current LP-solution). If we end up with any violated constraints, these are added to the LP-relaxation.

Otherwise, we start separating constraints of type (13), if $k \geq 3$, or (16), if $k = 2$. It is easy to see that these constraints can be separated in polynomial time: for a fixed combination of i_1, i_2, and w, we add a node j to W if and only if the current LP-values of $x_{i_1 j}$ and $x_{i_2 j}$ add up to more than the one of y_{jw}.

If we did not find any violated constraints of type (13) or (16), we separate the different kinds of orbit constraints: for $k \geq 3$, we search for violated constraints of type (10) or (11), which can be done in polynomial time by submodular function minimization. The same is true for the blossom constraints (14) for $k = 2$. If we still did not find any violated constraints, we start branching.

6.2 Primal Heuristics

Primal heuristics exploit the optimal solution of the current LP-relaxation in order to find feasible (but not necessarily optimal) solutions of the original ILP. In our algorithm, we use two heuristics working in an opposite way.

The first one starts with determining a symmetry of the given order k of the complete graph K_n that is close to the current (possibly fractional) values of the mapping variables. For example, we can traverse the mapping variables x_{ij} in descending order according to their value and let i be mapped to j if and only if no other node has been mapped to j before and i has not been mapped to any other node before. Once having fixed the node permutation, we can easily determine optimal values for all edge variables.

The second heuristic first fixes the edges of G' by rounding the values of all edge variables. Then the mapping variables x_{ij} are again traversed in descending order according to their value; the node i is mapped to j if and only if this contradicts neither earlier mapping decisions nor the fixing of edges in G'. Even if this strategy only yields a partially defined permutation of V, we can still use its value as a bound on the optimal solution of the ILP.

6.3 Runtime Results

Our branch & cut-algorithm for fuzzy symmetry detection is not fully developed yet. It is necessary and certainly possible to improve it significantly by future work. Nevertheless, we experimented with an implementation based on the polyhedral results obtained so far. In this evaluation, we searched for fuzzy reflections and fuzzy rotations of order n. The penalty for every modification of the graph was set to one. In the reflection case, any fixed node also involved a penalty of one, i.e., one edge modification was allowed to prevent one fixed node.

For all evaluations, we used an AMD 1400 MHz Athlon processor. Runtime results are given in CPU-seconds. Our implementation is based on ABACUS [1] in combination with CPLEX [7]. We used random simple undirected graphs for our experiments. Each pair of different nodes is adjacent by a probability of 1/4. Observe that random graphs are hard instances for fuzzy symmetry detection, since in general a lot of modifications are necessary to make them symmetric. For every number n of nodes, we tested 100 graphs in each evaluation.

Compared with our results for exact symmetry detection presented in [5], runtimes for fuzzy symmetry detection are rather disappointing. The results for reflection detection are displayed in Table 1. We list average and maximal runtime, number of subproblems, and number of LPs. Since we can use our algorithm heuristically by stopping it at any time, we also list the CPU-time needed to find an optimal solution—without knowing its optimality at this point.

Table 1. Results for fuzzy reflection detection

	runtime		opttime		#subprobs		#LPs	
n	avg	max	avg	max	avg	max	avg	max
8	0.04	0.32	0.00	0	4.0	25	13.2	79
9	0.13	1.58	0.01	1	8.8	57	28.9	201
10	0.53	3.15	0.04	2	20.7	95	68.8	317
11	2.01	14.46	0.43	5	48.4	263	160.9	894
12	11.92	75.08	1.86	49	167.5	931	567.8	3185

Table 1 shows that the runtime, the number of subproblems, and the number of LPs increase sharply already for small graphs. However, the time needed to find the optimal solution is much shorter than the total runtime, i.e., we know the optimal solution much earlier than the fact that it is optimal. We conclude that the primal heuristics used in our algorithm work well, whereas the cutting planes and separation algorithms must be improved in order to get a practically useful algorithm. In fact, a lot of violated cutting planes were usually found and added without changing the objective value of the optimal LP-solution.

The results displayed in Table 1 are much more homogeneous than those for exact symmetry detection; see [5]. This is even more evident if we split up the results by the optimal objective function value, i.e., by the minimial number of modifications and fixed nodes; see Table 2 for $n = 12$.

Table 2. Results for fuzzy reflection detection, $n = 12$

obj	% of insts	runtime avg	runtime max	opttime avg	opttime max	#subprobs avg	#subprobs max	#LPs avg	#LPs max
1	6	0.17	0.46	0.00	0	9.0	21	26.2	62
2	17	1.18	5.81	0.59	5	30.5	97	99.4	333
3	30	4.12	10.85	1.03	8	75.0	155	249.0	489
4	33	13.44	24.32	1.33	15	192.6	311	660.5	1031
5	13	40.72	60.22	7.77	49	511.0	791	1729.1	2526
6	1	75.08	75.08	0.00	0	931.0	931	3185.0	3185

Table 3. Results for fuzzy n-rotation detection

n	runtime avg	runtime max	opttime avg	opttime max	#subprobs avg	#subprobs max	#LPs avg	#LPs max
8	0.10	1.56	0.00	0	12.1	135	28.6	299
9	1.67	45.07	0.03	2	87.8	1239	232.6	3775
10	6.93	133.58	0.46	7	165.7	2251	541.4	8044
11	23.34	414.23	2.25	56	568.0	9867	1707.9	29759
12	305.64	15.0 %	35.60	711	2616.1	16785	8692.9	56342

Table 4. Results for fuzzy reflection detection, deleting edges only

n	runtime avg	runtime max	opttime avg	opttime max	#subprobs avg	#subprobs max	#LPs avg	#LPs max
8	0.02	0.14	0.00	0	3.1	15	9.0	38
9	0.05	0.24	0.00	0	5.9	29	17.5	59
10	0.10	0.35	0.00	0	8.6	33	24.0	70
11	0.35	1.53	0.05	1	16.7	53	48.7	156
12	0.84	4.39	0.06	2	28.8	111	85.0	327
13	4.68	20.13	1.23	11	85.5	273	260.3	846
14	8.19	26.94	1.22	8	122.7	371	380.1	1112
15	51.86	161.30	11.95	124	405.5	1149	1280.5	3623
16	79.16	219.41	10.05	107	554.9	1419	1791.2	4521

The results displayed in Table 2 reveal a strong connection between the hardness of an instance and its distance from being reflectional symmetric. This is good news, as the fuzzy symmetry detection algorithm is designed to draw nearly symmetric graphs.

Now consider the case $k = n$. As shown in Table 3, average runtimes for fuzzy rotation detection are even longer—by a factor of more than 25 for $n = 12$. For 15 instances, we could not even find a provably optimal solution within one hour of CPU-time. One reason may be the larger number of mapping variables in the fuzzy rotation ILP compared with the fuzzy reflection ILP.

To investigate the effect of the number of variables more closely, we finally evaluated a variant of the fuzzy reflection detection algorithm: we only allowed to delete edges but not to create them. In the corresponding ILP, we only needed an edge variable for each adjacent node-pair in the original graph. By construction of our test set, we could thus save about 3/4 of the edge variables. Table 4 shows that a smaller number of edge variables can decrease runtime significantly. Compared with the results given in Table 1, we observed an improvement of average runtime by a factor of more than 14 for $n = 12$.

In summary, if used as an exact algorithm, our branch & cut-method only works for small graphs yet. Its performance is much better for nearly symmetric graphs than for random graphs; restricting the set of allowed modifications also helps. We are optimistic that the runtime can be improved significantly by future work. In fact, for exact symmetry detection, we were able to accelerate our branch & cut-algorithm by a factor of up to 50 until now, compared with [4].

References

1. ABACUS – A Branch-And-CUt System. `www.informatik.uni-koeln.de/abacus`.
2. D. Abelson, S. Hong, and D. Taylor. A group-theoretic method for drawing graphs symmetrically. In M. Goodrich and S. Kobourov, editors, *Graph Drawing 2002*, volume 2528 of *LNCS*, pages 86–97. Springer-Verlag, 2002.
3. C. Buchheim. *An Integer Programming Approach to Exact and Fuzzy Symmetry Detection.* PhD thesis, Institut für Informatik, Universität zu Köln, 2003. Available at `kups.ub.uni-koeln.de/volltexte/2003/918`.
4. C. Buchheim and M. Jünger. Detecting symmetries by branch & cut. In P. Mutzel, M. Jünger, and S. Leipert, editors, *Graph Drawing 2001*, volume 2265 of *Lecture Notes in Computer Science*, pages 178–188. Springer-Verlag, 2002.
5. C. Buchheim and M. Jünger. Detecting symmetries by branch & cut. *Mathematical Programming, Series B*, 98:369–384, 2003.
6. H.-L. Chen, H.-I. Lu, and H.-C. Yen. On maximum symmetric subgraphs. In J. Marks, editor, *Graph Drawing 2000*, volume 1984 of *LNCS*, pages 372–383. Springer-Verlag, 2001.
7. CPLEX 7.0. `www.ilog.com/products/cplex`.
8. P. Eades and X. Lin. Spring algorithms and symmetry. *Theoretical Computer Science*, 240(2):379–405, 2000.
9. S. Hong and P. Eades. Drawing planar graphs symmetrically II: Biconnected graphs. Technical Report CS-IVG-2001-01, University of Sydney, 2001.
10. S. Hong and P. Eades. Drawing planar graphs symmetrically III: Oneconnected graphs. Technical Report CS-IVG-2001-02, University of Sydney, 2001.
11. S. Hong and P. Eades. Drawing planar graphs symmetrically IV: Disconnected graphs. Technical Report CS-IVG-2001-03, University of Sydney, 2001.
12. S. Hong, B. McKay, and P. Eades. Symmetric drawings of triconnected planar graphs. In *SODA 2002*, pages 356–365, 2002.
13. J. Manning. Computational complexity of geometric symmetry detection in graphs. In N. Sherwani, E. de Doncker, and J. Kapenga, editors, *First Great Lakes Computer Science Conference*, volume 507 of *LNCS*, pages 1–7. Springer-Verlag, 1991.
14. H. Purchase. Which aesthetic has the greatest effect on human understanding? In G. Di Battista, editor, *Graph Drawing '97*, volume 1353 of *LNCS*, pages 248–261. Springer-Verlag, 1998.

Barycentric Drawings of Periodic Graphs

Olaf Delgado-Friedrichs

University of Tübingen, WSI Computer Science, Sand 14, 72076 Tübingen, Germany

Abstract. We study barycentric placement of vertices in periodic graphs of dimension 2 or higher. Barycentric placements exist for every connected periodic graph, are unique up to affine transformations, and provide a versatile tool not only in drawing, but also in computation. Example applications include symmetric convex drawing in dimension 2 as well as determining topological types of crystals and computing their ideal symmetry groups.

1 Introduction

Periodic graphs provide convenient models for studying the topology of the interatomic structure of crystal species. Usually, atoms are represented by vertices, while bonds are represented by edges. Some crystals, however, are made from larger building blocks using linkers, in which case it is often more useful to model this meta-structure [OEL+00].

Apart from their relevance for chemistry, periodic graphs occur as the 1-skeleta of periodic tilings [DDH+99] and have applications in VLSI circuit design and integer programming [Orl84,CM91].

Although periodic graphs may be described in a finite way, algorithms for finite graphs do not usually carry over automatically. For example, precise practical methods for determining isomorphism of periodic graphs have been missing up until now, although much searched for by chemists and crystallographers [Kle87].

Several authors [OB92,TRR+97,DDH+99], have constructed many examples of potential crystal structures by means of combinatorially enumerating periodic graphs. In addition to the isomorphism problem, this gives rise to a demand for chemically reasonable 3-dimensional drawing algorithms.

Barycentric placements have been introduced by Tutte [Tut60,Tut63] for constructing convex drawings of "almost" 3-connected planar graphs. Here, an outer face is chosen and realized as an arbitrary convex polygon. Tutte showed that for any such choice a barycentric placement of the remaining vertices uniquely exists. If, moreover, the only pairs of vertices at which the graph can be cut are situated on different sides of the outer polygon, it induces an embedding with convex inner faces and the maximal possible symmetry which is compatible with the outer polygon.

We obtain a general existence and uniqueness result for connected, periodic graphs in general and a convexity result similar to Tutte's for 3-connected, 2-periodic, planar graphs. It should be noted that barycentric drawings have been

G. Liotta (Ed.): GD 2003, LNCS 2912, pp. 178–189, 2004.

employed by crystallographers for quite a while, but the author is not aware of any previous rigorous justification of that practice.

We show further that for a class of graphs which we call *stable*, a number of algorithmic problems can be solved in polynomial time using exact barycentric drawings as a tool.

In [Del01], barycentric drawings for periodic graphs are treated mathematically, with only a brief glance at computational issues. Here, we skip the more intricate mathematical details, while concentrating on the algorithmic and graphical aspects. The next section introduces periodic graphs. Section 3 defines barycentric drawings and gives some basic results. The two-dimensional case is treated in Section 4, and an algorithm for finding canonical forms and symmetries in Section 5. Section 6, finally, discusses some applications.

2 Periodic Graphs

A d-periodic graph is a simple undirected graph on which the group $\mathbb{Z}^d$ of d-dimensional integer vectors acts by automorphisms. The action is to be free, i.e., no vector may map any vertex onto itself, and there may be only finitely many orbits on the set of vertices as well as on the set of edges. Consequently, the set of vertices must be countable and the graph must be locally finite, i.e., only finitely many edges may be incident to any given vertex.

We interpret the result of applying an integer vector to a vertex or edge as a formal translation and write $v + s$ or $e + s$, respectively, where $s = (s_1, \ldots, s_d)$, and $v + (s + t) = (v + s) + t$ and so on.

An isomorphism between periodic graphs $G = (V, E)$ and $G' = (V', E')$ is an ordinary graph isomorphism, say f, which respects the $\mathbb{Z}^d$ action. This means that for each vector $s \in \mathbb{Z}^d$, there is another vector, s', such that $f(v + s) = f(v) + s'$ for any vertex $v \in V$. It is easy to see that the function $f^*: \mathbb{Z}^d \to \mathbb{Z}^d$ which maps every vector s to its corresponding s' must be linear and bijective, thus it can be represented by a $d \times d$ unimodular matrix (a matrix with integer entries and determinant 1 or -1).

A natural way to encode periodic graphs is the so-called vector or (labelled) orbit graph method [CHK84]: let n be the number of vertex orbits and m the number of edge orbits under the $\mathbb{Z}^d$ action. Choose arbitrary representatives $v_1, \ldots, v_n$ of the vertex orbits. Now each vertex $v \in V$ can be written uniquely as $v_i + s$ for some $i \in \{1, \ldots, n\}$ and some $s \in \mathbb{Z}^d$. For every edge orbit, there is a unique representative $e = (v_i, v_j + t)$, where either $i < j$ or else $i = j$ and the first nonzero entry of t is positive. Specifying all these edge representatives completely characterizes a periodic graph. Consequently, a periodic graph of dimension d with m edge orbits is given as an array of $m \times (d + 2)$ numbers, where the first two entries in each row are indices of vertex representatives, while the remaining d entries form a vector, which we call the shift vector for this edge, and by which the second vertex is to be shifted to obtain the second end of the edge.

Here are a few tiny examples:

$$\begin{matrix} 1\,1 & 1\,0 \\ 1\,1 & 0\,1 \end{matrix} \quad \text{and} \quad \begin{matrix} 1\,2 & 0\,0 \\ 1\,2 & 1\,0 \\ 1\,2 & 0\,1 \end{matrix}$$

encode the 1-skeleton of a face-to-face tiling by regular squares and the 1-skeleton of a face-to-face tiling by regular hexagons, respectively.

Their 3-dimensional analogues are

$$\begin{matrix} 1\,1 & 1\,0\,0 \\ 1\,1 & 0\,1\,0 \\ 1\,1 & 0\,0\,1 \end{matrix} \quad \text{and} \quad \begin{matrix} 1\,2 & 0\,0\,0 \\ 1\,2 & 1\,0\,0 \\ 1\,2 & 0\,1\,0 \\ 1\,2 & 0\,0\,1, \end{matrix}$$

the 1-skeleton of a face-to-face tiling by regular cubes and the atom-bond network of diamond.

A vector representation is naturally interpreted as a finite directed multigraph with edges labelled by shift vectors, where vertices, edges and their incidences represent vertex and edge orbits and their incidences in the original graph, hence the alternative term orbit graph.

For each periodic graph, there are infinitely many choices of vertex representatives. There are, moreover, infinitely many ways of choosing a basis for $\mathbb{Z}^d$, a choice which determines the actual coordinates used in specifying the shift vectors. An obvious improvement is to require that the set of representatives $v_1, \ldots, v_n$, or rather the subgraph induced by them, be connected, if possible. This leaves, up to translations, only finitely many choices for the set of vertex representatives. It is easy to see that a periodic graph is connected if and only if it has a connected set of vertex representatives and for any such set the induced set of shift vectors spans $\mathbb{Z}^d$ [CM91]. It should be possible to formulate additional conditions on the set of shift vectors in order to arrive at a finite set of permissible representations. Note, however, that the set of shift vectors need not contain an integral basis for $\mathbb{Z}^d$. Even in the 1-dimensional case, the "vectors" 2 and 3 form a counterexample.

In the following, we will assume that all periodic graphs under consideration are connected.

3 Barycentric Drawings and Isomorphisms

It is reasonable to require that a drawing of a periodic graph shall display the periodicity. In the following, we will restrict our attention to straight line drawings. Consequently, a drawing is determined by a vertex placement $p: \{v_1, \ldots, v_n\} \to \mathbb{R}^d$ and a linear, nonsingular function $p^*: \mathbb{Z}^d \to \mathbb{R}^d$, where the extension of p to V is given by $p(v_i + s) := p(v_i) + p^*(s)$.

A barycentric drawing is one which satisfies

$$p(v) = \frac{1}{|N(v)|} \sum_{w \in N(v)} p(w),$$

where $N(v)$ denotes the set of all vertices adjacent to v. This definition, quite obviously, is invariant under affine transformations, i.e., if p^* is changed or the whole graph is shifted by some vector, a barycentric drawing remains barycentric.

If p^* and the position of an arbitrary vertex are preassigned, say $p^* = \mathrm{Id}_d$ and $p(v_1) = \mathbf{0}$, there is a unique barycentric drawing satisfying these conditions. Consequently, barycentric drawings always exist for connected, periodic graphs and are unique up to affine transformations.

To see why, consider the "energy function"

$$E(p) := \sum_{vw \in E_0} d(p(v), p(w))^2,$$

where d denotes Euclidean distance and E_0 is a set of representatives for the edge orbits. It is straightforward to see that p is barycentric if and only if p, taken as a point in $\mathbb{R}^{nd}$, is critical for E, i.e., all partial derivatives vanish. Now consider

$$||p||^2 = |p(v_1)|^2 + \ldots + |p(v_n)|^2.$$

Because (V, E) is connected and $p(v_1) = \mathbf{0}$, the maximum edge length can be made arbitrarily large by just taking any p with $||p||^2$ large enough. Now E is a quadratic function, which by the above is non-degenerate in each direction, thus must have a unique critical point, a minimum (c.f. [RG96]).

This does not only show the existence and uniqueness of barycentric drawings, but also points to an aesthetically interesting property, namely that barycentric drawings minimize the sum of squared edge lengths among all periodic drawings with a given p^*.

In the following, we will always use p to denote the barycentric vertex placement with $p^* = \mathrm{Id}_d$ and $p(v_1) = \mathbf{0}$.

We call a graph stable if p is injective, thus vertices can be identified uniquely by their positions. This is a key to lowering the complexity of several computational problems dealing with periodic graphs. Where traversal algorithms are employed, it is usually enough, though, to require local stability, meaning that p need only be injective on $N(v)$ for each vertex v.

For display purposes, p can be conveniently approximated by iterative, force directed methods (c.f [Ead84,FR91]), where all forces are attractive, only exist between neighbors and are proportional in size to the respective edge lengths. In order to identify vertices by their positions, however, the linear system of equations induced by the barycentricity condition has to be solved exactly, i.e., using arbitrary precision rational arithmetic. This is necessary because in principle, vertices positions can get arbitrarily close without being identical.

Next, we claim that an isomorphism between periodic graphs induces an affine mapping between their barycentric drawings. To see this, let f be an isomorphism between graphs (V, E) and (V', E') with barycentric placements p and p', respectively. Define $g^*(s) := p'^*(f^*(p^{*-1}(s)))$ and consider new placements q_1 and q_2 on V, where $q_1(v) := p'(f(v))$ and $q_2(v) = g^*(p(v)) + p'(f(v_1)) - g^*(p(v_1))$. It is then an easy exercise to see that $q_1(v_1) = q_2(v_1) = p'(f(v_1))$ and $q_1(v+s) - q_1(v) = q_2(v+s) - q_2(v) = g^*(p^*(s))$ for any $s \in \mathbb{Z}^d$. Because f is an isomorphism and all the $\cdot^*$ functions are linear, both q_1 and q_2 are barycentric drawings, so they must be identical.

We state this as our first theorem:

Theorem 1. *If f is an isomorphism between connected periodic graphs (V, E) and (V', E') with barycentric vertex placements p and p', respectively, then there is an affine function φ such that $p'(f(v)) = \varphi(p(v))$ for all $v \in V$.*

Theorem 1 has a nice consequence regarding automorphisms of periodic graphs:

Theorem 2. *If (V, E) is a connected periodic graph, there is a linear function $p^*: \mathbb{Z}^d \to \mathbb{Z}^d$ such that for any barycentric placement satisfying $p(v+s) = p(v) + p^*(s)$ for all $v \in V$ and all $s \in \mathbb{Z}^d$ and any automorphism f of (V, E), there is an isometry φ of $\mathbb{R}^d$ with respect to the Euclidean metric, such that $p(f(v)) = \varphi(p(v))$ for all $v \in V$.*

What this means, essentially, is that the full automorphism group of a periodic graph can be realized as a crystallographic space group. There can be only finitely many conjugacy classes of automorphisms with respect to the $\mathbb{Z}^d$ action. Turning a finite set of affine maps into isometries by redefining the metric, or equivalently, the basis with respect to which they act, is a standard trick in mathematical crystallography, see for example [Sch80].

For stable periodic graphs, every nontrivial automorphism must move some vertex, so the mapping from automorphisms to isometries is injective. Consequently:

Corollary 1. *The automorphism group of any stable, connected, periodic graph is isomorphic to a crystallographic space group.*

For unstable graphs, some nontrivial automorphism may be mapped to the identity. As an example, see Figure 1, where any barycentric placement must map pairs of adjacent white vertices to the same position, so that the automorphism that just flips each such pair induces the identity on $\mathbb{R}^d$.

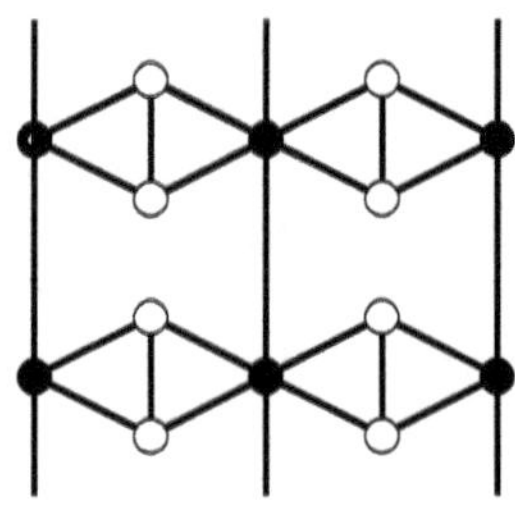

Fig. 1. An unstable 2-periodic graph.

4 Planar Graphs

Tutte's classical results on convex drawings of finite planar graphs [Tut60,Tut63] carry over to planar 2-periodic graphs. In particular, we have the following:

Theorem 3. *For a connected, 2-periodic, planar graph G, the following are equivalent:*

1. *G is 3-connected.*
2. *G has a convex embedding in the plane.*
3. *G has a convex embedding in the plane such that every automorphism of G is realized by a crystallographic symmetry of the embedding.*

Here, as usual, an embedding of a graph is a drawing in which the vertices and interiors of edges are all pairwise disjoint. Whenever the context is clear, we do not distinguish in terminology between vertices or edges and their images in a drawing.

A similar convexity criterion has been given by Thomassen [Tho80] for infinite planar graphs in general, but does not imply anything with respect to symmetries. Obviously, (3) implies (2), and, employing an easy lemma stating that any finite vertex cut in a periodic graph has exactly one infinite component, it is straightforward to see that (2) implies (1).

The implication (1) $\rightarrow$ (3) follows from a result in [Del01] (Theorem 20), namely:

Theorem 4. *The straight line drawing induced by any good, periodic vertex placement of a 2-periodic, 3-connected, planar graph is a convex embedding.*

Here, a vertex placement is called good if every vertex v is placed in the relative interior (c.f. [RG96]) of its neighbor's positions, defined as

$$\Big\{ \sum_{w \in N(v)} \lambda_w p(w) \,\Big|\, \sum_{w \in N(v)} \lambda_w = 1;\, \lambda_w > 0 \Big\}.$$

The proof, which we are going to sketch here briefly, was largely inspired by Richter-Gebert's treatment [RG96] of the classical Tutte Theorem.

The first step is to show that the neighbors of a given vertex v can never be placed on a common straight line h. For, assume otherwise and let, without loss of generality, h be horizontal. Consider the largest connected subgraph S of G containing v with all vertices placed on h. Then S, by periodicity, can not be the whole of G. Because of the 3-connectedness, and using Menger's theorem, there must be at least 3 vertices a, b and c in S with neighbors not on h. Because the placement is good, each of these has neighbors both above and below h. Using the periodicity again, one can show that those above h are pairwise connected by paths staying above h. Likewise, those below h are connected below h. But then we have 3 independent ways of interconnecting vertices a, b and c, one on, one above and one below h, which establishes a $K_{3,3}$ minor in G.

Consequently, in a good placement, every vertex falls inside the convex hull of its neighbors, which must be non-degenerate. The next step is to prove the theorem for triangulations without degenerate triangles. By summing up angles in two different ways, first around vertices and then inside triangles, and using an Euler characteristic argument, one shows that the triangles are all oriented consistently and that angles sum up to 2π around each vertex. The simple-connectedness of the plane then implies that there can be no overlaps between triangles.

Finally, the theorem is extended to the general case by first triangulating each facial cycle of the given graph without introducing new vertices. This can be done in such a way that the $\mathbb{Z}^d$ images of a facial cycle are triangulated consistently, making the resulting triangulated graph G' periodic again. If p is a good placement for G, it is also a good placement for G'. A perturbation argument is used to show that no degenerate triangles can occur. Consequently, p must induce a straight-line embedding of G' and thus of G. But because p is good, all interior facial angles occurring in the induced embedding of G are less than π, so all the faces are convex.

A nice feature of Theorem 4 is that it allows for some flexibility in the construction of drawings. Tutte's barycentric construction results in aesthetically pleasing pictures, but is also well-known for allowing a rather large variation in edge lengths, making some graph structures hard to recognize. This can be remedied to some extend by using a force-directed drawing method with the barycentric placement as its initial value. Such methods can be adjusted to support a more uniform distribution of edge lengths, for example by using attractive forces proportional to some power larger than 1 of the edge lengths. As long as the process is symmetric and carefully tuned as to keep the result of each iteration good, the final drawing will still be convex and display the full symmetry. The author has first employed this idea to polyhedra graphs, for which an analogue of Theorem 4 holds [RG96]. Figure 2 shows an example.

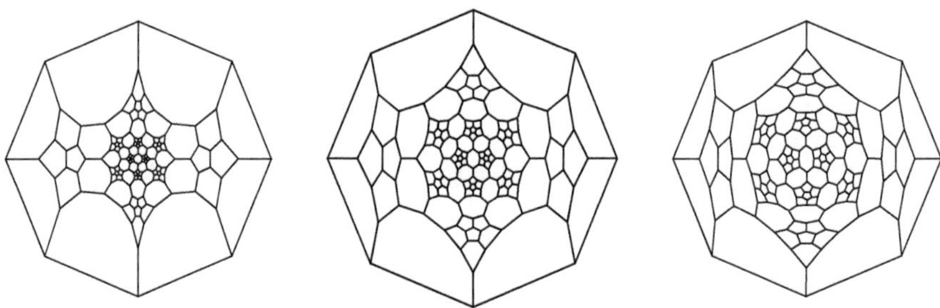

Fig. 2. Convex drawings of a large polyhedron graph. Left: barycentric drawing. Middle and right: variants with attractive forces proportional to the square and fourth power of the edge length.

A more general approach of relaxing the barycentric drawing would define an appropriate penalty function and use optimization methods such as simulated annealing or downhill simplex to search for a minimum. In Figure 3 we show three drawings of the same periodic graph, all derived from variations and relaxations of the barycenter construction.

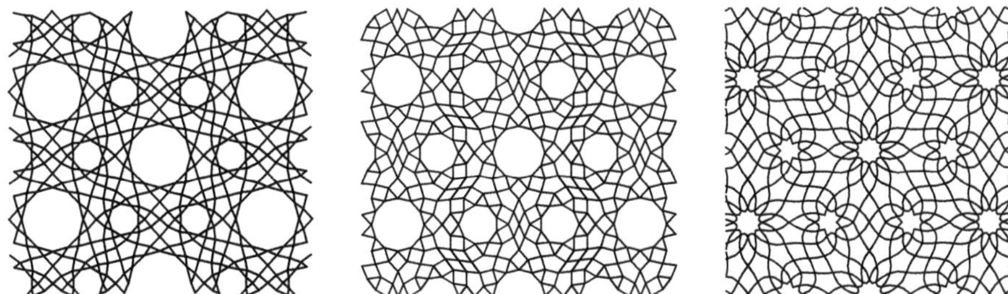

Fig. 3. Three drawings of a periodic planar graph, inspired by a Moorish pattern. Left: barycentric drawing. Middle: relaxed barycentric drawing with equal edge lengths. Right: vertices on mirror planes were spaced evenly, then those inside the fundamental triangle were placed using a barycentric drawing of the second derived subdivision.

5 Symmetries and Canonical Forms

Isomorphism of two structures can be tested in linear time when these are given in a canonical form, i.e., a representation that uniquely depends on the isomorphism type. Therefore, an efficient canonical forms algorithm is highly desirable.

A vector representation of a periodic graph as in Section 2 can be converted into a unique linear string of numbers by first sorting the rows lexicographically and then concatenating them into a single row. Our approach towards canonical forms is to efficiently construct a small characteristic set of vector representations and pick a minimal one with respect to the lexicographic order of their corresponding strings.

A straightforward way to obtain a finite set is by picking vertex representatives as they occur in, say, a breadth first traversal of the graph. To make the resulting set characteristic, every possible order of picking vertices has to be executed, which may easily lead to an exponentially large set of representations.

We use barycentric placements to arrive at a smaller set of traversals. Let p be a barycentric placement and recall that a graph is called locally stable if for each vertex v the function p is injective on the set $N(v)$ of v's neighbors. Then a traversal must pick these in lexicographical order of their coordinates as given by p, with respect to some appropriate basis. Because p is unique up to affine transformations, there is a unique breadth first traversal of this type for every choice of start vertex and coordinate basis.

It remains to pick a characteristic set of pairs, each consisting of a vertex and a basis. The term characteristic means that these must depend only on the isomorphism type of G, not on its original representation. A natural choice is an ordered d-tuple of edge vectors. There are $2m$ of these in a graph with m edge orbits, thus $(2m)^d$ potential bases, which can be readily determined from a given vector representation. Of course, only linearly independent tuples can be used, but by the d-periodicity and connectedness of G, at least one such exists. As a start vertex for the traversal, the first vertex of the first directed edge is used.

If during the traversal, a vertex is encountered which belongs to a new orbit, it is made the representative of this orbit and an edge with trivial shift vector is added to the new representation. If the orbit has already been seen, an edge with a non-trivial shift vector is added. These shift vectors are expressed in the basis used for the traversal, which is determined by the original tuple of edge vectors and in general is not an integral basis of the translation lattice. Thus, in the end, a basis for the integral span of the shift vectors must be computed in a deterministic way and the shift vectors expressed in terms of that basis. The complete algorithm is shown as Algorithm 5 below.

It is straightforward to extend the algorithm to computing the automorphisms as well. Indeed, whenever two d-tuples of directed edges lead to the same representation, an automorphism of the periodic graph G has been found and all automorphisms must occur in this way.

The runtime analysis of the Algorithm is complicated by the fact that exact rational numbers have to be used in all the calculations involving vertex positions. These arise from the solution of a linear system of equations $AX = B$, where A is an $n \times n$ integer matrix, B is an $n \times d$ integer matrix and X is an $n \times d$ rational matrix. Here, the i-row of A corresponds, of course, to the barycenter condition for the i-th vertex representative. It is not hard to show that the logarithm of the numerators and denominators of entries in X is $O(n \log n)$ if the entries in A and B are assumed to be constant in size, say, representable by machine words. In practice, we may then also assume that n is representable by a machine word, so the above boils down to $O(n)$. It turns out that the maximal representation size of any rational number occurring during the execution of the algorithm rises to $O(dn)$.

Ignoring the complexity of arithmetic for the moment, the main loop of Algorithm 5 runs in time $O(dm^{d+1} \log D)$, where m, once again, is the number of edge representatives and D is the maximal degree of a vertex. Clearly, the main loop is executed $(2m)^d$ times and the time spent per iteration is dominated by sorting the transformed shift vectors at each step of the traversal and accumulates to $O(dm \log D)$. Everything else, including the preparations before the execution of the main loop, takes time $O(dm)$.

It remains to consider the final steps after the main loop. In order to find a basis for the integral span of the shift vectors, we need to triangulate the matrix formed by these using column operations which only add integer multiples of columns to other columns. The problem here is that it may take several steps to eliminate a single entry, which may lead to an overall doubly exponential growth in some of the other entries. Clearly, by multiplying with the least common denominator of all entries, we can reduce the original problem to that of

Algorithm 5. Canonical Form.
Input

- A list E of edge representatives of the form (v, w, s) with $v, w \in \mathbb{N}$ and $s \in \mathbb{Z}^{d\times 1}$ (a d-dimensional column vector), representing a connected, locally stable, d-periodic graph G.
- A barycentric vertex placement p for G.

Output

- A list C of edge representatives, constituting a canonical form for G.

Set $A = [\,]$.
<u>**for**</u> (v, w, s) in E:
  Append $(v, w, (p_w + s) - p_v)$ and $(w, v, p_v - (p_w + s))$ to A.
<u>**for**</u> $v = 1$ to n:
  Let $S_v = [\,(w, s) \mid (v, w, s) \in A\,]$.

<u>**for**</u> each ordered d-tuple $((v_1, w_1, s_1), \ldots, (v_d, w_d, s_d))$ from A:
  Let $v = v_1$.
  Let B be the $d \times d$ matrix with columns $s_1, \ldots, s_d$.
  <u>**if**</u> B is singular:
    Skip to the next loop iteration.
  Set up an empty FIFO queue Q and enter v.
  Set $I_v = 1$, $q_v = \mathbf{0}$ and $k = 2$.
  Set $N = [\,]$.

  <u>**while**</u> Q is not empty:
    Pop v from Q.
    Set $T = [\,(w, B^{-1}s) \mid (w, s) \in S_v\,]$ and sort by second components.
    <u>**for**</u> (w, s) in T:
      <u>**if**</u> I_w is undefined:
        Set $I_w = k$, $q_w = q_v + s$ and increase k.
        Enter w into Q.
      Set $t = q_v + s - q_w$.
      <u>**if**</u> $I_v < I_w$ or ($I_v \equiv I_w$ and $t > \mathbf{0}$):
        Append (I_v, I_w, t) to N.

  <u>**if**</u> C is undefined or $N < C$:
    Set $C = N$.

Let B be a deterministic basis for the integral span of $[\,s \mid (v, w, s) \in C\,]$.
<u>**for**</u> (v, w, s) in C:
  Set $t = B^{-1}s$.
  <u>**if**</u> $v \equiv w$ and $t < \mathbf{0}$:
    Set $t = -t$.
  Replace (v, w, s) by (v, w, t) in C.
Sort and return C.

triangulating a matrix over the integers. Bachem and Kannan [BK79] were the first to present an polynomial time algorithm for this latter problem.

We conclude that, even after taking into account the extra cost involved in using precise rational number arithmetic, canonical forms for stable, connected, periodic graphs can be computed in polynomial time (assuming the dimension is taken to be constant) using barycentric vertex placement, and that two such graphs, given by their canonical forms, can be tested for isomorphism in time $O(md)$.

6 Applications

The methods discussed here have been implemented in the programming language Python. The program, called *Systre*, is available from the author upon request. It contains a library for dealing with crystallographic groups in Dimension 3, following the conventions of the so-called International Tables [Hah83]. Because crystal structures often display fewer translational symmetries than possible, an algorithm for finding additional potential translations has been incorporated as well. These can be recognized immediately as translational symmetries of the barycentric drawing. Systre computes both the ideal symmetry group and

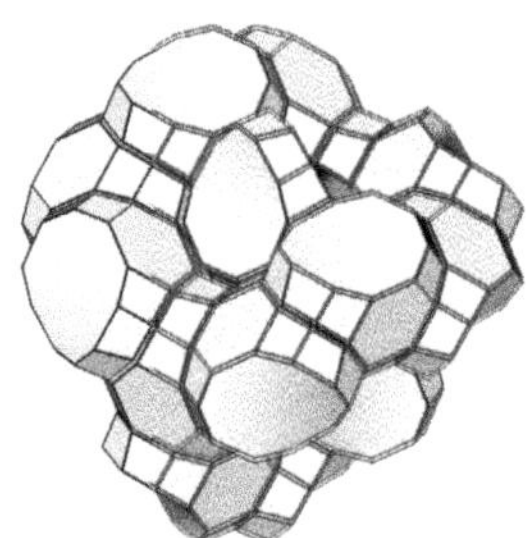

Fig. 4. A periodic tiling of 3-dimensional space with some non-planar faces.

canonical form of a given periodic graph. The latter can be used as a database index. Graphs are specified by vector representations or, alternatively, by atom positions and unit cell parameters. Systre then tries to deduce the bonds. A relaxed drawing, aiming for uniform edge lengths and a large total volume can be produced, using the barycentric drawing as a starting point.

In Section 4 we discussed barycentric placements for periodic, planar graphs and thus periodic tilings of the plane. Tilings of 3-dimensional space can be drawn similarly: first the 1-skeleton is embedded, using barycentric placement and force-models. Non-planar faces are then realized by first triangulating them, then barycentrically placing the inner vertices. This leads to nice, smooth-looking curved surfaces with the full symmetry of their respective boundaries. Figure 4 shows an example.

References

[BK79] A. Bachem and R. Kannan. Polynomial algorithms for computing the Smith and Hermite normal forms of an integer matrix. *SIAM Journal Computing*, 8:499–507, 1979.

[CHK84] S. J. Chung, Th. Hahn, and W.E. Klee. Nomenclature and generation of three-periodic nets: the vector method. *Acta Cryst.*, A40:42–50, 1984.

[CM91] Edith Cohen and Nimrod Megiddo. Recognizing properties of periodic graphs. In *Applied geometry and discrete mathematics*, volume 4 of *DIMACS Ser. Discrete Math. Theoret. Comput. Sci.*, pages 135–146. Amer. Math. Soc., Providence, RI, 1991.

[DDH+99] O. Delgado Friedrichs, A.W.M. Dress, D.H. Huson, J. Klinowski, and A.L. Mackay. Systematic enumeration of crystalline networks. *Nature*, 400:644–647, 1999.

[Del01] O. Delgado-Friedrichs. Equilibrium placement of periodic graphs and tilings. submitted, 2001.

[Ead84] P. Eades. A heuristic for graph drawing. *Congressus Numerantium*, 42:149–160, 1984.

[FR91] T. Fruchterman and E. Reingold. Graph drawing by force-directed placement. *Software—Practice and Experience*, 21(11):1129–1164, 1991.

[Hah83] T. Hahn, editor. *International Tables for Crystallography*, volume A. D. Reidel Publishing Company, Dordrecht, Boston, 1983.

[Kle87] W.E. Klee. The topology of crystal structures. *Z. Kristallogr.*, 179:67–76, 1987.

[OB92] M. O'Keeffe and N. E. Brese. Uninodal 4-connected 3d nets. I. Nets without 3- or 4-rings. *Acta Cryst.*, A48:663–669, 1992.

[OEL+00] M. O'Keeffe, M. Eddaoudi, Hailian Li, T. Reineke, and O. M. Yaghi. Frameworks for extended solids: Geometrical design principles. *J. Solid State Chem.*, 152(1):3–20, 2000.

[Orl84] James B. Orlin. Some problems on dynamic/periodic graphs. In *Progress in combinatorial optimization (Waterloo, Ont., 1982)*, pages 273–293. Academic Press, Toronto, ON, 1984.

[RG96] Jürgen Richter-Gebert. *Realization Spaces of Polytopes*. Springer Verlag, Berlin, 1996.

[Sch80] Rolf L. E. Schwarzenberger. *n-dimensional crystallography*, volume 41 of *Research Notes in Mathematics*. Pitman (Advanced Publishing Program), Boston, Mass., 1980.

[Tho80] C. Thomassen. Planarity and duality of finite and infinite graphs. *Journal of Combinatorial Theory*, Series B, 29:244–271, 1980.

[TRR+97] M.M.J. Treacy, K.H. Randall, S. Rao, J.A. Perry, and D.J. Chadi. Enumeration of periodic tetrahedral frameworks. *Z. Krist.*, 212:768–791, 1997.

[Tut60] W.T. Tutte. Convex representations of graphs. *Proc. London Math. Soc.* (3), 10:304–320, 1960.

[Tut63] W.T. Tutte. How to draw a graph. *Proc. London Math. Soc.*, 13:743–767, 1963.

Three-Dimensional Grid Drawings with Sub-quadratic Volume*

Vida Dujmović[1] and David R. Wood[2]

[1] School of Computer Science, McGill University, Montréal, Canada
vida@cs.mcgill.ca

[2] School of Computer Science, Carleton University, Ottawa, Canada
davidw@scs.carleton.ca

Abstract. A three-dimensional grid drawing of a graph is a placement of the vertices at distinct points with integer coordinates, such that the straight line-segments representing the edges are pairwise non-crossing. A $\mathcal{O}(n^{3/2})$ volume bound is proved for three-dimensional grid drawings of graphs with bounded degree, graphs with bounded genus, and graphs with no bounded complete graph as a minor. The previous best bound for these graph families was $\mathcal{O}(n^2)$. These results (partially) solve open problems due to Pach, Thiele, and Tóth [*Graph Drawing* 1997] and Felsner, Liotta, and Wismath [*Graph Drawing* 2001].

1 Introduction

A *three-dimensional straight-line grid drawing* of a graph, henceforth called a *3D drawing*, is a placement of the vertices at distinct points in $\mathbb{Z}^3$ (called *gridpoints*), such that the line-segments representing the edges are pairwise non-crossing. That is, distinct edges only intersect at common endpoints, and each edge only intersects a vertex that is an endpoint of that edge. In contrast to the case in the plane, it is well known that every graph has a 3D drawing. We are therefore interested in optimising certain measures of the aesthetic quality of such drawings.

The *bounding box* of a 3D drawing is the minimum axis-aligned box containing the drawing. If the bounding box has side lengths $X-1$, $Y-1$ and $Z-1$, then we speak of an $X \times Y \times Z$ drawing with *volume* $X \cdot Y \cdot Z$. That is, the volume of a 3D drawing is the number of gridpoints in the bounding box. This definition is formulated so that 2D drawings have positive volume. We are interested in 3D drawings with small volume, which is a widely studied problem [3,4,5,6,7,8,10, 11,12,14,18,24,25,27]. Three-dimensional graph drawings in which the vertices are allowed real coordinates have also been studied (see the references in [11]). The authors have also established bounds on the volume of three-dimensional *polyline* grid drawings, where bends in the edges are also at gridpoints [11]. Table 1 summarises the best known upper bounds on the volume of 3D drawings, including those established in this paper.

* Research supported by NSERC and FCAR.

G. Liotta (Ed.): GD 2003, LNCS 2912, pp. 190–201, 2004.

Table 1. Upper bounds on the volume of 3D drawings of n-vertex m-edge graphs.

graph family	volume	reference
arbitrary	$\mathcal{O}(n^3)$	Cohen *et al.* [5]
arbitrary	$\mathcal{O}(m^{4/3}n)$	Theorem 4
maximum degree Δ	$\mathcal{O}(\Delta mn)$	Theorem 3
constant maximum degree	$\mathcal{O}(n^{3/2})$	Theorem 10
constant chromatic number	$\mathcal{O}(n^2)$	Pach *et al.* [24]
constant chromatic number	$\mathcal{O}(m^{2/3}n)$	Theorem 6
no K_h-minor (h constant)	$\mathcal{O}(n^{3/2})$	Theorem 9
constant genus	$\mathcal{O}(n^{3/2})$	Theorem 8
constant tree-width	$\mathcal{O}(n)$	Dujmović and Wood [12]

Cohen *et al.* [5] proved that every graph has a 3D drawing with $\mathcal{O}(n^3)$ volume, and that this bound is asymptotically optimal for complete graphs K_n. Our edge-sensitive bounds of $\mathcal{O}(m^{4/3}n)$ and $\mathcal{O}(\Delta mn)$ are greater than $\mathcal{O}(n^3)$ in the worst case. It is an open problem whether there are edge-sensitive bounds that match the $\mathcal{O}(n^3)$ bound in the case of complete graphs.

Pach *et al.* [24] proved that graphs with constant chromatic number have 3D drawings with $\mathcal{O}(n^2)$ volume. For c-colourable graphs the actual bound is $\mathcal{O}(c^2n^2)$. Our edge-sensitive bound of $\mathcal{O}(m^{2/3}n)$ is an improvement on this result for graphs with constant chromatic number and $o\big(n^{3/2}\big)$ edges. Pach *et al.* [24] also proved an $\Omega(n^2)$ lower bound for the volume of 3D drawings of complete bipartite graphs $K_{n,n}$. This lower bound was generalised to all graphs by Bose *et al.* [3], who proved that every 3D drawing has volume at least $\frac{1}{8}(n+m)$.

Graphs with constant maximum degree have constant chromatic number, and by the result of Pach *et al.* [24], have 3D drawings with $\mathcal{O}(n^2)$ volume. Pach *et al.* [24] conjectured that graphs with constant maximum degree have 3D drawings with $o\big(n^2\big)$ volume. We verify this conjecture by proving that such graphs have 3D drawings with $\mathcal{O}(n^{3/2})$ volume.

The first $\mathcal{O}(n)$ upper bound on the volume of 3D drawings was established by Felsner *et al.* [14] for outerplanar graphs. This result was generalised by the authors for graphs with constant tree-width [12]. Improved constants have been obtained in a number of special cases [6,7,8,18]. Felsner *et al.* [14] proposed the following inviting open problem: does every planar graph have a 3D drawing with $\mathcal{O}(n)$ volume? In this paper we provide a partial solution to this question, by proving that planar graphs have 3D drawings with $\mathcal{O}(n^{3/2})$ volume. Note that $\mathcal{O}(n^2)$ is the optimal area for plane 2D grid drawings, and $\mathcal{O}(n^2)$ was the previous best upper bound on the volume of 3D drawings of planar graphs.

A graph H is a *minor* of a graph G if H is isomorphic to a graph obtained from a subgraph of G by contracting edges. The *genus* of a graph G is the minimum γ such that G can be embedded in the orientable surface with γ handles. Of course, planar graphs have genus 0 and no K_5-minor. A generalisation of our result for planar graphs is that every graph with constant genus or with no K_h-minor for constant h has a 3D drawing with $\mathcal{O}(n^{3/2})$ volume.

2 Track Layouts

We consider undirected, finite, and simple graphs G with vertex set $V(G)$ and edge set $E(G)$. The number of vertices and edges of G are respectively denoted by $n = |V(G)|$ and $m = |E(G)|$. A *vertex c-colouring* of G is a partition $\{V_i : 1 \leq i \leq c\}$ of $V(G)$, such that for every edge $vw \in E(G)$, if $v \in V_i$ and $w \in V_j$ then $i \neq j$. Each $i \in \{1, 2, \ldots, c\}$ is a *colour*, and each set V_i is a *colour class*. At times it will be convenient to write $\mathsf{col}(v) = i$ rather than $v \in V_i$. If G has a vertex c-colouring then G is *c-colourable*. The *chromatic number* of G, denoted by $\chi(G)$, is the minimum c such that G is c-colourable.

Let $\{V_i : 1 \leq i \leq c\}$ be a vertex c-colouring of a graph G. Let $<_i$ be a total order on each colour class V_i. Then each pair $(V_i, <_i)$ is a *track*, and $\{(V_i, <_i) : 1 \leq i \leq t\}$ is a *t-track assignment* of G. To ease the notation we denote track assignments by $\{V_i : 1 \leq i \leq c\}$ when the ordering on each colour class is implicit. An *X-crossing* in a track assignment consists of two edges vw and xy such that $v <_i x$ and $y <_j w$, for distinct colours i and j. A *t-track layout* of G consists of a t-track assignment of G with no X-crossing. The *track-number* of G, denoted by $\mathsf{tn}(G)$, is the minimum t such that G has a t-track layout[1]

Track layouts and track-number were introduced in [10,12] although they are implicit in many previous works [14,19]. Track layouts and 3D drawings are closely related, as illustrated by the following result by Dujmović *et al.* [10]. Also note that other authors have used track layouts to produce 3D drawings [6,8, 18], and there is a tight relationship between track layouts and another type of graph layout called a *queue layout* [27], which is a dual structure to a book embedding introduced by Heath *et al.* [19].

Theorem 1. [10] *Every n-vertex graph G with track-number* $\mathsf{tn}(G) \leq t$ *has a* $2t \times 4t \times 4t\lceil \frac{n}{t} \rceil$ *drawing, which has* $\mathcal{O}(t^2 n)$ *volume. Conversely, if a graph G has an* $X \times Y \times Z$ *drawing then G has track-number* $\mathsf{tn}(G) \leq 2XY$.

We have the following upper bounds on the track-number.

Lemma 1. *Let G be a graph with n vertices, maximum degree* Δ*, path-width p, tree-width w, genus* γ*, and with no* K_h*-minor. Then the track-number of G satisfies:* (a) $\mathsf{tn}(G) \leq p + 1$, (b) $\mathsf{tn}(G) \leq \mathcal{O}(6^{4^w})$, (c) $\mathsf{tn}(G) \leq 72w\Delta$, (d) $\mathsf{tn}(G) \in \mathcal{O}(\gamma^{1/2} n^{1/2})$, (e) $\mathsf{tn}(G) \in \mathcal{O}(h^{3/2} n^{1/2})$.

Proof. Part (a) is by Dujmović *et al.* [10]. Parts (b) and (c) are by the authors [12]. Gilbert *et al.* [16] and Djidjev [9] independently proved that G has a $\mathcal{O}(\gamma^{1/2} n^{1/2})$-separator, and thus has $\mathcal{O}(\gamma^{1/2} n^{1/2})$ path-width (see Bodlaender [2, Theorem 20(iii)]). Hence (d) follows from (a). Similarly (e) follows from the result by Alon *et al.* [1] that G has a $\mathcal{O}(h^{3/2} n^{1/2})$-separator. □

The next result is the fundamental contribution of this section.

Theorem 2. *Every graph G with m edges and maximum degree* Δ *has track-number* $\mathsf{tn}(G) \leq 14\sqrt{\Delta m}$.

[1] This definition of *track-number* is unrelated to that of Gyárfás and West [17].

To prove Theorem 2 we introduce the following concept. A vertex colouring is a *strong star colouring* if between every pair of colour classes, all edges (if any) are incident to a single vertex. That is, each bichromatic subgraph consists of a star and possibly some isolated vertices. The *strong star chromatic number* of a graph G, denoted by $\chi_{\mathrm{sst}}(G)$, is the minimum number of colours in a strong star colouring of G. Note that *star colourings*, in which each bichromatic subgraph is a forest of stars, have also been studied (see [23] for example). The *star chromatic number* of a graph G, denoted by $\chi_{\mathrm{st}}(G)$, is the minimum number of colours in a star-colouring of G. With an arbitrary order on each colour class in a strong star colouring, there is no X-crossing. Thus track-number $\mathsf{tn}(G) \leq \chi_{\mathrm{sst}}(G)$ for every graph G, and Theorem 2 is an immediate corollary of the next lemma.

Lemma 2. *Every graph G with m edges and maximum degree $\Delta \geq 1$ has strong star chromatic number $\chi_{\mathrm{sst}}(G) \leq 14\sqrt{\Delta m}$.*

The proof of Lemma 2 uses the weighted version of the Lovász Local Lemma [13].

Lemma 3. [22, p. 221] *Let $\mathcal{E} = \{A_1, \dots, A_n\}$ be a set of 'bad' events. Let $0 \leq p \leq \frac{1}{4}$ be a real number, and let $t_1, \dots, t_n \geq 1$ be integers. Suppose that for all $A_i \in \mathcal{E}$,*

(a) *the probability $\mathbf{P}(A_i) \leq p^{t_i}$,*
(b) *A_i is mutually independent of $\mathcal{E} \setminus (\{A_i\} \cup D_i)$ for some $D_i \subseteq \mathcal{E}$, and*
(c) $\displaystyle\sum_{A_j \in D_i} (2p)^{t_j} \leq \frac{t_i}{2}$.

Then with positive probability, no event in $\mathcal{E}$ occurs.

Proof (of Lemma 2). Let $c \geq 4$ be a positive integer to be specified later. Let $p = \frac{1}{c}$. Then $0 < p \leq \frac{1}{4}$. For each vertex $v \in V(G)$, randomly and independently choose $\mathsf{col}(v)$ from $\{1, 2, \dots, c\}$. For each edge $vw \in E(G)$, let A_{vw} be the *type-I* event that $\mathsf{col}(v) = \mathsf{col}(w)$. Let E' be the set of arcs $E' = \{(v, w), (w, v) : vw \in E(G)\}$. For each pair of arcs $(v, w), (x, y) \in E'$ with no endpoint in common, let $B_{(v,w),(x,y)}$ be the *type-II* event that $\mathsf{col}(v) = \mathsf{col}(x)$ and $\mathsf{col}(w) = \mathsf{col}(y)$.

We will apply Lemma 3 to obtain a colouring such that no type-I event and no type-II event occurs. No type-I event implies that we have a (proper) vertex colouring. No type-II event implies that no two disjoint edges share the same pair of colours; that is, we have a strong star colouring. For each type-I event A, $\mathbf{P}(A) = \frac{1}{c}$. Let $t_A = 1$. Then $\mathbf{P}(A) = p^{t_A}$. For each type-II event B, $\mathbf{P}(B) = \frac{1}{c^2}$. Let $t_B = 2$. Then $\mathbf{P}(B) = p^{t_B}$. Thus condition (a) of Lemma 3 is satisfied.

An event involving a particular set of vertices is dependent only on other events involving at least one of the vertices in that set. Each vertex is involved in at most Δ type-I events, and at most $2\Delta|E'| = 4\Delta m$ type-II events. A type-I event involves two vertices, and is thus mutually independent of all but at most 2Δ type-I events and at most $8\Delta m$ type-II events. A type-II event involves four vertices, and is thus mutually independent of all but at most 4Δ type-I events and at most $16\Delta m$ type-II events.

For condition (c) of Lemma 3 to hold we need $2\Delta\left(\frac{2}{c}\right)^1+8\Delta m\left(\frac{2}{c}\right)^2 \leq \frac{1}{2}$ for the type-I events, and $4\Delta\left(\frac{2}{c}\right)^1+16\Delta m\left(\frac{2}{c}\right)^2 \leq 1$ for the type-II events. It is a happy coincidence that these two equations are equivalent, and it is easily verified that $c = \lceil 4(\Delta + \sqrt{\Delta(1+4m)}\,)\rceil \geq 4$ is a solution. Thus by Lemma 3, with positive probability no type-I event and no type-II event occurs. Thus for every vertex $v \in V(G)$, there exists $\mathsf{col}(v) \in \{1,\ldots,c\}$ such that no type-I event and no type-II event occurs. As proved above such a colouring is a strong star colouring. Since $\Delta \leq \sqrt{\Delta m}$, the number of colours $c \leq \lceil 4(1+\sqrt{5})\sqrt{\Delta m}\rceil \leq 14\sqrt{\Delta m}$. □

Theorems 1 and 2 imply:

Theorem 3. *Every graph with n vertices, m edges and maximum degree Δ has a $\mathcal{O}((\Delta m)^{1/2}) \times \mathcal{O}((\Delta m)^{1/2}) \times \mathcal{O}(n)$ drawing with $\mathcal{O}(\Delta mn)$ volume.* □

We have the following corollary of Lemma 2.

Lemma 4. *Every graph G with m edges has strong star chromatic number $\chi_{\mathrm{sst}}(G) \leq 15m^{2/3}$.*

Proof. Let X be the set of vertices of G with degree greater than $\frac{1}{4}m^{1/3}$. Let H be the subgraph of G induced by $V(G) \setminus X$. Thus H has maximum degree at most $\frac{1}{4}m^{1/3}$. By Lemma 2, H has a strong star colouring with at most $14(\frac{1}{4}m^{1/3}m)^{1/2} = 7m^{2/3}$ colours. Now $|X| \leq 2m/(\frac{1}{4}m^{1/3}) = 8m^{2/3}$. By adding each vertex in X to its own colour class we obtain a strong star colouring of G with at most $15m^{2/3}$ colours. □

Since $\mathsf{tn}(G) \leq \chi_{\mathrm{sst}}(G)$, Lemma 4 implies that $\mathsf{tn}(G) \leq 15m^{2/3}$, and by Theorem 1 we have:

Theorem 4. *Every graph with n vertices and m edges has a $\mathcal{O}(m^{2/3}) \times \mathcal{O}(m^{2/3}) \times \mathcal{O}(n)$ drawing with $\mathcal{O}(m^{4/3}n)$ volume.* □

3 Sub-quadratic Volume Bounds

Vertex colourings [24] and track layouts [10] have previously been used to produce 3D drawings with small volume. In the following sequence of results we combine vertex colourings and track layouts to reduce the quadratic dependence on t in Theorem 1 to linear. This comes at the expense of a higher dependence on the chromatic number. However, in the intended applications the chromatic number will be constant, or at least will be independent of the size of the graph. The proof of the next lemma is a further generalisation of the 'moment curve' method for producing three-dimensional graph drawings [5,24,10].

Lemma 5. *Let G be a graph with a vertex c-colouring $\{V_i : 0 \leq i \leq c-1\}$, and a track layout $\{T_{i,j} : 0 \leq i \leq c-1, 1 \leq j \leq t_i\}$, such that each $T_{i,j} \subseteq V_i$. Then G has a $\mathcal{O}(c) \times \mathcal{O}(c^2 t) \times \mathcal{O}(c^5 tn')$ drawing, where $t = \max_i t_i$ and $n' = \max_{i,j} |T_{i,j}|$.*

Proof. Let p be the minimum prime such that $p \geq c$. Then $p < 2c$ by Bertrand's postulate. Let $v(i,j,k)$ denote the kth vertex in track $T_{i,j}$. Define

$$Y(i,j) = p(2\,i\,t + j) + (i^2 \bmod p), \text{ and}$$
$$Z(i,j,k) = p(20\,c\,i\,n' \cdot Y(i,j) + k) + (i^3 \bmod p) \ .$$

Position each vertex $v(i,j,k)$ at the gridpoint $(i, Y(i,j), Z(i,j,k))$, and draw each edge as a line-segment between its endpoints. Since $Y(i,j) \in \mathcal{O}(c^2\,t)$ and $Z(i,j,k) \in \mathcal{O}(c^3\,n' \cdot Y(i,j))$, the drawing is $\mathcal{O}(c) \times \mathcal{O}(c^2\,t) \times \mathcal{O}(c^5\,tn')$.

Observe that the tracks from a single colour class are within a distinct YZ-plane, each track occupies a distinct vertical line, and the Z-coordinates of the vertices within a track preserve the given ordering of that track. In addition, the Y-coordinates satisfy the following property.

Claim 1. *For all distinct colours i_1 and i_2 and for all $1 \leq j_1, j_2 \leq t$, we have $2c\,|\,(Y(i_1,j_1) - Y(i_2,j_2)\,|$ is greater than the Y-coordinate of any vertex.*

Proof. Without loss of generality $i_1 > i_2$. Observe that every Y-coordinate is less than $p(2(c-1)t+t)+p = p(2\,ct-t+1) \leq 2cpt$. Now $2c\,|\,Y(i_1,j_1)-Y(i_2,j_2)\,| > 2c\,|\,p(2\,i_1\,t+1) - p(2\,i_2\,t+t+1)\,| \geq 2cp\,|\,2(i_2+1)t - (2i_2t+t)\,| = 2cpt.$ □

We first prove that the only vertices each edge intersects are its own endpoints. It suffices to prove that if three tracks are collinear in the XY-plane then they are all from the same colour class. Loosely speaking, an edge does not pass through any track. Clearly two tracks from the same colour class are not collinear (in the XY-plane) with a third track from a distinct colour class. Thus we need only consider tracks $\{T(i_\alpha, j_\alpha) : 1 \leq i_\alpha \leq 3\}$ from three distinct colour classes $\{i_1, i_2, i_3\}$. Let R be the determinant,

$$R = \begin{vmatrix} 1 & i_1 & Y(i_1,j_1) \\ 1 & i_2 & Y(i_2,j_2) \\ 1 & i_3 & Y(i_3,j_3) \end{vmatrix} \ .$$

If the tracks $\{T(i_\alpha, j_\alpha) : 1 \leq i_\alpha \leq 3\}$ are collinear in the XY-plane then $R = 0$. However $Y(i,j) \equiv i^2 \pmod p$, and thus

$$R \equiv \begin{vmatrix} 1 & i_1 & i_1^2 \\ 1 & i_2 & i_2^2 \\ 1 & i_3 & i_3^2 \end{vmatrix} = \prod_{1 \leq \alpha < \beta \leq 3} (i_\alpha - i_\beta) \ \not\equiv 0 \pmod p \ ,$$

since $i_\alpha \neq i_\beta$, and p is a prime greater than any $i_\alpha - i_\beta$. Thus $R \neq 0$, and the tracks $\{T(i_\alpha, j_\alpha) : 1 \leq i_\alpha \leq 3\}$ are not collinear in the XY-plane. Hence the only vertices that an edge intersects are its own endpoints.

It remains to prove that there are no edge crossings. Consider two edges e and e' with distinct endpoints $v(i_\alpha, j_\alpha, k_\alpha)$, $1 \leq \alpha \leq 4$. (Clearly edges with a common endpoint do not cross.) Let $Y_\alpha = Y(i_\alpha, j_\alpha)$. Consider the determinant

$$D = \begin{vmatrix} 1 & i_1 & Y_1 & Z(i_1,j_1,k_1) \\ 1 & i_2 & Y_2 & Z(i_2,j_2,k_2) \\ 1 & i_3 & Y_3 & Z(i_3,j_3,k_3) \\ 1 & i_4 & Y_4 & Z(i_4,j_4,k_4) \end{vmatrix} \ .$$

If e and e' cross then their endpoints are coplanar, and $D = 0$. Thus it suffices to prove that $D \neq 0$. We proceed by considering the number $N = |\{i_1, i_2, i_3, i_4\}|$ of distinct colours assigned to the four endpoints of e and e'. Clearly $N \in \{2, 3, 4\}$.

Case $N = 4$: Since $Y_\alpha \equiv i_\alpha^2 \pmod p$ and $Z(i_\alpha, j_\alpha, k_\alpha) \equiv i_\alpha^3 \pmod p$,

$$D \equiv \begin{vmatrix} 1 & i_1 & i_1^2 & i_1^3 \\ 1 & i_2 & i_2^2 & i_2^3 \\ 1 & i_3 & i_3^2 & i_3^3 \\ 1 & i_4 & i_4^2 & i_4^3 \end{vmatrix} = \prod_{1 \leq \alpha < \beta \leq 4} (i_\alpha - i_\beta) \not\equiv 0 \pmod p ,$$

since $i_\alpha \neq i_\beta$, and p is a prime greater than any $i_\alpha - i_\beta$. Thus $D \neq 0$.

Case $N = 3$: Without loss of generality $i_1 = i_2$. It follows that $D = 5S_0 + S_1 + S_2 + S_3 + S_4$ where

$$\begin{aligned} S_0 &= 4cpn'(i_3 - i_1)(i_4 - i_1)(Y_2 - Y_1)(Y_3 - Y_4) \\ S_1 &= p(Y_2 - Y_1)\big(k_3(i_4 - i_1) - k_4(i_3 - i_1)\big) \\ S_2 &= p(i_4 - i_3)(k_2 Y_1 - k_1 Y_2) \\ S_3 &= p(k_2 - k_1)(Y_4(i_3 - i_1) - Y_3(i_4 - i_1)) \\ S_4 &= (Y_2 - Y_1)\big((i_3 - i_4)(i_1^3 \bmod p) - (i_3 - i_1)(i_4^3 \bmod p) + (i_4 - i_1)(i_3^3 \bmod p)\big) . \end{aligned}$$

If $Y_1 = Y_2$ then e and e' do not cross, since no three tracks from distinct colour classes are collinear in the XY-plane. Assume $Y_1 \neq Y_2$. If $i_3 < i_1 < i_4$ or $i_4 < i_1 < i_3$ then e and e' do not cross, simply by considering the projection in the XY-plane. Thus $i_1 < i_3, i_4$ or $i_1 > i_3, i_4$, which implies

$$(i_4 - i_1)(i_3 - i_1) > |i_4 - i_3| . \tag{1}$$

Claim 2. *If* $|S_0| \geq |S_1|$, $|S_0| \geq |S_2|$, $|S_0| \geq |S_3|$ *and* $|S_0| \geq |S_4|$ *then* $D \neq 0$.

Proof. To prove that $D = 5S_0 + S_1 + S_2 + S_3 + S_4$ is nonzero it suffices to show that $D' = \pm 5|S_0| \pm |S_1| \pm |S_2| \pm |S_3| \pm |S_4|$ is nonzero for all combinations of pluses and minuses. Consider $X = \pm|S_1| \pm |S_2| \pm |S_3| \pm |S_4|$ for some combination of pluses and minuses. Since $|S_1| \leq |S_0|, |S_2| \leq |S_0|, |S_3| \leq |S_0|$, and $|S_4| \leq |S_0|$, we have $-4|S_0| \leq X \leq 4|S_0|$. Since $S_0 \neq 0$, we have $5|S_0| + X \neq 0$ and $-5|S_0| + X \neq 0$. That is, all values of D' are nonzero. Therefore $D \neq 0$. □

Therefore, to prove that $D \neq 0$ it suffices to show that $|S_0| \geq |S_1|$, $|S_0| \geq |S_2|$, $|S_0| \geq |S_3|$ and $|S_0| \geq |S_4|$. We will use the following elementary facts regarding absolute values:

$$\begin{aligned} &\forall a_1, \ldots, a_k \in \mathbb{R} \quad |a_1 a_2 \ldots a_k| = |a_1||a_2| \cdots |a_k|, \text{ and} \\ &|a_1 + a_2 + \cdots + a_k| \leq |a_1| + |a_2| + \cdots + |a_k| \leq k \cdot \max\{|a_1|, |a_2|, \ldots, |a_k|\} . \end{aligned}$$

- First we prove that $|S_0| \geq |S_1|$. That is,

$$|4cpn'(i_3 - i_1)(i_4 - i_1)(Y_2 - Y_1)(Y_3 - Y_4)| \geq |p(Y_2 - Y_1)\big(k_3(i_4 - i_1) - k_4(i_3 - i_1)\big)| .$$

Hence, $|S_0| > |S_1|$ is implied if

$$\begin{aligned} & 2n'|i_3 - i_1||i_4 - i_1||Y_3 - Y_4| \geq |k_3(i_4 - i_1) - k_4(i_3 - i_1)| \;. \\ \Longleftarrow\; & 2n'|i_3 - i_1||i_4 - i_1||Y_3 - Y_4| \geq 2 \cdot \max\{|k_4(i_3 - i_1)|, |k_3(i_4 - i_1)|\} \;. \end{aligned}$$

Since $n' \geq k_3, k_4$ and $|Y_3 - Y_4| \geq 1$,

$$|S_0| > |S_1| \Longleftarrow |i_3 - i_1||i_4 - i_1| \geq \max\{|i_3 - i_1|, |i_4 - i_1|\} \;.$$

Thus $|S_0| \geq |S_1|$ since $|i_3 - i_1| \geq 1$ and $|i_4 - i_1| \geq 1$.

- Now we prove that $|S_0| \geq |S_2|$. That is,

$$|4cpn'(i_3 - i_1)(i_4 - i_1)(Y_2 - Y_1)(Y_3 - Y_4)| \geq |p(i_4 - i_3)(k_2 Y_1 - k_1 Y_2)| \;.$$

By (1) and since $|Y_2 - Y_1| \geq 1$, $|S_0| \geq |S_2|$ is implied if $|4cn'(Y_3 - Y_4)| \geq |k_2 Y_1 - k_1 Y_2|$. This in turn is implied if $|2cn'(Y_3 - Y_4)| \geq \max\{|k_2 Y_1|, |k_1 Y_2|\}$, which holds since $n' \geq k_1, k_2$ and $|2c(Y_3 - Y_4)| \geq \max\{Y_1, Y_2\}$ by Claim 1.

- Now we prove that $|S_0| \geq |S_3|$. That is,

$$|4cpn'(i_3 - i_1)(i_4 - i_1)(Y_2 - Y_1)(Y_3 - Y_4)| \geq |p(k_2 - k_1)\big(Y_4(i_3 - i_1) - Y_3(i_4 - i_1)\big)| \;.$$

Now $n' \geq |k_2 - k_1|$ and $|Y_2 - Y_1| \geq 1$. Thus $|S_0| \geq |S_3|$ is implied if

$$\begin{aligned} & |4c(i_3 - i_1)(i_4 - i_1)(Y_3 - Y_4)| \geq |Y_4(i_3 - i_1) - Y_3(i_4 - i_1)| \;. \\ \Longleftarrow\; & |2c(i_3 - i_1)(i_4 - i_1)(Y_3 - Y_4)| \geq \max\{|Y_4(i_3 - i_1)|, |Y_3(i_4 - i_1)|\} \;, \end{aligned}$$

which holds since $|2c(Y_3 - Y_4)| \geq \max\{Y_1, Y_2\}$ by Claim 1.

- Finally we prove that $|S_0| \geq |S_4|$. That is,

$$\begin{aligned} & |4cpn'(i_3 - i_1)(i_4 - i_1)(Y_2 - Y_1)(Y_3 - Y_4)| \geq \\ & \quad |(Y_2 - Y_1)\big((i_3 - i_4)(i_1^3 \bmod p) - (i_3 - i_1)(i_4^3 \bmod p) + (i_4 - i_1)(i_3^3 \bmod p)\big)| \;. \end{aligned}$$

Now $cn'|Y_3 - Y_4| \geq 1$. Thus $|S_0| > |S_4|$ is implied if

$$\begin{aligned} & |3p(i_3 - i_1)(i_4 - i_1)| \geq \\ & \quad |(i_3 - i_4)(i_1^3 \bmod p) - (i_3 - i_1)(i_4^3 \bmod p) + (i_4 - i_1)(i_3^3 \bmod p)| \\ \Longleftarrow\; & |3p(i_3 - i_1)(i_4 - i_1)| \geq \\ & \quad 3 \cdot \max\{|(i_3 - i_4)(i_1^3 \bmod p)|, |(i_3 - i_1)(i_4^3 \bmod p)|, |(i_4 - i_1)(i_3^3 \bmod p)|\} \\ \Longleftarrow\; & |(i_3 - i_1)(i_4 - i_1)| \geq \max\{|i_3 - i_4|, |i_3 - i_1|, |i_4 - i_1|\} \;, \end{aligned}$$

which holds by (1).

Case $N = 2$: Without loss of generality $i_1 = i_2 \neq i_3 = i_4$. If $Y_1 = Y_2$ and $Y_3 = Y_4$ then e and e' do not cross as otherwise there would be an X-crossing in the track layout. If $Y_1 = Y_2$ and $Y_3 \neq Y_4$ (or $Y_1 \neq Y_2$ and $Y_3 = Y_4$) then e

and e' do not cross, by considering the projection in the XY-plane. Thus we can assume that $Y_1 \neq Y_2$ and $Y_3 \neq Y_4$. It follows that

$$D = p(i_1 - i_3)\ \big(\ 5 \cdot 4cn'(Y_2 - Y_1)(Y_4 - Y_3)(i_3 - i_1) \\ + (k_1 - k_2)(Y_4 - Y_3) + (k_4 - k_3)(Y_2 - Y_1)\ \big)\ .$$

As in Claim 2, to show that $D \neq 0$ it suffices to show that

$$|\,4cn'(Y_2 - Y_1)(Y_4 - Y_3)(i_3 - i_1)\,| \geq |\,(k_1 - k_2)(Y_4 - Y_3)\,|\ , \qquad (2)$$

$$\text{and } |\,4cn'(Y_2 - Y_1)(Y_4 - Y_3)(i_3 - i_1)\,| \geq |\,(k_4 - k_3)(Y_2 - Y_1)\,|\ . \qquad (3)$$

Inequalities (2) and (3) hold since $n' > |k_1 - k_2|$ and $n' > |k_4 - k_3|$. □

Note that the constant 20 in the definition of $Z(i, j, k)$ in the proof of Lemma 5 is chosen to enable a relatively simple proof. It is easily seen that it can be reduced. The proof of the next lemma is based on an idea of Pach *et al.* [24] for balancing the size of the colour classes in a vertex colouring.

Lemma 6. *Let G be an n-vertex graph with a c-colouring $\{V_i : 0 \leq i \leq c-1\}$ and a track layout $\{T_{i,j} : 0 \leq i \leq c-1, 1 \leq j \leq t_i\}$, such that each $T_{i,j} \subseteq V_i$. Let $k = \sum_i t_i$ be the total number of tracks. Then G has a $\mathcal{O}(c) \times \mathcal{O}(ck) \times \mathcal{O}(c^4 n)$ drawing.*

Proof. Replace each track by tracks of size exactly $\lceil \frac{n}{k} \rceil$, except for at most one track of size at most $\lceil \frac{n}{k} \rceil$. Order the vertices within each track according to the original track, and consider the new tracks to belong to the same colour class as the original. Clearly no X-crossing is created. Within V_i there are now at most $t_i + |V_i|/\lceil \frac{n}{k} \rceil$ tracks. The total number of tracks is $\sum_i (t_i + |V_i|/\lceil \frac{n}{k} \rceil) \leq 2k$. For each colour class V_i, partition the set of tracks in V_i into sets of size exactly $\lceil \frac{2k}{c} \rceil$, except for one set of size at most $\lceil \frac{2k}{c} \rceil$. Consider each set to correspond to a colour. The number of colours is now at most $c + 2k/\lceil \frac{2k}{c} \rceil \leq 2c$. Applying Lemma 5 with $2c$ colours, $n' = \lceil \frac{n}{k} \rceil$, and $t = \lceil \frac{2k}{c} \rceil$, we obtain the desired drawing. □

Theorem 5. *Every c-colourable graph G with n vertices and track-number $\mathsf{tn}(G) \leq t$ has a $\mathcal{O}(c) \times \mathcal{O}(c^2 t) \times \mathcal{O}(c^4 n)$ drawing with $\mathcal{O}(c^7 tn)$ volume.*

Proof. Let $\{V_i : 0 \leq i \leq c-1\}$ be a c-colouring of G. Let $\{T_j : 1 \leq j \leq t\}$ be a t-track layout of G. For all $0 \leq i \leq c-1$ and $1 \leq j \leq t$, let $T_{i,j} = V_i \cap T_j$. Then $\{V_i : 0 \leq i \leq c-1\}$ and $\{T_{i,j} : 0 \leq i \leq c-1, 1 \leq j \leq t\}$ satisfy Lemma 6 with $k = ct$. Thus G has the desired drawing. □

In the case of bipartite graphs we have a simple proof of Theorem 5 with improved constants.

Lemma 7. *Every n-vertex bipartite graph G with track-number $\mathsf{tn}(G) \leq t$ has a $2 \times t \times n$ drawing.*

Proof. Let $\{A, B\}$ be the bipartition of $V(G)$. Let $\{T_i : 1 \leq i \leq t\}$ be a t-track layout of G. For each $1 \leq i \leq t$, let $A_i = T_i \cap A$ and $B_i = T_i \cap B$. Order each A_i and B_i as in T_i. Place the jth vertex in A_i at $(0, i, j + \sum_{k=1}^{i-1} |A_k|)$. Place the jth vertex in B_i at $(1, t-i+1, j + \sum_{k=1}^{i-1} |B_k|)$. The drawing is thus $2 \times t \times n$. Let A_iB_j be the set of edges with one endpoint in A_i and the other in B_j. There is no crossing between edges in A_iB_j and A_iB_j as otherwise there would be an X-crossing in the track layout. Clearly there is no crossing between edges in A_iB_j and A_iB_k for $j \neq k$. Suppose there is a crossing between edges in A_iB_j and A_kB_ℓ with $i \neq k$ and $j \neq \ell$. Without loss of generality $i < k$. Then the projections of the edges in the XY-plane also cross, and thus $j < \ell$. Hence the projections in the XZ-plane do not cross, and thus the edges do not cross. □

Lemma 4 with $\mathsf{tn}(G) \leq \chi_{\text{sst}}(G)$ and Theorem 5 imply:

Theorem 6. *Every c-colourable graph with n vertices and m edges has a $\mathcal{O}(c) \times \mathcal{O}(c^2 m^{2/3}) \times \mathcal{O}(c^4 n)$ drawing with $\mathcal{O}(c^6 m^{2/3} n)$ volume.* □

The next result is one of the main contributions of this paper.

Theorem 7. *Every planar graph with n vertices has a $\mathcal{O}(1) \times \mathcal{O}(n^{1/2}) \times \mathcal{O}(n)$ drawing with $\mathcal{O}(n^{3/2})$ volume.*

Proof. Planar graphs have $\mathcal{O}(n^{1/2})$ path-width (see [2]), and thus have $\mathcal{O}(n^{1/2})$ track-number by Lemma 1(a). The result follows from Theorem 5 since planar graphs are 4-colourable. □

The following generalisation of Theorem 7 follows from Lemma 1(d), Theorem 5, and the classical result of Heawood [20] that $\chi(G) \in \mathcal{O}(\gamma^{1/2})$.

Theorem 8. *Every n-vertex graph with genus γ has a $\mathcal{O}(\gamma^{1/2}) \times \mathcal{O}(\gamma^{3/2} n^{1/2}) \times \mathcal{O}(\gamma^2 n)$ drawing with $\mathcal{O}(\gamma^4 n^{3/2})$ volume.* □

The next generalisation of Theorem 7 for graphs with no K_h-minor follows from Lemma 1(e), Theorem 5, and the result independently due to Kostochka [21] and Thomason [26] that $\chi(G) \in \mathcal{O}(h \log^{1/2} h)$.

Theorem 9. *Every n-vertex graph with no K_h-minor has a $\mathcal{O}(h \log^{1/2} h) \times \mathcal{O}(h^{7/2} \log h \cdot n^{1/2}) \times \mathcal{O}(h^4 \log^2 h \cdot n)$ drawing with volume $\mathcal{O}(h^{17/2} \log^{7/2} h \cdot n^{3/2})$.*

Finally we consider the maximum degree Δ as a parameter. By the sequential greedy algorithm, G is $(\Delta + 1)$-colourable. Thus by Theorems 2 and 5 we have:

Theorem 10. *Every graph with n vertices, m edges, and maximum degree Δ has a $\mathcal{O}(\Delta) \times \mathcal{O}(\Delta^{5/2} m^{1/2}) \times \mathcal{O}(\Delta^4 n)$ drawing with $\mathcal{O}(\Delta^{15/2} m^{1/2} n)$ volume.* □

Graphs with constant Δ have $\mathcal{O}(n)$ edges. By Theorems 8, 9, and 10 we have:

Corollary 1. *Every n-vertex graph with constant genus, or with no K_h-minor for some constant h, or with constant maximum degree has a $\mathcal{O}(1) \times \mathcal{O}(n^{1/2}) \times \mathcal{O}(n)$ drawing with $\mathcal{O}(n^{3/2})$ volume.* □

We conclude with the following open problems: Does every graph have a 3D drawing with $\mathcal{O}(nm)$ volume? Does every graph with constant chromatic number have a 3D drawing with $\mathcal{O}(n\sqrt{m})$ volume? These bounds match the lower bounds for K_n and $K_{n,n}$, and would make edge-sensitive improvements to the existing upper bounds of $\mathcal{O}(n^3)$ and $\mathcal{O}(n^2)$, respectively. These edge-sensitive bounds would be implied by Theorems 1 and 5 should every graph have $\mathcal{O}(\sqrt{m})$ track-number. In turn, this bound on track-number would be implied should every graph have $\mathcal{O}(\sqrt{m})$ strong star chromatic number. As far as the authors are aware, a $\mathcal{O}(\sqrt{m})$ bound is not even known for star chromatic number. The best known bound in this direction is $\chi_{st}(G) \leq 11m^{3/5}$, which can be proved in a similar fashion to Lemma 4, in conjunction with the result of Fertin *et al.* [15] that $\chi_{st}(G) \leq \lceil 20\Delta^{3/2} \rceil$ (see [11]).

Acknowledgements. Thanks to Stefan Langerman for stimulating discussions. Thanks to Ferran Hurtado and Prosenjit Bose for graciously hosting the second author, whose research was completed at the Departament de Matemàtica Aplicada II, Universitat Politècnica de Catalunya, Barcelona, Spain.

References

1. Noga Alon, Paul Seymour, and Robin Thomas. A separator theorem for nonplanar graphs. *J. Amer. Math. Soc.*, 3(4):801–808, 1990.
2. Hans L. Bodlaender. A partial k-arboretum of graphs with bounded treewidth. *Theoret. Comput. Sci.*, 209(1-2):1–45, 1998.
3. Prosenjit Bose, Jurek Czyzowicz, Pat Morin, and David R. Wood. The maximum number of edges in a three-dimensional grid-drawing. In *Proc. 19th European Workshop on Computational Geometry*, pages 101–103, Germany, 2003. Univ. of Bonn.
4. Tiziana Calamoneri and Andrea Sterbini. 3D straight-line grid drawing of 4-colorable graphs. *Inform. Process. Lett.*, 63(2):97–102, 1997.
5. Robert F. Cohen, Peter Eades, Tao Lin, and Frank Ruskey. Three-dimensional graph drawing. *Algorithmica*, 17(2):199–208, 1996.
6. Emilio Di Giacomo. Drawing series-parallel graphs on restricted integer 3D grids. In these proceedings.
7. Emilio Di Giacomo, Giuseppe Liotta, and Stephen Wismath. Drawing series-parallel graphs on a box. In *Proc. 14th Canadian Conf. on Computational Geometry* (CCCG '02), pages 149–153. The Univ. of Lethbridge, Canada, 2002.
8. Emilio Di Giacomo and Henk Meijer. Track drawings of graphs with constant queue number. In these proceedings.
9. Hristo N. Djidjev. A separator theorem. *C. R. Acad. Bulgare Sci.*, 34(5):643–645, 1981.
10. Vida Dujmović, Pat Morin, and David R. Wood. Path-width and three-dimensional straight-line grid drawings of graphs. In Michael T. Goodrich and Stephen G. Kobourov, eds., *Proc. 10th Int'l Symp. on Graph Drawing* (GD '02), volume 2528 of *Lecture Notes in Comput. Sci.*, pages 42–53. Springer, 2002.
11. Vida Dujmović and David R. Wood. New results in graph layout. Tech. Report TR-2003-04, School of Computer Science, Carleton Univ., Ottawa, Canada, 2003.

12. VIDA DUJMOVIĆ and DAVID R. WOOD. Tree-partitions of k-trees with applications in graph layout. In HANS L. BODLAENDER, ed., *Proc. 29th Workshop on Graph Theoretic Concepts in Comput. Sci.* (WG'03), Lecture Notes in Comput. Sci. Springer, to appear.
13. PAUL ERDŐS and LÁSZLÓ LOVÁSZ. Problems and results on 3-chromatic hypergraphs and some related questions. In *Infinite and Finite Sets*, volume 10 of *Colloq. Math. Soc. János Bolyai*, pages 609–627. North-Holland, 1975.
14. STEFAN FELSNER, GIUSEPPE LIOTTA, and STEPHEN WISMATH. Straight-line drawings on restricted integer grids in two and three dimensions. In PETRA MUTZEL, MICHAEL JÜNGER, and SEBASTIAN LEIPERT, eds., *Proc. 9th Int'l Symp. on Graph Drawing* (GD '01), volume 2265 of *Lecture Notes in Comput. Sci.*, pages 328–342. Springer, 2002.
15. GUILLAUME FERTIN, ANDRÉ RASPAUD, and BRUCE REED. On star coloring of graphs. In ANDREAS BRANSTÄDT and VAN BANG LE, eds., *Proc. 27th Int'l Workshop on Graph-Theoretic Concepts in Computer Science* (WG '01), volume 2204 of *Lecture Notes in Comput. Sci.*, pages 140–153. Springer, 2001.
16. JOHN R. GILBERT, JOAN P. HUTCHINSON, and ROBERT E. TARJAN. A separator theorem for graphs of bounded genus. *J. Algorithms*, 5(3):391–407, 1984.
17. ANDRÁS GYÁRFÁS and DOUGLAS WEST. Multitrack interval graphs. In *Proc. 26th Southeastern Int'l Conf. on Combinatorics, Graph Theory and Computing*, volume 109 of *Congr. Numer.*, pages 109–116, 1995.
18. TORU HASUNUMA. Laying out iterated line digraphs using queues. In these proceedings.
19. LENWOOD S. HEATH, FRANK THOMSON LEIGHTON, and ARNOLD L. ROSENBERG. Comparing queues and stacks as mechanisms for laying out graphs. *SIAM J. Discrete Math.*, 5(3):398–412, 1992.
20. PERCY J. HEAWOOD. Map colour theorem. *Quart. J. Pure Appl. Math.*, 24:332–338, 1890.
21. ALEXANDR V. KOSTOCHKA. The minimum Hadwiger number for graphs with a given mean degree of vertices. *Metody Diskret. Analiz.*, 38:37–58, 1982.
22. MICHAEL MOLLOY and BRUCE REED. *Graph colouring and the probabilistic method*, volume 23 of *Algorithms and Combinatorics*. Springer, 2002.
23. JAROSLAV NEŠETŘIL and PATRICE OSSONA DE MENDEZ. Colorings and homomorphisms of minor closed classes. In BORIS ARONOV, SAUGATA BASU, JÁNOS PACH, and MICHA SHARIR, eds., *Discrete and Computational Geometry, The Goodman-Pollack Festschrift*, volume 25 of *Algorithms and Combinatorics*. Springer, 2003.
24. JÁNOS PACH, TORSTEN THIELE, and GÉZA TÓTH. Three-dimensional grid drawings of graphs. In GIUSEPPE DI BATTISTA, ed., *Proc. 5th Int'l Symp. on Graph Drawing* (GD '97), volume 1353 of *Lecture Notes in Comput. Sci.*, pages 47–51. Springer, 1997. Also in BERNARD CHAZELLE, JACOB E. GOODMAN, and RICHARD POLLACK, eds., Advances in discrete and computational geometry, volume 223 of *Contempory Mathematics*, pp. 251–255, Amer. Math. Soc., 1999.
25. TIMO PORANEN. A new algorithm for drawing series-parallel digraphs in 3D. Tech. Report A-2000-16, Dept. of Computer and Information Sciences, Univ. of Tampere, Finland, 2000.
26. ANDREW THOMASON. An extremal function for contractions of graphs. *Math. Proc. Cambridge Philos. Soc.*, 95(2):261–265, 1984.
27. DAVID R. WOOD. Queue layouts, tree-width, and three-dimensional graph drawing. In MANINDRA AGRAWAL and ANIL SETH, eds., *Proc. 22nd Foundations of Software Technology and Theoretical Computer Science* (FST TCS '02), volume 2556 of *Lecture Notes in Comput. Sci.*, pages 348–359. Springer, 2002.

Laying Out Iterated Line Digraphs Using Queues

Toru Hasunuma

Department of Computer Science
The University of Electro-Communications,
1-5-1 Chofugaoka, Chofu, Tokyo 182–8585 Japan
hasunuma@cs.uec.ac.jp

Abstract. In this paper, we study a layout problem of a digraph using queues. The queuenumber of a digraph is the minimum number of queues required for a queue layout of the digraph. We present upper and lower bounds on the queuenumber of an iterated line digraph $L^k(G)$ of a digraph G. In particular, our upper bound depends only on G and is independent of the number of iterations k. Queue layouts can be applied to three-dimensional drawings. From the result on the queuenumber of $L^k(G)$, it is shown that for any fixed digraph G, $L^k(G)$ has a three-dimensional drawing with $O(n)$ volume, where n is the number of vertices in $L^k(G)$. We also apply these results to particular families of iterated line digraphs such as de Bruijn digraphs, Kautz digraphs, butterfly digraphs, and wrapped butterfly digraphs.

1 Introduction

Let H be a graph. The vertex set and the edge set of H are denoted by $V(H)$ and $E(H)$, respectively. A *k-queue layout* of H consists of a linear ordering σ of the vertices, i.e., σ is a bijection from $V(H)$ to $\{1, 2, \ldots, |V(H)|\}$, and an assignment of each edge to one of k queues so that the set of edges assigned to each queue obeys a first-in/first-out discipline. (Consider scanning in left-to-right (ascending) order. If the left vertex of an edge e is encountered, e enters its assigned queue. If the right vertex of e is encountered, e exits from its assigned queue.) We can formally define such a discipline as follows; if edges $\{a, b\}, \{c, d\}$ are in the same queue such that $\sigma(a) \leq \sigma(b)$, $\sigma(c) \leq \sigma(d)$, and $\sigma(a) \leq \sigma(c)$, then one of the following inequalities holds:

- $\sigma(a) \leq \sigma(b) \leq \sigma(c) \leq \sigma(d)$
- $\sigma(a) \leq \sigma(c) \leq \sigma(b) \leq \sigma(d)$

The minimum number of queues required for a queue layout of H is the *queuenumber* of H, and denoted by $\mathsf{qn}(H)$.

As a dual concept of a queue layout, a stack layout is known. A *k-stack layout* of H consists of a linear ordering σ of the vertices, and an assignment of each edge to one of k stacks so that the set of edges assigned to each stack obeys a last-in/first-out discipline. (If the left vertex of an edge e is encountered, e is pushed into its assigned stack, and if the right vertex of e is encountered, e is

G. Liotta (Ed.): GD 2003, LNCS 2912, pp. 202–213, 2004.

popped out of its assigned stack.) That is, if $\{a, b\}, \{c, d\}$ are in the same stack such that $\sigma(a) \leq \sigma(b)$, $\sigma(c) \leq \sigma(d)$, and $\sigma(a) \leq \sigma(c)$, then one of the following inequalities holds:

- $\sigma(a) \leq \sigma(b) \leq \sigma(c) \leq \sigma(d)$
- $\sigma(a) \leq \sigma(c) \leq \sigma(d) \leq \sigma(b)$

The minimum number of stacks required for a stack layout of H is the *stacknumber* of H, and denoted by $\mathsf{sn}(H)$. (A stack layout is equivalent to a book-embedding. In the study of book-embeddings, pagenumber is used instead of stacknumber.) Queue and stack layouts of a digraph are similarly defined as those of the underlying graph of the digraph. (Note that our definitions for digraphs are different from queue and stack layouts defined in [19,20] in which all arcs must have the same direction with respect to the vertex ordering.)

Queue and stack layouts are motivated by several area of computer science [5,9,18,21,25,27,29,30,31]. In particular, such layouts of interconnection networks have applications to the DIOGENES approach, proposed by Rosenberg [27], to fault-tolerant processors array. Also, Wood [30,31] have shown that queue and stack layouts can be applied to three-dimensional drawings. For queue layouts, he proved that every graph G with n vertices from a proper minor-closed family has an $O(1) \times O(1) \times O(n)$ three-dimensional straight-line grid drawing if and only if G has $O(1)$ queuenumber.

Until now, queue and/or stack layouts have been studied for many graph classes:

- Stacknumber: complete graphs [4], complete bipartite graphs [24], butterfly graphs [12], trees, grids, X-trees [5], hypercubes [5,22], de Bruijn digraphs, Kautz digraphs, shuffle-exchange graphs [16], planar graphs [32], genus-g graphs [23], bandwidth-k graphs [28], k-trees [13], iterated line digraphs [14].
- Queuenumber: complete graphs, complete bipartite graphs, trees, grids, unicyclic graphs, X-trees, binary de Bruijn graphs, butterfly graphs (all in [21]), k-tree [6,26,31],

In this paper, we treat iterated line digraphs and study queue layouts for the class. Let G be a digraph. The vertex set and the arc set of G are denoted by $V(G)$ and $A(G)$, respectively. The *line digraph* $L(G)$ of G is defined as follows. The vertex set of $L(G)$ is the arc set of G, i.e., $V(L(G)) = A(G)$. The vertex (u, v) is a predecessor of every vertex of the form (v, w), i.e., $A(L(G)) = \{((u, v), (v, w)) \mid (u, v), (v, w) \in A(G)\}$. When we regard "$L$" as an operation on digraphs, the operation is called the *line digraph operation*. The *k-iterated line digraph* $L^k(G)$ of G is the digraph obtained from G by iteratively applying the line digraph operation k times. Iterated line digraphs have many desirable properties for interconnection networks of massively parallel computer such as bounded degree, small diameter, and high connectivity. In fact, the class of iterated line digraphs contains several well-known interconnection networks such as de Bruijn digraphs, Kautz digraphs, butterfly digraphs and wrapped butterfly digraphs.

We present upper and lower bounds on the queuenumber of an iterated line digraph $L^k(G)$. In particular, our upper bound depends only on G and is independent of the number of iterations k. As corollaries, upper and lower bounds on the queuenumbers of de Bruijn digraphs, Kautz digraphs, butterfly digraphs, and wrapped butterfly digraphs, some of which are generalizations of previously known results, are obtained. Also, it is shown that for any fixed digraph G, every digraph with n vertices in $\{L^k(G) \mid k \geq 1\}$ has an $O(1) \times O(1) \times O(n)$ three-dimensional straight-line grid drawing.

This paper is organized as follows. In Section 2, we present upper and lower bounds on the queuenumber of $L^k(G)$. In Section 3, we apply the results to specific families of iterated line digraphs. In Section 4, we consider three-dimensional drawings of iterated line digraphs.

2 Queue Layouts of Iterated Line Digraphs

We assume in this paper that a digraph may have loops but not multiple arcs, since $L^k(G)$ has no multiple arcs for all $k \geq 1$ even if G has multiple arcs. Let G be a digraph. If $(x, y) \in A(G)$, then we say that x is a *predecessor* of y, y is a *successor* of x, x is the *tail* of (x, y), and y is the *head* of (x, y). Two arcs such that the head of one arc is the tail of the other arc, are *successive arcs.* For $v \in V(G)$, let $\Gamma_G^+(v)$ denote the set of successors of v in G, and let $A_G^+(v)$ denote the set of arcs with tail v in G. Also let $\delta^+(G) = \min_{v \in V(G)} |A_G^+(v)|$ and $\Delta^+(G) = \max_{v \in V(G)} |A_G^+(v)|$. Analogously, $\Gamma_G^-(v)$, $A_G^-(v)$, $\delta^-(G)$, and $\Delta^-(G)$ are defined.

2.1 An Upper Bound

We first define a restricted queue layout, and then present an algorithm to construct such a restricted queue layout of $L(G)$, based on a restricted queue layout of G, while preserving the number of queues.

For a queue layout of a digraph and an arc $e = (u, v)$ of the digraph, if $\sigma(u) < \sigma(v)$ (resp. $\sigma(v) < \sigma(u)$), then we say that e has right-direction (resp. left-direction). For a loop, we consider that it has no direction. We say that an arc with right-direction and an arc with left-direction have opposite directions.

As a restricted queue layout, we define a tree-queue layout as follows.

Definition 1. *A* tree-queue layout *of a digraph G is a queue layout of G such that for arcs assigned to the same queue, the following two conditions hold.*

- *any two arcs with the same head have opposite directions.*
- *any successive arcs have the same direction except for a loop.*

The tree-queuenumber *of G, denoted by $\mathsf{tqn}(G)$, is the minimum number of queues required for a tree-queue layout of G.*

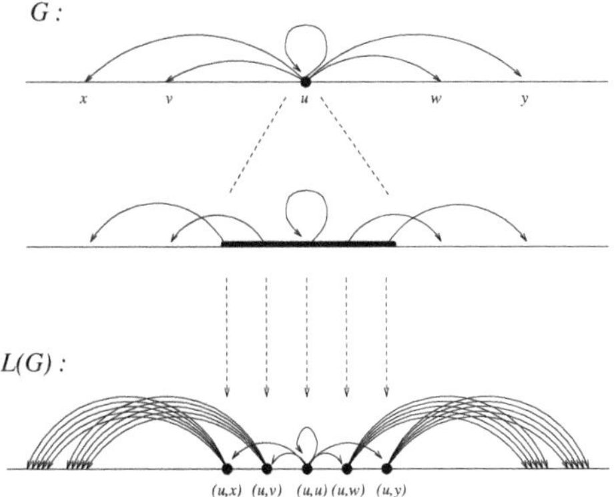

Fig. 1. Algorithm for laying out $L(G)$.

Since a tree-queue layout is a restricted version of a queue layout, it holds that $\mathsf{qn}(G) \leq \mathsf{tqn}(G)$. On the other hand, given a k-queue layout, we can construct a tree-queue layout as follows. For each queue, we divide the set of arcs in the queue according to the direction, and assign arcs with the same direction to the same queue. A loop is assigned to either queue. Next, for each queue, we assign arcs with the same head in the queue to distinct queues. Thus, a layout using at most $2k\Delta^-(G)$ queues is obtained. The resulting layout is a tree-queue layout. Therefore, $\mathsf{tqn}(G) \leq 2\Delta^-(G)\mathsf{qn}(G)$.

The terminology "tree-queue" named after the structural property that the set of arcs assigned to each queue induces a nearly disjoint union of rooted trees, where "nearly" means that two trees may have common leaves and the root of a tree may have a loop.

We will show that the tree-queuenumber of G is an upper bound on the tree-queuenumber of $L^k(G)$.

Theorem 1. $\mathsf{tqn}(L^k(G)) \leq \mathsf{tqn}(G)$ *for all* $k \geq 1$.

Proof. Given a tree-queue layout of G, we present an algorithm to construct a tree-queue layout of $L(G)$ with the same number of queues. For convenience, we consider a line and regard the vertex ordering as a placement of vertices on the line. A vertex x is placed on the left of another vertex y iff $\sigma(x) < \sigma(y)$. We write $x <_l y$ if x is placed on the left of y, and $x \leq_l y$ if x is not placed on the right of y. Note that x may equal to y when $x \leq_l y$.

1. For each vertex u of G, we first regard all the tails of arcs in $A^+_G(u)$ as being different and then stretch the point corresponding to u on the line so that any two arcs in $A^+_G(u)$ obey the first-in/first-out discipline. Then, we assign each arc in $A^+_G(u)$ to the point of its tail (see Figure 1). More formally, the vertices of $L(G)$ are placed on the line in the following way:
   a) If $u <_l w$, then $(u,a) <_l (w,b)$ for any $a \in \Gamma^+_G(u)$, $b \in \Gamma^+_G(w)$.

b) Suppose that $(u,v),(u,w) \in A(G)$.
   i. If $v <_l u <_l w$, then $(u,v) <_l (u,w)$.
   ii. If $(u,u) \in A(G)$ and $v <_l u <_l w$, then $(u,v) <_l (u,u) <_l (u,w)$.
   iii. If $u <_l v <_l w$, then $(u,v) <_l (u,w)$.
   iv. If $w <_l v <_l u$, then $(u,w) <_l (u,v)$.

According to this placement of vertices on the line, we define the vertex ordering of $L(G)$.

2. For each arc $((u,v),(v,w))$ of $L(G)$, we assign it to the queue to which (u,v) is assigned in the tree-queue layout of G.

In what follows, we show that the above algorithm correctly produces a tree-queue layout of $L(G)$. We first show that any two arcs in each queue obey the first-in/first-out discipline. Let $((u,v),(v,w)),((x,y),(y,z)) \in A(L(G))$ such that they are assigned to the same queue. This means that (u,v) and (x,y) are in the same queue in the tree-queue layout of G. If $u = x$, then clearly $((u,v),(v,w))$ and $((x,y),(y,z))$ obey the first-in/first-out discipline, by our algorithm. Then suppose that $u \neq x$. Without loss of generality, we can assume that $u <_l x$. Thus, $(u,v) <_l (x,y)$ (by 1-(a) in the algorithm). Now we assume that $((u,v),(v,w))$ and $((x,y),(y,z))$ do not obey the first-in/first-out discipline, i.e., one of the following inequalities holds:

Case 1: $(u,v) <_l (x,y) \leq_l (y,z) <_l (v,w)$
Case 2: $(y,z) <_l (v,w) \leq_l (u,v) <_l (x,y)$
Case 3: $(u,v) <_l (y,z) \leq_l (x,y) <_l (v,w)$
Case 4: $(y,z) <_l (u,v) \leq_l (v,w) <_l (x,y)$

Assume that Case 1 holds. By our algorithm, we can see that $u <_l x \leq_l y \leq_l v$. Since two arcs with the same head must have opposite directions, $y \neq v$. Hence $u <_l x \leq_l y <_l v$. However, this contradicts the first-in/first-out discipline. Therefore, Case 1 does not hold. Similarly, Case 2 does not hold. Next, consider Case 3, and assume that it holds. Then we obtain that $u \leq_l y \leq_l x \leq_l v$. If $u \neq y$ and $x \neq v$, then (u,v) and (x,y) contradicts the first-in/first-out discipline. Thus, it must be that $u = y$ or $x = v$. Suppose that $x = y$, i.e., (x,y) is a loop. If $x = v$, then (x,y) and (u,v) have the same head but not opposite directions. Thus, $u = y$. However, by our algorithm (1-(b)-(ii)), (x,y) must be placed on the left of (u,v), which contradicts the situation in Case 3. Suppose that $x \neq y$, i.e., (x,y) is not a loop. Then (u,v) and (x,y) are successive arcs with opposite directions, which contradicts a tree-queue layout of G. Hence, Case 3 does not hold. It can be similarly shown that Case 4 does not hold. Therefore, our algorithm correctly produces a queue layout of $L(G)$.

Next, we show that the queue layout of $L(G)$ is a tree-queue layout. Suppose that $((a,u),(u,v)),((b,u),(u,v)) \in A(L(G))$ such that these two arcs are assigned to the same queue. From our algorithm, (a,u) and (b,u) must be in the same queue in the tree-queue layout of G such that their directions are opposite. Note that neither (a,u) nor (b,u) is a loop. Since the directions of $((a,u),(u,v))$ and $((b,u),(u,v))$ are the same as those of (a,u) and (b,u), respectively, $((a,u),(u,v))$ and $((b,u),(u,v))$ have opposite directions.

Next suppose that $((u,v),(v,w))$ and $((v,w),(w,x))$ are in the same queue. Then (u,v) and (v,w) are in the same queue in the tree-queue layout of G. Thus, (u,v) and (v,w) have the same direction, or (u,v) is a loop. If (u,v) is not a loop, then $((u,v),(v,w))$ and $((v,w),(w,x))$ have the same direction. Suppose that (u,v) is a loop. Without loss of generality, we can assume that $u = v <_l w$. Then by our algorithm (1-(b)-(ii)), $(u,v) <_l (v,w) <_l (w,x)$. Therefore, $((u,v),(v,w))$ and $((v,w),(w,x))$ have the same direction.

Consequently, the layout of $L(G)$ obtained by our algorithm is a tree-queue layout with the same number of queues in the tree-layout of G. By applying the algorithm iteratively, we have $\mathsf{tqn}(L^k(G)) \le \mathsf{tqn}(L^{k-1}(G)) \le \cdots \le \mathsf{tqn}(G)$. □

Corollary 1. $\mathsf{qn}(L^k(G)) \le \mathsf{tqn}(G)$ *for all* $k \ge 1$.

It can be easily checked that $\delta^-(G) \le \mathsf{tqn}(G) \le |V(G)|$. (Thus, for the complete digraph K_n°, it holds that $\mathsf{tqn}(K_n^\circ) = n$.) Therefore, we have the following corollary.

Corollary 2. $\mathsf{qn}(L^k(G)) \le |V(G)|$ *for all* $k \ge 1$.

2.2 A Lower Bound

It has been shown in [21] that in a queue layout of a graph with n vertices, one queue has at most $2n-3$ edges. Thus, a general lower bound on the queuenumber of a graph is obtained as follows.

Theorem 2. [21] *Let G be a graph with n vertices. Then* $\mathsf{qn}(G) \ge \left\lceil \frac{|E(G)|}{2n-3} \right\rceil$.

The number of arcs in $L^k(G)$ cannot be expressed in a simple form in general. Besides, we need to consider not only the number of arcs, but also the numbers of loops and cycles of length two (symmetric arcs) in a digraph in order to count edges in the underlying graph. Then, we use a structural property of a line digraph.

For a vertex v with no loop in G, $L(G)$ has a complete bipartite digraph corresponding to v, since $A_G^-(v) \cup A_G^+(v)$ forms a complete bipartite digraph with partite sets of size $|A_G^-(v)|$ and $|A_G^+(v)|$ in $L(G)$. The queuenumber of a complete bipartite graph $K_{m,n}$ with partite sets of size m and n has been determined in [21] to be $\min\{\lceil \frac{m}{2} \rceil, \lceil \frac{n}{2} \rceil\}$,

If G has only loops, then $\Delta^-(G) = \Delta^+(G) = 1$. On the other hand, if G has an arc which is not a loop and $\max\{\delta^-(G), \delta^+(G)\} > 0$, then there is a vertex with no loop in $L^k(G)$ for all $k \ge 1$. Now suppose that G has an arc which is not a loop and $\max\{\delta^-(G), \delta^+(G)\} > 0$. Let v be a vertex with no loop in $L^k(G)$. It can be easily checked that each vertex in $L^k(G)$ corresponds to a walk of length k in G. Suppose that v corresponds to a walk $(v_1, v_2, \ldots, v_{k+1})$ in G. Since $|A_{L^k(G)}^-(v)| = |A_G^-(v_1)|$ and $|A_{L^k(G)}^+(v)| = |A_G^+(v_{k+1})|$, $L^{k+1}(G)$ contains a complete bipartite digraph with partite sets of size $|A_G^-(v_1)|$ and $|A_G^+(v_{k+1})|$. Thus, $\mathsf{qn}(L^{k+1}(G)) \ge \min\{\frac{|A_G^-(v_1)|}{2}, \frac{|A_G^+(v_{k+1})|}{2}\}$. Therefore, the following theorem holds.

Theorem 3. $\mathsf{qn}(L^k(G)) \geq \min\{\lceil \frac{\delta^-(G)}{2} \rceil, \lceil \frac{\delta^+(G)}{2} \rceil\}$ *for all* $k \geq 2$.

Suppose that G has a loop. Let $\mathsf{diam}(G)$ denote the diameter of G. (When G is not strongly connected, we define $\mathsf{diam}(G)$ as ∞.) Then for all $k > 2\mathsf{diam}(G)$, there is a walk of length k from a vertex with indegree $\Delta^-(G)$ to a vertex with outdegree $\Delta^+(G)$ passing through a loop. Thus, the following theorem holds. (Note that it trivially holds for the case that G has only loops, and the case that G is not strongly connected.)

Theorem 4. *Let G be a digraph with a loop. Then,*
$\mathsf{qn}(L^k(G)) \geq \min\{\lceil \frac{\Delta^-(G)}{2} \rceil, \lceil \frac{\Delta^+(G)}{2} \rceil\}$ *for all* $k > 2\mathsf{diam}(G)$.

3 Queue Layouts for Specific Classes

3.1 De Bruijn and Kautz Digraphs

The *de Bruijn digraph* $B(d, D)$ can be defined as $L^{D-1}(K_d^\circ)$ where K_d° is the complete digraph with d vertices [11]. The *Kautz digraph* $K(d, D)$ can be also defined as $L^{D-1}(K_{d+1}^*)$, where K_{d+1}^* is the complete symmetric digraph with $d+1$ vertices. These digraphs are representative interconnection networks in iterated line digraphs, and have many nice properties (see [3]). As a direct consequence of Corollary 2 and Theorem 3, upper and lower bounds on $\mathsf{qn}(B(d, D))$ and $\mathsf{qn}(K(d, D))$ are obtained. (These results can be also obtained as corollaries of a result on the queuenumbers of generalized de Bruijn and Kautz digraphs [15].)

Proposition 1. $\lceil \frac{d}{2} \rceil \leq \mathsf{qn}(B(d, D)) \leq d$ $(D \geq 3)$.

Proposition 2. $\lceil \frac{d}{2} \rceil \leq \mathsf{qn}(K(d, D)) \leq d + 1$ $(D \geq 2)$.

3.2 Butterfly Digraphs

The butterfly graph is one of the most popular interconnection networks. The *k-ary butterfly digraph* $b(k, r)$ is a directed version of the k-ary butterfly graph, and can be defined as an iterated line digraph. As an origin digraph to which the line digraph operation is applied, there are two digraphs.

The *Kronecker product* of two digraphs G_1 and G_2, denoted by $G_1 \otimes G_2$, is defined as follows:

$$\begin{cases} V(G_1 \otimes G_2) = V(G_1) \times V(G_2), \\ A(G_1 \otimes G_2) = \{((u_1, u_2), (v_1, v_2)) \mid (u_1, v_1) \in A(G_1) \text{ and } (u_2, v_2) \in A(G_2)\}. \end{cases}$$

Then it holds that $b(k, r) \cong L^{r-1}(K_k^\circ \otimes P_{2r})$, where P_{2r} is a directed path with $2r$ vertices [2].

A *complete k-ary out-tree* is an out-tree such that every non-leaf vertex has outdegree k, and every path from the root to a leaf has the same length. A *complete k-ary in-tree* is an in-tree obtained from a complete k-ary out-tree by reversing the orientations of arcs. Let $X(k, r)$ denote the digraph obtained from the complete k-ary in-tree of depth r and the complete k-ary out-tree of depth r by identifying their roots. Then it holds that $b(k, r) \cong L^r(X(k, r))$ [17].

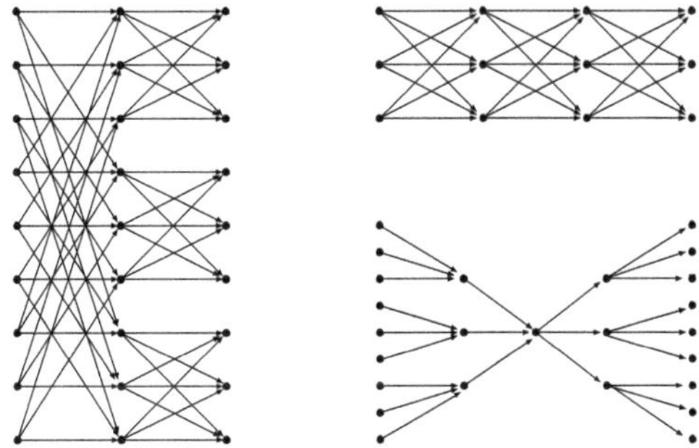

Fig. 2. $b(3,2)$, $K_3^\circ \otimes P_4$, and $X(3,2)$.

Proposition 3. $\lceil \frac{k}{2} \rceil \leq \mathsf{qn}(b(k,r)) \leq \min\left\{k, \lfloor \frac{k}{2} \rfloor + 2\right\}$.

Proof. The k-ary butterfly graph contains a complete bipartite graph $K_{k,k}$. Thus, $\lceil \frac{k}{2} \rceil$ is a lower bound on $\mathsf{qn}(b(k,r))$. For an upper bound, we first consider a tree-queue layout of $K_k^\circ \otimes P_{2r}$. Let $V(K_k^\circ) = \{v_1, v_2, \ldots, v_k\}$ and $V(P_{2r}) = \{w_1, w_2, \ldots, w_{2r}\}$ such that $(w_i, w_{i+1}) \in A(P_{2r})$ for $1 \leq i < 2r$. Then $V(K_k^\circ \otimes P_{2r}) = \{(v_i, w_j) \mid 1 \leq i \leq k, 1 \leq j \leq 2r\}$. As a vertex-ordering of $K_k^\circ \otimes P_{2r}$, we employ the lexicographical ordering with respect to the indices j and i of (v_i, w_j). (In the following, we omit the notation σ for the vertex-ordering.)

$$(v_1, w_1) < \cdots < (v_k, w_1) < (v_1, w_2) < \cdots < (v_{k-1}, w_{2r}) < (v_k, w_{2r}).$$

Let $A_t = \{((v_t, w_p), (v_i, w_{p+1})) \mid 1 \leq i \leq k, 1 \leq p < 2r\}$ for $t = 1, 2, \ldots, k$. We prepare k queues and assign the arcs in A_t into the t-th queue. It can be easily checked that this layout is a tree-queue layout.

Next we consider a tree-queue layout of $X(k,r)$. It is not difficult to see that the complete k-ary out-tree of depth r has a 1-tree-queue layout. For the complete k-ary in-tree of depth r, we prove by induction on the depth that it has a $(\lfloor \frac{k}{2} \rfloor + 1)$-tree-queue layout. Let $T(k,r)$ denote the complete k-ary in-tree of depth r. Also, let $V(T(k,r)) = \{(i,j) \mid 1 \leq i \leq k^j, 0 \leq j \leq r\}$, where j denotes the depth of a vertex (i,j). Note that $(1,0)$ is the root of $T(k,r)$. Suppose that $r = 1$. We order the vertices as follows. $(1,1) < \cdots < (\lfloor \frac{k}{2} \rfloor, 1) < (1,0) < (\lfloor \frac{k}{2} \rfloor + 1, 1) < \cdots < (k,1)$. Then we assign arcs $((i,1),(1,0))$ and $((\lfloor \frac{k}{2} \rfloor + i, 1), (1,0))$ to the i-th queue $(1 \leq i \leq \lfloor \frac{k}{2} \rfloor)$. When k is odd, we assign the arc $((k,1),(1,0))$ to the $(\lfloor \frac{k}{2} \rfloor + 1)$-th queue. Clearly, this layout is a tree-queue layout. Now assume that $T(k,r)$ has a tree-queue layout using $\lfloor \frac{k}{2} \rfloor + 1$ queues. By adding k arcs to each leaf of $T(k,r)$, $T(k,r+1)$ is obtained. For each leaf of $T(k,r)$, we do the similar manipulation to the case of $r = 1$. Let v be a leaf. Also let (v,w) be the arc with tail v. About half of $\Gamma^-_{T(k,r+1)}(v)$ is placed on the left of v, and the remaining half is placed on the right of v. Then we assign two arcs with opposite directions in $A^-_{T(k,r+1)}(v)$ to each queue. Here, if k is even, then we assign no arc to the queue to which (v, w) is assigned. If k is odd, then we assign one arc with the same direction as (v, w) to the queue to which (v, w) is assigned, and use the other queues for the remaining arcs. (Thus, if (v, w) has right (resp. left) direction, then $\lfloor \frac{k}{2} \rfloor + 1$ vertices in $\Gamma^{-}_{T(k,r+1)}(v)$ are placed on the left (resp. right) of v.) After such an assignment of arcs, we precisely set the order position of each tail while preserving the tree-queue layout style. Such a setting is always possible, since the underlying graph of a digraph induced by the set of arcs assigned to each queue is a disjoint union of paths such that for any vertex x with degree two in the graph, one neighbor is on the left of x and the other is on the right of x. Therefore, $T(k, r)$ has a tree-queue layout with $\lfloor \frac{k}{2} \rfloor + 1$ queues, and thus $X(k, r)$ has a tree-queue layout with $\lfloor \frac{k}{2} \rfloor + 2$ queues. □

3.3 Wrapped Butterfly Digraphs

The k-ary wrapped butterfly digraph $wb(k, r)$, $r \geq 3$, can be defined as $L^{r-1}(K^{\circ}_{k} \otimes C_r)$, where C_r is a directed cycle of length r [2].

Proposition 4. $\lceil \frac{k}{2} \rceil \leq \mathsf{qn}(wb(k, r)) \leq 2k$.

Proof. Similarly to $b(k, r)$, $\lceil \frac{k}{2} \rceil$ is a lower bound on $\mathsf{qn}(wb(k, r))$. Let $V(K^{\circ}_{k}) = \{v_1, v_2, \dots, v_k\}$ and $V(C_r) = \{w_1, w_2, \dots, w_r\}$ such that $(w_i, w_{i+1}) \in A(C_{2r})$ for $1 \leq i < r$ and $(w_r, w_1) \in A(C_{2r})$. Similarly to the first case of butterfly digraphs, we order the vertices, and assign the arcs into k queue, except for the arcs $((v_i, w_r), (v_j, w_1))$, $1 \leq i, j \leq k$. We prepare other k queues and assign the arcs in $\{((v_t, w_r), (v_j, w_1)) \mid 1 \leq j \leq k\}$ into the $(k + t)$-th queue. We can easily check that the resulting layout is a tree-queue layout. □

4 Three-Dimensional Drawings of Iterated Line Digraphs

A three-dimensional drawing treated here is a *three-dimensional straight-line grid drawing* of a graph. The vertices are represented by distinct points in $\mathbf{Z}^3$. Each edge are represented by a line-segment between its end-vertices such that edges only intersect at common end-vertices, and an edge only intersects a vertex which is an end-vertex of the edge. If a three-dimensional drawing is contained in an axis-aligned box with side length $X - 1$, $Y - 1$, $Z - 1$, then the drawing is called an $X \times Y \times Z$ drawing with volume $X \cdot Y \cdot Z$. A three-dimensional straight-line grid drawing of a digraph is similarly defined.

Dujmović et al. [8] introduced a track layout. A *k-track assignment* of a graph G consists of a partition $V_1, V_2, \dots, V_k$ of $V(G)$ such that each V_i is an independent set (i.e., if $\{u, v\} \in E(G)$ such that $u \in V_i$ and $v \in V_j$, then $i \neq j$) and a linear ordering σ_i of each V_i. Each pair of V_i and σ_i is called a track. An *X-crossing* in a track assignment consists of two edges $\{v, w\}$ and $\{x, y\}$ such that $v, x \in V_i$ and $w, y \in V_j$, where $i \neq j$, and $\sigma_i(v) < \sigma_i(x)$, $\sigma_j(y) < \sigma_j(w)$. A k-track assignment with no X-crossing is called a *k-track layout.* The *track-number* of G, denoted by $\mathsf{tn}(G)$, is the minimum number of tracks required for a track layout of G. Dujmović et al. proved the following.

Theorem 5. [8] *If G has a k-track layout, then G has a $k \times 2k \times 2kn'$ three-dimensional drawing, where n' is the maximum number of vertices in a track.*

The star chromatic number of a graph is the minimum number of colors required for a vertex-coloring of the graph such that each bichromatic subgraph is a disjoint union of stars. Based on the star chromatic number and the queuenumber, Wood [31] presented an upper bound on the track-number. (The upper bound has been improved in [7].)

Theorem 6. [7] *Let G be a graph with star chromatic number $\chi_{st}(G) \leq c$, and queuenumber $\mathsf{qn}(G) \leq q$. Then $\mathsf{tn}(G) \leq c(2q+1)^{c-1}$.*

In Section 2, we have shown that for any fixed digraph G, the queuenumber of every $L^k(G)$ is at most the tree-queuenumber of G. Also the underlying graph of $L^k(G)$ has bounded star chromatic number, since the maximum outdegree and indegree of $L^k(G)$ are equal to those of G, respectively, and it has been shown that graphs with bounded maximum degree have bounded star chromatic number [1,10]. Thus, we have the following theorem.

Theorem 7. *For any fixed digraph G, every iterated line digraph of G has an $O(1) \times O(1) \times O(n)$ three-dimensional drawing.*

Corollary 3. *For any fixed $d \geq 2$, $B(d, D)$ and $K(d, D)$ have three-dimensional drawings with $O(n)$ volume.*

For $b(k,r)$ and $wb(k,r)$, the origin digraphs to which the line digraph operation is applied, depend on r. However, their maximum degree and upper bounds on the queuenumber depend only on k. Thus, we also have the following corollary.

Corollary 4. *For any fixed $k \geq 2$, $b(k,r)$ and $wb(k,r)$ have three-dimensional drawings with $O(n)$ volume.*

For $B(2, D)$ and $b(2, r)$, their three-dimensional drawings with $O(n)$ volume was previously shown in [31].

We can show that the number of vertices in G is an upper bound on the track-number of $L^k(G)$, if we restrict ourselves to digraphs with no loop and no symmetric arcs.

Theorem 8. *Let G be a digraph with no loop and no symmetric arcs. Then $\mathsf{tn}(L^k(G)) \leq |V(G)|$ for all $k \geq 1$.*

Proof. Let $V(G) = \{v_1, v_2, \ldots, v_p\}$. We order the vertices of G in any manner. Then we prepare p queues, and assign the arcs in $A^+_G(v_i)$ to the i-th queue. This trivial queue layout is indeed a tree-queue layout. Also this layout corresponds to a p-track layout by considering that each track V_i consists of only v_i.

By applying the algorithm in Theorem 1, we can obtain a tree-queue layout of $L(G)$ using the same number of queues. We show that this layout also corresponds to a p-track layout, where each track $V_i = A_G^+(v_i)$ and the linear ordering of each V_i follows the vertex ordering of the queue layout. Since G has no loop, there is no arc between the vertices in each track V_i. Thus, the queue layout corresponds to a p-track assignment. Since G has no symmetric arcs, for any V_i and V_j $(i \neq j)$, all arcs between V_i and V_j are in the same queue. Also all vertices in each track are consecutive with respect to the vertex ordering σ of the queue layout of $L(G)$, i.e., if for some $u \in V_i$ and $v \in V_j$, $\sigma(u) < \sigma(v)$, then for every $x \in V_i$ and $y \in V_j$, $\sigma(x) < \sigma(y)$. Therefore, there is no X-crossing in the track assignment.

Similarly, we can see that the tree-queue layout of $L^k(G)$ obtained by applying the algorithm iteratively, corresponds to a p-track layout, where each track V_i consists of the vertices corresponding to walks of length k starting from v_i in G, and the linear ordering of V_i follows the vertex ordering of the tree-queue layout of $L^k(G)$. □

Corollary 5. *Let G be a digraph with no loop and no symmetric arcs. Also let $p = |V(G)|$. Then $L^k(G)$ has a $p \times 2p \times 2pn'$ three-dimensional drawing, where n' is the maximum number of vertices in a track.*

References

1. N. Alon, C. McDiarmid, and B. Reed, Acyclic coloring of graphs, *Random Structures & Algorithms* 2 (1991) 277-288.
2. J.-C. Bermond, E. Darrot, O. Delmas, and S. Perennes, Hamiltonian circuits in the directed wrapped butterfly network, *Discrete Applied Math.* 84 (1998), 21–42.
3. J.-C. Bermond and C. Peyrat, "De Bruijn and Kautz networks: A competition for the hypercube?," Hypercube and distributed computers, F. André, J.P. Verjus (Editors), North Holland, Amsterdam, 1989, pp.279–293.
4. F. Bernhart and P.C. Kainen, The book thickness of a graph, *J. Combin. Theory* Ser. B 27 (1979), 320–331.
5. F.R.K. Chung, F.T. Leighton, and A.L. Rosenberg, Embedding graphs in books: a layout problem with application to VLSI design, *SIAM J. Algebraic Discrete Methods* 8 (1987), 33–58.
6. V. Dujmović and D.R. Wood, Tree-Partitions of k-trees with applications in graph layouts, Proc. of WG'03, Lecture Notes in Comput. Sci., Springer, to appear.
7. V. Dujmović and D.R. Wood, New results in graph layout, Tech. Report TR-2003-04, School of Computer Science, Carleton University, Canada, 2002.
8. V. Dujmović, P. Morin, and D.R. Wood, Path-width and three-dimensional straight-line grid drawings of graphs, Proc. of GD'02, Lecture Notes in Comput. Sci. 2528, pp. 42–53, Springer, 2002.
9. S. Even and A. Itai, "Queues, stacks and graphs," Theory of Machines and Computations, Z. Kohavi and A. Paz (Editors), Academic Press, New York, 1971, pp. 71–86.
10. G. Fertin, A. Raspaud, and B. Reed, On star coloring of graphs, Proc. of WG'01, Lecture Notes in Comput. Sci. 2204, pp.140–153, Springer, 2001.
11. M.A. Fiol, J.L.A. Yebra, and I. Alegre, Line digraph iterations and the (d,k) digraph problem, *IEEE Trans. Comput.* 33 (1984) 400–403.

12. R.A. Games, Optimal book embeddings of the FFT, Benes, and barrel shifter networks, *Algorithmica* 1 (1986), 233–250.
13. J.L. Ganley and L.S. Heath, The pagenumber of k-trees is $O(k)$, *Discrete Applied Math.* 109 (2001), 215–221.
14. T. Hasunuma, Embedding iterated line digraphs in books, *Networks* 40 (2002) 51–62.
15. T. Hasunuma, Queuenumbers and stacknumbers of generalized de Bruijn and Kautz digraphs, submitted.
16. T. Hasunuma and Y. Shibata, Embedding de Bruijn, Kautz and shuffle-exchange networks in books, *Discrete Applied Math.* 78 (1997), 103–116.
17. T. Hasunuma and Y. Shibata, Containment of butterflies in networks constructed by the line digraph operation, *Inform. Process. Lett.* 61 (1997), 25–30.
18. L.S. Heath, F.T. Leighton, and A.L. Rosenberg, Comparing queues and stacks as mechanisms for laying out graphs, *SIAM J. Discrete Math.* 5 (1992)
19. L.S. Heath, S.V. Pemmaraju, and A.N. Trenk, Stack and queue layouts of directed acyclic graphs I, *SIAM J. Comput.* 28 (1999) 1510–1539.
20. L.S. Heath and S.V. Pemmaraju, Stack and queue layouts of directed acyclic graphs II, *SIAM J. Comput.* 28 (1999) 1588–1626.
21. L.S. Heath and A.L. Rosenberg, Laying out graphs using queues, *SIAM J. Comput.* 21 (1992) 927–958.
22. M. Konoe, K. Hagiwara, and N. Tokura, On the pagenumber of hypercubes and cube-connected cycles, *IEICE Trans.* J71-D (1988), 490–500 (in Japanese).
23. S.M. Malitz, Genus g graphs have pagenumber $O(\sqrt{g})$, *J. Algorithms* 17 (1994), 85–109.
24. D.J. Muder, M.L. Weaver, and D.B. West, Pagenumber of complete bipartite graphs, *J. Graph Theory* 12 (1988), 469–489.
25. S.V. Pemmaraju, Exploring the powers of stacks and queues via graph layouts, Ph.D thesis, Virginia Polytechnic Institute and State University, Virginia, USA, 1992.
26. S. Rengarajan and C.E. Veni Madhavan, Stack and queue number of 2-trees, Proc. of COCOON 95, Lecture Notes in Comput. Sci. 959, pp.203–212, Springer, 1995.
27. A.L. Rosenberg, The Diogenes approach to testable fault-tolerant arrays of processors, *IEEE Trans. Comput.* C-32 (1983), 902–910.
28. R.P. Swaminathan, D. Giriaj, and D.K. Bhatia, The pagenumber of the class of bandwidth-k graphs is $k-1$, *Inform. Process. Lett.* 55 (1995), 71–74.
29. R.E. Tarjan, Sorting using networks of queues and stacks, *J. Assoc. Comput. Mach.*, 19 (1972) 341–346.
30. D.R. Wood, Bounded degree book embeddings and three-dimensional orthogonal graph drawing, Proc. of GD'01, Lecture Notes in Comput. Sci. 2265, pp.312–327, Springer, 2002.
31. D.R. Wood, Queue layouts, tree-width, and three-dimensional graph drawing, Proc. of FSTTCS'02, Lecture Notes in Comput. Sci. 2556, pp.348–359, Springer, 2002.
32. M. Yannakakis, Embedding planar graphs in four pages, *J. Comput. System Sci.* 38 (1989), 36–67.

Track Drawings of Graphs with Constant Queue Number*

Emilio Di Giacomo[1] and Henk Meijer[2]

[1] Università degli Studi di Perugia, Perugia, Italy.
digiacomo@diei.unipg.it
[2] Queen's University, Kingston, Ontario, Canada.
henk@cs.queensu.ca

Abstract. A k-track drawing is a crossing-free $3D$ straight-line drawing of a graph G on a set of k parallel lines called *tracks*. The minimum value of k for which G admits a k-track drawing is called the *track number* of G. In [9] it is proved that every graph from a proper minor closed family has constant track number if and only if it has constant queue number. In this paper we study the track number of well-known families of graphs with small queue number. For these families we show upper bounds and lower bounds on the track number that significantly improve previous results in the literature. Linear time algorithms that compute track drawings of these graphs are also presented and their volume complexity is discussed.

1 Introduction and Overview

The problem of computing a drawing of a graph with small area/volume has received a lot of attention in the graph drawing literature during the last decade.

Felsner et al. [5] initiated the study of integer grids consisting of parallel grid lines, called *tracks*. In particular, they focus on the *box* and the 3-*prism*. A box is a grid consisting of four parallel lines, one grid unit apart from each other and a 3-prism uses three non-coplanar parallel lines. It is shown that all outerplanar graphs can be drawn on a 3-prism where the length of the lines is $O(n)$. This result gives the first algorithm to compute a crossing-free straight-line 3D drawing with linear volume for a non-trivial family of planar graphs. Moreover it is shown that there exist planar graphs that cannot be drawn on the prism and that even a box does not support all planar graphs.

Dujmović et al. [2] show that if a graph G admits a drawing Γ on a a grid ϕ consisting of a constant number of parallel lines, then G has a linear volume upper bound. This result suggests that the focus of the research should be on minimizing the number of tracks in a restricted integer grid, independent of the length of the tracks themselves. The *track number* $tn(G)$ of a graph G is the minimum number of tracks that is required to compute such a drawing.

* Research partially supported by "Progetto ALINWEB: Algoritmica per Internet e per il Web", MIUR – PRIN. We thank Giuseppe Liotta for useful discussions on the subject of this paper.

G. Liotta (Ed.): GD 2003, LNCS 2912, pp. 214–225, 2004.

Wood [9] shows a relationship between $tn(G)$ and another well-studied graph parameter, the *queue number* $qn(G)$ (i.e. the minimum number of queues in a queue layout of G [8]). He proves that every graph from a proper minor closed family has constant track number if and only if it has constant queue number. By the result of Wood all families of graphs whose queue number is known to be constant (for example series-parallel graphs, Halin graphs, Benes networks, arched leveled planar graphs, X-trees, unicyclic graphs), have a three dimensional straight-line grid drawing with linear volume. A recent result by Dujmović and Wood [4] shows that linear volume can also be achieved for graphs with bounded tree width. We observe however that value of the track number (and hence of the volume) deriving from the results above are often very large. In [3] it is shown that $tn(G) \leq c(2qn(G)+1)^{c-1}$ where c is the star chromatic number of G ($c \geq 3$ for any graph). Therefore, even for the (apparently innocent) family of planar graphs whose queue number is 1 we obtain an upper bound for the track number of at least 27 and by using the technique in [2] a drawing with a bound on the volume of $54 \times 59 \times 59\lceil \frac{n}{27} \rceil$. From the observation above it is natural to ask whether it is possible to reduce the bounds on track number and volume for graphs with constant queue number.

In this paper we present new lower and upper bounds on the track number (and hence on the volume) of some families of graphs that are known to have constant queue number. Our main contributions can be listed as follows.

- The family of graphs with queue number one (i.e. arched leveled planar graphs) are proved to have track number at least four and at most five. A drawing algorithm is presented for arched leveled planar graphs that gives a volume bound of $3 \times 3 \times n$.
- The track number of X-trees (that have queue number 2) is proved to be three and a volume bound of $2 \times 2 \times \frac{4(n+1)}{7}$.
- A lower bound of three and an upper bound of four is presented for Halin Graphs. A volume bound of $2 \times 2 \times n$ is also shown.

The drawing algorithm described in order to prove the above results all have $O(n)$ time complexity and use integer arithmetic.

The remainder of of the paper is organized as follows. In Section 2 some definitions and results about track and queue layout are recalled. The bounds on the track number and on the volume of arched leveled planar graphs are proved in Section 3. The track number of X-trees and Halin Graphs is studied in Section 4. For reasons of space some detail are omitted.

2 Preliminaries

Let $G = (V, E)$ be a graph. A *track assignment* of G consists of a partition $\{t_i \mid i \in I \subseteq \mathbb{N}\}$ of V, and of a total ordering $<_i$ of the vertices in each set t_i. Each set t_i is called a *track-set*. An *overlap* in a track assignment consists of three vertices u, v, and w such that they are in the same track-set t_i, there exists the edge (u, w) and $u <_i v <_i w$. An *X-crossing* in a track assignment consists of two edges (u_0, v_0) and (u_1, v_1) such that u_0 and u_1 are on the same track-set t_i, v_0 and v_1 are on another track-set t_j $(i \neq j)$ with $u_0 <_i u_1$ and $v_1 <_j v_0$. Figure 1(a) shows an example of track assignment for a graph G. Vertices v_1, v_5 and v_2 form an overlap, as well as vertices v_3, v_6 and v_4. Edges (v_5, v_4), (v_2, v_3) form an X-crossing. Another X-crossing is formed by edges (v_6, v_8) and (v_4, v_7).

A *track layout* is a track assignment with no overlaps and no X-crossings. Figure 1(b) shows an example of a track layout of the graph of Figure 1(a). A track layout with k track-sets is also called a k-track layout. The *track number* of a graph G, denoted by $tn(G)$, is the minimum k such that G has a k-track layout[1]. A set of k track-sets is also called a *k-prism*.

In the rest of the paper a track layout will be specified by assigning to each vertex v two numbers: $track(v)$ is an integer that denotes the index of the track-set containing v; $order(v)$ is an integer that denotes the ordering of v in $track(v)$. We say that $u <_i v$ if $track(u) = track(v) = i$ and $order(u) < order(v)$. We shall sometimes simplify the notation and write $u < v$ instead of $u <_i v$.

A *track* is a set of 3D grid points on straight line of infinite length. We always assume that a track is parallel to the x-axis, thus a track is the set of all the grid points having the same y- and z-coordinate. We denote as (x, Y, Z) a track whose points have y-coordinate Y and z-coordinate Z. A *track drawing* of a graph G on k tracks, also called k-track drawing, is a 3D straight-line crossing-free grid drawing of G such that each vertex of G is drawn on one of k tracks. The drawing in Figure 1(c) is a track drawing of the graph of Figure 1(a). The k tracks of a k-track drawing will be denoted as $T_0, T_1, \ldots, T_{k-1}$. A *strip* σ_{ij} is the portion of plane delimited by tracks T_i and T_j. A strip is assumed to be an open set. Two strips σ_{ij} and σ_{hl} are called *crossing strips* if they cross each other. If a drawing of a graph is contained in an axis-aligned box with side lengths $X - 1$, $Y - 1$ and $Z - 1$ then we say that the volume of the drawing is $X \times Y \times Z$.

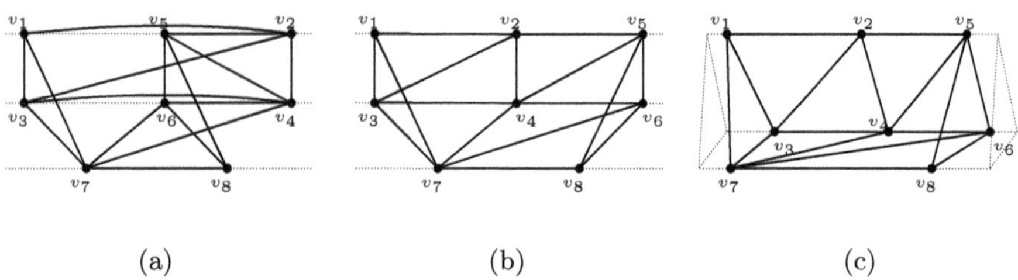

Fig. 1. (a) A track assignment of a graph G. (c) A track layout of G. (d) A track drawing of G.

[1] The concepts of track assignment and track layout are introduced also in [2], where it is assumed that no edge has both vertices in a same track-set.

In [2] a general technique is described for computing a track drawing of a graph G from a track layout of G. In order to reduce the volume of the drawing we use a different technique to obtain a track drawing from a track layout. The obtained drawing have neither overlaps nor X-crossings, because there is no overlap and no X-crossing in a track layout. The only possible crossings are crossings between two edges on a pair of crossing strips. Let $e_0 = (u_0, v_0)$ and $e_1 = (u_1, v_1)$ be two edges of G on two crossing strips σ_{ij} and σ_{hl}. If e_0 and e_1 cross each other, then the four points representing u_0, v_0, u_1 and v_1 are co-planar, i.e. the following equation is satisfied:

$$\begin{vmatrix} 1 & 1 & 1 & 1 \\ x(u_0) & x(u_1) & x(v_0) & x(v_1) \\ y(u_0) & y(u_1) & y(v_0) & y(v_1) \\ z(u_0) & z(u_1) & z(v_0) & z(v_1) \end{vmatrix} = 0$$

The substitution of the y- and z-coordinates of each vertex in the equation above gives a condition on the x-coordinates of the vertices that must be satisfied in order to have a crossings between e_0 and e_1. Thus it is sufficient to prove that the equation has no solution in order to prove that e_0 and e_1 do not cross each other. We call this equation *co-planarity equation* of e_0 and e_1.

A *queue layout* [8] of a graph G consists of a linear ordering λ of the vertices of G, and a partition of the edges of G into queues, such that no two edges in the same queue are *nested* with respect to λ. In other words there are no edges $e_0 = (u_0, v_0)$ and $e_1 = (u_1, v_1)$ such that e_0 and e_1 are assigned to the same queue and $u_0 < u_1 < v_1 < v_0$ in λ. A queue layout with q queues is also called a *q-queue layout.* The *queue number* of a graph G, denoted by $qn(G)$, is the minimum q such that G has a q-queue layout.

The following theorem [3] relates the track number of a graph to the queue number of a graph.

Theorem 1. *[3] Let G be a graph with star chromatic number $\chi_{st}(G) \le c$, and queue number $qn(G) \le q$. Then G has a t-track layout, where $t \le c(2q+1)^{c-1}$.*

3 Graphs with $qn = 1$

Graphs with queue number equal to 1 are characterized in [8], where it is shown that they are planar graphs and admit a leveled planar embedding. For this reason they are called *arched leveled planar graphs.* We first give a lower bound on the track number and then we present an upper bound that improves the result in [9]. We start with a basic lemma that is used to prove the lower bound.

Lemma 1. *Let G_0 be the graph in Figure 2(a). Then in any 3-track layout, vertices u and v are in different track-sets, and at least one of the vertices w_i $(i = 0, \ldots, 3)$ is on the third track-set.*

Lemma 2. *Let G be the graph in Figure 2(b). Then $qn(G) = 1$ and $tn(G) \ge 4$.*

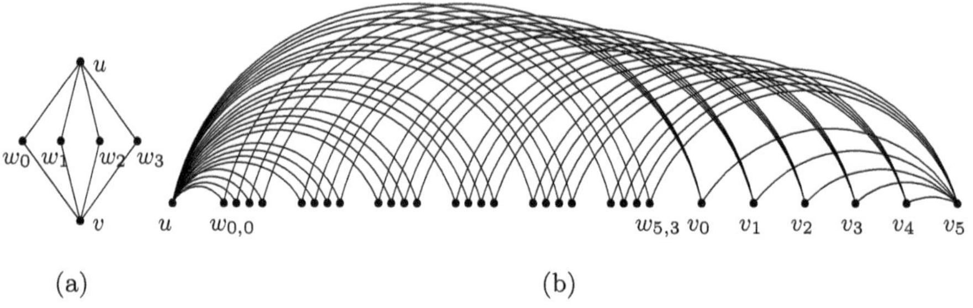

Fig. 2. (a) Graph G_0 of Lemma 1. (b) Graph G with $qn(G) = 1$ and $tn(G) = 4$.

Proof. The proof that G has queue number 1 follows from Figure 2(b) where a 1-queue layout of G is given. We prove that it does not admit a track layout on 3 track-sets. Suppose for a contradiction that there exists a 3-track layout of G with track-sets t_0, t_1 and t_2, and let t_0 be the track-set containing u. By Lemma 1 vertices v_i $(0 \leq i \leq 5)$ must be in a track-set different from t_0. Let t_1 be the track-set containing v_5. Two cases are possible:

1. There exist three distinct vertices v_i, v_j and v_k $(0 \leq i, j, k \leq 4)$ that are in t_1. In this case two of the three edges (v_5, v_i), (v_5, v_j), and (v_5, v_k) form an overlap.
2. There exist three distinct vertices v_i, v_j and v_k, that are in t_2. By Lemma 1, there exist three vertices $w_{i,a}$, $w_{j,b}$ and $w_{k,c}$ $(0 \leq a, b, c \leq 3)$ in t_1 adjacent to u and to v_i, v_j and v_k respectively. Assume without loss of generality that $v_i < v_j < v_k$ in t_2. It follows that $w_{i,a} < w_{j,b} < w_{k,c}$ in t_1, because otherwise there would be an X-crossing between edges $(v_i, w_{i,a})$, $(v_j, w_{j,b})$ and $(v_k, w_{k,c})$. If $v_5 < w_{i,a}$ then edges (v_5, v_j) and $(v_i, w_{i,a})$ form an X-crossing. If $w_{i,a} < v_5 < w_{j,b}$ then edges (v_5, v_k) and $(v_j, w_{j,b})$ form an X-crossing. If $w_{j,b} < v_5 < w_{k,c}$ then edges (v_5, v_i) and $(v_j, w_{j,b})$ form an X-crossing. Finally, if $v_5 > w_{k,c}$ then edges (v_5, v_j) and $(v_k, w_{k,c})$ form an X-crossing.

It follows that $tn(G) \geq 4$. □

In order to prove the upper bound on the track number of arched leveled planar graphs we first describe a block decomposition of a connected graph G with $qn(G) = 1$. Then we will describe how to assign the vertices to 5 track-sets. Suppose we have a 1-queue layout of G where the linear ordering of the vertices is $\lambda = v_0, v_1, \ldots, v_{n-1}$. We say that $v_i < v_j$ if $i < j$. A vertex v is called a *special cut vertex* if there are no edges (u, w) with $u < v < w$. A block is a subset of consecutive vertices in λ. The first vertex of a block B_i is called *source vertex* and is denoted as s_i. The last vertex of a block B_i is called *sink vertex* and is denoted as t_i. All the other vertices are called *internal vertices* of B_i. The *block decomposition* is defined as follows: (i) Block B_0 consists of the vertices $v_0, v_1, \ldots, v_j$, where v_j is such that there exists the edge (v_0, v_j) and there is no edge (v_0, v_h) with $v_j < v_h$. In other words j is the largest index such that there is an edge (v_0, v_j). (ii) If the sink vertex t_{i-1} of block B_{i-1} is not a special cut vertex then block B_i consists of the vertices $v_k, v_{k+1}, \ldots, v_j$ with the following properties: there is an edge (v_k, v_j); $v_k < t_{i-1}$; there is no edge (v_g, v_h) with $v_g < t_{i-1}$ and $v_j < v_h$; there is no edge (v_h, v_j) with $v_h < v_k$. In other words j is the largest index such that there is an edge from a vertex smaller than t_{i-1} to v_j and k is the smallest index such that there is an edge (v_k, v_j). (iii) If the sink vertex t_{i-1} of block B_{i-1} is a special cut vertex then block B_i consists of the vertices $v_k, v_{k+1}, \ldots, v_j$, where $v_k = t_{i-1}$, and where v_j is such that there exists an edge (t_{i-1}, v_j) and there is no edge (t_{i-1}, v_h) with $h > j$. In other words j is the largest index such that there is an edge from t_{i-1} to v_j.

An example of a block decomposition is illustrated in Figure 3.

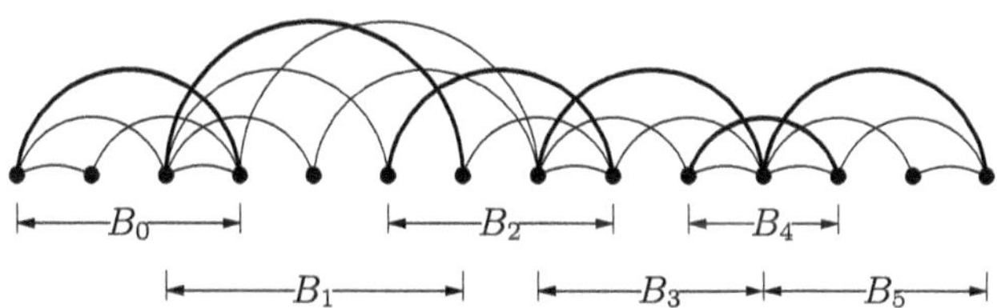

Fig. 3. A block decomposition of an arched leveled planar graph.

Algorithm 1 computes a 5-track layout of an arched leveled planar graph G.

ALPTRACKLAYOUT(G)
Input: 1-queue layout of connected graph G with $\lambda = v_0, v_1, \ldots, v_{n-1}$
Output: A 5-track layout of G.
Let t_0,t_1,t_2,t_3 and t_4 be five track-sets;
Let $B_0, \ldots, B_{k-1}$ be a block decomposition of G;
$track(s_0) \leftarrow 0$;
for $i = 0$ **to** $k-1$
  $t \leftarrow track(s_i)$;
  foreach internal vertex v of B_i not assigned to any track-set
    $track(v) \leftarrow (t+1) \mod 5$;
  $track(t_i) \leftarrow (t+2) \mod 5$;
for $i = 0$ **to** $n-1$
  $order(v_i) = i$;

Algorithm 1: Algorithm ALPTRACKLAYOUT()

Lemma 3. *Let G be a connected graph with $qn(G) = 1$. Let $\lambda = v_0, v_1, \ldots, v_{n-1}$ be the linear ordering of a 1-queue layout of G. The track assignment computed by Algorithm 1 is such that for all edges (v_g, v_h) with $g < h$, either $track(v_h) = (track(v_g) + 1) \mod 5$ or $track(v_h) = (track(v_g) + 2) \mod 5$.*

Lemma 4. *Let G be a graph with $qn(G) = 1$, then $tn(G) \leq 5$.*

Proof. We first assume that G is connected. Suppose we have a 1-queue layout of G with $\lambda = v_0, v_1, \ldots, v_{n-1}$. We prove that Algorithm 1 computes a track layout of G. We have no overlaps since there are no edges (v_i, v_j) with $track(v_i) = track(v_j)$. Also we have no X-crossing. Consider two edges (v_g, v_h) and (v_i, v_j) such that v_g and v_i are in a track-set a and v_h and v_j are in another track-set b. By Lemma 3, there are two cases: (i) if $b = (a+1) \mod 5$ or $b = (a+2) \mod 5$, then $v_g < v_h$ and $v_i < v_j$; (ii) if $b = (a-1) \mod 5$ or $b = (a-2) \mod 5$, then $v_g > v_h$ and $v_i > v_j$. Consider case (i), and assume without loss of generality that $v_g < v_i$. If $v_j < v_h$ then $v_g < v_i < v_j < v_h$ in λ, but this is not possible, because there would be two nested edges in the 1-queue layout. Hence $v_h < v_j$ and therefore there is no X-crossing. With analogous arguments also case (ii) can be considered. □

Lemma 4 proves that every arched leveled planar graph G has track number at most 5. The following theorem summarize the results above and describe how to produce a 5-track drawing with volume $3 \times 3 \times n$.

Theorem 2. *Every arched leveled planar graph has track number at most 5 and admits a 5-track drawing with volume at most $3 \times 3 \times n$ that can be computed in $O(n)$ time. Also, there exists an arched leveled planar graph G such that $tn(G) \geq 4$.*

Proof. The bounds on the track number follow from Lemma 2 and 4. We describe now how to compute a track drawing with volume $3 \times 3 \times n$. Consider the five tracks $(x, 2, 1)$, $(x, 0, 1)$, $(x, 0, 0)$, $(x, 2, 0)$ and $(x, 1, 2)$ and denote them as T_0,T_1,T_2,T_3, and T_4, respectively. Compute a track layout with five track-sets using Algorithm 1. Let n_0, n_1, n_2, n_3 and n_4 be the number of vertices in track-set t_0, t_1, t_2, t_3 and t_4, respectively. Draw the vertices assigned to track-set t_i on track T_i according to the total order defined in the track-set, so that they occupy x-coordinates from $\sum_{j=0}^{i-1} n_j$ to $\sum_{j=0}^{i} n_j$ $(i = 0, 1, 2, 3, 4)$. We prove that the drawing has no crossing. Overlaps and X-crossings are not possible, because there is no overlap and no X-crossing in the track layout.

A crossing is possible only between edges that are on crossing strips. There are five pairs of such strips: (1) σ_{02} and σ_{13}; (2) σ_{01} and σ_{24};(3) σ_{01} and σ_{34}; (4) σ_{02} and σ_{34};(5) σ_{13} and σ_{24}. Let σ_{ij} and σ_{hl} be the two crossing strips of one of the five cases above and let e_1 and e_2 be two edges on σ_{ij} and σ_{hl}, respectively. Denote as x_i, x_j, x_h and x_l the x-coordinates of the vertices on tracks T_i, T_j, T_h and T_l. The co-planarity equations for each of the five cases are: (1) $x_0 + x_2 = x_1 + x_3$ (2) $x_0 + 3x_1 = 2x_2 + 2x_4$ (3) $3x_0 + x_1 = 2x_3 + 2x_4$ (4) $4x_0 + x_2 = 3x_3 + 2x_4$ (5) $4x_1 + x_3 = 3x_2 + 2x_4$. Since $x_0 < x_1 < x_2 < x_3 < x_4$, none of the above equations has a solution, and therefore no crossing is possible.

The drawing algorithm consider a vertex per time and executes a constant number of operations for each vertex. The time complexity is then $O(n)$. □

Theorem 2 shows that every arched leveled planar graph admits a 5-track layout. However, there are specific classes of graphs that have queue number 1 and that admit a track layout with less than 5 track-sets. One of such class is that of square meshes. The following Theorem hold. Proof is omitted for reasons of space.

Theorem 3. *Every square mesh has track number at most 3 and admits a 3-track drawing with volume at most* $2 \times 2 \times \frac{n+2\sqrt{n}}{3}$ *that can be computed in* $O(n)$ *time. Also, there exists a square mesh* G *such that* $tn(G) \geq 3$.

4 Graphs with $2 \leq qn \leq 3$

In this section we study the track number of graphs with queue number equal to 2 or queue number greater than 2 and less than 3. X-trees have queue number 2 [8]. The following theorem, whose proof is omitted for reasons of space, hold.

Theorem 4. *Every* X*-tree has track number at most 3 and admits a 3-track drawing with volume at most* $2 \times 2 \times \frac{4}{7}(n+1)$ *that can be computed in* $O(n)$ *time. Also, there exists an* X*-tree* G *such that* $tn(G) \geq 3$.

Halin Graphs [7] have queue number greater than 2 and less than 3 [6]. A *Halin graph* is a graph such that: (i) every vertex of G has degree greater or equal to 3;(ii) G can be decomposed into a spanning tree T of G and a cycle C through the leaves of T; (iii) G has a planar embedding in which C is the boundary of the external face.
T is called the *characteristic tree* of G and C is called the *adjoint cycle* of G. Figure 4 shows a Halin graph.

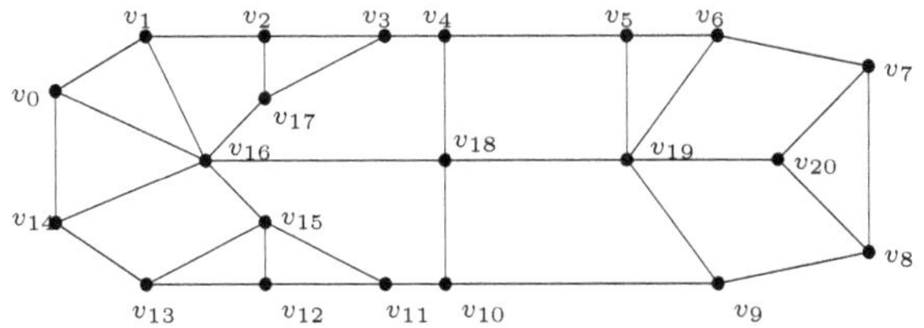

Fig. 4. A Halin graph.

It is known that a graph admits a 2-track layout only if it is outerplanar [1]. Since Halin graphs are not outerplanar (every outerplanar graph has at least on vertex of degree two) a lower bound on the track number of Halin graphs is 3.

We now describe an algorithm to compute a 4-track layout of a Halin graph. An *external path* of an embedded rooted ordered tree T is the path $\pi = \pi_l \cup \pi_r$, where π_l is the path from the leftmost leaf of T to the root of T and π_r is the path from the root of T to the rightmost leaf of T. Let v be a vertex in an external path π. If v has children that are not in π then every subtree rooted at a child of v not in π is called a *dangling subtree* of π.

Let G be a Halin graph. Assume that G is embedded in the plane such that it is planar and its adjoint cycle C is the exterior face. Let T be the characteristic tree of G. T inherits its embedding from G. Arbitrarily choose one of the non-leaf vertices of T as the root. A *level decomposition* of T is an assignment of a level to each vertex v of T that is defined as follows (see Figure 5): (i) all the vertices on the external path of T are given level 0; (ii) Let π be an external path of level i. For any dangling subtree T' of π, the vertices on the external path of T' are given level $i+1$.

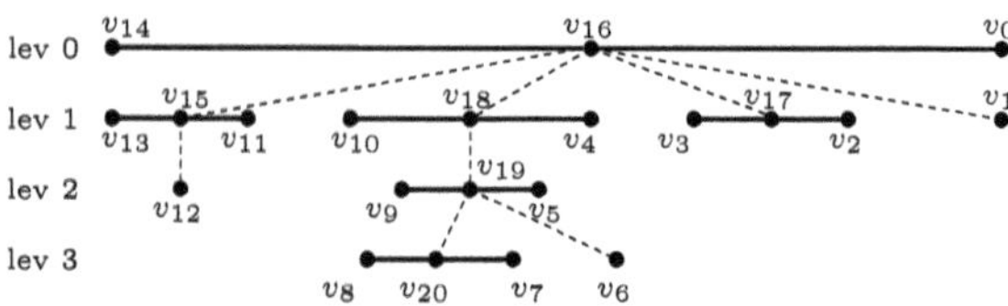

Fig. 5. A level decomposition of the characteristic tree of the Halin graph in Figure 4.

Let π be an external path of any level. Let $T_0, T_1, \ldots, T_{h-1}$ be the dangling subtrees of π. We define a *natural ordering* of the dangling subtrees as follows: (i) $T_0, T_1, \ldots, T_{h-1}$ are ordered from left to right according to their parents order in π; (ii) dangling subtrees that have the same parent are ordered from left to right. A consequence of the level decomposition is the following.

Property 1. Let G be a Halin graph and let π be an external path of any level with at least one dangling subtree. Let $T_0, T_1, \ldots, T_{h-1}$ be the natural order of the dangling subtrees of π. Then in the adjoint cycle C of G: (i) the leftmost leaf of π is adjacent to the leftmost leaf of T_0; (ii) the rightmost leaf of π is adjacent to the rightmost leaf of T_{h-1}; (iii) the rightmost leaf of T_j is adjacent to the leftmost leaf of T_{j+1} $(j = 0, \ldots, h-2)$.

Since the embedding of G is not changed when T is rooted, and since the boundary of the external face of G is C, then there exist an edge of C that connects the leftmost leaf of T to the rightmost leaf of T. The edge connecting the leftmost leaf of T and the rightmost leaf of T is called the *long edge*. Let π be an external path without dangling subtrees; if π consists of three vertices u, v and w in this order and u and w are leafs of T, then u and w are connected by an edge of C and this edge is called a *short edge*. The long edge and the short edges are generically called *overlapping edges*.

We now describe an algorithm to compute a 3-track layout of a Halin graph without overlapping edges. Later we will describe how the overlapping edges can be added back to the track layout using a fourth track-set. A Halin graph after the deletion of the overlapping edges is called a *reduced Halin graph*.

```
RHTrackLayout(G)
  Input: An embedded reduced Halin graph G
  Output: A 3-track layout of G.
Let T be the characteristic tree of G, rooted at any vertex r;
Q ← new queue();
Q.enqueue(r);
ord ← 0;
while Q is not empty
   v ← Q.dequeue();
   if v = r then t ← 0;
           else t ← (track(parent(v)) + 1)  mod 3;
   Let T' be the subtree rooted at v;
   Let π = v_0, v_1, ..., v_{h-1} be the external path of T';
   for i = 0 to h - 1
      track(v_i) ← t;
      order(v_i) ← ord;
      ord ← ord + 1;
      Let w_0, w_1, ..., w_{k-1} be the children of v_i not in π ordered
      from left to right;
      for j = 0 to k - 1
         Q.enqueue(w_j);
```

Algorithm 2: Algorithm RHTrackLayout()

Lemma 5. *Let G be a reduced Halin graph. Algorithm 2 computes a 3-track layout of G.*

Proof. We prove that the track assignment computed by Algorithm 2 has no overlaps nor X-crossing. The edges having both vertices in a same track-set are either edges of an external path or edges of the adjoint cycle connecting leafs of the same level. The edges of the external paths do not overlap since the two vertices of each edge are consecutive in a track-set. Let π be an external path of level i and let $T_0, T_1, \ldots, T_{h-1}$ be the natural ordering of the dangling subtrees of π. By Property 1 the edges of the adjoint cycle connecting two leafs of the same level $i+1$ connect the rightmost leaf of T_j to the leftmost leaf of T_{j+1}. Since Algorithm 2 lays out $T_0, T_1, \ldots, T_{h-1}$ according to their natural order, then the rightmost leaf of T_j and the leftmost leaf of T_{j+1} are consecutive in a track-set and therefore they do not overlap.

Let $e_0 = (u_0, v_0)$ and $e_1 = (u_1, v_1)$ be two edges having the two vertices in two different track-sets and assume that u_0 and u_1 are in the same track-set with $u_0 < u_1$. Edges e_0 and e_1 are either edges connecting the roots of dangling trees to their parents or edges of the adjoint cycle connecting leafs at consecutive levels. If u_0 and u_1 are in different external path π_0 and π_1, then u_0 is adjacent to a vertex (a leaf or the root) of a dangling subtree of π_0 and u_1 is adjacent to a vertex (a leaf or the root) of a dangling subtree of π_1. The vertices of the dangling subtrees of π_0 precede the vertices of the dangling subtree of π_1 in each track-set. Therefore $v_0 < v_1$ and an X-crossing is not possible. Assume u_0 and

u_1 be in the same external path π. If u_0 is the leftmost vertex of π then v_0 is the leftmost leaf of the first dangling subtree T_0 of π (Property 1). It follows that $v_0 < v_1$ and an X-crossing is not possible also in this case. If u_1 is the rightmost vertex of π then v_1 is the is the rightmost leaf of the last dangling subtree T_{h-1} of π (Property 1). It follows that $v_0 < v_1$ and an X-crossing is not possible. If u_0 is not the leftmost vertex of π and u_1 is not the rightmost vertex of π, then they are adjacent to the roots of two dangling subtrees of π. Since $u_0 < u_1$ and since the dangling subtree are laid out according to their natural order, then $v_0 < v_1$ and an X-crossing is not possible. □

Notice that the track layout of the Reduced Halin Graph is such that the vertices of each overlapping edge are both in the same track-set. The track layout of the reduced Halin graph of the graph in Figure 4 is shown in Figure 6. The long edge (v_0, v_{14}) and the short edges (v_2, v_3) and (v_7, v_8) would create an overlap if considered in the track layout.

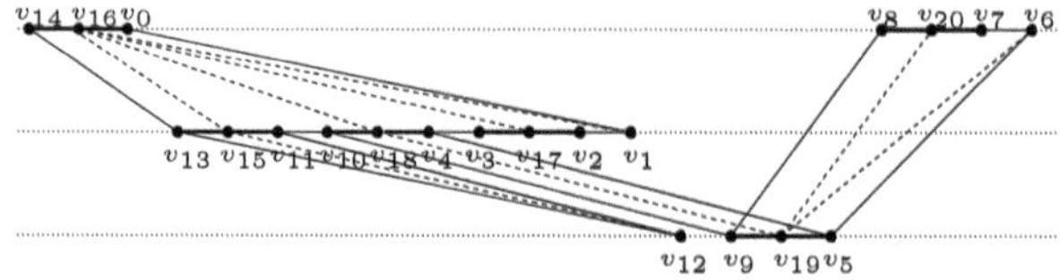

Fig. 6. A track layout of the reduced Halin graph of the graph in Figure 4.

Lemma 6. *Let G be a Halin graph and let G' the corresponding reduced Halin graph. Let $\gamma(G')$ be a 3-track layout of G' computed by Algorithm 2 and let $e_0 = (u_0, v_0)$ and $e_1 = (u_1, v_1)$ be any pair of overlapping edges. The two edges e_0 and e_1 do not have any vertex in common, and if their vertices are in the same track-set then either $u_0 < u_1 < v_0 < v_1$ or $u_1 < u_0 < v_1 < v_0$.*

By Lemma 6 it is easy to see that a track layout of a Halin graph can be computed from a 3-track layout of a reduced Halin graph by adding a new track-set and moving the left vertex of each overlapping edge in the new track-set.

Lemma 7. *Let G be a Halin graph, then $tn(G) \leq 4$.*

Proof. Let $\gamma(G)$ the 3-track layout of the reduced Halin graph G' of G computed by Algorithm 2. Denote as t_0,t_1, and t_2 the three track-set used and consider a new track-set denoted as t_3. For each overlapping edge $e = (u, v)$ assume that $u < v$ and set $track(u) = 3$, i.e. for each overlapping edge one of the vertices is assigned to the new track-set. This change of track-set do not introduce an overlap since no edge has both vertices in track-set t_4. Also, no X-crossing is introduced. Namely, consider the the pair of track-sets consisting of track t_4 and of any of the other three track-sets (denote this track-set as t_i). The edges whose vertices are in this pair are the overlapping edges whose vertices were originally in t_i. Let $e_0 = (u_0, v_0)$ and $e_1 = (u_1, v_1)$ be two of these edges and assume that u_0

and u_1 are in track-set t_3 with $u_0 < u_1$. By Lemma 6 we have $u_0 < v_0 < u_1 < v_1$ in track-set t_i in $\gamma(G')$. Since $order(u_0)$, $order(v_0)$, $order(u_1)$, and $order(v_1)$ are not changed, we have $u_0 < u_1$ in track-set t_3 and $v_0 < v_1$ in track-set t_i. So there is no X-crossing. □

Theorem 5. *Every Halin graph has track number at most 4 and admits a 4-track drawing with volume at most* $2 \times 2 \times n$ *that can be computed in* $O(n)$ *time. Also, for every Halin graph* $tn(G) \geq 3$.

Proof. The results about the bounds on track number follows from the fact that Halin graphs are not outerplanar and by Lemma 7. We prove the result about volume. Consider the four tracks $(x, 0, 0)$, $(x, 1, 0)$, $(x, 0, 1)$ and $(x, 1, 1)$, and denote them as T_0,T_1,T_2, and T_3, respectively. Compute a 4-track layout on four track-sets as described above . Let n_0, n_1, n_2 and n_3 be the number of vertices on track-set t_0, t_1, t_2 and t_3, respectively. Draw the vertices assigned to track-set t_i on track T_i according to the total order defined in the track-set, so that they occupy x-coordinates from $\sum_{j=0}^{i-1} n_j$ to $\sum_{j=0}^{i} n_j$ $(i = 0, 1, 2, 3, 4)$. We prove that the drawing has no crossing. Overlap and X-crossings are not possible, because there is no overlap and no X-crossing in the track layout. A crossing is then possible only between edges on crossing strips. There are only two such strips σ_{02} and σ_{13}. The co-planarity equation of an edge on σ_{02} and an edge on σ_{13} is $x_1 + x_3 = x_0 + x_2$. Since $x_0 < x_1 < x_2 < x_3$ then the equation has no solution, i.e. a crossing is not possible. It follows that the drawing is a track drawing with volume $2 \times 2 \times n$. The drawing algorithm consider a vertex per time and executes a constant number of operations for each vertex. The time complexity is therefore $O(n)$. □

References

1. S. Cornelsen, T. Schank, and D. Wagner. Drawing graphs on two and three lines. In *Proc. GD 2002*, volume 2528 of *LNCS*, pages 31–41. Springer-Verlag, 2002.
2. V. Dujmović, P. Morin, and D. Wood. Pathwidth and three-dimensional straight line grid drawings of graphs. In *Proc. GD 2002*, volume 2528 of *LNCS*, pages 42–53. Springer-Verlag, 2002.
3. V. Dujmović and D. Wood. New results in graph layout. Technical Report TR-03-04, School of Computer Science, Carleton University, 2003.
4. V. Dujmović and D. Wood. Tree-partitions of k-trees with application in graph layout. In *Proc. WG 2003*, LNCS. Springer-Verlag, to appear.
5. S. Felsner, G.Liotta, and S. Wismath. Straight line drawings on restricted integer grids in two and three dimensions. In *Proc. GD 2001*, volume 2265 of *LNCS*, pages 328–342. Springer-Verlag, 2001.
6. J. L. Ganley. Stack and queue layouts of Halin graphs. 1995, manuscript.
7. R. Halin. Studies in minimally connected graphs. In *Combinatorial Mathematics and its Applications*, pages 129–136. New York, Academic Press, 1971.
8. L. S. Heath and A. L. Rosenberg. Laying out graphs using queues. *SIAM J. Computing*, 21:927–958, 1992.
9. D. Wood. Queue layouts, tree-width, and three-dimensional graph drawing. In *Proc. FSTTCS 2002*, volume 2556 of *LNCS.*, pages 348–359. Springer-Verlag, 2002.

3D Visibility Representations of Complete Graphs

Jan Štola

Department of Applied Mathematics, Charles University
Malostranské nám. 25, Prague, Czech Republic
Jan.Stola@mff.cuni.cz

Abstract. This paper continues the study of 3D visibility representations of complete graphs where vertices are represented by equal convex polygons lying in planes parallel to the xy-plane. Edges correspond to the z-parallel visibility among these polygons.
We give several bounds on the size of the largest complete graph that has a 3D visibility representation with particular properties. Namely we improve the best known lower bound for representations by regular n-gons from $\lfloor \frac{n+1}{2} \rfloor + 2$ to $n+1$ and the upper bound from 2^{2^n} to $\binom{6n-3}{3n-1} - 3$.

1 Introduction

In this paper we study visibility representations of graphs. Two-dimensional variants of these representations were intesively studied in the past because of their connection to the VLSI routing and circuit board layout [8], [10]. Particularly the rectangle visibility graphs (axis-aligned rectangles in the plane with axis-aligned visibility segments among them) received wide attention (see for example [5] and [7]) in the last decade.

It is natural to generalize these representations into three dimensions. Two-dimensional Euclidean sets act as vertices there. The sets are situated in the planes parallel to the xy-plane. Two vertices are connected if and only if they can see each other in the direction that is orthogonal to their planes.

Fekete et al. study representations by squares and rectangles. In [6] and [9] it is shown that K_7 can be represented by equal squares but K_8 cannot. Similarly K_{22} has a representation by rectangles of different shapes but K_{56} has not. Alt, Godeau and Whitesides [1] prove that there doesn't exist a convex polygon P such that any complete graph has a visibility representation by shifted copies of P. Particularly K_m for $m > 2^{2^n}$ cannot be represented by convex n-gons. They also describe a representation of any graph on n vertices by (possibly different) polygons with at most $2n$ sides. Babilon et al. [2] are interested in the maximum size of a clique represented by regular n-gons and present a lower bound $\lfloor \frac{n+1}{2} \rfloor + 2$.

Cobos et al. [4] extend visibility representations even into the higher dimensions and characterize the dimensionality (e.g. the lowest dimension in which a representation of the given graph exists) of several classes of graphs.

G. Liotta (Ed.): GD 2003, LNCS 2912, pp. 226–237, 2004.

We continue in the study of the maximum size of a complete graph with a visibility representation where vertices are represented by shifted copies of a given (open) polygon. It can be easily shown (see for example [1]) that a complete graph of an arbitrary size can be represented by any non-convex polygon. So, our attention is restricted to convex polygons only. Therefore the term polygon means convex polygon in the sequel.

Let $s(P)$ denote the maximum size of a complete graph with a visibility representation by copies of the polygon P. Denote by R_n a regular n-gon. It was shown in [1] and [2] that $\lfloor \frac{n+1}{2} \rfloor + 2 \leq s(R_n) \leq 2^{2^n}$. We improve both sides of this estimate.

2 Coordinates

Let's remind the notion of polygon coordinates in a visibility representation (introduced in [2]). Firstly note that because of the openness of the polygons used in the representation we can expect that no two projections (into the plane $z = 0$) of sides of the polygons lie in the same line. Let P be a projection of some polygon into the plane $z = 0$. Denote by $\boldsymbol{u}_i$, $i \in \{1, \ldots, n\}$ the unitary vector orthogonal to the i-th side of P. We choose the orientation of $\boldsymbol{u}_i$ such that $\boldsymbol{u}_i$ aims out of P.

Note that the definition of $\boldsymbol{u}_i$ doesn't depend on the selection of P because all polygons in the representation are shifted copies of the same polygon.

Now consider projections of i-th sides of all polygons from the representation. The *i-th coordinate* of the polygon P is the integer number that denotes the order (in the direction of $\boldsymbol{u}_i$) of the projection of the i-th side of P among the projections of i-th sides of the other polygons.

The polygon coordinates allow us to recognize configurations that cannot appear in a representation of a complete graph.

Lemma 1. [2] *Let P, Q, R be n-gons whose i-th coordinates are monotone for every $i \in \{1, \ldots, n\}$. If the z-coordinates of those polygons are also monotone then P cannot see R along the z-axis.*

3 Relationships

It is important to understand that individual coordinates of polygons are not independent.

Lemma 2. *Let P and Q be polygons from a visibility representation. Let $(p_1, \ldots, p_n)$ and $(q_1, \ldots, q_n)$ be their coordinates. There exist $i, j \in \{1, \ldots, n\}$, $i < j$ such that one of the following conditions is satisfied:*

- *$p_k < q_k$ for $k \in \{i, \ldots, j-1\}$ and $p_k > q_k$ otherwise*
- *$p_k > q_k$ for $k \in \{i, \ldots, j-1\}$ and $p_k < q_k$ otherwise*

Proof. Let Q' be a projection of Q into the plane of P. Denote by $\boldsymbol{u}$ the translation vector from P to Q' and by l_k the line that goes through the origin and is parallel to the k-th side of P. Let $\boldsymbol{u}_k$ be the vector from the definition of polygon coordinates. The line l_k (that is orthogonal to $\boldsymbol{u}_k$) divides the plane into two half planes. We call them positive and negative depending on the fact whether $\boldsymbol{u}_k$ points into them or not. It is clear that $p_k < q_k$ if and only if $\boldsymbol{u}$ points into the positive half plane.

Now consider all lines l_k together. The polygon P is convex. Therefore one of the following two possibilities must occur.

- There exist i and j such that $\boldsymbol{u}$ lies in the positive half planes that correspond to lines l_k, $i \leq k < j$ and in the negative half planes that correspond to the other lines.
- There exist i and j such that $\boldsymbol{u}$ lies in the negative half planes that correspond to lines l_k, $i \leq k < j$ and in the positive half planes that correspond to the other lines.

And that is a reformulation of what we want to prove. □

Definition 1. *Let P and Q be polygons in a visibility representation. Denote by i and j the numbers from the previous lemma. We say that P has a relationship (i, j) to Q if the first condition is satisfied. Otherwise we say that P has to Q a relationship (j, i). Finally the relationship between P and Q is the relationship of the polygon with the smaller z-coordinate to the other one.*

Consider relationships in a representation by the given polygon P. Each relationship corresponds to one part of the plane divided by cuts (going through the origin) that are parallel to the individual sides of P. For example we have $2n$ relationships for a general n-gon or for a regular $2n$-gon. We denote this number by $r(P)$.

Now we can reformulate Lemma 1. in terms of relationships.

Lemma 3. *Let P, Q, R be polygons from a visibility representation. If P has to Q the same relationship as Q has to R then if Q lies (according to the z-coordinate) between P and R then P cannot see R.*

4 Upper Bounds

In this section we present several upper bounds on the size of a complete graph with a visibility representation with particular properties. Firstly we improve the upper bound for $s(R_n)$ by using the notion of relationships.

Theorem 1. *If P is a polygon then $s(P) \leq \lfloor e.r(P)! \rfloor$.*

Proof. We proceed by induction on $r(P)$. Let's suppose that there are $k + 1$ relationships in the given visibility representation. Divide polygons into sets

$M_1, \ldots, M_{k+1}$ according to their relationship to the polygon with the maximum z-coordinate.

If r is a relationship that corresponds to the set M_i then there are no two polygons in M_i with this relationship. Otherwise these two polygons together with the one with the maximum z-coordinate would fulfill presumptions of Lemma 3. Hence, there are at most k types of relationships among polygons in M_i. Therefore $|M_i| \leq \lfloor e.k! \rfloor$ and there is at most $1 + (k+1)\lfloor e.k! \rfloor = \lfloor e.(k+1)! \rfloor$ polygons in the whole representation.

It remains to prove the simplest case $k = 1$. If there is only one type of relationship in the representation then there can be at most $2 = \lfloor e.1! \rfloor$ polygons. Otherwise we would be able to apply Lemma 3. to any triple of polygons. □

Later in the paper we derive an estimate with a better asymptotic behavior than the previous one but the previous estimate of $s(R_n)$ gives the best known upper bounds for n equal to 5, 6, 8 and 10.

For the sake of simplicity we work in the rest of this paper with objects in different parallel planes as if they are in the same plane. From the formal point of view such a description should be understood as an operation on projections into a common plane and projections of a result back to the individual planes.

In the rest of this section we denote by X_i the first vertex of the i-th polygon from the representation. We also say that polygons $P_1, \ldots, P_k$ are in z-order if they are ordered according to their increasing z-coordinate.

Lemma 4. *Let polygons $P_1, \ldots, P_k$ (in z-order) form a visibility representation of K_k. If $X_1 \ldots X_k X_1$ form a boundary of a convex polygon then $k \leq r(P_1) + 2$.*

Proof. Denote by $\boldsymbol{u}_i$ the vector $X_{i+1} - X_i$. $X_1X_2 \ldots X_kX_1$ is a boundary of a convex polygon. Therefore $(\boldsymbol{u}_i)_{i=1}^{k-1}$ is a sequence of vectors that rotate around the origin by less than 2π.

The vector $\boldsymbol{u}_i$ is clearly a translation vector from P_i to P_{i+1}. This vector determines the relationship between P_i and P_{i+1}. Because of the rotation around the origin and Lemma 3. there cannot be two equal relationships between polygons P_i and P_{i+1}, with one exception – the relationship between P_1 and P_2 could be the same as the relationship between P_{k-1} and P_k. Hence $k \leq r(P_1) + 2$. □

Regular $(2n+1)$-gons can have two types of relationships (i, j): $(j - i) \bmod (2n+1)$ could be either n or $n+1$. For $(4n+1)$-gons the first situation corresponds to the pull of the upper polygon by a vertex and the second to the pull by an edge. The opposite holds for $(4n+3)$-gons. That is why we call such relationships edge-relationships and vertex-relationships.

Definition 2. *Let V be a vertex of a regular $(2n+1)$-gon P with the center S_P. Denote by P' the copy of P shifted by the vector S_PV. We call the relationship of P to P' V-relationship (or the relationship V).*

Let E be an edge of P. If we replace V in the previous paragraph by the center of E we get the definition of *E-relationship* (the *relationship E*). The following two lemmas give us the basic properties of these relationships.

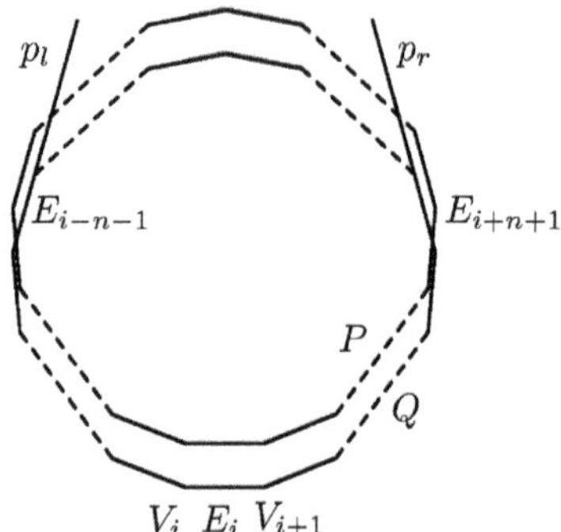

Lemma 5. *Let P, Q, R (in z-order) be regular $(4n+1)$-gons from a visibility representation. Denote by V_i (E_i) the i-th vertex (edge) of P. If P and Q have E_i-relationship and Q and R have V_{i+1}-relationship then P and R cannot see each other.*

Proof. Let's rotate the whole situation such that the vector V_iV_{i+1} is parallel to the vector $(1,0)$. Denote by p_l (p_r) the line that contains E_{i-n-1} (E_{i+n+1}) – the edge going from the vertex with the smallest (biggest) x-coordinate that goes away from E_i. It is clear that P and R cannot see each other if the translation vector between Q and R is lying between p_r and p_l, but the translation vector that corresponds to V_{i+1}-relationship has to lie between $V_{i+n+2}V_{i+n+1}$ and $V_{i-n}V_{i-n+1}$. □

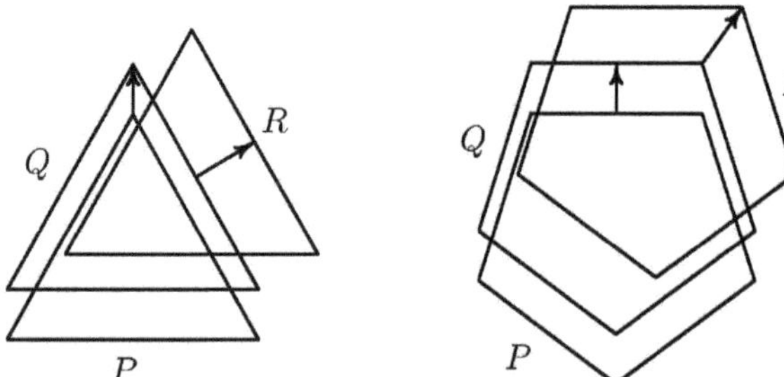

Lemma 6. *Let P, Q, R (in z-order) be regular $(4n+3)$-gons from a visibility representation. Denote by V_i (E_i) the i-th vertex (edge) of P. If P and Q have V_i-relationship and Q and R have E_i-relationship then P and R cannot see each other.*

Proof. We proceed in the same way as in the previous proof. □

Lemma 4. gives an upper bound $2n + 2$ for regular $2n$-gons. Using Lemmas 5. and 6. we can prove the same bound also for regular $(2n + 1)$-gons.

Lemma 7. *Let regular* $(2n + 1)$*-gons* $P_1, \ldots, P_k$ *(in* z*-order) form a visibility representation of* K_k*. If the centers of the polygons form a boundary of a convex polygon then* $k \leq 2n + 2$.

Proof. We prove the lemma for $(4n + 1)$-gons. The proof for $(4n + 3)$-gons is similar.

Using the same argument as in the proof of Lemma 4. we can show that there cannot be the same relationships between polygons P_i and P_{i+1}. Again with one exception – the relationship between P_1 and P_2 could be the same as the relationship between P_{k-1} and P_k.

Without loss of generality we can assume that the vertices and edges are numbered such that the relationship between P_1 and P_2 is V_1 or E_1 and that the sequence of relationships between P_i and P_{i+1} is a subsequence of $V_1 E_1 V_2 \ldots E_m V_1$ or $E_1 V_2 E_2 \ldots E_m V_1 E_1$, where $m = 4n + 1$. Due to Lemma 5. we know that this sequence doesn't contain any pair $E_i V_{i+1}$.

Hence, it is sufficient to show that a subsequence avoiding mentioned pair that is chosen

- from $E_1 V_2 \ldots E_m V_1$ has length at most m,
- from $V_1 E_1 V_2 \ldots E_m V_1$ has length at most $m + 1$.

We proceed by induction, for $m = 1$ the claim holds because we have to leave at least one relationship from the sequences $E_1 V_2$ and $V_1 E_1 V_2$.

Let's consider what can remain from the first two relationships in the first case. If we omit E_1 then by the second case the length of the subsequence is at most m. If we omit V_2 we get the first case for a shorter sequence and again the length at most m.

In the second case we have to consider the first three relationships. If we omit V_1 then by the first case the length of the subsequence is at most m. If we omit E_1 then we get the second case for a shorter sequence and the length at most $m + 1$. Finally if we omit V_2 we get again the length at most $m + 1$ by the first case. □

We demanded that corresponding points of polygons lie in a convex position and that polygons are ordered according to their z-coordinates. Let's see how the situation changes when we remove the second requirement.

Lemma 8. *Let polygons* $P_1, \ldots, P_k$ *form a visibility representation of* K_k*. If* $X_1 \ldots X_k X_1$ *form a boundary of a convex polygon then* $k \leq 3r(P_1)$.

Proof. Let $\boldsymbol{u}_i$ has the same meaning as in the proof of Lemma 4. Denote by r_i the relationship that corresponds to the translation vector $\boldsymbol{u}_i$. $(\boldsymbol{u}_i)_{i=1}^{k}$ is a sequence of vectors that rotate around the origin by less than 2π. Therefore the relationships that occur multiple times in the sequence (r_i) must have all occurrences adjacent (after possible renumbering).[1]

If the sequence is longer than $3r(P_1)$ then in the sequence (r_i) there must exist a constant segment of length at least 4. This segment corresponds to the sequence of at least 5 polygons with the same relationships. It is always possible to choose from those polygons a subsequence of length 3 with monotone z-coordinates. According to Lemma 3. polygons with this property cannot lie in a representation of a complete graph. Therefore $k \leq 3r(P_1)$. □

Again we can use Lemmas 5. and 6. to improve the estimate for $(2n+1)$-gons.

Lemma 9. *Let regular $(2n+1)$-gons $P_1, \ldots, P_k$ form a visibility representation of K_k. If the centers of the polygons form a boundary of a convex polygon then $k \leq 3(2n+1)$.*

Proof. We proceed in the same way as in the previous proof to the definition of the sequence $(r_i)_{i=1}^k$. There we have to utilize the fact that some relationships cannot be adjacent. We show the proof only for $(4n+1)$-gons because the argument for $(4n+3)$-gons is similar.

There cannot be the relationship E_i immediately followed by the relationship V_{i+1} or V_i in the sequence (r_i). Without loss of generality we can assume that vectors $\boldsymbol{u}_i$ turn from E_i to V_{i+1}, otherwise we can take the opposite numbering of polygons.

Let's group the relationships into $4n+1$ groups such that there are relationships E_i and V_{i+1} in the i-th group. Members from the same group are (after possible renumbering) adjacent in the sequence (r_i). If the sequence is longer than $3(4n+1)$ then there must be a segment of length at least 4 whose members belong into the i-th group for some $i \in \{1, \ldots, 4n+1\}$. By the construction the relationships E_i are before V_{i+1} in this segment. The segment corresponds to the sequence of at least 5 polygons from which we can chose a subsequence of 3 polygons with monotone z-coordinates. These three polygons have one type of relationship or E_i-relationship followed by V_{i+1}-relationship. According to Lemma 3. or 5. such polygons cannot lie in a representation of a complete graph. □

If we remove the requirement for a convex position of corresponding points of polygons we get a general estimate of $s(P)$.

Theorem 2. $s(P) \leq \binom{6r(P)-3}{3r(P)-1} - 3$.

Proof. Due to the upper bound for the Erdős-Szekeres problem presented in [3] there exist among any $\binom{6r(P)-3}{3r(P)-1} - 2$ points in the plane at least $3r(P)+1$ points in a convex position.

Choose an arbitrary point of P. If we have a representation with more than $\binom{6r(P)-3}{3r(P)-1} - 3$ polygons then there exists a group of at least $3r(P)+1$ of them

[1] It could happen that a part of the occurrences lies at the start of the sequence and a part at the end.

such that their points that correspond to the chosen point are in a convex position. According to Lemma 8. such a group cannot appear in a representation of a complete graph. □

Particularly we improved the best known upper bound for $s(R_n)$.

Corollary 1. $s(R_n) \leq \binom{6n-3}{3n-1} - 3 \approx 2^{6n}$.

Proof. For even n this result is a consequence of the previous theorem. If n is odd then we can repeat the previous proof and use Lemma 9. instead of Lemma 8. at the end. □

The last exponential estimate doesn't look very tight after the linear estimates from the rest of the paper. Below we show that its exponential behavior could not be improved by the used technique. So far, the only forbidden configuration was a triple of polygons where the middle one blocked the visibility between the other two polygons. There exist representations of an exponential size that don't contain this configuration.

Lemma 10. *Let P, Q, R (in z-order) be regular n-gons from a visibility representation. If the relationship between P and Q is different from the relationship between Q and R and*

- *for $n = 4k + 1$ the relationship between P and Q is a vertex-relationship, or it is an E-relationship but the relationship between Q and R is not a V-relationship such that $V \in E$,*
- *for $n = 4k + 3$ the relationship between P and Q is an edge-relationship, or it is a V-relationship but the relationship between Q and R is not an E-relationship such that $V \in E$,*

then polygon P can see R provided that the translation vectors among P, Q and R are sufficiently small.

Proof. Let (i, j) be the relationship between P and Q. If the translation vector from P to Q is sufficiently small then the i-th side of P intersects the $(i-1)$-th side of Q and the j-th side of Q intersects the $(j-1)$-th side of P. Denote by p_k the line that contains the k-th side of Q. If the translation vector from Q to R lies in the opposite half plane given by the line p_{i-1} (respectively p_j) than polygon Q then P and R can see each other in the neighborhood of the $(i-1)$-th (resp. the j-th) side of Q.

Before we continue we must understand how to find for the given translation vector $\boldsymbol{u}$ all vectors that correspond to the same relationship. It is clear from the proof of Lemma 2. that those vectors are the ones between the sides that form the smallest positive and the smallest negative angle with $\boldsymbol{u}$.

With this information we can rewrite the first paragraph into the words of relationships. The situation is simple for $2n$-gons because for their relationships (i, j) we have $(j - i) \equiv n \bmod 2n$. Hence, the $(i-1)$-th and j-th side form the smallest angle and translation vectors between them correspond to one relationship – the relationship between P and Q. If Q and R have a different relationship then P can see R provided that the translation vectors are small.

We can use a similar argument for $(4n+1)$-gons and a vertex-relationship or for $(4n+3)$-gons and an edge-relationship between P and Q.

The remaining situations are slightly more complicated because there are two sides between the $(i-1)$-th and the j-th side of Q. The translation vectors between those sides correspond to the relationships $V_l E_l V_{l+1}$ (for $(4n+1)$-gons) or $E_{l-1} V_l E_l$ (for $(4n+3)$-gons). Therefore P can see R if Q and R don't have a relationship adjacent to the relationship between P and Q. □

Theorem 3. *There exists a visibility representation with 4^n (resp. 4^{n+1}) regular $2n$-gons (resp. $(2n+1)$-gons) where every triple of polygons alone form a representation of K_3.*

Proof. Let's choose n adjacent relationships for $2n$-gons or relationships $V_1, \dots, V_{n+1}$ for $(2n+1)$-gons. It is important that all corresponding translation vectors lie in one half plane.

We proceed by induction on the number of chosen relationships. If we have one relationship we place 4 polygons in a direction given by the chosen relationship and set their z-coordinates equal to 2, 1, 4 and 3 subsequently.

Let's suppose that we have a construction for k relationships. The $(k+1)$-th relationship corresponds to the translation vectors between two half lines p_l and p_r. Consider projections (to the plane $z=0$) of the centers of all polygons from the construction. Draw two tangents to the convex hull of the projections – one parallel to p_l and one parallel to p_r. Now we place a copy of the construction for k relationships to the opposite quadrant determined by the tangents. We repeat the process three times.

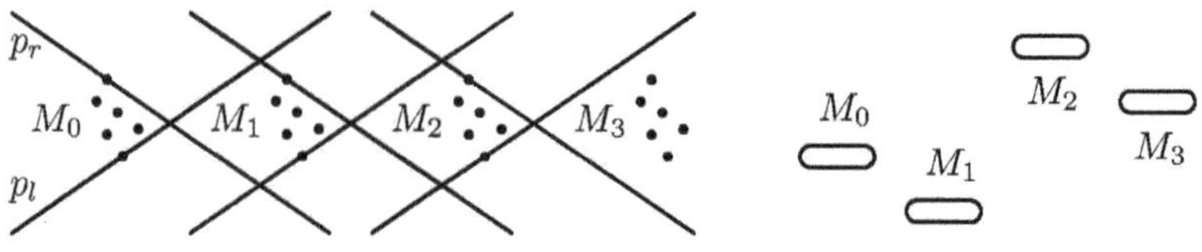

We have four copies of 4^k polygons. It is important that polygons from the different copies have $(k+1)$-th relationship. Now we place all polygons from the first copy below and from the third copy above the original polygons. The second copy is placed above all other polygons. Finally we proportionally shorten the distances among the centers of the polygons.

Let P, Q, R (in z-order) be polygons from the construction. If P and Q don't have the same relationship as Q and R then they can see each other because of the previous lemma and the set of relationships considered. The mentioned relationships cannot be the same because we place individual copies in all steps such that there are no three polygons (in z-order) with $(k+1)$-th relationships among them. □

5 Lower Bound

We improved the best known upper bound for $s(R_n)$ in the previous section. Now we concentrate on the lower bound.

Theorem 4. $s(R_n) \geq n+1$.

Proof. By [2] we have $s(R_3) \geq 14$. So, we can assume $n \geq 4$.

At first we describe a representation of K_n. Later we add the $(n+1)$-th polygon to obtain a representation of K_{n+1}.

Let P be a regular n-gon whose vertices v_i lie on the unitary circle centered at the origin e.g. $v_i = \left(\cos\frac{2\pi}{n}i, \sin\frac{2\pi}{n}i\right)$. Polygons P_i are created by 1/2 long shifts of P in the direction of $\boldsymbol{u}_i = \left(\cos\left(\frac{2\pi}{n}i+\alpha\right), \sin\left(\frac{2\pi}{n}i+\alpha\right)\right)$, where α is an angle whose value is determined at the end of the proof. We claim that P_i moved into planes $z = i$ form the desired representation of K_n.

It can be easily verified that we can obtain P_{i+1} by a rotation of P_i around the origin by $\frac{2\pi}{n}$. Hence, P_0 can see P_k if and only if P_l can see P_{k+l}, $k \in \{1,\ldots,n-1\}$, $l \in \{0,\ldots,n-1-k\}$. Therefore it is sufficient to show that P_0 can see all P_k, $k \in \{1,\ldots,n-1\}$.

Denote by S_i the center of the polygon P_i. We claim that P_0 and P_k can see each other through the intersection of their inscribed circles, particularly through the intersection X that is in the opposite part (than points S_i) of the plane determined by the angle S_0OS_k – O denotes the origin.

To prove that P_0 and P_k can see each other through X it is sufficient to show that X is not inside circles circumscribed to the polygons P_j, $0 < j < k$, e.g. to show that it is at least 1 unit far from their centers.

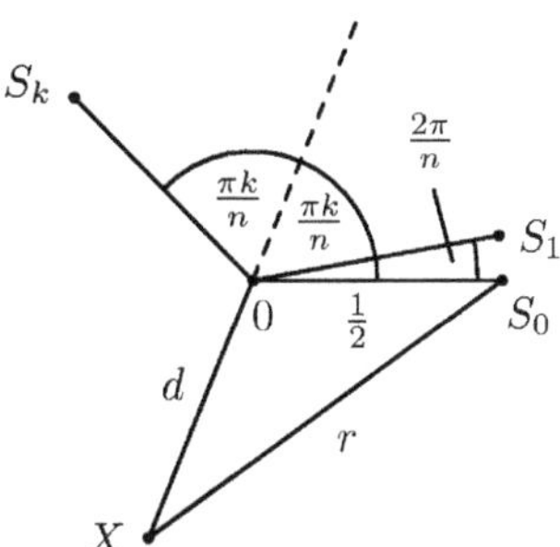

Let r denote the radius of the circle inscribed to P. We have $r > 1/2$ because $r = \cos\frac{\pi}{n}$ and $n \geq 4$. So, there exists an intersection of the inscribed circles outside the angle S_0OS_k and the point X is correctly defined. The nearest points from $S_1, \dots, S_{k-1}$ are S_1 and S_{k-1}, both equally far. Therefore it remains to prove the inequality $|XS_1| > 1$.

Let $d = |XO|$. According to the cosines laws we have

$$|XS_1|^2 = d^2 + \frac{1}{4} - d\cos\left(\frac{n-k+2}{n}\pi\right)$$

$$r^2 = d^2 + \frac{1}{4} - d\cos\left(\frac{n-k}{n}\pi\right)$$

Hence, it is sufficient to show

$$1 < |XS_1|^2 = r^2 + d\left(\cos\left(\frac{n-k}{n}\pi\right) - \cos\left(\frac{n-k+2}{n}\pi\right)\right) =$$

$$= \cos^2\frac{\pi}{n} + 2d\sin\frac{\pi}{n}\sin\left(\frac{k-1}{n}\pi\right),$$

for $k \in \{1, \dots, n-1\}$. We can ignore the case $k = 1$ because two adjacent polygons in this representation can see each other and we can write $\sin\frac{k-1}{n}\pi \geq \sin\frac{\pi}{n}$ for the rest. Our inequality can be rewritten to

$$1 < \cos^2\frac{\pi}{n} + 2d\sin^2\frac{\pi}{n} \equiv 0 < (2d-1)\sin^2\frac{\pi}{n} \equiv d > \frac{1}{2}\ .$$

The estimate $\sin\frac{k-1}{n}\pi \geq \sin\frac{\pi}{n}$ is sharp for $k \neq 2$. Hence, for $k \neq 2$ it suffices to show $d \geq 1/2$. The inequality $d < 1/2$ leads to the following contradiction:

$$\cos^2\frac{\pi}{n} = r^2 = d^2 + \frac{1}{4} + d\cos\frac{k\pi}{n} < \frac{1}{2} + \frac{1}{2}\cos\frac{2\pi}{n} = \cos^2\frac{\pi}{n}\ .$$

We have proved that $d \geq 1/2$. It remains to solve the case $k = 2$ and $d = 1/2$. This happens when the circles inscribed into the polygons P_0 and P_2 intersect exactly on the circle circumscribed to the polygon P_1. If there is not a vertex of the polygon P_1 in the intersection then polygons P_0 and P_2 still can see each other. We achieve this by the mysterious angle α from the beginning of the proof.

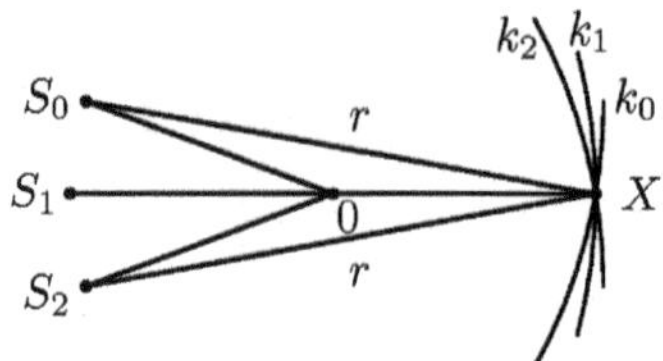

The described problematic situation could happen only when the vector $\boldsymbol{u}_1$ has the same direction as one of the vectors $-\boldsymbol{v}_i$. So, we have to choose the vector α to avoid a finite number of those critical values.

Now we have a representation of K_n. We must add a new polygon to create a representation of K_{n+1}. We claim that such polygon could be a copy of P_0 shifted to the plane $z = n$. We could consider such a polygon to be a natural successor of P_{n-1} in our construction – it can be created by a shift of P by a vector $\boldsymbol{u}_0 = \boldsymbol{u}_n = (\cos(\frac{2\pi}{n}n + \alpha), \sin(\frac{2\pi}{n}n + \alpha))$. From this point of view it is clear that P_n can see P_k, $k \in \{1, \ldots, n-1\}$ if and only if P_{n-1} can see P_{k-1} which is true. The only problem could be the visibility between P_0 and P_n.

For simplicity we rotate the whole situation such that the translation vector $\boldsymbol{u}_n$ is parallel to the vector $(1, 0)$. The polygons P_0 and P_n can see each other through the neighborhood of the vertex with the biggest x-coordinate. There cannot be a polygon that would break such a visibility because its translation vector $\boldsymbol{u}_i$ would have to have bigger x-coordinate than the vector $\boldsymbol{u}_n$ which is impossible because we rotated the situation such that this vector is parallel to $(1, 0)$. □

6 Conclusion

We derived several bounds for $s(P)$ for representations with particular properties and improved the lower bound for $s(R_n)$ from $\lfloor \frac{n+1}{2} \rfloor + 2$ to $n + 1$ and the upper bound from 2^{2^n} to $\binom{6n-3}{3n-1} - 3 \approx 2^{6n}$. In Theorem 3. we showed that the exponential upper bound cannot be improved if we study only visibility in triples of polygons (e.g. if we look for one polygon that blocks the visibility between another two polygons).

References

1. H. Alt, M. Godau, S. Whitesides (1998): Universal 3-dimensional visibility representations for graphs, *Comput. Geom.* **9**, 111–125
2. R. Babilon, H. Nyklová, O. Pangrác, J. Vondrák (1999): Visibility representations of complete graphs, *Proc. Graph Drawing* **99**, 333–341
3. G. Tóth, P. Valtr (1998): Note on the Erdős-Szekeres theorem, *Discrete Comput. Geom.* **19(3)**, 457–459
4. F.J. Cobos, J.C. Dana, F. Hurtado, A. Márquez, F. Mateos (1995): On a visibility representation of graphs, *Proc. Graph Drawing* **95**, 152–161
5. P. Bose, A. Dean, J. Hutchinson, T. Shermer (1996): On rectangle visibility graphs, *Proc. Graph Drawing* **96**, 25–44
6. S.P. Fekete, M.E. Houle, S. Whitesides (1995): New results on a visibility representation of graphs in 3D, *Proc. Graph Drawing* **95**, 234–241
7. J.P. Hutchinson, T. Shermer, A. Vince (1999): On representations of some thickness-two graphs, *Comput. Geom.* **13(3)**, 161–171
8. R. Tamassia, I.G. Tollis (1986): A unified approach to visibility representations of planar graphs, *Discrete and Computational Geometry*, 1:321–341
9. P. Bose, H. Everett, S. Fekete, A. Lubiw, H. Meijer, K. Romanik. T. Shermer, S. Whitesides (1993): On a visibility representation for graphs in three dimensions, *Proc. Graph Drawing* **93**, 38–39
10. A. Dean, J. Hutchinson (1997): Rectangle-visibility representations of bipartite graphs, *Discrete Appl. Math.* **75(1)**, 9–25

Drawing Series-Parallel Graphs on Restricted Integer 3D Grids*

Emilio Di Giacomo

Dipartimento di Ingegneria Elettronica e dell'Informazione,
Universitá degli Studi di Perugia, Perugia, Italy
digiacomo@diei.unipg.it

Abstract. A k-track drawing is a crossing-free $3D$ straight-line grid drawing of a graph G on a set of k parallel lines called *tracks*. The minimum value of k for which G admits a k-track drawing is called the *track number* of G. In the existing literature a lower bound of five and an upper bound of fifteen are known for the track number of series-parallel graph. In this paper we reduce this gap for a large subclass of series-parallel graph for which the lower bound remains five but we show an upper bound of eight. We also describe a linear time drawing algorithm that computes a $3D$ straight-line grid drawing of these graphs in volume $4 \times 4 \times 2n$.

1 Introduction

In this paper we study the problem of computing 3D straight-line crossing-free grid drawings of series-parallel graphs with linear volume. Many recent papers devoted to the study of this problem have been presented(see, e.g., [5,6,7,8,9]).

The first subquadratic volume bound for series-parallel graphs was presented in [6] where it is proved that every series-parallel graph admits a drawing in $O(n \log^2 n)$ volume. In [9] Wood gives the first $O(n)$ volume bound for series-parallel graphs. The hidden multiplicative constant in the volume bound provided in [9] is in the order of 10^{16}.

Both papers [6,9] study the problem of computing small-volume 3D drawings by using the following approach, introduced by Felsner Liotta and Wismath [8]. A grid consisting of k parallel lines, called *tracks*, is considered. A drawing of a graph G on such a grid is called a *k-track drawing* of G and the minimum value of k such that G admits a k-track drawing is called the *track number* of G. In [6] it is proved that if a graph has constant track number then it admits a drawing with linear volume. This result suggests that in order to reduce the volume bound one can minimize the track number.

Recent results about track number are due to Dujmović and Wood [7] who proved that k-trees have constant track number, and therefore linear volume bound. As a special case, Dujmović and Wood [7] refine the track number and

* Research supported in part by "Progetto ALINWEB: Algoritmica per Internet e per il Web", MIUR Programmi di Ricerca Scientifica di Rilevante Interesse Nazionale.

G. Liotta (Ed.): GD 2003, LNCS 2912, pp. 238–246, 2004.

volume bounds of 2-trees (every 2-tree is a series-parallel graph and every series-parallel graph can be augmented to become a 2-tree) and show that the track number of a series-parallel graph is at most 18; as a consequence they show that a series-parallel graph admits a 3D straight-line grid drawings of volume at most $36 \times 37 \times 37\lceil \frac{n}{18} \rceil$.

Very recently, Di Giacomo, Liotta and Meijer [4] improved the results in [7] by presenting new upper bounds on the track number of k-trees; as for series-parallel graph it is proved a track number of at most 15. The corresponding volume described in [4] is $30 \times 31 \times 31\lceil \frac{n}{15} \rceil$. A lower bound of five for the track number of series-parallel graphs is presented in [5].

This paper is motivated by the natural question of reducing the gap between the lower and the upper bound on the track number of series-parallel graphs (i.e. a $5 \div 15$ gap). Our main results can be listed as follows:

- We study a large subclass of series-parallel graphs, called *flat series-parallel graphs*, for which an upper bound of eight on the track number is proved. The lower bound for this class is five.
- We present a linear time drawing algorithm that computes 3D straight-line grid drawings of flat series-parallel graphs in volume $4 \times 4 \times 2n$.

We remark that the class of flat series-parallel graphs have been the subject of investigation in the literature for their interest in book-embedding and VLSI applications (see, e.g., [1]).

2 Preliminaries

A *series-parallel digraph* (also called as an *SP-digraph* in the following) is a directed planar graph recursively defined as follows [2]: (i) A directed edge $e = (s,t)$ is an SP-digraph; s is the *source pole* and t is the *sink pole* of the SP-digraph. (ii) Let G' and G'' be two SP-digraphs, whose source poles are s' and s'', respectively, and whose sink poles are t' and t'', respectively; the digraph G obtained by identifying t' with s'' is also an SP-graph; s' is the *source pole* of G and t'' is the *sink pole* of G. (iii) Let G' and G'' be two SP-digraphs, whose sources are s' and s'', respectively, and whose sinks are t' and t'', respectively; the digraph G obtained by identifying s' with s'' and t' with t'' is also an SP-digraph; $s' = s''$ is the *source pole* and $t' = t''$ is the *sink pole* of G. The source pole and the sink pole of an SP-digraph G are also called the *poles* of G. The undirected graph underlying an SP-digraph is a *series-parallel graph*, also called SP-graph.

Let G be an SP-digraph. A *split pair* of G is either a separation pair or a pair of adjacent vertices of G. A *split component* of G with respect to a split pair $\{s,t\}$ is either an edge (s,t) or a maximal subgraph C of G such that C is an st-graph [2] and $\{s,t\}$ is not a split pair of C.

We briefly recall the definition of the *SPQ-tree* of G, also called the *decomposition tree* of G (see [3] for a complete definition). A SPQ-tree is an ordered rooted tree T whose nodes are of three types: S,P, or Q. A S-node corresponds to a series composition of blocks, and it has a child for each block; A P-node corresponds to a parallel composition of split components with respect to a separation pair and it has a child for each split component; A Q-node corresponds to a single edge and it has no child. The *pertinent digraph* of a node μ of T is the subgraph of G whose decomposition tree is the tree rooted at μ. We call *poles* of a node μ of T the poles of the pertinent digraph of μ. Also, to simplify the description of our algorithms we shall distinguish the S-nodes having only Q-nodes as children from the others. Namely, an S-node whose children are all Q-nodes will be called an *S^*-node.*

Two parallel components of an SP-digraph G are said to be *nested* if they share a pole and they are not in series. Let v be a vertex of G, the *depth* of v is the number of nested parallel components having v has a pole. An SP-digraph *has depth k* if each pole has depth at most k. An SP-digraph of depth 1 is said to be *flat*. An SP-graph G is *flat* if there exists an st-orientation [2] of G such that the resulting SP-digraph is flat.

A *track* is a set of 3D grid points on a straight line of infinite length. We always assume that a track is parallel to the x-axis, thus a track is the set of all the grid points having the same y- and z-coordinate. We denote as (x, Y, Z) a track whose points have y-coordinate Y and z-coordinate Z. A *k-track drawing* of a graph G is a 3D straight-line crossing-free grid drawing of G such that each vertex of G is drawn on one of k tracks. The minimum value of k such that G admits a k-track drawing is called the *track number* of G.

An *8-prism* is the 3D integer grid consisting of the eight tracks $T_0 = (x, 0, 2)$, $T_1 = (x, 0, 1)$, $T_2 = (x, 3, 2)$, $T_3 = (x, 3, 1)$, $T_4 = (x, 1, 0)$, $T_5 = (x, 1, 3)$, $T_6 = (x, 2, 0)$, and $T_7 = (x, 2, 3)$. A *strip* σ_{ij} is the portion of plane delimited by tracks T_i and T_j. The 8 tracks of an 8-prism can be grouped in two subgrids: the first one, called *even ingot*, consists of the four tracks with even numbers; the second one, called *odd ingot*, consists of the four tracks with odd numbers. A track drawing on the eight tracks of an 8-prism will be called *8-prism drawing* in the following. Also, an 8-prism drawing will be specified by assigning to each vertex two numbers: $track(v)$ is an integer that denotes the track to which v is assigned; $x(v)$ is the x-coordinate assigned to v. Finally, $|G|$ will denote the number of vertices of a given digraph G.

3 8-Prism Drawings of Flat SP-Digraphs

In this section we present an algorithm that computes an 8-prism drawing of a flat SP-graph. Since a flat st-orientation of a flat SP-graph can be computed in polynomial time we focus on flat SP-digraphs.

Let G be a flat SP-digraph and let T be the SPQ-tree of G. A *level numbering* of T is a numbering of each node μ of T denoted by $level(\mu)$ and computed as follows. The root ρ of T has $level(\rho) = 0$. For each non-root node μ, if μ is an S-node then $level(\mu) = level(parent(\mu)) + 1$, else $level(\mu) = level(parent(\mu))$.

Figure 1 depicts an SP-digraph G and the level numbering of its decomposition tree. Using the level numbering of T we assign a number $level(v)$ to each

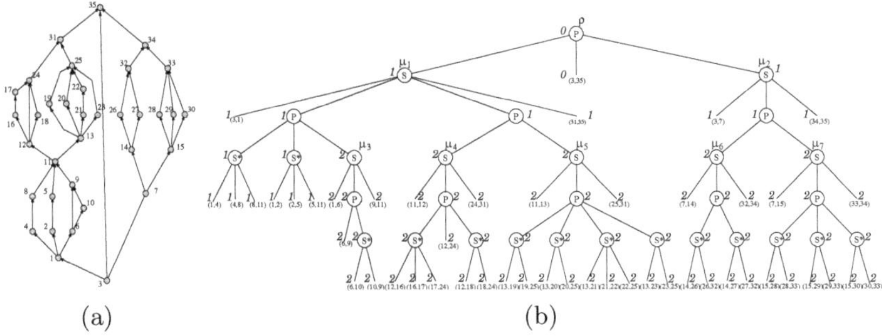

Fig. 1. (a) An SP-digraph G and (b) the level numbering of its decomposition tree.

vertex v of G, called *level of* v. Such number is the minimum among the levels of all the nodes of T having v as a pole. We call *jumping edges* those edges whose endvertices are assigned to different levels. Notice that the level numbering is such that the level number changes in correspondence of the S-nodes.

Lemma 1. *Let G be a flat SP-digraph, let T be the decomposition tree of G, and let μ be an S-node of T such that $level(\mu) = j$ $(j > 0)$. The leftmost child and the rightmost child of μ are both Q-nodes and the edges associated with them are both jumping edges.*

Let μ be either an S-node or the root node of T and let $level(\mu) = j$, we call *j-digraph of μ* the subgraph G^j_μ of G_μ, induced by those vertices having level number j. Notice that a j-digraph may not be connected . The subtree of T_μ consisting of all nodes with level umber equal to j except, for the non-root node, the leftmost child and the rightmost child is called *j-tree rooted at μ* and is denoted as T^j_μ. A *j-component* is the union of all the j-digraph of level j. By Lemma 1, each j-digraph G^j_μ $(j > 0)$ is connected to vertices of level $j - 1$ by two jumping edges, corresponding to the rightmost child and the leftmost child of μ.

We describe now how to compute a drawing of a single j-digraph. Each j-digraph is drawn on one of the two ingots defined by the 8-prism. The four tracks of the ingot are denoted as T_s, T_t, T_0 and T_1 and are arranged as shown in Figure 3. Also, we refer to T_s as the *source track* and to T_t as the *sink track*. Figure 2 shows Algorithm DRAWJDIGRAPH() and Figure 3 is an example of a resulting drawing. In the algorithm, μ is an S-node or the root of T and $\nu_0, \ldots, \nu_h$ are the children of μ in T^j_μ. For a child ν_i $(i = 0, \ldots, h)$, the source pole is s_{ν_i} and the sink pole is t_{ν_i}.

Lemma 2. *Let G be a flat SP-digraph, whose decomposition tree is T. Let μ be an S-node of T or the root of T and let $level(\mu) = j$. Let G^j_μ be the j-digraph of μ. Algorithm* DRAWJDIGRAPH*() computes an ingot-drawing of G^j_μ in $O(|G^j_\mu|)$ time.*

DrawJDigraph(G_μ^j,x_s)

Input: The j-digraph G_μ^j of an S-node or of the root μ of T and a starting x-coordinate x_s;

Output: A crossing-free straight-line grid drawing of G_μ^j on the ingot defined by T_s, T_t, T_0 and T_1;

$track(s_{\nu_0}) := T_s$; $x(s_{\nu_0}) := x_s$; $\Delta_x := 1$;
if ν_0 is a P-node having S^*-nodes as children
  let $\mu_0, \ldots, \mu_\xi$ be the children of ν_0 that are S^*-nodes;
  for $g = 0$ **to** ξ
    let $\{(s_{\nu_0}, v_1), (v_1, v_2), \ldots, (v_m, t_{\nu_0})\}$ be the pertinent digraph of μ_g;
    for $y = 1$ **to** m
      $track(v_y) := T_t$; $x(v_y) := x_s + \Delta_x$; $\Delta_x := \Delta_x + 1$;
$track(t_{\nu_0}) := T_0$; $x(t_{\nu_0}) := x_s + \Delta_x$; $\Delta_x := \Delta_x + 1$;
for $i = 1$ **to** $h - 1$
  if ν_i is a P-node having S^*-nodes as children
    let $\mu_0, \ldots, \mu_\xi$ be the children of ν_i that are S^*-nodes;
    for $g = 0$ **to** ξ
      let $\{(s_{\nu_i}, v_1), (v_1, v_2), \ldots, (v_m, t_{\nu_i})\}$ be the pertinent digraph of μ_g;
      for $y = 1$ **to** m
        $track(v_y) := T_t$; $x(v_y) := x_s + \Delta_x$; $\Delta_x := \Delta_x + 1$;
  $\alpha := i \mod 2$; $track(t_{\nu_i}) := T_\alpha$; $x(t_{\nu_i}) := x_s + \Delta_x$; $\Delta_x := \Delta_x + 1$;
if ν_h is a P-node having S^*-nodes as children
  let $\mu_0, \ldots, \mu_\xi$ be the children of ν_h that are S^*-nodes;
  for $g = 0$ **to** ξ
    let $\{(s_{\nu_h}, v_1), (v_1, v_2), \ldots, (v_m, t_{\nu_h})\}$ be the pertinent digraph of μ_g;
    for $y = 1$ **to** m
      $\alpha := h \mod 2$; $track(v_y) := T_\alpha$; $x(v_y) := x_s + \Delta_x$; $\Delta_x := \Delta_x + 1$;
$track(t_{\nu_h}) := T_t$; $x(t_{\nu_h}) := x_s + \Delta_x$;

Fig. 2. Algorithm DrawJDigraph()

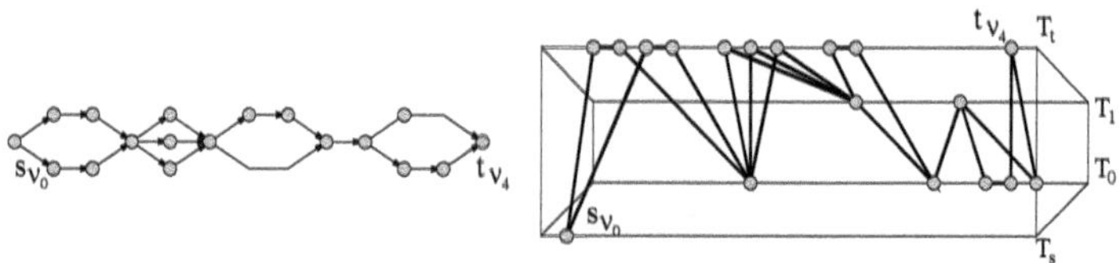

Fig. 3. A j-digraph and its drawing on an ingot.

Proof. We prove that the drawing computed by Algorithm DrawJDigraph() is crossing-free. Suppose that two edges $e_1 = (u_1, v_1)$ and $e_2 = (u_2, v_2)$ cross each other. Since all the vertices have different x-coordinates, a crossing is possible only if one of the endvertices of e_2 has x-coordinate between $x(u_1)$ and $x(v_1)$ or one of the endvertices of e_1 has x-coordinate between $x(u_2)$ and $x(v_2)$. There are three types of crossings:

1. Crossing on a track (overlap). The only edges having both endvertices on the same track are the edges of the simple paths connecting the poles of a P-child of μ. The endvertices of these edges have consecutive x-coordinates (see rows 4–8, 12–15, 19–22), thus a crossing of this type is not possible.

2. Crossing between edges lying on a strip. Without loss of generality, suppose that e_1 has endvertices with non-consecutive x-coordinates. At least one of the endvertices of e_1 is a pole of a P-child ν_i of μ. We consider only the case in which u_1 coincides with the source pole s_{ν_i} of ν_i, the other cases are analogous. The vertex v_1 can be either the first vertex of any simple path which connects $u_1 = s_{\nu_i}$ to t_{ν_i}, or it may coincide with t_{ν_i}. If $v_1 = t_{\nu_i}$ then the vertices having x-coordinates between $x(u_1)$ and $x(v_1)$ are the vertices of the simple paths connecting $u_1 = s_{\nu_i}$ to $v_1 = t_{\nu_i}$; no one of these vertices is either on the track of u_1 or on the track of v_1, thus edge e_2 cannot have an endvertex with x-coordinate between u_1 and v_1 and a crossing is not possible. If $v_1 \neq t_{\nu_i}$ then the vertices having x-coordinates between $x(u_1)$ and $x(v_1)$ are a subset of the vertices of the simple paths connecting $u_1 = s_{\nu_i}$ to t_{ν_i}; these vertices are on the same track of v_1. The edges having one of these vertices as an endvertex, and lying on the same strip as e_1, have u_1 as the other endvertex. The edge e_2 must be one of these edges. Edges e_1 and e_2 cannot cross because they have a common endvertex.

3. Crossing between edges lying on two strips. For such a crossing, edges e_1 and e_2 must lie on two different strips which cross each other along a straight line which is not one of the four tracks T_s, T_t, T_0, and T_1. The only two strips which cross in such a way are the strip $\sigma_{s,1}$, and the strip $\sigma_{t,0}$. On the strip $\sigma_{s,1}$ we have no edges. Namely, the only vertex on track T_s is the source pole s_{ν_0} of ν_0 and it can be connected either to vertices on track T_t or to a vertex on track T_0. It follows that a crossing is not possible.

Since the algorithm visit a vertex at a time and executes a constant number of operations for each vertex, its time complexity is $O(|G^j_\mu|)$. □

Algorithm DRAWJDIGRAPH() draws a single j-digraph on an ingot. The drawing of a j-component $G(j)$ can be obtained by two different algorithms called GRAINEDDRAWING() and COUNTERGRAINEDDRAWING(), respectively. Both algorithms have two main steps: in the first step a drawing is computed by Algorithm DRAWJDIGRAPH() for each j-digraph $G^j_{\mu_i} \in G(j)$, and all these drawings are consecutively arranged on the four tracks, according to the left to right order of the nodes μ_i $(i = 1, \ldots, n_j)$ in T. In the second step vertices on each track are shifted. More precisely, Algorithm GRAINEDDRAWING() shifts vertices so that: (i) all vertices on track T_0 precede all vertices on track T_1, which precede all vertices on track T_s, which precede all vertices on track T_t; (ii) the vertices on track T_0 and T_1 have odd coordinates, while the vertices on T_s and T_t have even coordinates. Algorithm COUNTERGRAINEDDRAWING() shifts vertices so that: (i) all vertices on track T_1 precede all vertices on track T_0, which precede all vertices on track T_s, which precede all vertices on track T_t; (ii) the vertices on track T_0 and T_1 have odd coordinates, while the vertices on T_s and T_t have even coordinates. We call a drawing produced by Algorithm GRAINEDDRAWING()

grained and a drawing produced by Algorithm CounterGrainedDrawing() *counter-grained.*

Lemma 3. *Let G be a flat SP-digraph, with decomposition tree T, and let $G(j) = \{G^j_{\mu_1}, \ldots, G^j_{\mu_{n_j}}\}$ be the j-component of level j. Then the Algorithm* GrainedDrawing*() and the Algorithm* CounterGrainedDrawing*() compute an ingot-drawing of $G(j)$ in $O(|G(j)|)$ time.*

To conclude the description of our technique we present now an algorithm, called 8PrismDrawing(), that draws all the j-components $G(j)$ ($j = 0, \ldots, n_{max}$) of an SP-digraph G on an 8-prism. The different j-components of G are considered consecutively from $G(n_{max})$ to $G(0)$. If j is odd $G(j)$ is drawn on the odd ingot and its drawing is a grained drawing. If j is even $G(j)$ is drawn on the even ingot and its drawing is a counter-grained drawing. Moreover the source track of $G(j)$ coincides with track T_a, where $a = j \mod 4$; as a consequence the sink track of the drawing coincides with track T_b, where $b = (j + 2) \mod 4$. Figure 4 depicts the arrangement of the j-components on the 8-prism.

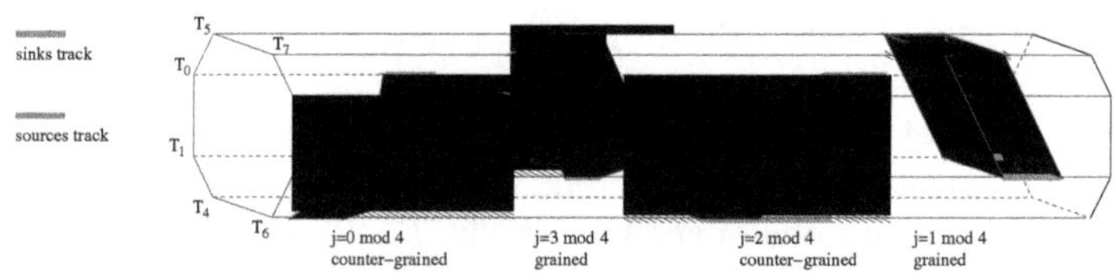

Fig. 4. The arrangement of the drawings of the j-components on the 8-prism.

In order to prove that the jumping edges connecting the j-digraph of different levels can be added to the drawing computed by Algorithm 8PrismDrawing() without creating crossings, we recall some properties of jumping edges. Each j-digraph G^j_μ ($j > 0$) is connected to vertices of level $j - 1$ by two jumping edges, corresponding to the rightmost child and leftmost child of μ. We call *core poles* the poles of a j-digraph and *simple poles* the poles that are not core poles. Thus, given a jumping edge, the endvertex with the larger level number is a core pole of a j-digraph G^j_μ. The other endvertex can be either a core pole or a simple pole of a $(j - 1)$-digraph. If a jumping edge connect two core poles, then these two core poles are of the same type, i.e. they are either both source core pole or both sink core poles. The following lemma proves that the drawing obtained by Algorithm 8PrismDrawing() is crossing-free.

Lemma 4. *Let G be a flat SP-digraph. Algorithm* 8PrismDrawing*() computes an 8-prism drawing of G.*

Proof. The drawing of a single j-component $G(j)$ ($0 \leq j \leq n_{max}$) computed either by Algorithm GrainedDrawing() or by Algorithm CounterGrained-Drawing() is crossing free by Lemma 3.

We show that the jumping edges do not cross each other and do not cross any other edge. As pointed out the jumping edges connect either source (sink) core pole to source (sink) core pole, or simple poles to core poles. Thus, the jumping edges are drawn on strips $\sigma_{1,2}$, $\sigma_{0,3}$, $\sigma_{0,1}$, $\sigma_{2,3}$, $\sigma_{0,5}$, $\sigma_{0,7}$, $\sigma_{2,5}$, $\sigma_{2,7}$, $\sigma_{1,4}$, $\sigma_{1,6}$, $\sigma_{3,4}$, and $\sigma_{3,6}$. First, we prove that jumping edges lying on the same strip do not cross each other. Let $e_1 = (u_1, v_1)$ and $e_2 = (u_2, v_2)$ be two jumping edges such that $track(u_1) = track(u_2)$ and $track(v_1) = track(v_2)$. It is either $level(u_1) = level(v_1)+1$ and $level(u_2) = level(v_2)+1$ or $level(v_1) = level(u_1)+1$ and $level(v_2) = level(u_2)+1$. Assume without loss of generality that $level(v_1) = level(u_1) + 1$ and $level(v_2) = level(u_2) + 1$. It follows that v_1 is the core pole of a j-digraph $G^j_{\mu_1}$ and that v_2 is the core pole of a h-digraph $G^h_{\mu_2}$. Consider the case when $j \neq h$, and assume $j < h$. Since v_1 and v_2 are on the same track then h is at least $j + 2$ and therefore $level(u_1) < level(v_1) < level(u_2) < level(v_2)$. It follows that $x(u_1) < x(v_1) < x(u_2) < x(v_2)$ and a crossing is not possible. If $j = h$ then $G^j_{\mu_1}$ and $G^h_{\mu_2}$ are in the same j-component $G(j)$. Since $x(v_1) < x(v_2)$ then μ_1 is to the left of μ_2 in the SPQ-tree. If μ_1 and μ_2 have the same parent ν then $u_1 = u_2$ and the two edges can not cross since they have a common vertex. If μ_1 and μ_2 have different parents ν_1 and ν_2, then since μ_1 is to the left of μ_2, it follows that ν_1 is to the left of ν_2, and therefore $x(u_1) < x(u_2)$. A crossing is not possible either in this case.

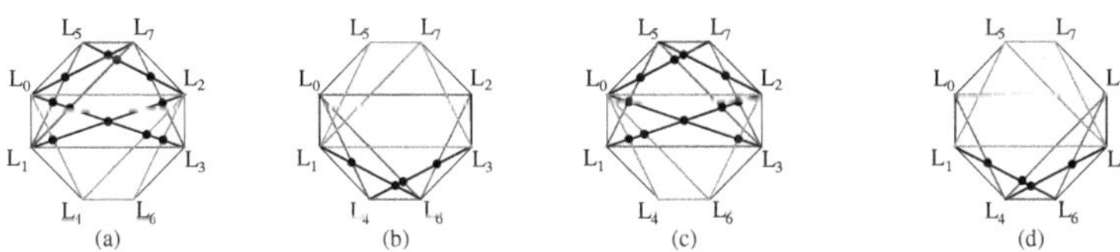

Fig. 5. The possible crossings of the jumping edges.

Now we prove that the jumping edges do not cross any other edge on different strips. Depending on the value of j the jumping edges between levels j and $j+1$ are on different strips and therefore the possible crossings are different. Figure 5 shows all the possible cases depending on the value of j. In Figure 5(a) the drawing on the even ingot has level $j \mod 4 = 0$ and the drawing on the odd ingot has level $(j-1) \mod 4 = 3$. In Figure 5(b) the drawing on the odd ingot has level $j \mod 4 = 3$ and the drawing on the odd ingot has level $(j-1) \mod 4 = 2$. In Figure 5(c) the drawing on the even ingot has level $j \mod 4 = 2$ and the drawing on the odd ingot has level $(j-1) \mod 4 = 1$. In Figure 5(d) the drawing on the odd ingot has level $j \mod 4 = 1$ and the drawing on the even ingot has level $j - 1 \mod 4 = 0$. The black dots represent the 30 possible crossings.

Let $e_1 = (u_1, v_1)$ and $e_2 = (u_2, v_2)$ be two edges; if they cross each other, the four endvertices are co-planar. The co-planarity condition of the four vertices can be expressed by means of an equation on the x-, y- and z-coordinates of the vertices and since the y- and z- coordinates are determined by the choice of the tracks, the equation reduces to a co-planarity condition on the x-coordinates. For each of the 30 cases the x-coordinates assigned to vertices are such that the equation above is never satisfied. This is a consequence of the fact that some vertices have even x-coordinates and some vertices have odd x-coordinates, and of the relative position of the vertices induced by the grained and counter-grained drawings. The analysis of the 30 cases is straightforward but tediousand is omitted here for reasons of brevity. □

Theorem 1. *Let G be a flat series-parallel digraph with n vertices. Then G has track number eight and there exists an algorithm that computes an 8-track-drawing of G in $O(n)$ time and with at most $4 \times 4 \times 2n$ volume.*

Proof. The drawing computed by Algorithm 8PrismEmbedder() is crossing free by Lemma 4. For what concern the time complexity of the algorithm: the SPQ-tree T and the level numbering of G can be computed in $O(n)$ time. The drawing of each j-component $G(j)$ can be computed in $O(|G(j)|)$ time by Lemma 3. It follows that Algorithm 8PrismEmbedder() has time complexity $O(n)$. For what concerns the volume of the drawing, we have that: (i) each vertex has a different x-coordinate; (ii) the grained or counter-grained drawing of each j-component $G(j)$ requires a portion of the 8-prism of length $2|G(j)|-1$, in order to guarantee that some vertices have even x-coordinates and some other have odd x-coordinates. The length of the drawing is then $\sum_{j=0}^{n_{max}}(2|G(j)|-1) = 2n - n_{max}$, and the overall volume is at most $4 \times 4 \times 2n$. □

References

1. F. R. K. Chung, F. T. Leighton, and A. Rosenberg. Embedding graphs in books: A layout problem with applications to VLSI design. *SIAM J on Alg. and Disc. Methods*, 8:33–58, 1987.
2. G. Di Battista, P. Eades, R. Tamassia, and I. G. Tollis. *Graph Drawing*. Prentice Hall, Upper Saddle River, NJ, 1999.
3. G. Di Battista and R. Tamassia. On-line maintenance of triconnected components with SPQR-trees. *Algorithmica*, 15(4):302–318, 1996.
4. E. Di Giacomo, G. Liotta, and H. Meijer. 3D straight-line drawings of k-trees. Submitted for publication.
5. E. Di Giacomo, G. Liotta, and S. K. Wismath. Drawing series-parallel graphs on a box. In *Proc. CCCG 2002*, 2002.
6. V. Dujmović, P.Morin, and D. Wood. Pathwidth and three-dimensional straight line grid drawings of graphs. In *Proc. GD 2002*, volume 2528 of *LNCS*, pages 42–53. Springer-Verlag, 2002.
7. V. Dujmović and D. R. Wood. Tree-partitions of k-trees with application in graph layout. In *Proc. WG 2003*, LNCS. Springer-Verlag, to appear.
8. S. Felsner, G.Liotta, and S. K. Wismath. Straight line drawings on restricted integer grids in two and three dimensions. In *Proc. GD 2001)*, volume 2265 of *LNCS*, pages 328–342. Springer-Verlag, 2001.
9. D. R. Wood. Queue layouts, tree-width, and three-dimensional graph drawing. In *Proc. FSTTCS '02*, volume 2556 of *LNCS*, pages 348–359. Springer-Verlag, 2002.

Nearly Optimal Three Dimensional Layout of Hypercube Networks

Tiziana Calamoneri and Annalisa Massini

Department of Computer Science, University of Rome "La Sapienza" - Italy, via Salaria 113, 00198 Roma, Italy.
massini@dsi.uniroma1.it

Abstract. In this paper we consider the three-dimensional layout of hypercube networks. Namely, we study the problem of laying hypercube networks out on the three-dimensional grid with the properties that all nodes are represented as rectangular slices and lie on two opposite sides of the bounding box of the layout volume. We present both a lower bound and a layout method providing an upper bound on the layout volume of the hypercube network.

Keywords: Hypercube Network, Three Dimensional Layout, VLSI layout volume.

1 Introduction

The layout of interconnection networks has important cost and performance implications for multiprocessors and parallel and distributed systems based on components. Thus, there is currently renewed interest in finding efficient VLSI layouts for various interconnection networks (see for instance [2,4,9,13,14,15,16]).

Recent hardware advances have allowed three-dimensional circuits to have a cost low enough to make them commonly available. For this reason three-dimensional layouts of interconnection topologies on rectilinear grids are becoming of wide interest. The importance of efficiently representing interconnection networks in three dimensions has already been stated in the 80's by Rosenberg [11], and the most relevant aims are to shorten wires and to save in material (i.e. to minimize volume). The three-dimensional layout problem is related both with the study of the VLSI layout problem for integrated circuits and with the study of algorithms for drawing graphs. Indeed, the tie between VLSI layout studies and theoretical graph drawing is very strong since laying out a network on a grid is equivalent to orthogonally drawing the underlying graph, except for the fact that wire crossings are allowed. Nevertheless, almost all the results known in the literature about three-dimensional grid drawing of graphs are valid for very general graphs (see, the proceedings of the past editions of the *Graph Drawing* Symposia and, for instance, [1,5,12]) and therefore they do not work efficiently for structured and regular graphs such as the most commonly used interconnection topologies, either the produced volume is too large (in order of

G. Liotta (Ed.): GD 2003, LNCS 2912, pp. 247–258, 2004.

magnitude or in the multiplicative constant), or the length of each wire is not kept of the same order of magnitude and sufficiently small, or the shape of nodes is not feasible to represent a processor, a switch, etc.. Hence, usually, lower and upper bounds on the layout volume of specific interconnection topologies must be established and the layout technique is as more good as these bounds are closer; of course, the layout method should try to highlight all the interesting properties of the network and to make as fast as possible the flow through it by shortening the maximum wire length.

The hypercube network has been widely considered as a network for parallel computing for its good parameters, e.g. regularity, logarithmic diameter, fault tolerance. A further parameter distinguishing 'good' networks is whether they can be laid out compactly in the orthogonal grid. This issue has already been studied in two dimensions [6]. In this paper we study the three-dimensional layout of the hypercube networks with the property that nodes lie on two planes, corresponding to opposite sides of the bounding box, partitioned into two equally sized sets, and they are represented as rectangular slices. The model of keeping all nodes on the surface of the bounding box is motivated by two main reasons: (i) it decreases the problems due to power consumption and concomitant heat generation and (ii) it makes easier the connection of the topologies to other part of the architecture, as the wires are necessarily connected to nodes. We present upper and lower bounds for the layout volume of the hypercube networks that are very close, although not coinciding (in fact they differ by a factor of $\log^{1/2} N$). The maximum wire-length of the presented layout is $O(\log N\sqrt{N})$.

The rest of this paper is organized as follows: in Section 2 we give some preliminary definitions and recall some useful previous results. Section 3 is the core of the work and we present both a lower bound and a layout method providing an upper bound on the layout volume of the hypercube network. Finally, in Section 4 we address some conclusions and list some interesting open problems.

2 Preliminaries

In this section we recall some useful definitions and state some preliminary results.

A *hypercube* Q_n of dimension n is an interconnection network with $N = 2^n$ nodes, each labeled with an n-bit binary string. Two nodes are connected by an edge if their labels differ in exactly one bit.

It is easy to see that a Q_n can be constructed by considering two copies of Q_{n-1}, adding an edge for each pair of equally labeled nodes and completing the labels of nodes by adding one bit (0 in the first copy and 1 in the second copy of Q_{n-1}). In the following we will call *homonym nodes* in Q_n two nodes connected by this inductive step, i.e. two nodes whose labels differ only in the leftmost bit.

By iterating this recursive construction, we deduce that Q_n can be constructed starting from 2^{n-4} copies of Q_4, adding some edges and readjusting the labels. In the following we will extensively use this observation and we will clarify the reason why Q_4 is chosen.

Another interesting property of Q_n we will exploit in the following is its bipartiteness; a feasible partition of the nodes can be obtained by dividing nodes having an even number of 0s in their label from nodes having an odd number of 0s in their label. The bipartiteness property is particularly evident when Q_n is visualized as a lattice (see Fig. 1).

Finally, observe that the hypercube Q_n is a regular graph of degree n, so it is not possible to represent nodes as dots for each $n > 6$.

A *three-dimensional grid layout* $\Gamma(\mathcal{G})$ of an interconnection network $\mathcal{G}$ is a mapping of $\mathcal{G}$ to the three-dimensional grid such that nodes are mapped to grid-nodes and edges are mapped to independent grid-paths satisfying the following conditions:

- distinct grid-paths are edge-disjoint (then at most three paths can cross at a grid-node);
- grid-paths that share an intermediate grid-node must cross at that node (that is 'knock-knee' paths [8] are not allowed);
- a grid-path may touch no mapped node, except at its end-points.

If $\Gamma(\mathcal{G})$ can be enclosed in a $H \times W \times L$ three-dimensional box, the *volume* of $\Gamma(\mathcal{G})$ is the product $H \times W \times L$.

We define the $a \times b$-3D double channel routing as a slight modification of the concept of k-3D channel routing introduced in [3]. To this aim, we need to recall the following definition:

Definition 1. *A* k-channel routing *involves a bidimensional grid and two sets* S *and* S' *each consisting of* k *nodes to be connected by a 1-1 function.* S *and* S' *are arranged onto opposite sides of the grid.*

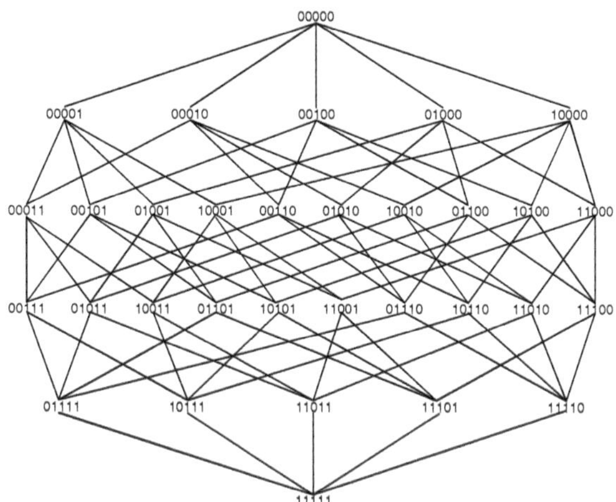

Fig. 1. Q_5 represented as a lattice.

The grid involved in any k-channel routing is not greater than $(k+1)\times(\frac{3}{2}k+2)$ and S and S' lie on the shortest sides, as shown in [10].

Definition 2. *An* $a\times b$-3D double channel routing *involves a three-dimensional grid (the channel) and two sets S and S', each containing $a\cdot b$ nodes, arranged on the grid nodes of opposite faces of dimension $a\times b$, to be connected by a 1-1 function f. Function f associates to a node (x,y) of S a node (x',y') of S' such that $x'=g(x)$ and $y'=h(y)$, where functions g and h are two-dimensional $a-$ and $b-$channel routings, respectively.*

In the following, for the sake of brevity, when no confusion arises, we refer to the $a\times b$-3D double channel routing as *3D channel routing.*

The following theorem states the necessary volume for any 3D channel routing. In this preliminary version of the paper we omit its proof.

Theorem 1. *A three-dimensional grid of size $(a+1)\times(b+1)\times max\{\frac{3}{2}a+2,\frac{3}{2}b+2\}$ is enough to realize a 3D channel routing.*

3 Three Dimensional Layout of Hypercube Network

In this section we discuss the problem of laying the hypercube network out; in particular, we determine lower and upper bounds for the volume.

3.1 Lower Bound

In order to compute a lower bound on the layout volume of the hypercube network, we can use the general formula provided in [3]:

$$\text{lower bound on the layout volume of } \mathcal{H} \geq (\sqrt{BW(\mathcal{H})}-1)^3$$

where $\mathcal{H}$ is a general network and $BW(\mathcal{H})$ is its bisection width.

It is well known that the bisection width of Q_n is $\frac{N}{2}$ [7], so a lower bound for the layout volume for Q_n is $\Omega(N^{\frac{3}{2}})$. One can object that this lower bound does not take into account the fact that, when $n>6$, nodes cannot be represented as dots (i.e. with null area), but it is necessary to use a space (either area or volume) proportional to the degree for each of them. Furthermore, the previous formula assumes that the nodes can be distributed anywhere inside the volume of the layout. On the contrary, our aim is to find a three-dimensional layout of Q_n with some special properties; namely, all nodes are represented as rectangular slices with area proportional to their degree, they must lie on two planes, corresponding to opposite sides of the bounding box, and they are partitioned into two equally sized sets. Under these assumptions, we can better specify the value of the lower bound on the volume as $\Omega(N^{3/2}\log^{1/2}N)$.

We compute this lower bound by exploiting a proof technique presented in [11] for one-active-level layouts. Let W,L and H be the dimensions of the bounding box of any three-dimensional layout of Q_n with nodes partitioned into two equally sized sets and laid out on two opposite sides. Without loss of generality, assume the nodes lie on the sides of dimension $W \times L$ of the bounding box. Since each side contains $\frac{N}{2}$ nodes, each of degree $\log N$, it is straightforward to see that $W \times L = \Omega(\frac{N}{2} \log N)$. Two cases can occur, according to the fact that $H \geq L, W$ or not.

If $H \geq L, W$, then $H \geq (\frac{N}{2} \log N)^{1/2}$. It follows that the volume must be $\Omega((\frac{N}{2})^{3/2} \log^{3/2} N)$, which is even greater than we claim.

If H is not the maximum dimension of the bounding box, we can consider a plane π orthogonal to the sides of the bounding box containing nodes and parallel to the pair of parallel sides of the bounding box of dimension $L \times H$. We place π in such a way that π cuts the layout into two equal portions. The intersection of the bounding box with plane π has area $L \times H$ and π partitions each set of nodes into two equally sized sets leaving $\frac{N}{2}$ nodes to its left ($\frac{N}{4}$ on the upper face and $\frac{N}{4}$ on the lower face of the bounding box) and $\frac{N}{2}$ nodes to its right ($\frac{N}{4}$ on the upper face and $\frac{N}{4}$ on the lower face of the bounding box). Hence, π cuts a number of edges at least equal to the bisection width of Q_n, i.e. $\frac{N}{2}$. It follows that the area of π is at least $\frac{N}{2}$, i.e. $H \times L \geq \frac{N}{2}$. Since $L \geq N^{1/2} \log^{1/2} N$ we have that the volume is $\Omega(N^{3/2} \log^{1/2} N)$.

The previous considerations lead to the following:

Theorem 2. *A lower bound for a 3D grid layout of Q_n is $\Omega(N^{3/2} \log^{1/2} N)$, if the layout has the following properties:*

- *all nodes are represented as rectangular slices with area proportional to their degree,*
- *all nodes lie on two planes, corresponding to opposite sides of the bounding box,*
- *nodes are partitioned into two equally sized sets.*

3.2 Layout of Q_n

In this subsection we provide a method to lay the hypercube network out for each $n \geq 4$ (indeed, for $n = 3$ the layout is trivial).

In our layout we choose to represent nodes as rectangular slices of area $1 \times \lceil \frac{n}{2} - 1 \rceil$ (see Fig. 2). Each dot at integer coordinates on the slice is a pin where one wire is connected. Observe that when n is odd, the last column of each slice has only one active pin, whereas when n is even, all pins of the slice are active (see Fig. 2).

As we highlighted in the previous section, the basic building box of our result is the layout of Q_4, so we show it, before generalizing the construction to Q_n.

Since the degree of nodes in Q_4 is 4, we represent each node as a unit area slice and we arrange these slices in two rectangles of dimension 3×7, as shown in Fig. 3. Our layout puts nodes onto opposite sides of the bounding box, and all edges are routed inside. Hence, we exploit the property that Q_4 is bipartite and put on one side (let it be the *upper side*) all nodes having an even number of 0s in their label, and on the opposite side (let it be the *lower side*) the remaining nodes (see Fig. 3).

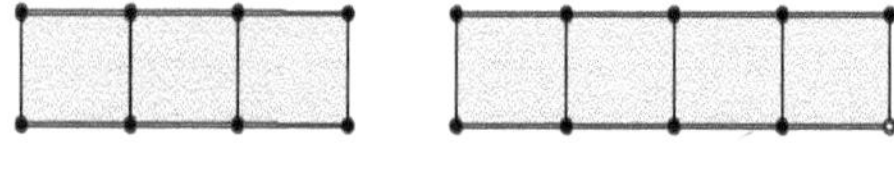

Fig. 2. Slices representing nodes of degree n; a. n even; b. n odd; the white dot is not active.

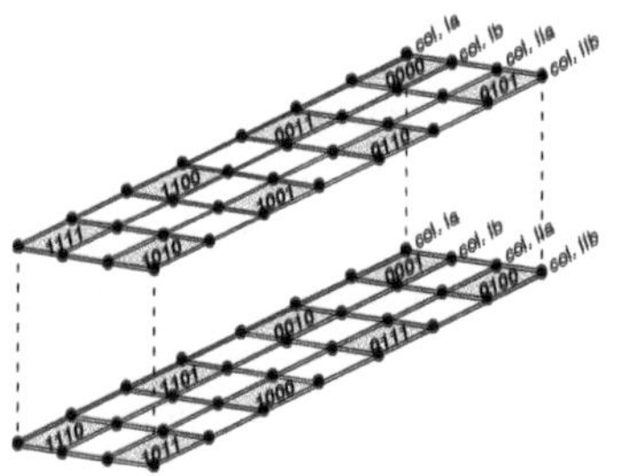

Fig. 3. Placement of Q_4's nodes on a three-dimensional grid.

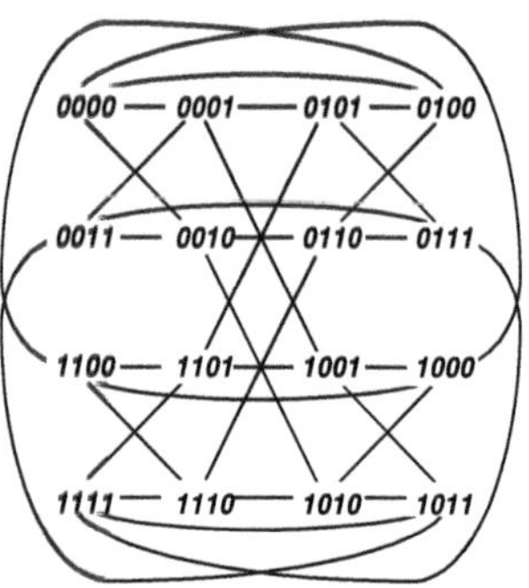

Fig. 4. Q_4 represented as a butterfly-like network.

Furthermore, we want to use a 3D channel routing to arrange all edges on the grid. To this aim, we need to assign to each node a position in such a way that the rules of 3D channel routing (cf. Def. 2) are maintained; namely, we have to map columns of the upper side to columns of the lower side. In other words, all nodes of each column i on the upper side must be mapped to nodes of a column j on the lower side. The way how to arrange nodes on the rectangle is highlighted from the representation of Q_4 shown in Fig. 4. Indeed, we can map the leftmost node column of Fig. 4 to the first column of the upper side (see Fig. 3), the second node column in Fig. 4 in the first column of the lower side, and so on. Of course, we use pins in the same subcolumn (either a or b) to layout edges outgoing from each node and going towards right (left) in Fig. 4.

Such an assignment respects the rules of a 3D channel routing, and a possible final three-dimensional layout is sketched in Fig. 5, where the 4-channel routings inside the bent planes are not drawn for the sake of clarity. The volume of the built layout for Q_4 is $8 \times 4 \times 5$.

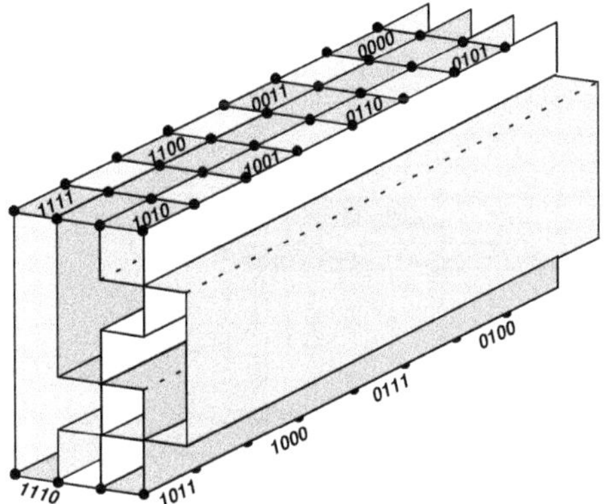

Fig. 5. 3D layout of Q_4, where the 4-channel routings inside the bent planes are not drawn.

In order to generalize our layout to each $n \geq 4$, we exploit the layout of Q_4 and the recursive definition of Q_n. Namely, given a hypercube Q_n, we will arrange it on the three-dimensional grid as 2^{n-4} copies of hypercubes of dimension 4 plus a set of edges. We claim that this set of edges constitutes a 3D channel routing.

We give the construction and the proof of our claim by induction on n.

Base. The claim is trivially true when $n = 4$.

Inductive Hypothesis. $Q_r, 4 \leq r < n$, can be arranged on the grid as 2^{r-4} copies of Q_4s plus a set of edges constituting a 3D channel routing.

Inductive Step. Q_n can be obtained by considering two copies of Q_{n-1} and connecting with an edge each pair of homonym nodes. In view of the inductive hypothesis, the claim is true for Q_{n-1}. So, we can represent Q_n on the three-dimensional grid by putting two copies of 3D layouts of Q_{n-1}s one beside the other. Since the copies of Q_{n-1}s consist of 2^{n-5} copies of Q_4s each by inductive hypothesis, Q_n will consist of 2^{n-4} copies of Q_4s. It remains to prove that the edges connecting homonym pairs plus the existing 3D channel routings inside the two copies of Q_{n-1} constitute a new 3D channel routing.

Let us distinguish two different cases according to the parity of n:

- n **odd**
  Arrange the 3D layouts of Q_{n-1}s one beside the other but swapping upper and lower sides in one of them, in such a way that homonym nodes lie on opposite sides of the bounding box (see Fig. 6.a).

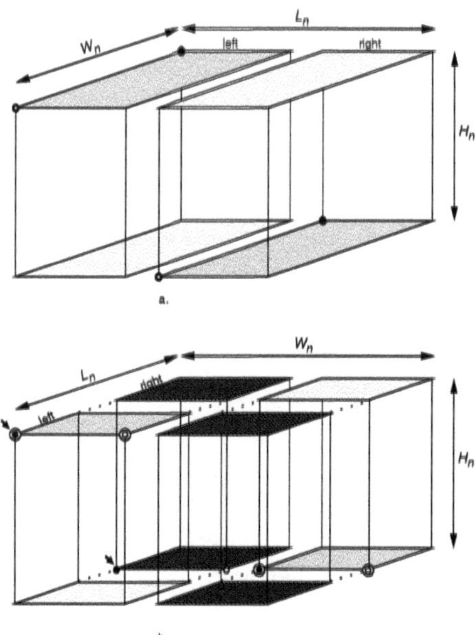

Fig. 6. Construction of Q_n starting from two copies of Q_{n-1}s; a. n odd; b. n even. Double circled nodes represent homonym nodes in the two copies of Q_{n-1}s. Simply circled nodes are homonym nodes of those in the upper part of the layout at the previous step, i.e. in the two copies of Q_{n-2}s inside the same copy of Q_{n-1}.

Observe that the set of edges connecting the copies of Q_4s is a 3D channel routing. Indeed, it is constituted by a set of edges (inside the two copies of Q_{n-1}s) and by a set of edges connecting homonym nodes of the two copies of Q_{n-1}s. The first set forms two disjoint 3D channel routings by inductive hypothesis; the second set connects homonym nodes located on the upper side of the left (right) Q_{n-1} with nodes located on the lower side of the right (left) Q_{n-1} (for an example, see nodes in Fig. 6.a).
Each node has the last column constituted by only one active pin and it corresponds to the edge that must be still laid out. Hence, the second set is completely disjoint from the other one and it has even stronger properties than a 3D channel routing, namely it ordinately connects nodes of the same row with nodes on the same row and nodes on the same column with nodes on the same column.
From the previous observations it follows that these three 3D channel routings form all together a unique 3D channel routing.

– n **even**
In this case each node has all columns constituted by two pins and the lower pin in the last column corresponds to the edges to be still laid out (see Fig. 2). The upper pin in the last column corresponds to edges laid out in the previous step (when connecting two copies of Q_{n-2}s in a Q_{n-1}). Hence,

when we route the edges connecting homonym nodes using the lower pins, we must take into account the existing 3D channel routing using the upper pins. Again, arrange the 3D layouts of Q_{n-1}s one beside the other, swapping upper and lower sides in one of them and swapping also the order of its two copies of Q_{n-2}s inside, as shown in Fig. 6.b.
This arrangement guarantees that the rules of the 3D channel routing are respected by the set of all edges connecting the copies of Q_4s. Actually, this set is constituted by a set of edges inside the two copies of Q_{n-1}s and a set of edges connecting homonym nodes of the two copies of Q_{n-1}s. The first set forms two disjoint 3D channel routings by inductive hypothesis; inside this set, we want to highlight the subset of edges outgoing from the upper pin of the last column of each node. In view of the 3D channel routing of the previous step $n-1$ (odd), they connect each node in general position (i,j) of the upper face, left (right) side of each copy of Q_{n-1} with its homonym in position (i,j) of the lower face, right (left) side of the same copy (for an example, see nodes pointed by the arrows in Fig. 6.b). Now consider the second set: it connects each node in general position (i,j) of the upper face, left (right) side of the first copy, with its homonym in position (i,j) of the lower face, right (left) side of the second copy by means of the lower pin of the last column (for an example, see double circled nodes in Fig. 6.b). So all connections coming out from the last column of the slice go to the same column and we have a 3D channel routing.

3.3 Upper Bound

Starting from the construction of the layout of Q_n presented in the previous subsection, we deduce an upper bound on the volume of the layout of Q_n.

Theorem 3. *There exists a three-dimensional layout of Q_n having volume $\Theta(N^{3/2}\log N)$.*

Proof. Let W_n, L_n and H_n be the three dimensions of the bounding box of the layout of Q_n we presented in the previous subsection. We define some recurrence equations in order to compute these values for each $n \geq 4$.

If $n = 4$ we have already obtained $W_4 = 8, L_4 = 4$ and $H_4 = 5$ (see Fig. 5), and we use these values as base of the recurrence.

As our layout method is conceived, the equations are different according to n odd or even.

If n is **odd**, we have:

W_n: The width of the layout of Q_n is the same as the width of Q_{n-1} (see Fig. 6.a), hence $W_n = W_{n-1}$.

L_n: $L_n = 2L_{n-1} + \sqrt{N/2}$ as we use two copies of Q_{n-1}s and we need to enlarge the slices constituting nodes by one column. The contribution of $\sqrt{N/2}$ follows from the way of positioning the slices.

H_n: The height is increased by the 3D channel routing introduced to connect the two copies. As this 3D channel routing has $a = b = \sqrt{N/2}$, its contribution to the height is $\frac{3}{2}\sqrt{N/2} + 2$, hence $H_n = H_{n-1} + \frac{3}{2}\sqrt{N/2} + 2$.

If n is **even**, we have:

W_n: $W_n = 2W_{n-1}$ as we use two copies of Q_{n-1}s.
L_n: The length of the layout of Q_n is the same as the length of Q_{n-1} (see Fig. 6.b), hence $L_n = L_{n-1}$. Indeed, the newly inserted edges do not introduce any new contribution to this dimension.
H_n: Also in this case the height is increased by the 3D channel routing introduced to connect the two copies and $a = 2\sqrt{N/2}$ and $b = \sqrt{N/2}$. Nevertheless, the contribution to the height is due only to the connections outgoing from the lower pins of the last columns in the slices. Indeed, the contribution of the other connections has been already computed in H_{n-1}. Hence, we have to add only the difference between the contributions of the two 3D channel routings, i.e. $H_n = H_{n-1} + \frac{3}{2}\sqrt{N/2}$.

By solving the previous recurrence equations, we get:

$$W_n = \Theta(\sqrt{N})$$
$$L_n = \Theta(\sqrt{N}\log N)$$
$$H_n = \Theta(\sqrt{N})$$

both when n is even and when n is odd.
It follows that the volume is $\Theta(N^{\frac{3}{2}}\log N)$.

Observe that this upper bound differs from the lower bound only by a $\log^{1/2} N$ factor.

Lemma 1. *The 3D layout of Q_n presented in the previous section has maximum wire length $O(\sqrt{N}\log N)$.*

Proof. It is easy to see that the maximum wire length can be computed considering the worst case in the 3D channel routing inside the layout.

As highlighted in Fig. 7.a, each plane constituting the 3D channel routing is at most $H_n + 2L_n - 1$ long. Inside this plane (see Fig. 7.b), the longest wire is $H_n + 2L_n - 1 + 2W_n - 1$.

From this formula and from the values of H_n, W_n and L_n found in the previous proof, the claim follows.

Observe that in the two-dimensional layout of Q_n, the maximum wire length is $\Theta(N)$ [6], so this 3D layout saves both in material and in wire-length.

We conclude this subsection by summarizing all the results in the following theorem:

Theorem 4. *The method described in the previous subsections provides a 3D grid layout of Q_n in $\Omega(N^{3/2}\log N)$ volume, with $O(\sqrt{N}\log N)$ maximum wire length, if the layout has the following properties:*

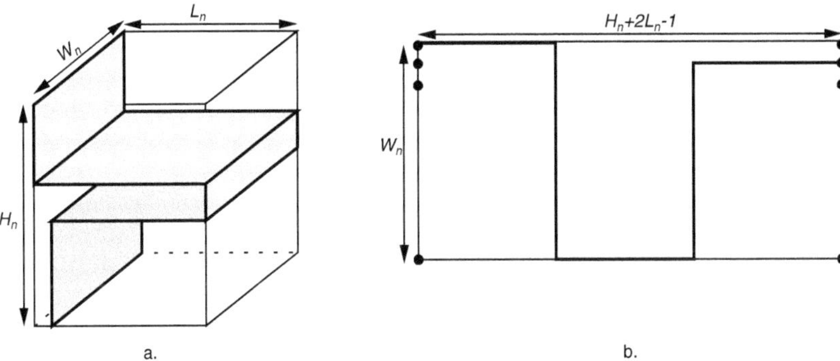

Fig. 7. a. A plane of the 3D channel routing inside the layout of Q_n and b. a wire inside the same plane.

- *all nodes are represented as rectangular slices with area proportional to their degree,*
- *all nodes lie on two planes, corresponding to opposite sides of the bounding box,*
- *nodes are partitioned into two equally sized sets.*

4 Conclusions and Open Problems

In this paper we have considered the three-dimensional layout of hypercube networks. Namely, we have studied the problem of laying hypercube networks out on the three-dimensional grid with the properties that all nodes are represented as rectangular slices and lie on two opposite sides of the bounding box of the layout volume. We have presented upper and lower bounds for the layout volume of the hypercube networks that are very close, although not coinciding (in fact they differ by a factor of $\log^{1/2} N$). The maximum wire-length of the presented layout is $O(\log N\sqrt{N})$.

Two open problems arise from this work. The first problem derives from the not coinciding upper and lower bounds, and consists in understanding which bound is tight. Secondly, it would be interesting to study the more general 3D layout problem, i.e. the problem where nodes are not constrained to be on two opposite sides of the bounding box but wherever inside the volume. In this case, it is possible to consider also other shapes for nodes (i.e. boxes). Concerning this problem, note that it is quite easy to find a $O(N^{3/2})$ volume 3D layout of CCC_n network (i.e. the network obtained from Q_n by substituting each Q_n's node with a cycle of length n) from the technique used in [3]. We have wondered if it is possible to deduce a 3D layout of Q_n from the layout of CCC_n with the same volume exploiting the definition of CCC_n. This does not seem possible, because the nodes constituting the cycles are not clustered in the layout.

References

1. T. Biedl, T. Thiele, D.R. Wood: Three-Dimensional Orthogonal Graph Drawing with Optimal Volume. *Proc. Graph Drawing (GD'00)*, pp. 284–295, 2001.
2. T. Calamoneri, A. Massini: An Optimal Layout of Multigrid Networks. *Information Processing Letters* , **72**, pp. 137-141, 1999.
3. T. Calamoneri, A. Massini: Optimal three-dimensional layout of interconnection networks. *Theoretical Computer Science*, 255, pp. 263–279, 2001.
4. A. DeHon: Compact, Multilayer Layout for Butterfly Fat-Tree. *ACM Symp. on Parallel Algorithms and Architectures (SPPA)*, 2000.
5. P. Eades, A. Symvonis, S. Whitesides: Two Algorithms for Three Dimensional Orthogonal Graph Drawing. *Proc.Graph Drawing '96 (GD '96)*, Lectures Notes in Computer Science 1190, Springer-Verlag ,1996, pp 139–154.
6. R.I. Greenberg and L. Guan: On the area of hypercube layouts. *Information Processing Letters*, 84, pp. 41–46, 2002.
7. F.T. Leighton: *Introduction to Parallel Algorithms and Architectures: Arrays, Trees and Hypercubes*, Morgan Kaufmann Publ., San Mateo, CA, 1992.
8. K. Melhorn, F.P. Preparata and M. Sarrafzadeh: Channel routing in knock-knee mode: simplified algorithms and proofs. *Algorithmica* 1, 213-221, 1986.
9. S. Muthukrishnan, M. Paterson, S.C. Sahinalp and T. Suel: Compact Grid Layouts of Multi-Level Networks. *31st ACM Symp. on Theory of Computing*, 1999.
10. R.Y. Pinter, On routing two-point nets across a channel, *19th ACM-IEEE Design Automation Conf.* (1982) 894–902.
11. A.L Rosenberg: Three-Dimensional VLSI: A Case Study. *Journal of the ACM*, 30(3), 1983, pp 397–416.
12. D.R. Wood: A New Algorithm and Open Problems in Three-Dimensional Orthogonal Graph Drawing. *Proc. 10th Australasian Work. Combinatorial Algorithms, (AWOCA)*, pp. 157–167, 1999.
13. T.Yamada and S. Ueno: On Three-Dimensional Layout of De Bruijn Networks. *2002 IEEE International Symposium on Circuits and Systems (ISCAS '02)*, Vol. III, pp.779-782, 2002.
14. T.Yamada and S. Ueno: On Three-Dimensional Layout of Pyramid Networks. *2002 IEEE Asia Pacific Conference on Circuits and Systems (APCCAS '02)*, 2002.
15. C.-H. Yeh, B. Parhami, E.A. Varvarigos and H. Lee: VLSI Layout and Packaging of Butterfly Networks. *ACM Symp. on Parallel Algorithms and Architectures (SPPA)*, pp. 196–205, 2000.
16. C.-H. Yeh, E.A. Varvarigos and B. Parhami: Multilayer VLSI Layout for Interconnection Networks. *2000 Int.l Conference on Parallel Processing (ICPP)*, pp. 33–40, 2002.

Graph Embedding with Minimum Depth and Maximum External Face
(Extended Abstract)

Carsten Gutwenger[1] and Petra Mutzel[2]

[1] Stiftung caesar
Ludwig-Erhard-Allee 2, D-53175 Bonn, Germany
[2] Vienna University of Technology
Favoritenstr. 9–11 E186, A-1040 Vienna, Austria

Abstract. We present new linear time algorithms using the SPQR-tree data structure for computing planar embeddings of planar graphs optimizing certain distance measures. Experience with orthogonal drawings generated by the topology-shape-metrics approach shows that planar embeddings following these distance measures lead to improved quality of the final drawing in terms of bends, edge length, and drawing area. Given a planar graph, the algorithms compute the planar embedding with

1. the minimum depth among the set of all planar embeddings of G,
2. the external face of maximum size among the set of all planar embeddings of G,
3. the external face of maximum size among the set of all embeddings of G with minimum depth.

1 Introduction

A *combinatorial embedding* Γ of a planar graph G is defined as a clockwise ordering of the incident edges for every vertex with respect to a crossing-free drawing of G in the plane. An equivalent definition of a combinatorial embedding is an ordered list of the boundary edges for each face. A combinatorial embedding Γ together with a given *external face* $f \in \Gamma$ is called a *planar embedding* (Γ, f).

A successful approach for generating orthogonal drawings of general graphs is the *topology-shape-metrics approach* [18,2,6]. Here, in a first step, the crossing structure of the graph is computed. The second step takes this topology as input and produces an orthogonal representation fixing the shape of the drawing. Finally, the third step determines the lengths of the horizontal and vertical edge segments. The goal of the three steps is to minimize the number of crossings, the number of bends, and the total edge length, respectively.

Algorithms for the second step typically deal with planar graphs (i.e., the planarized graphs arising in step 1). Such algorithms, in general, need as input a planar embedding (e.g., the bend minimization algorithm by Tamassia [17]). Fortunately, a planar embedding of a planar graph can be computed in linear

G. Liotta (Ed.): GD 2003, LNCS 2912, pp. 259–272, 2004.

time (see [14,5]). However, the choice of the embedding typically has a big impact on the quality of the resulting drawing, and the number of possible embeddings of a planar graph may be exponential in the size of the graph. Since optimizing aesthetic criteria like the number of bends is NP-hard in general [10], finding criteria for planar embeddings that lead to good drawings in practice, as well as efficient algorithms to compute such planar embeddings is desired.

Several authors have studied the problem of computing planar embeddings which are optimal with respect to some distance measures. Bienstock and Monma [4] have suggested polynomial time algorithms for computing planar embeddings that minimize the distance measures *radius*, *width*, *outerplanarity*, and *face depth*. They also showed that it is NP-complete to test whether a planar graph has an embedding with dual diameter bounded above by an input number.

Exact algorithms for the NP-hard problem of computing planar embeddings that minimize the number of bends of orthogonal planar graphs have been suggested by Liotta et al. [13] and by Mutzel and Weiskircher [15]. Mutzel and Weiskircher [15] have suggested an exact algorithm for the NP-hard problem of computing a linear cost function on the face cycles. However, all the exact algorithms only work well for graphs with up to 80 edges.

In this paper we give linear time algorithms based on the SPQR-tree data structure for computing planar embeddings of planar graphs optimizing various distance measures. Experience in the graph drawing community has shown that planar embeddings following these distance measures lead to improved quality of the final drawing in terms of bends, edge length, and drawing area [1,3,9,16, 20].

The first distance measure is the depth of a planar embedding introduced by Pizzonia and Tamassia [16]. They have suggested an algorithm for a restricted version of computing a minimum depth embedding of a planar graph in which the embeddings of the biconnected components are given and fixed. Our algorithm finds the minimum depth embedding over the class of all planar embeddings without any restriction.

The *depth* of a planar embedding is a measure of the topological nesting of the biconnected components of G in (Γ, f). For the formal definition, we need to introduce some more terms. A *block* of G is a biconnected component of G. Given a connected graph, the *BC-tree* $\mathcal{B}$ of G has vertices for each block and each cut vertex of G. There is an edge between a block vertex b associated with a block B and a cut vertex v if $v \in B$. BC-trees are rooted at an arbitrary block of G, and the edges are directed from parent to child (e.g., an edge (v, w)); sometimes they will be considered as undirected (e.g., denoted as $\{v, w\}$). Given a planar embedding (Γ, f) of a planar graph $G = (V, E)$ with face set F, the *dual graph* $G^* = (V^*, E^*)$ is constructed as follows: $V^* = F$ and E^* contains an edge $\{f_i, f_j\}$ for each $e \in E$ such that e is on the boundary of both f_i and f_j. An *extended dual BC-tree* of a given planar embedding (Γ, f) is either the BC-tree of the dual graph G^* in the case that the edges on the external face do not belong to a single block in G; then, we define r as the (cut) vertex $v_f \in V^*$ associated

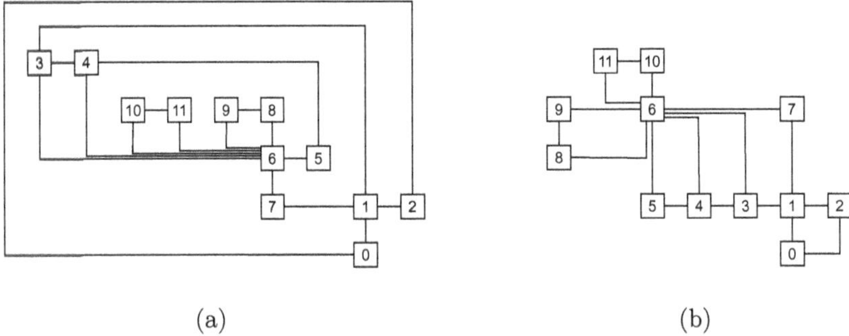

Fig. 1. Two drawings of the same graph given in [16] (computed by the GDToolkit system [11]). The planar embedding in (a) has depth 5, while the one in (b) has depth 1.

with f in G. Otherwise, all edges bordering the external face belong to a unique block B in the dual BC-tree containing the external face vertex $v_f \in V^*$. Then, the *extended dual BC-tree* is the dual BC-tree extended by an additional vertex r and the edge (r, b), where b is the block vertex associated with B. Now we are ready to introduce the formal definition of the *depth* of a planar embedding:

For a planar embedding (Γ, f) of a connected planar graph G, the *depth* is defined as the height of the extended dual BC-tree with respect to (Γ, f) rooted at r.

The example shown in Figure 1 has already been provided by Pizzonia and Tamassia in order to justify their approach for minimizing the depth. Each of the two drawings has bends and area optimized for its embedding (computed by the GDToolkit system [11]). The planar embedding in Figure 1(a) has depth 5, while the one in Figure 1(b) has depth 1. Obviously, the drawing in Figure 1(b) is much easier to read and to understand than the one in Figure 1(a).

Figure 2 shows two drawings of the same graph realizing different planar embeddings. Again, both drawings have the minimal number of bends with respect to their planar embeddings. The drawing in Figure 2(b) looks much more compact than the drawing in Figure 2(a). We observe that the graph is biconnected and hence the depth of any planar embedding is 1. On the other hand, the external face in Figure 2(b) is bordered by 9 edges and is much larger than in Figure 2(a) with only 3 edges.

Also the two embeddings in Figure 1 differ in the number of edges contained in the external face. Figure 1(a) has 3 edges bordering the external face, while the better drawing shown in Figure 1(b) has 15. This goes along with our observation that a higher number of edges on the external face leads to improved layout quality.

Hence, the second distance measure investigated in this paper is the length of the external face cycle (also called, the *size of a face*) in the planar embed-

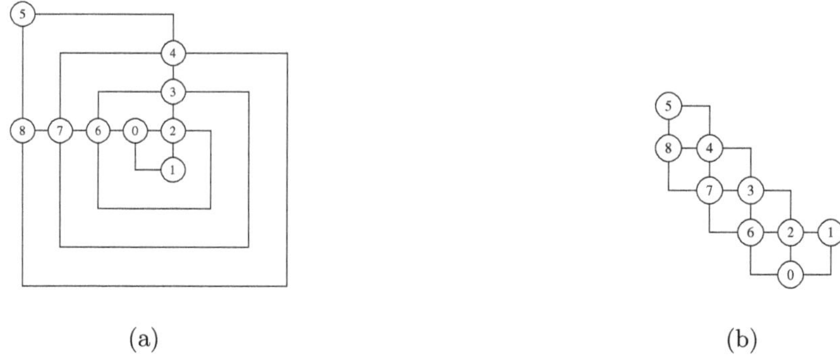

Fig. 2. Two bend-minimal orthogonal drawings of a graph with different planar embeddings.

ding. Our algorithm computes the planar embedding with the external face of maximum size among the set of all possible planar embeddings.

The third distance measure considered in this paper is a combination of the first two distance measures: we search the planar embedding with the external face of maximum size among all planar embeddings with minimum depth. We conjecture that this measure provides, in general, the *best* planar embeddings among the considered distance measures in our paper, and in the literature, leading to improved planar layouts.

The rest of the paper is organized as follows. Section 2 provides the required graph-theoretical background. For technical reasons, we describe the algorithm for computing a planar embedding with maximum external face first (see Section 3), before we discuss the two remaining algorithms in Section 4. Because of space restrictions, we have omitted some of the proofs.

2 Preliminaries

SPQR-trees [8,7] basically represent the decomposition of a biconnected graph into its triconnected components. Let G be a biconnected graph. A *split pair* of G is either a separation pair or a pair of adjacent vertices. A *split component* of a split pair $\{u,v\}$ is either an edge (u,v) or a maximal subgraph C of G such that $\{u,v\}$ is not a split pair of C. Let $\{s,t\}$ be a split pair of G. A *maximal split pair* $\{u,v\}$ of G with respect to $\{s,t\}$ is such that, for any other split pair $\{u',v'\}$, vertices u, v, s, and t are in the same split component.

Let $e=(s,t)$ be an edge of G, called the *reference edge*. The SPQR-tree $\mathcal{T}$ of G with respect to e is a rooted ordered tree whose nodes are of four types: S, P, Q, and R. Each node μ of $\mathcal{T}$ has an associated biconnected multi-graph, called the *skeleton* of μ. Tree $\mathcal{T}$ is recursively defined as follows:

Trivial Case: If G consists of exactly two parallel edges between s and t, then $\mathcal{T}$ consists of a single Q-node whose skeleton is G itself.

Parallel Case: If the split pair $\{s,t\}$ has at least three split components $G_1,\ldots,G_k$, the root of $\mathcal{T}$ is a P-node μ, whose skeleton consists of k parallel edges $e=e_1,\ldots,e_k$ between s and t.

Series Case: Otherwise, the split pair $\{s,t\}$ has exactly two split components, one of them is e, and the other one is denoted with G'. If G' has cut-vertices $c_1,\ldots,c_{k-1}$ $(k\geq 2)$ that partition G into its blocks $G_1,\ldots,G_k$, in this order from s to t, the root of $\mathcal{T}$ is an S-node μ, whose skeleton is the cycle $e_0,e_1,\ldots,e_k$, where $e_0=e$, $c_0=s$, $c_k=t$, and $e_i=(c_{i-1},c_i)$ $(i=1,\ldots,k)$.

Rigid Case: If none of the above cases applies, let $\{s_1,t_1\},\ldots,\{s_k,t_k\}$ be the maximal split pairs of G with respect to $\{s,t\}$ $(k\geq 1)$, and, for $i=1,\ldots,k$, let G_i be the union of all the split components of $\{s_i,t_i\}$ but the one containing e. The root of $\mathcal{T}$ is an R-node, whose skeleton is obtained from G by replacing each subgraph G_i with the edge $e_i=(s_i,t_i)$.

Except for the trivial case, μ has children $\mu_1,\ldots,\mu_k$, such that μ_i is the root of the SPQR-tree of $G_i\cup e_i$ with respect to e_i $(i=1,\ldots,k)$. The endpoints of edge e_i are called the *poles* of node μ_i. Edge e_i is said to be the *virtual edge* of node μ_i in skeleton of μ and of node μ in skeleton of μ_i. We call node μ the *pertinent node* of e_i in skeleton of μ_i, and μ_i the *pertinent node* of e_i in skeleton of μ. The virtual edge of μ in skeleton of μ_i is called the reference edge of μ_i.

Let μ_r be the root of $\mathcal{T}$ in the decomposition given above. We add a Q-node representing the reference edge e and make it the parent of μ_r so that it becomes the new root.

Let e be an edge in *skeleton*(μ) and ν the pertinent node of e. Deleting edge $\{\mu,\nu\}$ in $\mathcal{T}$ splits $\mathcal{T}$ into two connected components. Let $\mathcal{T}_\nu$ be the connected component containing ν. The *expansion graph* of e (denoted with *expansion*(e)) is the graph induced by the edges that are represented by the Q-nodes in $\mathcal{T}_\nu$. We further introduce the notation *expansion*$^+(e)$ for the graph *expansion*$(e)\cup e$.

Replacing a skeleton edge e by its expansion graph is called *expanding* e. The *pertinent graph* of a tree node μ results from expanding all edges in *skeleton*(μ) except for the reference edge of μ and is denoted with *pertinent*(μ). Hence, if e is a skeleton edge and ν its pertinent node, then *expansion*$^+(e)$ equals *pertinent*(ν). If v is a vertex in G, a node in $\mathcal{T}$ whose skeleton contains v is called an *allocation node* of v.

BC-trees and SPQR-trees can be constructed in linear time and use only linear space (see [19,12,8]). They are used to encode all embeddings of a planar connected graph G. Denote with $\mathcal{T}_\mathcal{B}$ the SPQR-tree of a block $B\in G$. The embeddings of B are in one-to-one correspondence with the embeddings of the skeletons of $\mathcal{T}_\mathcal{B}$. An embedding of B is basically constructed by replacing skeleton edges with their expansion graphs while preserving the embedding. Once all blocks are embedded, an embedding of G is constructed by several applications of the following procedure. Let Γ' be an embedding consisting of a connected union of some blocks. We want to insert a further embedded block Γ_B with vertex

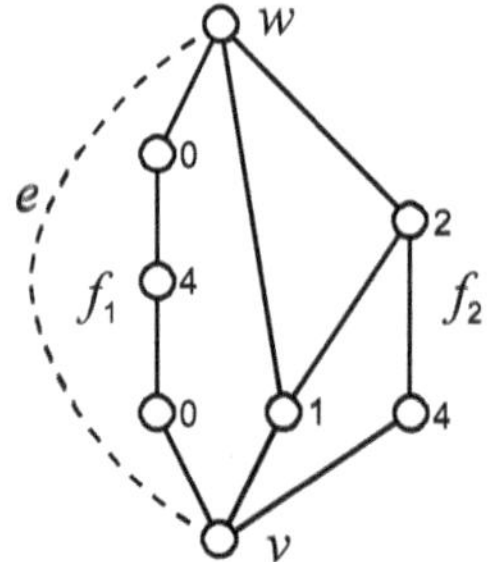

Fig. 3. An embedding of *expansion*$^+(e)$ with given vertex lengths. Assuming that e, v, w have length 0, face f_1 has size 8 and f_2 has size 9, therefore edge e has component length ≥ 9.

$c \in B \cap \Gamma'$ into Γ'. Then, we have to choose a face containing c as external face of Γ_B and insert Γ_B into a face of Γ' containing c.

3 Planar Embeddings with Maximum External Face

Let $G = (V, E)$ be a planar connected graph without self-loops. We first consider biconnected graphs and present linear-time algorithms for finding an embedding with maximum external face and for computing the size of a maximum external face containing a prescribed vertex v for every vertex $v \in V$.

3.1 Biconnected Graphs

Let $B = (V_B, E_B)$ be a block of G and $\mathcal{T}_B$ its SPQR-tree. We associate a non-negative length with each vertex $v \in B$ and each skeleton edge in $\mathcal{T}_B$. The length of an edge in B is simply 1. The size of a face f is defined as $\sum_{e \in f} \mathit{length}(e) + \sum_{v \in f} \mathit{length}(v)$.

Consider an edge $e = (v, w)$ in a skeleton graph and let Γ_e be an embedding of *expansion*$^+(e)$ such that Γ_e has a face f^* containing e of maximum size among all embeddings of *expansion*$^+(e)$. We call such an embedding an embedding of *expansion*$^+(e)$ with *maximum length* and define the *component length* of e to be $\mathit{size}(f^*) - \mathit{length}(e) - \mathit{length}(v) - \mathit{length}(w)$ (compare also Figure 3).

The general idea of the algorithm is as follows: Let S be a skeleton for which we have chosen an embedding Γ_S. In order to extend Γ_S to an embedding Γ of B, we have to choose embeddings of the graphs *expansion*$^+(e)$ for all edges $e \in S$. Each face f_S in Γ_S corresponds to a face f_Γ in Γ in which each skeleton edge $e \in f_S$ is replaced by a path p_e on the external face of its expansion graph (see Figure 4). We call this *expanding face* f_S to face f_Γ. Vice versa, for each face f'_Γ in Γ, we can find a face in a skeleton that can be expanded to f'_Γ. Face f_Γ is made as large as possible by embedding each expansion graph of an edge in f_S, such that the *length* of the path p_e, which we define as the number of edges

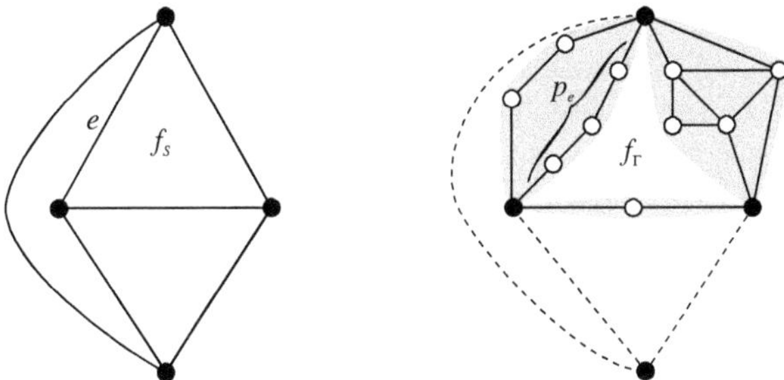

Fig. 4. A face f_S in the skeleton of an R-node (left) and the resulting face f_Γ after expanding all edges in f_S (right).

in p_e plus the lengths of the interior vertices on p_e, is the component length of e. We get the following Lemma:

Lemma 1. *Let B be a planar biconnected graph with vertex lengths and $\mathcal{T}_B$ its SPQR-tree. Then the following holds:*

(i) Each skeleton face f can be expanded to a face f' such that the size of f' is the sum of the lengths of the vertices in f plus the component lengths of all edges in f.
(ii) There exists an internal node $\mu \in \mathcal{T}_B$ and an embedding Γ_μ of skeleton(μ) such that there is a face f in Γ_μ that can be expanded to a maximum face f^* of B. All edges in f are expanded to expansion graphs with maximum length.

We can compute the component length of all non-reference skeleton edges by a bottom-up traversal of $\mathcal{T}_\mathcal{B}$. Let e be an edge in *skeleton*(μ) and ν the pertinent node of e. Denote with e_r the reference edge of ν and with ℓ the sum of the length of the two poles of μ. If we assume that the length of all edges in *skeleton*(ν) except for e_r are set to their component length, the component length of e is

Q-node: 1
S-node: size of a face in *skeleton*(ν) minus ℓ
P-node: length of the longest edge different from e_r in *skeleton*(ν)
R-node: size of the largest face containing e_r in *skeleton*(ν) minus ℓ

Once we have set the length of all non-reference skeleton edges to their component lengths, we can compute the component length of the reference edges by a top-down traversal of $\mathcal{T}_\mathcal{B}$. Since $\mathcal{T}_\mathcal{B}$ is rooted at a Q-node, the component length of the reference edge of its only son is 1. Consider now a node $\mu \in \mathcal{T}_\mathcal{B}$ and let S be the skeleton of μ. We assume that the length of each edge in S is set to its component length. We want to compute the component length of the reference edge of each child ν of μ. We distinguish three cases according to the type of μ:

S-node: Let L be the sum of the length of all vertices and edges in S. The component length of the reference edge of ν is L minus the length of $e_{S,\nu}$ minus the lengths of the two vertices incident to $e_{S,\nu}$.

P-node: The component length of the reference edge of ν is the length of the longest edge in S different from $e_{S,\nu}$.

R-node: Let f be the largest face in S containing $e_{S,\nu}$. The component length of the reference edge of ν is the size of f minus the length of $e_{S,\nu}$ minus the lengths of the two vertices incident to $e_{S,\nu}$.

According to Lemma 1, we need to find an embedding Γ_μ of a skeleton with a face f of maximum size among all possible embeddings of skeletons. This can be achieved by inspecting all skeletons S. If S is the skeleton of an R-node, S has only two embeddings which are a mirror of each other. The largest face we can produce is simply a largest face in any embedding of S. If S is the skeleton of a P-node, say a bundle of parallel edges $e_1, \dots, e_k$, we can form any face consisting of two edges e_i and e_j with $i \neq j$. Hence, the largest face we can produce consists of the two longest edges in S. If S is the skeleton of an S-node, S has only a single embedding consisting of two equally sized faces.

This shows that we can find a tree node μ and produce a skeleton face $f \in \Gamma_\mu$ that can be expanded to a maximum face of B. According to Lemma 1, all edges in f have to be expanded to expansion graphs with maximum length, which can be achieved by recursively traversing the tree nodes involved.

Theorem 1. *Let $B = (V_B, E_B)$ be a biconnected planar graph. Then, there exists an $O(|V_B| + |E_B|)$-time algorithm that computes a planar embedding of B with maximum external face.*

In order to compute the size of a maximum external face containing a prescribed vertex v, we have to consider all allocation nodes of v in $\mathcal{T}_B$. Suppose we have precomputed the size of all skeleton faces in S- and R-nodes, as well as the length of the two longest edges within a P-node. Then, we can compute the size of a maximum face containing v very efficiently, i.e. in time $O(n_v + m_v)$, where n_v is the number of allocation nodes of v in $\mathcal{T}_B$ and m_v is the number of skeleton edges in R-nodes incident to representatives of v. Since the size of $\mathcal{T}_B$ including all skeleton graphs is linear in the size of B, we know that $\sum_{v \in B}(n_v + m_v)$ is linear in the size of B and we get the following lemma.

Lemma 2. *There exists an $O(|V_B| + |E_B|)$-time algorithm that computes the size of a maximum face containing v for every vertex $v \in V_B$.*

3.2 Connected Graphs

Let $\mathcal{B}$ be the BC-tree of G. Consider a block B and a cut-vertex c in G with $c \in B$. If we delete edge $\{c, B\}$ in $\mathcal{B}$, $\mathcal{B}$ is split into two connected components, one component $\mathcal{B}_B$ containing B. We denote the graph induced by the edges in all blocks contained in $\mathcal{B}_B$ with $G_{c,B}$.

The idea of the algorithm is as follows (compare Figure 5): Suppose we have fixed an embedding Γ_0 of a block B_0. In order to extend Γ_0 to an embedding of G, we have to find an embedding $\Gamma_{c,B}$ with c on its external face for each graph $G_{c,B}$ with $c \in B$, $B \neq B_0$, and place $\Gamma_{c,B}$ into an adjacent face of c in

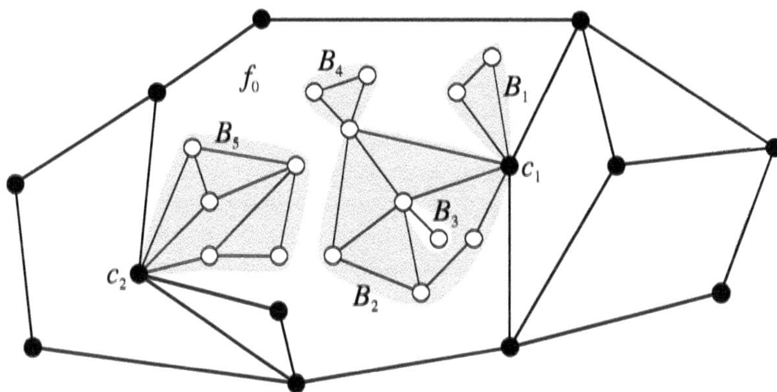

Fig. 5. A fixed embedding Γ_0 of B_0 (thick edges), where the graphs G_{c_1,B_1}, G_{c_1,B_2} and G_{c_2,B_5} have been placed into face f_0.

Γ_0. For a fixed face f_0 in Γ_0, we can enlarge f_0 as much as possible if we place all $\Gamma_{c,B}$ with $c \in f_0$ into f_0. We denote with $smf_B(c)$ the sum of the sizes of the maximum faces containing c of all $G_{c,B'}$ with $c \in B'$, $B' \neq B$. If c is not a cut-vertex in G, then $smf_B(c)$ is 0. We will use the value $smf_B(c)$ as the vertex length of c in B_0 and apply the procedures developed in the previous section. The following lemma states the crucial results on which the correctness of our algorithm is based.

Lemma 3. *(i) Let B be a block of G and define the length of a vertex v in B as $smf_B(v)$. If f_B is a face in an arbitrary embedding Γ_B of B, then there exists an embedding Γ of G with a face f, such that the size of f (disregarding vertex lengths) equals the size of f_B. Face f results from embedding each $G_{c,B'}$ with $c \in B$, $B' \neq B$ with a maximum external face containing c and placing it into f_B.*

(ii) Let f be a face in an embedding Γ of G. Then, there exists a block B of G, a length ℓ_v for each vertex $v \in B$ with $\ell_v \leq smf_B(v)$, and an embedding Γ_B of B with a face f_B such that the size of f (disregarding vertex lengths) equals the size of f_B.

Theorem 2. *For each block B of G, let $smf_B(v)$ be the length of a vertex $v \in B$. Let B_{max} be the block having the embedding Γ_{max} with the largest maximum face f_{max} among all blocks. Then, we can extend Γ_{max} to a planar embedding of G with maximum external face f. Face f results from embedding each $G_{c,B'}$ with $c \in B_{max}$ and $B' \neq B_{max}$ with a maximum external face containing c and placing it into f_{max}.*

Proof. According to Lemma 3(i), we will find face f with the required size in an embedding Γ of G. Lemma 3(ii) shows that a maximum face in G cannot be larger than f, hence Γ is an embedding of G with a face f of maximum size. □

Algorithm 1: Planar embedding with maximum external face.

```
MaximumFace (graph G)
    Compute BC-tree 𝓑 of G (rooted at any block) and SPQR-tree T_B for each block B ∈ 𝓑.
    let B_r be the root block of 𝓑
    forall v ∈ B_r do
        length_{B_r}(v) := Σ_{(v,B)∈𝓑} ConstraintMaxFace(B, v)
    od
    (B, ℓ) := MaximumFaceRec(B_r)
    Embed G by expanding a maximum face in B
end

ConstraintMaxFace (block B, vertex c)
    forall v ∈ B, v ≠ c do
        length_B(v) := Σ_{(v,B')∈𝓑} ConstraintMaxFace(B', v)
    od
    length_B(c) := 0
    cstrLength(B, c) := size of a maximum face in B containing c
    return cstrLength(B, c)
end

MaximumFaceRec (block B)
    (B*, ℓ*) := (B, size of a maximum face in B)
    forall (B, c) ∈ 𝓑 with ex. (c, B') ∈ 𝓑 do
        cstrLength(B, c) := size of a maximum face in B containing c
        L := Σ_{{B',c}∈𝓑} cstrLength(B', c)
        forall c → B' ∈ 𝓑 do
            length_{B'}(c) := L − cstrLength(B', c)
            ℓ' := MaximumFaceRec(B')
            if ℓ' > ℓ* then (B*, ℓ*) := (B', ℓ')
        od
    od
    return (B*, ℓ*)
end
```

Algorithm 3.2 shows how to compute a planar embedding with maximum external face. Variable $length_B(c)$ stores the length of vertex c in block B, and $cstrLength(B, c)$ is set to the size of a maximum face of $G_{c,B}$ containing c. Function ConstraintMaxFace(B, c) computes a maximum face in $G_{c,B}$ containing c and is called for each edge $(c, B) \in \mathcal{B}$. Function MaximumFaceRec recursively traverses $\mathcal{B}$ from top to bottom. When MaximumFaceRec is called for block B, the length of each vertex $v \in B$ is already set to $smf_B(v)$. Since the function finally returns the block B^* with the largest maximum face of size ℓ^*, this is the block B_{max} in Theorem 2. For embedding each $G_{c,B}$ with $c \in B^*$ and $B \neq B^*$ with a maximum external face containing c, $\mathcal{B}$ has to be recursively traversed starting at block B^*.

Theorem 3. *Let $G = (V, E)$ be a planar connected graph.* MaximumFace(G) *computes a planar embedding of G with maximum external face in time $O(|V| + |E|)$.*

All algorithms presented in this section can easily be generalized to graphs with predefined non-negative edge lengths, in particular to edges with length 0.

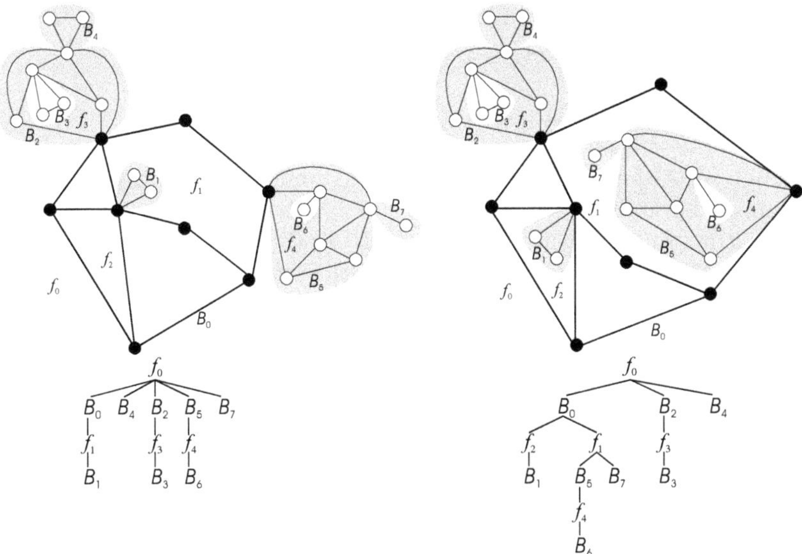

Fig. 6. Two different extensions of an embedding Γ_0 of a block B_0 with their corresponding block-cutface trees.The left extension has depth 3 whereas the right one has depth 5.

4 Planar Embeddings with Minimum Depth

Let $G = (V, E)$ be a planar graph. Consider a block B with planar embedding Γ_B. An *extension* of Γ_B denotes a planar embedding of G which results from embedding all graphs $G_{c,B'}$ with $c \in B$ and $B' \neq B$ so that c lies on the external face and placing them into a face of Γ_B.

Let $m_{c,B'}$ be the minimum depth of a planar embedding of $G_{c,B'}$ with c on the external face. We define further

$$\begin{aligned} m_B(c) &:= \max\{0\} \cup \{m_{c,B'} \mid c \in B', B' \neq B\} \\ m_B &:= \max_{c \in B} m_B(c) \\ M_B &:= \{c \in B \mid m_B(c) = m_B\} \end{aligned}$$

Then, the minimum depth of an extension of an embedding of B is m_B if there is an embedding of B in which all vertices in M_B lie in a common face, and $m_B + 2$ otherwise (compare Figure 6). The following lemma shows the relationship between minimizing the depth and maximizing the external face.

Lemma 4. *Set the length of all edges in B to 0 and the length of a vertex v in B to 1 if $v \in M_B$, and 0 otherwise. Let (Γ^*, f^*) be a planar embedding of B with maximum external face f^*. Then, the minimum depth of an extension of an embedding of B is*

$$\begin{array}{ll} m_B & \textit{if size}(f^*) = |M_B| \\ m_B + 2 & \textit{otherwise} \end{array}$$

The problem of finding a planar embedding of a graph $G_{c,B}$ with minimum depth such that c is on the external face can be tackled the same way. Based on this result, algorithm `MinimumDepth` proceeds similar to algorithm `MaximumFace` for connected graphs.

First, the values $m_{c,B}$ for all edges (c, B) in the BC-tree $\mathcal{B}$ are computed by a bottom-up traversal of $\mathcal{B}$. The length of each vertex $v \neq c$ in B is set according to Lemma 4 and a maximum face in B is computed that contains c.

Afterwards, $\mathcal{B}$ is traversed from top to bottom. When a block $B = (V_B, E_B)$ is visited, the values $m_{c,B'}$ are already computed for each cut-vertex c in G that is contained in B. According to Lemma 4, we set the vertex lengths in B and compute the minimum depth of an extension of B by finding a maximum face in B. Before we proceed with the descendants of B in tree $\mathcal{B}$, we have to compute the values $m_{c,B}$ for all edges $(B, c) \in \mathcal{B}$. We distinguish two cases:

$M_B = \{c_1, \ldots, c_k\}$ with $k \geq 2$: We compute a maximum face in block B containing c using the vertex lengths we have already assigned.

$M_B = \{c\}$: In this case, $m_2 = \max_{v \in V_B, v \neq c} m_B(v)$ is less than m_B and we cannot reuse the vertex lengths. However, this case can occur at most once for a block which allows us to invest $O(|B|)$ running time. We calculate new vertex lengths according to $M_2 = \{c \in V_B \setminus \{v\} \mid m_B(c) = m_2\}$ and find a maximum face in B containing c using these vertex lengths.

Since `MaximumFace` and `MinimumDepth` proceed in a similar fashion, it is possible to combine both algorithms to a new algorithm `MinDepthMaxFace` that computes a planar embedding with maximum external face among all planar embeddings with minimum depths. This can be achieved by using a pair (d, ℓ) as length attribute, where the first component d refers to the $0-1$ length attribute used in `MinimumDepth` and ℓ refers to the length attribute used in `MaximumFace`. The linear order on these tuples is defined componentwise, i.e.

$$(d, \ell) > (d', \ell') \iff d > d' \text{ or } (d = d' \text{ and } \ell > \ell')$$

Each time a maximum face in a block has to be computed, we first determine a maximum face according to the linear order defined above. If the resulting maximum face has size (d^*, ℓ^*) and d^* is the number of cut-vertices we want to place in a common face (e.g. $d^* = |M_B|$ as in Lemma 4), we also have found a planar embedding with a maximum face among all planar embeddings with minimum depth. Otherwise, the value d^* is irrelevant since all extensions will have the same depth and there might be planar embeddings with a larger external face. Therefore, we call the procedure for finding a maximum face in a block again, this time respecting only the second component ℓ of the length attribute.

Theorem 4. *Let $G = (V, E)$ be a planar connected graph.*

i) Algorithm `MinimumDepth`(G) *computes a planar embedding of* G *with minimum depth.*

ii) Algorithm `MinDepthMaxFace`(G) *computes a planar embedding of* G *with maximum external face among all planar embeddings of* G *with minimum depth.*

References

[1] C. Batini, E. Nardelli, and R. Tamassia. A layout algorithm for data-flow diagrams. *IEEE Trans. Soft. Eng.*, SE-12(4):538–546, 1986.

[2] C. Batini, M. Talamo, and R. Tamassia. Computer aided layout of entity relationship diagrams. *The Journal of Systems and Software*, 4:163–173, 1984.

[3] C. Batini, M. Talamo, and R. Tamassia. Computer aided layout of entity-relationship diagrams. *Journal of Systems and Software*, 4:163–173, 1984.

[4] D. Bienstock and C. L. Monma. On the computational complexity of embedding planar graphs to minimize certain distance measures. *Algorithmica*, 5(1):93–109, 1990.

[5] N. Chiba, T. Nishizeki, S. Abe, and T. Ozawa. A linear algorithm for embedding planar graphs using PQ-trees. *J. Computer and System Sciences*, 30:54–76, 1985.

[6] G. Di Battista, P. Eades, R. Tamassia, and I. G. Tollis. *Graph Drawing.* Prentice Hall, 1998.

[7] G. Di Battista and R. Tamassia. On-line maintanance of triconnected components with SPQR-trees. *Algorithmica*, 15:302–318, 1996.

[8] G. Di Battista and R. Tamassia. On-line planarity testing. *SIAM J. Comput.*, 25(5):956–997, 1996.

[9] Didimo and Liotta. Computing orthogonal drawings in a variable embedding setting. In K. Chwa, editor, *Algorithms and Computation (Proc. ISAAC '98)*, volume 1533 of *LNCS*, pages 79–88. Springer-Verlag, 1998.

[10] A. Garg and R. Tamassia. On the computational complexity of upward and rectilinear planarity testing. *SIAM J. Computing*, 31(2):601–625, 2002.

[11] Graph drawing toolkit: An object-oriented library for handling and drawing graphs. http://www.dia.uniroma3.it/ gdt.

[12] C. Gutwenger and P. Mutzel. A linear time implementation of SPQR trees. In J. Marks, editor, *Graph Drawing (Proc. GD 2000)*, volume 1984 of *LNCS*, pages 77–90. Springer-Verlag, 2001.

[13] G. Liotta, F. Vargiu, and G. Di Battista. Orthogonal drawings with the minimum number of bends. In *Proceedings of the 6th Canadian Conference on Computational Geometry*, pages 281–286. University of Saskatchewan, 1994.

[14] K. Mehlhorn and P. Mutzel. On the embedding phase of the Hopcoft and Tarjan planarity testing algorithm. *Algorithmica*, 16(2):233–242, 1996.

[15] P. Mutzel and R. Weiskircher. Bend minimization in orthogonal drawings using integer programming. In O. H. Ibarra and L. Zhang, editors, *Proceedings of the Eighth Annual International Conference on Computing and Combinatorics (COCOON 2002)*, volume 2387 of *LNCS*, pages 484–493. Springer-Verlag, 2002.

[16] M. Pizzonia and R. Tamassia. Minimum depth graph embedding. In M. Paterson, editor, *Algorithms – ESA 2000*, volume 1879 of *LNCS*, pages 356–367. Springer-Verlag, 2000.

[17] R. Tamassia. On embedding a graph in the grid with the minimum number of bends. *SIAM J. Comput.*, 16(3):421–444, 1987.
[18] R. Tamassia, G. Di Battista, and C. Batini. Automatic graph drawing and readability of diagrams. *IEEE Trans. Syst. Man Cybern.*, SMC-18(1):61–79, 1988.
[19] R. E. Tarjan. Depth-first search and linear graph algorithms. *SIAM J. Comput.*, 1(2):146–160, 1972.
[20] R. Weiskircher. *New Applications of SPQR-Trees in Graph Drawing*. PhD thesis, Universität des Saarlandes, 2002.

More Efficient Generation of Plane Triangulations

Shin-ichi Nakano[1] and Takeaki Uno[2]

[1] Gunma University, Kiryu 376-8515, Japan
nakano@cs.gunma-u.ac.jp
http://www.msc.cs.gunma-u.ac.jp/~nakano/index.html
[2] National Institute of Informatics, Tokyo 101-8430, Japan
uno@nii.jp,
http://research.nii.ac.jp/~uno/

Abstract. In this paper we give an algorithm to generate all biconnected plane triangulations having exactly n vertices including exactly r vertices on the outer face. The algorithm uses $O(n)$ space in total and generates such triangulations without duplications in $O(rn)$ time per triangulation, while the previous best algorithm generates such triangulations in $O(r^2n)$ time per triangulation.

1 Introduction

Generating all graphs with some properties without duplications has many applications, including unbiased statistical analysis[M98]. Many algorithms to solve these problems are already known [A96,B80,M98,W86]. Many nice textbooks have been published on the subject[G93,KS98].

In this paper we wish to generate all biconnected plane triangulations having exactly n vertices including exactly r vertices on the outer face. Such triangulations play an important role in many algorithms, including graph drawing algorithms [CN98,FPP90,S90]. A graph G is *biconnected* if removing any vertex leaves G connected. An embedded planar graph is called *a plane triangulation* if each inner face has exactly three edges on its boundary.

Recently, we proposed an algorithm to generate all biconnected "based" plane triangulations having exactly n vertices including exactly r vertices on the outer face[LN01]. A *based* plane triangulation means a plane triangulation with one designated "base" edge on the outer face. Two based plane graphs are said to be isomorphic if they are isomorphic and the base edges correspond to each other in the bijection. For instance, four biconnected based plane triangulations are shown in Fig. 1, where the base edges are depicted by thick lines. Note that, however, those based triangulations are isomorphic as non-based plane triangulations. The algorithm for based plane triangulations uses $O(n)$ space in total and runs in constant time per output while the previous best algorithm generates such triangulations in $O(n^2)$ time per triangulation[A96]. The algorithm does not output entire triangulations but the difference from the previous triangulation.

G. Liotta (Ed.): GD 2003, LNCS 2912, pp. 273–282, 2004.

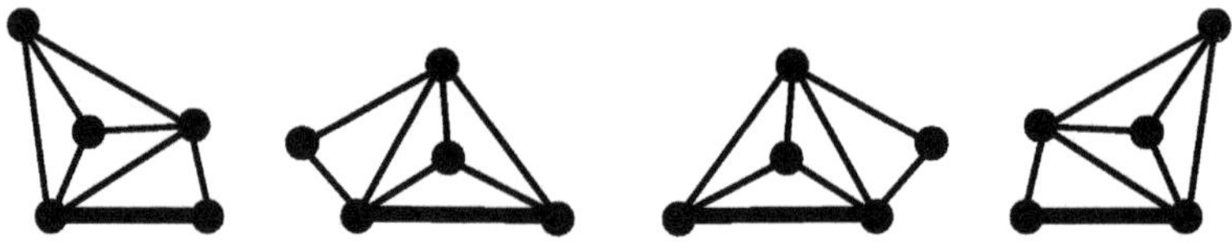

Fig. 1. Biconnected based plane triangulations.

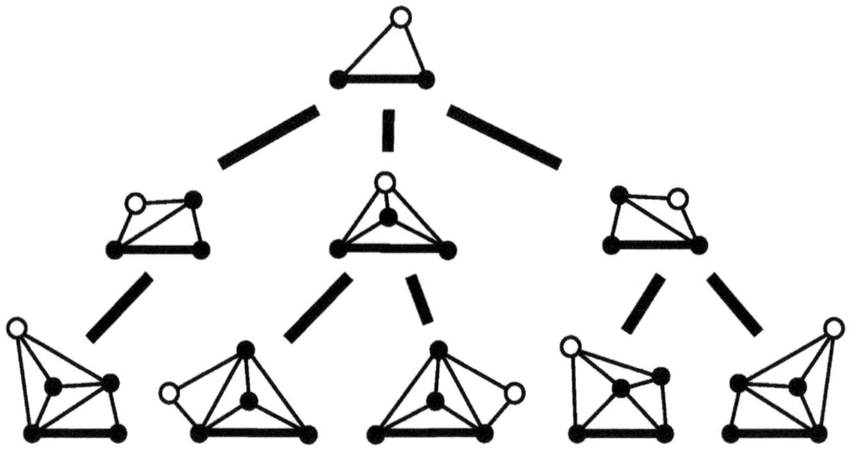

Fig. 2. The genealogical tree for $n = 5$ and $r = 4$.

The algorithm of [LN01] generates all based triangulations as follows. Given n and r, we first define a tree T such that the inner vertices of T correspond to the biconnected based plane triangulations having at most $n-1$ vertices including at most $n-r$ inner vertices, the leaves of T correspond to the biconnected based plane triangulations having exactly n vertices including exactly r vertices on the outer face, and the edges of T correspond to some relation between the biconnected based plane triangulations. T is called the *genealogical tree,* and the genealogical tree for $n = 5$ and $r = 4$ is shown in Fig. 2. In the figure, we can observe that if we remove a vertex depicted by a white circle from a biconnected based plane triangulation, then we can have the "parent" biconnected based plane triangulation. Also we can prove the number of vertices of T is within 3 times the number of leaves of T. Since the size of T is huge in general, thus we cannot construct whole part of T at once. However, we can simply traverse T in $O(1)$ time per edge of T by partially constructing T. We need only $O(n)$ space in total. And on the traversal we can find all the vertices of T, which correspond to all the triangulations.

By modifying the algorithm, we can also generate without duplications all biconnected (non-based) plane triangulations having exactly n vertices including exactly r vertices on the outer face in $O(r^2n)$ time per triangulation on average

[LN01]. If we maintain output triangulations in a database to avoid the duplication, then the space complexity will be exponential. So to obtain an efficient algorithm, we have to check the occurrence of duplication without any database, hence this is a non-trivial modification. Another algorithm with $O(n^2)$ time per triangulation is also claimed in [M98] without detail but using a complicated theoretical linear-time plane graph isomorphism algorithm [HW74], while the algorithm in [LN01] is simple and does not need the isomorphism algorithm.

The strategy of the algorithm in [LN01] is as follows. The genealogical tree T has many biconnected based plane triangulations which are distinct as based plane triangulations, but isomorphic as (non-based) plane triangulations. The only difference is the choice of the base edge on the outer face. See Fig. 1. On the traversal of T, we have to output exactly one biconnected (non-based) plane triangulations for each isomorphic class. By giving a unique sequence of letters for each biconnected based plane triangulations, we can define a representative triangulation among each isomorphic class as the triangulation having the lexicographically-first sequence of letters. The algorithm in [LN01] needs $O(rn)$ time computation at each leaf v of T to decide whether the sequence of letters for the based triangulation corresponding to v is the lexicographically-first one among the isomorphic class, and only in such case the based plane triangulation is output as the representative plane triangulation. Otherwise the based triangulation is not output. For each output triangulation, T may contain r isomorphic ones corresponding to the r choices of the base edge. Thus, the amortized time complexity of the algorithm is $O(r^2n)$ per output triangulation.

In this paper we improve the running time of the algorithm in [LN01] as follows. We define a new unique sequence of letters for each biconnected based plane triangulation. Given a biconnected based plane triangulation, the new sequence of letters needs less computation to decide whether the sequence of letters for the based plane triangulation is the lexicographically-first one among the isomorphic class. Our algorithm needs only $O(n)$ time computation at each leaf of T. Again, for each output triangulation, T may contain r isomorphic ones corresponding to the r choices of the base edge. Thus, the amortized time complexity of our algorithm is $O(rn)$ per triangulation.

The rest of the paper is organized as follows. Section 2 defines our new sequence of letters. Section 3 gives a new algorithm to find the lexicographically first sequence. Finally Section 4 is a conclusion.

2 Sequence of Letters

In this section, given a biconnected based plane triangulation G with the base edge e, we define a sequence of letters. Then we show that the sequence of letters has enough information to re-construct G. This means that we give a unique name to each biconnected based plane triangulation. We need some definitions first.

For a graph, a *cut* is a set of vertices whose removal results in a disconnected graph or a single-vertex graph K_1. A graph is said to be *biconnected* if no cut

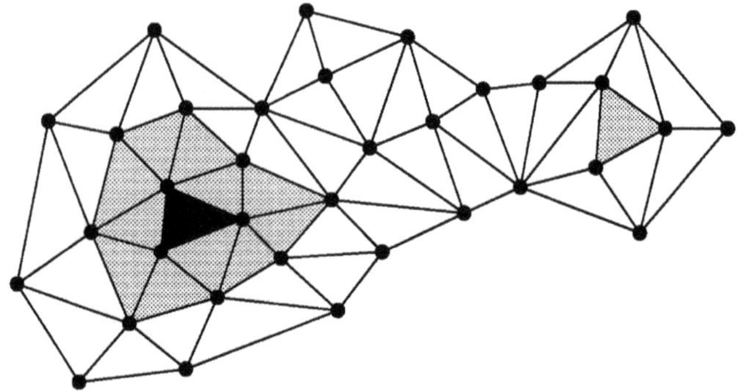

Fig. 3. Graph $G_1 = G$: the gray area is G_2, and the black area is G_3.

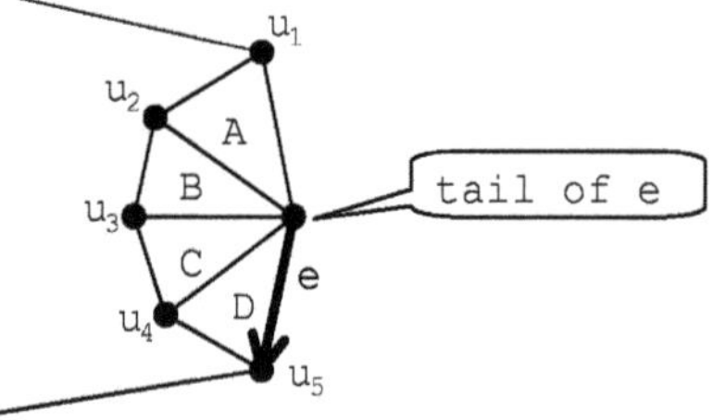

Fig. 4. An edge e and its tail: triangle sequence of e is (A, B, C, D). For triangle C, $v_1(C) = u_4$ and $v_2(C) = u_3$.

is composed of a single vertex. A graph is *planar* if it can be embedded in the plane so that no two edges intersect geometrically except at a vertex to which they are both incident. A *plane* graph is a planar graph with a fixed planar embedding. A plane graph divides the plane into connected regions called *faces*. The unbounded face is called *the outer face*, and other faces are called *inner faces*. A plane graph is called *plane triangulation* if each inner face includes exactly three edges. Without loss of generality, we assume that the vertices are on the general position, so that no triangle is "empty." An *isomorphism* from a graph G to H is a bijection $f : V(G) \to V(H)$ such that $(u, v) \in E(G)$ if and only if $(f(u), f(v)) \in E(H)$. We say G is isomorphic to H if there is an isomorphism from G to H. Two plane graphs are said to be *isomorphic* if there is an isomorphism preserving the cyclic ordering of edges incident to each vertex.

A triangle of G is a *boundary triangle* of G if it has at least one vertex on the outer face. Let $G_1 = G$, and $G_i, i > 1$ be the graph obtained from G_{i-1} by removing all vertices included in only boundary triangles, i.e. the graph induced by non-boundary triangles of G_{i-1} (see Fig. 3). For an edge e of the outerface, the *tail* of e is the endpoint of e followed by the other endpoint of e in clockwise

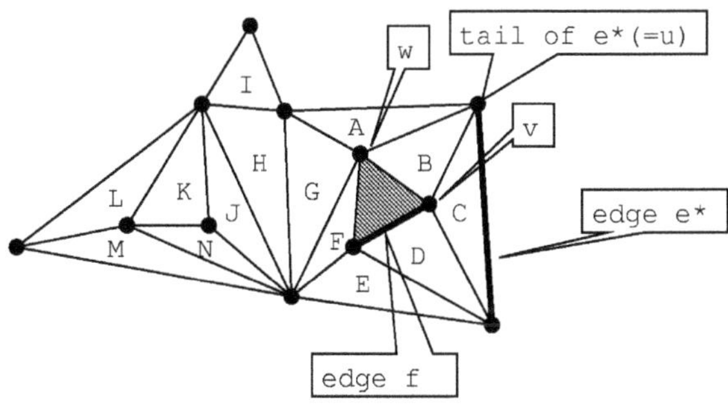

Fig. 5. seq_1 =(A,B,C, C,D,E, E,F,G,H,J,N,M M,L, L,K,J,H,I, I, I,H,G,A). By setting the triangle sequence (A, B, C) of e^* to the top, this circular sequence becomes a sequence. The outer triangle of edge f is D.

order. We define the *triangle sequence* of e by the sequence of triangles incident to the tail of e in counter-clockwise order (see Fig. 4).

Let seq_1 be the circular sequence obtained by connecting the triangle sequence of the edges of the outer face of G_1 in clockwise order (see Fig. 5). Note that this is a circular sequence so it has no end. For an edge e on the outer face of $G_i, i \geq 2$, we define the *outer triangle* of e by the boundary triangle of G_{i-1} including e (see Fig. 5). Let seq_i be the circular sequence obtained by connecting the triangle sequence of the edges on the outer face of G_i in the order of their outer triangles in seq_{i-1} (see Fig. 6). Any outer triangle appears just once in seq_{i-1}, and other boundary triangles appear in seq_i at least once and at most three times. For instance in Fig. 5, H appears in seq_1 three times and F appears once. Each appearance of a triangle in seq_i is called an *occurence.*

For two occurrence t_1 and t_2 of seq_i, we define the *distance from t_1 to t_2* by the number of occurrences from t_1 to t_2 in seq_i (see Fig. 5). Note that the distance is zero if t_2 is next to t_1. If t_1 and t_2 are the same occurrence, we define the distance from t_1 to t_2 by the number of occurrences in seq_i minus one. For instance in Fig. 5, the distance from the occurrence corresponding to the first A to the occurrence corresponding to the first C is 1, and the distance from the occurrence corresponding to the last A to the occurrence corresponding to first A is 0.

For each occurrence t we assign a sequence of three integers $l(t)$ as follows. Assume that t is in the triangle sequence of edge e, u is the tail of e, and u, v, and w are the three vertices on the boundary of t, and they appear clockwise in this order. Let $d(u)$ be the distance from u' to u, where u' is the last occurence corresponding to t. Let $d(v)$ be the distance from v' to v, where v' is the last occurence corresponding to a triangle containing v. Let $d(w)$ be the distance from w' to w, where w' is the last occurence corresponding to a triangle containing v. We set $l(t) = (d(u), d(v), d(w))$.

For example, see Fig.5. The traingle sequence of e^* has an occurrence corresponding to triangle B. Now $d(u) = 24$ since seq_i has 25 occurrences, and

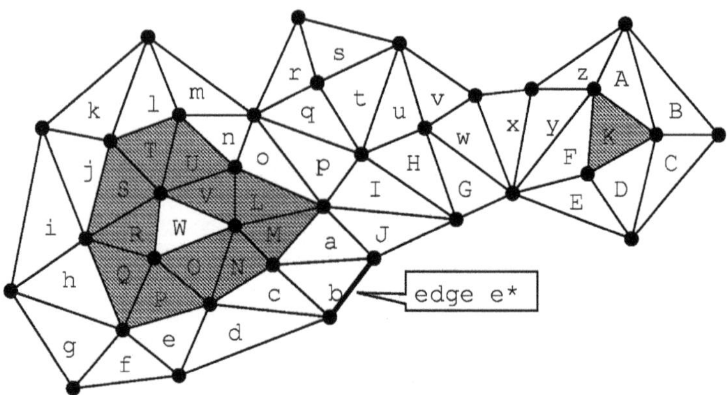

Fig. 6. seq_1 is circular sequence (b,c,d, d,e,f, f,g, g,h,i,..., G,H,I,J, J,a,b). The order of outer triangles of edges of G_1 is (a,c,e,h,j,l,n,o,A,D,F), hence seq_2 is (M,N, N,O,P, P,Q, Q,R,S, S,T, T,U, U,V,L, K, K, K, L,M). By setting the triangle sequence of e^* to the top of the sequence, seq_1 becomes a sequence (J,a,b, b,c,d, d,e,f,...,I,J), and seq_2 becomes a sequence (L,M,M,N,N,...,K,K,K).

has only one occurence corresponding to B, $d(v) = 21$ since from D to B there are 21 occurences, and $d(w) = 0$ since from A to B there are zero occurences. Note that A, B and D mean the occurrences corresponding to triangle A, B, and D, respectively. Thus $l(B) = (24, 21, 0)$. Similarly for the first occurrence C, $l(C) = (23, 20, 0)$.

For two occurences t_1 and t_2, we say $l(t_1)$ is *smaller* than $l(t_2)$ if $l(t_1)$ is lexicographically smallerthan $l(t_2)$, and we write $l(t_1) < l(t_2)$.

We can observe that the first integer of $l(t)$ is 0 iff t is the first triangle in the triangle sequence of some edge. Hence, the following property holds.

Property 1. If occurence t_1 is the first triangle of the corresponding triangle sequence and occurence t_2 is not the first triangle of the corresponding triangle sequence, then $l(t_1) < l(t_2)$. □

For any edge e on the outer face of G, we can obtain an ordinal letter sequence from the circular sequences $seq_1, seq_2, \ldots$ as follows. First we cut the circular sequence seq_1 to a sequence so that the triangle sequence of e appears first (see Fig. 5). By using the order of seq_{i-1}, seq_i can be considered as a sequence (see Fig. 6). We connect these sequences so that the sequence obtained from seq_i follows the sequence obtained from seq_{i-1}, and replace each occurrence t by $l(t)$. We define the *letter sequence* $L(e)$ of e by the sequence of letters. We denote the ith letter of $L(e)$ by $l^i(e)$.

Given letter sequence L, we can uniquely re-construct the biconnected plane triangulation G based at e such that e is the base edge of G, and $L(e) = L$ as follows.

First, we consider the graph G' induced by the boundary triangles of G_1. If G is not equal to G', then there exists at least one boundary triangle appearing in

seq_1 only once. From the first number of the letter of the occurrence, we can get the length of seq_1. The occurrences such that the first number of their letters are 0 are the tops of the triangle sequences. Thus, we can get the number of edges on the outer face, and the length of the triangle sequence of each edge. From the first number of the letters, we can know which triangles of distinct triangle sequences are identical. From the second and the third numbers of the letters, we can know the order of triangles incident to a vertex. Therefore, we can construct the graph G' uniquely.

Suppose that we have constructed the graph induced by the boundary triangles of $G_1, .., G_{k-1}$. In the similar way to the above, we can construct the graph induced by the boundary triangles of G_k. By connecting the graphs we constructed, we obtain the graph induced by the boundary triangles of $G_1, .., G_k$. Therefore, G can be constructed uniquely from the letter sequence.

Thus, for any biconnected based plane triangulation with base edge e, the letter sequence $L(e)$ is defined uniquely and given $L(e)$, and we can re-construct a unique based biconnected plane triangulation with base edge e. Therefore, we gave a unique name to each biconnected based plane triangulation.

3 Algorithm for Finding the Minimum Letter Sequence

We have defined a unique letter sequence for each biconnected based plane triangulation in Section 2. In this section, we define a unique letter sequence for each biconnected (non-based) plane triangulation, and show an algorithm for computing the letter sequence in short time.

A biconnected plane triangulation G corresponds to many biconnected "based" plane triangulations G_e, since we can choose any edge on the outer face as the base edge e. So intuitively each biconnected (non-based) plane triangulation G has many names which correspond to the choice of the base edge. We are going to find e^* which is the lexicographically minimum name $L(e^*)$ among all over $L(e)$, and define the name of G as $L(e^*)$.

For two sequences L_1 and L_2, we say that L_1 is smaller than L_2 iff the first $k-1$ letters of L_1 and L_2 are the same, and kth letter of L_1 is smaller than L_2. Note that k can be 1. Let e^* be the edge of the outer face of G whose letter sequence is the lexicographically minimum. In this section, we describe an algorithm for finding the edges whose letter sequences are equal to $L(e^*)$.

For an edge e on the outer face of G, we define $m(e)$ by the number such that $l^i(e) = l^i(e^*)$ if $i < m(e)$ and $l^i(e) > l^i(e^*)$ if $i = m(e)$. We define $m(e)$ by $+\infty$ if $L(e) = L(e^*)$. We are going to find the edges e with $m(e) = +\infty$ by comparing the ith letter of the letter sequences iteratively. $l^i(e^*)$ is the minimum letter among $l^i(e)$ with $m(e) \geq i$. We can see that if $m(e) \geq i$ and $l^i(e) \neq l^i(e^*)$, then $m(e) = i$, and we do not care about e in the $(i+1)$th iteration.

In this way, if we check each occurrence of seq_i to get its letter only once, the computation time is $O(n)$. However, if some occurrences are checked more than once, then the computation time may be up to $O(n^2)$. We will explain how we can save computation time in this case.

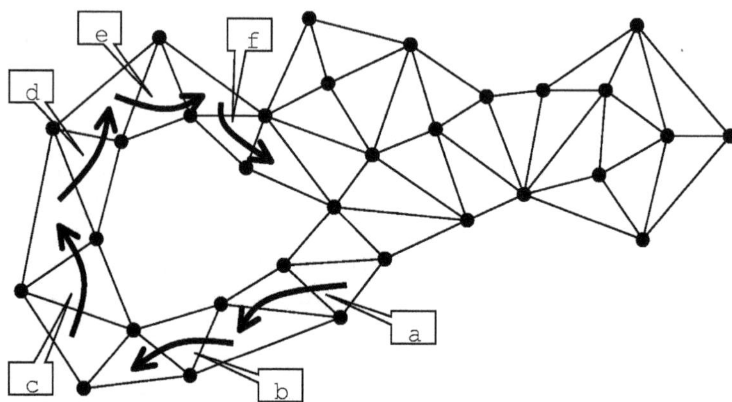

Fig. 7. a overlaps with b. Edges c, d, e, and f compose a chain, and the tail of the chain is f. The length of the chain is 4, and the length of the chain (a, b) is 2, hence only c remains.

In the ith iteration, for an edge e and e' of the outer face satisfying $m(e) \geq i$, if there exists h satisfying $h < i, h < m(e')$, and the occurrences giving $l^i(e)$ and $l^h(e')$ are on the same position in seq_j for some j, we say that e *overlaps* with e'. Let z_i be the sum of the cardinalities of $seq_1, ..., seq_i$. We define z_0 by 0. Note that $h = z_{j-1} + 1$ if e does not overlap with e' in the $(i-1)$th iteration.

Suppose that edges $e_1, ..., e_k$ overlap with $r(e_1), ..., r(e_k)$ in the ith iteration and $z_q < i \leq z_{q+1}$. From Property 1, if an edge e satisfying $m(e) \geq i$ overlaps with no edge, $m(e) = i$.

First, we consider the case that no $r(e_j)$ is in $\{e_1, ..., e_k\}$. Let $m^* = \max\{m(r(e_1)), ..., m(r(e_k))\}$. We can see that $m(e_j) = +\infty$ only if $m(r(e_j)) = m^*$, and $m(e_j) = i + m(r(e_j)) - z_q - 1$ if $m(r(e_j)) < m^*$. Hence, we can go to $(i + m^* - z_q)$th iteration. The skipped iterations are not necessary to be executed.

Second, we consider the case that some $r(e_j)$ are in $\{e_1, ..., e_k\}$. We define the *chain* of e_j by $\{e_j = e_{j_1}, e_{j_2}, ..., e_{j_h}\}$ such that $r(e_{j_p}) = e_{j_{p+1}}$, and $r(e_{j_h})$ is not in $\{e_1, ..., e_k\}$ or e_{j_h} overlaps with no edge. We denote e_{j_h} by $tail(e_j)$ (see Fig. 7). Let $d(e_j)$ be the cardinality of the chain. We consider the case later that the chain is circular, i.e., any $r(e_j)$ is in $\{e_1, ..., e_k\}$.

Let $d^* = \max\{d(e_1), ..., d(e_k)\}$. We can see that $m(e_j) = (i - z_q - 1) \times (d(e_j) - 1) + m(tail(e_j))$ if $d(e_j) < d^*$. Hence, $m(e_j) = +\infty$ only if $d(e_j) = d^*$, and we can go to $((i - z_q - 1) \times (d^* - 1) + z_q)$th iteration.

Finally, we consider the case that the chain is circular, i.e. any e_j overlaps with an edge of $e_1, ..., e_k$. In this case, a chain includes all $e_1, ..., e_k$. Hence, the first z_{q+1}th letters of $L(e_1), ..., L(e_k)$ are the same, and we can go to $(z_{q+1} + 1)$th iteration.

We describe this algorithm as a procedure below.

1. Compute seq_j for each j
2. Compute the letters of all occurrences in seq_j for each j
3. $i := 1, F :=$ edges of the outer face
4. Set q so that $z_q < i \leq z_{q+1}$
5. **For each** $e \in F$ satisfying $l^i(e) < \max_{f \in F} l^i(f)$
6. $m(e) := i$ and remove e from F
7. $i := i + 1$
8. **If** any edge of F overlaps with an edge of F **then**
9. $i := z_{q+1} + 1$
10 **Else if** an edge of F overlaps with an edge of F **then**
11. **For each** $e \in F$ satisfying $d(e) < \max_{f \in F} d(f)$
    $m(e) := (i - z_q - 1) \times (d(e) - 1) + m(tail(e)) - 1$ and remove e from F
12. $i := ((i - z_q - 1) \times (\max_{f \in F} d(f) - 1) + z_q)$
13. **Else if** an edge of F overlaps with an edge not in F **then**
14. **For each** $e \in F$ satisfying $m(r(e)) < \max_{f \in F} m(r(f))$
    $m(e) := i - z_q + m(r(e)) - 1$ and remove e from F
15. $i := i - z_q + \max_{f \in F} m(r(f))$
16. **End if**
17. **If** i is less than the length of the letter sequences **and** $|F| > 1$ **then go to** 4.
18. **Output** all edges of F

The time complexity of this algorithm is proportional to the length of a letter sequence plus the number of overlaps. Since the length of a letter sequence is $O(n)$, we estimate the upper bound of the overlaps.

Consider the the ith iteration of the algorithm. For each $e \in F$, one of the following three cases occurs.

(1) e overlaps with no other edge.
(2) e overlaps with another edge e', and $e' = tail(f)$ for no f.
(3) e overlaps with another edge e', and $e' = tail(f)$ for some f.

If (3) occurs, $e' = tail(f)$ can never hold again for any f, hence (3) will never occur for e'. If (2) occurs, then e' can never be overlapped. Hence the number of the overlaps is at most twice the length of the letter sequence, and is $O(n)$.

By replacing the $O(rn)$ time computation at each leaf of the genealogical tree T in [LN01] to the $O(n)$ time computation above, we can improve the running time of the algorithm. We have the following theorem.

Theorem 1. *All the biconnected plane triangulations having n vertices including r vertices on the outer face can be generated in $O(rn)$ time per triangulation.* □

4 Conclusion

In this paper we have given an algorithm to generate all biconnected plane triangulations without duplication. Our idea is to define a unique sequence of letters for each biconnected based plane triangulation, and by finding the lexicographically minimum among all the bases of a biconnected plane triangulations, we can efficiently decide for each whether the based triangulation should be output as a biconnected plane triangulation.

References

[A96] D. Avis, *Generating rooted triangulations without repetitions*, Algorithmica, 16, (1996), pp.618–632.

[B80] T. Beyer and S. M. Hedetniemi, *Constant time generation of rooted trees*, SIAM J. Comput., 9, (1980), pp.706–712.

[CN98] M. Chrobak and S. Nakano, *Minimum-width grid drawings of plane graphs*, Computational Geometry: Theory and Applications, 10, (1998), pp.29–54.

[FPP90] H. de Fraysseix, J. Pach and R. Pollack, *How to draw a planar graph on a grid*, Combinatorica, 10, (1990), pp.41–51.

[G93] L. A. Goldberg, *Efficient algorithms for listing combinatorial structures*, Cambridge University Press, New York, (1993).

[HW74] J. E. Hopcroft and J.K. Wong, *Linear time algorithm for isomorphism of planar graphs*, Proc. of 6th STOC, (1974), pp.172–184.

[KS98] D. L. Kreher and D. R. Stinson, *Combinatorial algorithms*, CRC Press, Boca Raton, (1998).

[LN01] Z. Li and S. Nakano, *Efficient generation of plane triangulations without repetitions*, Proc. ICALP2001, LNCS 2076, (2001), pp.433–443.

[M98] B. D. McKay, *Isomorph-free exhaustive generation*, J. of Algorithms, 26, (1998), pp.306–324.

[S90] W. Schnyder, *Embedding planar graphs on the grid*, Proc. 1st Annual ACM-SIAM Symp. on Discrete Algorithms, San Francisco, (1990), pp.138–148.

[W86] R. A. Wright, B. Richmond, A. Odlyzko and B. D. McKay, *Constant time generation of free trees*, SIAM J. Comput., 15, (1986), pp.540–548.

Planar Embeddings of Graphs with Specified Edge Lengths*

Sergio Cabello[1], Erik D. Demaine[2], and Günter Rote[3]

[1] Institute of Information and Computing Sciences, Universiteit Utrecht, P.O.Box 80.089, 3508 TB Utrecht, The Netherlands.
sergio@cs.uu.nl
[2] MIT Laboratory for Computer Science, 200 Technology Square, Cambridge, MA 02139, USA.
edemaine@mit.edu
[3] Institut für Informatik, Freie Universität Berlin, Takustraße 9, 14195 Berlin, Germany.
rote@inf.fu-berlin.de

Abstract. We consider the problem of finding a planar embedding of a (planar) graph with a prescribed Euclidean length on every edge. There has been substantial previous work on the problem without the planarity restrictions, which has close connections to rigidity theory, and where it is easy to see that the problem is NP-hard. In contrast, we show that the problem is tractable—indeed, solvable in linear time on a real RAM—for planar embeddings of planar 3-connected triangulations, even if the outer face is not a triangle. This result is essentially tight: the problem becomes NP-hard if we consider instead planar embeddings of planar 3-connected infinitesimally rigid graphs, a natural relaxation of triangulations in this context.

1 Introduction

Given a graph and the lengths of the edges from a planar straight-line drawing of a (planar) graph, can we reconstruct the drawing? When is this reconstruction unique? Can we recognize realizable length assignments? These three problems have an extensive history, having been studied in the fields of computational geometry [8,14,26,27], rigidity theory [7,16,18], sensor networks [5,25], and structural analysis of molecules [1,9,17]. The reconstruction problem arises frequently when only distance information is known about a given structure, such as the atoms in a protein [1,9,17] or the nodes in an ad-hoc wireless network [5, 24,25]. A reconstruction is always unique and easy-to-compute for a complete graph of (exact) distances, or any graph that can be "shelled" by incrementally locating nodes according to the distances to three noncollinear located neighbors

* The research by the first author has been supported by Cornelis Lely Stichting and by a Marie Curie Fellowship of the European Community programme "Combinatorics, Geometry, and Computation" under contract number HPMT-CT-2001-00282. The third author is partially supported by the National Science Foundation under grant number ITR ANI-0205445.

G. Liotta (Ed.): GD 2003, LNCS 2912, pp. 283–294, 2004.

(Fig. 1). More interesting is that such graphs include visibility graphs [8] and segment visibility graphs [14]. In general, however, the reconstruction problem is NP-hard [27], even in the strong sense [26]. The uniqueness of a reconstruction in the generic case (in 2D) was recently shown to be testable in polynomial time by a simple characterization related to generic rigidity [16,18], but this result has not yet lead to efficient algorithms for actual reconstruction in the generic case.

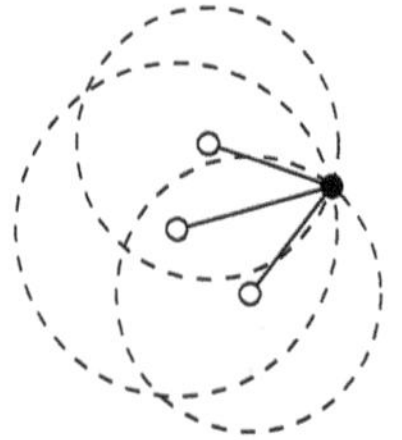

Fig. 1. Locating a vertex from the distances to three located neighbors.

Planar embeddings. We consider a variation on this basic problem of reconstruction from distances: the graph is planar and the planar straight-line drawing must be a planar embedding (edges not incident in the graph should not be incident in the plane). Our problem is then, given a planar graph with prescribed lengths on the edges, to construct a planar embedding of the graph that adheres to the specified edge lengths, and determine whether this embedding is unique, or determine that no such embedding exists.

Applications. The restriction to planar embeddings makes sense in many applications, for example when the underlying structure we want to reconstruct is known to be planar. Another application specifically in the context of graph drawing is the generation of linear cartograms. A *cartogram* is a map in which the size of each entity is proportional to some value associated with the entity [4]. *Area cartograms* are the most common example, in which the area of each region is proportional to some function of the region, e.g., its population. In *linear cartograms*, we want to display a network in such a way that the length of a connection is related to some characteristic of the connection. In common maps, this length is correlated (through a planar projection of the sphere) with the length of the connection in the real world. However, we may be interested in showing, e.g., the traveling time for each connection, or the traffic on each connection. The construction of such a map can be modeled by defining the length of each edge appropriately and trying to realize the graph with these edge lengths. In real-life applications, we would also like to keep some resemblance with the original network, and so we may restrict where the vertices of the graph can be embedded. However, as we will see, the problem is already hard without this restriction.

Our results. We prove the following main results:

1. Even for planar 3-connected graphs, deciding planar embeddability with unit edge lengths is strongly NP-hard, even when the embeddings are guaranteed to be infinitesimally rigid. [1] This improves upon results in [13], where the

[1] Infinitesimal rigidity is a strong form of rigidity, stating that no first-order motion of the vertices preserves the lengths of the edges to the first order. See e.g. [15] for formal definitions.

hardness is shown for 2-connected graphs with unit edge lengths and for 3-connected graphs with arbitrary edge lengths. Another (aesthetic) difference is that our reduction is directly from planar 3-SAT, rather than using a synthetic problem as a bridge. See Section 3.
2. For planar 3-connected graphs, we can decide in $O(|V|)$ time whether there is a planar embedding with specified edge lengths in which only the outermost face is not a triangle. Furthermore, such an embedding is always unique up to rigid motions (translations, rotations, and reflections), and can be constructed in $O(|V|)$ time. More generally, we can find planar embeddings in which the triangular faces form a connected family of cells and the nontriangular faces form a forest of cells. See Section 2.

These results give a fairly precise division between tractable and intractable forms of planar embedding with specified edge lengths. To our surprise, triangles seem to play a more fundamental role than other rigid structures, despite the close connections between rigidity and embedding with specified edge lengths [7,16,18]. Other than visibility graphs [8,14] and dense graphs [1], our results are the first positive results for efficient embeddings of (special) graphs with specified edge lengths.

Model of computation. Even the simple task of embedding a triangle with given side lengths in the plane and computing coordinates for the vertices involves square roots. Thus, we do not know whether our embedding problems belong to NP, and for our algorithmic results we have to assume the real RAM model of computation [23] which supports constant-time exact arithmetic operations ($+$, $-$, $\times$, $\div$, $\sqrt{}$) on real numbers. This model is customary in computational geometry. On the other hand, our NP-hardness result is in the standard Turing machine model, and therefore must be careful to ensure that the input lengths can be encoded in a number of bits that is polynomial in the input graph size.

2 Triangulated Graphs

Let G be a 3-connected planar graph. By Whitney's Theorem, G has only one topological embedding into the 2-dimensional sphere, or equivalently the faces in any planar embedding of G are always induced by the same cycles [12, Chapter 6]. In particular, all embeddings of G have the same dual graph G^*, and once we have fixed the outer face, the topological embedding into the plane is completely determined. This is the basic ingredient for the following result:

Theorem 1. *If $G = (V, E)$ is a 3-connected graph with specified edge lengths, we can decide in $O(|V|)$ time on a real RAM whether there is a planar embedding such that all faces are triangles, with the possible exception of the outer face.*

Proof. Consider any planar embedding of G, which can be computed in $O(|V|)$ time [11]. If two or more faces are not triangles, then we can decide that the desired realization is not possible because of Whitney's theorem. If exactly one face is not a triangle, that face must be the outer face in the desired realization.

If all faces are triangles, all the longest edges have to be part of the outer face, which gives us at most two candidates T and T' for the outer face. If T is the outer face, then T' must fit inside T while sharing the common edge, and vice versa. This test leaves us with at most one candidate for the outer face f_{ext}.

All nodes in $G^* \setminus f_{ext}$ correspond to triangular faces. We pick a node f_0 in this graph, and compute coordinates for the vertices of the corresponding triangle that realize the triangle edge lengths. Now we visit all nodes in $G^* \setminus f_{ext}$ using breadth-first search from f_0. When visiting a node f_i, two options arise:

1. If all vertices of the face f_i have already been assigned coordinates, we check that all the edges in f_i have the specified edge lengths.
2. If some vertex of the face f_i has not been assigned coordinates, we know that the other two vertices u, v of f_i participate in another face f_j that has been already visited, and so they have already been assigned coordinates. We can compute the coordinates of the third vertex using the specified edge lengths and the restriction that f_i and f_j must lie on opposite sides of the line segment uv by Whitney's Theorem.

At the end, every edge in the graph has been checked whether it satisfies the specified edge length, including the lengths of the edges of the outer face f_{ext}. In the process, we visited each face once, and we spent constant time per face, so, overall, the embedding process takes $O(|V|)$ time.

We need to check that the realization that we constructed is indeed planar, to avoid situations like the ones depicted in Fig. 2. A simple plane sweep would do this in $O(|V| \log |V|)$ time. To get linear time, we first construct a triangulation of the whole plane: We enclose all points in a large triangle T and triangulate the area between T and the boundary of the outer face f_{ext}. To do this, we insert an edge from an extreme vertex of V to a corner of T and triangulate the resulting simple polygon in linear time [6]. Under the assumption that the original embedding was planar, we obtain a graph which is a triangulation of T and is embedded in the plane without crossings. On the other hand, if the original embedding contains crossings, the triangulation algorithm will either (i) terminate in error, or (ii) it will produce a subdivision of T which is topologically consistent but whose embedding contains crossings. Topological consistency means that the two triangle faces incident to an edge are embedded on different sides of the edge, except for the edges of T where the other triangle is embedded inside T. The planarity condition (ii) for a subdivision can be tested in linear time [10]. □

If *all* faces are triangles and we only want to *test* embeddability without constructing coordinates, it suffices to use the following result.

Lemma 1. *Let G be a triangulated planar graph with a designated outer triangle T and an embedding which embeds the two triangles incident to an edge on different sides of the edge, except for the edges of T. Then the realization is planar if and only if, for all vertices $v \notin T$, the sum of the angles that are incident to v is 360 degrees.*

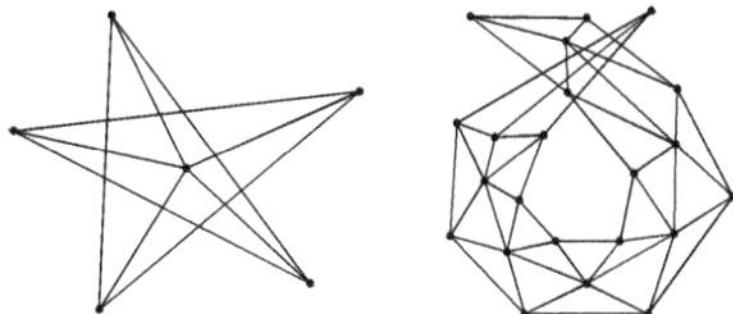

Fig. 2. These examples show that we need to check that the embedding is indeed planar.

It is clearly necessary that the angle sum is 360 degrees for all interior vertices v. To see that this property is sufficient, imagine that the faces (triangles) that we constructed are made of cardboard. The question becomes whether we can arrange these pieces of cardboard, respecting the vertex adjacencies, without overlap in the plane. This question appears in some proofs of the Koebe-Andreyev-Thurston circle packing theorem [22, Chapter 8], and a positive answer is easily shown by induction on the number of vertices: the triangles adjacent to an interior vertex can be packed to form a star-shaped polygon P, which is obviously planar; then you remove that vertex and retriangulate the polygon P.

For graphs of bounded degree, this can be tested in polynomial time in the classical Turing machine model, with rational edge lengths as inputs, using separation bounds for algebraic computations, see [3,2,20]. We may also allow square roots of rationals as inputs. (Otherwise, it will be difficult to come up with interesting examples of realizable graphs with rational edge lengths.) For general graphs, this algorithm is singly-exponential in the degree.

On the other hand, observe that, in the proof of the Theorem 1, we have only used that the coordinates of the vertices can be computed by considering the triangular faces in an appropriate order. To get this property, we only need that each vertex of G is incident to a triangular face, and that the set of triangular faces is connected in the dual graph G^*. Therefore, we can weaken the hypothesis on G as follows: the subgraph of G^* induced by vertices of degree 3 is connected, and the subgraph of G^* induced by vertices of degree larger than 3 is a forest. Once we have fixed the outer face, these hypotheses would be enough to prove the result.

3 NP-Hardness

To show the NP-hardness of our problem, we reduce from the P3-SAT (planar 3-satisfiability) problem, which is strongly NP-complete [21]. In an instance of P3-SAT, we are given a planar bipartite graph whose nodes on one side of the bipartition represent the variables $v_1, \ldots, v_n$, and whose nodes on the other side represent the clauses $C_1, \ldots, C_m$, and edges connect each clause to the three variables it contains. Moreover, the variables can be arranged on a horizontal line, and the three-legged clauses be drawn such that all edges lie either above or below this line; and the graph can be drawn on a rectangular grid of polynomial size as shown in Fig. 3, left [19].

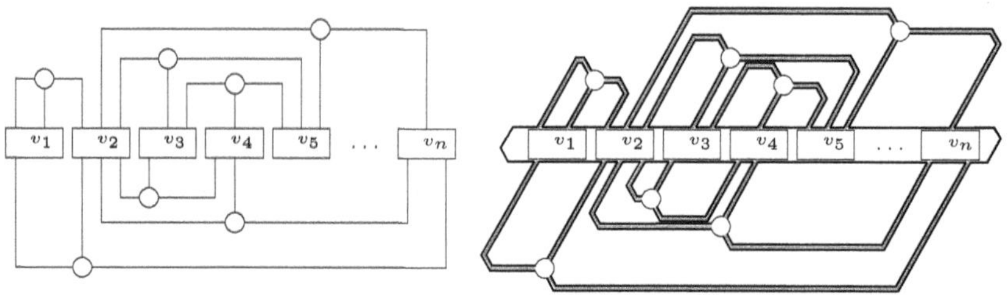

Fig. 3. Left: example of a planar 3-satisfiability instance. The variables can be arranged on a straight line, and the clauses are represented as a vertex with three orthogonal edges leaving from it and one bend in each edge. Right: High-level sketch of NP-hardness reduction. We will make a rigid 3-connected structure that resembles the bold lines.

The high-level workings of the reduction are as follows. We slant the grid into a hexagonal grid to get angles that are multiples of 60 degrees. This slant will allow us to make all lengths one. Furthermore, we modify the drawing so that all the corners have angles of 120 degrees, and the three edges arriving at a clause form angles of 120 degrees; see Fig. 3, right. We make a rigid structure that will leave a tunnel for each edge connecting a variable with a clause. A variable will be represented by a rigid structure that has two different realizations, representing the truth assignment of the variable. The value of the literal will be transmitted to the clause through the tunnel corresponding to the edge, and we will represent the clause by a structure that can be realized if and only if at least one of the literals is true. Furthermore, each of the lines in the figure will be represented by a rigid 3-connected bar, like a "thick" line. This will be the basic trick to make the whole graph 3-connected as well.

The construction relies on three basic rigid structures that are depicted on Fig. 4, and that we explain in the following. In all cases, the grey regions represent 3-connected, rigid structures which are fixed. Firstly, in Fig. 4A, the edges p_1q_1 and p_2q_2 have the same length, and so do p_1p_2 and q_1q_2. Under these conditions, in any realization of this structure, the edges p_1p_2 and q_1q_2 have to be parallel. Secondly, in Fig. 4B, if the vertices p_1 and p_2, marked with squares, are fixed, then the vertex marked with a circle has two possible positions, q and q'. This is so because the distance between this vertex and p_1 and p_2 is fixed, and therefore it has to be placed at the intersection of two circles centered at p_1 and p_2. Finally, in Fig. 4C, there is a 3-connected structure that allows q to rotate around p.

Theorem 2. *Deciding planar embeddability of a planar 3-connected graph with specified edge lengths is NP-hard.*

Proof. We have already described the general idea, so it only remains to describe the gadgets that are used. For the tunnels, we need a structure that allows us

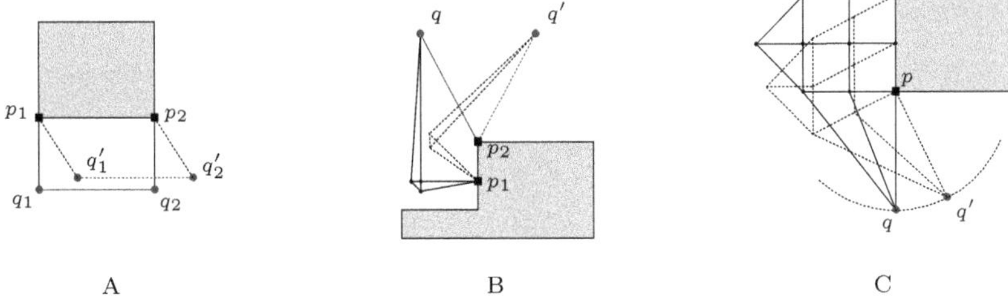

Fig. 4. Assume that the grey regions are rigid and fixed. A. The segments p_1p_2 and q_1q_2 are parallel in any realization. B. The vertex q can only be realized in two positions. C. How to make rotations while keeping 3-connectedness.

to fix the relative positions of both sides of the tunnel, while transmitting the value of the literal through the tunnel. The value will be either true or false, so we need a structure that allows two realizations.

In Fig. 5A the *holder* gadget is shown. Consider the upper half of it. Observe that the two points that are marked with big dots, p_1, p_2, and the two points that are marked with squares, q_1, q_2, represent a situation like shown in Fig. 4A. Therefore, the bar that supports q_1, q_2 is always parallel to the one that supports p_1, p_2, and the point q_2 is always vertically above point q. The points q, q_2 and p_2, implement the idea shown in Fig. 4B, and so p_2 has only two possible placements with respect to q, q_2. Overall, this implies that the upper half of Fig. 5A can be realized in two ways.

The holder gadget can be realized in four different ways: two of them keep the relative position of both sides of the tunnel (A and B), while two of them would move them (C and D). We can concatenate two of these gadgets with one *bend*, as shown in Fig. 5E, in such a way that the realizations C and D are not possible. Thus, the two sides of the tunnel are connected in a (globally) rigid way. We define the *transmitter* to be the bar that is inside the tunnel, because it will transmit the truth value of the literal from the variable to the clause. Below we will discuss the possible realizations of the transmitter.

The structure that we have described is 3-connected, and so we can construct a rigid 3-connected structure, as shown in Fig. 3, right, where the distance between the upper and the lower part will be defined later on by the height of the variables. The sides of the tunnels taken together form a rigid structure in which the transmitters and the variables can move: If a tunnel contains a bend, its two sides can be connected rigidly by two holders, as in Fig. 5E. One can check that the sides of a tunnel without a bend are always connected to a tunnel with a bend, and therefore are also immobile. (Or we could introduce two bends in a zigzag way to make an otherwise straight tunnel rigid in its own right.)

We still have to discuss how the variables, the transmitter, and the clauses work.

For each variable we repeat the structure of the upper half of the holder gadget, but with a thicker bar (*variable-bar*) inside; see Fig. 6. Consider the

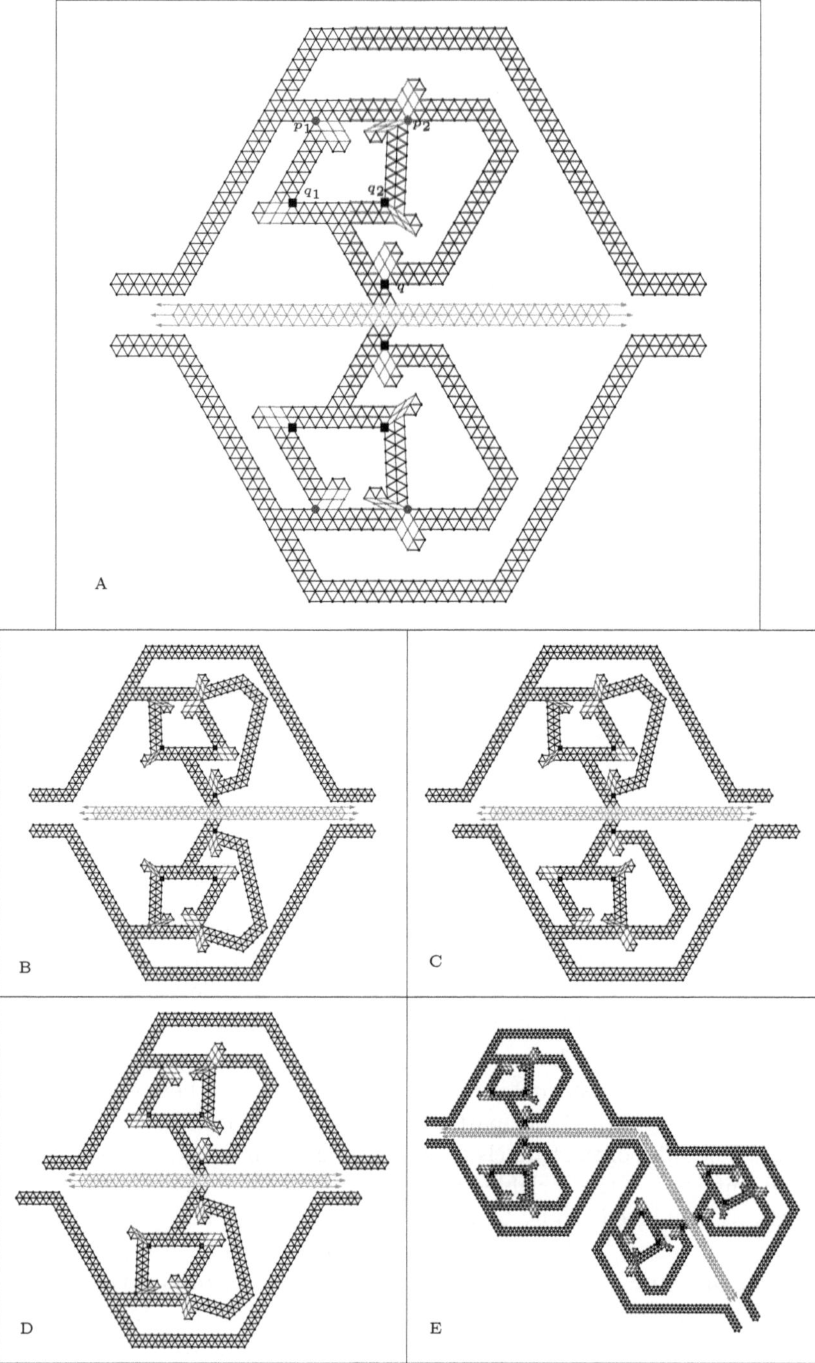

Fig. 5. A. The holder gadget. A–D. Four possible positions of the holder. E. Two holders connected by a bend. This rigidly connects the sides of the tunnel.

realization of the structure assigned to true. On the sides of the variable-bar that are facing the tunnels, for each literal that is not negated, we place an indentation on it that prolongates the tunnel of the literal. For the literals that are negated, we place such an indentation on the part of the variable bar that faces the tunnel in the "false" realization of the structure; see Fig. 6. We have to make the variable-bar large enough that tunnels for all occurrences of each variable can be accommodated on its sides. (In Fig. 6, there are three tunnels on each side.)

The graph that we have constructed so far is 3-connected and rigid. Furthermore, whenever a literal is true, the transmitter bar inside the tunnel can be pushed towards the variable-bar. Furthermore, we can transmit this "pushing information", or pressure, through the tunnel, and also through the corners using Fig. 5E, so that it can be used at the clause.

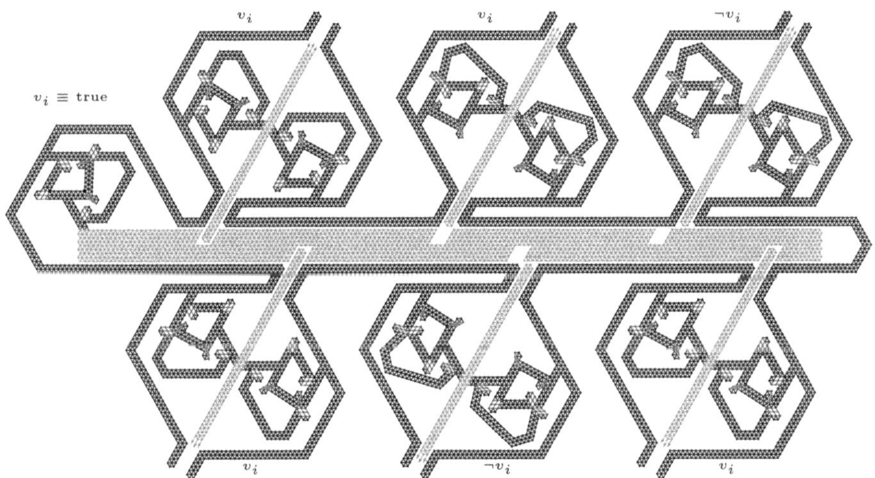

Fig. 6. A variable assigned to true.

Our next goal is to design a *clause checker* that is realizable if and only if one of the three transmitters can be pushed towards its variable. It turns out to be easier to solve the reverse problem: a clause that is realizable if and only if one of the transmitters is pushed towards the clause. Therefore, we design a pushing-inverter which we place on each tunnel just before the clause. It is described in Fig. 7, where its two possible realizations are displayed. The inverter is 3-connected, and inverts pressure towards the clause into pressure towards the variable, and vice versa.

Finally, a clause is described in Fig. 8, with its relevant realizations. The big dots in each literal indicate the two possible positions for the end of the transmitter, and the one that is closer to the center indicates that the literal is true (pushing towards the clause). In all cases, the position of the big dot in

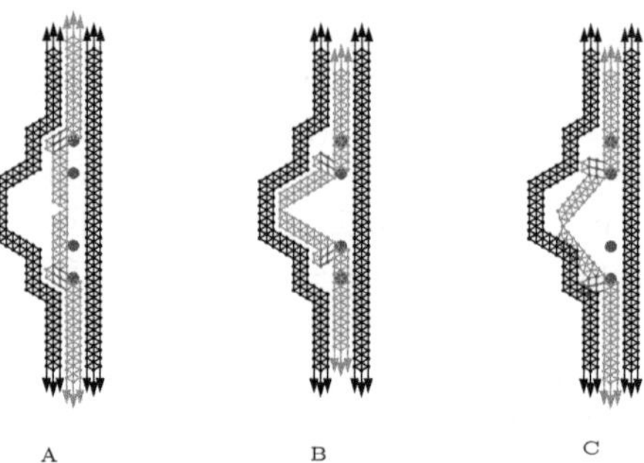

Fig. 7. An inverter. A–B are realizable, but C is not.

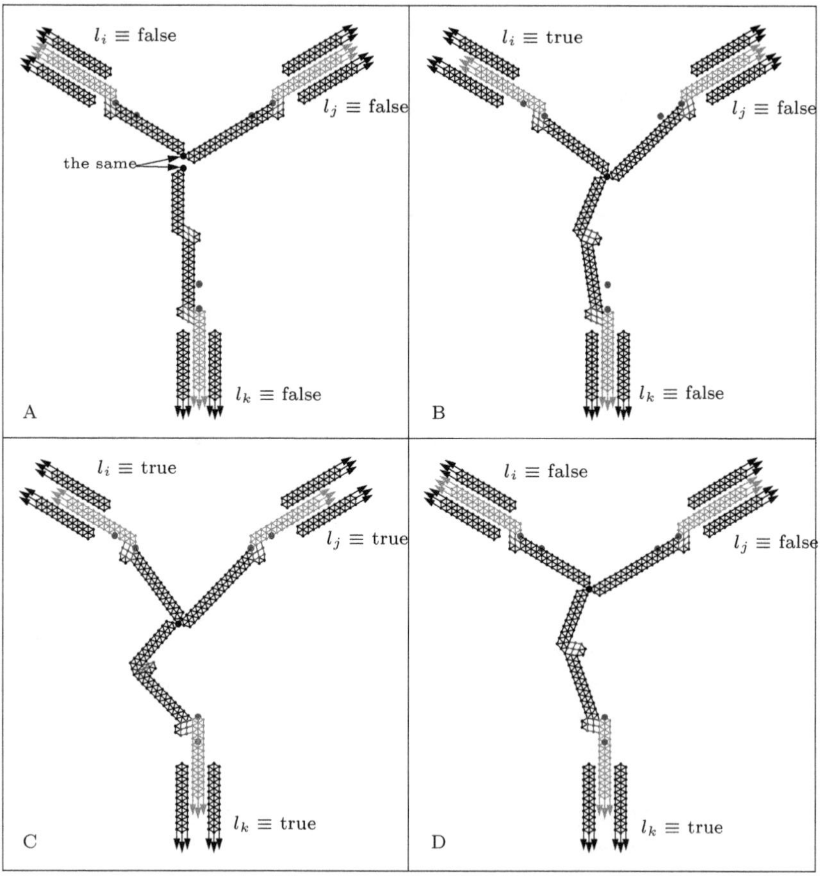

Fig. 8. A clause checker. The situation in A is not realizable, but the ones in B–D are realizable.

the center is completely determined by the values of l_i and l_j. When all l_i, l_j, l_k are false, then, the big dot in the center is too far from l_k to be realizable; see Fig. 8A. In the other cases, it is always realizable; see Fig. 8B–D for some cases.

To conclude, we summarize the argument why a realization of the graph corresponds to a satisfying truth assignment. The clause checker can be realized if and only if at least one transmitter is at the position closer to the clause checker. This can only be the case if, at the variable side of the corresponding inverter, the transmitter is pushed away from the clause checker. This pushing is transmitted through all bends and holders to the variable wheels. It follows that the literal must be true.

One can check by inspection that the clause-gadget is 3-connected, and therefore the whole construction is 3-connected. Furthermore, the lengths of the constructed graph are one because the vertices and edges lie on a hexagonal grid. The grid has polynomial size. Therefore, we are using a polynomial number of edges and bits, and so the reduction can be done in polynomial time. □

Observe that, when the graph is realizable, the realization is infinitesimally rigid. In other words, its vertices cannot be infinitesimally perturbed in a way that preserves the edge lengths to the first order. This condition is stronger than rigidity, and implies that the underlying graph is generically rigid [15]. Therefore, the problem remains NP-hard even when we know that the graph is generically rigid.

The 3-SAT problem is NP-hard even if each variable occurs at most 6 times, and this property is maintained in the reduction from 3-SAT to P3-SAT. Therefore, the faces that participate in the variable gadget have bounded degree. By filling the free space between the tunnels and on the outside by a triangulation, we can make sure that all the faces have bounded degree. Therefore, the problem remains NP-hard even if we assume bounded face degree.

Acknowledgements. We would like to thank the anonymous referee who brought [10,13] to our attention.

References

1. B. Berger, J. Kleinberg, and T. Leighton. Reconstructing a three-dimensional model with arbitrary errors. In *Proc. 28th Annu. ACM Sympos. Theory Comput.*, pages 449–458, May 1996.
2. C. Burnikel, R. Fleischer, K. Mehlhorn, and S. Schirra. A strong and easily computable separation bound for arithmetic expressions involving radicals. *Algorithmica*, 27(1):87–99, 2000.
3. C. Burnikel, S. Funke, K. Mehlhorn, S. Schirra, and S. Schmitt. A separation bound for real algebraic expressions. In *Algorithms — ESA 2001, 9th Annual European Symposium, Proceedings*, volume 2161 of *Lecture Notes in Computer Science*, pages 254–265. Springer-Verlag, 2001.
4. J. Campbell. *Map Use and Analysis.* McGraw-Hill, Boston, 4th edition, 2001.
5. S. Čapkun, M. Hamdi, and J. Hubaux. GPS-free positioning in mobile ad-hoc networks. In *Proceedings of the 34th Hawaii International Conference on System Sciences*, pages 3481–3490, January 2001.

6. B. Chazelle. Triangulating a simple polygon in linear time. *Discrete Comput. Geom.*, 6(5):485–524, 1991.
7. R. Connelly. On generic global rigidity. In P. Gritzman and B. Sturmfels, editors, *Applied Geometry and Discrete Mathematics: The Victor Klee Festschrift*, volume 4 of *DIMACS Series in Discrete Mathematics and Theoretical Computer Science*, pages 147–155. AMS Press, 1991.
8. C. Coullard and A. Lubiw. Distance visibility graphs. *Internat. J. Comput. Geom. Appl.*, 2(4):349–362, 1992.
9. G. M. Crippen and T. F. Havel. *Distance Geometry and Molecular Conformation.* John Wiley & Sons, 1988.
10. O. Devillers, G. Liotta, F. P. Preparata, and R. Tamassia. Checking the convexity of polytopes and the planarity of subdivisions. *Comput. Geom. Theory Appl.*, 11:187–208, 1998.
11. G. Di Battista, P. Eades, R. Tamassia, and I. G. Tollis. *Graph Drawing.* Prentice Hall, Upper Saddle River, NJ, 1999.
12. R. Diestel. *Graph Theory.* Springer-Verlag, New York, 2nd edition, 2000.
13. P. Eades and N. Wormald. Fixed edge length graph drawing is NP-hard. *Discrete Appl. Math.*, 28:111–134, 1990.
14. H. Everett, C. T. Hoàng, K. Kilakos, and M. Noy. Distance segment visibility graphs. Manuscript, 1999. `http://www.loria.fr/~everett/publications/distance.html`.
15. J. Graver, B. Servatius, and H. Servatius. *Combinatorial Rigidity.* American Mathematical Society, 1993.
16. B. Hendrickson. Conditions for unique graph realizations. *SIAM J. Comput.*, 21(1):65–84, February 1992.
17. B. Hendrickson. The molecule problem: Exploiting structure in global optimization. *SIAM J. on Optimization*, 5:835–857, 1995.
18. B. Jackson and T. Jordán. Connected rigidity matroids and unique realizations of graphs. Manuscript, March 2003.
19. D. E. Knuth and A. Raghunathan. The problem of compatible representatives. *SIAM J. on Discrete Mathematics*, 5(3):422–427, Aug. 1992.
20. C. Li and C. Yap. A new constructive root bound for algebraic expressions. In *Proc. 12th Annu. ACM–SIAM Sympos. Discrete Algorithms*, pages 496–505, 2001.
21. D. Lichtenstein. Planar formulae and their uses. *SIAM J. Comput.*, 11(2):329–343, 1982.
22. J. Pach and P. Agarwal. *Combinatorial Geometry.* John Wiley & Sons, New York, NY, 1995.
23. F. P. Preparata and M. I. Shamos. *Computational Geometry: An Introduction.* Springer-Verlag, 3rd edition, Oct. 1990.
24. N. B. Priyantha, A. Chakraborty, and H. Balakrishnan. The Cricket location-support system. In *Proceedings of 6th Annual International Conference on Mobile Computing and Networking*, pages 32–43, Boston, MA, August 2000.
25. C. Savarese, J. Rabaey, and J. Beutel. Locationing in distributed ad-hoc wireless sensor networks. In *Proceedings of the International Conference on Acoustics, Speech, and Signal Processing*, pages 2037–2040, Salt Lake City, UT, May 2001.
26. J. B. Saxe. Embeddability of weighted graphs in k-space is strongly NP-hard. In *Proc. 17th Allerton Conf. Commun. Control Comput.*, pages 480–489, 1979.
27. Y. Yemini. Some theoretical aspects of position-location problems. In *Proc. 20th Annu. IEEE Sympos. Found. Comput. Sci.*, pages 1–8, 1979.

BGPlay: A System for Visualizing the Interdomain Routing Evolution*
(Long Demo)

Giuseppe Di Battista, Federico Mariani, Maurizio Patrignani, and Maurizio Pizzonia

Università degli studi "Roma Tre"
Dipartimento di Informatica e Automazione
Via della Vasca Navale 79
00146 Roma, Italy

Abstract. In this paper we describe the visual interface of BGPlay, an on-line service for the visualization of the behavior and of the instabilities of the Internet routing at the autonomous system level.
A graph showing only connections among autonomous systems is not enough to convey all the information needed to fully understand the routing and its changes. BGPlay provides specifically tailored techniques and algorithms to show the routing at specific instants of time and to animate its changes. The system obtains routing data from well known on-line archives of routing information constantly kept up-to-date.

1 Introduction

The Internet is administratively partitioned into networks, called *Autonomous Systems* (*AS*), where each AS is under a single administrative authority. Usually, each Internet Service Provider (ISP) controls one or more ASes.

Roughly speaking, at the AS level the Internet routing is described by a collection of sequences of ASes, called *AS-paths*. An AS-path $AS1, AS2 \dots, ASn$ says that the packets directed to ASn and originated by some device in $AS1$ should traverse $AS2$, $AS3$, etc. in the specified order. The AS-paths that give the full routing status of the Internet at a certain moment can be "merged" into a graph, called a *routing graph*.

To understand our goals, consider the Network Operating Center (NOC) of some ISP and suppose that the NOC wants to know what are the paths that the packets currently follow to reach ASx which is operated by the NOC. BGPlay aims at showing the portion of the routing graph that describes how the traffic

* Work partially supported by European Commission - Fet Open project COSIN – COevolution and Self-organization In dynamical Networks – IST-2001-33555, by "Progetto ALINWEB: Algoritmica per Internet e per il Web", MIUR Programmi di Ricerca Scientifica di Rilevante Interesse Nazionale, and by "The Multichannel Adaptive Information Systems (MAIS) Project", MIUR Fondo per gli Investimenti della Ricerca di Base.

G. Liotta (Ed.): GD 2003, LNCS 2912, pp. 295–306, 2004.

flows to ASx from a selected set of ASes which are considered, in some sense, "representative" of the entire Internet.

Further, suppose that the NOC is interested in understanding how the routing evolved during a specific time interval. This can be important for several reasons: to determine faults in the pipes surrounding ASx, to check the consistency of the routers configurations, or even to monitor the behavior of the partners of the considered ISP with the purpose to verify whether they fulfill the commercial agreements. BGPlay aims at showing all the routing changes "around" ASx that occurred during the prescribed time interval.

Instabilities and faults of interdomain routing have been the subject of recent research in the networking area (see for example [17,12,10,16]). Also, several works address the problem of effectively visualizing network related graphs [11, 13,20,5,1]

In this paper we describe the visual interface of BGPlay [7], which runs at `http://www.dia.uniroma3.it/~compunet/bgplay`. The purpose of BGPlay is the visualization of the behavior and of the instabilities of the Internet routing at the AS level.

BGPlay has a client-server architecture. The user performs a query on a Web browser (the client side) specifying an IP prefix and a time interval. The server identifies the AS that contains the given prefix and extracts information that describes the routing evolution in the given interval from the Routing Information Service (RIS) [2] of the RIPE and from a local mirror of the Oregon Route View (ORV) project [3]. RIS and ORV are large and well known repositories of routing information.

All the visualization problems are addressed on the client side by means of a java applet. The user interface exposed by the BGPlay client can visualize the routing graph detected at any instant of the specified time interval. The BGPlay visualization puts in evidence ASes interconnections and which AS-paths are active on that interconnections in a given instant of time.

The paper is organized as follows. In Section 2 we give basic interdomain routing concepts. In Section 3 we introduce the concept of *routing graph*, discuss the requirements for its visualization, and argue about the unfeasibility of some possible visualization approaches. In Section 4 we show the techniques adopted in BGPlay for the visualization of the routing graph. In Section 5 we show how BGPlay animates the evolution of the routing. Section 6 shows a typical session with BGPlay.

2 Networking Background

In the Internet each host is identified by an IP address (32 bits, usually written in the dotted notation, e.g. 193.204.161.48). An IP prefix identifies a set of (contiguous) IP addresses having the same leftmost n bits, with $0 \leq n \leq 32$. Such IP prefix is usually indicated attaching a $/n$ at the end of the prefix (e.g. 193.204.0.0/15 indicates a prefix 15 bits long) [22]. Routing in the Internet is prefix based (like for telephone call routing). Since a prefix identifies a set of addresses, it implicitly identifies a set of hosts having such addresses. In the following the term *prefix* is used to denote both a set of addresses and a set of hosts, the distinction will be clear by the context.

An *Autonomous System* (*AS*) is a portion of the Internet under a unique administrative authority. In the Internet each AS is identified by an integer number. ASes cooperate in order to ensure good connectivity service to their customers but are competitors from an commercial point of view [14,15].

Traffic starting from an AS and directed to a specific prefix traverses an ordered set of ASes (*AS-path*). The configuration of such paths on the routing devices is too complicated to be manually performed. Hence, ASes exchange routing information with other ASes by means of a routing protocol called Border Gateway Protocol (*BGP*) [19,21]. Such a protocol is based on a distributed architecture where *border routers* that belong to distinct ASes exchange the information they know about reachability of prefixes. Two border routers that directly exchange information are said to perform a *peering session*, and the ASes they belong to are said to be *adjacent*.

Each router stores information about routing into its *routing information base* (*RIB*). The RIB is a table where each line is a pair $\langle$prefix, AS-path$\rangle$ meaning that a certain prefix is reachable through the associated AS-path. Such pairs are called *routes*. The main purpose of BGP is to allow the routers to exchange the routes they know. Since RIBs may be huge, the BGP process running on a router sends to its peers the full RIB only when a peering session is set up. During regular operation only updates are sent.

A BGP *update* is either a route *announcement* or a route *withdrawal*. An announcement conveys the following information: “through me you can reach a certain prefix traversing a certain AS-path”. A withdrawal nullifies a previously communicated route related to a specified prefix. In other words a withdrawal means “you can no longer reach this prefix through me”.

The receiver of an update may or may not modify its routing table depending on whether the router knows routes which are considered better or not. If the router modifies the routing table, the update is propagated to its peers.

Routes related to a certain prefix “born” within an AS called the *originator* of the prefix. Then, routes are propagated to adjacent ASes, which prepend their AS identifiers to the AS-path of the route and propagate it again by means of route announcements.

3 Effective Routing Graph Visualization

The purpose of BGPlay is to provide a graphical representation of a portion of the Internet routing at a given instant of time and of the changes that affect routing over time. In this section we describe the requirements we considered and show some negative preliminary results.

We define a *collector-peer* as an AS that has a BGP peering with ORV or with RIS collectors. The collector-peers are the vantage points we use to inspect the routing in the Internet and are determined by the current configuration of

RIS and ORV. The set of collector-peers we are currently using is easy to identify by just looking at the ASes appearing as leftmost ASes in at least one AS-path among the AS-paths provided by ORV and RIS.

We focus the attention on a specific prefix that is in turn contained in one AS, the *target AS*. The *routing status* at a given time for that prefix gives for each collector-peer the AS-path representing the route chosen at that time by that collector-peer to reach the prefix. Such a status is effectively represented with a *routing graph*. A routing graph is a graph in which each vertex is an AS and edges are the pairs of ASes that appear consecutively in at least one of the AS-paths.

We have identified the following requirements for the visualization of the routing status:

- The attention of the user should mainly be focused on the target AS.
- An AS should appear in the drawing at a geometric distance from the target AS that is roughly proportional to the number of (AS-)hops separating them.
- For each visualized AS, the AS-path used to reach the target must be fully identifiable in the drawing, even if traversing edges that are traversed by other paths. Observe that this requirement would be easy to meet if the graph was a tree; in fact in this case there is just one tree path from each AS to the target AS. The presence of cycles makes the problem more complex.

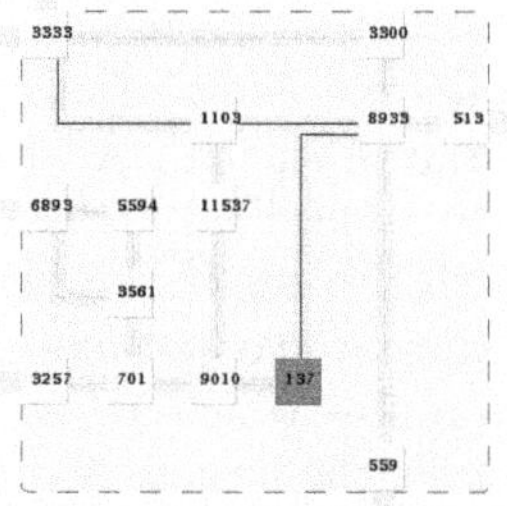

Fig. 1. In a first prototype of BGPlay the drawing standard used was orthogonal. The target AS (the dark node) was arbitrarily placed. The topological distance between AS137 and AS559 is two, while they are located very close in the drawing.

The choice of the drawing standard is biased by the first two requirements. Preliminary experiments we performed in this field have shown how the orthogonal drawing standard is not suitable for visualizing the routing graph. In fact, automatic layout tools in order to effectively compact the drawing may put the target AS in an arbitrary position and geometric distance among nodes does not convey any information about the topological distance (see Figure 1).

In the current version of BGPlay we adopt the straight-line drawing standard. For such standard there exist several layout algorithms that can fulfill the cited requirements (see Section 4).

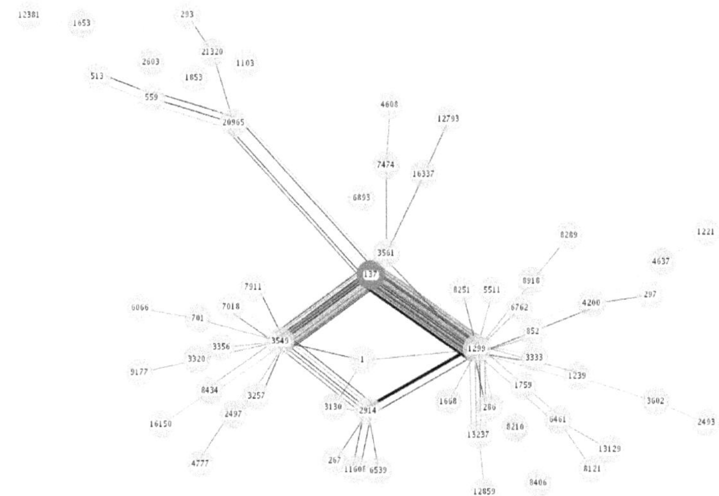

Fig. 2. In a first prototype of BGPlay all AS-paths were separately displayed. Paths were hardly distinguishable.

The third requirement is much more complex. Our first approach was to represent each AS-path separately and with a distinct color. A first prototype using this approach demonstrated that the number of paths to be shown was, usually, too high for the paths to be easily recognized due to the closeness of the paths and to the limited number of colors distinguishable on a monitor by human eyes (see Figure 2). The problem remains hard even using the techniques presented in [8] for drawing with fat edges.

Hence, we decided to adopt a different approach, described in Section 4.

4 Visualization Algorithms

In this section we describe the main algorithmic problems that we had to solve in order to design and implement BGPlay.

The third requirement stated in Section 3, that is that each AS-path must be easy to identify, is the one that was more difficult to meet. As we noted, if a set of AS-paths forms a tree, then such a set can be actually represented as a tree in an arbitrary color, satisfying the requirement. In fact, since each pair of vertices of a tree are joined by a single path, there would be no doubt about the sequence of ASes from the target AS to each collector-peer.

However, the routing graph is very often not a tree. Hence, our strategy is to partition the AS-paths into sets such that the graph obtained by merging the AS-paths of the same set is acyclic and can be unambiguously represented as a tree with a specific color. More formally, we have to solve the following problem.

Problem: TREE PARTITION
Instance: A set of AS-paths $\mathcal{P}$, such that each AS-path $p \in \mathcal{P}$ starts from a common AS_x.
Target: Find a partition for $\mathcal{P}$ in k sets such that (i) the graph induced by the AS-paths in the same set is acyclic (ii) every partition for $\mathcal{P}$ respecting (i) has at least k colors (that is, the number of sets is minimum).

Unfortunately, the TREE PARTITION problem is NP-hard. In order to prove this, following a standard technique, we show the NP-hardness of the corresponding decision problem, defined as follows.

Problem: K-TREE PARTITION
Instance: A positive integer K and a set of AS-paths $\mathcal{P}$, such that each AS-path $p \in \mathcal{P}$ starts from a common AS_x.
Question: Does a partition for $\mathcal{P}$ in K sets exist such that the graph induced by the AS-paths in the same set is acyclic?

We prove that K-TREE PARTITION is NP-hard by reducing the GRAPH K-COLORABILITY to it. Recall that GRAPH K-COLORABILITY is an NP-complete problem defined as follows.

Problem: GRAPH K-COLORABILITY
Instance: A graph $G = (V, E)$ and a positive integer $K < |V| + 1$.
Question: Is G K-colorable, i.e., does a coloring of the vertices of G in K colors exist such that adjacent vertices have different colors?

Theorem. 1 *The problem* K-TREE PARTITION *is NP-hard.*

Proof. We reduce GRAPH K-COLORABILITY to K-TREE PARTITION. Given an instance of GRAPH K-COLORABILITY we construct the corresponding instance of K-TREE PARTITION as follows. For each vertex v of $G(V, E)$ we introduce an AS-path p_v. At the beginnig of the construction all AS-paths have length one, and contain the same AS_x. Then, for each edge $(u, v) \in E$, we append to the two AS-path p_u and p_v the same $AS_{(u,v)}$. It is easy to show that the construction of the K-TREE PARTITION instance can be made in polynomial time, and that a solution for the instance of the GRAPH K-COLORABILITY problem exists iff a solution for the original instance of the K-TREE PARTITION does. □

Because of Theorem 1, in BGPlay, to solve TREE PARTITION, we used the following greedy algorithm which runs in polynomial time.

The sets of AS-paths are denoted $S_0, S_1, \ldots, S_{m-1}$, where the current number of sets is m.

1. A compatibility matrix is computed in which each AS-path is compared with each other. Two AS-paths are incompatible if and only if they form a cycle.
2. Only one empty set exists at the beginning: $m = 1$ and $S_0 = \phi$
3. foreach AS-path p

3.1 foreach set i from 0 to $m-1$
   – if p is compatible with all paths in S_i then accommodate p into S_i, skip the rest of this cycle and continue with the next path, otherwise try the next set S_{i+1} if it exists.

3.2 if p has not been accommodated in any of the available sets add a new set S_m, initialize $S_m = \{p\}$ and increment m.

For the correctness of the algorithm it is crucial to observe that all the paths have a common endpoint (target AS). The worst case time complexity of the algorithm described above is $O(n^2)$ where n is the number of paths (considered bounded in length). In fact, the number of compatibility tests performed for each AS-path is at most n.

An edge traversed by more than one tree is displayed using as many lines as the number of trees traversing that edge, where each line is colored with the color of the corresponding tree. Fig. 3 shows a drawing produced by BGPlay. It is about a prefix announced by AS702. Three levels of gray are used. AS-paths 16150 8434 3549 702 and 1853 1239 702 are drawn in gray. The other paths are grouped in two trees drawn in black and and light-gray. Only dashed lines are used since, as we shall see in Section 6, BGPlay uses solid lines to highlight paths involved in routing changes.

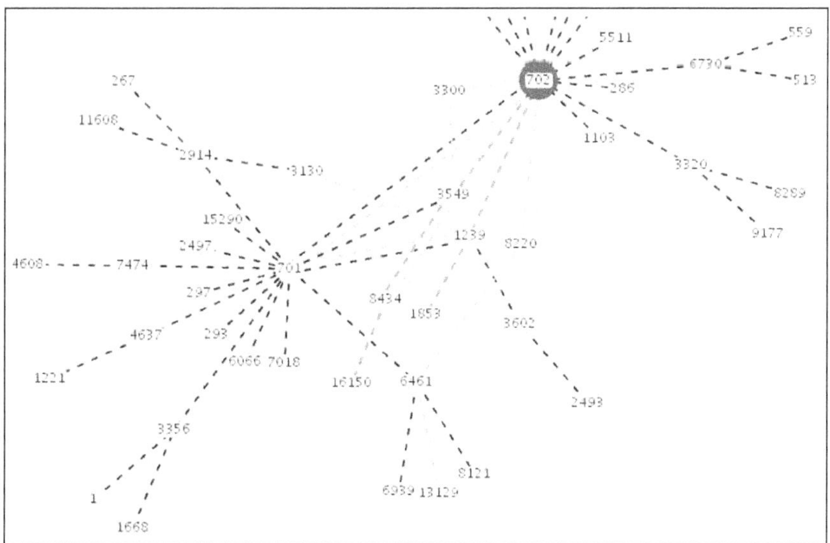

Fig. 3. A drawing produced by BGPlay. Three levels of gray are used. AS-paths 16150 8434 3549 702 and 1853 1239 702 are drawn in mid-gray. The other paths are grouped in two trees drawn respectively in black and light-gray.

The drawing layout is computed by using a spring embedder [6]. The target AS is constrained to be in the center of the drawing area. The computation is performed for a fixed number of steps that are much larger of the steps needed in the average case by a drawing to reach equilibrium.

What we said up to now concerns the visualization of the routing graph at a given instant of time. However, even for a single user query, BGPlay needs to visualize several routing graphs, each corresponding to the status of the routing at a specific instant.

Obviously, the visualization of the routing status before and after a BGP update occurred is sufficient to convey the information of the update. However, it is essential for two consecutive visualizations to be similar for not changing too much the users "mental map" [9,18].

We address the above problem by computing, before any visualization or animation takes place, the layout on a graph that is the union of all the AS-paths that appear in any routing graph at any instant of the selected interval. When the animation takes place, at any instant we display only the edges that belong to the current routing graph.

This trick ensures that consecutive animation steps show drawings that are very similar, since node positions are the same. Further, nodes involved in route changes (see Section 5) are most of the times placed near. In fact, two AS-paths involved in a routes change begin and end with the same AS, hence, the spring embedder tends to draw them closely.

5 Animation of the Routing Changes

Obviously, the visualization of the routing status before and after a BGP update occurred is sufficient to convey the information of the update. However, it is essential for two consecutive visualizations to be "similar" [9,18]. Further, the routing change should be apparent to the user.

A *routing history* is given by a starting routing status and a sequence of routing changes. From a routing history it is possible to reconstruct the routing status at each instant covered by the routing history.

A routing change happening at instant t can be one of the following types.

New route. A collector-peer that is not connected at time right before t with the target AS now acquires connectivity using a specified AS-path.

Route change. A collector-peer which is connected at time right before t with the target AS using AS-path p changes its connectivity by using AS-path q with $q \neq p$.

Route withdrawal. A collector-peer which is connected at time right before t with the target AS, looses its connectivity.

Further, it is interesting to consider the following event that is not a routing change but can be a symptom of network problems.

Route re-announcement. A collector-peer is connected at time t with the target AS and an announcement is received for the same route as if the route was new.

The collector-peers are partitioned in stable and unstable peers according to the following rules:

- if during the period of time covered by the considerd routing history a collector-peer is continuously connected with the target AS using the same AS-path, that collector-peer is considered stable,
- in all other cases the collector-peer is considered unstable.

To each stable collector-peers (and its stable AS-paths) a color is assigned according to the greedy algorithm described in Section 3. Distinct colors are assigned to all unstable collector-peers.

The union of all AS-paths that appear in the considered routing history forms a graph G in which the nodes are the involved ASes. A layout of the nodes of G is performed by using a spring-embedder algorithm as in Section 3. Such a layout remains unchanged during all the animation.

Animation is performed in the following way.

- AS-paths related to stable collector-peers are drawn dashed with their colors as in Section 3 and are not affected by any animation.
- AS-paths for new routes appear in their regular position (according to the layout of G). The attention of the user is caught by making the thickness of new AS-paths pulsing.
- Route changes are animated by means of a polyline morphing from the old AS-path to the new AS-path.
- Route withdrawals catch the attention of the user by a thickness pulse of the involved AS-path before it disappears.
- Route re-announcements catch the attention of the user by a thickness pulse of the involved AS-path.

The polyline morphing used to show route changes could be realized in several ways (see for example [4]). In BGPlay we adopted a rather simple technique which is illustrated in Figure 4. Let p be the polyline representing the old AS-path and p' be the polyline representing the new AS-path. A bijection between points of p and points of p' is defined in the following way. Polylines p and p' are consistently oriented incoming the target AS. Points in p are mapped with real numbers x in $[0 \dots 1]$ preserving the ordering. The same operation is performed on p'. Morphing between points which have the same value of x is performed linearly both in space and in time. In our case the two extremes of a polyline are always the same, namely the target AS and the collector-peer. The figure shows also the shape of the polyline in three intermediate instants. Morphing time is approximately set to one second.

6 Monitoring the Routes of the Traffic Incoming a Given AS with BGPlay

In this section we show how a user (e.g. the NOC of an ISP) can exploit the BGPlay service to monitor the evolution of the routing around an AS manager by the ISP.

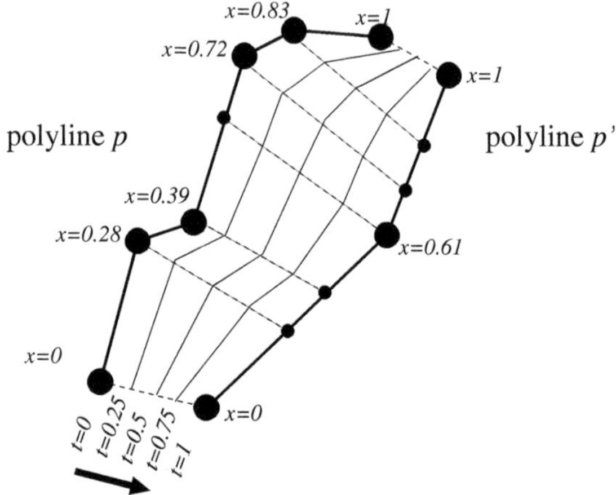

Fig. 4. Route changes are animated using a linear polyline morphing.

Welcome to Flap Viever v.2.0 (beta)
This tool is designed to show the instabilities in BGP routing of a specific prefix

Insert here the prefix that you want to explore:

Prefix: 193 . 0 . 0 . 0 / 21

Time Interval:

Select the time interval of your query:
- to have Route Views data your starting time must be less than a week ago
- to have Ris data the query starting time must be less than three months ago

Starting ti...

Date (DD/MM/YY... 21 / 5 / 2003 Time (hh:mm:... 18 : 0 : 00

Ending time:

Date (DD/MM/YY... 23 / 5 / 2003 Time (hh:mm:... 18 : 0 : 00

Data Sources:

Choose the Data Sources that you want to "use" in your query. RIS data are taken from on line user interface, while Route Views data come from local cache

RIS RRC00, RIS RRC01, RIS RRC02, RIS RRC03, RIS RRC04, RIS RRC05, RIS RRC06, RIS RRC07, RIS RRC08, Route Views

Clear Canc... OK

Fig. 5. The starting form.

Suppose that the NOC is interested in monitoring the routing evolution concerning the prefix 193.0.0.0/21 from May 21, 2003 to May 23, 2003, because in that period some network instabilities have been somehow perceived. She/He fills the form of Fig. 5. Observe that the user is filling out the form selecting all the available sources of information (checkbox RRC00, RRC01, etc.).

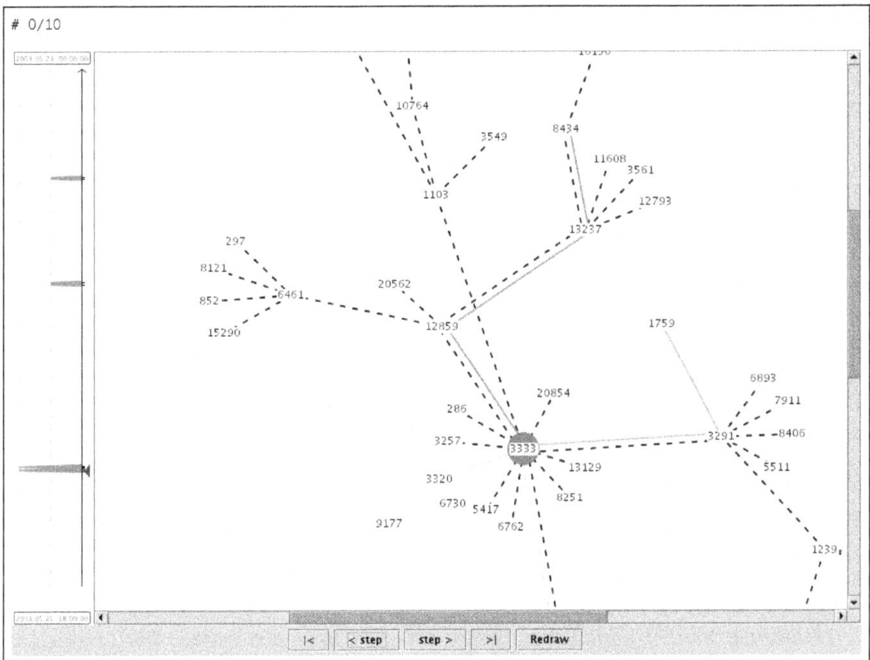

Fig. 6. A routing graph shown by BGPlay.

BGPlay shows (Fig. 6) a routing graph that represents the routing status at the time of the first event in the selected time interval. Observe that the AS originating the prefix is 3333 and it is placed in the center of the window. The paths that represents stable routes are drawn dashed while the paths associated to unstable routes are drawn solid.

The buttons on the bottom allow the user to move through the sequence of events that happened in the specified time interval. Both forward and backward moves are possible. A time panel (on the left) shows the distribution of the routing events in the interval and an arrow indicates the time of the event just displayed.

Now, the user steps through the events. BGPlay shows the the evolution of the routing. The status bar (on top) indicates the event identifier, a timestamp, and the type of the event (see Section 5). Depending on the type, additional information is displayed:

New route. The AS-path of the new route.
Route change. Both the old and the new AS-path.
Route withdrawal. The AS-path that is no longer valid.
Route re-announcement. The AS-path that is re-announced.

References

1. Hermes. http://www.dia.uniroma3.it/~hermes/.
2. Routing Information Service of the RIPE (RIS). http://www.ripe.net/ripencc/pub-services/np/ris/.
3. University of Oregon RouteViews project. http://www.routeviews.org.
4. S. Bespamyatnikh. An optimal morphing between polylines. *International Journal of Computational Geometry & Applications*, 12(3):217–228, 2002.
5. A. Carmignani, G. Di Battista, W. Didimo, F. Matera, and M. Pizzonia. Visualization of the high level structure of the internet with hermes. *J. of Graph Algorithms and Applications*, 6(3):281–311, 2002.
6. G. Di Battista, P. Eades, R. Tamassia, and I. G. Tollis. *Graph Drawing*. Prentice Hall, Upper Saddle River, NJ, 1999.
7. G. Di Battista, F. Mariani, M. Patrignani, and M. Pizzonia. Archives of bgp updates: Integration and visualization. In *Proceedings of IPS 2003, International Workshop on Inter-domain Performance and Simulation*, pages 123–129, 2003. online.
8. C. A. Duncan, A. Efrat, S. G. Kobourov, and C. Wenk. Drawing with fat edges. *Lecture Notes in Computer Science*, 2265:162–177, 2002.
9. P. Eades, R. F. Cohen, and M. L. Huang. Online animated graph drawing for web navigation. In G. Di Battista, editor, *Graph Drawing (Proc. GD '97)*, volume 1353 of *Lecture Notes Comput. Sci.*, pages 330–335. Springer-Verlag, 1997.
10. L. Gao and J. Rexford. Stable internet routing without global coordination. In *Measurement and Modeling of Computer Systems*, pages 307–317, 2000.
11. R. Govindan and H. Tangmunarunkit. Heuristics for internet map discovery. In *IEEE INFOCOM 2000*, pages 1371–1380, Tel Aviv, Israel, March 2000.
12. T. Griffin and G. T. Wilfong. An analysis of BGP convergence properties. In *SIGCOMM*, pages 277–288, 1999.
13. B. Huffaker, D. Plummer, D. Moore, and k claffy. Topology discovery by active probing. Technical report, Cooperative Association for Internet Data Analysis - CAIDA, San Diego Supercomputer Center, University of California, San Diego, 2002.
14. G. Huston. Interconnection, peering and settlements – part 1. *Internet Protocol Journal*, 2(1):2–16, 1999.
15. G. Huston. Interconnection, peering and settlements – part 2. *Internet Protocol Journal*, 2(2):2–23, 1999.
16. C. Labovitz, A. Ahuja, A. Bose, and F. Jahanian. Delayed internet routing convergence. In *SIGCOMM*, pages 175–187, 2000.
17. C. Labovitz, G. R. Malan, and F. Jahanian. Internet routing instability. *IEEE/ACM Transactions on Networking*, 6(5):515–528, 1998.
18. K. Misue, P. Eades, W. Lai, and K. Sugiyama. Layout adjustment and the mental map. *J. Visual Lang. Comput.*, 6(2):183–210, 1995.
19. Y. Rekhter. A border gateway protocol 4 (BGP-4). IETF, RFC 1771.
20. N. Spring, R. Mahajan, and D. Wetherall. Measuring isp topologies with rocketfuel. In *Proceedings of ACM/SIGCOMM '02*, Aug. 2002.
21. J. W. Stewart. *BGP4: Inter-Domain Routing in the Internet.* Addison-Wesley, Reading, MA, 1999.
22. A. S. Tanenbaum. *Computer Networks.* Prentice-hall International, Inc., 1996. ISBN: 0-13-394248-1.

GraphEx: An Improved Graph Translation Service

Stina Bridgeman

Computer Science Department, Colgate University, Hamilton, NY 13346 USA
sbridgeman@mail.colgate.edu

Abstract. The Internet-based translation service GraphEx automatically converting between different graph formats, making it easier for users to take advantage of the full variety of graph drawing tools. GraphEx builds on the prototype translation component of the Graph Drawing Server [2], improving the handling of information mismatch problems and adding support for user-defined formats.

1 Introduction

The development of a wide variety of graph drawing tools and libraries has led to the creation of many different formats for representing graph-structured information. The simplest formats contain only a list of edges to describe the combinatorial structure of the graph, while the most sophisticated allow labels, geometry, and other attributes to be associated with the graph structure. The variety of graph formats can make it difficult for people to take full advantage of the available graph drawing technology because of the effort required to translate existing graph data to the required format(s).

One solution is to develop a common graph interchange format flexible enough to support application-specific data. Efforts in this direction include GML [4], GraphXML [3], GXL [5], and GraphML [1]. These formats specify a core representation for the combinatorial structure of the graph and a framework for adding arbitrary attributes. Well-behaved applications are expected to quietly ignore (and possibly preserve) unknown attributes.

However, problems remain: there are multiple competing interchange formats, applications already in existence still use their own specialized formats, and graph-structured data arises in many domains and a standard format in one domain may be unfamiliar in another. Furthermore, implementors of a new tool may define their own format because the existing formats do not support the features they need or are too complex to parse, or because they are not aware of a suitable format. As a result, there is a need for a graph translation service which can automatically convert between different graph formats.

GraphEx (for *G*raph *Ex*change) improves on the graph translation component of the Graph Drawing Server [2] by reducing unnecessary loss of information and by supporting user-defined formats. A key component in reducing information loss is the incorporation of a merging step to restore attributes lost as a result of the format required by the desired drawing algorithm.

G. Liotta (Ed.): GD 2003, LNCS 2912, pp. 307–313, 2004.

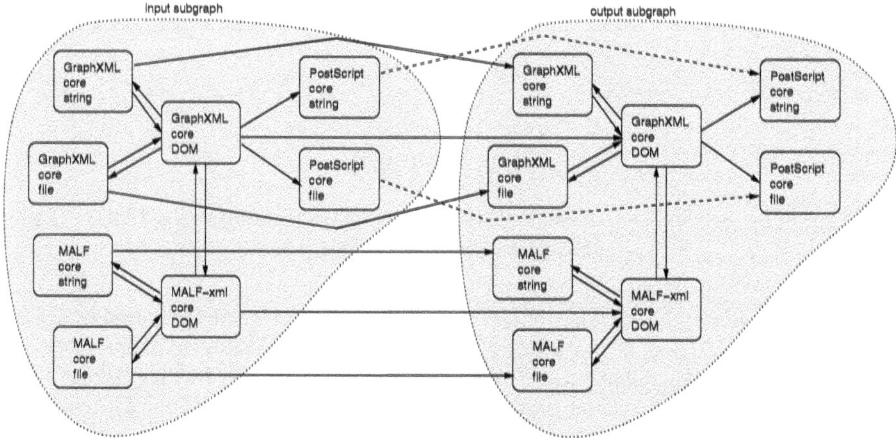

Fig. 1. An extended translation graph. Dashed lines indicate dummy merging edges.

Section 2 gives an overview of issues to be addressed by a general-purpose graph translator, section 3 presents the architecture of GraphEx, and section 4 summarizes the contributions of GraphEx and directions for future work.

2 Issues in Graph Format Translation

One of the primary issues facing a graph format translator is the fact that nearly any pair of graph formats will be incompatible in at least one aspect.

Structural mismatch refers to incompatibility of the graph structures which can be represented by the format: format A supports hierarchical graphs or multiedges, for example, while format B does not. In this case translation is impossible unless the original graph can somehow be encoded in a way that is compatible with the new format.

Information mismatch addresses incompatibility of the attributes used to specify information beyond the graph structure: format A supports attributes that format B does not, or format A places restrictions on the legal values of attributes which are different from B's requirements. A certain amount of information loss is inevitable when this occurs.

Due to the large number of formats in existence, it is not practical to require one-step translation filters for every pair of formats. A solution is to use translation sequences, so that converting from format F_0 to format F_n may involve multiple intermediate formats F_1, ..., F_2, ..., F_{n-1}. Unnecessary information loss can result if an attribute supported by both F_0 and F_n is not supported by one of the intermediate formats F_1, ..., F_{n-1}.

Two translation sequences are involved if a drawing algorithm is used: one to translate from the user's input format F_0 to the algorithm's input format F_k, and one to translate from the algorithm's output format F_{k+1} to the user's output format F_n. In addition to within-sequence information loss, unnecessary information loss can occur if an attribute supported by both F_0 and F_n is not supported by F_k or F_{k+1}. Furthermore, information can be lost if there is a pair of formats F_i and F_j such that F_i is converted to F_j in the first translation sequence, F_j is converted to F_i in the second translation sequence, and the translations are asymmetric i.e. if a particular attribute x of F_i is converted to attribute y of F_j in the translation $F_i \to F_j$, but a different attribute z of F_j is converted to attribute x of F_i in the translation $F_j \to F_i$.

Another issue results from extensible formats such as GML and GraphML: since users can define their own attributes, a translator cannot be guaranteed to understand all of the attributes present in a graph description. However, users often want the output from a drawing algorithm in the same format as their original input, and it is desirable for the extra application-specific attributes to be preserved. The matter is complicated by *fragile attributes* whose values may be altered the algorithm, as these attributes should *not* be preserved. Most formats do not provide an automatic way to identify fragile attributes; GML, which solves the problem by stipulating a naming convention, is one exception.

3 System Architecture

3.1 The Translation Graph

GraphEx extends the Graph Drawing Server's idea of a *translation graph*, a directed graph which describes all of the format conversions supported by the service. Figure 1 shows an example of an extended translation graph.

The translation graph is partitioned into two subgraphs, the "input subgraph" and the "output subgraph".

The input and output subgraphs each contain a vertex for each format. A format is described by a name, a version, and a form. The version primarily applies to extensible formats, and distinguishes between different attribute sets. The form distinguishes between different ways a graph in a given format may be represented: "a graph in GML format" may refer to a filename, URL, string, parsed object, etc., each of which may require different handling.

The formats supported by GraphEx can be classified into two groups: *graph-oriented* formats such as GML and GraphML, and *diagram-oriented* formats such as PostScript and GIF. Graph-oriented formats contain an explicit representation of the graph structure, while diagram-oriented formats contain only a graphical representation of a drawing; the graph structure cannot be easily extracted from a diagram-oriented format. Diagram-oriented formats can generally only be translated to other diagram-oriented formats.

Edges in the translation graph correspond to translation or merging functions, and are directed from the input format to the output format.

"Translation" refers to anything which changes the form or format of the input data — a parser is considered to be a translation because it changes the form from a filename to a parsed object, and a function downloading a remote URL is a translation because it converts the URL into a local file. Translation edges only connect vertices within a subgraph and are duplicated in the input and output subgraphs.

Merging operations are used in the second sequence of a two-sequence translation to restore attributes of the sequence one input graph which are lost as a result of formats required by the drawing algorithm. The output of the merging step contains all of the non-fragile attributes from the original input graph plus attributes present in the current graph from the second translation sequence. Currently all unknown attributes are copied; an extension will allow the user to specify which attributes should not be copied.

Merging requires establishing a correspondence between the vertices and edges of two separate but structurally identical graphs. This is made more difficult because not all graph formats support unique vertex identifiers, even fewer support unique edge identifiers, and some formats with unique identifiers do not guarantee that the identifiers remain associated with the same object if the graph is processed in some way. Merging establishes a correspondence by attempting to match vertices according to their unique identifiers, if available. If this is not successful, the graph structure is used to determine a partial correspondence which is then refined further, if necessary, by attempting to match attributes present in both graphs. Once the vertices have been matched, a similar procedure is applied to match edges.

Merging edges connect vertices in the input subgraph with vertices in the output subgraph and are directed from the input subgraph to the output subgraph. Every pair of same-format vertices in the input and output subgraphs must be connected by a merging edge; dummy edges are included in cases where merging is not supported (e.g. diagram-oriented formats) or not implemented. To reduce the amount of work required to support merging, it is generally expected that merging will only occur between graphs in the same format; however, additional merges could be incorporated into the translation graph without modifications.

3.2 Performing Translations and Handling Mismatch

A translation sequence is found by finding the shortest path from the input format in the input subgraph to the output format in the output subgraph. Since the only connections between these two components are merging edges, the shortest path is guaranteed to contain exactly one merging edge.

GraphEx supports two options for dealing with formats which are structurally incompatible. One option is to simply not define translations between such formats, resulting in a disconnected translation graph. If the formats are truly incompatible, a "translation not possible" error is a reasonable outcome.

A second possibility is to recognize that translating from format A to incompatible format B is only impossible if those features of format A which are incompatible with format B are used. For example, if format A supports multiedges but format B does not, the translation only needs to fail if the particular graph to be converted actually contains multiedges. GraphEx allows these "partial translators" as long as the translation function generates a warning or error if the graph cannot be translated.

(The possibility of encoding the original graph G as some other graph G' which can be represented in the target format is ignored, as such an encoding is typically application-specific and thus is not appropriate for a general-purpose translator. Encoding functions can be added as algorithms to the Graph Drawing Server, if desired, and can be explicitly accessed by the user in that way.)

Unnecessary information loss due to information mismatch can be avoided by providing translations between every pair of formats, or by defining a single "most powerful format" $\mathcal{F}$ and ensuring that $\mathcal{F}$ is the only intermediate format in every translation sequence. Unfortunately, both of these solutions are impractical: both require a significant amount of work to realize, and there is always a possibility that a new or user-defined format will contain a feature not supported by $\mathcal{F}$. Furthermore, the second scheme limits the degree to which existing translation filters can be incorporated into the service because a specific set of translations is required.

Instead, GraphEx tries to minimize unnecessary information loss through edge weights and clever structuring of the translation graph.

Translation edges are assigned weights according to the quality of the translation: completely compatible formats where no attributes are added or lost have the lowest weight, followed by translations which add attributes and finally those which involve a loss of information. This ensures that the best translation sequence is chosen if there is more than one possible translation sequence for a given pair of formats.

It is also envisioned that the translation graph will evolve to contain a small core of expressive formats (such as those intended as exchange formats), with each application-specific format linked to one of the core formats. The intention is that translating from an application-specific format to one of the core formats would not involve any information loss, and that translations between core formats would involve little or no information loss.

Dummy merging edges always have weight 0 since it is not possible to merge these formats. The weights of other merging edges depend on the translation scenario.

In a single-sequence translation or in the first sequence of a two-sequence translation, merging is not necessary and thus all merging edges are assigned a weight of 0 so they do not influence the rest of the translation process.

For the second sequence of a two-sequence translation, it is also necessary to consider the translation required to convert the input for the first sequence to the input format for the merging step. To ensure that merging happens at the best possible time, merging edges are assigned a weight corresponding to the cost of this conversion. If such a translation is not possible, the merging edge is assigned a weight of ∞.

3.3 User-Defined Formats

The purpose of GraphEx is to allow a user to access a variety of graph drawing algorithms without having to worry about format translations, but a user cannot take advantage of the system if her data is not already in one of the formats supported by the service. A new feature of GraphEx is its ability to allow users to install new formats, translations, and merges, which are then added to the translation graph and can be automatically and transparently in future drawing requests. For security, the new items are only available to the user who installed them.

Currently the user must provide implementations of the translations required to link the new format to an existing format, including any necessary parsers, but future plans include simplifying and automating the process to reduce the burden on the user.

3.4 Implementation Details

The core GraphEx service is implemented in Java; it provides an interface accessible via Java RMI. Java was chosen because of its portability — a GraphEx server can be placed on a wide variety of hosts so that it is not necessary to port existing translation filters to different environments. Java RMI also affords several advantages: it automatically handles the marshalling and unmarshalling of entire objects sent between hosts, and it supports dynamic code loading allowing Java classes to be distributed at runtime as needed throughout the system. Furthermore, RMI can easily be layered over SSL to provide secure communications between client and server.

GraphEx also supports user authentication and allows individual translations to be restricted to certain users.

There is a great deal of flexiblity as to how individual translation filters are implemented: they can be Java classes, standalone executables, shell scripts, C or C++ libraries, XSLT transforms (for XML-based formats), or anything else that can be called from a Java program. Every translation is associated with a wrapper class which hides details of the invocation of the translation filter and provides a common interface to the translation service for all translations.

A web-based frontend for GraphEx is under development. This allows the client to send a request to a webserver which then uses servlets to contact the translation service to process the request.

4 Conclusions and Future Work

In summary, GraphEx provides an Internet-accessible graph format translation service. It improves on the original translation service of the Graph Drawing Server [2] by reducing unnecessary loss of information and adding support for user-defined formats and additional security. Future work includes development of a web-based frontend and a more powerful interface to facilitate the specification of user-defined formats.

References

1. U. Brandes, M. Eiglsperger, I. Herman, M. Himsolt, and M. S. Marshall. GraphML progress report: Structural layout proposal. In *Proc. 9th Intl. Symp. Graph Drawing (GD '01)*, volume 2265 of *Lecture Notes Comput. Sci.*, pages 501–512. Springer-Verlag, 2002.
2. S. Bridgeman, A. Garg, and R. Tamassia. A graph drawing and translation service on the WWW. *Internat. J. Comput. Geom. Appl.*, 9(4/5):419–446, 1999.
3. Ivan Herman and M. Scott Marshall. GraphXML — an XML-based graph description format. In Joe Marks, editor, *Proc. 8th Intl. Symp. Graph Drawing (GD 2000)*, volume 1984 of *Lecture Notes Comput. Sci.*, pages 52–62. Springer-Verlag, 2000.
4. M. Himsolt. GML: Graph modelling language. Manuscript, Universität Passau, Innstraße 33, 94030 Passau, Germany, 1996. `http://infosun.fmi.uni-passau.de/Graphlet/GML/`.
5. A. Winter. Exchanging graphs with GXL. In Petra Mutzel, Michael Jünger, and Sebastian Leipert, editors, *Proc. 9th Intl. Symp. Graph Drawing (GD '01)*, volume 2265 of *Lecture Notes Comput. Sci.*, pages 485–500. Springer-Verlag, 2002.

A Constrained, Force-Directed Layout Algorithm for Biological Pathways

Burkay Genc and Ugur Dogrusoz

Computer Engineering Department and
Center for Bioinformatics, Bilkent Univ., Ankara 06800, Turkey

Abstract. We present a new elegant algorithm for layout of biological signaling pathways. It uses a force-directed layout scheme, taking into account directional and regional constraints enforced by different molecular interaction types and subcellular locations in a cell. The algorithm has been successfully implemented as part of a pathway integration and analysis toolkit named PATIKA, and results with respect to computational complexity and quality of the layout have been found satisfactory.

1 Introduction

As graphical user interfaces have improved, and more state-of-the-art software tools have incorporated visual functions, interactive graph editing and diagramming facilities have become important components in visualization systems [4]. Biology is no exception. In order to make useful deductions about a cell, an inherently complex multi-body system, one needs to consider cellular pathways as an interconnected network rather than separate linear signal routes.

There has been a few studies done specifically for layout of biological pathways as well, focusing on metabolic pathways. Karp and Paley [6] proposed a divide-and-conquer algorithm to identify a number of pre-determined subtopologies such as paths, cycles, and trees so that different layout approaches may be applied on each part. Becker and Rojas [1] improve this approach by supplementing a special force-directed layout algorithm and additional layout heuristics.

PATIKA [3], a pathway database and tool, is mainly intended for signaling pathways whose underlying graph structure can be arbitrarily more complicated and irregular than that of metabolic pathways.

In this paper, we introduce an efficient and powerful layout algorithm devised for pathway graphs as defined by PATIKA ontology [2]. It is based on the spring force directed layout algorithm [5] with regional constraints. It also uses a similar idea to magnetic fields of Sugiyama [7] but employs per edge fields to enforce edge orientation constraints, which are allowed to adaptively change during layout.

2 Pathway Model

The structure of pathway graphs highly depend on the type of the pathways and the ontology used to represent the biological phenomenon. We assume the basics

G. Liotta (Ed.): GD 2003, LNCS 2912, pp. 314–319, 2004.

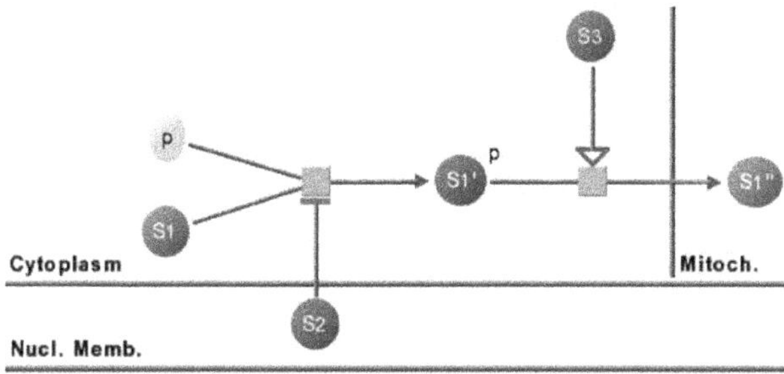

Fig. 1. An example illustrating the basics of the assumed ontology. The states, transitions, and interactions (substrates such as the one with source $S1$, products such as the one with target $S1'$, and effectors such as the one with source $S2$) are represented with ovals, rectangles, and lines of varying types, respectively, and cellular compartments are separated by orthogonal lines.

of the ontology described in [3,2], which represents a cellular process in the form of a directed graph called *pathway graph* (Figure 1). Usually the pathway graphs representing signaling pathways do not possess the uniform properties that those representing metabolic pathways do.

3 Layout Algorithm

We have chosen to use a force-directed layout algorithm with constraints to satisfy the criteria of the specific underlying model as well as the general conventions in pathway graph drawings. Basically, it is a virtual dynamic system in which nodes are assumed to have a certain "mass", connected via "springs" of a pre-specified desired length. Thus each node in a pathway graph is applied both *spring* and *node-to-node repulsion* forces. Spring forces include *relativity constraint* forces that are applied on each substrate, product, activator or inhibitor node to align the corresponding edge to lie towards the left, right, top or bottom of the associated transition, respectively. Furthermore, each horizontal (vertical) compartment separator is part of this physical system, on which the rest of the system can apply forces, moving them in only vertical (horizontal) direction. We also assume "gravitational" forces on compartment separators, disallowing a compartment to unnecessarily expand (Figure 2). Thus the optimal layout is regarded as the state of this system in which total energy is minimal.

The layout algorithm is split into three *major phases*, each of which alternates between odd and even-numbered *minor phases*. The first major phase is mainly for unscrambling the pathway graph with the help of high repulsion force ranges and the concept of *pulsing*. This is achieved by expanding the graph to a much larger area in a new minor phase compared to the previous one, and vice versa.

The second phase is where each edge adapts a best orientation for itself with the concept of "maturity". As an edge stays in a certain orientation (e.g., left-to-

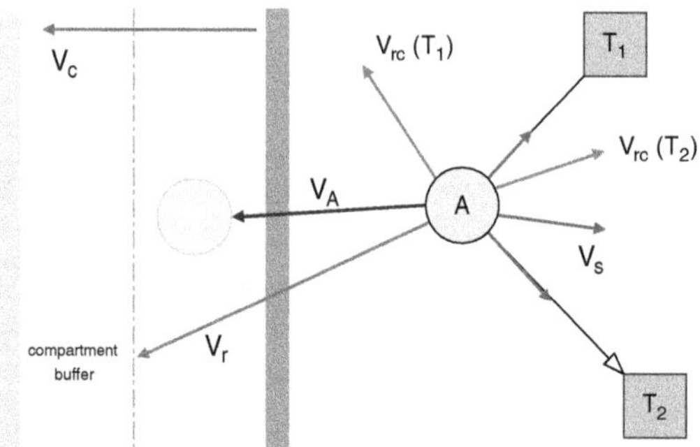

Fig. 2. An example showing various types of forces on a state A (V_s, V_r, and V_{rc}: spring, repulsion, and relativity constraint forces, respectively) and a compartment separator. Both move towards left by total forces V_A and V_c, respectively.

right) over consecutive iterations, its maturity is increased; and after a certain period, it "adapts" this orientation.

The last major phase is the stabilization phase, where all forces are pulled down to a minimum level, and pulsing and adaptive layout are disabled. In this phase compartments are also allowed to shrink.

The following method is used for calculating the relativity constraint forces acting over an edge. The method is clearly of $\Theta(1)$ time complexity.

```
algorithm ApplyRCF(Edge e)
    {u,v} = {source,target} node of e
    if this is an adaptive layout then
        if we are at major phase 2 then
            Increment maturity of e
            if e is mature and orientation is not satisfied then
                Change orientation of e as appropriate
    Calculate V_rc on e according to its orientation
    Split the force into components: V_rc^x and V_rc^y
    Update {u,v}.sf.x and {u,v}.sf.y by V_rc^x and V_rc^y, resp.
```

The next method is of $\Theta(|E|)$ and calculates the general spring forces acting on each edge using $F_s = (\lambda - edgeLength)^2/\eta$, where λ is the ideal edge length and η is the elasticity constant of the edge.

```
algorithm ApplySpringForces(Graph G = (V,E))
    for e ∈ E do
        {u,v} = {source,target} node of e
        Calculate the spring force V_s acting on e
        Split the force into components: V_s^x and V_s^y
        Update {u,v}.sf.x and {u,v}.sf.y by V_s^x and V_s^y, resp.
        call ApplyRCF(e)
```

Node-to-node repulsion forces are calculated using the formula $F_m = \alpha/(d_x^2 + d_y^2)$, where α is the repulsion constant and d_x and d_y are the differences in x and y coordinates of the two repulsing nodes, respectively.

algorithm APPLYMASSFORCES(*Graph* $G = (V, E)$)
(1) Create empty set S of layout nodes
(2) **for** $u \in V$ **do**
(3) Insert u into S
(4) **for** $v \in V - S$ **do**
(5) **if** u and v are in repulsion range **then**
(6) Calculate repulsion force V_r acting on u and v
(7) Split the force into components: V_r^x and V_r^y
(8) Update $\{u, v\}.rf.x$ and $\{u, v\}.rf.y$ by V_r^x and V_r^y, resp.

Steps 6-8 are handled in $\Theta(1)$ steps executed a total of maximum $O(|V|^2)$ times. However, since a node pair affect each other only when they are below a certain geometric distance, the average complexity is expected to be lower.

The following method controls the compartment constraints.

algorithm CHECKCOMPARTMENTRULES(*Graph* $G = (V, E)$)
(1) **for** $u \in V$ **do**
(2) Calculate $newX$, $newY$, $newRx$ and $newRy$ based on old coordinates and V_s^x, V_s^y, V_r^x and V_r^y values of u
(3) **if** u is a *state* **then**
(4) **if** compartment bounds are violated by $newRx$ or $newRy$ **then**
(5) **if** compartment resizing is enabled **then**
(6) Resize compartment of u
(7) **else**
(8) Alter V_r^x, V_r^y so as to keep u within compartment borders
(9) **if** compartment bounds are violated by $newX$ or $newY$ **then**
(10) Alter V_s^x, V_s^y so as to keep u within compartment borders
(11) Increment *error* by V_r^x, V_r^y, V_s^x and V_s^y of u
(12) Update coordinates of $u.x$ and $u.y$ with $newX$ and $newY$, resp.

Step 6 might require displacement of all nodes taking $O(|V|)$ time to complete. The compartments are normally resized no more than once or twice per iteration. Thus, the overall time complexity is $O(|V|^2)$ in the worst case and $O(|V|)$ on the average.

The main layout algorithm is as follows:

algorithm LAYOUT()
(1) Set *step* to 0
(2) **if** an incremental layout is to be done **then**
(3) Increment *step* to second major phase
(4) **else**
(5) Set *repulsionRange* to *MAX_REPULSION_RANGE*
(6) **while** *step* $\leq$ *MAX_ITERATION_COUNT* **do**
(7) **if** entering second major phase **then**
(8) Set *repulsionRange* to *desiredRange* for second major phase
(9) Set *error* to 0
(10) **call** APPLYSPRINGFORCES()
(11) **if** in an odd minor phase **or** in third major phase **then**
(12) **call** APPLYMASSFORCES()
(13) **call** CHECKCOMPARTMENTRULES()

```
    if in third major phase and step mod shrinkPeriod == 0 then
        Shrink all compartments from all sides as much as possible
    if error <ERROR_THRESHOLD then
        Jump to next minor phase by adjusting step
        if in third major phase then
            Immediately finish layout
    Increment step by 1
```

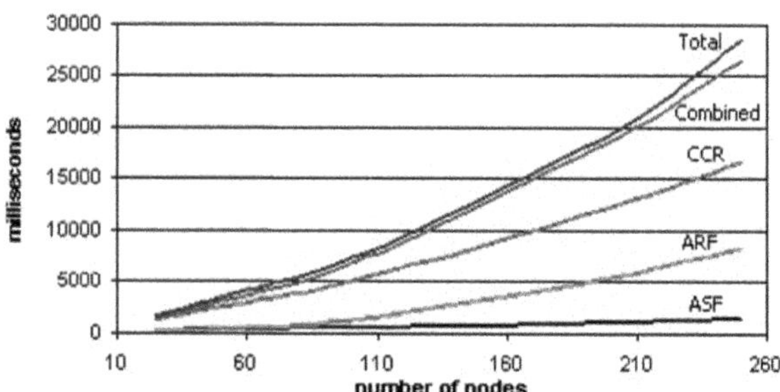

Fig. 3. Graph size vs. execution time.

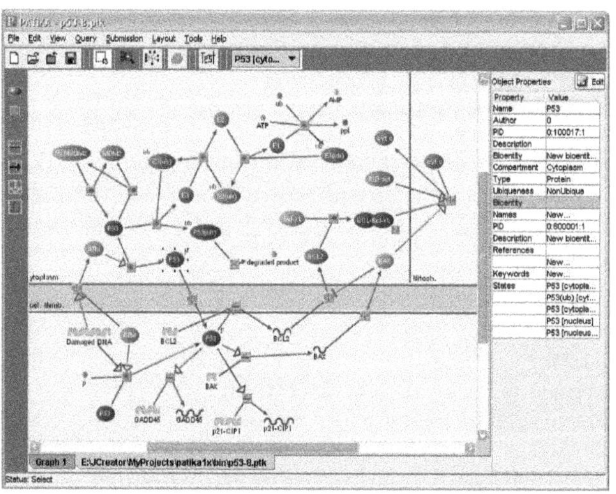

Fig. 4. An example layout for the p53 pathway.

The first and second major phases only differ in the amount of repulsion range considered when calling ApplyMassForces. For the odd-numbered minor phases and first two major phases the overall worst-case time complexity of each layout iteration is $O(|E| + |V|^2 + |V|) = O(|V|^2)$ for sparse graphs. For the even-numbered phases this is reduced to $O(|E| + |V|)$. In the third major phase, the repulsion forces are always calculated; additionally, a shrink operation is performed at certain periods yielding an overall complexity of $O(|V|^2)$ for sparse graphs. In the worst case if we assume that all phases are executed to the end and all node pairs are considered for repulsion calculations, the overall time complexity is $O(K \cdot |V|^2)$ over a total of K iterations needed for minimizing the total energy of the system.

4 Implementation and Results

The algorithm described above has been implemented and tested within the PATIKA pathway editor [3]. For each test a random graph is generated and all nodes are randomly assigned a compartment. The number of edges per graph is chosen to be linear in the number of nodes as in a typical pathway graph. For similar reasons one in every 20 edges or so are added as a back edge to form a new cycle.

Figure 3 shows the run time behavior of each layout component with increasing number of nodes. It is clear that the time spent inside the ApplySpringForces method is linear with respect to the number of nodes as expected. Execution time of the algorithm is affected by other parameters of the algorithm as suggested by the theoretical analysis.

The quality of the layout is found to be acceptable in terms of general graph drawing criteria (e.g., discovering symmetries, minimizing edge crossings) as well as pathway graph drawing conventions (Figure 4).

References

1. M. Y. Becker and I. Rojas. A graph layout algorithm for drawing metabolic pathways. *Bioinformatics*, 17:461–467, 2001.
2. E. Demir, O. Babur, U. Dogrusoz, A. Gursoy, A. Ayaz, G. Gulesir, G. Nisanci, and R. Cetin-Atalay. An ontology for collaborative construction and analysis of cellular pathways. To appear in Bioinformatics, 2003.
3. E. Demir, O. Babur, U. Dogrusoz, A. Gursoy, G. Nisanci, R. Cetin-Atalay, and M. Ozturk. PATIKA: An integrated visual environment for collaborative construction and analysis of cellular pathways. *Bioinformatics*, 18(7):996–1003, 2002.
4. U. Dogrusoz, Q. Feng, B. Madden, M. Doorley, and A. Frick. Graph visualization toolkits. *IEEE Computer Graphics and Applications*, 22(1):30–37, January/February 2002.
5. T. M. J. Fruchterman and E. M. Reingold. Graph drawing by force-directed placement. *Software Practice and Experience*, 21(11):1129–1164, 1991.
6. P. D. Karp and S. Paley. Automated drawing of metabolic pathways. In *Third International Conference on Bioinformatics and Genome Research*, pages 225–238, Tallahassee, Florida, June 1994.
7. K. Sugiyama and K. Misue. A simle and unified method for drawing graphs: Magnetic-spring algorithm. In R. Tamassia and I. Tollis, editors, *Graph Drawing (Proc. GD '94)*, volume 894 of *Lecture Notes in Computer Science*, pages 364–375. Springer-Verlag, 1995.

Intersection-Free Morphing of Planar Graphs*

Cesim Erten[1], Stephen G. Kobourov[1], and Chandan Pitta[2]

[1] Department of Computer Science
University of Arizona
cesim,kobourov@cs.arizona.edu

[2] Department of Electrical and Computer Engineering
University of Arizona
chandanp@ece.arizona.edu

Abstract. Given two different drawings of a planar graph we consider the problem of morphing one drawing into the other. We designed and implemented an algorithm for intersection-free morphing of planar graphs. Our algorithm uses a combination of different techniques to achieve smooth transformations: rigid morphing, compatible triangulations, as well as morphing based on interpolation of the convex representations of the graphs. Our algorithm can morph between drawings with straight-line segments, bends, and curves. Our system is implemented in Java and available as an applet at `http://gmorph.cs.arizona.edu`.

1 Introduction

Morphing refers to the process of transforming one shape (the source) into another (the target). Morphing is widely used in computer graphics, animation, and modeling; see a survey by Gomes *et al* [13]. In planar graph morphing we would like to transform a given source graph to another target graph. A smooth transformation of one graph into another can be useful for numerous problems from graph drawing [5,18]. In particular, when dealing with dynamic graphs and graphs that change through time, it is crucial to preserve the mental map of the user. Thus, it is important to minimize the changes to the drawing and to create a smooth transition between consecutive drawings.

In this paper we consider the problem of morphing between two drawings, D_s and D_t, of the same planar graph $G = (V, E)$. We assume that both drawings realize the same embedding of G, have the same outer-face, and are intersection-free. The source drawing D_s and the target drawing D_t can be straight-line drawings, or drawings with bends and curves. The positions of the vertices in the two drawings may be different (as long as the embedding is the same in both). The main objective is to find a morph that preserves planarity throughout the transformation. Secondary objectives include obtaining simple and smooth trajectories for the vertices (and bends) and preserving drawing invariants throughout the transformation. Preservation of drawing invariants refers to the continuity of

* This work is partially supported by the NSF under grant ACR-0222920.

G. Liotta (Ed.): GD 2003, LNCS 2912, pp. 320–331, 2004.

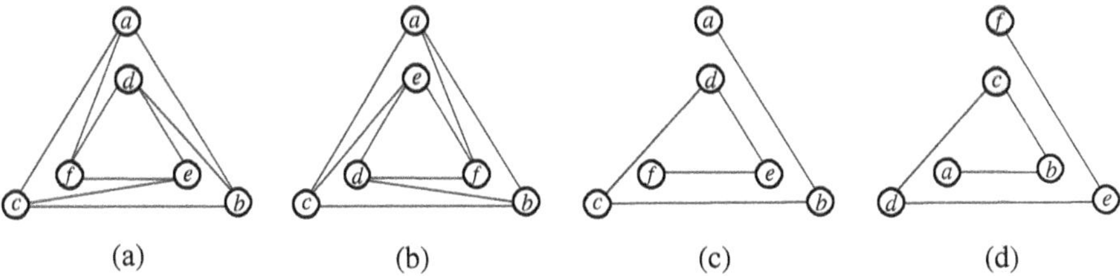

Fig. 1. The drawing from part (a) cannot be morphed into the drawing of part (b) while preserving the edge lengths. In particular, edges $(e, f), (b, d), (c, e)$ will have to shorten and lengthen if we are to preserve planarity. The drawing from part (c) cannot be morphed into the drawing of part (d) using linear trajectories while avoiding crossings.

the change: for example, a particular edge may shorten throughout the process but it should not shorten and then lengthen repeatedly. Similarly, the morph should avoid shrinking and growing of the graph faces.

We designed and implemented an algorithm for morphing planar graphs which preserves planarity throughout the transformation. It is easy to see that if we want to preserve planarity, some edges may have to lengthen and shorten; see Fig. 1(a-b). Similarly, linear trajectories cannot always be achieved; see Fig. 1(c-d). Thus, our algorithm yields smooth trajectories and preserves edge lengths whenever possible.

2 Previous Work

Morphing has been extensively studied in graphics, animation, modeling and computational geometry, e.g., morphing 2D images [2,14,25], polygons and poly-lines [3,12,20,21,22], 3D objects [15,17] and free form curves [19].

Graph morphing, refers to the process of transforming a given graph G_1 into another graph G_2. Early work on this problem includes a result by Cairns in 1944 [4] who shows that if G_1 and G_2 are maximally planar graphs with the same embedding, then there exists a non-intersecting morph between them. Later, Thomassen [26] showed that if G_1 and G_2 are isomorphic convex planar graphs with the same outer face, then there exists a non-intersecting morph between them that preserves convexity.

A naive approach to morphing one graph to another is *linear morphing*, where all the vertices move in a straight line at constant velocity from their positions in the source drawing to their final positions in the target drawing [8,16]. This is the simplest form of morphing but it may result in poor animation as all the trajectories may intersect at a common point, thus shrinking the drawing to a point on the way from the source to the target; see Fig. 2. Another problem with linear morphing is that intermediate graphs may have self-intersections even thought the source and the target are non-crossing; see Fig. 3.

Friedrich and Eades [10] present a graph animation technique based on rigid motion and linear interpolation. In the rigid motion stage the trajectories of the

Fig. 2. Linear morphing can result in degenerate intermediate drawings.

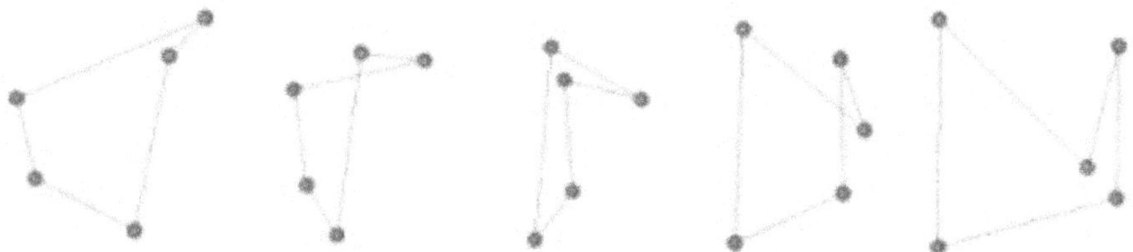

Fig. 3. Linear morphing can create crossings.

vertices are computed by an affine linear transformation. As a result, the source and target vertices are aligned as close as possible. In the linear interpolation stage the vertices travel on straight-line trajectories. While the rigid motion leads to smooth animations, in the interpolation stage, crossings may occur, even if the source and target are intersection-free. Friedrich and Houle [11] modify the algorithm in [10] by clustering groups of nodes that share similar motions in order to create better animations.

Graph morphing is also related to the problem of compatible triangulations. This problem arises when it is necessary to find isomorphic triangulations of two point sets on n vertices, or of two n-sided polygons. Aronov *et al* [1] show that it is always possible to create isomorphic triangulations, provided that $O(n^2)$ additional points (Steiner points) are created. Given two compatible triangulations with the same convex boundaries, Floater and Gotsman [7] and Surazhsky and Gotsman [24] show how to morph between them using convex representation of triangulations using barycentric coordinates, originally described by Tutte in 1963 [27]. A generalization of the same approach is used in [14] for morphing simple planar polygons, while guaranteeing that the intermediate polygons are also simple.

3 Algorithm Overview

We assume that the source drawing D_s and the target drawing D_t are intersection-free, have the same outer-face, and their underlying graphs are isomorphic. If the two drawings are isomorphic but the outer-face is different then there does not exists a transformation that preserves planarity throughout the process. If the graphs are not isomorphic, then nodes and edges that are not in

MAIN ALGORITHM
1. compute trajectories based on rigid motion
2. introduce "bend" vertices
3. compatibly triangulate all faces
4. compute trajectories based on convex representations

Fig. 4. Summary of the algorithm.

the intersection of the two graphs can be faded in and out as in earlier systems [9, 10].

Our algorithm for intersection-free morphing of planar graphs has four distinct stages. In the first stage the two drawings are aligned using 2-D transformations consisting of translation, rotation, scaling and shearing. That is, we move the source drawing as close as possible to the destination drawing as a rigid object in space. In the second stage we introduce "bend" vertices on all edges with bends. In the case of curvilinear drawings, we approximate the curves by piecewise linear curves. For every edge in the graph, we ensure that the same number of bend vertices are introduced in both D_s and D_t. In the third stage we identify all faces and compatibly triangulate all corresponding pairs of faces. In this process, we introduce additional triangulation vertices (Steiner points), internal to the faces. There are at most $O(k^2)$ Steiner points, where k is equal to the number of vertices together with the number of bends. In the fourth stage we compute trajectories for all vertices (including the bend vertices) based on convex graph representations and using interpolation of the matrices that represent the two graphs. The four steps of the algorithm are summarized in Fig. 4 and illustrated through an example in Fig. 5.

In the following sections we discuss steps 1, 3 and 4 in detail, leaving out step 2 as it is quite straight-forward.

4 Computing Trajectories Using Rigid Motion

4.1 Affine Matrix of Transformations

The rigid motion in 2-D can be associated with a natural interpolation of four transformations over time: translation, rotation, scaling and shearing. All these transformations can be accommodated by an affine matrix, which can be considered a 2×2 matrix, appended with a translation row:

$$\begin{bmatrix} c_{11} & c_{12} & 0 \\ c_{21} & c_{22} & 0 \\ t_x & t_y & 1 \end{bmatrix}$$

Then a point (x, y) which can be represented with the vector $[x \; y \; 1]$ and multiplied on the right by the matrix, is transformed into (x', y') using the

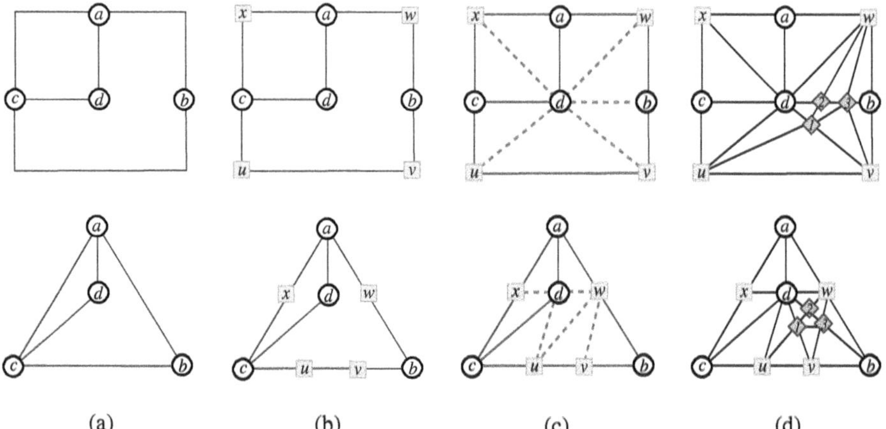

Fig. 5. Part (a) shows the graphs to be morphed: the source is on the top and target is on the bottom. Part (b) shows the addition of the bend vertices u, v, w, x (shown as squares). Part (c) shows the independent triangulation of both graphs (dashed edges). Part (d) shows the compatible triangulation with three Steiner points $1, 2, 3$ (shown as diamonds).

linear equations:

$$x' = c_{11}x + c_{21}y + t_x$$
$$y' = c_{12}x + c_{22}y + t_y$$

Given a point $p_s = (x_s, y_s)$ in D_s and the corresponding target point $p_t = (x_t, y_t)$ in D_t, we want p'_s, the resulting point after the transformations being applied on p_s, to be as close as possible to the target point p_t. Thus, to align the two drawings as best as possible we minimize the sum of squares of such pairwise distances:

$$\sum_{p_s \in D_s} dist^2(p'_s, p_t)$$

where *dist* is the Euclidean distance between two points. Minimizing this sum can be realized by setting the derivative with respect to c_{ij}, t_x, t_y to zero and solving the resulting equations which can be done in linear time.

4.2 Linear Interpolation of the Affine Matrix

Once we find the affine matrix of transformations, M, it is straight-forward to perform a linear interpolation in order to obtain the sequence of matrices throughout the morph in the rigid motion stage: $(1 - t) \times I + t \times M$, where I is the identity matrix, gives us a natural interpolation throughout time. Once again, the linear interpolation can lead to degeneracies, such as the collapse of the drawing to a single point [23]. Consider, for example, a square as D_s and

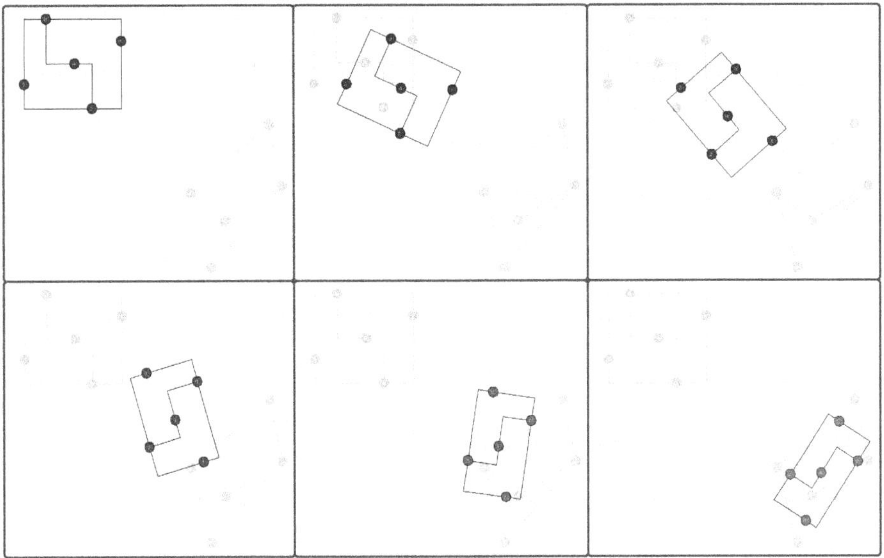

Fig. 6. D_s, at the top-left corner is aligned to D_t, at the bottom-right corner using rigid motion. The transformations include translation, rotation, scaling and shearing. Both D_s and D_t are opaquely drawn in all images.

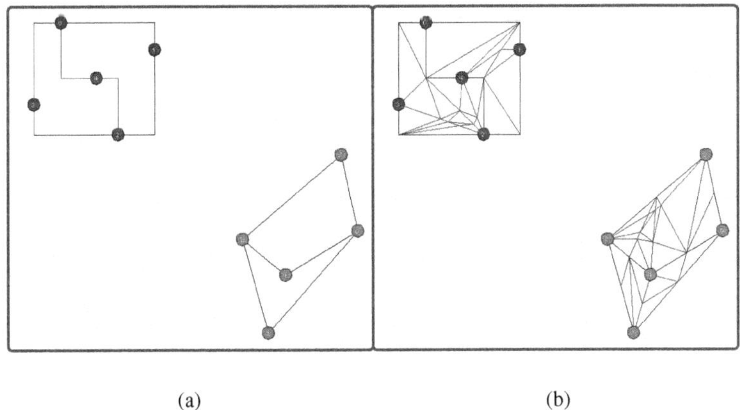

(a) (b)

Fig. 7. Part (a) shows D_s and D_t and part (b) shows the compatible triangulation.

the same square rotated 180° around the center as D_t. If we perform the linear interpolation from the identity matrix to the rotation matrix the square collapses into a point in the center. Fortunately, rotation is the only rigid transformation that is distorted by matrix interpolation and we can extract the rotation from a given affine matrix of transformations in constant time [23]. Once rotation is extracted from the affine matrix M, the linear interpolation of M does not introduce degeneracies, and the rotation can be applied separately by a linear interpolation of the rotation angle.

When aligning the two drawings using rigid motion we only use the vertices in D_s and D_t. Fig. 6 shows the snapshots when rigid motion is applied to D_s (an orthogonal drawing) to align it with D_t (a straight-line drawing). It is also possible to first introduce all bend vertices and then align D_s and D_t, using all vertices (including the bend vertices). This will correspond to swapping steps 1 and 2 of our algorithm; see Fig. 4. The main reason for aligning the drawings based only on the original graph vertices is that the placement of the bend vertices can be quite arbitrary, as long as the bends are added in the right order along the original graph edges.

5 Compatible Triangulation of the Faces

After introducing all the bend vertices in both D_s and D_t we proceed to the third stage of the algorithm and compatibly triangulate all matching pairs of faces in D_s and D_t. Once we have the embedding of the drawing D_s, i.e., the clockwise order of the edges around each vertex in D_s, it is easy to identify the faces. We make each edge bi-directed and traverse through the directed edges each time following a neighboring edge in the clockwise order. This traversal continues until all the edges are traversed in which case we have all the faces identified. A face in D_s and D_t is a simple polygon if the graph is biconnected, or possibly a polygonal subdivision. We first consider the simple polygon case and then the polygonal subdivision case.

Given two corresponding polygons P_1 and P_2, the goal then is to compatibly triangulate the two polygons, i.e. triangulate them in such a way that the resulting triangulations are isomorphic. In general, it is not always possible to compatibly triangulate two simple polygons. However, if we allow the introduction of extra vertices (Steiner points) then we can always find a compatible triangulation, using $O(k^2)$ Steiner points, where k is the number of vertices in each polygon. We use the algorithm of [1] to construct compatible triangulations. First we independently triangulate P_1 and P_2 in $O(k)$ time. Then we overlay the two triangulations on a newly created convex polygon P with k vertices. This overlay introduces intersection points between the triangulation edges of P_1 and those of P_2. These intersections are the Steiner points and it is easy to see that there are at most $O(k^2)$ of them, since every triangulation edge of P_1 can cross at most $O(k)$ triangulation edges of P_2. The overlay of the two triangulations can create faces with more than 3 edges. Fortunately, all these faces are convex and can be easily triangulated by selecting a vertex and adding all the needed chordal edges. The resulting full triangulation is a compatible triangulation of both P_1 and P_2. Fig. 7 shows the compatible triangulations of the D_s (orthogonal edges) and D_t (straight-line edges). If the graph underlying D_s and D_t is biconnected, then the above approach for compatible triangulation of polygons can be applied to all matching pairs of faces. If the underlying graph is not bi-

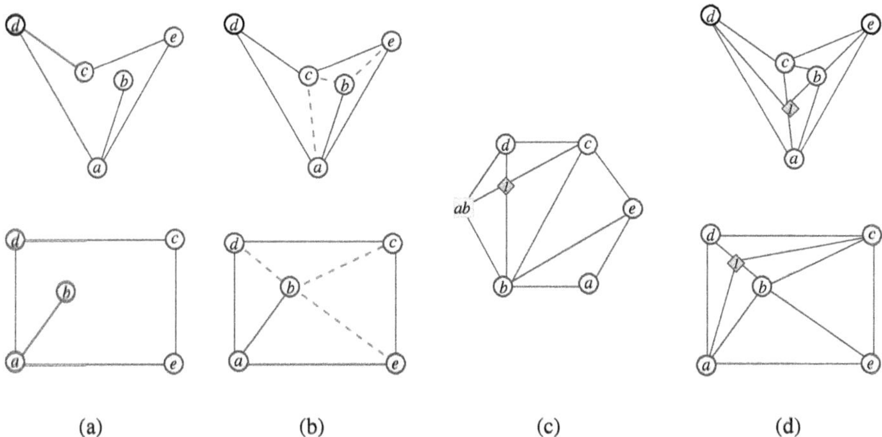

Fig. 8. Dealing with non-simple polygons; (a) two non-simple faces, P_1 (top) and P_2 (bottom); (b) independent triangulation of P_1 and P_2; (c) overlay of triangulations on P. Note that triangulation edge (a, c) from P_1 is replaced with (ab, c) in the overlay; (d) the compatible triangulation of P_1 and P_2.

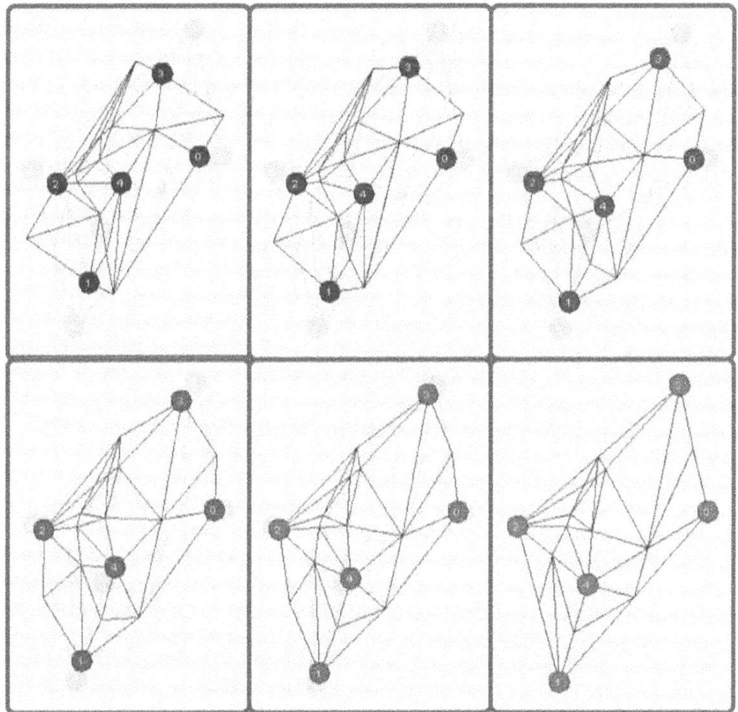

Fig. 9. Convex representation morph. D_t is opaquely drawn in all images.

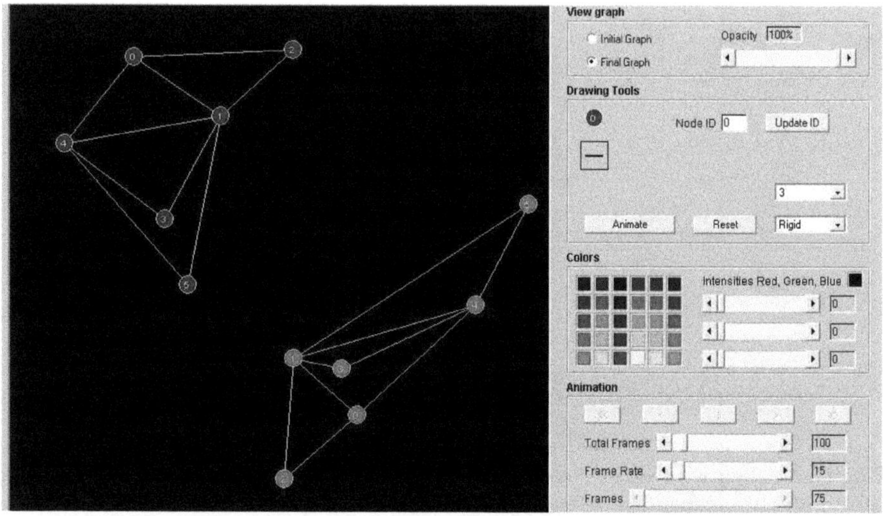

Fig. 10. A snapshot of the morphing system interface.

connected, additional complications arise. In this case the polygonal subdivision P, constructed from a particular face in the graph, may not be simple, as some vertex may be repeated; see Fig. 8. This problem can be overcome as follows. The triangulations of P_1 and P_2 are obtained independently as before. However, we might need to add extra vertices to P, corresponding to each repeated vertex, before overlaying the triangulations on P. Special care must be taken while overlaying the triangulation edges connected to such vertices. Let a be such a vertex, as in Fig. 8. Each repetition must be the result of an edge (a, b) that is traversed in both directions while constructing the face. Denote each such repeated vertex with the corresponding edge, i.e. ab. Let (a, c) be a triangulation edge in P_1 (or P_2) that follows (a, b) in the counter clockwise order. The triangulation edge (a, c) in the top drawing of Fig. 8(b) is such an edge. While overlaying the triangulations in P we create an edge between vertices c and ab, rather than c and a, see Fig. 8(c). We then overlay the resulting triangulation on P_1 and P_2 as before.

6 Computing Trajectories Using Convex Representations

In 1963 Tutte proposed the following *barycentric mapping* to generate straight line drawing of a 3-connected planar graph G: Given an embedding of G, we map the outer face of G onto a convex polygon. Then the locations of interior vertices are determined by their barycentric coordinates:

$$u_i = \sum_{j \in N(i)} \lambda_{ij} \times u_j, \qquad \sum_{j \in N(i)} \lambda_{ij} = 1,$$

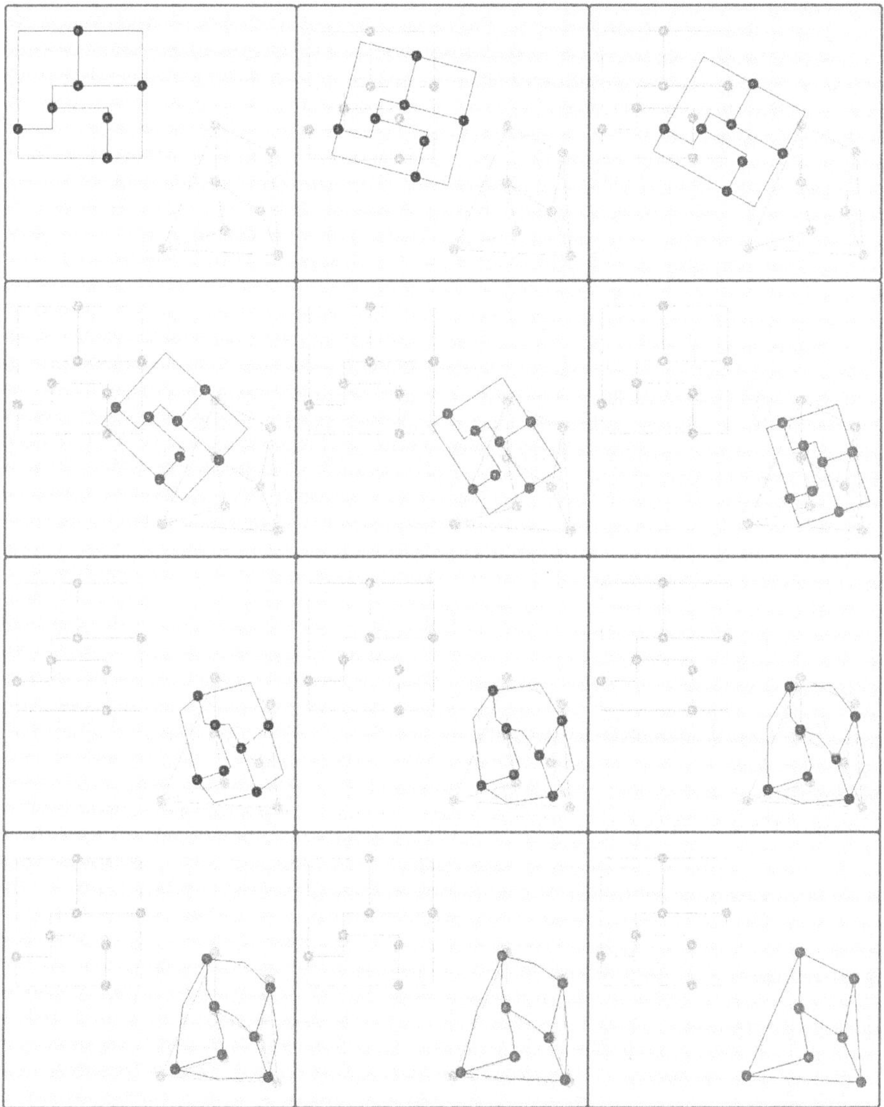

Fig. 11. A complete morph from D_s at the top-left corner, to D_t at the bottom-right.

where λ_{ij} is called a barycentric coordinate of u_i with respect to u_j and $N(i)$ is the set of neighbors of u_i. In Tutte's mapping, $\lambda_{ij} = 1/d_i$, where d_i is the degree of u_i.

Floater and Gotsman [7] applied the idea to morphing compatible triangulations. The basic idea is to obtain a barycentric representation of source/target triangulations as $n \times n$ matrices (λ_{ij} is the entry at row i and column j), call them M_s and M_t respectively, and apply a linear interpolation from M_s to M_t, $(1-t) \times M_s + t \times M_t$. Since throughout the interpolation each resulting matrix is a barycentric representation the sequence of graphs obtained from these matrices are all planar and the morphing is intersection-free [7].

In order to find proper λ_{ij} values that depend continuously and smoothly on the neighbors of u_i we use the mean value coordinates described in [6]:

$$\lambda_{ij} = \frac{w_{ij}}{\sum_{j \in N(i)} w_{ij}}, \qquad w_{ij} = \frac{tan(\alpha_{ij-1}/2) + tan(\alpha_{ij}/2)}{dist(v_i, v_j)},$$

where α_{ij} is the angle between the segments $v_j v_i$ and $v_i v_{j+1}$. Fig. 9 shows the morph to D_t after computing the trajectory using convex representation.

Note that this approach assumes D_s and D_t share the same outer face, i.e. the outer face vertices are located at exactly the same locations. In our general setting for planar graphs this usually will not be the case. To handle this problem we embed D_s and D_t inside the same triangle T. It remains to connect an outer face vertex with a vertex of T in D_s and D_t. A simple way to do this is to pick a vertex v_s in D_s that is visible from one of the triangle vertices, t_i. Connect v_s and t_i with a straight-line segment. Find a vertex v_t in D_t that is visible from t_i. Create a path from v_s to v_t in D_t following the outer face edges and connect the path to t_i.

7 System Implementation

We have implemented our morphing algorithm using Java; see Fig. 10 for a snapshot of the system. An applet for this implementation and graph morphing movies can be found at, `http://gmorph.cs.arizona.edu`. Fig. 11 shows the complete morphing sequence of two different drawings of the same graph.

References

1. B. Aronov, R. Seidel, and D. Souvaine. On compatible triangulations of simple polygons. *CGTA: Computational Geometry: Theory and Applications*, 3:27–35, 1993.
2. T. Beier and S. Neely. Feature-based image metamorphosis. In E.E. Catmull, editor, *Computer Graphics (SIGGRAPH '92 Proceedings)*, volume 27, pages 35–42, July 1992.
3. S. Bespamyatnikh. An optimal morphing between polylines. *IJCGA: International Journal of Computational Geometry and Applications*, 12(3):217–228, 2002.
4. S. S. Cairns. Deformations of plane rectilinear complexes. *American Math. Monthly*, 51:247–252, 1944.
5. G. Di Battista, P. Eades, R. Tamassia, and I. G. Tollis. *Graph Drawing: Algorithms for the Visualization of Graphs.* Prentice Hall, Englewood Cliffs, NJ, 1999.
6. M. Floater. Mean value coordinates. *Comp. Aided Geom. Design*, 20:19–27, 2003.
7. M. Floater and C. Gotsman. How to morph tilings injectively. *Journal of Computational and Applied Mathematics*, 101:117–129, 1999.

8. C. Friedrich. The ffgraph library. Technical Report 9520, Universität Passau, 1995.
9. C. Friedrich and P. Eades. The Marey graph animation tool demo. In *Proceedings of the 8th Symposium on Graph Drawing (GD)*, pages 396–406, 2000.
10. C. Friedrich and P. Eades. Graph drawing in motion. *Journal of Graph Algorithms and Applications*, 6(3):353–370, 2002.
11. C. Friedrich and M. E. Houle. Graph drawing in motion II. In *Proceedings of the 9th Symposium on Graph Drawing (GD)*, pages 220–231, 2001.
12. E. Goldstein and C. Gotsman. Polygon morphing using a multiresolution representation. In *Graphics Interface '95*, pages 247–254, 1995.
13. J. Gomes, L. Darsa, B. Costa, and D. M. Vello. *Warping and Morphing of Graphical Objects*. Morgan Kaufmann, 1999.
14. C. Gotsman and V. Surazhsky. Guaranteed intersection-free polygon morphing. *Computers and Graphics*, 25(1):67–75, Feb. 2001.
15. T. He, S. Wang, and A. Kaufman. Wavelet-based volume morphing. In *Proceedings of the Conference on Visualization*, pages 85–92, 1994.
16. M. L. Huang and P. Eades. A fully animated interactive system for clustering and navigating huge graphs. In *Proceedings of the 6th Symposium on Graph Drawing (GD)*, pages 374–383, 1998.
17. J. F. Hughes. Scheduled Fourier volume morphing. *Computer Graphics*, 26(2):43–46, July 1992.
18. M. Kaufmann and D. Wagner. *Drawing graphs: methods and models*, volume 2025 of *Lecture Notes in Computer Science*. Springer-Verlag, 2001.
19. T. Samoilov and G. Elber. Self-intersection elimination in metamorphosis of two-dimensional curves. *The Visual Computer*, 14:415–428, 1998.
20. T. W. Sederberg, P. Gao, G. Wang, and H. Mu. 2D shape blending: An intrinsic solution to the vertex path problem. In *Computer Graphics (SIGGRAPH '93 Proceedings)*, volume 27, pages 15–18, 1993.
21. T. W. Sederberg and E. Greenwood. A physically based approach to 2-D shape blending. In E. E. Catmull, editor, *Computer Graphics (SIGGRAPH '92 Proceedings)*, volume 26, pages 25–34, July 1992.
22. M. Shapira and A. Rappoport. Shape blending using the star-skeleton representation. *IEEE Computer Graphics and Applications*, 15(2):44–50, Mar. 1995.
23. K. Shoemake and T. Duff. Matrix animation and polar decomposition. In *Proceedings of Graphics Interface '92*, pages 258–264, May 1992.
24. V. Surazhsky and C. Gotsman. Controllable morphing of compatible planar triangulations. *ACM Transactions on Graphics*, 20(4):203–231, Oct. 2001.
25. A. Tal and G. Elber. Image morphing with feature preserving texture. *Computer Graphics Forum*, 18(3):339–348, Sept. 1999. ISSN 1067-7055.
26. C. Thomassen. Deformations of plane graphs. *J. Combin. Theory Ser. B*, 34:244–257, 1983.
27. W. T. Tutte. How to draw a graph. *Proc. London Math. Society*, 13(52):743–768, 1963.

Fixed Parameter Algorithms for ONE-SIDED CROSSING MINIMIZATION Revisited

Vida Dujmović[1], Henning Fernau[2,3], and Michael Kaufmann[2]

[1] McGill University, School of Computer Science,
3480 University St., Montreal, QC H3A 2A7, Canada,
vida@cs.mcgill.ca

[2] Universität Tübingen, WSI für Informatik, Sand 13,
72076 Tübingen, Germany
fernau / mk@informatik.uni-tuebingen.de

[3] The University of Newcastle, School of Electr. and Computer Science,
University Drive, Callaghan, NSW 2308, Australia
fernau@newcastle.edu.au

Abstract. We exhibit a small problem kernel for the problem ONE-SIDED CROSSING MINIMIZATION which plays an important role in graph drawing algorithms based on the Sugiyama layering approach. Moreover, we improve on the search tree algorithm developed in [5] and derive an $\mathcal{O}(1.4656^k + kn^2)$ algorithm for this problem, where k upperbounds the number of tolerated crossings of straight lines involved in the drawings of an n-vertex graph. Relations of this graph-drawing problem to the algebraic problem of finding a weighted linear extension of an ordering similar to [7] are exhibited.

1 Introduction and Problem Definition

We consider the following problem k-ONE-SIDED CROSSING MINIMIZATION, called k-OSCM for short in this paper: Given: A bipartite graph $G = (V_1, V_2, E)$ and a linear order $<_1$ on V_1. Parameter: k Question: Is there a linear order $<$ on V_2 such that, when the vertices from V_1 are placed on a line (also called layer) L_1 in the order induced by $<_1$ and the vertices from V_2 are placed on a second layer L_2 (parallel to L_1) in the order induced by $<$, then drawing straight lines for each edge in E will introduce no more than k edge crossings.

We denote by OSCM the optimization version (that is, non-parameterized version) of this problem. (k-)OSCM is the key procedure in the well-known layout framework for layered drawings commonly known as the Sugiyama algorithm. After the first phase (the assignment of the vertices to layers), the order of the vertices within the layers has to be fixed such that the number of the corresponding crossings of the edges between two adjacent layers is minimized. Finally, the concrete position of the vertices within the layers is determined according the previously computed order.

The crossing minimization step, although most essential in the Sugiyama approach, is an NP-complete problem. The most commonly used method is the

G. Liotta (Ed.): GD 2003, LNCS 2912, pp. 332–344, 2004.

layer-by-layer sweep where, starting from $i = 1$, the order for L_i is fixed and the order for L_{i+1} that minimizes the number of crossings amongst the edges between layers L_i and L_{i+1} is determined.

After increasing index i to the maximum layer index, we turn around and repeat this process from the back with decreasing indices. In each step, an OSCM problem has to be solved. Unfortunately, this seemingly elementary problem is NP-complete [6], even for sparse graphs [8]. Jünger and Mutzel [7] transformed the OSCM problem to a linear ordering problem which they solve with the branch and cut method.

This elementary graph drawing problem attracted several researchers from the area of fixed-parameter tractable (FPT) algorithms [2]. The first approaches to the more general variant of this problem have been published by Dujmović *et al.* in [3] and [4]. The last one has been greatly improved by Dujmović and Whitesides [5] who achieved an $\mathcal{O}(1.6182^k n^2)$ algorithm for this problem.

In this paper, we derive an $\mathcal{O}(1.4656^k + kn^2)$ algorithm, which significantly lowers the constants involved in the exponential search tree algorithm part.

Moreover, we exhibit a small problem kernel for this problem, which has not been done before.

More specifically, this means that, with the aid of some reduction rules, we can arrive at an instance of k-OSCM with the number of vertices involved being bounded by $3k(k + \frac{1}{2})$.

2 Basics

We start with some formalities. The number of vertices of a graph $G = (V, E)$ is denoted by $n = |V|$. For each vertex $v \in V$, let $N(v)$ denote the set of vertices adjacent to v and let $\deg(v) = |N(v)|$ denote the degree of v in G. The subgraph of G induced by the set of vertices $V' \subseteq V$ is denoted by $G[V']$. A bipartite graph $G = (V_1, V_2, E)$ together with linear orderings on V_1 and on V_2 is also called a *drawing* of G. This formulation implicitly assumes a drawing of G where the vertices of V_1 and V_2 are drawn on two (virtual) horizontal lines, the line L_1 corresponding to V_1 being above the line L_2 corresponding to V_2. If $u < v$ for two vertices on L_1 or on L_2, we will also say that u is *to the left* of v. A linear order on V_2 that minimizes the number of crossings subject to the fixed linear order of V_1 is called an *optimal ordering* and the corresponding drawing of G is called an *optimal drawing*. Since the positions of isolated vertices in any drawing are irrelevant, we disregard isolated vertices of the input graph G in what follows.

If $|V_2| = 2$, then there are only two different drawings. This gives us the useful notion of a *crossing number*. Given a bipartite graph $G = (V_1, V_2, E)$ with $|V_2| > 1$, for any two distinct vertices $a, b \in V_2$, define c_{ab} to be the number of crossings in the drawing of $G[\{a, b\} \cup (N(a) \cup N(b))]$ when $a < b$ is assumed. Furthermore, for any $a \in V_2$ with $\deg(v) > 0$, let l_a be the leftmost neighbour of a on L_1, and r_a be the rightmost neighbour of a. We call two vertices $a, b \in V_2$ *interfering* or *unsuited* if there exists some $x \in N(a)$ with $l_b < x < r_b$, or there exists some $x \in N(b)$ with $l_a < x < r_a$. Otherwise, they are called *suited*.

Observe that, for $\{a, b\}$ suited, $c_{ab} \cdot c_{ba} = 0$. Dujmović and Whitesides have shown that, in any optimal ordering $<$ of the vertices of V_2, we find $a < b$ if $r_a \leq l_b$.

This means that all suited pairs appear in their *natural ordering.*

In the next section we will need the following general notation. A directed graph (digraph) obtained from an undirected graph G by replacing every edge $\{u, v\}$ by the two arcs (u, v) and (v, u) is denoted by $D(G)$. The undirected graph obtained from a digraph G by putting an edge $\{u, v\}$ whenever there is an arc (u, v) in G is denoted by $U(G)$. Finally, the graph complement of G is denoted by G^c.

It is important to note that we only deal with simple (di-)graphs, i.e., graphs having no (self-)loops or multiple edges. The set of arcs of a digraph G is denoted by $A(G)$.

3 Getting a Small Kernel

In this section, we give a drastic reduction of the complexity of k-OSCM. Recall that a partial order is an irreflexive, asymmetric and transitive relation. A partially ordered set (or poset), denoted by $P = (V, A)$, is a set V taken together with a partial order A on it. For two distinct elements $a, b \in V$, if either $a < b$ or $b < a$ in A, then the pair $\{a, b\}$ is *comparable* in P; else the pair is *incomparable.* A partial order is linear if every pair of elements of V is comparable. If $A \subseteq V \times V$ is a partial order on V, then we call any partial order $L \supseteq A$ a *completion of* A. More algebraically speaking, it can also be called a (weighted) *linear extension.* As is well known, every poset $P(V, A)$ can be represented as a directed acyclic digraph with vertex set V and arc relation A which is a partial order. For simplicity, we will equate posets and directed acyclic graphs in the following, so we refer to posets as *ordered* or *linear digraphs.*

Consider the following problem k-WEIGHTED COMPLETION OF AN ORDERING (k-WCO):

Given: An ordered digraph $P = (V, A)$ and a cost function c mapping $A(D([U(P)]^c))$ into the nonnegative integers; by setting c to zero for arcs in $A(D(U(P)))$, we can interpret the domain of c as $V(P) \times V(P)$. Parameter: k
Question: Is there a selection A' of arcs from $A(D([U(P)]^c))$ such that the transitive closure $(A' \cup A(P))^+$ is a linear order and $\sum\{c(a) \mid a \in (A' \cup A(P))^+\} \leq k$?

It is quite obvious that OSCM can be solved with the help of WCO; the linear order which is aimed at is the permutation of the vertices on the second layer which minimizes the number of crossings involved, so that the crossing number c_{ab} is the cost of the arc (a, b) in the digraph model. Since it is easy to see that WCO is nondeterministically solvable in polynomial time, and since Eades and Wormald [6] have shown that OSCM is NP-hard, we can immediately deduce:

Lemma 1. *WCO is NP-complete.*

Jünger and Mutzel [7] gave an ILP-formulation of the OSCM problem as a linear ordering problem which they could solve with a branch and cut method.

In contrast, our algorithm for OSCM (both the kernelization and the search tree part) can be seen as growing larger and larger parts of the linear order of the vertices of the second layer we are aiming at. When settling the ordering between a and b, we also say that we are *committing* $a < b$ or $b < a$.

Dujmović and Whitesides have shown that, for each instance of OSCM, suited pairs appear in their natural ordering in any optimal drawing. That justifies the following first reduction rule: RR1: For every suited pair of vertices $\{a, b\}$ from V_2 with $c_{ba} > 0$, commit $a < b$.

Furthermore, consider a set $S \subseteq V_2$ such that all the vertices in S are adjacent to the exact same set of neighbours in V_1. It is simple to observe that arbitrarily permuting the vertices of S in any drawing cannot affect the total number of crossings. This observation justifies the following second reduction rule:
RR2: For every pair of vertices $\{a, b\}$ from V_2 with $N(a) = N(b)$, (arbitrarily) commit $a < b$.

Notice that if $c_{ab} = c_{ba} = 0$ then (disregarding isolated vertices) $\deg(a) = \deg(b) = 1$ and $N(a) = N(b)$. Therefore, after applying these two reduction rules to an instance of k-OSCM, we obtain the ordered digraph $P = (V_2, A)$, where a pair of vertices is incomparable in P only if both of the crossing numbers of that pair are nonzero. Hence, we can assume that c is mapping each arc of $A(D([U(P)]^c))$ onto a positive integer. We call the corresponding specialized problem k-POSITIVE COMPLETION OF AN ORDERING (k-PWCO).

Our reasoning up to now shows that this special case is also NP-hard. Exhaustively applying the following two reduction rules shows that k-PWCO has a small linear-size problem kernel.

RRLO1:

If v is a vertex in ordered digraph $P = (V, A)$ of maximal degree, that is, the sum of the indegrees and outdegrees of v equals $|V| - 1$, then remove v and consider the instance $(P[V - v], k)$.

To see the soundness of this rule, observe that both P and $P[V - v]$ have transitive arc relations. After having applied RRLO1 exhaustively, there is, for each vertex $v \in V$, another vertex $v' \in V(P)$ such that the pair $\{v, v'\}$ is incomparable in P. By the positivity assumption, committing $v < v'$ or $v' < v$ will cost at least one "unit" and will, furthermore, at best maximize the degree of v and v'. By an inductive argument, it follows that the ordered digraph P in a RRLO1-reduced instance of k-PWCO cannot have more than $2k$ vertices.

In an ordered digraph $P = (V, A)$, call two incomparable vertices $u, v \in V$

- *independent with respect to* w if $\{u, w\}$ and $\{v, w\}$ are comparable in P;
- *dependent with respect to* w if $\{u, w\}$ or $\{v, w\}$ are incomparable in P;
- *independent* (*in* P) if $\{u, v\}$ are independent with respect to all $w \in V \setminus \{u, v\}$;
- *dependent* (*in* P) if $\{u, v\}$ are dependent with respect to some $w \in V \setminus \{u, v\}$;
- *transitive with respect to* w if either $\{u, w\}$ or $\{v, w\}$ is comparable but not both.

Being transitive with respect to w means that committing $v < u$ or $u < v$ commits by transitivity $v < w$, $w < v$, $u < w$ or $w < u$.

RRLO2: If the vertices u and v are independent in a given ordered digraph $P = (V, A)$ and if $c((u, v)) < c((v, u))$, then reduce the problem instance to $((V, A \cup \{(u, v)\}), k - c((u, v))$.

The soundness of RRLO2 is obvious. In fact, this rule is generalizable:

RRLOq for any fixed $q > 1$: For each connected component $C \subseteq V$ of $[U(P)]^c$ with $|C| \leq q$, solve PWCO optimally on $P[C]$.

The reduced instance will see the orderings between all pairs of vertices from C settled, and the parameter will be reduced accordingly. Note that this implies that all the vertices of C are isolated in the reduced $[U(P)]^c$ and can thus be deleted by the RRLO1. This new rule yields a kernel of size $k\frac{q+1}{q}$, since after exhaustive application of this rule and RRLO1, each connected component of $[U(P)]^c$ has at least $(q+1)$ vertices and thus at least q edges. Correspondingly, at least q arcs with a weight of at least one per arc have to be added per component, therefore there are at most k/q components.

In other words, a kernel size of k can be "approximated" with arbitrary precision. This type of behaviour has not been found in other parameterized algorithms yet. However, this approximation comes at a price: Solving this problem on graphs with q vertices can only be done in exponential time (assuming P$\neq$NP). Still, arbitrary constants q or $q = \log k$ are possible choices for polynomial-time kernelization algorithms. It is therefore desirable to find easy optimality criteria for small values of q (as $q = 2$).

The soundness of the reduction rule RRLOq is readily seen.Assume that C is a connected component of $[U(P)]^c$. Consider a vertex $x \notin C$. Since C is a connected component in $[U(P)]^c$, x is comparable with each $y \in C$ (otherwise, y and x would be connected in the complement graph). Hence, C can be portioned into $C_\ell = \{y \in C \mid y < x\}$ and $C_r = \{y \in C \mid x < y\}$. Since $C_\ell < x < C_r$, either $C_\ell = \emptyset$ or $C_r = \emptyset$; otherwise, there would be no connection between vertices of C_ℓ and vertices of C_r in the complement graph. Hence, x will never be ordered "in-between" two vertices from C. Therefore, we can conclude:

Theorem 1. *Fix some $1 < \alpha \leq 1.5$. Then, each instance of k-PWCO admits a problem kernel of size αk.*

Considering k-OSCM in a slightly more general fashion, namely, allowing some pairs of V_2 to be already committed as part of the input, we can conclude:

Corollary 1. *Fix some $1 < \alpha \leq 1.5$. Then, each instance of k-OSCM can be reduced to an instance $(P = (V_2, A), k)$ of k-PWCO, with $|V_2| \leq \alpha k$.*

Proof. Applying RRLO1 and RRLO2 (and RRLOq) to k-OSCM exhaustively results in a partial order on V_2 where the crossing numbers of the incomparable pairs are positive. Therefore, this partial order on V_2, together with the positive crossing numbers of incomparable pairs comprises an instance of k-PWCO, which by Theorem 1 has the kernel of size $|V_2| \leq \alpha k$. □

This has not yet given us a problem kernel for the original problem, since $|V_1|$ (and hence $|E|$) has not yet been bounded by a function of k.

With that aim, consider the following simple reduction rule

RRlarge: If $c_{ab} > k$ then do: if $c_{ba} \leq k$ then commit $b < a$ else return NO.

This is clearly a sound rule. Moreover notice, that if a vertex $a \in V_2$ has $\deg(a) > 2k+1$, then for every vertex $b \in V_2$, $c_{ab} > k$ or $c_{ba} > k$ are true. Therefore, after performing RRlarge exhaustively in combination with RRLO1, all the vertices of V_2 of degree larger than $2k+1$ will have been removed from the OSCM-instance. This yields:

Theorem 2. *For some* $1 < \alpha \leq 1.5$, *k-OSCM admits a problem kernel* $G = (V_1, V_2, E)$ *with* $|V_1| \leq (2k+1)|V_2| (\leq 3k(k+\frac{1}{2}))$, $|V_2| \leq \alpha k$, *and* $|E| \leq (2k+1)|V_2|$.

The following algorithm summarizes how to obtain a problem kernel:

Algorithm 1 (Kernelization for k-OSCM).

Compute the crossing numbers c_{ab} for all pairs (a,b).
If $k < \sum_{a,b \in V_2} \min(c_{ab}, c_{ba})$ then answer NO.
If $k \geq \sum_{a,b \in V_2} \max(c_{ab}, c_{ba})$ then YES; output an arbitrary linear order.
Apply reduction rules RR1, RR2 (and RR3).
Apply reduction rules RRLO1, RRLO2, and RRlarge exhaustively.

The rule RR3 is derived in Section 6, as it is needed by the search tree part of our algorithm. For now, while considering kernelization only, we can disregard this reduction rule. Since computing the crossing numbers is the predominant computational part (taking time $\mathcal{O}(kn^2)$ even when combined with RRlarge according to Dujmović and Whitesides), we conclude that the kernelization preprocessing takes time $\mathcal{O}(kn^2)$. As a result of this kernelization, the vertices of V_2 are partially ordered. Our search tree algorithm presented in the next section can cope with such "pre-set" instances of k-OSCM. Therefore, we will obtain running time of the form $\mathcal{O}(f(k) + kn^2)$ for the k-OSCM problem.

4 A General Overview of the Search Tree Algorithm

Our search tree for the OSCM problem slightly deviates from the one constructed by Dujmović and Whitesides in their paper. Hence, we give some details in what follows. Consider two distinct vertices $a, b \in V_2$ with $c_{ab} = i$ and $c_{ba} = j$. Then, the pair $\{a, b\}$ is also called an *i/j pattern*.

As is the standard practice when developing search tree based FPT algorithms, each node of a search tree is associated with a problem instance. In the case of k-OSCM, we let an instance consist of an ordered digraph $P = (V_2, A)$ and an integer k'. P contains the arcs corresponding to all the pairs of vertices of V_2 committed thus far, and k' gives the remaining number of allowed edge crossings, that is, $k' = k - \sum_{(v,u) \in A} c_{vu}$. Initially (in the root node) the arcs of P are determined by the kernelization. In each node of the search tree, if k' is negative, then the node returns NO, otherwise we look for a pair of dependent vertices $\{a, b\}$ that form an i/j pattern such that either $i+j \geq 4$, or $i = 2$, $j = 1$ and $\{a, b\}$ is transitive. If such a pair $\{a, b\}$ is found we branch the problem instance into two new problems instances. In one we commit $a < b$ and in the other we commit $b > a$; consequently, in each problem instance we update the ordered digraph P and lower the parameter k' accordingly. (By adding (v, w) to A and then computing transitive closure $(A \cup (v, w))^+$ we get the updated ordered digraph $P = (V_2, (A \cup (v, w))^+)$.) If no pair as described above is found, then we commit all the remaining incomparable pairs in P deterministically. The details and the correctness of this approach will be discussed in Section 7.

The running time of FPT algorithms is dominated by the part exponential in parameter k. In our case, that part is bounded by the size of the search tree.

We now analyse that size. Firstly, observe that each internal node of our search tree has two branches. If one branch lowers the parameter k' by b_1, and the other by b_2, we denote the corresponding *branching vector* by (b_1, b_2). Since all i/j patterns with $i = 0$ or $j = 0$ are already committed (by the reduction rules RR1 and RR2 during the kernelization), each internal node of the search tree has $b_1 > 0$ and $b_2 > 0$. Furthermore, each node that branches on a transitive 2/1 pattern commits by transitivity an extra i/j pattern where $i, j > 0$. Thus each internal node has $b_1 + b_2 \geq 4$. Therefore, in the worst case, each node of our search tree has a branching vector $(2, 2)$ or $(3, 1)$. The size $s(k)$ of the search tree obeying these branching vectors can be deduced as follows:

- The recurrence corresponding to the $(2, 2)$ branching vector is $s(k) = 2s(k-2) + \mathcal{O}(1)$. Solving this recurrence gives $s(k) < 1.4143^k$.
- The recurrence corresponding to the $(1, 3)$ branching vector is $s(k) = s(k-1) + s(k-3) + \mathcal{O}(1)$. Solving this recurrence gives $s(k) < 1.4656^k$.

Thus in the worst case, the size of our search tree is less than 1.4656^k. This is in contrast to the search tree with a $(2, 1)$ branching vector derived by Dujmović and Whitesides that gives $s(k) < 1.6181^k$.

In a node of our search tree that does not branch (that is, in a leaf), all the incomparable pairs form 1/1 and 2/1 patterns. To be able to commit these pairs deterministically, we study the structural properties of such patterns in the next section. In Section 7, we finally give a detailed algorithm for the k-OSCM problem and the proof of its correctness.

5 Some Structural Properties of 2/1 and 1/1 Patterns

We start this section with the following known and useful equality [1, Chapter 9]:

$$c_{ab} + c_{ba} = \deg(a)\deg(b) - |N(a) \cap N(b)|. \tag{1}$$

This implies that if $\deg(b) \leq \deg(a)$ then

$$(\deg(a) - 1)\deg(b) \leq c_{ab} + c_{ba} \leq \deg(a)\deg(b). \tag{2}$$

We now study 1/1 and 2/1 patterns. As a convention, we will label vertices from the first layer by (decorated) letters x, y, z and vertices from the second layer by letters a, b, c. For clarity, we will draw neighbours of a as non-filled circles and neighbours of b as filled-in circles.

Lemma 2. *In the case of 1/1 or 2/1 patterns, and if* $c_{ba} = 1$, *we must find the situation depicted in the figure below.*

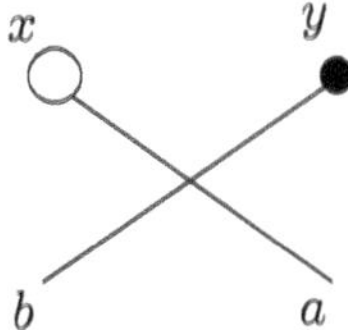

Moreover, each remaining neighbour of a *(if any) must be to the right of* y *(or* y *itself), while each remaining neighbour of* b *(if any) must be to the left of* x *(or* x *itself). (Otherwise,* $c_{ba} > 1$.*)*

Theorem 3. *If* $c_{ab} = c_{ba} = 1$, *then there are two basically different situations (that can be obtained by enhancing the situation sketched in Lemma 2):*

1. a *and* b *are each adjacent to* x *and* y *only. In other words,* $\deg(a) = \deg(b) = 2$ *and* $N(a) = N(b)$, *as illustrated in the figure below.*

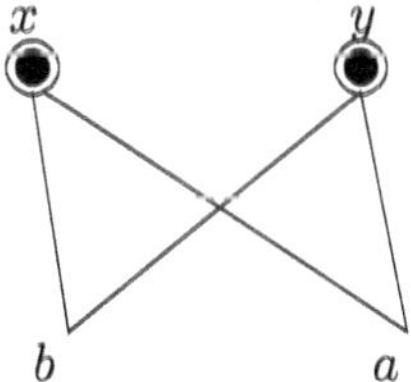

2. *Two subcases arise: (a) If* $\deg(b) = 1$, *then* a *has (besides* x*) another neighbour* x' *to the right of* y. *In addition to* x *and* x', a *may only be adjacent to* y. *(b) If* $\deg(a) = 1$, *then* b *has (besides* y*) another neighbour* y' *to the left of* x. *In addition to* y *and* y', b *may only be adjacent to* x. *Both situations are illustrated in the figure below.*

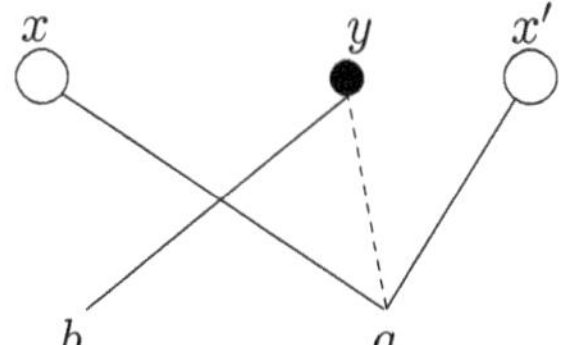

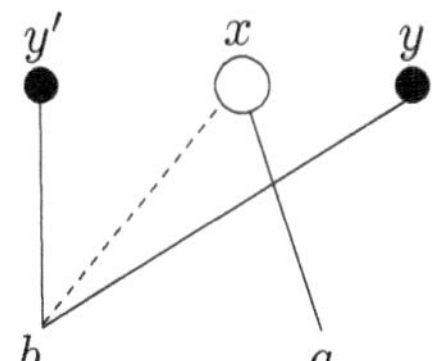

Proof. By inequality (2) either $\deg(a) = \deg(b) = 2$, or one of a, b has degree one. Consider first the case where $\deg(a) = \deg(b) = 2$.

Then by equality (1), $N(a) = N(b)$ as otherwise $c_{ab} + c_{ba} \geq 3$. Consider now the case where $\deg(b) = 1$. Then a has (besides x) exactly one neighbour x' strictly to the right of y as otherwise either the pair $\{a, b\}$ is suited or $c_{ab} > 1$.

Vertex a cannot have a neighbour strictly to the left of y as otherwise $c_{ab} > 1$. Thus in addition to x and x', a can only be adjacent to y. The remaining case where $\deg(a) = 1$ is symmetric.

□

Theorem 4. *Let the pair $\{a, b\}$ form a $1/2$ pattern, that is, $c_{ab} = 1$ and $c_{ba} = 2$. Then there are two basically different situations:*

1. *In the following figures, neither a nor b can have any other neighbours.*

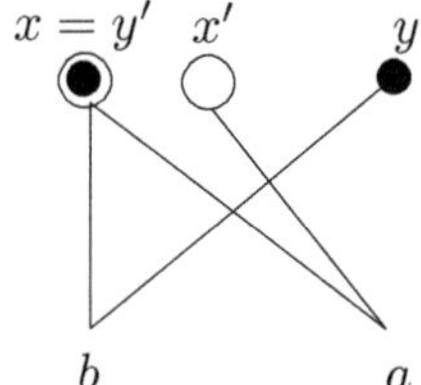

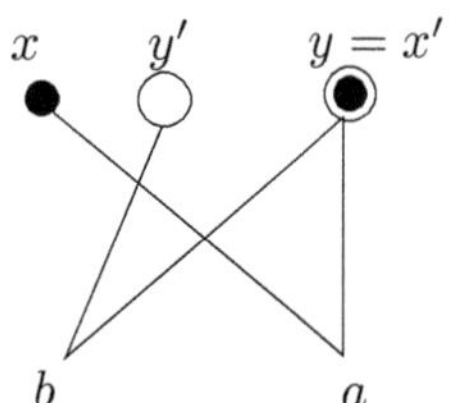

2. *In the following figure (on the left-hand side), in addition to x, x' and x'', vertex a may only be adjacent to y, while b has no other neighbours. The figure on the right-hand side can be symmetrically interpreted.*

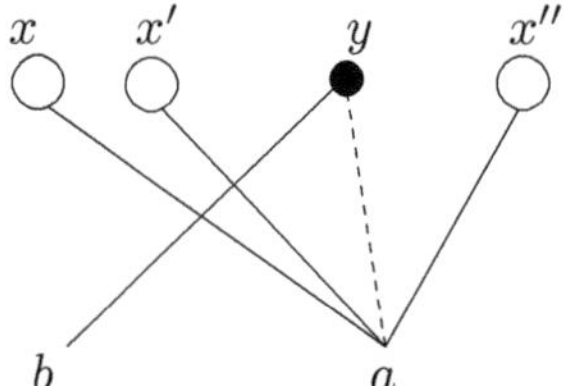

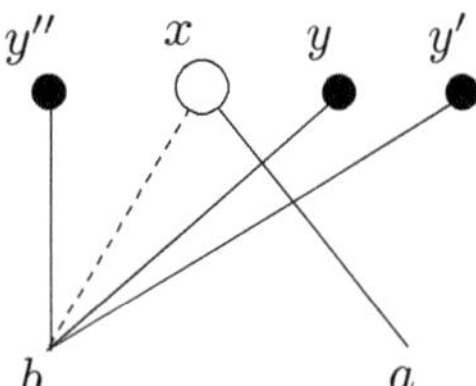

Proof. By inequality (2) either $\deg(a) = \deg(b) = 2$, or one of a, b has degree one. Consider first the case where $\deg(a) = \deg(b) = 2$ and let x' and y' denote the remaining neighbours of a and b, respectively. Having both $x' \neq y$ and $y' \neq x$ is not possible as otherwise by equality (1), $c_{ab} + c_{ba} = 4$. Having both $x' = y$ and $y' = x$ gives the first case of the previous theorem.

Therefore, if $x' = y$ then $x < y' < y$ as otherwise, $\{a, b\}$ is suited or $c_{ba} = 1$. On the other hand, if $y' = x$, then $x < x' < y$ as otherwise, $\{a, b\}$ is suited or $c_{ba} = 1$.

Consider now the case where $\deg(b) = 1$. The case where $\deg(a) = 1$ is symmetric. In addition to x, a has exactly one neighbour x' strictly to the left of y as otherwise, $c_{ba} \neq 2$. Vertex a has exactly one neighbour x'' strictly to the right of y as otherwise, either the pair $\{a, b\}$ is suited or $c_{ab} > 1$. Thus in addition to x, x' and x'', a can only be adjacent to y.

□

6 A Reduction Rule for a 2/1 Pattern

The reduction rule RR2 presented in Section 3 resolves the situation described in the first case of Theorem 3. Unfortunately, the second case of Theorem 3 does not admit a similar simple resolution. Let us now reconsider the 2/1 patterns identified in Theorem 4. The second kind of the pattern described in that theorem is in a sense similar to (but more complicated than) the second pattern of Theorem 3 and hence admits no deterministic solution.

Consider the first case in Theorem 4 with a having a neighbour distinct from x and y. Let I denote any drawing where $b < a$ and let II denote the drawing obtained from drawing I by swapping a and b. Let the total number of crossings in I and II be denoted by c_I and c_{II} respectively.

Consider the following figure and tables.

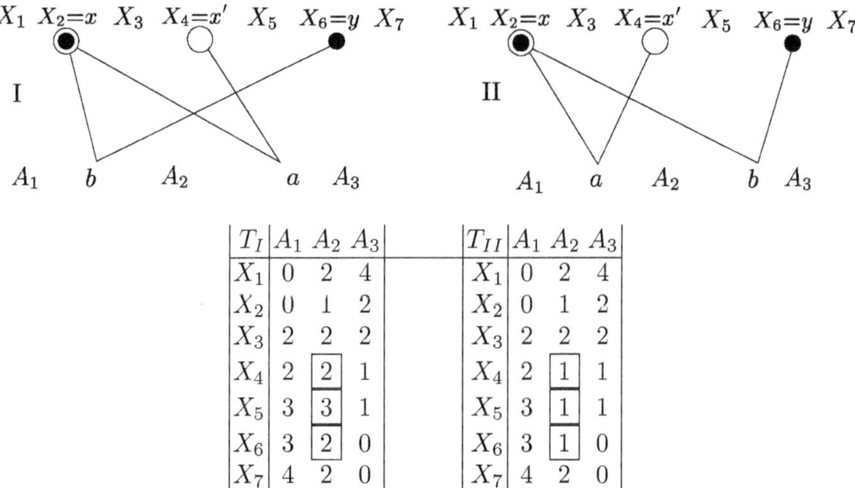

T_I	A_1	A_2	A_3
X_1	0	2	4
X_2	0	1	2
X_3	2	2	2
X_4	2	[2]	1
X_5	3	[3]	1
X_6	3	[2]	0
X_7	4	2	0

T_{II}	A_1	A_2	A_3
X_1	0	2	4
X_2	0	1	2
X_3	2	2	2
X_4	2	[1]	1
X_5	3	[1]	1
X_6	3	[1]	0
X_7	4	2	0

These tables read as follows: if there are m_{ij} edges connecting vertices from X_i with vertices from A_j, then there are $T_x(X_i, A_j)m_{ij}$ crossings between these m_{ij} edges and the edges shown in the above sketches for case $x \in \{I, II\}$. It is clear that the columns labelled A_1 and A_3 are identical in both tables: swapping a and b can only affect the relative order of vertices in A_2 compared to a and b. The boxed entries show the only differences between the tables. It follows that

$$c_I - c_{II} = c_{ba} - c_{ab} + m_{42} + 2m_{52} + m_{62} > 0,$$

since $m_{42} + 2m_{52} + m_{62} \geq 0$ and $c_{ba} - c_{ab} = 1$. This implies that whenever this situation from Theorem 4 arises, any optimal drawing shows $a < b$.

The first case in Theorem 4 where b has a neighbour distinct from x and y is symmetric; the same analysis applies. This justifies the next reduction rule:

RR3: In the situations described in the first case of Theorem 4, always let $a < b$.

Since in every optimal drawing all the pairs of this type appear in their "cheaper ordering", we apply this reduction rule at the very beginning of our algorithm, that is we add this rule to the kernelization as shown in Algor. 1.

Having this reduction rule is instrumental in bounding the size of the search tree, as will become clear from the two lemmas in the next section.

7 The Algorithm for the OSCM Problem

We are now ready to present the parameterized algorithm we suggest for OSCM:

Algorithm 2 (A parameterized algorithm for OSCM).

```
- kernelize
In a node of a search tree with instance (P,k') do

  0: if k' < 0 return NO.
1: Apply RRLO1, RRLO2 and RRlarge exhaustively.
2: if in P there is an incomparable i/j pattern {a,b} such that i+j ≥ 4
 branch on a < b and b < a; update P and k' accordingly
3: else if in P there is a dependent 2/1 pattern {a,b}
branch on a < b and b < a; update P and k' accordingly
4: else
//the remaining 2/1 patterns are independent; thus, RRLO2 applies
resolve all remaining 1/1 patterns arbitrarily
update P and k' accordingly; if k' < 0 return NO else YES.
```

The correctness of this algorithm is clear from the analysis presented in the previous sections. What remains is to analyse the branching vectors of the internal nodes in the search tree associated with the algorithm. As required by the analysis presented in Section 3, we now show that the branching vector (b_1, b_2) in each internal node has $b_1 + b_2 \geq 4$ and $b_1, b_2 > 0$.

Lemma 3 (Main Lemma). *Let $\{a, b\}$ be a pair of dependent vertices forming 2/1 pattern in step 3 of the algorithm. Then $\{a, b\}$ is a transitive pair.*

Proof. Since $\{a, b\}$ is dependent in P, there must be a vertex c such that $\{a, c\}$ or $\{b, c\}$ are incomparable in P. It suffices to show that one of these two pairs are comparable in P. In the step 3 of the algorithm, the only remaining incomparable pairs are 1/1 patterns of the second type in Theorem 3 and 2/1 patterns of the second type in Theorem 4. Therefore, let without loss of generality $\deg(a) = 3$ (or $\deg(a) = 4$) and $\deg(b) = 1$. If $\deg(c) \geq 2$ then by the inequality (2), $c_{ac} + c_{ca} \geq 4$ and thus the ordering of $\{a, c\}$ is settled either by RR1 or by the step 2 of the algorithm. Therefore, $\deg(c) = 1$. In that case, the pair $\{b, c\}$ forms a $0/i$ pattern and its ordering is settled by either RR1 or RR2. □

A direct consequence of this lemma is the following.

Lemma 4. *The branching vector (b_1, b_2) in each internal node of the search tree has $b_1 + b_2 \geq 4$ and $b_1, b_2 > 0$.*

Proof. Based on the reduction rules RR1 and RR2, in each node of a search tree all the incomparable (and dependent) patterns i/j have $i > 0$ and $j > 0$, thus in each node $b_1, b_2 > 0$. Furthermore, all the nodes branched in the step 2, have $i + j \geq 4$ and thus have $b_1 + b_2 \geq 4$. The remaining nodes are branched in step 3. Each such node branches on a dependent 2/1 pattern $\{a, b\}$. By the previous lemma, committing either $a < b$ or $b < a$ determines, without loss of generality, the ordering of a pair $\{a, c\}$. Having been incomparable at step 3 initially, the pair $\{a, c\}$ has $c_{ac}, c_{ca} \geq 1$. Therefore, we have $b_1 + b_2 \geq 4$ in this case, too. □

8 Conclusions

In this paper, we present a new search tree based parameterized algorithm for the k-OSCM problem. Due to possible rekernelizations [9], we may deduce:

Theorem 5. *The problem of k-OSCM can be solved in time* $\mathcal{O}(1.4656^k + kn^2)$.

It remains to be determined whether further progress is possible, especially in further lowering the base of the exponent in the running time of the search tree algorithm. Our present case analysis shows at two places a branching behaviour which matches the upper bound stated in the previous theorem, namely when branching at 3/1 and at 2/1 patterns . A detailed analysis of 3/1 patterns may be worthwhile, but will probably be very tedious, requiring considerations of all the possible combinations.

Let us finally remark that even with the possibly more difficult problem k-PWCO, we are able to derive a search tree algorithm with a better branching behaviour than Dujmović and Whitesides did for k-OSCM. Since the result is of certain interest on its own, we will state it in the following.

Theorem 6. *k-PWCO can be solved in time* $\mathcal{O}(1.5175^k + kn^2)$.

In fact, the worst case branching vector in our algorithm is $(2, 4, 4, 4)$.

Acknowledgments. We are grateful for hints of F. Rosamond, P. Shaw and M. Suderman on draft versions of this paper.

References

1. Di Battista, G., Eades P., Tamassia R. and I. G. Tollis, *Graph Drawing: Algorithms for the Visualization of Graphs*, Prentice Hall, 1999.
2. R. Downey and M. Fellows. *Parameterized Complexity*, Springer, 1999.
3. V. Dujmović, M. Fellows, M. Hallet, M. Kitching, G. Liotta, C. McCartin, N. Nishimura, P. Ragde, F. Rosemand, M. Suderman, S. Whitesides,and D. Wood, An efficient fixed parameter approach to two-layer planarization. In P. Mutzel and M. Jünger, eds., *Graph Drawing GD'01*, LNCS 2265, pages 1-15. Springer, 2001.
4. V. Dujmović, M. Fellows, M. Hallet, M. Kitching, G. Liotta, C. McCartin, N. Nishimura, P. Ragde, F. Rosemand, M. Suderman, S. Whitesides,and D. Wood, On the parameterized complexity of layered graph drawing, In F. Meyer auf der Heide, ed., *European Symposium on Algorithms ESA'01*, LNCS 2161, pages 488–499. Springer, 2001.

5. V. Dujmović and S. Whitesides. An efficient fixed parameter tractable algorithm for 1-sided crossing minimization. In M. T. Goodrich and S. G. Kobourov, eds., *Graph Drawing GD'02*, LNCS 2528, pages 118–129. Springer, 2002.
6. P. Eades and N. C. Wormald. Edge crossings in drawings of bipartite graphs. *Algorithmica*, 11:379–403, 1994.
7. M. Jünger and P. Mutzel. 2-layer straightline crossing minimization: performance of exact and heuristic algorithms. *J. Graph Algorithms Appl.*, 1:1–25, 1997.
8. X. Munoz, W. Unger, and I. Vrto, One sided crossing minimization is NP-hard for sparse graphs. In P. Mutzel and M. Jünger, eds., *Graph Drawing GD'01*, LNCS 2265, pages 115–123. Springer, 2001.
9. R. Niedermeier and P. Rossmanith. A general method to speed up fixed-parameter-tractable algorithms. *Information Processing Letters*, 73:125–129, 2000.
10. Sugiyama K., S. Tagawa and M. Toda. Methods for visual understanding of hierarchical system structures, *IEEE Transactions on Systems, Man and Cybernetics*, 11:109-125, 1981.

Experiments with the Fixed-Parameter Approach for Two-Layer Planarization

Matthew Suderman and Sue Whitesides*

School of Computer Science, McGill University
3480 University Street, Room 318
Montreal, Quebec, H3A 2A7 CANADA
{msuder,sue}@cs.mcgill.ca

Abstract. We present computational results of an implementation based on the fixed parameter tractability (FPT) approach for biplanarizing graphs. These results show that the implementation can efficiently minimum biplanarizing sets containing up to about 18 edges, thus making it comparable to previous integer linear programming approaches. We show how our implementation slightly improves the theoretical running time to $O(6^{\mathsf{bpr}(G)} + |G|)$. Finally, we explain how our experimental work predicts how performance on sparse graphs may be improved.

1 Introduction

A layered drawing of a graph G is a 2-dimensional drawing of G in which each vertex is placed on one of several parallel lines called *layers*, and each edge is drawn as a straight line between its end-vertices. In this paper, we also require that the end-vertices of each edge lie on different layers. Layered graph drawings [16] have applications in visualization [1], DNA mapping [17], and VLSI layout [11]. In this paper, we consider 2-layer drawings which are of fundamental importance in the "Sugiyama" approach to multi-layer graph drawing [14].

One of the most studied criteria for obtaining "good" drawings of graphs is to minimize the number of edge crossings in the drawing. For 2-layer drawings, the minimum number of crossings is called the *bipartite crossing number* of a graph. Unfortunately, the 2-LAYER CROSSING MINIMIZATION problem, of deciding whether or not the bipartite crossing number of a given graph is at most a given constant $k \geq 0$, is NP-Hard [6]. Furthermore, in practice, exact solutions are practical for up to only 15 vertices per layer, and heuristics are extremely inaccurate for sparse graphs [8].

Some recent experimental evidence suggests that, reducing the number of edges creating crossings produces "better" drawings in some cases than minimizing the number of crossings [13]. In other words, this motivates a strategy of removing a minimum number of edges so that the resulting graph can be drawn without crossings (and then possibly re-inserting the removed edges). A graph is

* Research supported by an NSERC operating grant and an FQRNT scholarship.

G. Liotta (Ed.): GD 2003, LNCS 2912, pp. 345–356, 2004.

biplanar if it admits a planar 2-layer drawing. A set of edges whose removal from a graph G makes it biplanar is called a *biplanarizing set* for G. The *biplanarizing number* of a graph G, denoted by $\mathsf{bpr}(G)$, is the size of the minimum biplanarizing set for G. Thus, the 2-Layer Planarization problem is: given a graph G and an integer $k \geq 0$, determine whether or not $\mathsf{bpr}(G) \leq k$. This problem was first studied by Tomii *et al.* [15], who showed that it is NP-Complete. Therefore the optimization problem of finding the biplanarizing number of a graph is NP-Hard. Interestingly, Mutzel [12, 13] reports much better results for integer linear programming solutions that find the biplanarizing number of a graph than those that find its bipartite crossing number. Thus, [12, 13] provide two compelling reasons to study 2-Layer Planarization and its corresponding optimization problem.

Biplanarity has been studied from the parametric complexity view. A problem with input size n and parameter k is said to be *Fixed-Parameter Tractable*, or in the class FPT, if it can be solved in $O(f(k) \cdot n^{\alpha})$ time, for some function f and constant α (see Downey and Fellows [2]). Dujmović *et al.* [3] describe such an algorithm for solving the 2-Layer Planarization problem that runs in time $O(k \cdot 6^k + |G|)$.

In this paper, we describe an implementation based on the algorithm of [3]. Our implementation finds a biplanarizing set of size $\mathsf{bpr}(G)$. We also present experimental evidence showing that the FPT approach to biplanarization is of more than theoretical interest. In particulare, our results show that our implementation can be used in practice to efficiently find minimum biplanarizing sets containing up to about 18 edges. Furthermore, we show that its running time is roughly comparable to the well-studied integer linear programming approach. We show that our implementation runs in time $O(6^k + |G|)$, a slight improvement on the theoretical bound of [3].

Finally, we predict, on the basis of our experimental results, that a further variation of our implementation, described in Section 5, will be able to efficiently planarize sparse graphs with biplanarization numbers much larger than 18.

The rest of the paper is organized as follows. The next section defines several terms and presents previous work. Section 3 describes our implementation and its running time. Section 4 presents computational results for our implementation and compares them to those of Mutzel in [12, 13]. Finally, Section 5 describes a further variation of our implementation for sparse graphs.

2 Preliminaries

In this paper, each graph $G = (V, E)$ is simple and undirected, but not necessarily connected. A *leaf* is a vertex with exactly one neighbor, and we use $\deg'_G(v)$ (or $\deg'(v)$ when the context is clear) to denote the number of non-leaf neighbors of a vertex v with respect to G. Any graph that can be transformed into a path by removing all its leaves is a *caterpillar*. This unique path is called the *spine* of the caterpillar. The 2-*claw* is the smallest tree that is not a caterpillar. It

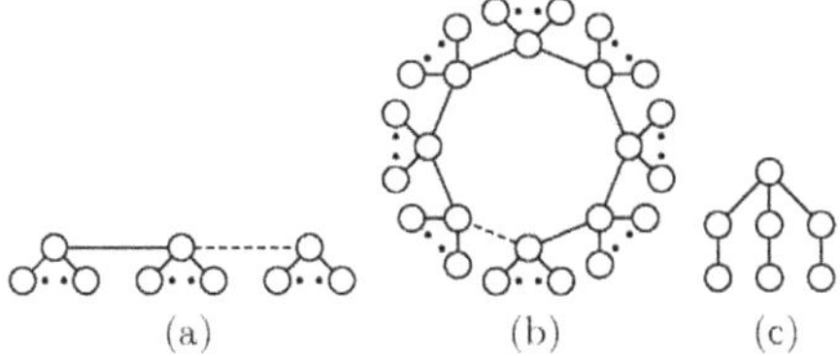

Fig. 1. (a) Caterpillar, (b) Wreath, (c) 2-Claw

consists of a vertex called the *root* that has three neighbors, and each neighbor is additionally adjacent to a leaf.

Lemma 1 ([5, 7, 15]). *Let G be a graph. The following are equivalent:*

1. *G is biplanar;*
2. *G is a forest of caterpillars;*
3. *G is acyclic and contains no 2-claw; and*
4. *The graph obtained from G by deleting all leaves is a forest and contains no vertex with degree three or greater.*

Let $P = v_1 \dots v_p$ be a simple path of length at most two in a graph G. If $\deg'_G(v_1) \geq 3$, $\deg'_G(v_p) = 1$, and the remaining vertices v_i have $\deg'_G(v_i) = 2$, then the subgraph induced by the vertices of P and the neighbors of vertices $v_2, \dots, v_p$ is called a *pendant caterpillar* of G. This pendant caterpillar is said to be *connected* to the graph at v_1, its *connection point*. If, instead, we have $\deg'_G(v_p) \geq 3$, then the subgraph induced by the vertices of P and the neighbors of vertices $v_2, \dots, v_{p-1}$ is called an *internal caterpillar* of G. This internal caterpillar is said to be *connected* to G at vertices v_1 and v_p, its *connection points*. If an internal caterpillar is a path, then it is also called an *internal path.*

Any graph that can be transformed into a cycle C by removing all of its leaves is called a *wreath.* The edges of C are called the *cycle edges* and C is called the *wreath cycle* of the wreath. A wreath subgraph is called a *pendant wreath* if exactly one of the vertices v has a neighbor outside of the wreath, and v is on the wreath cycle[1]. The wreath subgraph is said to be *connected* to the rest of the graph at v, its *connection point.* A *pendant triangle* is a pendant wreath composed of three edges, all cycle edges. A *middle edge* of a pendant triangle is any edge whose end-vertices have degree equal to 2.

3 Algorithm Implementation

We begin by recalling the following lemma whose proof we include in order to describe our implementation.

[1] Note: a wreath component may be regarded as a pendant wreath by thinking of one of its leaves as outside the wreath subgraph.

Lemma 2 ([3]). *If there exists a vertex v in a graph G such that $deg'_G(v) \geq 3$, then v belongs to a 2-claw or a 3- or 4-cycle in G.*

Proof. Let w_1, w_2, w_3 be three distinct non-leaf neighbors of v, and let x_1, x_2, x_3 be neighbors of w_1, w_2, w_3, respectively, that are distinct from v. If $x_i = w_j$ for some i and j, then vw_jw_i is a 3-cycle. On the other hand, if $x_i \neq w_j$ for each i and j but $x_i = x_j$ for some $i \neq j$, then $vw_ix_iw_j$ is a 4-cycle. If neither of these is true, then vertices $v, w_1, w_2, w_3, x_1, x_2, x_3$ form a 2-claw rooted at v. □

We call a 2-claw or a 3- or 4-cycle in a graph a *forbidden structure.*

One approach for producing FPT algorithms is the *method of bounded search trees* [2]. The basic idea is to exhaustively search for a solution to the problem in a tree whose size is bounded by a function of the problem parameter. For intuition, we give a basic bounded search tree algorithm for solving the 2-Layer Planarization problem. Both our implementation and the algorithm of [3] elaborate on this basic idea.

We construct the search tree recursively, beginning at the root. To each node, we associate a subgraph H of G; for the root node, we have $H = G$. For each non-leaf node, we also associate a forbidden structure S in H. By Lemma 1, at least one edge in S is in every biplanarizing set of H; consequently, the current node has $|S|$ children, one corresponding to each edge in S. The subgraph associated with each child is obtained by removing an edge in S from H. A node is a leaf if its subgraph H is obtained from G by removing more than k edges, or does not contain any forbidden structures. In the second case, we have, by Lemma 2, that every vertex v in H has $\deg'_H(v) \leq 2$; in other words, each connected component in H is either a caterpillar or a wreath. By Lemma 1, any minimum biplanarizing set H contains exactly one cycle edge from each wreath in H. Thus, a leaf node represents a *yes*-instance to the problem if its subgraph H does not contain any forbidden structures, and the sum of the edges removed from H to obtain G together with the number of wreaths in H is at most k. The corresponding biplanarizing set is simply the edges removed from G to obtain H together with one cycle edge from each wreath in H.

The resulting tree has at most $O(6^k)$ nodes because, first of all, each node has at most 6 children, and secondly, each non-root node corresponds to an edge removal so the height of the tree is at most k. Constructing this tree naively requires $O(|G|)$ time at each node; therefore, we have an $O(6^k \cdot |G|)$ time algorithm for solving the 2-Layer Planarization problem.

Although this is enough to prove that the 2-Layer Planarization problem is Fixed-Parameter Tractable, the running time can be further improved. In fact, an $O(k \cdot 6^k + |G|)$ time algorithm is given in [3], roughly by reducing the graph to a "kernel" of size $O(k)$ so that at most $O(k)$ time is needed at each node.

In looking for a convenient implementation, we discovered that we could further reduce the running time to $O(6^k + |G|)$. We obtained this reduction by finding a way to determine, at each search tree node, whether or not its subgraph H contains a forbidden structure, and, if so, to exhibit one such structure, all in constant time. In addition, instead of handling component wreaths only at leaf nodes, we handle them as soon they are created by an edge-removal.

It is possible to find a forbidden structure in constant time by maintaining the list F of vertices with $\deg' \geq 3$, and, for each vertex v, the list $f(v)$ of edges incident on v that correspond to the non-leaf neighbors of v. We construct a forbidden structure in constant time as in the proof of Lemma 2. We first select the first vertex v in F, then the first three edges (v, w_1), (v, w_2) and (v, w_3) in $f(v)$, and, for each w_i, an incident edge other than (v, w_i). If these six edges induce a cycle, then we have found either a 3- or 4-cycle; otherwise, we have a 2-claw rooted at v. If F is empty, then, we are at a leaf node in the tree.

The following lemmas show that we can update these lists after each edge removal in constant time. If (v, w) is an edge, then $\mathrm{nl}(v, w)$ denotes the other neighbor $w' \neq w$ of v if $\deg(v) = 2$; otherwise, $\mathrm{nl}(v, w) = v$.

Lemma 3. *Let $e = (v_0, v_1)$ be an edge in a graph G. Let F be the set of vertices in G with $deg' \geq 3$, and let F' the set of vertices in $G \setminus e$ with $deg' \geq 3$. Then:*

$$F' \subseteq F \subseteq F' \cup \{nl(v_0, v_1), nl(v_1, v_0)\}.$$

Proof. After removing edge e, only vertices whose $\deg'$ decreases to 2 are removed from F. In other words, these vertices either lose a neighbor or one of their neighbors becomes a leaf. Thus, the value of $\deg'$ may change only for v_0, v_1, $\mathrm{nl}(v_0, v_1)$ and $\mathrm{nl}(v_1, v_0)$ when e is removed. If we have $v_0 \neq \mathrm{nl}(v_0, v_1)$, then $\deg(v_0) = 2$ so $v_0 \notin F$ or F'. Symmetric comments apply to v_1. □

The proof of the next lemma is similar so it is omitted.

Lemma 4. *Let $e = (v_0, v_1)$ be an edge in a graph G, and let f be the mapping from each vertex in G to the set of incident edges corresponding to its non-leaf neighbors. Similarly, let f' be the mapping from each vertex in G to the set of incident edges corresponding to its non-leaf neighbors in $G \setminus e$.*

Then, $f'(w) = f(w)$ for each vertex w in $E(G) \setminus \{v_0, nl(v_0, v_1), v_1, nl(v_1, v_0)\}$, and, otherwise:

- $f(v_0) \subseteq f'(v_0) \cup \{(v_0, v_1)\}$,
- $f(nl(v_0, v_1)) \subseteq f'(nl(v_0, v_1)) \cup \{(v_0, nl(v_0, v_1))\}$,
- $f(v_1) \subseteq f'(v_1) \cup \{(v_0, v_1)\}$, *and*
- $f(nl(v_1, v_0)) \subseteq f'(nl(v_1, v_0)) \cup \{(v_1, nl(v_1, v_0))\}$.

As mentioned earlier, we handle component wreaths at each node in the tree rather than leaving them for the leaf nodes. Furthermore, we detect and planarize each wreath in constant time. As a result, we cannot expect to detect a component wreath by traversing each of its member vertices. Instead, we rely on pointers called *cheaters*. Cheaters link the first and last vertices on the spine of every internal caterpillar. For convenience, we will think of a pendant wreath as an internal caterpillar with one connection point.

Suppose that the subgraph H of a node contains no component wreaths but that, for some edge $e = (v_0, v_1)$ in H, $H \setminus e$ contains at least one. By Lemma 3, $v_0' = \mathrm{nl}(v_0, v_1)$ or $v_1' = \mathrm{nl}(v_1, v_0)$ belongs to the wreath, and, for each v_i' in the wreath, we have $\deg'_{H \setminus e}(v_i') = 2$ and $\deg'_H(v_i') > 2$. All vertices in a component wreath have $\deg' \leq 2$, so, of all vertices in the wreath subgraph in H, only v_0' and v_1' have $\deg_H' > 3$. Thus, the wreath subgraph in H is composed of zero or more internal caterpillars and possibly vertices v_0' and v_1' acting as connection points for these internal caterpillars. Thus, removing edge e creates a component wreath if and only if some v_i' is the single connection point for an internal caterpillar in H and $\deg'_{H\setminus e}(v_i') = 2$, or both v_0' and v_1' are the connection points for two internal caterpillars and $\deg'_{H\setminus e}(v_0') = \deg'_{H\setminus e}(v_1') = 2$. In both cases, we planarize the wreath by removing any edge in $f(v_i')$.

Having handled component wreaths using cheaters, we now show how to efficiently update cheaters whenever an edge is removed. If a new internal caterpillar is created by an edge removal, then the edge removal decreases the $\deg'$ of a vertex v down to two. If v is a connection point for two internal caterpillars P_1 and P_2 before the edge removal, then the new internal caterpillar is the concatenation of P_1 and P_2. If v is a connection point for only one internal caterpillar P, then the new internal caterpillar is the concatentation of P with v and its leaf neighbors. Otherwise, the new internal caterpillar is composed only of v and its leaf neighbors. In each of these three cases, it is a simple matter to update the cheaters in constant time after an edge removal using existing cheaters.

Thus, we have shown how to explore the bounded search tree in $O(6^k + |G|)$ time. The resulting algorithm Bounded Search Tree for solving 2-LAYER PLANARIZATION is given below. We assume that set F and the map f are correctly initialized for the graph G in $O(|G|)$ time before the algorithm is executed.

Algorithm Bounded Search Tree (graph G; vertex-set F; map f; integer k)

1. **if** $F = \emptyset$ **then return** YES;
2. **else if** $k = 0$ **then return** NO;
3. **else**
   a) $S \leftarrow$ a 2-claw, 3-cycle or 4-cycle in G using F and f;
   b) **for each** edge $(x, y) \in S$ **do**
      i. Remove (x, y) from G and planarize any resulting component wreaths while updating F, f and the cheaters. Let P be the set of removed edges;
      ii. **if** Bounded Search Tree($k - |P|$) **then return** YES;
      iii. **else**
         Add edges in P back to G and undo the resulting changes to F, f and the cheaters;
4. **return** NO;

We have following result:

Lemma 5. *Given a graph G and integer k, algorithm* Bounded Search Tree *determines if* bpr $(G) \leq k$ *in $O(6^k + |G|)$ time.*

We use the algorithm Bounded Search Tree to find the minimum biplanarizing set for a graph by repeatedly executing Bounded Search Tree, intially with some lower bound value for k, and then increasing k until the algorithm returns YES.

Thus, we can find the minimum biplanarizing set for a graph in $O(6^0 + 6^1 + \ldots + 6^{\mathsf{bpr}(G)} + |G|) = O(6^{\mathsf{bpr}(G)} + |G|)$ time. Variable k is initially set to either $\Phi(G)/2$ or $|E| - |V| + 1$. As defined in [3], the potential function $\Phi(G)$ denotes the following sum:

$$\sum_{v \in V(G)} \max\{\deg'(v) - 2, 0\}.$$

Lemma 6 ([3]). *For every graph G, $\mathsf{bpr}(G) \geq \frac{\Phi(G)}{2}$.*

The value of $\mathsf{bpr}(G)$ equals $|E| - |V| + 1$ whenever G has a spanning caterpillar. This is best possible for $\mathsf{bpr}(G)$ for any graph G. Thus, we obtain:

Theorem 7. *Given a graph G, there exists an algorithm that finds a minimum biplanarizing set for G in $O(6^{\mathsf{bpr}(G)} + |G|)$ time.*

In our implementation, we include three other improvements to the algorithm Bounded Search Tree described above. We obtain the first improvement by slightly generalizing the method for planarizing component wreaths at each search tree node, which we described above, to planarize all pendant wreaths as well. The generalization can be applied in constant time after each edge removal.

The second improvement is based on the observation that, if a vertex is the root of more than one 2-claw, then planarizing these 2-claws one-at-a-time could result in exploring unnecessary branches of the search tree. For example, if a vertex v has exactly four non-leaf neighbors, and each neighbor has degree equal to two, then v is the root of $\binom{4}{3}$ 2-claws. The original algorithm would typically choose one of these 2-claws, branch on it, and then, on each branch, branch on the 2-claw remaining at v. This results in trying to planarize the 2-claws rooted at v in 36 different ways. Some of these branches are redundant because there are exactly 24 different ways to planarize these 2-claws with two edge removals. Avoiding these redundant branches could lead to a substantially shorter running time when many vertices in the graph are the roots of more than one 2-claw. In fact, we end up with a branching factor of $\sqrt(24) < 5$ rather than 6.

The redundant branches are due to deleting the same set of edges but in a different order. For example, if the children of node N correspond to the edges $e_1, \ldots, e_6$ of a 2-claw, then the search subtree corresponding to removing e_1 may explore the possibility of also removing edge e_2, and, similarly, the search subtree corresponding to removing e_2 may explore the possibility of also removing edge e_1. Entirely exploring both subtrees would be redundant because, if the graph remaining at node N could have been planarized by removing both e_1 and e_2, then that solution node appears in both subtrees. Thus, it would be more efficient to completely explore the subtree corresponding to e_1, and then explore only parts of the subtree corresponding to e_2 that do not involve removing e_1. We avoid these redundant parts by marking e_1 as *tried* after we have explored its subtree and failed to find a solution. If we fail to find a solution at a descendant of node N, then, just before backtracking from N, we remove the mark on e_1. In general, then, we mark an edge as *tried* as soon as we have explored its subtree and then remove that mark whenever we backtrack away from its parent node. The extra overhead is clearly constant per node so the resulting algorithm runs in $O(6^{\mathsf{bpr}(G)} + |G|)$ time.

The third improvement is based on the observation that we can avoid exploring subtrees that correspond to removing a *candidate* edge from the graph. As defined in [4], an edge is called a *candidate* for removal if it is the middle edge of an internal 3-path or triangle, or it does not belong to an internal 3-path or triangle and one end-vertex has $\deg' > 2$ and the other has $\deg > 1$. We use $\mathcal{K}$ to denote the set of candidate edges, so a *canonical biplanarizing set* is a biplanarizing set that is a subset of $\mathcal{K}$. The following lemma shows that there is always a minimum biplanarizing set that is canonical.

Lemma 8 ([4]). *If T is a biplanarizing set for a graph G, then there exists a canonical biplanarizing set T^* of G such that $|T^*| \leq |T|$.*

One can easily test whether or not an edge is a candidate for removal in constant time, so the algorithm still runs in $O(6^{\mathsf{bpr}(G)} + |G|)$ time.

4 Computational Results

We implemented the algorithm described in the previous section using the Java programming language. We compiled the program using the byte-code compiler from the Java SDK version 1.4.1 from Sun Microsystems. We ran the experiments using their byte-code interpreter on a 1004.542 MHz Pentium III computer with 901156 Kb RAM running Debian Linux version 2.4.18. Actual running times depend on many factors such as the speed and architecture of the computer, other processes running in the background, the quality of the implementation, and the choice of implementation language (Java programs are generally 2-3 times slower than similar C++ programs); therefore, we also recorded the sizes of the search trees explored. These values depend only on the input graph and the algorithm. We include the running times in our results only to give a rough idea of how long the implementation takes to planarize a graph.

We applied our implementation to the bipartite graphs from the Stanford Graphbase [10] that were used in the experiments of Mutzel [12,13]. The results of our experiments are shown alongside the results of Mutzel [13] in Table 1. Each row in the table corresponds to the average values from applying the algorithm to 100 different graphs. From Mutzel's ILP experiments, we included both the running time and the average guarantee of the solution value (Gar), i.e. $\frac{UpBound - Sol}{UpBound} \times 100$ where Sol denotes the number of edges in a biplanar subgraph of G having the most edges among the biplanar subgraphs found, and UpBound denotes the value determined by the linear programming relaxation when the time limit of 300 seconds expired. It turns out that for the graphs investigated, the initial value for k, $\max(\frac{\phi(G)}{2}, |E|-|V|+1)$, is quite close to $\mathsf{bpr}(G)$. For denser graphs we would expect this value to be closer to $\mathsf{bpr}(G)$ than for sparser graphs because of the higher probability of have a spanning caterpillar. However, on

Table 1. Results for bipartite graphs with $|V_i|$ vertices per bipartition and $|E|$ edges.

		Mutzel ILP [13]		FPT			
$\|V_i\|$	$\|E\|$	Gar	Time ($\leq$ 300s)	Time ($\leq$ 600s)	Steps	bpr	Success
20	20	0.00	0	0	5	1	100
20	25	0.00	0	0	8	1.5	100
20	30	0.00	0	0	25	3	100
20	35	0.00	1	0	90	4.9	100
20	40	0.00	6	0	595	7.7	100
20	45	0.03	26	0	1,829	10.7	100
20	50	0.67	100	2	53,416	14.3	100
20	55	0.53	81	41	1,767,872	18.2	96
20	60	0.37	56	-	-	-	-
20	65	0.32	54	-	-	-	-
20	70	0.13	26	-	-	-	-
20	75	0.13	22	-	-	-	-
20	80	0.03	12	-	-	-	-
20	85	0.10	20	-	-	-	-
20	90	0.02	8	-	-	-	-
20	95	0.00	4	-	-	-	-
20	100	0.00	4	-	-	-	-
20	40	0.00	6	0	495	7.4	100
30	60	0.13	49	1	10,559	11.3	100
40	80	0.55	150	9	243,760	15.6	100
50	100	1.45	253	43	1,281,694	19.4	97

average the biplanarizing number was at most one more than our initial value for k. Thus, even our intial lower bound which could be calculated in $O(|G|)$ time had a good average guarantee of the solution value.

It would not be very meaningful to directly compare our running times to those of Mutzel because of environment differences. More specifically, they were originally reported in 1996 in [12] so the computers used were much slower than the ones we used, and the implementation language was C++ using ABACUS [9].

It is, however, meaningful to compare the shapes of the $|E|$ versus running time graphs. In the first 17 rows of Table 1, we see that the FPT implementation is quite efficient up to $|E| = 55$, finding exact solutions to all input graphs. After $|E| = 55$, the FPT implementation was able to obtain exact solutions to only a few input graphs for maximum time 10 minutes per graph. The ILP implementation, on the other hand, demonstrates poorest performance at $|E| = 50$. However, after $|E| = 50$, it improves as $|E|$ approaches 100.

Thus, we see that these two different approaches may be complimentary. It appears that, whereas FPT approaches tend to be efficient on sparse graphs, ILP approaches tend to be efficient on dense graphs. This is because FPT algorithms have running times like $O(f(k) \cdot n^{\alpha})$; therefore, they will be efficient when $\mathsf{bpr}(G)$ is small, that is, the graph is sparse. ILP algorithms using branch-and-cut, on the other hand, find an optimal solution by repeatedly finding approximate solutions that close in on an optimal solution. For planarization, this means that an ILP algorithm begins with a biplanarizing set of size between $\mathsf{bpr}(G)$ and $|E|-|V|+1$, and then find increasingly smaller biplanarizing sets until one of size $\mathsf{bpr}(G)$ is found. For dense graphs, the probability that $\mathsf{bpr}(G) = |E| - |V| + 1$ is high, so the ILP algorithm begins with a solution close to the optimal.

Designing and implementing FPT solutions is still quite new as compared to integer linear programming, especially for graph drawing algorithms. Consequently, further work on FPT algorithms is almost sure to yield improvements. In the next section, we describe some future possibilities.

5 Future Work

One way that we might improve the performance of our implementation on larger sparse graphs is to integrate divide-and-conquer into the algorithm. For example, planarizing two subgraphs each with $\mathsf{bpr}= k$ as a single graph could use up to $O(6^{2k} + |G|) = O(36^k + |G|)$ time. If, on the other hand, it is possible to planarize them separately, then we would use $O(2 \cdot 6^k + |G|)$ time. Clearly, the second option is preferrable. In sparse graphs, we would expect this to be often possible.

Certainly, if a graph is disconnected, then the minimum biplanarizing set for the whole graph is simply the union of the minimum biplanarizing sets for the connected components. We can, however, do slightly better than this by dividing the graph into p-components. A p-component of a graph is a maximal connected subgraph consisting of biconnected components that are connected by internal paths of length at most three, and the internal caterpillars that connect this subgraph to other p-components. Notice that two p-components are not necessarly disjoint since they may share a single internal caterpillar. The following lemma shows that each p-component can be planarized separately. The proof is not trivial but omitted.

Lemma 9. *If H_i, $1 \leq i \leq p$, are the p-components of a graph G, and M_i are their minimum canonical biplanarizing sets, then $\bigcup M_i$ is a minimum canonical biplanarizing set for G and $M_i \cap M_j = \emptyset$ for each $i \neq j$.*

Lemma 9 suggests a divide-and-conquer variation of the algorithm: divide the graph into p-components, and then planarize each p-component individually. In fact, we are able to do better than this by planarizing in such a way as to break a larger p-component into smaller p-components, and then to planarize each of them. One strategy for breaking up a p-component is to branch on forbidden structures containing cut vertices, which we would expect to find in sparse graphs.

A slight complication of this variation of the algorithm is that, when planarizing a p-component, we are using a bounded search tree so we have bounded the number of edge removals by some parameter k. Thus, if we break the p-component C into smaller child p-components, then we must somehow divide the parameter k for C into smaller parameters for each child p-component. The problem is that we do not know what parameter to assign each child without knowing the size of its minimum biplanarizing set. We solve this problem by initializing the parameter for each child to some lower bound. We re-apply the algorithm to the child, increasing its parameter until we find its minimum biplanarizing set. If the sum of the parameters for the children ever becomes greater than the parameter for their parent, then we realize that the way we divided the parent p-component into smaller p-components will not yield a biplanarizing set matching the parent's parameter. In response, we immediately backtrack to the point in the search tree where we disconnected the parent p-component and continue searching from there.

The extra work of computing the p-components and determining if the current p-component C has been broken into smaller p-components can all be done in $O(\mathsf{bpr}(C))$ time. To compute sub-p-components, we simply apply a modification of the algorithm fory finding biconnected components in a graph. We apply the algorithm to the at most $O(\mathsf{bpr}(C))$ vertices having three or more non-leaf neighbors, skipping over internal caterpillars during graph traversal using the cheater pointers described in the previous section.

Straight-forward but tedious analysis shows that the running time of this variation of the algorithm runs in $O(6^k + |G|)$ time. It remains to be seen how well this approach will work in practice. We expect that the running time of the algorithm will differ polynomially with respect the size of sparse graphs with uniform density.

6 Conclusion

We have described the implementation and computational results of an algorithm inspired by parameterized complexity. We have shown that for computing the minimum biplanarizing sets, this algorithm has both practical as well as theoretical value. Furthermore, we have presented experimental evidence showing that it is roughly comparable with the well-studied field of linear programming for finding practical solutions to NP-hard problems. Finally, we have described one possible way to dramatically improve on the experimental results presented in this paper.

In the future, we would plan to obtain computational results from the algorithm based on p-components. We plan to perform further experiments with graphs from sources other than the Stanford Graphbase, such as from DNA-mapping applications.

One of the limitations of our implementation is that it obtains only exact solutions. We plan to investigate using FPT algorithms for finding approximation solutions.

References

1. Battista, G.D., Eades, P., Tamassia, R., Tollis, I.G.: Graph Drawing: Algorithms for the Visualization of Graphs. Prentice-Hall (1999)

2. Downey, R.G., Fellows, M.R.: Parametrized Complexity. Springer-Verlag (1999)
3. Dujmović, V., Fellows, M.R., Hallett, M.T., Kitching, M., Liotta, G., McCartin, C., Nishimura, N., Ragde, P., Rosamond, F.A., Suderman, M., Whitesides, S., Wood, D.R.: A fixed-parameter approach to two-layer planarization. In Mutzel, P., Jünger, M., Leipert, S., eds.: Graph Drawing, 9th International Symposium (GD 2001). Volume 2265 of Lecture Notes in Computer Science., Springer-Verlag (2001) 1–15
4. Dujmović, V., Fellows, M.R., Hallett, M.T., Kitching, M., Liotta, G., McCartin, C., Nishimura, N., Ragde, P., Rosamond, F.A., Suderman, M., Whitesides, S., Wood, D.R.: A fixed-parameter approach to two-layer planarization, manuscript (2003)
5. Eades, P., McKay, B., Wormald, N.: On an edge crossing problem. In: Proceedings of the 9th Australian Computer Science Conference, Australian National University (1986) 327–334
6. Garey, M.R., Johnson, D.S.: Crossing number is NP-complete. SIAM Journal of Algebraic Discrete Methods **4** (1983) 312–316
7. Harary, F., Schwenk, A.: A new crossing number for bipartite graphs. Utilitas Mathematica **1** (1972) 203–209
8. Jünger, M., Mutzel, P.: 2-layer straightline crossing minimization: performance of exact and heuristic algorithms. Journal of Graph Algorithms and Applications **1** (1997) 1–25
9. Jünger, M., Thienel, S.: The ABACUS-system for branch and cut and price algorithms in integer programming and combinatorial optimization. In: Software–Practice and Experience. Volume 30. (2000) 1324–1352
10. Knuth, D.: The Stanford GraphBase: A Platform for Combinatorial Computing. ACM Press, Addison-Wesley Publishing Company (1993)
11. Lengauer, T.: Combinatorial Algorithms for Integrated Circuit Layout. Wiley (1990)
12. Mutzel, P.: An alternative method to crossing minimization on hierarchical graphs. In North, S.C., ed.: Graph Drawing, Symposium on Graph Drawing (GD '96). Volume 1190 of Lecture Notes in Computer Science., Springer-Verlag (1996) 318–333
13. Mutzel, P.: An alternative method to crossing minimization on hierarchical graphs. SIAM Journal of Optimization **11** (2001) 1065–1080
14. Sugiyama, K., Tagawa, S., Toda, M.: Methods for visual understanding of hierarchical system structures. IEEE Transactions on Systems, Man, and Cybernetics **11** (1981) 109–125
15. Tomii, N., Kambayashi, Y., Yajima, S.: On planarization algorithms of 2-level graphs. Papers of tech. group on elect. comp., IECEJ **EC77-38** (1977) 1–12
16. Warfield, J.N.: Crossing theory and hierarchy mapping. IEEE Transactions on Systems, Man, and Cybernetics **7** (1977) 502–523
17. Waterman, M.S., Griggs, J.R.: Interval graphs and maps of DNA. Bulletin of Mathematical Biology **48** (1986) 189–195

Characterizing Families of Cuts That Can Be Represented by Axis-Parallel Rectangles*

Ulrik Brandes[1], Sabine Cornelsen[2], and Dorothea Wagner[3]

[1] Fakultät für Mathematik & Informatik, Universität Passau.
brandes@algo.fmi.uni-passau.de
[2] Dipartimento di Ingegneria Elettrica, Università dell'Aquila.
cornelse@inf.uni-konstanz.de
[3] Fakultät für Informatik, Universität Karlsruhe.
dwagner@ira.uka.de

Abstract. A drawing of a family of cuts of a graph is an augmented drawing of the graph such that every cut is represented by a simple closed curve and vice versa.

We show that the families of cuts that admit a drawing in which every cut is represented by an axis-parallel rectangle are exactly those that have a cactus model that can be rooted such that edges of the graph that cross a cycle of the cactus point to the root. This includes the family of all minimum cuts of a graph. The proof also yields an efficient algorithm to construct a drawing with axis-parallel rectangles if it exists.

1 Introduction

A cut of a graph is a partition of its vertex set into two non-empty subsets. In a drawing of a graph, it is therefore natural to represent a cut by a closed curve partitioning the plane into two regions containing one subset each.

When a set of cuts is drawn in this way, the curves can intersect so that their union might contain closed curves that represent other cuts not contained in that set. We are interested in families of cuts that can be drawn without creating confusing non-members. In particular, we consider the problem of drawing families of cuts such that every member is represented by an axis-parallel rectangle and vice versa.

Our main result is a characterization of all families of cuts that can be represented by axis-parallel rectangles, namely those that can be modeled by a cactus containing a certain, rootable, node. These include the important family of all minimum cuts of a graph [4]. We give an algorithm to test this property. Provided a set of cuts of a graph with n vertices and m edges is represented by a cactus model, the test works in $\mathcal{O}(nm)$ time. We also show how to construct a drawing with axis-parallel rectangles if one exists. The construction works in two

* Research partially supported by the DFG under grant BR 2158/1-1 and WA 654/13-1. It was also partially supported by the Human Potential Program of the EU under contract no HPRN-CT-1999-00104 (AMORE Project).

G. Liotta (Ed.): GD 2003, LNCS 2912, pp. 357–368, 2004.

steps. First, we solve the problem for families of minimal cuts of a planar connected graph. Then, we use similar planarization techniques as in [3] to extend the result to general graphs.

The paper is organized as follows. In Sect 2, we define drawings of families of cuts. In Sect. 3, we introduce the cactus model and show that its existence is necessary for the existence of a drawing with axis-parallel rectangles. Using hierarchically clustered graphs (briefly introduced in Sect. 4), we give an additional necessary condition for families of cuts that admit a drawing with axis-parallel rectangles in Section 5 and show how to test it. To show that the conditions are also sufficient, we construct a drawing in Section 6.

2 Drawing Families of Cuts

Throughout this paper, let $G = (V, E)$ denote a simple, connected undirected graph with n vertices and m edges. A drawing $\mathcal{D}$ of G maps vertices on distinct points in the plane and edges on simple curves connecting the drawings of their incident vertices. The interior of an edge must not contain the drawing of a vertex. A drawing is planar, if edges do not intersect but in common end points. Let $S, T \subseteq V$. With $G(S)$ we denote the subgraph of G that is induced by S and with $E(S, T)$ we denote the set of edges that are incident to a vertex in S and a vertex in T. A *cut* of G is a partition $C = \{S, \overline{S}\}$ of the vertex set V into two non-empty subsets S and $\overline{S} := V \setminus S$. We say that the cut $\{S, \overline{S}\}$ is induced by S. The edges in $E(C) = E(S, \overline{S})$ are the *cut-edges* of C. A *minimum cut* is a cut with the minimum number of cut-edges among all cuts of G. A *minimal cut* is a cut $\{S, \overline{S}\}$ that is inclusion-minimal, i.e. there is no cut $\{T, \overline{T}\}$ with $E(T, \overline{T}) \subsetneq E(S, \overline{S})$. Note that both, $G(S)$ and $G(\overline{S})$, are connected if $\{S, \overline{S}\}$ is a minimal cut of the connected graph G.

A *drawing of a cut* $C = \{S, \overline{S}\}$ of G in a drawing $\mathcal{D}(G)$ is a simple closed curve γ, such that

- γ *separates* S and $\overline{S}$, i.e. the drawings of edges and vertices in $G(S)$ and $G(\overline{S})$, respectively, are in different connected regions of $\mathbb{R}^2 \setminus \gamma$, and
- $|\mathcal{D}(e) \cap \gamma| = 1$ for $e \in E(C)$, i.e. the drawing of a cut-edge crosses the drawing of the cut exactly once.

Let $\mathcal{C}$ be a set of cuts of a graph G. A mapping $\mathcal{D}$ is a *(planar) drawing* of a graph G and a family of cuts $\mathcal{C}$ of G, if

1. $\mathcal{D}(G)$ is a (planar) drawing of G and
2. $\mathcal{D}(C)$ is a drawing of C in $\mathcal{D}(G)$ for every cut $C \in \mathcal{C}$, and
3. every simple closed curve $\gamma \subseteq \bigcup_{C \in \mathcal{C}} \mathcal{D}(C)$ is a drawing of some cut in $\mathcal{C}$.

Note that the third condition eliminates any potential ambiguity regarding which cuts are in the family.

A drawing $\mathcal{D}$ of a graph and a family $\mathcal{C}$ of cuts is a *drawing with axis-parallel rectangles*, if every cut is drawn as an axis-parallel rectangle – more precisely, if every simple closed curve in $\bigcup_{C \in \mathcal{C}} \mathcal{D}(C)$ is an axis-parallel rectangle. See Fig. 1(a) for a planar drawing with axis-parallel rectangles of the set of all minimum and minimum+1 cuts of a graph.

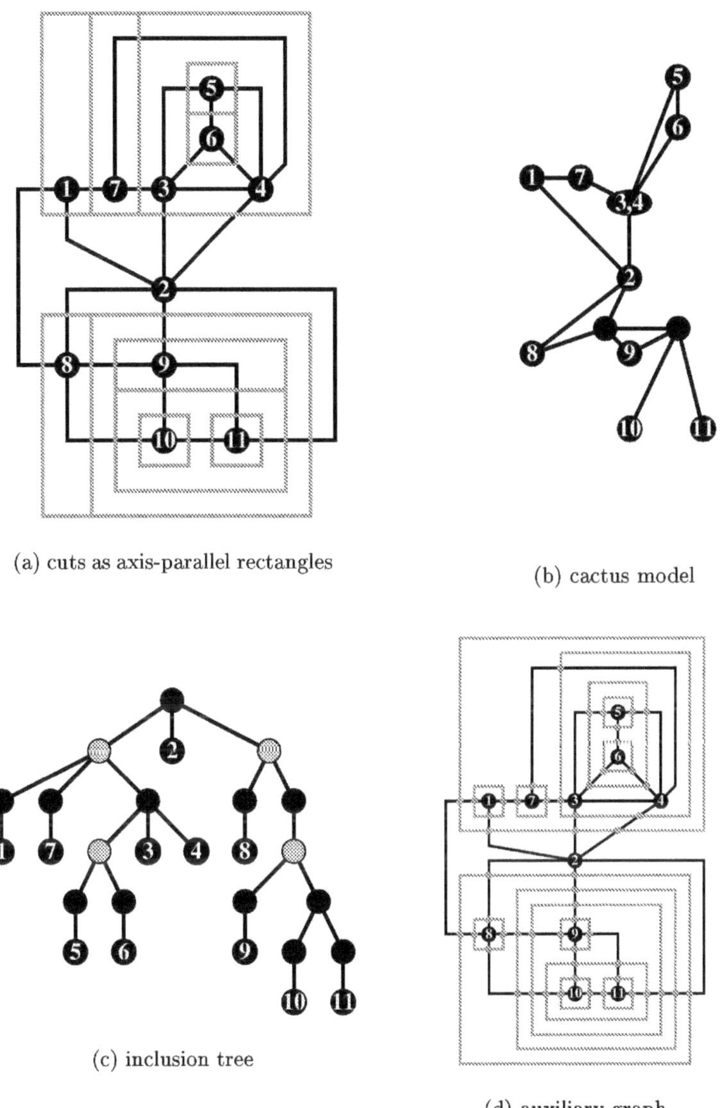

(a) cuts as axis-parallel rectangles

(b) cactus model

(c) inclusion tree

(d) auxiliary graph

Fig. 1. A drawing of the family of all minimum and minimum+1 cuts of a graph with axis-parallel rectangles. In the cactus model of the cuts, φ is indicated by node labels, and cycle-replacement nodes in the inclusion tree constructed from the cactus are shown in grey. The auxiliary graph is shown in a c-planar drawing of the hierarchical clustering represented by the inclusion tree.

3 Necessity of a Cactus Model

A *cactus* is a connected graph in which every edge belongs to at most one cycle. A *cactus model* for a set $\mathcal{C}$ of cuts of a graph $G = (V, E)$ is a pair $(\mathcal{G}, \varphi)$ that consists of a cactus $\mathcal{G} = (\mathcal{V}, \mathcal{E})$ and a mapping $\varphi : V \to \mathcal{V}$ such that $\mathcal{C}$ is *modeled* by the minimal cuts of $\mathcal{G}$, i.e.,

$$\mathcal{C} = \{\{\varphi^{-1}(S), \varphi^{-1}(\overline{S})\}; \{S, \overline{S}\} \text{ is a minimal cut of } \mathcal{G}\}.$$

To avoid confusion, the vertices of the cactus $\mathcal{G}$ are called *nodes*. A node $\nu \in \mathcal{V}$ is called *empty* if $\varphi^{-1}(\nu) = \emptyset$.

An important family of cuts that has a cactus model is the set of all minimum cuts of a connected graph [4]. A cactus model for the minimum and minimum+1 cuts of the graph in Fig. 1(a) is given in Fig. 1(b).

While not every family of cuts has a cactus model, we show that only those that do can have a drawing with axis-parallel rectangles. Two cuts $\{S, \overline{S}\}$ and $\{T, \overline{T}\}$ *cross*, if and only if the four *corners* $S \cap T$, $S \setminus T$, $T \setminus S$, and $\overline{S \cup T}$ are non-empty. The four cuts induced by the four corners of two crossing cuts, respectively, are called *corner cuts*. The cut induced by $(S \setminus T) \cup (T \setminus S)$ is called the *diagonal cut*.

Theorem 1 ([5]). *A set $\mathcal{C}$ of cuts of the graph G can be modeled by a cactus if and only if, for any two crossing cuts $\{S, \overline{S}\}$ and $\{T, \overline{T}\}$ in $\mathcal{C}$,*

- *the four corner cuts are in $\mathcal{C}$, and*
- *the diagonal cut is not in $\mathcal{C}$.*

If a cactus model exists, there is always one with $O(n)$ nodes.

The properties of crossing cuts in this characterization are implied by overlapping axis-parallel rectangles.

Lemma 1. *If a set of cuts has a drawing with axis-parallel rectangles, it has a cactus model.*

Proof. Let $\mathcal{D}$ be a drawing of a set $\mathcal{C}$ of cuts with axis-parallel rectangles and suppose that $\mathcal{C}$ contains crossing cuts $\{S, \overline{S}\}$ and $\{T, \overline{T}\}$. There are essentially the seven cases indicated in Fig. 2 for the drawings of two crossing cuts by axis-parallel rectangles. Since the cases in Fig. 2(a) contain simple closed curves that are not axis-parallel rectangles, only the case in Fig. 2(b) needs to be considered.

Let $D_S, D_T \subseteq \mathbb{R}^2$ be the rectangular regions bounded by $\mathcal{D}(\{S, \overline{S}\})$ and $\mathcal{D}(\{T, \overline{T}\})$, respectively. Then regions $D_S \cap D_T$, $D_S \setminus D_T$, $D_T \setminus D_S$, and $D_S \cup D_T$ are bounded by axis-parallel rectangles in $\mathcal{D}(\{S, \overline{S}\}) \cup \mathcal{D}(\{T, \overline{T}\})$. These are drawings of the four corner cuts of $\{S, \overline{S}\}$ and $\{T, \overline{T}\}$. Hence, they are in $\mathcal{C}$.

On the other hand, suppose the diagonal cut C induced by $(S \setminus T) \cup (T \setminus S)$ is in $\mathcal{C}$ and let D_C be the rectangular region bounded by its drawing $\mathcal{D}(C)$. Without loss of generality, suppose that $\mathcal{D}(S) \subset D_S$ and $\mathcal{D}(T) \subset D_T$. Either D_C contains $\mathcal{D}(G(S \setminus T))$ and $\mathcal{D}(G(T \setminus S))$, but not $\mathcal{D}(G(S \cap T))$, or it contains $\mathcal{D}(G(\overline{(S \setminus T) \cup (T \setminus S)}))$, but not $\mathcal{D}(G(S \setminus T \cup T \setminus S))$. In the first case, the drawing contains a simple closed curve bounding the region $D_C \cap D_S \cap D_T$ (thus inducing an empty cut), and in the second case the union of the three rectangles contains a simple closed curve that is not an axis-parallel rectangle. □

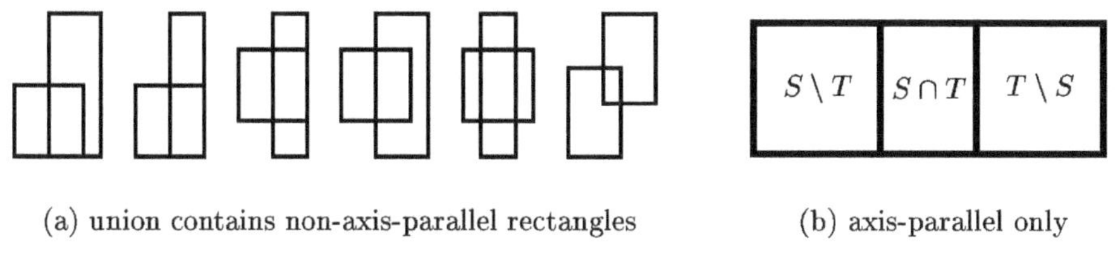

(a) union contains non-axis-parallel rectangles (b) axis-parallel only

Fig. 2. Drawings of two crossing cuts $\{S, \overline{S}\}$ and $\{T, \overline{T}\}$ with axis-parallel rectangles.

4 Cactus-Induced Hierarchical Clusterings

Since a cactus model is necessary for a drawing to exist, we can make use of a transformation originally developed for drawing the particular family of all minimum cuts of a planar connected graph [1]. Given a cactus model $(\mathcal{G}, \varphi)$ of a family $\mathcal{C}$ of G, we use the tree $T = \mathcal{T}(\mathcal{G}, \varphi)$ that is constructed as follows.

1. Choose a node r of $\mathcal{G}$ as root.
2. Replace each cycle c of $\mathcal{G}$ by a star, i.e. delete every edge of c, add a new empty node ν_c – called *cycle-replacement node* of c – to $\mathcal{G}$, and for every node ν of c, add an edge $\{\nu_c, \nu\}$ to $\mathcal{G}$.
3. For each $v \in V$, add v to the vertex set of $\mathcal{G}$ and add an edge $\{v, \varphi(v)\}$ to $\mathcal{G}$.

The construction is illustrated in Fig. 1(c) and yields a triple (G, T, r) of

- a graph $G = (V, E)$,
- a tree T, and
- an inner vertex r of T
- where the set of leaves of T is exactly V.

This is the hierarchically clustered graph model introduced by Feng et al. [6]. G is called the *underlying graph* and T the *inclusion tree* of (G, T, r). Inner vertices of T are called *nodes*. We denote the tree T rooted at r by (T, r). Each node ν of T represents the *cluster* $V_r(\nu)$ of leaves in the subtree of (T, r) rooted at ν. Note that $\{\{V_r(\nu), \overline{V_r(\nu)}\};\ \nu \neq r \text{ node of } \mathcal{T}(\mathcal{G}, \varphi)\}$ equals the subset $\mathcal{C}_{\|}$ of *non-crossing cuts* of $\mathcal{C}$, i.e. the set of cuts in $\mathcal{C}$ that do not cross any other cut in $\mathcal{C}$. A *c-planar drawing* of a hierarchically clustered graph (G, T, r) consists of

- a planar drawing of the underlying graph G and
- an inclusion representation[1] of the rooted tree (T, r) such that
- each edge crosses the boundary of the drawing of a node of T at most once.

[1] In an inclusion representation of a rooted tree (T, r), each node of T is represented by a simple closed region bounded by a simple closed curve. The drawing of a node or leaf ν of T is contained in the interior of the region representing a node μ of T if and only if μ is contained in the path from ν to r in T. The drawings of two nodes μ and ν are disjoint if neither μ is contained in the path from ν to r nor ν is contained in the path from μ to r in T.

Note that the vertices of G are the leaves of T and thus have the same drawing. Also note, that a c-planar drawing of $(G, \mathcal{T}(\mathcal{G}, \varphi), r)$ contains a planar drawing of the set of all non-crossing cuts $\mathcal{C}_{\|}$. A hierarchically clustered graph (G, T, r) is *c-planar* if it has a c-planar drawing. It is *completely connected*, if each cluster and the complement of each cluster induces a connected subgraph of G. A hierarchically clustered graph with planar underlying graph does not have to be c-planar. However, a completely connected hierarchically clustered graph is c-planar if only the underlying graph is planar [2].

According to the construction in [1,9], we associate an auxiliary graph $G_{\mathcal{D}}$ with a c-planar drawing $\mathcal{D}$ of a hierarchically clustered graph. Let V' be the set of points, in which drawings of edges and boundaries of drawings of clusters intersect. Then the vertex set of $G_{\mathcal{D}}$ is $V \cup V'$. The edge set of $G_{\mathcal{D}}$ contains two types of edges. For an edge $e = \{v, w\}$ of G, let $v_1, \ldots, v_k$ be the points in $\mathcal{D}(e) \cap V'$ in the order they occur in the drawing of e from v to w. Then $G_{\mathcal{D}}$ contains the edges $\{v, v_1\}, \{v_1, v_2\}, \ldots, \{v_k, w\}$. Let $\nu \neq r$ be a node of T. Let $v_1, \ldots, v_k$ be the points in $\partial\mathcal{D}(\nu) \cap V'$ in the order they occur in the boundary $\partial\mathcal{D}(\nu)$ of the drawing of ν. Then $G_{\mathcal{D}}$ contains the edges $\{v_1, v_2\}, \ldots, \{v_{k-1}, v_k\}, \{v_k, v_1\}$. The cycle $v_1, \ldots, v_k$ of $G_{\mathcal{D}}$ is called the *boundary cycle* of ν. (To avoid loops and parallel edges, additional vertices of degree two may be inserted into boundary cycles). See Fig. 1(d).

5 Towards a Characterization

Another necessary condition for families of cuts that have a drawing with axis-parallel rectangles depends also on the edges in the graph. Let $(\mathcal{G}, \varphi)$ be the cactus model of a set $\mathcal{C}$ of cuts of a graph G. For a cycle $c : \nu_1, \ldots, \nu_k$ in $\mathcal{G}$ let $V_i := V_{\nu_c}(\nu_i), i = 1, \ldots, k$. An edge $\{v, w\}$ of G *crosses cycle* c if there are $1 \leq i, j \leq k$ such that

$$v \in V_i, w \in V_j, \text{ and } i - j \not\equiv \pm 1 \bmod k.$$

If $(\mathcal{G}, \varphi)$ is the cactus of all minimum cuts of G, then no edge of G crosses a cycle of $\mathcal{G}$. In general, it depends on the edges that cross a cycle of the cactus model, whether a set of cuts has a drawing with axis-parallel rectangles. More precisely, if $\mathcal{C}$ has a drawing $\mathcal{D}$ with axis-parallel rectangles and e crosses the cycle c then there exists an $i \in \{1, \ldots, k\}$ such that e is incident to a vertex in V_i and the drawing of $\overline{V_i}$ is contained in the simple region bounded by $\mathcal{D}(V_i, \overline{V_i})$. This statement is further formalized in the next lemma.

Lemma 2. *A family of cuts that has a cactus model $(\mathcal{G}, \varphi)$ has a drawing with axis-parallel rectangles only if the root r of $\mathcal{T}(\mathcal{G}, \varphi)$ can be chosen such that*

(R) *each edge of G that crosses a cycle c of $\mathcal{G}$ is incident to a vertex in $V \setminus V_r(\nu_c)$.*

Proof. Suppose that the set of cuts modeled by $(\mathcal{G}, \varphi)$ has a drawing $\mathcal{D}$ with axis-parallel rectangles. Let r be a node of $T = \mathcal{T}(\mathcal{G}, \varphi)$ such that for every node $\nu \neq r$ of T the set $V_r(\nu)$ is contained in the simple region bounded by

$\mathcal{D}(\{V_r(\nu), \overline{V_r(\nu)}\})$. Using the fact that $\overline{V_r(\mu)} = V_\mu(r)$ for two adjacent nodes r and μ of T, it can be shown that such a node r exists.

Let c be a cycle of $\mathcal{G}$ and let e be an edge of G that crosses c. Let $\nu_1, \dots, \nu_k$ be the descendents of ν_c such that $\nu_1, \dots, \nu_k$ is a path in c. Suppose that e is not incident to a vertex in $V \setminus V_r(\nu_c)$. Then $e \in E(V_r(\nu_i), V_r(\nu_j))$ for some $1 < i+1 < j \le k$. Let $S = \bigcup_{\ell=1}^{i+1} V_r(\nu_\ell)$ and $T = \bigcup_{\ell=i+1}^{k} V_r(\nu_\ell)$. Then, on one hand, $e \in E(S \setminus T, T \setminus S)$. But, on the other hand, $\{S, \overline{S}\}$ and $\{T, \overline{T}\}$ are two crossing cuts that are modeled by $(\mathcal{G}, \varphi)$ such that S and T are contained in the simple closed region D_S and D_T bounded by $\mathcal{D}(\{S, \overline{S}\})$ and $\mathcal{D}(\{T, \overline{T}\})$, respectively. Hence, we have again the situation as indicated in Fig. 2(b). By the definition of drawings of cuts, the drawing of the edges and vertices of $G(S \cup T)$ has to be contained inside $D_S \cup D_T$ and the drawing of the edges and vertices of $\mathcal{D}(G(\overline{S \cap T}))$ may not intersect $D_S \cap D_T$. This implies that $E(S \setminus T, T \setminus S) = \emptyset$. □

In the following, a node of the inclusion tree $T = \mathcal{T}(\mathcal{G}, \varphi)$ is called *rootable*, if it fulfills Condition (R) of the previous lemma. We give an algorithm for finding all rootable nodes of the inclusion tree $T = \mathcal{T}(\mathcal{G}, \varphi)$. We assume that the size of $\mathcal{G}$ is in $\mathcal{O}(n)$.

We use a node and edge array CROSSED on T to store the information about edges of G that cross a cycle of the cactus. Let c be a cycle of $\mathcal{G}$. CROSSED(ν_c) is true if and only if there is an edge of G that crosses c. CROSSED($\{\nu_c, \nu\}$) is true for an edge $\{\nu_c, \nu\}$ of T if and only if CROSSED(ν_c) is true and each edge that crosses c is incident to a vertex in $V_{\nu_c}(\nu)$. For all other edges and nodes of T CROSSED is false. Clearly, it can be tested in linear time, which cycles are crossed by an edge. Hence, CROSSED can be computed in $\mathcal{O}(mn)$ time. The next lemma is a reformulation of Condition (R).

Lemma 3. *A node r of T is rootable if and only if for each node ν of T with* CROSSED(ν) = TRUE *there exists an adjacent node μ of ν with* CROSSED$\{\nu, \mu\}$ = TRUE *such that r is contained in the subtree of (T, ν) rooted at μ.*

Hence, we can obtain the possible roots for constructing a drawing with axis-parallel rectangles of $\mathcal{C}$ – possibly including some cycle-replacement nodes – by deleting for all marked nodes ν the subtrees of (T, ν) rooted at those adjacent nodes μ of ν for which $\{\nu, \mu\}$ is not marked. Proceeding first from the leaves to an arbitrary root and then from this root to the leaves of T, this can be done in time $\mathcal{O}(n)$.

By constructing a drawing, we show in the next section that the conditions in Lemma 2 are also sufficient.

6 The Drawing

Let $(\mathcal{G}, \varphi)$ continue to be a cactus model of a set $\mathcal{C}$ of cuts of the graph G and let $T = \mathcal{T}(\mathcal{G}, \varphi)$ be the inclusion tree constructed in Sect. 4. We assume that the root r of T is a rootable node. We show how to construct a drawing with axis-parallel rectangles for $\mathcal{C}$. In a first step, we consider the case that $\mathcal{C}$ is a set of minimal cuts of a connected planar graph. Using planarization, we generalize the result to general sets of cuts of not necessarily planar graphs.

6.1 Planar Graphs

In this section, we assume that $\mathcal{C}$ is a set of minimal cuts of a planar connected graph G. We show how the ideas for drawing the set of all minimum cuts of a planar graph [1] yield a drawing with axis-parallel rectangles of $\mathcal{C}$. We briefly sketch the general construction and explain in more details the parts that differ from the case of drawings for the set of minimum cuts. The construction starts with a c-planar drawing $\mathcal{D}$ of the hierarchically clustered graph (G, T, r). Such a c-planar drawing always exists: Since $\mathcal{C}$ is a set of minimal cuts of a connected graph it follows that (G, T, r) is completely connected and hence c-planar.

Let $c : \nu_1, \ldots, \nu_\ell$ be a cycle of $\mathcal{G}$. Suppose that ν_c is a descendent of ν_ℓ in the tree T rooted at r. Let $V_i = V_{\nu_c}(\nu_i), i = 1, \ldots, \ell$ and let $V_0 = V_\ell$. Let $i \in \{1, \ldots, \ell-1\}$ and let $e_1, \ldots, e_k$ be the sequence of edges incident to a vertex in V_i in the cyclic order around V_i. Then $E(V_i, V_{i-1})$ and $E(V_i, V_{i+1})$ are non-empty subsequences of $e_1, \ldots, e_k$, i.e. suppose $e_1 \in E(V_i, V_{i-1})$ and $e_k \notin E(V_i, V_{i-1})$, then there are indices $1 < k_1 \leq k_2 < k_3 \leq k$ such that

$$\underbrace{e_1, \ldots, e_{k_1}}_{E(V_i,V_{i-1})}, \underbrace{e_{k_1+1}, \ldots, e_{k_2}}_{E(V_i,V_\ell)}, \underbrace{e_{k_2+1}, \ldots, e_{k_3}}_{E(V_i,V_{i+1})}, \underbrace{e_{k_3+1}, \ldots, e_k}_{E(V_i,V_\ell)} . \tag{1}$$

Further, $E(V_\ell, V_1)$ and $E(V_\ell, V_{\ell-1})$ are non-empty subsequence of the edges around $V \setminus V_\ell$. This observation can be proved in the following two steps.

1. Since G is connected and $\mathcal{C}$ is a set of minimal cuts, it follows that the sets $E(V_i, V_{i-1}), E(V_i, V_{i+1}), i = 1, \ldots, \ell - 1$ are non-empty.
2. It follows from Step 1 and the c-planarity that $E(V_i, V_{i-1}), E(V_i, V_{i+1}), i = 1, \ldots, \ell - 1$ are subsequences of the cyclic sequence of edges around V_i and that $E(V_\ell, V_1), E(V_\ell, V_{\ell-1})$ are subsequences of the cyclic sequence of edges around $V \setminus V_\ell$.

Besides, let $e \in E(V_i, V_j)$ for some $i, j \in \{1, \ldots, \ell\}, i \neq j$. Then c-planarity implies that the two vertices in the auxiliary graph $G_\mathcal{D}$ that represent the intersection of e with the boundary of ν_i and ν_j (or the boundary of ν_c, if $i = \ell$ or $j = \ell$), respectively, are adjacent. Hence the situation in the auxiliary graph $G_\mathcal{D}$ is as indicated in Fig. 3a. For $i \in \{1, \ldots, \ell - 1\}$, let $p_i^\pm$ be the path in $G_\mathcal{D}$ that is induced by the intersection of the edges in $E(V_i, V_{i\pm 1})$ with the boundary of ν_i. Similarly, let p_0^+ (p_ℓ^-) be the path in $G_\mathcal{D}$ that is induced by the intersection of the edges in $E(V_\ell, V_1)$ ($E(V_\ell, V_{\ell-1})$) with the boundary of ν_c.

A planar drawing of $\mathcal{C}$ with axis-parallel rectangles can now be obtained as follows. First, for each i, paths p_i^+ and p_{i+1}^- are united to one path. The parts of the boundary-cycle of ν_c that are not in p_0^+ or p_ℓ^- are removed. The result can be seen in Fig. 3b. Finally, at each end of each thus united path p_i^+/p_{i+1}^-,

$i = 1, \ldots, \ell - 2$, an additional vertex is inserted as indicated in Fig. 3c. In the following, we refer to these vertices as *cycle-path end-vertices*. Let G' be the graph that results from $G_{\mathcal{D}}$ by doing this for every cycle of $\mathcal{G}$. Applying a min-cost flow approach for quasi-orthogonal drawings [10,8,7,11] with some restriction on the flow to G' yields a drawing of $\mathcal{C}$ with axis-parallel rectangles. There are two restrictions on the flow necessary

1. the flow over a boundary edge from outside a boundary cycle into the inside of a boundary cycle is zero
2. the flow from a cycle-path end-vertex into the inside of a boundary cycle is minimum, i.e. there is a rectangle.

It can be shown that there is always a feasible flow for the thus restricted flow network.

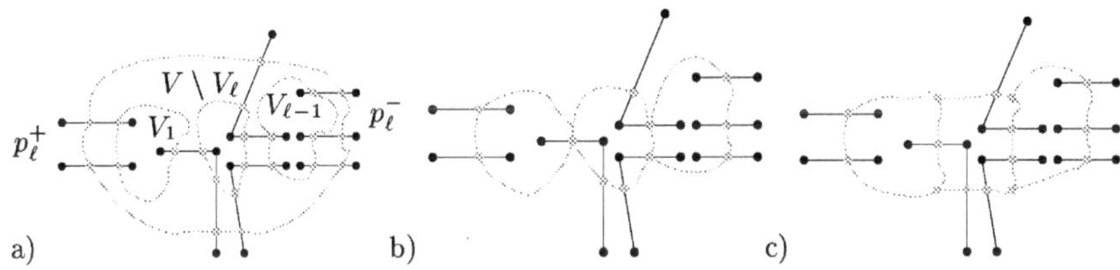

Fig. 3. Constructing a drawing for a set of cuts from a c-planar drawing.

6.2 General Graphs

In this section, we extend the results on planar drawings for families of minimal cuts to not necessarily planar drawings. Similar to the method described in [3] for hierarchically clustered graphs, the idea for the construction uses planarization techniques. Recall that $(\mathcal{G}, \varphi)$ is a cactus model for a family of cuts of G. We assume again that the root r of $T = \mathcal{T}(\mathcal{G}, \varphi)$ is rootable.

1. Let $G_P = (V, E_P)$ be a planar connected graph on the same set of vertices as G, such that $\mathcal{G}$ models a set of minimal cuts and each edge of G_P that crosses a cycle c of $\mathcal{G}$ is incident to a vertex in $V \setminus V_r(\nu_c)$.
2. Construct the auxiliary graph G'_P for $(G_P, \mathcal{G}, \varphi)$ as it is described in Sect. 6.1.
3. Construct a planar graph G', by inserting the remaining edges of G into G'_P such that each edge crosses the boundary cycle of a cut at most once.
4. Apply an algorithm for quasi-orthogonal drawings with the restrictions indicated in Sect. 6.1 to G'.
5. Delete the edges that do not correspond to original edges in G.

Remarks to Step 1. It is not always possible to choose the graph G_P with the properties required in Step 1 as a subgraph of G. The following lemma shows, however, that at least some graph with the properties required in Step 1 exists.

Lemma 4. *Let $(\mathcal{G}, \varphi)$ be a cactus model for a set of cuts of a graph with vertex set V. Then there exists a planar connected graph H with vertex set V such that*

1. *$\mathcal{G}$ models a set of minimal cuts of H and*
2. *no edge of H crosses a cycle of $\mathcal{G}$.*

Proof. Let μ be a node of T that is adjacent to a leaf v. In a first step, we start with an empty graph. Proceeding from bottom to top of the tree T rooted at μ, we can construct a graph $H_0 = (V, E_0)$ such that

1. for any node $\nu \neq \mu$ the subgraph of H_0 induced by $V_\mu(\nu)$ is a path,
2. no edge of H_0 crosses a cycle of $\mathcal{G}$, and
3. $E_0(V_\mu(\nu), \overline{V_\mu(\nu)}) = \emptyset$ if ν is adjacent to μ.

In a second step, for each node ν of T that is adjacent to μ let v_1 and v_2 be the two vertices of degree one or let $v_1 = v_2$ be the vertex of degree zero in the subgraph of H_0 induced by $V_\mu(\nu)$, respectively. Add edges $\{v, v_1\}$ and $\{v, v_2\}$ to H_0. Finally, for all leaves $w \neq v$ of T that are adjacent to r, add an edge $\{v, w\}$ to H_0.

It is now easy to see that no edge of the thus constructed graph H crosses a cycle of $\mathcal{G}$ and that $\mathcal{G}$ models a set of minimal cuts. H is a cactus and hence it is planar and connected. □

Let $H = (V, E_H)$ be the graph from the previous lemma and let G_P be a maximal planar subgraph of $(V, E \cup E_H)$ such that H is a subgraph of G_P. Then G_P is planar and connected, $\mathcal{G}$ models a set of minimal cuts of G_P and r is rootable. Hence, G_P fulfills the properties required in Step 1.

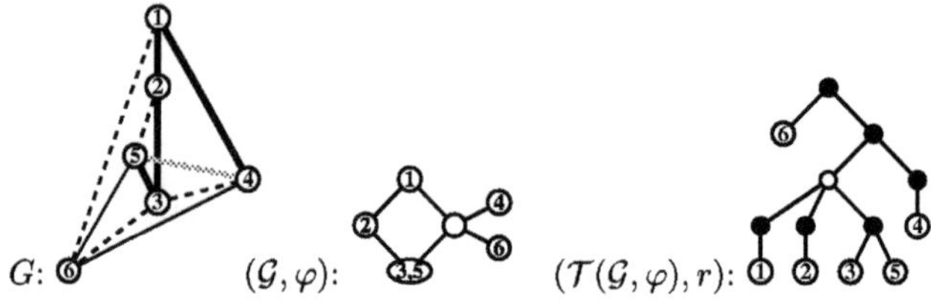

Fig. 4. A non-planar graph G and a cactus model. Deleting the grey edge yields a maximal planar subgraph. Solid thick black edges induce a spanning path in every cluster $S \neq V$ of $(G, \mathcal{T}(\mathcal{G}, \varphi), r)$. All solid black edges indicate the graph H constructed in the proof of Lemma 4.

Remarks to Step 3. We use an extention of planarization-techniques that is similar to the method introduced by Di Battista et al. [3] for drawing non-planar hierarchically clustered graphs. Edges in G that are not represented in G_P are routed iteratively through the dual graph of G'_P, replacing every crossing of an edge and a dual in the rout by a vertex of degree four. Suppose, we want to insert an edge $\{v, w\}$. Let F be the set of faces of G'_P. Let $E^*_{\{v,w\}}$ be the following set of arcs. For every edge e of G'_P, let $f_1, f_2 \in F$ be the faces that are incident to e. Then $E^*_{\{v,w\}}$ contains the *dual arcs* $(f_1, f_2)_e$ and $(f_2, f_1)_e$. Further, $E^*_{\{v,w\}}$ contains the arcs $(f, x), (x, f)$, $x \in \{v, w\}, f \in F$, f incident to x. There are no other arcs contained in $E^*_{\{v,w\}}$. In general, any simple path in the *extended dual* $G^*_{\{v,w\}} = (\{v, w\} \cup F, E^*_{\{v,w\}})$ from v to w can be used as a rout for the edge $\{v, w\}$. To achieve that the drawing of $\{v, w\}$ does not cross a boundary cycle twice, we use a restricted version of the extended dual. Let $v = \nu_1, \nu_2 \ldots, \nu_\ell = w$ be the path in T between v and w. Let $k \in \{2, \ldots, \ell - 1\}$ be such that ν_{k-1} and ν_{k+1} are descendants of ν_k. For each boundary edge e of G'_P,

- if e is contained in the drawing γ of $\{V_r(\nu_i), \overline{V_r(\nu_i)}\}, i = 2, \ldots, k-1$, delete the dual arc of e that is directed from the outside of γ to the inside of γ from $E^*_{\{v,w\}}$.
- if e is contained in the drawing γ of $\{V_r(\nu_i), \overline{V_r(\nu_i)}\}, i = k+1, \ldots, \ell - 1$, delete the dual arc of e that is directed from the inside of γ to the outside of γ from $E^*_{\{v,w\}}$.
- if e is contained in the drawing of a cut $\{V_r(\nu), \overline{V_r(\nu)}\}$ for a node $\nu \neq \nu_i, i - 2, \ldots, k-1, k+1, \ldots, \ell - 1$ of T, delete both dual arcs of e from $E^*_{\{v,w\}}$.

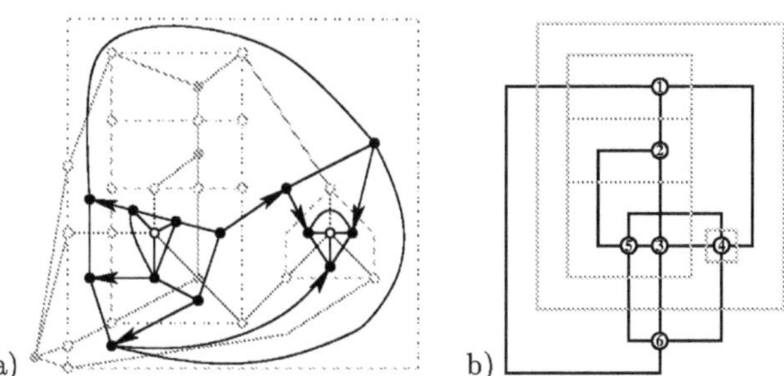

Fig. 5. a) Black edges show a connected component of the restricted extended dual for adding the edge $\{5, 4\}$ from 5 to 4 of the graph in Fig. 4. Bidirected edges are indicated by simple curves without arrows. b) A drawing with axis-parallel rectangles for the family of cuts modeled by $(\mathcal{G}, \varphi)$ in Fig. 4.

The thus restricted dual (see Fig. 5 for an illustration) guarantees that any path from v to w crosses any boundary cycle at most once. On the other hand, there is a path from v to w in the restricted dual if and only if either ν_k is not a cycle-replacement node or ν_k is a cycle-replacement node and ν_{k-1}, ν_{k+1} are adjacent in $\mathcal{G}$. Hence, by Condition (R), there is a path from v to w. Summarizing, we have shown the following characterization

Theorem 2. *A family $\mathcal{C}$ of (minimal) cuts of a (planar) graph G has a (planar) drawing with axis-parallel rectangles if and only if*

1. *$\mathcal{C}$ has a cactus model $(\mathcal{G}, \varphi)$ and*
2. *the root r of $\mathcal{T}(\mathcal{G}, \varphi)$ can be chosen such that each edge that crosses a cycle c is incident to a vertex in $V \setminus V_r(\nu_c)$.*

Provided a cactus model is given, it can be tested in time $\mathcal{O}(nm)$ whether a drawing with axis-parallel rectangles exists.

References

1. U. Brandes, S. Cornelsen, and D. Wagner. How to draw the minimum cuts of a planar graph. In J. Marks, editor, *Proceedings of the 8th International Symposium on Graph Drawing (GD 2000)*, volume 1984 of *Lecture Notes in Computer Science*, pages 103–114. Springer, 2001.
2. S. Cornelsen and D. Wagner. Completely connected clustered graphs. To appear in *Proceedings of the 29th International Workshop on Graph Theoretic Concepts in Computer Science (WG 2003)*.
3. G. Di Battista, W. Didimo, and A. Marcandalli. Planarization of clustered graphs. In M. Jünger and P. Mutzel, editors, *Proceedings of the 9th International Symposium on Graph Drawing (GD 2001)*, volume 2265 of *Lecture Notes in Computer Science*, pages 60–74. Springer, 2002.
4. Y. Dinitz, A. V. Karzanov, and M. Lomonosov. On the structure of a family of minimal weighted cuts in a graph. In A. Fridman, editor, *Studies in Discrete Optimization*, pages 290–306. Nauka, 1976. (in Russian).
5. Y. Dinitz and Z. Nutov. A 2-level cactus model for the system of minimum and minimum+1 edge–cuts in a graph and its incremental maintenance. In *Proceedings of the 27th Annual ACM Symposium on the Theory of Computing (STOC '95)*, pages 509–518. ACM, The Association for Computing Machinery, 1995.
6. Q. Feng, R. F. Cohen, and P. Eades. Planarity for clustered graphs. In P. Spirakis, editor, *Proceedings of the 3rd European Symposium on Algorithms (ESA '95)*, volume 979 of *Lecture Notes in Computer Science*, pages 213–226. Springer, 1995.
7. U. Fößmeier and M. Kaufmann. Drawing high degree graphs with low bend numbers. In F. J. Brandenburg, editor, *Proceedings of the 3rd International Symposium on Graph Drawing (GD '95)*, volume 1027 of *Lecture Notes in Computer Science*, pages 254–266. Springer, 1996.
8. G. W. Klau and P. Mutzel. Quasi orthogonal drawing of planar graphs. Technical Report MPI-I-98-1-013, Max-Planck-Institut für Informatik, Saarbrücken, Germany, 1998. Available at http://data.mpi-sb.mpg.de/internet/reports.nsf.
9. D. Lütke-Hüttmann. Knickminimales Zeichnen 4-planarer Clustergraphen. Master's thesis, Universität des Saarlandes, 1999. (Diplomarbeit).
10. R. Tamassia. On embedding a graph in the grid with the minimum number of bends. *SIAM Journal on Computing*, 16:421–444, 1987.
11. R. Tamassia, G. Di Battista, and C. Batini. Automatic graph drawing and readability of diagrams. *IEEE Transactions on Systems, Man and Cybernetics*, 18(1):61–79, 1988.

Convex Drawing for c-Planar Biconnected Clustered Graphs

Hiroshi Nagamochi[1] and Katsutoshi Kuroya[2]

[1] Toyohashi University of Technology, Toyohashi, Aichi 441-8580, Japan
naga@ics.tut.ac.jp
[2] Hitachi Advanced Digital, Inc. 292 Yoshida-cho, Totsuka-ku, Yokohama 244, Japan

Abstract. In a graph, a cluster is a set of vertices, and two clusters are said to be non-intersecting if they are disjoint or one of them is contained in the other. A clustered graph is a graph with a set of non-intersecting clusters. In this paper, we assume that the graph is planar, each non leaf cluster has exactly two child clusters in the tree representation of non-intersecting clusters, and each cluster induces a biconnected subgraph. Then we show that such a clustered graph admits a drawing in the plane such that (i) edges are drawn as straight line segments with no crossing between two edges, and (ii) the boundary of the biconnected subgraph induced by each cluster is convex polygon.

1 Introduction

A *clustered graph* $C = (G, T)$ consists of an undirected graph $G = (V, E)$ and a rooted tree $T = (\mathcal{V}, \mathcal{E})$ such that each node $c \in \mathcal{V}$, called a *cluster*, corresponds to a subset of V, denoted by $V(c)$. For a cluster c, $Ch(c)$ denotes the set of children of c, and $parent(c)$ denotes the parent of c if c is not the root. We assume that, for each non-leaf node c, it holds $V(c) = \cup_{c' \in Ch(c)} V(c')$. (Notice that a leaf cluster c may contain more than one vertex.) The subgraph of G induced by $V(c)$ for a cluster c is denoted by $G(c)$.

A clustered graph can be used to draw graphs with large size such as WWW connection graphs or VLSI schematics. The vertex set of such a graph is clustered to display a part of the graph [4]. On the other hand, there are already clustered graphs in applications such as statistics (e.g. [9]) and linguistics (e.g., [2]). Drawing clustered graphs in an understandable way is important to visualize these structures. See [1,12] for recent developments in graph drawings. In this paper, we consider how to draw clustered graphs nicely in the plane.

In a *drawing* of a clustered graph $C = (G, T)$, graph G is drawn as points and curves as usual. For each node c of T, the cluster is drawn as a simple closed region R_c that contains the drawing of $G(c)$, such that:

(1) the regions for all child clusters of c are completely contained in the interior of R_c;
(2) the regions for all other clusters are completely contained in the exterior of R_c;

G. Liotta (Ed.): GD 2003, LNCS 2912, pp. 369–380, 2004.

(3) if there is an edge e between two vertices of $V(c)$ then the drawing of e is completely contained in R_c.

We say that the drawing of edge e and region R have an *edge-region crossing* if the drawing of e crosses the boundary of R more than once. A drawing of a clustered graph is *compound planar* (c-planar, for short) if there are no edge crossings or edge-region crossings. If a clustered graph C has a c-planar drawing then it is called *c-planar* [8]. Fig. 1(a) shows an example of a c-planar drawing of a c-planar clustered graph. It is known that, if each cluster induces a connected subgraph, then testing whether a given clustered graph C has a c-planar drawing or not can be done in linear time [3]. However, the complexity status of the testing problem is unknown (see [10] for a recent progress on this issue). In this paper, we are given a clustered graph equipped with a c-planar drawing.

One of the fundamental questions in planar clustered graph drawing is: does every c-planar clustered graph admit a planar drawing such that edges are drawn as straight-line segments and clusters are drawn as convex polygons? The question has been solved affirmatively by Eades et al. [5,6,7]. To determine the xy-coordinates of the vertices in a c-planar graph which has been embedded in the plane, their method first computes y-coordinates of vertices based on an extended numbering of the st-numbering, and then determines an adequate x-coordinate of each vertex. The idea behind this is that after fixing the y-coordinates of the vertices any two disjoint clusters are separable by a horizontal line (a line parallel with the x-axis) no matter how their x-coordinates will be determined; x-coordinates can be determined so as to draw each edge as a straight line segment without taking into account the cluster structure any more. As a result, in the obtained drawing, clusters are arranged in a special way. Fig. 1(b) shows an output of their method applied to the clustered graph in Fig. 1(a). The characteristic of their drawing may be favorable to some purpose, but is rather disadvantageous to obtain a drawing in which clusters are required to be packed compactly in the plane. In this paper, we propose a new way of drawing c-planar clustered graphs. Our method is based on a divide-and-conquer approach without relaying on any special numbering on vertices, and may produce a drawing that has no structure biased in a certain direction.

2 Preliminaries

For a c-planar clustered graph $C = (G, T)$, we assume that the embedding of G of a c-planar drawing of C is given, and C is stored in an $O(n)$ space as follows, where n denotes the number of vertices in G.

A graph $G = (V, E)$ is represented by a set of adjacency lists $L(v)$ for all vertices $v \in V$, and in an adjacency list $L(v)$, all the edges incident to v appear in the same order that they appear around v in clockwise order (that is how the embedding of G in the drawing is represented). Each edge $e = (v, u)$ is equipped pointers that indicate the cells for e in the lists $L(v)$ and $L(u)$. Also we assume that, for two nodes $c, c' \in T$, their least common ancestor, denoted by $\mathrm{LCA}(c, c')$, can be found in $O(1)$ time after an $O(n)$ time preprocessing is

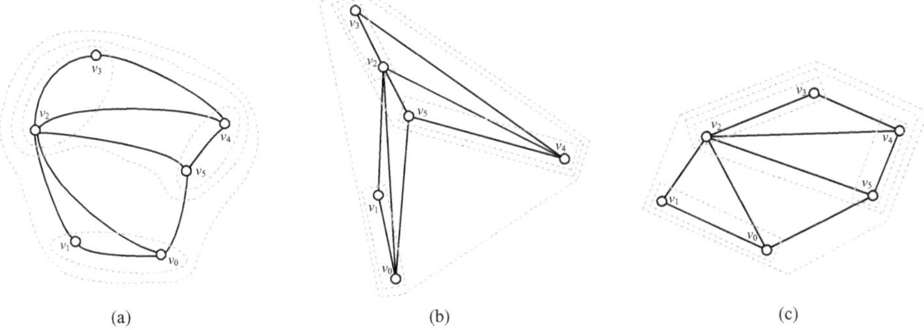

Fig. 1. Three drawings of a clustered graph C, which has five clusters $c_1 = \{v_0, v_1, v_2, v_3, v_4, v_5\}$, $c_2 = \{v_0, v_1\}$, $c_3 = \{v_2, v_3, v_4, v_5\}$, $c_4 = \{v_2, v_3\}$ and $c_5 = \{v_4, v_5\}$ each surrounded by a broken line: (a) a c-planar drawing, (b) a straight line cluster drawing, and (c) a c-planar convex cluster drawing.

applied to a rooted tree $T = (\mathcal{V}, \mathcal{E})$ [11]. We assume that in a clustered graph $C = (G, T)$, every non-leaf node of tree T has at least two children. Hence the size of $T = (\mathcal{V}, \mathcal{E})$ is $O(|\mathcal{V}| + |\mathcal{E}|) = O(n)$.

Definition 1. *For a c-planar clustered graph $G = (G, T)$, a drawing of G that satisfies the followings is called a* c-planar straight line cluster drawing *of C.*

- *Each vertex is drawn as a point and each edge is drawn as a straight line segment between two points drawn for its end vertices.*
- *For each cluster c, let the region R_c of c be the convex hull of the points drawn for the vertices in $V(c)$.*
- *For each cluster c, the edges in $G(c)$ are drawn in the interior of R_c, and the vertices and edges in $G - V(c)$ are drawn in the exterior of R_c.*
- *There are no edge crossings or edge-region crossings.*

Storing a region R_c for a cluster c in a given drawing may take $\Omega(|V(c)|)$ space, and just computing all regions would take $\Omega(n^2)$ space and time (even though computing the points for the vertices in G in a c-planar straight line cluster drawing may be done in linear time). In the above definition, we do not need explicitly store a data structure for the region R_c for each cluster c, which can be obtained by the convex hull of the points drawn for the vertices in $V(c)$. For example, the region R_c for cluster $c = \{v_0, v_1, \ldots, v_5\}$ is given by the convex hull with corner points v_0, v_1, v_3, v_4.

A clustered graph $C = (G, T)$ is a *connected clustered graph* if each cluster induces a connected subgraph of G. Eades et al. [7] proved the following result.

Theorem 1. [7] *Let $C = (G, T)$ be a c-planar connected clustered graph. Then there always exists a c-planar straight line cluster drawing of C, and such a drawing of C with n vertices can be constructed in $O(n)$ time.* □

Definition 2. *For a c-planar clustered graph $G = (G, T)$, a c-planar straight line clustered drawing of G is called a* c-planar convex cluster drawing *of C if the following holds for each cluster c, whose region R_c is defined to be the convex hull of the points drawn for the vertices in $V(c)$.*

- *The boundary of the embedding of $G(c)$ is a strict convex polygon (i.e., a convex polygon such that the internal angle at each corner is less than π).*

A connected clustered graph $C = (G, T)$ is a *biconnected clustered graph* if each cluster with more than two vertices induces a biconnected subgraph of G. In this paper, we show the next result.

Theorem 2. *Let $C = (G, T)$ be a c-planar biconnected clustered graph such that T is a binary tree. Then there always exists a c-planar convex cluster drawing of C, and such a drawing of C with n vertices and n' clusters can be constructed in $O(n + n' \log^2 n')$ time.* □

The clustered graph in Fig. 1(a) is a c-planar biconnected cluster graph, but the drawing in Fig. 1(b) is not a convex cluster drawing. Fig. 1(c) shows a convex cluster drawing of the c-planar biconnected cluster graph C in Fig. 1(a). Note that in a convex cluster drawing one can indicate the region R_c for a particular cluster c by emphasizing the boundary of $G(c)$ (say, by allocating a different color to the edges in the boundary), without drawing the boundary of R_c.

For a clustered graph $C = (G, T)$ in Theorem 2, we introduce some terminology. Fix an embedding of G in a c-planar drawing of C. For each cluster c, let $\text{ex}(c)$ denote the set of vertices on the boundary of the outer face of $G(c)$. When $\text{ex}(c)$ is represented by its members $\{v_1, v_2, \ldots, v_n\}$, we assume that $v_1, v_2, \ldots, v_n$ appear in clockwise order along the boundary of c.

Let c be a non-root cluster c with $|V(c)| \geq 3$, and c' and c'' be its child clusters of c. The set of edges commonly used in the boundaries of $G(c)$ and $G(c')$ forms a single path, and hence the set of edges that are used in the boundary of $G(c')$ but not in that of $G(c)$ forms a single path, say P_c. The set of vertices in P_c is denoted by $\lambda(c')$. The end vertices of P_c are denoted by $q_1(c')$ and $q_2(c')$, where we assume that P_c is a path that goes from $q_1(c')$ to $q_2(c')$ in clockwise order along the boundary of $G(c')$. We call these vertices $q_1(c')$ and $q_2(c')$ *joint vertices* of c'. The edge set $\{(v, w) \in E | v \in \lambda(c'), w \in \lambda(c'')\}$ is denoted by $\Delta(c)$ (see Fig. 2).

In what follows, for notational simplicity, a point embedded from a vertex v in a drawing of a clustered graph may be also denoted by v. For two points p_1 and p_2, the line segment between them is denoted by p_1p_2, and the line passing through them is denoted by $\ell(p_1, p_2)$.

2.1 Basic Transformations

In this subsection, we review some operations that can transform a c-planar convex cluster drawing into another c-planar convex cluster drawing. We use the following five types of operations.

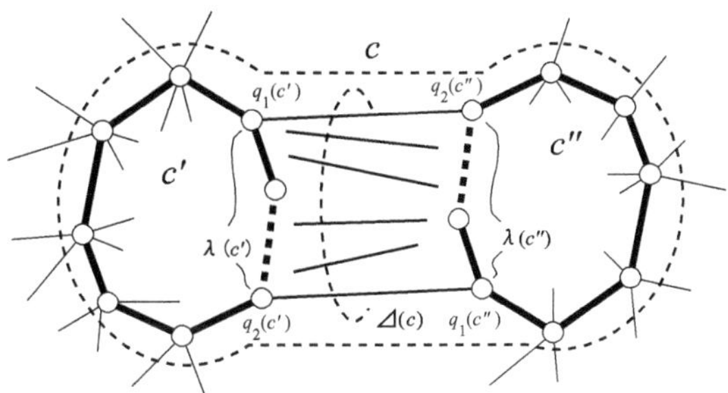

Fig. 2. Definition of $\Delta(c)$ of a cluster c, $\lambda(c')$, $q_1(c')$ and $q_2(c')$ for a child cluster c' of c'' (thick lines show cycles $\mathrm{ex}(c')$ and $\mathrm{ex}(c'')$ for the two child clusters).

Let $p = (p_x, p_y)$ be a point with an x-coordinate p_x and a y-coordinate p_y in the plane.

- Translation with respect to a vector $\boldsymbol{a} = (a_x, a_y)$: Move p to a new point $p' = (p + a_x, p + a_y)$.
- Rotation with respect to a real θ and a reference point r: Rotate p in clockwise order by the angle θ around r.
- Scaling with respect to a reference point r and $\gamma > 0$: Scale the line segment between r and p by factor γ fixing the end point r, i.e., move p to a new point $p' = (r_x + \gamma \cdot (p_x - r_x), r_y + \gamma \cdot (p_y - r_y))$.
- One-dimensional scaling with respect to a reference line ℓ, and a real $\gamma > 0$: Let r^p be the point on ℓ that is closest to p (i.e., $pr^p \perp \ell$), and move p to a new point $p' = (r^p_x + \gamma(p_x - r^p_x), r^p_y + \gamma(p_y - r^p_y))$.
- Shearing with respect to a reference line ℓ with a head and a tail, and a real $\gamma > 0$: Let $\boldsymbol{a}$ be the unit vector in the direction from the tail to the head of ℓ, and h be the distance of p from ℓ. Then move p to a new point $p' = (p_x + \gamma \cdot h \cdot a_x,\ p_y + \gamma \cdot h \cdot a_y)$ if p is on the left side with respect to ℓ with the head upward, or to a new point $p' = (p_x - \gamma \cdot h \cdot a_x,\ p_y - \gamma \cdot h \cdot a_y)$ otherwise.

For a point p in the plane, let $f(p)$ denote the point obtained by an operation f in the above. Any operation f in the above is an affine transformation, by which a given point $p = (p_x; p_y)$ is projected to a point $f(p) = (p'_x, p'_y)$ by $(p'_x, p'_y)^t := A \cdot (p_x, p_y)^t + b$ for a 2×2 matrix A and a vector b.

For a set P of points, let $f(P)$ denote the set $\{f(p) \mid p \in P\}$. We see that for a set P of points on a line segment p_1p_2, $f(P)$ is a set of a line segment $f(p_1)f(p_2)$. Also, we see that for a set P of corner points of a convex polygon, $f(P)$ is a set of corner points of some convex polygon, where the points $f(p) \in f(P)$ appear in clockwise order in the same order that they appear around P. This also implies that, after any transformation, no two edges create a corner point, i.e., no two edges cross. From these observations, we have the next.

Lemma 1. *Let $\mathcal{D}$ be a c-planar convex cluster drawing of a c-planar clustered graph C. Then the drawing $\mathcal{D}'$ obtained from $\mathcal{D}$ by applying any of the above five transformations is also a c-planar convex cluster drawing of C.* □

Note that a point $p'' = (p''_x, p''_y)$ obtained from a given point p by a sequence of operations $f_1, f_2, \ldots, f_k$ is given by $(p''_x, p''_y)^t := A'' \cdot (p_x, p_y)^t + b''$ for adequate matrix A'' and vector b''.

2.2 Drawing for c-Planar Biconnected Clustered Graphs

In the next section, we prove that every c-planar biconnected clustered graph $C = (G, T)$ such that T is a binary tree admits a c-planar convex cluster drawing. Our proof is algorithmic and can be implemented to run in polynomial time.

We construct a drawing of a given c-planar cluster graph $C = (G, T)$ by a divide-and-conquer approach. Supposing that, for a non-leaf cluster c, c-planar convex cluster drawings $\mathcal{D}(c')$ and $\mathcal{D}(c'')$ for its child clusters c' and c'' are obtained, we consider how to combine them to obtain a c-planar convex cluster drawing $\mathcal{D}(c)$ for their parent cluster c, where we may transform the two drawings if necessary. However, as shown in Fig. 3(a), in general two c-planar convex cluster drawings cannot be transformed by any of the above five operations to get a desired drawing. To overcome this, we impose a technical constraint on c-planar convex cluster drawings.

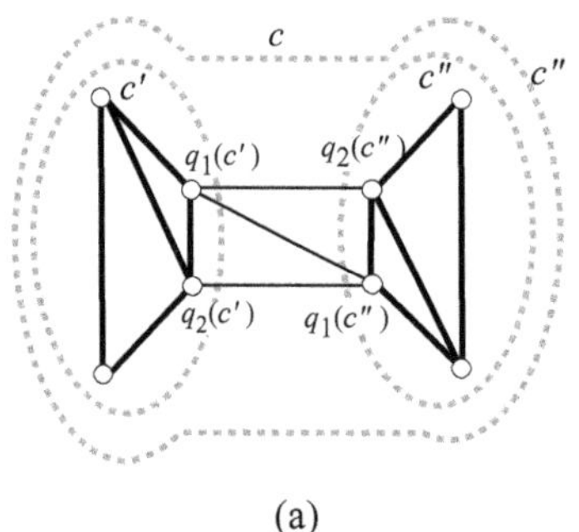

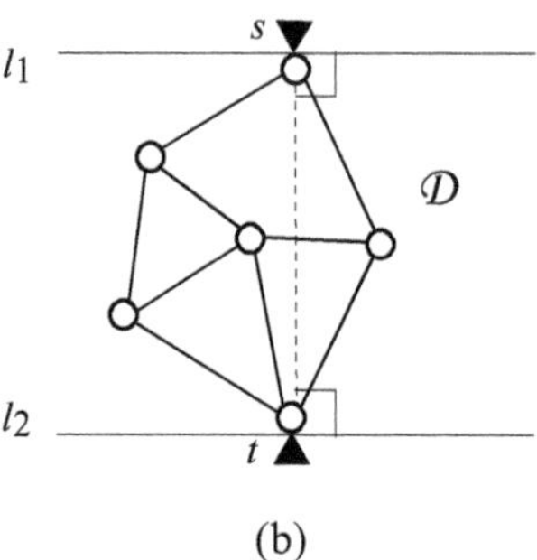

Fig. 3. (a) c-planar convex cluster drawings for child clusters c' and c'', depicted by thick lines, which cannot be combined into a c-planar convex cluster drawing for the parent cluster c; (b) a supported c-planar convex cluster drawing with support vertices s and t.

Let $\mathcal{D}$ be a c-planar convex cluster drawing of a clustered graph C. We say that a line *supports* $\mathcal{D}$ if it passes through a corner point of the boundary and all other vertices are situated in one of the half planes divided by the line. We also say that two parallel lines ℓ_1 and ℓ_2 ($\ell_1 \neq \ell_2$) *support* $\mathcal{D}$ if each of the lines supports the drawing, where the two corner points s and t (resp., their corresponding vertices in V) are called *support corners* of the drawing (resp., *support vertices*) of the clustered graph. We further say that ℓ_1 and ℓ_2 *properly support* $\mathcal{D}$ if $\ell_1 \perp st$.

Definition 3. *A c-planar convex cluster drawing is called a* supported c-planar convex cluster drawing *if there are two parallel lines* ℓ_1 *and* ℓ_2 *that properly support the drawing.*

Fig. 3(b) illustrates an example of a supported c-planar convex cluster drawing Now we formulate the following problem:

Problem 1. Input: A c-planar clustered graph $C = (G, T)$ with a planar graph G embedded in the plane and a binary tree T, and two distinct vertices s and t on the boundary of G.
Output: A supported c-planar convex cluster drawing $\mathcal{D}(C)$ of C with support vertices s and t.

For a leaf c in T, which has no child cluster, a supported c-planar convex cluster drawing $\mathcal{D}(c)$ of $G(c)$ can be easily constructed by a conventional convex drawing algorithm [13]. Also if $G(c)$ has at most three vertices, then such a drawing $\mathcal{D}(C)$ is trivially obtained. In the next section, we prove that the problem can be solved by a divide-and-conquer method.

3 Divide and Conquer

Let $C = (G, T)$ be a c-planar clustered graph with a planar graph G and a binary tree T. The clustered graph induced from C by a cluster c is denoted by $C(c) = (G(c), T(c))$, where $T(c)$ is a subtree of T rooted at node c.

Let c_r denote the root cluster of C. We choose two arbitrary vertices $\{s_r, t_r\}$ on the boundary of the outer face of G as a set $S(c_r)$ of the support vertices of C. To obtain a supported c-planar convex cluster drawing $\mathcal{D}(C)$ of C with support vertices s_r and t_r, we apply the next recursive procedure to (c_r, s_r, t_r).

DRAW(c, s, t)
if c is a leaf in T or $|V(c)| \leq 3$ holds **then**
  Return a supported c-planar convex cluster drawing $\mathcal{D}(c)$ of $C(c)$;
else
  Let c' and c'' be the two child clusters of c;
  Determine adequate pairs of support vertices $S(c') = \{s', t'\}$ and $S(c'') = \{s'', t''\}$ for induces clustered graphs $C(c')$ and $C(c'')$;
  Call DRAW(c', s', t') and DRAW(c'', s'', t'') to compute supported c-planar convex cluster drawings $\mathcal{D}(c')$ of $C(c')$ and $\mathcal{D}(c'')$ of $C(c'')$;
  Transform $\mathcal{D}(c')$ and $\mathcal{D}(c'')$ into another c-planar convex cluster drawings respectively by affine transformations $f_{c''}$ and $f_{c''}$ (which are obtained by sequences of basic transformations) before combining them into a supported c-planar convex cluster drawing $\mathcal{D}(c)$ of $C(c)$;

Return $\mathcal{D}(c)$.
/* end if */

3.1 Dividing Phase

In the dividing phase of our divide-and-conquer procedure DRAW, support vertices of a cluster are determined from the support vertices of its parent cluster in the following way. Let c be a cluster and $S(c) = \{s, t\}$ be the pair of support vertices. We choose the sets of support vertices $S(c') = \{s', t'\}$ and $S(c'') = \{s'', t''\}$ of its child clusters c' and c''. We distinguish the next two cases.

Case-1. One of $G(c')$ and $G(c'')$ contains both support vertices in $S(c)$; $S(c) \subseteq V(c')$ is assumed without loss of generality (see Fig. 4(a)). Define the pairs $S(c')$ and $S(c'')$ of support vertices of c' and c'' by

$$s' := s, \quad t' := t, \quad s'' := q_2(c'') \text{ and } t'' := q_1(c'').$$

Case-2. One of the support vertices in $S(c)$ belongs to $G(c')$ and the other $G(c'')$ (see Fig. 4(b)). Define the pairs $S(c')$ and $S(c'')$ of support vertices of c' and c'' by

$$s' := q_1(c'), \quad t' := q_2(c'), \quad s'' := q_2(c'') \text{ and } t'' = q_1(c'').$$

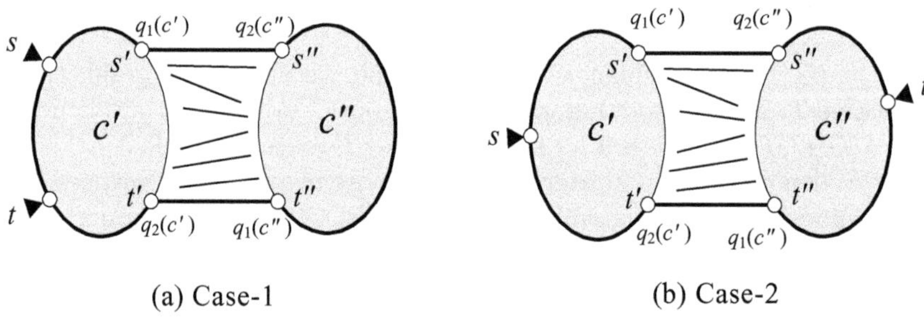

Fig. 4. Illustration for support vertices, where (a) and (b) indicate Cases-1 and 2, respectively.

3.2 Combining Phase

In this subsection, we consider the combining phase of DRAW. Suppose that, for two child clusters c' and c'' of a cluster c, we have obtained their supported c-planar convex cluster drawings $\mathcal{D}(c')$ and $\mathcal{D}(c'')$ with the support vertices which have been determined during the dividing phase. We may transform each of the drawings $\mathcal{D}(c')$ and $\mathcal{D}(c'')$ by some of the basic transformations.

In what follows, we further assume that $|V(c')| \geq 3$ and $|V(c'')| \geq 3$ (the case of $|V(c')| \leq 2$ or $|V(c'')| \leq 2$ can be treated with a slight modification of the subsequent argument). It suffices to show that two drawings for child clusters can be combined so as to meet the following three conditions:

(i) The boundary formed for $G(c)$ is a strict convex polygon.
(ii) Edges in $\Delta(c)$ (i.e., those between $G(c')$ and $G(c'')$) are drawn by line segments without creating any intersection with each other or with the boundaries of $G(c')$ and $G(c'')$.
(iii) There exits a pair of parallel lines that support the drawing for $G(c)$.

In (iii), the parallel lines do not necessarily *properly* support the drawing since we can always make them properly support the drawing by applying a shearing operation with respect to one of the lines.

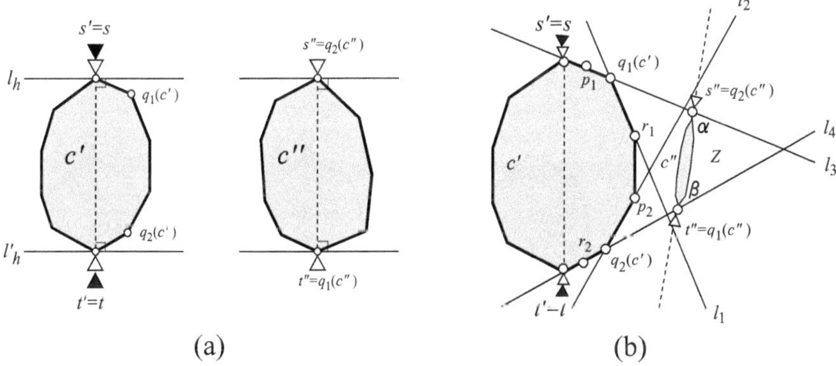

Fig. 5. (a) Given supported c-planar convex cluster drawings $\mathcal{D}(c')$ and $\mathcal{D}(c'')$ in Case 1, and (b) Combining supported c-planar convex cluster drawings $\mathcal{D}(c')$ and $\mathcal{D}(c'')$ in Case 1.

We fix the drawing $\mathcal{D}(c')$ in the xy-plane so that its supporting lines ℓ_h and ℓ'_h are parallel with the x-axis (hence the line segment $s't'$ is parallel with the y-axis). Although the other drawing $\mathcal{D}(c'')$ may be transformed before being combined with $\mathcal{D}(c')$, we temporarily fix $\mathcal{D}(c'')$ so that its supporting lines are parallel with the x-axis. We assume without loss of generality that ℓ_h and ℓ'_h pass through point s' and t', respectively, and that the y-coordinate of s' (resp., $q_1(c')$, s'', $q_2(c'')$) is larger than that of t' (resp., $q_2(c')$, t'', $q_1(c'')$).

Let $p_1, r_1 \in \text{ex}(c')$ (resp., $p_2, r_2 \in \text{ex}(c')$) be the vertices adjacent to $q_1(c')$ (resp., $q_2(c')$) such that these vertices appear along the boundary of $\mathcal{D}(c')$ in the order of $p_1, q_1(c'), r_1, p_2, q_2(c')$ and r_2. Define lines $\ell_1 = \ell(q_1(c'), r_1)$, $\ell_2 = \ell(q_2(c'), p_2)$, $\ell_3 = \ell(p_1, q_1(c'))$ and $\ell_4 = \ell(r_2, q_2(c'))$, where for a technical reason we set $\ell_3 = \ell_h$ if $q_1(c') = s'$ (resp., $\ell_4 = \ell'_h$ if $q_1(c') = t'$).

The other drawing $\mathcal{D}(c'')$ will be situated in a region Z which is defined below distinguishing Cases-1 and 2.

Case-1 (See Fig. 5(a).) The Z is set to be the region that is above lines ℓ_1 and ℓ_4 and below lines ℓ_2 and ℓ_3. We now show how to put $\mathcal{D}(c'')$ inside the Z. Choose an internal point α (resp., β) of the line segment induced by Z from ℓ_3 (resp., ℓ_4).

By applying translation, rotation and scaling operations, we put $\mathcal{D}(c'')$ in Z in such a way that $q_2(c'')$ and $q_1(c'')$ fall on the points α and β, respectively. Let $\mathcal{D}_1$ be the drawing transformed from $\mathcal{D}(c'')$ in this way. Finally we apply to $\mathcal{D}_1$ a one-dimensional scaling with respect to line $\ell(\alpha,\beta)$ and a sufficiently small real γ so that all points (except for $q_2(c'')$ and $q_1(c'')$) in the drawing $\mathcal{D}_2$ transformed from $\mathcal{D}_1$ are situated properly inside Z (see Fig.5(b)). Then a drawing $\mathcal{D}(c)$ for the parent cluster c is set to be the union of $\mathcal{D}(c')$ and $\mathcal{D}_2$. It is not difficult to see that the resulting drawing $\mathcal{D}(c)$ for $C(c)$ satisfies all conditions (i),(ii) and (iii) for supporting lines ℓ_h and ℓ'_h.

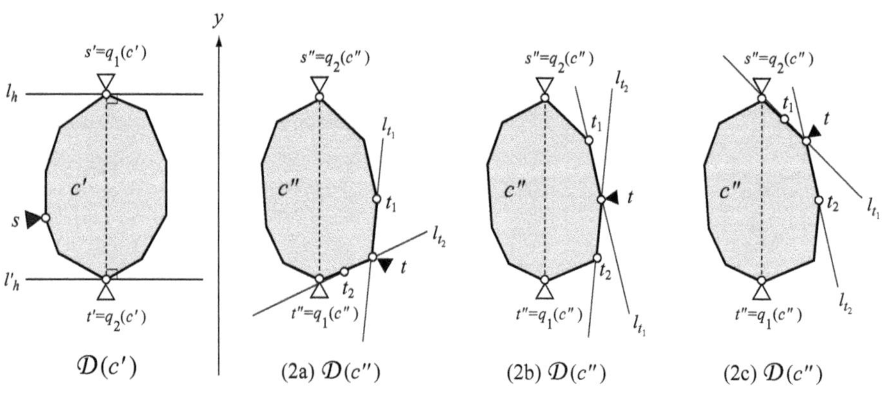

Fig. 6. Illustration for supported c-planar convex cluster drawings $\mathcal{D}(c')$ and $\mathcal{D}(c'')$ in Case 2, where (2a), (2b) and (2c) show three subcases for $\mathcal{D}(c'')$.

Case-2 (See Fig. 6.) Assume that two support vertices s and t in $S(c)$ belong to c' and c'', respectively. Let s_1 and s_2 be the vertices in $ex(c')$ that appear respectively before and after s when we visit the boundary of $G(c')$ in clockwise order. Let ℓ_{s_i} $(i = 1, 2)$ be the line $\ell(s, s_i)$.

Let t_1 and t_2 be the vertices in $ex(c'')$ that appear respectively before and after t when we visit the boundary of $G(c'')$ in clockwise order, and let $\ell_{t_i} = \ell(t, t_i)$ $(i = 1, 2)$. Then we define lines ℓ^{ref} and ℓ^{tan} by distinguishing the following three subcases, where $g(\ell)$ for a line ℓ denotes the gradient of ℓ.

(2a) $g(\ell_{t_1}) > 0$ or $\ell_{t_1} \parallel s''t''$ (see Fig. 6(2a)). Then define ℓ^{ref} (resp., ℓ^{tan}) to be a line parallel with ℓ_{s_1} (resp., ℓ_{s_2}).
(2b) $g(\ell_{t_2}) > 0$ and $g(\ell_{t_1}) < 0$ (see Fig. 6(2b)). Then define ℓ^{ref} (resp., ℓ^{tan}) to be a line parallel with ℓ_{s_1} (resp., ℓ_{s_2}).
(2c) $g(\ell_{t_2}) < 0$ or $\ell_{t_2} \parallel s''t''$ (see Fig. 6(2c)). Then define ℓ^{ref} (resp., ℓ^{tan}) to be a line parallel with ℓ_{s_2} (resp., ℓ_{s_1}).

We consider case (2a) (the rest of the cases can be treated similarly). Let α (resp., β) be the intersection of ℓ_{ref} and ℓ_h (resp., ℓ_{ref} and ℓ'_h). We choose ℓ_{ref} so that both α and β are strictly above ℓ_1 and below ℓ_2. Let $\ell'_1 = \ell(q_1(c'), \beta)$ and $\ell'_2 = \ell(q_2(c'), \alpha)$. Then the Z is defined to be the region that is above ℓ'_1 and ℓ'_h and below ℓ'_2 and ℓ_h.

We now show how to put the drawing $\mathcal{D}(c'')$ within the region Z. We first transform $\mathcal{D}(c'')$ by applying translation, rotation and scaling operations so that points $q_2(c'')$ and $q_1(c'')$ fall on α and β, respectively. We next apply one-dimensional scaling operation to the resulting drawing $\mathcal{D}_1$ so that all the points in the drawing except for $\alpha = q_2(c'')$ and $\beta = q_1(c'')$ are properly contained in Z and ℓ^{tan} has a gradient between those of ℓ_{s_1} and ℓ_{s_2} (see Fig. 7). Since line $\ell^{ref} = \ell(\alpha, \beta)$ is parallel with ℓ_{s_2}, line $\ell_{t_1} = \ell(t, t_1)$ will have the gradient between those of ℓ_{s_1} and ℓ_{s_2} if $\mathcal{D}(c'')$ gets enough close to line segment $\alpha\beta$. Let $\mathcal{D}_2$ be the resulting drawing.

As observed in Case 1, it is not difficult to see that the drawing $\mathcal{D}(c)$ obtained by combining $\mathcal{D}(c')$ and $\mathcal{D}_2$ meets conditions (i), (ii) and (iii) with lines ℓ_{s_2} and ℓ^{tan} in $\mathcal{D}_2$.

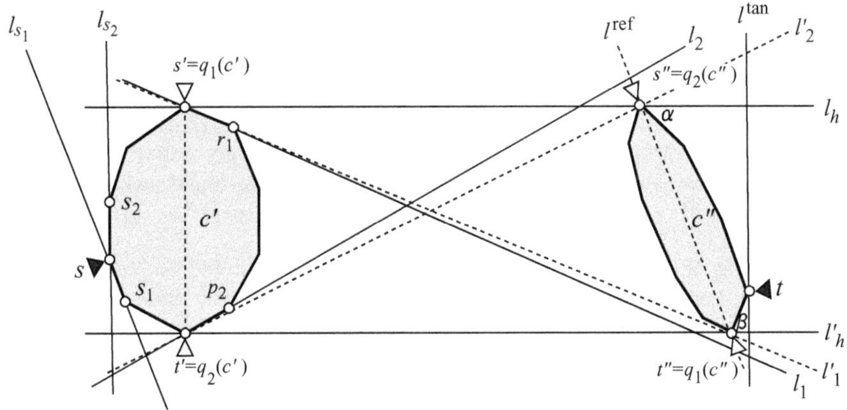

Fig. 7. Combining supported c-planar convex cluster drawings $\mathcal{D}(c')$ and $\mathcal{D}(c'')$ in Case 2.

Each iteration in the dividing and combining phases of DRAW can be executed in $O(n)$ time. There are $O(|\mathcal{V}|) = O(n')$ such iterations. A naive implementation of DRAW takes $O(n'n)$ time. This can be reduced to $O(n+n' \log^2 n')$. Note that, to determine $f_{c'}$ and $f_{c''}$ in an iteration of the combining phase, we only need to know the positions of joint nodes of c' and c'' and their neighbors on ex(c') and ex(c''). All such nodes can be identified in linear time before resorting procedure DRAW. The position of such a node, say $q_1(c')$ in drawing $\mathcal{D}(c')$ can be obtained by $f_{\langle c_1, c' \rangle}(q_1(c'))$ after computing the synthesized affine transformation

$$f_{\langle c_1, c' \rangle} \equiv f_{c_k} \circ f_{c_{k-1}} \circ \cdots \circ f_{c_2} \circ f_{c_1},$$

where c_1 is the leaf cluster containing $q_1(c')$ and $c_1, c_2, \ldots, c_k, c_{k+1}(= c')$ are the clusters that appear in this order from c to c' in T. The computation can be executed in $O(n' \log^2 n')$ time by a technique of pointer jumping and a decomposition of T into disjoint paths (the detail is omitted). Once the affine transformations f_c, $c \in \mathcal{V} - \{c_r\}$ have been determined, the final drawing $\mathcal{D}(c_r)$ can be constructed by transforming the points in each leaf cluster c' by $f_{\langle c', c_r \rangle}$ in $O(n + n' \log^2 n')$ time.

References

1. G. D. Battista, P. Eades, R. Tamassia, and I. G. Tollis, *Graph drawing*, Prentice-Hall, 1999.
2. V. Batagelj, D. Kerzic and T. Pisanski, Automatic clustering of languages, *Computational Linguistics*, 18(3) 339–352, 1992.
3. E. Dahlhaus, Linear time algorithm to recognize clustered planar graphs and its parallelization (extended abstract), *Latin America symposium on theoretical informatics* (LATIN'98), Lecture Notes in Computer Science, 1380, 239–248, 1998.
4. C. A. Duncan, M. T. Goodrich, and S. G. Kobourov, Planarity-preserving clustering and embedding for large planar graphs, *Graph Drawing* (GD'99), Lecture Notes in Computer Science, 1731, 186–196, 1999.
5. P. Eades, Q.-W. Feng and H. Nagamochi, Drawing clustered graphs on an orthogonal grid, *Journal of Graph Algorithms and Application*, 3(4), 3–29, 1999.
6. P. Eades and Q.-W. Feng, Drawing clustered graphs on orthogonal grid, *Graph Drawing* (GD'97), Lecture Notes in Computer Science, 1353, 146–157, 1997.
7. P. Eades, Q.-W. Feng, X. Lin, and H. Nagamochi, Straight-line drawing algorithms for hierarchical graphs and clustered graphs, *Technical Report Newcastle University* (submitted to a journal), 1999.
8. Q.-W. Feng, R.-F. Cohen and P. Eades, Planarity for clustered graphs, *Algorithms* (ESA'95), Lecture Notes in Computer Science, 979, 213–226, 1995.
9. E. Godehardt, *Graphs as Structural Models*, Advances in System Analysis 4, Vieweg, 1998.
10. C. Gutwenger, M. Jünger, S. Leipert, P. Mutzel, M. Percan and R. Weiskircher, Advances in c-planarity testing of clustered graphs (extended abstract), *Graph Drawing* (GD 2002), Lecture Notes in Computer Science, 2528, 220–235, 2002.
11. D. Harel and R. Tarjan, Fast algorithms for finding nearest common ancestors, *SIAM J. Computing*, 13, 338–355, 1984.
12. M. Kaufmann and D. Wagner, Drawing graphs, *Lecture Notes in Computer Science*, 2025, Springer, 2001.
13. T. Nishizeki and N. Chiba, *Planar Graphs: Theory and Algorithms*, North-Holland Mathematics Studies 140/32, 1988.

Layout of Directed Hypergraphs with Orthogonal Hyperedges
(Extended Abstract)*

Georg Sander

ILOG Deutschland GmbH, Ober-Eschbacher Str. 109,
61352 Bad Homburg, Germany
sander@ilog.fr

Abstract. We present a layout algorithm for directed hypergraphs. A hypergraph contains hyperedges that have multiple source and target nodes. Hyperedges are drawn with orthogonal segments. Nodes are organized in layers, so that for the majority of hyperedges the source nodes are placed in a higher layer than the target nodes, similar to traditional hierarchical layout [8,11]. The algorithm was implemented using ILOG JViews[10] for a project that targeted electrical signal visualization.

1 Introduction

While classical graphs deal with edges between pairs of nodes (or vertices), a hypergraph deals with hyperedges between more than two nodes. In mathematical terms, a directed hypergraph $G = (N, E)$ contains a set N of nodes and a set $E \subseteq \mathcal{P}(N) \times \mathcal{P}(N)$ of hyperedges. A hyperedge $e = (S, T)$ has source nodes $S \subseteq N$ and target nodes $T \subseteq N$.

In the literature, various drawing strategies are described for *undirected* hypergraphs [5,6,7]. To lay out *directed* hyperedges, some graph drawing tools use traditional polyline graph layout techniques applied on regular edges, and simulate the hyperedge by overlapping the start line segments of all regular edges that represent the hyperedge [9,10]. A disadvantage of this approach is that the paths of the simulated hyperedge branches out very early, so that the hyperedge drawings are unnecessary complex. We show in this article a better solution.

In a project for the automobile industry, we developed a display tool for schematics of automotive communication networks. To show different views to such a communication network and to show different levels of details we provided three different layout algorithms. This paper sketches the foundations of one of these algorithms which draws so-called *function net diagrams* of a communication network. The nodes in a function net diagram represent control units for, e.g., outside mirrors, turn signals, automatic break assistants, or horns. Connections between the nodes represent electric signals that are emitted from a single

* The full article is available via `ftp://ftp.ilog.fr/private/ILOG.de/rnd/gsander/public/hypergraph.ps.gz`

G. Liotta (Ed.): GD 2003, LNCS 2912, pp. 381–386, 2004.

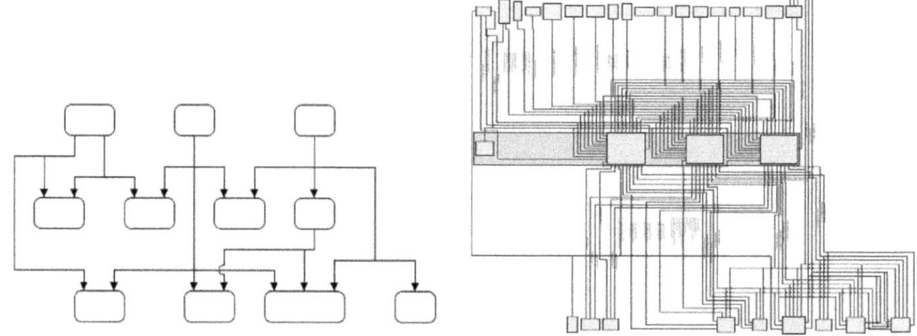

Fig. 1. Hypergraph with orth. hyperedges (left: schematic, right: real world diagram)

function and are received by one or several other functions. For example, a signal containing distance information is submitted to a park distance control unit.

An electrical signal is essentially a single source hyperedge: it starts at the electrical component that produces the signal, and it ends at all components that need the signal as input. The drawing conventions for these diagrams include that the hyperedge is drawn by a set of orthogonal line segments so that there is a path along the segments from the source node to each target node. Each path towards one target node should share many segments with the paths to the other target nodes. (Fig. 1). The customer requirements include that the components (nodes) are organized in horizontal layers. Additional goals are to balance the diagram and to avoid unnecessary crossings, segment overlaps and bends.

The traditional hierarchical layout [8,11] has similar layout objectives; however, it is designed for classical graphs and does not work for hypergraphs. In the following, we assume that the reader is already familiar with the details of traditional hierarchical layout. We sketch the new additions that are necessary to convert the traditional hierarchical layout into an algorithm suitable for hypergraphs. Even though the project used single source hyperedges, our algorithm can be applied to general directed hyperedges as well.

2 Grid Representation

The first goal is to position the nodes on a preliminary two-dimensional grid. Each node consumes one grid cell, i.e., it obtains an integer grid coordinate (i, j) where i is the column number and j is the row number of the grid. The grid representation easily allows the routing of long hyperedge segments that span several columns or several rows without bends because the nodes are aligned to the grid cells so that there is free space between rows and columns (Fig. 2). In order to produce a more balanced drawing, the layout will dissolve the preliminary grid in a later step.

Traditional hierarchical layout deals with layers, which corresponds to rows in this grid. For each hyperedge, all source nodes should be in a higher layer

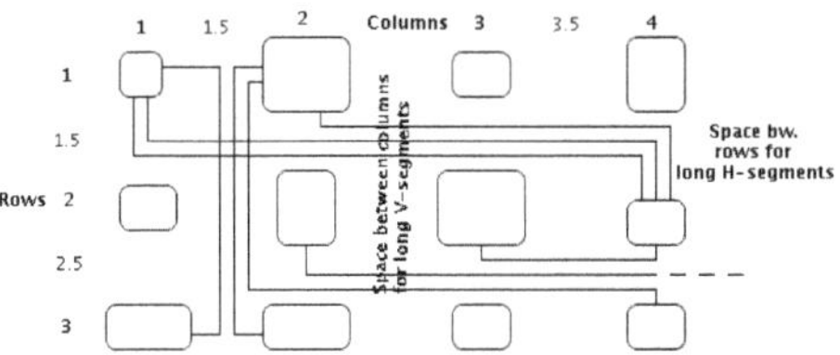

Fig. 2. Grid representation of the hypergraph

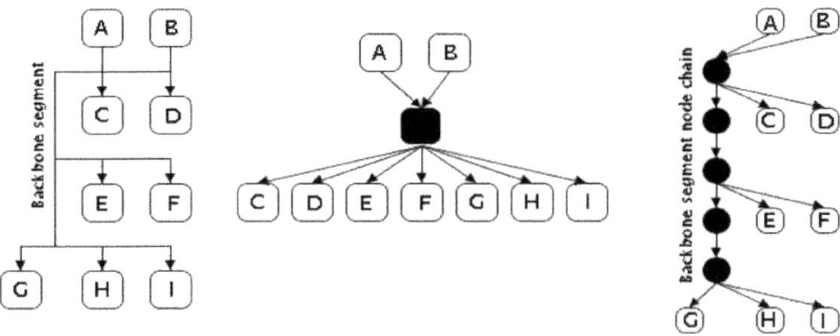

Fig. 3. Left: hypergraph with 1 hyperedge, mid: layering graph, right: cross.red.graph

than all target nodes (source-target condition). Therefore, we translate the hypergraph into a layering graph. Besides hypergraph nodes, each hyperedge e has a representative node n_e in the layering graph. For each hyperedge e, regular edges are added to the layering graph from the source nodes to n_e and from n_e to the target nodes (Fig. 3 middle). The layering graph is an appropriate representation of the source-target relationship of the hyperedges. If the layering graph is cyclic, then some hyperedges must violate the source-target condition. Since the layering graph is a regular graph, we use the standard heuristic to make the graph acyclic and to calculate a ranking of the nodes [1]. From the ranking, the layer number (row number) can be obtained by eliminating all ranks that contain only hyperedge nodes n_e but no regular nodes.

In order to calculate the column number, we need an appropriate relative ordering of the nodes within the layers. This can be considered as preliminary crossing reduction, because a good ordering ensures that later a routing with only few crossings is possible. Each hyperedge that spans several layers must have a backbone segment from which the segments can branch out that connect the end nodes. We translate the hypergraph into a crossing reduction graph (Fig. 3 right). The crossing reduction graph differs from the layering graph in the representation of the hyperedges:

- We duplicate the number of layers: each node at layer j corresponds to a node at layer $2j$ in the crossing reduction graph. The odd layer numbers are reserved for nodes representing hyperedges.
- The backbone segment of a hyperedge is represented by a chain of nodes in the crossing reduction graph that spans from $j_{\min} + 1$ to $j_{\max} - 1$, where

$j_{\min}$ and $j_{\max}$ is the minimal and maximal layer number of an end node of the hyperedge in the crossing reduction graph. All nodes of the chain are sequentially connected by regular edges in the crossing reduction graph.
- Each source node n of the hyperedge at layer j is connected by a regular edge to the node of the backbone segment chain at layer $j+1$.
- Each target node of the hyperedge at layer j is connected to the node of the backbone segment chain at layer $j-1$.

The crossing reduction graph is a proper hierarchy, hence the standard techniques [8] for crossing reduction in hierarchical layout can be applied to obtain the relative ordering of nodes within levels.

3 Creating the Hyperedge Segments

Once all nodes have integer row and column coordinates, we create the hyperedge segments so that they run between the node grid coordinates. All horizontal segments that must be placed in the free space between row j and $j+1$ get the preliminary coordinate $j+0.5$, and similarly with the vertical segments.

For each hyperedge, first the vertical backbone segment is created, and then the horizontal segments that run between the rows are connected as needed to the backbone segment. For each hyperedge, only maximally one horizontal row segment per row is needed. Finally, the row segments are connected to the source and target nodes of the hyperedge via vertical end segments. The preliminary coordinate assignment for all segments is straight forward, except for the backbone segment. We use a heuristic for the preliminary coordinate of the backbone segment. The heuristic has the goal to reduce the number of segment crossings. The full article illustrates the details. As result, all hyperedge segments are created, but segments of different hyperedges may overlap, since only segment coordinates 0.5, 1.5, 2.5, etc, are used.

4 Disentangling Segment Coordinates

To resolve the segment overlaps, all segments that run between the same grid row or grid column are collected. For instance, all horizontal segments at coordinate $k+0.5$ are collected in a set S_k^H. The goal is to spread them to different coordinates between k and $k+1$. The distribution of the segments influences the number of segment crossings (Fig. 4). A simple yet efficient solution of the problem based on sifting is illustrated in [3]. We sketch here quickly a different, more sophisticated solution: To minimize the number of crossings, we generate the segment crossing graph for S_k^H. Each segment of S_k^H corresponds to a node in the segment crossing graph. For each pair of segments $s_1, s_2 \in S_k^H$, we calculate the number of crossings C_1 if s_1 has a lower coordinate than s_2, and C_2 if s_2 has a lower coordinate than s_1, If $C_1 < C_2$, we add an edge between s_1 and s_2 with cost $C_2 - C_1$, otherwise we add an edge between s_2 and s_1 with cost $C_1 - C_2$.

If the segment crossing graph is acyclic, the segment crossing graph can be sorted topologically. If the final segment coordinates respect this topological

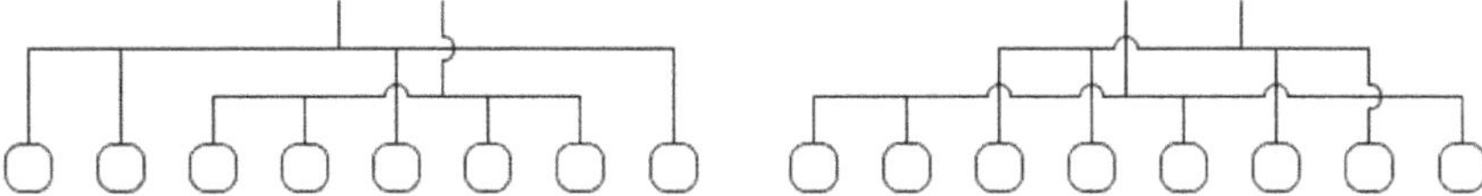

Fig. 4. Two distributions of the same segments. Left: 2 crossing, right: 5 crossings

ordering, then the number of crossings is minimal and no segments overlap. However, usually, the segment crossing graph contains cycles. In this case, some edges must be removed from the segment crossing graph to break the cycles, using one of the standard techniques for this problem [1].

Finally, we calculate a ranking of the segments from the acyclic segment crossing graph. Segments of S_k^H with rank $r \in \{1, \ldots, r_{\max}\}$ get the coordinate $k + r/(r_{\max} + 1)$, hence are spread between k and $k+1$. After this step is done for all sets S_k^H of horizontal segments, the same step is done for all sets S_k^V of vertical segments.

5 Balancing

Nodes aligned to a grid waste space. Furthermore, the layout is not yet well balanced, because the column numbers of the nodes were calculated from the ordering of the nodes in the crossing reduction graph, and the crossing reduction graph does not take any balance criteria into account. The requirements of the application included that the nodes are organized in layers but not on a grid. Hence we dissolve the grid now. We keep the rows (layers) but remove the columns, by allowing the nodes to be shifted within the layers to any non-integer coordinate. The balancing rules require for instance that a source node is centered above the target nodes (Fig. 5).

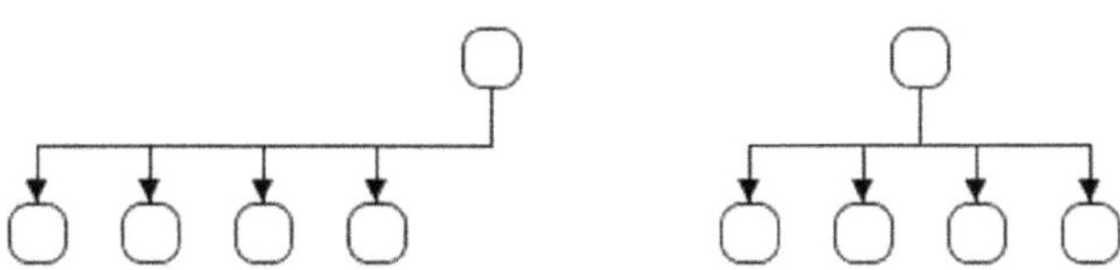

Fig. 5. Left: unbalanced situation. Right: source node is balanced.

In [8], we introduced the pendulum method to balance a traditional hierarchical layout. This method is extended for hypergraphs. The central question of the method is how far a node or segment can be shifted without overlapping the neighbored objects. In a classical, layered graph, this question is trivial to answer. In a hypergraph, it is more difficult, because long vertical segments may be influenced by nodes on different levels. Therefore, a visibility graph [2] is constructed. Edges in the visibility graph indicate the neighbor relationship between nodes and segments. The modification of the pendulum method with the visibility graph is described in the full article.

6 Conclusion and Acknowledgment

A variant of the layout algorithm was implemented by using the ILOG JViews Component Suite in Java [10]. As with the traditional hierarchical layout, many subproblems of the hypergraph layout are NP complete [4], and heuristics are used. The layout speed is sufficient, however it is slightly slower than traditional hierarchical layout, due to the fact that the treatment of hyperedges is more complex than the treatment of regular edges.

The layout algorithm was developed in a project funded by BMW. We thank the project team at ILOG: F. Baumann, X. Loiseau, C. Sirvain, F. Stork, and J.P. Vandara. Furthermore the close collaboration with Mr. Schumm and Dr. Schuller (both BMW) is appreciated.

References

1. G. Di Battista, P. Eades, R. Tamassia, and I. G. Tollis. *Graph Drawing.* Prentice-Hall, Inc., New Jersey, 1999.
2. G. Di Battista and R. Tamassia. Algorithms for plane representations of acyclic digraphs. *Theoret. Comput. Sci.*, 61:175–198, 1988.
3. T. Eschbach, W. Günther, and B. Becker. Crossing reduction for orthogonal circuit visualization. In *Proc. International Conference on VLSI, Las Vegas*, pages 107–113. CSREA Press, 2003.
4. M. R. Garey and D. S. Johnson. Computers and intractability: A guide through the theory of NP-Completeness. *W. H. Freeman, New York, NY*, 1979.
5. H. Gropp. The drawing of configurations. In *Proc. Symposium on Graph Drawing, GD'95*, pages 267–276. Springer, LNCS 1027, 1996.
6. D. S. Johnson and H. Pollak. Hypergraph planarity and the complexity of drawing venn diagrams. *Journal of Graph Theory*, 11(3):309–325, 1987.
7. E. Mäkinen. How to draw a hypergraph. *International Journal of Computer Mathematics*, 34:177–185, 1990.
8. G. Sander. Graph layout through the VCG tool. In *Proc. DIMACS International Workshop on Graph Drawing, GD'94*, pages 194–205. Springer, LNCS 894, 1995.
9. G. Sander. A fast heuristic for hierarchical Manhattan layout. In *Proc. Symposium on Graph Drawing, GD'95*, pages 447–458. Springer, LNCS 1027, 1996.
10. G. Sander and A. Vasiliu. The ILOG JViews graph layout module. In *Proc. Symposium on Graph Drawing, GD 2001*, pages 438–439. Springer, LNCS 2265, 2002.
11. K. Sugiyama, S. Tagawa, and M. Toda. Methods for visual understanding of hierarchical systems. *IEEE Trans. Sys. Man, and Cybernetics*, SMC 11(2):109–125, 1981.

No-Bend Orthogonal Drawings of Subdivisions of Planar Triconnected Cubic Graphs
(Extended Abstract)

Md. Saidur Rahman, Noritsugu Egi, and Takao Nishizeki

Graduate School of Information Sciences, Tohoku University,
Aoba-yama 05, Sendai 980-8579, Japan.
{saidur,egi}@nishizeki.ecei.tohoku.ac.jp
nishi@ecei.tohoku.ac.jp

Abstract. A plane graph is a planar graph with a fixed embedding. In a no-bend orthogonal drawing of a plane graph, each vertex is drawn as a point and each edge is drawn as a single horizontal or vertical line segment. A planar graph is said to have a no-bend orthogonal drawing if at least one of its plane embeddings has a no-bend orthogonal drawing. In this paper we consider a class of planar graphs, called subdividions of planar triconnected cubic graphs, and give a linear-time algorithm to examine whether such a planar graph G has a no-bend orthogonal drawing and to find one if G has.

1 Introduction

An *orthogonal drawing* of a plane graph G is a drawing of G with the given embedding such that each vertex is mapped to a point, each edge is drawn as a sequence of alternate horizontal and vertical line segments, and any two edges do not cross except at their common end [RN02,RNN02,RNN99,S84,T87]. A *bend* is a point where an edge changes its direction in a drawing. If G has a vertex of degree five or more, then G has no orthogonal drawing. On the other hand, if G has no vertex of degree five or more, that is, the maximum degree Δ of G is at most four, then G has an orthogonal drawing, but may need bends. Some plane graphs have an orthogonal drawing without bends, in which each edge is drawn by a single horizontal or vertical line segment. We call such a drawing a *no-bend drawing*. Figure 1(a) depicts a no-bend drawing of the plane graph in Fig. 1(b). Rahman *et al.* [RNN02] obtained a necessary and sufficient condition for a plane graph G with $\Delta \leq 3$ to have a no-bend drawing, and gave a linear-time algorithm to find a no-bend drawing if G has.

We say that *a planar graph G has a no-bend drawing* if at least one of the plane embeddings of G has a no-bend drawing. Figures 1(b), (c) and (d) depict three of the different plane embeddings of the same planar graph G. Among them only the embedding in Fig. 1(b) has a no-bend drawing as illustrated in Fig. 1(a). Thus the *planar* graph G has a no-bend drawing. It is a NP-complete problem to examine whether a planar graph G with $\Delta \leq 4$ has a no-bend drawing

G. Liotta (Ed.): GD 2003, LNCS 2912, pp. 387–392, 2004.

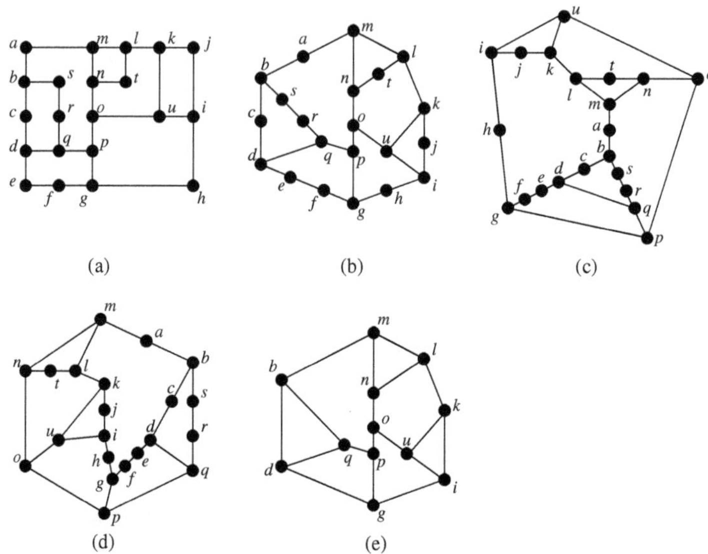

Fig. 1. A no-bend drawing (a), and three different embeddings (b), (c) and (d) of the same graph which is a subdivision of the triconnected cubic planar graph in (e).

[GT01]. However, for a planar graph G with $\Delta \leq 3$, Di Battista *et al.* [DLV98] gave an $O(n^5 \log n)$ time algorithm to find an orthogonal drawing of G with the minimum number of bends. Hence, by their algorithm, one can examine in time $O(n^5 \log n)$ whether a planar graph with $\Delta \leq 3$ has a no-bend drawing.

In this paper we consider a class of planar graphs, called subdivisions of planar triconnected cubic graphs, and give a linear-time algorithm to examine whether such a planar graph G has a no-bend drawing and to find a no-bend drawing if G has. The graph in Fig. 1(b) is obtained from the planar triconnected cubic graph in Fig. 1(e) by inserting vertices of degree two into some edges, and hence is a subdivision of the cubic graph.

The rest of the paper is organized as follows. Section 2 describes some definitions and presents preliminary results. Section 3 presents a necessary and sufficient condition for a subdivision G of a planar triconnected cubic graph to have a no-bend drawing; the condition leads to a linear-time algorithm. Finally Section 4 is a conclusion.

2 Preliminaries

In this section we give some definitions and present preliminary results.

Let $G = (V, E)$ be a connected simple graph with vertex set V and edge set E. The *degree* $d(v)$ of a vertex v is the number of neighbors of v in G. We call a vertex of degree two in G a *2-vertex* of G. A graph G is called *cubic* if $d(v) = 3$ for every vertex v. For $V' \subseteq V$, $G - V'$ denotes a graph obtained from

G by deleting all vertices in V' together with all edges incident to them. For a subgraph G' of G, we denote by $G - G'$ the graph obtained from G by deleting all vertices in G'. The *connectivity* $\kappa(G)$ of a graph G is the minimum number of vertices whose removal results in a disconnected graph or a single-vertex graph K_1. We say that G is *k-connected* if $\kappa(G) \geq k$.

Subdividing an edge (u, v) of a graph G is the operation of deleting the edge (u, v) and adding a path $u(= w_0), w_1, w_2, \cdots, w_k, v(= w_{k+1})$ passing through new 2-vertices $w_1, w_2, \cdots, w_k, k \geq 1$. A graph G is said to be a *subdivision* of a graph G' if G is obtained from G' by subdividing some of the edges of G'. A subdivision of a triconnected cubic graph is biconnected, and the degree of any vertex is either 2 or 3.

Let $P = w_0, w_1, w_2, \cdots, w_{k+1}, k \geq 1$, be a path of a graph G such that $d(w_0) \geq 3, d(w_1) = d(w_2) = \cdots = d(w_k) = 2$, and $d(w_{k+1}) \geq 3$. Then we call the subpath $P' = w_1, w_2, \cdots, w_k$ of P a *chain* of G, and we call vertices w_0 and w_{k+1} the *supports* of the chain P'. If G is a subdivision of a triconnected graph, then any 2-vertex of G is contained in exactly one of the chains of G.

Let G be a planar biconnected graph, and let Γ be a plane embedding of G. The contour of a face of Γ is a cycle of G, and is simply called a *face* or a *facial cycle* of Γ. We denote by $F_o(\Gamma)$ the *outer face* of Γ. For a cycle C of Γ, we call the plane subgraph of Γ inside C (including C) the *inner subgraph* $\Gamma_I(C)$ for C, and call the plane subgraph of Γ outside C (including C) the *outer subgraph* $\Gamma_O(C)$ for C. Any face of Γ is either in $\Gamma_I(C)$ or in $\Gamma_O(C)$. An edge of G is called a *leg* of C if it is incident to exactly one vertex of C and located outside C. A cycle C of Γ is called a *k-legged cycle* if C has exactly k legs in Γ and there is no edge which joins two vertices on C and is located outside C. An edge of G is called a *hand* of C if it is incident to exactly one vertex of C and located inside C. A cycle C of Γ is called a *k-handed cycle* if C has exactly k hands in Γ and there is no edge which joins two vertices on C and is located inside C. Both the set of k legs of a k-legged cycle and the set of k hands of a k-handed cycle in G correspond to a "cutset" of k edges in G.

Rahman *et al.* [RNN02] obtained a necessary and sufficient condition for a plane graph G with $\Delta \leq 3$ to have a no-bend drawing, and gave a linear time algorithm to find a no-bend drawing of G if G has, as follows.

Lemma 1. [RNN02] *A plane embedding Γ of a planar biconnected graph G with $\Delta \leq 3$ has a no-bend drawing if and only if Γ satisfies the following three conditions:*

(a) *there are at least four 2-vertices on the outer face $F_o(\Gamma)$;*
(b) *every 2-legged cycle in Γ contains at least two 2-vertices; and*
(c) *every 3-legged cycle in Γ contains at least one 2-vertex.*

Furthermore one can examine in linear time whether Γ satisfies the conditions above, and one can find a no-bend drawing of Γ in linear time if Γ has.

Although the results above for a plane graph is known, it is difficult to examine whether a planar graph has a no-bend drawing or not, since a planar graph may have an exponential number of plane embeddings in general. However, the following fact is known for subdivisions of planar triconnected graphs.

Fact 2 [NC88] *Let G be a subdivision of a planar triconnected graph. Then there is exactly one embedding of G for each face embedded as the outer face. Furthermore, for any two plane embeddings Γ and Γ' of G, any facial cycle of Γ is a facial cycle of Γ'.* ■

Fact 2 implies that a subdivision G of a planar triconnected graph has an $O(n)$ number of embeddings, one for each chosen outer face. Examining by the linear algorithm in [RNN02] whether the three conditions in Lemma 1 hold for each of the $O(n)$ embeddings, one can examine in time $O(n^2)$ whether the planar graph G has a no-bend drawing. In Section 3 we present a necessary and sufficient condition for G to have a no-bend drawing. Our condition leads a linear-time algorithm to examine whether G satisfies the condition and to find a no-bend drawing if G has.

We have the following lemma and facts on a subdivision of a planar triconnected cubic graph, which will be useful in Section 3.

Lemma 3. [RNG02] *Let G be a subdivision of a planar triconnected cubic graph, and let Γ be an arbitrary plane embedding of the planar graph G. Then the following* (a) and (b) *hold.*

(a) *For any 2-legged cycle C of Γ, the set of all vertices not in $\Gamma_I(C)$ induces a chain of G on $F_o(\Gamma)$.*
(b) *For any chain P on $F_o(\Gamma)$, the outer face of the plane graph $\Gamma - P$ is a 2-legged cycle in Γ.* ■

A cycle of Γ violating Condition (b) or (c) in Lemma 1 is called a *bad cycle* of Γ: a 2-legged cycle is *bad* if it contains at most one 2-vertex; a 3-legged cycle is *bad* if it contains no 2-vertex. Let C be a 3-handed cycle of a plane embedding Γ of G, let F be a face in $\Gamma_I(C)$, and let Γ' be a plane embedding of G in which F is embedded as $F_o(\Gamma')$. Then C is a 3-legged cycle of Γ'. If C contains no 2-vertex, then C is a bad cycle of Γ'. We thus call a 3-handed cycle C of Γ a *bad 3-handed cycle* of Γ if it contains no 2-vertex, otherwise we call C a *good 3-handed cycle.*

3 No-Bend Drawings of Planar Graphs

We obtain the following necessary and sufficient condition for a subdivision G of a planar triconnected cubic graph to have a no-bend drawing.

Theorem 1. *Let G be a subdivision of a planar triconnected cubic graph, and let Γ be an arbitrary plane embedding of G. Then the planar graph G has a no-bend drawing if and only if Γ has a face F satisfying the following conditions* (i)–(v)*:*

(i) *there are at least four 2-vertices on F;*

(ii) *F is contained in $\Gamma_I(C)$ for any bad 3-legged cycle C in Γ;*
(iii) *F is contained in $\Gamma_O(C)$ for any bad 3-handed cycle C in Γ;*
(iv) *if there is exactly one chain P on F, then the face F' which contains P and is different from F contains at least two 2-vertices which are not on P;and*
(v) *if there are exactly two chains P_1 and P_2 on F, and one of them, say P_1, contains exactly one vertex, then the face F' which contains P_2 and is different from F contains at least one 2-vertex which is not on P_2.*

Proof. Necessity: Assume that G is a subdivision of a planar triconnected cubic graph and has a no-bend drawing. Then G has a plane embedding Γ' which has a no-bend drawing. Let F be the face of Γ such that $F = F_o(\Gamma')$. Then we can show that F satisfies (i)–(v), as follows.

(i) Since Γ' has a no-bend drawing, by Lemma 1(a) there are at least four 2-vertices on $F_o(\Gamma') = F$.

(ii) Suppose for a contradiction that F is not contained in $\Gamma_I(C)$ for a bad 3-legged cycle C in Γ. Then F is contained in $\Gamma_O(C)$. One can observe that C is a bad 3-legged cycle of Γ', and hence by Lemma 1 Γ' does not have a no-bend drawing, a contradiction.

(iii) Suppose for a contradiction that F is not contained in $\Gamma_O(C)$ for a bad 3-handed cycle C in Γ. Then F is contained in $\Gamma_I(C)$. One can observe that C is a bad 3-legged cycle of Γ', and hence by Lemma 1 Γ' does not have a no-bend drawing, a contradiction.

(iv) Assume that there is exactly one chain P on F. By Lemma 3(b) Γ' has a 2-legged cycle $C = F_o(\Gamma' - P)$. Γ' does not have any 2-legged cycle other than C; otherwise, by Lemma 3(a) $F_o(\Gamma')$ has a chain other than P, and hence $F = F_o(\Gamma')$ would have two chains. Thus Γ' has exactly one 2-legged cycle C. Since Γ' has a no-bend drawing, by Lemma 1(b) C contains at least two 2-vertices. Since F has exactly one chain P, all 2-vertices on $F_o(\Gamma')$ are on P and hence the maximal subpath Q_1 of C that is on $F_o(\Gamma')$ contains no 2-vertex. Thus all 2-vertices on C are contained in the maximal subpath Q_2 of C that is not on $F_o(\Gamma')$. Clearly Q_2 is on F' and is not on P. Hence F' contains at least two 2-vertices which are not on P.

(v) Assume that F has exactly two chains P_1 and P_2 and P_1 contains exactly one vertex. Since $F_o(\Gamma') = F$ contains at least four 2-vertices by Lemma 1(a), P_2 contains at least three 2-vertices. By Lemma 3(b) Γ' has two 2-legged cycles $C_1 = F_o(\Gamma' - P_1)$ and $C_2 = F_o(\Gamma' - P_2)$. By Lemma 3(a) Γ' does not have any 2-legged cycle other than C_1 and C_2. By Lemma 1(b) C_2 contains at least two 2-vertices. The maximal subpath Q_1 of C_2 that is on $F_o(\Gamma')$ contains exactly one 2-vertex, and hence the maximal subpath Q_2 of C_2 that is not on $F_o(\Gamma')$ contains at least one 2-vertex. Clearly Q_2 is on F' and is not on P_2, and hence F' contains at least one 2-vertex which is not on P_2.

Sufficiency: Assume that Γ has a face F satisfying Conditions (i)–(v). Let Γ' be an embedding of G such that $F = F_o(\Gamma')$. We can prove that that Γ' satisfies Conditions (a)–(c) in Lemma 1 and hence Γ' has a no-bend orthogonal drawing. The detail of the proof is omitted in this extended abstract. $\mathcal{Q.E.D.}$

Theorem 1 immediately leads to a linear algorithm to examine whether a subdivision G of a planar triconnected cubic graph has a no-bend drawing and to find a no-bend drawing of G if it exists, as follows. Let Γ be an arbitrary plane embedding of G. Using a method similar to one in [RNN00,RNG02], one can examine in linear time whether Γ has a face F satisfying the conditions in Theorem 1. If Γ has such a face F, a plane embedding Γ' such that $F = F_o(\Gamma')$ can be obtained in linear time [NC88]. Finding a no-bend drawing of Γ' takes linear time [RNN02].

4 Conclusions

In this paper we presented a necessary and sufficient condition for a subdivision G of a planar triconnected cubic graph to have a no-bend drawing, and gave a linear-time algorithm to examine whether G satisfies the condition and to find a no-bend drawing if G has.

It is left as a future work to find a linear-time algorithm to find a no-bend drawing for a larger class of planar graphs.

References

[DLV98] G. Di Battista, G. Liotta and F. Vargiu, *Spirality and optimal orthogonal drawings*, SIAM J. Comput., 27(6), pp. 1764–1811, 1998.

[GT01] A. Garg and R. Tamassia, *On the computational complexity of upward and rectilinear planarity testing*, SIAM J. Comput., 31(2), pp. 601–625, 2001.

[NC88] T. Nishizeki and N. Chiba, *Planar Graphs: Theory and Algorithms*, North-Holland, Amsterdam, 1988.

[RN02] M. S. Rahman and T. Nishizeki, *Bend-minimum orthogonal drawings of plane 3-graphs*, Proc. of WG '02, Lect. Notes in Computer Science, 2573, pp. 265–276, 2002.

[RNG02] M. S. Rahman, T. Nishizeki, and S. Ghosh, *Rectangular drawings of planar graphs*, Proc. of GD '02, Lect, Notes in Computer Science, 2528, pp. 244–255, 2002, also Journal of Algorithms, to appear.

[RNN00] M. S. Rahman, S. Nakano and T. Nishizeki, *Box-rectangular drawings of plane graphs*, Journal of Algorithms, 37, pp. 363–398, 2000.

[RNN02] M. S. Rahman, M. Naznin and T. Nishizeki, *Orthogonal drawings of plane graphs without bends*, Proc. of Graph Drawing '01, Lect. Notes in Computer Science, Springer-verlag, 2265, pp. 392–406, 2002, also Journal of Graph Alg. and Appl., to appear.

[RNN99] M.S. Rahman, S. Nakano and T. Nishizeki, *A linear algorithm for bend-optimal orthogonal drawings of triconnected cubic plane graphs*, Journal of Graph Alg. and Appl., http://www.cs.brown.edu/publications/jgaa/, 3(4), pp. 31–62, 1999.

[S84] J. Storer, *On minimum node-cost planar embeddings*, Networks, 14, pp. 181–212, 1984.

[T87] R. Tamassia, *On embedding a graph in the grid with the minimum number of bends*, SIAM J. Comput., 16, pp. 421–444, 1987.

Radial Level Planarity Testing and Embedding in Linear Time*

(Extended Abstract)

Christian Bachmaier, Franz J. Brandenburg, and Michael Forster

University of Passau, 94030 Passau, Germany
{bachmaier,brandenb,forster}@fmi.uni-passau.de

Abstract. Every planar graph has a concentric representation based on a breadth first search, see [21]. The vertices are placed on concentric circles and the edges are routed as curves without crossings. Here we take the opposite view. A graph with a given partitioning of its vertices onto k concentric circles is k-radial planar, if the edges can be routed monotonic between the circles without crossings. Radial planarity is a generalisation of level planarity, where the vertices are placed on k horizontal lines. We extend the technique for level planarity testing of [18,17,15,16,12,13] and show that radial planarity is decidable in linear time, and that a radial planar embedding can be computed in linear time.

1 Introduction

The display of hierarchical structures is an important issue in automatic graph drawing. Directed acyclic graphs (DAGs) and ordered trees are usually drawn such that the vertices are placed on horizontal levels, and the edges are drawn as straight lines or as y-monotone polylines. This technique is used by the Sugiyama algorithm, the most common algorithm for drawing DAGs [6,20]. After the calculation of a level assignment, the algorithm tries to minimise the number of edge crossings. In the best case there are no crossings at all, and the graph is level planar. Level planarity has been investigated intensively, and there are linear time algorithms both for the test of k-level planarity and for the computation of an embedding.

A *k-level graph* $G = (V, E, \phi)$ is an undirected graph with a level assignment $\phi\colon V \to \{1, 2, \dots, k\}$, $1 \leq k \leq |V|$, that partitions the vertex set into $V = V^1 \dot\cup V^2 \dot\cup \cdots \dot\cup V^k$, $V^j = \phi^{-1}(j)$, $1 \leq j \leq k$, such that $\phi(u) \neq \phi(v)$ for each edge $(u, v) \in E$. A k-level graph is *proper* if $|\phi(u) - \phi(v)| = 1$ for each edge $(u, v) \in E$. The level planarity problem [7,12,17] can be formulated as follows: Is it possible to draw a level graph G in the Cartesian plane such that all vertices $v \in V^j$ of the j-th level are placed on a single horizontal line $l_j = \{ (x, j) \mid x \in \mathbb{R} \}$ and the edges are drawn as strictly y-monotone curves without crossings. For a planar

* This research has been supported in part by the Deutsche Forschungsgemeinschaft, grant BR 835/9-1.

G. Liotta (Ed.): GD 2003, LNCS 2912, pp. 393–405, 2004.

k-level drawing with the above restrictions, an embedding of the graph has to be computed. Level planar embeddings are characterized by linear orderings $\leq_j$ of the vertices in each V^j, $1 \leq j \leq k$, which is the order of the vertices from left to right.

Note that we do not consider the problem of finding a level planar or radial planar embedding for graphs without a given levelling. Heath and Rosenberg [14] have shown that the levelling problem is NP-hard for proper graphs. In the non-proper case, i. e., with arbitrarily long edges, each planar graph has a k-level planar levelling. This follows for example from the planar grid drawings of de Fraysseix et al. [5]. There neither the number of levels nor the length of the edges is taken into account, e. g., for a minimisation.

We generalise level planarity to radial level planarity or short radial planarity. In contrast to the above, the vertices are not drawn on k horizontal lines, but on k concentric circle lines $l_j = \{ (j \cos\theta, j \sin\theta) \mid \theta \in [0, 2\pi) \}$, $1 \leq j \leq k$. A k-level graph is *radial k-level planar* if there are orderings $\leq_j$ of the vertices on each radial level such that edges are drawn as strictly monotone curves from inner to outer levels without crossings.

The transformation of a level planar embedding to a radial planar embedding can be obtained by connecting the ends of each level and thus forming concentric level circles. This allows the insertion of some additional edges connecting the end of one level with the beginning of another. These *cut edges* cross an imaginary ray from the centre of the concentric levels to infinity through the points where the connected fronts and backs of the levels meet. There are two directions for routing cut edges around the centre, clockwise and counterclockwise. As an extension to level planar embeddings, *radial planar embeddings* need additional information about cut edges and their direction. Figure 1(b) shows a radial planar drawing of the graph in Fig. 1(a) which is not level planar. The edge $(1, 6)$ crosses the imaginary ray and thus is a clockwise cut edge, following its implicit direction from lower to higher levels. Obviously, a radial planar graph is level planar, if there are no cut edges.

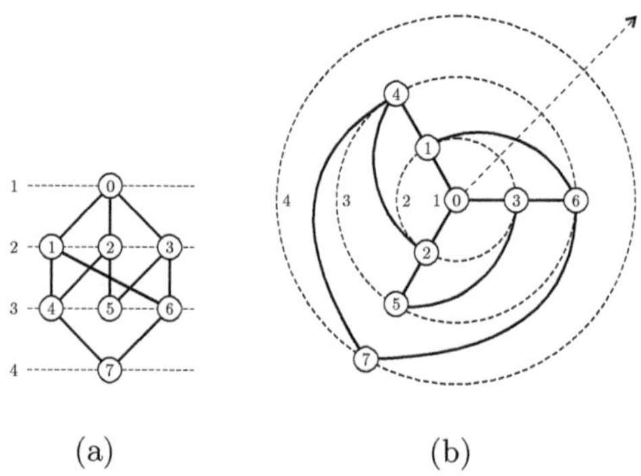

Fig. 1. A radial planar graph with a radial planar drawing

The more general approach of Dujmović et al. [9] leads to a linear time fixed paramter tractable algorithm for detecting radial planar graphs for a fixed number of levels. We give a practical algorithm based on the level planarity test of Leipert et al. [15,17,16,18] that improves this result to $\mathcal{O}(|V|)$ time for an arbitrary number of (non empty) levels.

Without loss of generality we only consider simple graphs without self loops and parallel edges. Input graphs with $|E| > 3|V| - 6$ are rejected as not radial planar because radial planar graphs are planar by definition. For more details and proofs we refer to the long version [1] of this paper.

2 Related Work

The basis of our algorithm is the linear time algorithm of Leipert et al. [18,17,15, 16] for level planarity testing and embedding, which is in turn based on previous work of Heath and Pemmaraju [12,13] and Di Battista and Nardelli [7]. As the vertex addition method for graph planarity tests [19,10], this algorithm makes heavy use of the *PQ-tree* data structure introduced by Booth and Lueker [3]. See [1, Sect. 2] for an introduction. Leipert's algorithm proceeds level by level and stores the admissible permutations of the vertices on each level efficiently in a set of PQ-trees. Also a level embedding can be computed in linear time. First, the input graph is augmented to an *st*-graph $G_{st} = (V_{st}, E_{st})$ and then an *st*-embedding is computed by the algorithm of Chiba et al. [4]. With this *st*-embedding, a level embedding is constructed by an ordered depth first search (DFS). See [18,17,15,16,12,13,1] for details.

Theorem 1 (Leipert et al. [18,17,15,16]). *There is an $\mathcal{O}(|V|)$ time algorithm for testing level planarity and computing a level planar embedding.*

3 Radial Level Planarity Test

In this section we describe our linear time algorithm for level planarity testing, using a new data structure, PQR-trees. This theorem is our main result. For its proof see [1].

Theorem 2. *There is an $\mathcal{O}(|V|)$ time algorithm for testing radial k-level planarity.*

3.1 Concepts

Level planarity and radial planarity look similar, but there are some essential differences. Leipert's algorithm heavily depends on the fact that a level graph is level planar if and only if each connected component is level planar. Therefore it suffices to test each connected component for level planarity. This is no more true for radial planarity, as Fig. 2 explains.

Clearly a graph is radial planar, if it consists only of level planar components, because it is level planar. Hence, we must consider those components that are radial planar but not level planar. Therefore we introduce the concept of a ring:

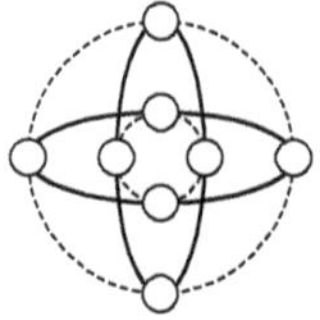

Fig. 2. A non-radial planar graph consisting only of radial planar connected components

Definition 1. *A* ring *is a biconnected component of a level graph which is radial planar but not level planar. A level graph containing a ring is called a* ring graph.

A priori it is not clear whether a biconnected component is a ring. We will see later how rings are detected. Nevertheless we investigate some interesting properties of rings first. The graph in Fig. 3(a) consists of four biconnected components, where the single ring is indicated by the darker shading. A component can be nested in another and here this is even necessary for a planar drawing. This only happens if the "outer" component is a ring. Moreover rings are not related to cycles. In fact every biconnected component with at least three vertices contains a cycle, but whether it is a ring depends on the levelling. If vertex 14 was on level 1, this graph would not contain a ring, because according to the ray in Fig. 3(b) there are no cut edges.

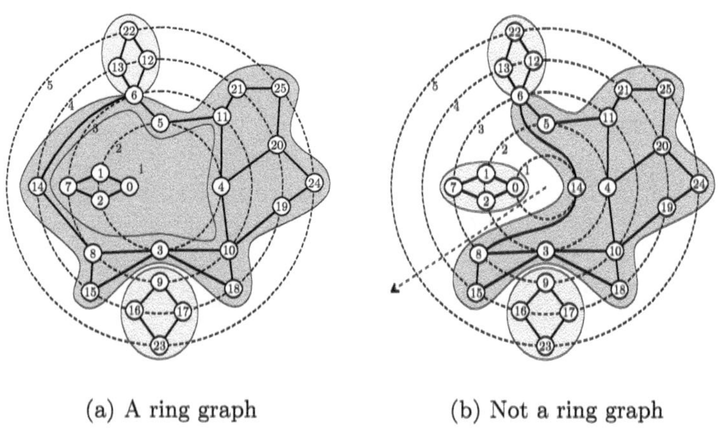

(a) A ring graph (b) Not a ring graph

Fig. 3. Rings depend on the levelling

Rings make the difference between level planarity and radial planarity. If a level graph does not contain a ring, it is radial planar if and only if it is level planar. Moreover this is equivalent to each connected component being level planar (or radial planar). Hence, if a graph does not contain a ring, we can use Leipert's level planarity test algorithm to test for radial planarity. For ring graphs, the algorithm has to be extended. Before we describe how our algorithm stores the admissible permutations of the vertices on each circle in the next section, we discuss some more properties of rings. Clearly, in any radial planar embedding of a ring graph the centre of the concentric levels lies in an inner face, the *centre face*. The nesting of rings is determined by some characterising parameters.

Definition 2. *Let G be a k-level graph containing a ring R. The minimum and maximum level with vertices of R are denoted by α_R and δ_R. These values are independent of the embedding. Let β_R be the maximum level with a vertex of the centre face in any radial planar embedding. Analogously, define γ_R as the minimum level with a vertex of the outer face of R in any radial planar embedding. See Fig. 4 for an example.*

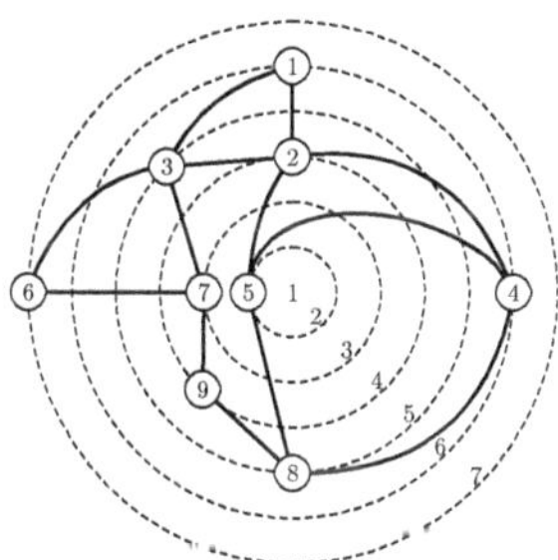

Fig. 4. Extreme levels of a ring. $\alpha_R = 2$, $\beta_R = 6$, $\gamma_R = 3$, $\delta_R = 7$

Lemma 1. *Let G be a level graph consisting of two disjoint rings R and S. G is radial planar if and only if R and S are radial planar and*

$$\alpha_S > \gamma_R \wedge \beta_S > \delta_R \quad \text{or} \quad \alpha_R > \gamma_S \wedge \beta_R > \delta_S$$

In other words, given appropriate embeddings of R and S, R must fit into the centre face of S or vice versa and cannot be placed side by side, as Leipert's algorithm would do. Because β_R and γ_R depend on the embedding, it is clear that for leaving maximum space for placing other components the embedding has to be calculated such that β_R is maximised and γ_R is minimised. Such an embedding is called *level optimal*. A priori it is not clear that there is a level optimal embedding for every ring. But our algorithm constructs such an embedding, see [1].

3.2 R-Nodes

Our goal is to extend Leipert's algorithm to test for radial planarity. The basic ideas of that algorithm can be adopted. The input graph is traversed in a top down sweep, which now becomes a "wavefront" sweep from the centre. The processed part of the graph is represented by a collection of trees, which is denoted by $\mathcal{T}$. For dealing with rings, we introduce a new data structure *PQR-trees*. PQR-trees store the admissible edge permutations of radial planar graphs. PQR-trees are not related to SPQR-trees, used for incremental planarity testing [8].

PQR-trees are based on PQ-trees, but contain a new "R" node type for the rings. Because of Lemma 1 it suffices that R-nodes occur only as the root of a PQR-tree. As usual [3], P-nodes represent all permutations of its children and Q-nodes the given children list or its reverse. *R-nodes* are similar to Q-nodes, but they have some new properties representing the differences between rings and other biconnected components. An R-node is drawn as an elliptical ring and as usual P-nodes and Q-nodes are drawn as circles and rectangles, respectively. The admissible operations on an R-node are reversion, i. e., inverting the iteration direction of its children in the same way as for Q-nodes, and *rotation* which is new. Since rings always contain the centre, it is possible to rotate a ring. This corresponds to rotating the graph around the centre, and is done by moving a subsequence of R-node children from the beginning of the children list to its end, or vice versa, while maintaining the relative order of the moved children. On a circular list this happens implicitly. R-nodes can be implemented, for example, with the improved symmetric list data structure [2]. Insertions, reversions, and rotations can be done in constant time, which is crucial for the linear running time of the test.

3.3 New Templates

The main operation on PQ(R)-trees is REDUCE [3]. By the application of several templates REDUCE ensures that a given subset of *pertinent* PQ-leaves occurs consecutively in any of the stored permutations. In addition to the eleven templates of PQ-trees we need twelve new templates for PQR-trees.

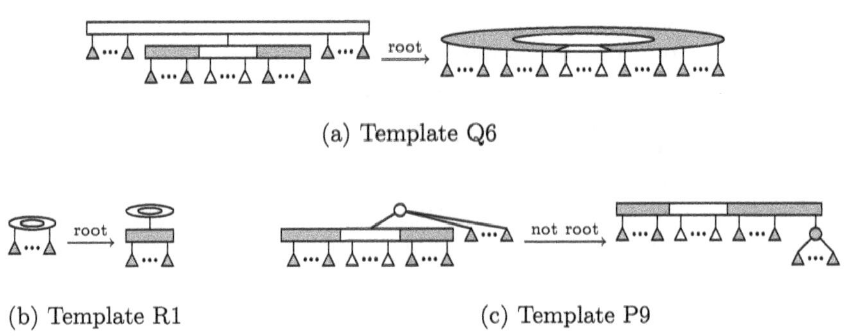

(a) Template Q6

(b) Template R1

(c) Template P9

Fig. 5. Some of the new templates

Figure 5 shows some of them. A complete description is given in [1]. As an example, template Q6 demonstrates some key elements of the new templates. In addition to empty, partial, and full Q-nodes, there are *boundary partial* Q-nodes where all pertinent children are at the beginning or at the end separated by at least one empty child. Instead of this boundary partial child, there can be some empty children, optionally bounded by partial Q-nodes. This is handled by another template. Further it shows that there are templates which can only be applied to the root of a PQR-tree. This is different from the restriction that some PQ-tree templates may only be applied to the root of the pertinent subtree (pert-root). Finally it is illustrated how an R-node is created. Because the full children cannot be made consecutive it is not possible to apply any of the existing Q-templates, i. e., the biconnected component is no more level planar. Hence, R-nodes are created only by necessity, i. e., if newly encountered edges change the represented biconnected component from level planar into a ring.

Now that there are some templates which create R-nodes, it is necessary to treat R-nodes on the left side of a template. Therefore we introduce templates R0–R3. It may be necessary to rotate the R-node before applying an R-template. The R-templates are the straightforward transformations of the respective Q-node templates with the exception of R1 (see Fig. 5(b)), where a pseudo Q-node is introduced for technical reasons. This preserves the information that the PQR-tree represents a ring component and makes it possible to compute a value minML in order to know what fits "below" this round component. The single *meet level* at the root is set to minML. The meet level ML of two adjacent siblings with a Q-node or R-node as their parent describes how much space is left between them. Other components can be nested at this position only if they start on a higher level.

Definition 3. *For an R-node X with children* $X_1, X_2, \ldots X_t$ *define*

$$\text{minML} = \min\{\, \text{ML}(X_i, X_{i+1}) \mid 1 \leq i \leq t, X_{t+1} = X_1 \,\}.$$

The presence of boundary partial Q-nodes forces us to provide the remaining templates P7–P9, Q5, Q7, and R4. See template P9 in Fig. 5(c) as an example. If a P-node has only full children with the exception of one boundary partial Q-node, the full children are grouped by a new P-node which is inserted into the Q-node. It is both admissible to place it at the front or at the back of the Q-node. The difference is only, whether the edges represented by the descendant leaves become cut edges later.

The templates are exactly designed to cover all cases occurring in PQR-trees while traversing a radial planar graph. In addition they maintain the invariant that once an R-node is created, it is preserved until its host PQR-tree is deleted. The following vertex addition step performed by a PQ(R)-tree operation REPLACE_PERT stays the same as in Leipert's algorithm.

3.4 Merge Operations on PQR-Trees

If there are PQ-leaves with the same label in more than one PQ-tree, these PQ-trees have to be merged according to merge conditions described in [1,18,

17,15,16,12,13]. Whether a PQ-tree T can be merged into another one depends on its *low indexed level* LL(T), the lowest level with vertices of the represented component. A PQR-tree with a lower LL is said to be *higher* than a *smaller* one with a greater LL. The merge conditions remain essentially the same in PQR-trees if no R-node occurs. Because of Lemma 1, merge condition E, i. e., placing two PQ-trees next to each other with a new Q-root, cannot be applied if one of the merge candidates has an R-root. As a consequence a merge operation may fail, contrary to the non-radial case, where condition E always is admissible if no other condition fits. For PQR-trees with an R-root we have to provide two additional merge conditions. The root of the smaller PQR-tree must not be an R-node. If the root of the higher PQR-tree is an R-node, condition B and C collapse to the new condition $\mathrm{C^R}$,because R-nodes can be rotated such that a merge can always be done in its interior. Similarly, if the root of the source pattern of condition D is an R-node we obtain $\mathrm{D^R}$,see [1].

3.5 Merge of Processed Non-rings

In level planarity testing, separate components can be placed side by side without violating planarity. This is not necessarily true here. Consider a component of the input graph G containing at least one ring. The other components detected so far must fit into some inner face of the ring or its outer face. We first consider the case that these other components contain no rings. For the efficient execution of the necessary additional checks the algorithm maintains a variable minLL.

Definition 4. minLL $= \min\{\,\mathrm{LL}(T) \mid T$ *is a completely processed PQR-tree with no R-root*$\}$. *A* completely processed PQR-tree *is a PQR-tree representing a component of the graph not having any vertices on the current or on higher levels. If there is no such* T, *then* minLL $= \infty$.

As soon as a PQR-tree T is identified as completely processed, minLL is updated by minLL $= \min\{\mathrm{minLL}, \mathrm{LL}(T)\}$. All processed PQR-trees are discarded as in Leipert's test algorithm. It suffices to check if the component C of the completely processed PQR-tree T starting at the lowest level fits into an internal face. For all other processed (non-ring) components there is enough space to embed them in the same face as C. Whenever inner faces are closed and there is a processed PQR-tree with an R-root, i. e., minLL $< \infty$, we have to check whether C can be included. We use the same mechanism as Leipert uses for v-singular components, i. e., those whose PQR-tree has exactly one leaf. If it fits, we set minLL $= \infty$. Otherwise we need not care whether another processed component smaller as C fits, for which its PQR-tree has been discarded. These will still fit later when a face for C is found. If no such face can be found, the graph is not radial level planar anyway. Recall that a processed PQR-tree with an R-node as root can never be included this way.

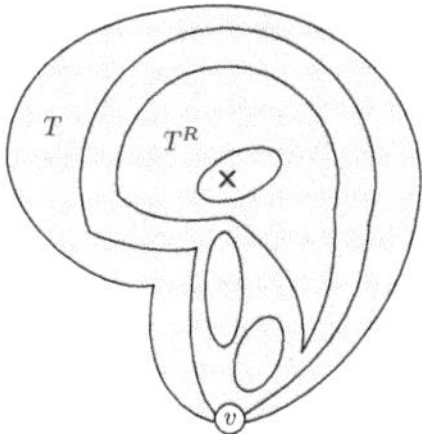

Fig. 6. Merge of rings

3.6 Merge of Processed Rings

The algorithm maintains the invariant that at any time there is at most one PQR-tree T^R containing an R-node. If another PQR-tree T gets an R-root during the reduction of a *link vertex* v, we proceed as follows: If there is a PQR-tree T^R with an R-node as root, the algorithm checks if T^R is completely processed. Otherwise G is not radial planar except if T^R is v-singular, which is solved as for level planarity. Afterwards the algorithm checks whether minML is small enough that T fits "below" T^R and the PQR-tree with the smallest low indexed level minLL and all others fit "between" T and T^R. If one of the checks fails, G is not radial planar as Lemma 1 states.

3.7 Completion

If at termination of the test algorithm there is no PQR-tree T^R representing a ring graph, the graph is radial planar, because all other left PQR-trees can be placed side by side. Otherwise, if no other trees occurred after T^R was detected, i. e., if minLL $= \infty$, the graph is radial planar, too. It remains to check whether the other PQR-trees fit below T^R, i. e., if minML $<$ minLL, otherwise, G is not radial planar.

4 Radial Planar Embedding

Leipert et. al. also have presented an algorithm for computing a level planar embedding for a level planar graph. This algorithm can be extended to compute a radial planar embedding for a radial planar graph by using PQR-trees. Again there are some differences that need attention.

Theorem 3. *There is an $\mathcal{O}(|V|)$ time algorithm for computing a radial planar embedding of a radial k-level planar graph.*

4.1 Embedding the Edges

As already mentioned in the introduction, we not only have to compute a vertex ordering $\leq_j$ on each level j but also the edge routing. It suffices to designate cut edges while calculating the *st*-embedding. Since cut edges are defined as those crossing the imaginary ray, R-nodes save the current position of this ray with a new child, an ignored PQ-leaf similar to the direction indicators of [4]. It is called *ray indicator* and labelled with \$. All templates involving an R-node must be aware of the ray indicator. If the pertinent sequence contains the ray indicator, this means that the edges represented by one part of the sequence must cross the ray. Thus they are moved over the ray indicator and are marked as cut edges, while the edges represented by the other part are non-cut edges.

4.2 Augmentation to an *st*-Graph

G is augmented to an *st*-graph G_{st} by a top down and a subsequent bottom up sweep of a slightly modified level planarity test. Each sink is connected to an appropriate vertex v on a higher level. Therefore the corresponding leaves in the PQR-trees are not deleted but replaced with *sink indicators*, which are ignored by most parts of the algorithm. Nodes having only ignored children are also ignored. Since v is not known before it is encountered, we do not discard processed PQR-trees, but save them in a list $\mathcal{T}^*$, similar to the list $\mathcal{T}$ of *active* PQR-trees. At most one PQR-tree $T^R \in \mathcal{T} \cup \mathcal{T}^*$ has an R-root. The necessary augmentations occur if an outer ring or a face is closed at a link vertex v.

Templates that create R-nodes must be applied to a node X different from the root if all predecessors of X have only one non-ignored child left. Then all PQ-leaves that are not descendants of X are sink indicators and consequently are connected to the link vertex. Thus X becomes the new R-root according to the applied template.

4.3 Computation of an Upward *st*-Embedding

If the graph has been augmented to an *st*-graph, we can generate a planar *st*-embedding. Leipert's algorithm achieves this by using the planar embedding algorithm of Chiba et al. [4] and a subsequent ordered DFS. Chiba's algorithm calculates a planar embedding for general planar graphs, where an edge (s, t) ensures that s and t lie in the same face. Such an edge may violate radial planarity if the graph contains a ring. Fortunately, in our case the algorithm of Chiba also works without an *st*-edge because the levels and the faces of s and t are already fixed. It only has to be extended with the same new templates as the radial planarity test. The final DFS of Chiba's algorithm must be omitted, because cut edges may lead to an incorrect result. Instead we use the upward *st*-embedding $\mathcal{E}_u$, generated as an intermediate result, i. e., the order of the incoming edges for each vertex.

4.4 Computation of a Radial Embedding

Leipert's algorithm uses an ordered DFS to compute an ordering of the vertices for each level. Again we need an extension. In level planar embeddings there are

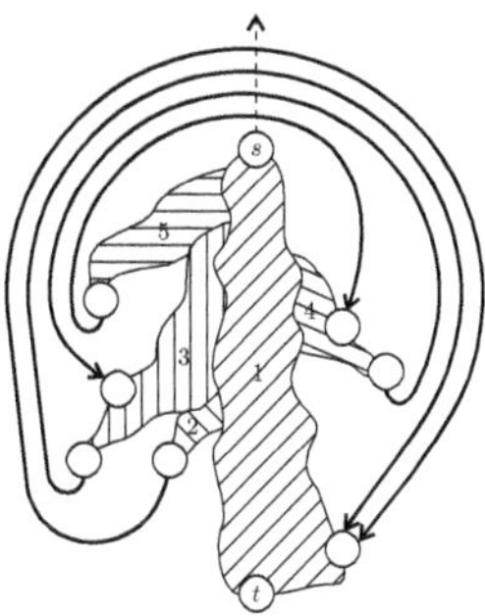

Fig. 7. Successive and ordered attachments of faces to the sides of the trunk

no cut edges. Hence, the order of the incoming edges in $\mathcal{E}_u$ leads directly to the order of the source vertices in a level embedding $\mathcal{E}_l$. Due to the presence of cut edges this is not true in the radial case. Instead, the cut edges have to be treated in a special way. A cut edge cannot lie between two non-cut edges in the ordered adjacency list of any vertex v in $\mathcal{E}_u$. So cut edges occur only at the front or only at the back of the adjacency list. This leads to two different types of cut edges according to their position in the adjacency list. A cut edge is called *clockwise* with respect to $\mathcal{E}_u$ if it occurs at the end of the incoming adjacency list of its target vertex, and *counterclockwise* otherwise.

Our algorithm starts with an ordered DFS from t in $\mathcal{E}_u$ using no cut edges. All vertices with at least one incoming cut edge are placed into a Queue Q. Every other newly detected vertex v is inserted at the end of its level $\phi(v)$. Afterwards an ordered DFS is started from all vertices $q \in Q$, in turn using only unvisited vertices. Clockwise cut edges are traversed from right to left and their respective source vertex w is inserted at the beginning of its level $\phi(w)$. Counterclockwise cut edges are traversed from left to right and their respective source vertex w is inserted at the end of its level $\phi(w)$. Now some unused non-cut edges and therefore some unvisited vertices can be reachable from w. They are inserted at the same end of the level lists as w. Note that source vertices of newly detected cut edges are treated in the same manner as above and are inserted into Q. The algorithm terminates when all vertices have been processed. Figure 7 illustrates the algorithm. Since no vertex v in $\mathcal{E}_u$ can be the target of both clockwise and counterclockwise cut edges, these edges do not cross and thus the algorithm calculates a correct embedding.

5 Conclusion

We have presented a new algorithm for detecting radial planarity of a k-level graph in linear time. For this we have enhanced the PQ-tree data structure of [3] with a new node type, R-nodes, representing a ring component of the graph. Further we have described the templates for our PQR-trees. We can compute a radial level planar embedding within the same linear time bounds as for planar embeddings [4] and level planar embeddings [18,15,16]. To check the practicability of our algorithm we have realised a prototypical implementation in C++ using the Graph Template Library [11].

References

[1] C. Bachmaier, F. J. Brandenburg, and M. Forster. Radial level planarity testing and embedding in linear time. Technical Report MIP-0303, University of Passau, June 2003.
[2] C. Bachmaier and M. Raitner. Improved symmetric lists. Submitted for publication, May 2003.
[3] K. S. Booth and G. S. Lueker. Testing for the consecutive ones property, interval graphs, and graph planarity using PQ-tree algorithms. *Journal of Computer and System Sciences*, 13:335–379, 1976.
[4] N. Chiba, T. Nishizeki, S. Abe, and T. Ozawa. A linear algorithm for embedding planar graphs using PQ-trees. *Journal of Computer and System Sciences*, 30:54–76, 1985.
[5] H. de Fraysseix, J. Pach, and R. Pollack. How to draw a planar graph on a grid. *Combinatorica*, 10:41–51, 1990.
[6] G. Di Battista, P. Eades, R. Tamassia, and I. G. Tollis. *Graph drawing: Algorithms for the Visualization of Graphs.* Prentice Hall, 1999.
[7] G. Di Battista and E. Nardelli. Hierarchies and planarity theory. *IEEE Transactions on Systems, Man, and Cybernetics*, 18(6):1035–1046, 1988.
[8] G. Di Battista and R. Tamassia. On-line planarity testing. *SIAM Journal on Computing*, 25(5):956–997, 1996.
[9] V. Dujmović, M. Fellows, M. Hallett, M. Kitching, G. Liotta, C. McCartin, N. Nishimura, P. Ragde, F. Rosamond, M. Suderman, S. Whitesides, and D. R. Wood. On the parameterized complexity of layered graph drawing. In F. Meyer auf der Heide, editor, *Proc. European Symposium on Algorithms, ESA 2001*, volume 2161 of *LNCS*, pages 488–499. Springer, 2001.
[10] S. Even. *Algorithms*, chapter 7, pages 148–191. Computer Science Press, 1979.
[11] GTL. Graph Template Library. `http://www.infosun.fmi.uni-passau.de/GTL/`. University of Passau.
[12] L. S. Heath and S. V. Pemmaraju. Recognizing leveled-planar dags in linear time. In *Proc. Graph Drawing '95*, volume 1027 of *LNCS*, pages 300–311. Springer, 1996.
[13] L. S. Heath and S. V. Pemmaraju. Stack and queue layouts of directed acyclic graphs: Part II. *SIAM Journal on Computing*, 28(5):1588–1626, 1999.
[14] L. S. Heath and A. L. Rosenberg. Laying out graphs using queues. *SIAM Journal on Computing*, 21(5):927–958, 1992.
[15] M. Jünger and S. Leipert. Level planar embedding in linear time. In *Proc. Graph Drawing '99*, volume 1731 of *LNCS*, pages 72–81. Springer, 1999.
[16] M. Jünger and S. Leipert. Level planar embedding in linear time. *Journal of Graph Algorithms and Applications*, 6(1):67–113, 2002.
[17] M. Jünger, S. Leipert, and P. Mutzel. Level planarity testing in linear time. In *Proc. Graph Drawing '98*, volume 1547 of *LNCS*, pages 224–237. Springer, 1998.
[18] S. Leipert. *Level Planarity Testing and Embedding in Linear Time.* Dissertation, Mathematisch-Naturwissenschaftliche Fakultät der Universität zu Köln, 1998.

[19] A. Lempel, S. Even, and I. Cederbaum. An algorithm for planarity testing of graphs. In P. Rosenstiehl, editor, *Theory of Graphs, International Symposium, Rome, July 1966*, pages 215–232. Gordon and Breach, 1967.
[20] K. Sugiyama, S. Tagawa, and M. Toda. Methods for visual understanding of hierarchical system structures. *IEEE Transactions on Systems, Man, and Cybernetics*, 11(2):109–125, 1981.
[21] J. D. Ullman. *Computational Aspects of VLSI*, chapter 3.5, pages 111–114. Computer Science Press, 1984.

An Improved Approximation to the One-Sided Bilayer Drawing

Hiroshi Nagamochi

Toyohashi University of Technology, Toyohashi, Aichi 441-8580, Japan
naga@ics.tut.ac.jp

Abstract. Given a bipartite graph $G = (V, W, E)$, a bilayer drawing consists of placing nodes in the first vertex set V on a straight line L_1 and placing nodes in the second vertex set W on a parallel line L_2. The one-sided crossing minimization problem asks to find an ordering of nodes in V to be placed on L_1 so that the number of arc crossings is minimized. In this paper, we prove that there always exits a solution whose crossing number is at most 1.4664 times of a well-known lower bound that is obtained by summing up $\min\{c_{uv}, c_{vu}\}$ over all node pairs $u, v \in V$, where c_{uv} denotes the number of crossings generated by arcs incident to u and v when u precedes v in an ordering.

1 Introduction

Given a bipartite graph $G = (V, W, E)$, a bilayer drawing consists of placing nodes in the first vertex set V on a straight line L_1 and placing nodes in the second vertex set W on a parallel line L_2. The problem of minimizing the number of crossings between arcs in a bilayer drawing was first introduced by Harary and Schwenk [5,6]. The one-sided crossing minimization problem asks to find an ordering of nodes in V to be placed on L_1 so that the number of arc crossings is minimized (while the ordering of the nodes in W on L_2 is given and fixed).

The problem has many applications such as VLSI layouts [11] and hierarchical drawings [1]. However, the two-sided and one-sided problems are shown to be NP-hard by Garey and Johnson [4] and by Eades and Wormald [3], respectively. Muñoz et al. [10] have proven that the one-sided problem remains to be NP-hard even for sparse graphs such as forests of 4-stars. Recently Dujmović and Whitesides [2] have given an $O(\phi^c \cdot n^2)$ time algorithm to the one-sided problem, where c is the number of crossings to be checked, $n = |V| + |W|$ and $\phi = \frac{1+\sqrt{5}}{2}$, thus showing that the problem is Fixed Parameter Tractable.

There are several heuristics that deliver theoretically or empirically good solutions. The so-called barycenter heuristic finds an $O(\sqrt{n})$-approximation solution or a $(d-1)$-approximation solution, where d is the maximum degree of nodes in the free side V (see [8] for the analysis). Eades and Wormald [3] proposed a simple and theoretically better heuristic, the median heuristic which delivers a 3-approximation solution. They also prove that the performance guarantee of the median heuristic approaches to 1 if graphs become dense. Yamaguchi and

G. Liotta (Ed.): GD 2003, LNCS 2912, pp. 406–418, 2004.

Sugimoto [13] gave a 2-approximation algorithm if $d \leq 4$. For all the known performance guarantees of these heuristics are based on a conventional lower bound that is obtained by summing up $\min\{c_{uv}, c_{vu}\}$ over all node pairs $u, v \in V$, where c_{uv} denotes the number of crossings generated by arcs incident to u and v when u precedes v in an ordering. An extensive computational experiment of several heuristics including the above two has been conducted by Jünger and Mutzel [7] and by Mäkinen [9]. Jünger and Mutzel [7] reported that most of the heuristics gave good solutions whose crossing numbers are nearly equal to the lower bound. However the theoretically best estimation to the gap between the optimal and the lower bound is 3 due to the heuristic by Eades and Wormald [3].

In this paper, we prove that there always exists a solution whose crossing number is at most 1.4664 times of the lower bound. Our argument is based on a probabilistic analysis, which provides a polynomial randomized algorithm that delivers a solution whose average number of crossings is at most 1.4664 times of the optimal.

2 Preliminaries

Let $G = (V, W, E)$ be a bipartite graph with a partition V and W of a node set. Assume that G has no isolated node. Let π denote a permutation of $\{1, 2, \ldots, |V|\}$ and σ denote a permutation of $\{1, 2, \ldots, |W|\}$. A pair of π and σ defines a *bilayer drawing* of G in the plane in such a way that, for two parallel horizontal lines L_1 and L_2, the nodes in V (resp., in W) are arranged on L_1 (resp., L_2) according to π (resp., σ) and each arc is depicted by a straight line segment joining the end-nodes, where the directions for traversing L_1 and L_2 are taken as the same one (see Fig. 1(a)). In a bilayer drawing (π, σ) of G two arcs $(v, w), (v', w') \in E$ intersect properly (or create a *crossing*) if and only if $(\pi(v) - \pi(v'))(\sigma(w) - \sigma(w'))$ is negative. In this paper, we consider the following problem.

One-sided Crossing Minimization: Given a bipartite graph $G = (V, W, E)$ and a permutation σ on W, find a permutation π on V that minimizes the number of crossings in a bilayer drawing (π, σ) of G.

Since the permutation σ on $W = \{1, 2, \ldots, |W|\}$ is fixed, we assume throughout the paper that $\sigma(i) = i$ for all $i \in W$. For each node u in G, let $\Gamma(u)$ denote the set of nodes adjacent to u, and let $d_u = |\Gamma(u)|$. For two nodes $u, v \in V$, let $\Delta_{uv} = |\Gamma(u) \cap \Gamma(v)|$. The *crossing number* c_{uv} for an ordered pair of two nodes $u, v \in V$ is the number of crossing generated by an arc incident to u and an arc incident to v when $\pi(u) < \pi(v)$ holds in a bilayer drawing (π, σ). (Fig. 1(b) shows the crossing numbers in the graph in Fig. 1(a).) It is a simple matter to see that

$$d_u d_v = c_{uv} + c_{vu} + \Delta_{uv}, \quad \min\{c_{uv}, c_{vu}\} \geq \frac{\Delta_{uv}(\Delta_{uv} - 1)}{2}.$$

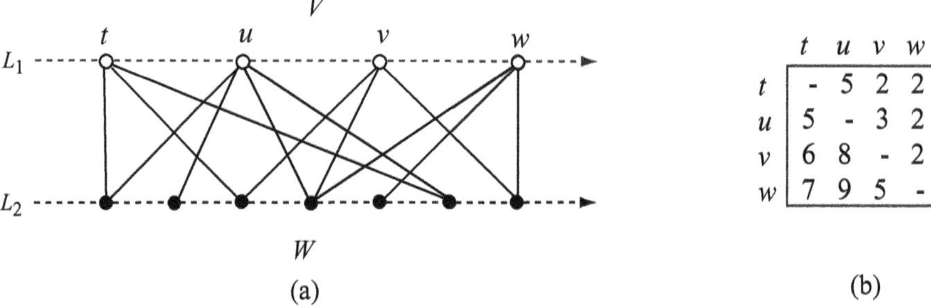

	t	u	v	w
t	-	5	2	2
u	5	-	3	2
v	6	8	-	2
w	7	9	5	-

Fig. 1. (a) A bilayer drawing of a bipartite graph. (b) Crossing numbers for each pair of nodes in the top layer.

For a permutation π on V, let

$$cross(u, v; \pi) := \begin{cases} c_{uv} & \text{if } \pi(u) < \pi(v), \\ c_{vu} & \text{otherwise.} \end{cases}$$

Define

$$cross(\pi) := \sum_{u,v \in Y : \pi(u) < \pi(v)} c_{uv} = \sum_{u,v \in V} cross(u, v; \pi).$$

The optimal to the problem is denoted by $opt = \min\{cross(\pi) \mid \text{permutation } \pi$ on $V\}$. For $LB = \sum_{u,v \in V} \min\{c_{uv}, c_{vu}\}$, it holds

$$opt \geq LB.$$

In this paper, we prove the next result.

Theorem 1. *For a bipartite graph $G = (V, W, E)$ with a permutation σ on W, there exists a permutation π on V such that $cross(\pi) \leq 1.4664LB$.* □

Note that the one-sided crossing minimization is a purely combinatorial problem in the sense that the number of crossings is determined by a permutation π, not by the actual positions of nodes in the layers. However, in this paper, we convert the problem into a geometric problem to derive Theorem 1. For this, we first introduce a geometric representation that illustrates how two sets $\Gamma(u)$ and $\Gamma(v)$ determine crossing numbers c_{uv} and c_{vu} in a bipartite graph G.

Rectangles that we treat here are axis-parallel in the xy-coordinate, and they are denoted by the coordinates of the lower-left corner and the upper-right corner, where the x-coordinate increases in the right direction and the y-coordinate increases in the upward direction. For example, $[(0,0),(1/2,1)]$ represents the square with four corners $(0,0),(0,1),(1/2,0)$ and $(1/2,1)$.

Let S denote the square $[(0,0),(1,1)]$. For a connected region R in S, we may use R to denote the sets of points in the region R, and let $a(R)$ denote the area size of R. For two points $b, b' \in S$, a line segment connecting b and b' is denoted by bb'. A part of the boundary of a region R may be called an *edge* if it is a line segment. For a line segment (or an edge) e, its length is denoted by $\ell(e)$. We say that edge e *overlaps* with another edge e' if the intersection of e and e' is a line segment of a positive length.

A path P between points $(0,0)$ and $(1,1)$ in S is called *monotone* if none of the x- and y-coordinates of the point on P decreases when we traverse points on P from $(0,0)$ to $(1,1)$ (in general a monotone path is not necessarily piecewise linear).

For two integers $d, d' \geq 1$, the square $S = [(0,0),(1,1)]$ is called *(d,d')-sliced* if it is sliced by $(d-1)$ horizontal line segments and $(d'-1)$ vertical segments so that these line segments give rise to $d \times d'$ congruent rectangles. Each of such rectangles is called a *block*, which has four edges.

We represent the positions of nodes in $\Gamma(u)$ and $\Gamma(v)$ in the permutation σ by using the unit square S in the xy-coordinate. Let $\Gamma(u) = \{u'_1, u'_2, \dots, u'_{d_u}\}$ and $\Gamma(v) = \{v'_1, v'_2, \dots, v'_{d_v}\}$. For an ordered pair (u,v) of nodes in V, we consider $d_u d_v$ blocks in the (d_u, d_v)-sliced square S. We denote these blocks by

$$bl(i,j) = [(\tfrac{j-1}{d_v}, \tfrac{i-1}{d_u}), (\tfrac{j}{d_v}, \tfrac{i}{d_u})],\ 1 \leq i \leq d_u \text{ and } 1 \leq j \leq d_v$$

We let $bl(i,j)$ correspond to a pair of arcs (u, u'_i) and (v, v'_j). Note that arcs (u, u'_i) and (v, v'_j) create a crossing in a permutation π with $\pi(u) < \pi(v)$ or $\pi(u) > \pi(v)$ if $u'_i \neq v'_j$, but generate no crossing in any permutation π otherwise. We call a block $bl(i,j)$ with $u'_i \neq v'_j$ an *up-block* if arcs (u, u'_i) and (v, v'_j) creates a crossing in a permutation π with $\pi(u) < \pi(v)$ and an *down-block* otherwise. We call a block $bl(i,j)$ with $u'_i = v'_j$ a *neutral-block*. Observe that the number of up-blocks (resp., down-blocks and neutral-blocks) is equal to c_{uv} (resp., c_{vu} and $\Delta_{uv} = \Delta_{vu}$). We here partition the set of these blocks into two groups UP and DWN as follows (where a neutral-block may be split into two half blocks in the partitioning).

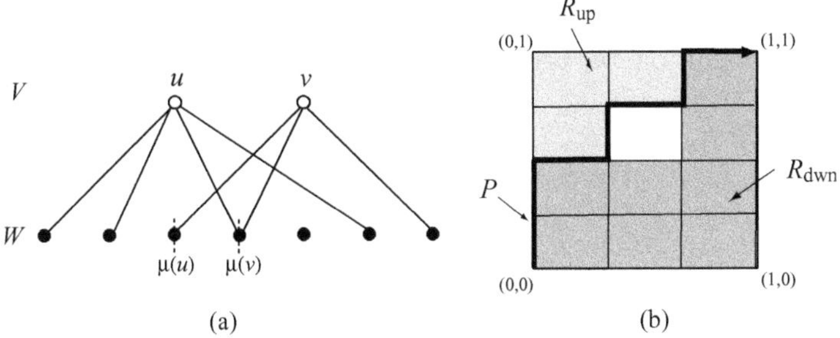

Fig. 2. (a) Two nodes u and v in the top layer, where $c_{uv} = 3$ and $c_{vu} = 8$. (b) A (u,v)-path P of a $(4,3)$-sliced square S in the case of (i).

Definition 1. *For each node $u \in V$, where $\Gamma(u) = \{w_1, w_2, \ldots, w_{d_u}\} \subseteq W$ $(w_1 < w_2 < \cdots < w_{d_u})$, we define the* median index $\mu(u)$ *of its neighbors by*

$$\mu(u) := \begin{cases} w_{\frac{d_u+1}{2}} & \text{if } d_u \text{ is odd,} \\ \frac{1}{2}(w_{\frac{d_u}{2}} + w_{\frac{d_u}{2}+1}) & \text{if } d_u \text{ is even.} \end{cases}$$

(i) *If $\mu(u) < \mu(v)$, then let UP be the set of all up-blocks, and DWN be the set of down-blocks and neutral-blocks (see Fig. 2).*
(ii) *If $\mu(u) > \mu(v)$, then let UP be the set of all up-blocks and neutral-blocks, and DWN be the set of down-blocks (see Fig. 3).*
(iii) *If $\mu(u) = \mu(v)$, then split each neutral-block $[p, q]$ into two parts by the line segment pq, and put the upper-left part into UP and the other in DWN. Then put all up-blocks in the UP, and all down-blocks in the DWN.*

The set of all points in the blocks in UP forms a connected region, which we denoted by R_{up}. Similarly R_{dwn} is defined by DWN. □

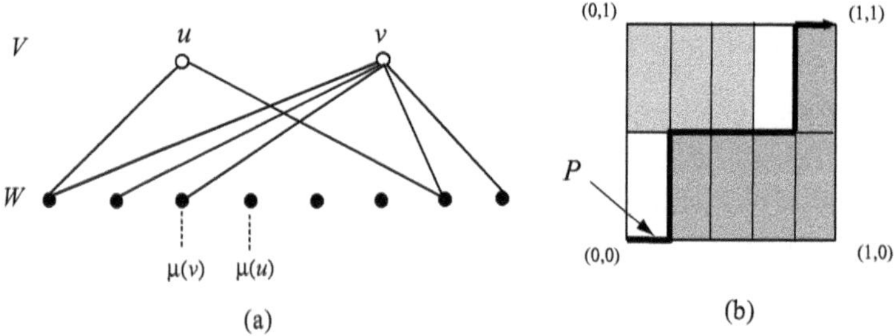

Fig. 3. (a) Two nodes u and v in the top layer. (b) A (u, v)-path P of a $(2, 5)$-sliced square S in the case of (ii).

From the definition, we observe the next property.

Lemma 1. *Let R_{up} and R_{dwn} be the regions defined for an ordered pair of nodes u and v in V. Then there is a monotone path P that separates S into R_{up} and R_{dwn}, and it holds*

$$a(R_{up}) = \begin{cases} \frac{c_{uv}}{d_u d_v} & \text{if } \mu(u) < \mu(v), \\ \frac{c_{uv} + \frac{\Delta_{uv}}{2}}{d_u d_v} & \text{if } \mu(u) = \mu(v), \\ \frac{c_{uv} + \Delta_{uv}}{d_u d_v} & \text{if } \mu(u) > \mu(v). \end{cases}$$

Moreover, R_{up} contains point $(0.5, 0.5)$ if $\mu(u) \geq \mu(v)$. □

Such a path P in the lemma is called the *(u, v)-path* with respect to G and σ.

Lemma 2. *Let $u, v \in V$ be two nodes in (G, σ) such that $1 \leq c_{uv} < c_{vu}$ and $c_{uv} < 2\Delta_{uv}$. Then $\mu(u) < \mu(v)$ holds unless u and v satisfies one of the following conditions (1), (2) and (3):*

$$c_{uv} = 3,\ c_{vu} = 4,\ d_u = d_v = 3 \text{ and } \Delta_{uv} = 2, \tag{1}$$

$$c_{uv} = 3,\ c_{vu} = 5,\ \{d_u, d_v\} = \{2, 5\} \text{ and } \Delta_{uv} = 2 \text{ (see Fig. 3)}, \tag{2}$$

$$c_{uv} = 5,\ c_{vu} = 7,\ \{d_u, d_v\} = \{3, 5\} \text{ and } \Delta_{uv} = 3. \tag{3}$$

Proof. Omitted. □

We close this section by showing some technical lemmas.

Lemma 3. *For constants $a > 0, b, c > 0$ and d such that $ad - bc \geq 0$, function $f(x) = (ax+b)(\frac{1}{cx+d} - 2)$ takes the maximum $\frac{1}{c}(\sqrt{a} - \sqrt{2(ad - bc)})^2$ over x with $cx + d > 0$.* □

Lemma 4. *For four positive constants a, b, c and d with $\frac{b}{a} < d \leq \frac{1}{\sqrt{2c}}$, function $f(x) = (ax - b)^2(\frac{1}{cx^2} - 2)$ $(\frac{b}{a} < x \leq d)$ takes the maximum at $x = \min\{d, (\frac{b}{2ac})^{\frac{1}{3}}\}$.* □

3 Algorithm and Analysis

Let $\theta : V \to [0, 1]$ be a function from V to the set of reals in $[0, 1]$, where $\theta(u)$ is called the *real key* of node u. Given a real-key function θ, we construct a permutation π_θ of $\{1, 2, \ldots, |V|\}$ by the next procedure.

PERMUTE($\theta; \pi_\theta$): Step 1. For each node $u \in V$, compute $j = \lceil \theta(u) d_u \rceil$, and define an *integer key* $\kappa(u)$ of u by

$$\kappa(u) := w_j \text{ for the } j\text{-th neighbor } w_j \in \Gamma(u),$$

where $\Gamma(u) = \{w_1, w_2, \ldots, w_{d_u}\}$ $(w_1 < w_2 < \cdots < w_{d_u})$.

Step 2. Sort nodes $u \in V$ in the lexicographical order with respect to $(\kappa(u), \mu(u))$, where the ties among nodes u with the same key $(\kappa(u), \mu(u))$ are broken randomly. We denote by π_θ the resulting permutation of $\{1, 2, \ldots, |V|\}$. □

We see the following important property.

Lemma 5. *For two nodes $u, v \in V$, let R_{up} and R_{dwn} be the regions in Definition 1. Then for a given real-key function θ, $\pi_\theta(u) < \pi_\theta(v)$ if point $(\theta(u), \theta(v))$ is inside R_{dwn} and $\pi_\theta(u) > \pi_\theta(v)$ if point $(\theta(u), \theta(v))$ is inside R_{up}.* □

A *scheme* based on which we choose a real-key function θ probabilistically is defined by a set of tuples of reals $\mathcal{S} = \{(s_i, t_i, p_i) \mid i = 1, 2, \ldots, h\}$, such that $0 < s_i \leq t_i < 1$ and $0 \leq p_i$ for $i = 1, 2, \ldots, h$ and $\sum_{1 \leq i \leq h} p_i = 1$, where we call each (s_i, t_i, p_i) a *subscheme*. Given a scheme $\mathcal{S}$, we choose a real-key function θ in the following manner.

RANDOM-KEY$(\mathcal{S}; \theta)$**:**
Step 1. Choose a subscheme $(s_i, t_i, p_i) \in \mathcal{S}$ with probability p_i.

Step 2. For each node $u \in V$, choose a real key $\theta(u)$ from $[s_i, t_i]$ uniformly. □

We denote by $E_{\mathcal{S}}[cross(u, v; \pi_\theta)]$ and $E_{\mathcal{S}}[cross(\pi_\theta)]$ respectively the expectations of $cross(u, v; \pi_\theta)$ and $cross(\pi_\theta)$ over all real-key functions θ resulting from RANDOM-KEY. In this paper, we prove the next result.

Theorem 2. *There is a scheme $\mathcal{S}$ such that $E_{\mathcal{S}}[cross(\pi_\theta)] \leq 1.4664LB$.* □

By the linearity of expectations, if we have a constant $\alpha \geq 1$ such that

$$E_{\mathcal{S}}[cross(u, v; \pi_\theta)] \leq \alpha \min\{c_{uv}, c_{vu}\}, \quad u, v \in V,$$

then it holds $E_{\mathcal{S}}[cross(\pi_\theta)] \leq \alpha LB$.

In the rest of this paper, we fix two nodes $u, v \in V$, and analyze $E_{\mathcal{S}}[cross(u, v; \pi_\theta)]$ for a given scheme $\mathcal{S}$. Without loss of generality we assume that $1.46c_{uv} < c_{vu}$ and $d_u \leq d_v$ (the case of $\max\{c_{uv}, c_{vu}\} < 1.46 \min\{c_{uv}, c_{vu}\}$ needs no special consideration to prove Theorem 2, and we have $d_u \leq d_v$ by renaming u and v after reversing the permutation σ). Note that none of (1) and (3) holds since $1.46c_{uv} < c_{vu}$. Moreover, we can assume that $c_{uv} \geq 1$ since otherwise (i.e., $c_{uv} = 0$) $\pi_\theta(u) < \pi_\theta(v)$ holds in any permutation π_θ computed by PERMUTE due to the comparison of $\mu(u)$ and $\mu(v)$.

For a given scheme $\mathcal{S}$ and a region $R \subseteq S$, let $p_{\mathcal{S}}(R)$ denote the probability that point $(\theta(u), \theta(v))$ falls inside R. By Lemma 5, we observe the next formula.

Lemma 6. $E_{\mathcal{S}}[cross(u, v; \pi_\theta)] = p_{\mathcal{S}}(R_{dwn})c_{uv} + p_{\mathcal{S}}(R_{up})c_{vu}$. □

We are ready to derive an important inequality.

Lemma 7. *Assume that $1 \leq c_{uv} < c_{vu}/1.46$ holds. Then it holds*

$$\frac{E_{\mathcal{S}}[cross(u, v; \pi_\theta)]}{\min\{c_{uv}, c_{vu}\}} \leq \begin{cases} 1 + p_{\mathcal{S}}(R_{up})(\frac{1}{a(R_{up})} - 2) & \text{if } \mu(u) < \mu(v), \\ 1 + p_{\mathcal{S}}(R_{up})(\frac{1.5}{a(R_{up})} - 2.5) & \text{if } \mu(u) \geq \mu(v) \end{cases}$$

unless (2) *holds.*

Proof. By Lemma 6, we get

$$\frac{E_{\mathcal{S}}[cross(u,v;\pi_\theta)]}{\min\{c_{uv},c_{vu}\}} = \frac{p_{\mathcal{S}}(R_{dwn})c_{uv}+p_{\mathcal{S}}(R_{up})c_{vu}}{c_{uv}}$$
$$= \frac{(1-p_{\mathcal{S}}(R_{up}))c_{uv}+p_{\mathcal{S}}(R_{up})(d_ud_v-c_{uv}-\Delta_{uv})}{c_{uv}} = 1+p_{\mathcal{S}}(R_{up})(\frac{d_ud_v-\Delta_{uv}}{c_{uv}}-2).$$

First consider the case of $\mu(u)<\mu(v)$. By Lemma 1, we have $a(R_{up}) = \frac{c_{uv}}{d_ud_v}$. Hence

$$\frac{d_ud_v-\Delta_{uv}}{c_{uv}}-2=\frac{1}{c_{uv}}(\frac{c_{uv}}{a(R_{up})}-\Delta_{uv})-2\le\frac{1}{a(R_{up})}-2.$$

Next consider the case of $\mu(u)\ge\mu(v)$. By Lemma 1, we have $a(R_{up})\le\frac{c_{uv}+\Delta_{uv}}{d_ud_v}$. Since $1\le c_{uv}<c_{vu}$ holds but (2) does not holds for the u and v, Lemma 2 implies $\Delta_{uv}\le c_{uv}/2$. Then

$$\frac{d_ud_v-\Delta_{uv}}{c_{uv}}-2\le\frac{1}{c_{uv}}(\frac{c_{uv}+\Delta_{uv}}{a(R_{up})}-\Delta_{uv})-2=\frac{c_{uv}+\Delta_{uv}(1-a(R_{up}))}{c_{uv}a(R_{up})}-2$$
$$\le\frac{c_{uv}+\frac{1}{2}c_{uv}(1-a(R_{up}))}{c_{uv}a(R_{up})}-2=\frac{1.5}{a(R_{up})}-2.5.$$

This completes the proof. □

We wish to find a scheme $\mathcal{S}$ that minimizes $\max_{u,v\in V}\frac{E_{\mathcal{S}}[cross(u,v;\pi_\theta)]}{\min\{c_{uv},c_{vu}\}}$ (even though finding such an $\mathcal{S}$ analytically seems a hard problem). For this, we consider the set of all monotone paths P for a given scheme $\mathcal{S}$. Let P be an arbitrary monotone path between points $(0,0)$ and $(1,1)$ in a unit square S (not necessarily a (u,v)-path for particular nodes $u,v\in V$). Define $R_{up}(P)$ and $R_{dwn}(P)$ be the regions obtained by splitting S with P, where we assume that $R_{up}(P)$ is above $R_{dwn}(P)$. Let

$$\beta(\mathcal{S},P):=\begin{cases} p_{\mathcal{S}}(R_{up}(P))(\frac{1}{a(R_{up}(P))}-2) & \text{if } (0.5,0.5)\notin R_{up}(P),\\ p_{\mathcal{S}}(R_{up}(P))(\frac{1.5}{a(R_{up}(P))}-2.5) & \text{if } (0.5,0.5)\in R_{up}(P),\end{cases}$$

and $\beta(\mathcal{S}):=\max\{\beta(\mathcal{S},P)\mid \text{monotone path } P\}$. Given a scheme $\mathcal{S}$, a monotone path P from $(0,0)$ to $(1,1)$ in a unit square S is called $\mathcal{S}$-*maximal* if $\beta(\mathcal{S},P)=\beta(\mathcal{S})$.

Since the choice of monotone paths P is relaxed, we obtain $E_{\mathcal{S}}[cross(\pi_\theta)]\le(1+\beta(\mathcal{S}))LB$. (Recall that $(0.5,0.5)\in R_{up}$ holds if $\mu(u)\ge\mu(v)$ by Lemma 1.) Therefore, to prove Theorem 2, it suffices to show that there exists a scheme $\mathcal{S}$ such that $\beta(\mathcal{S})<0.4664$ (provided that the case of (2) is treated separately to prove Theorem 2).

4 A Scheme $\mathcal{S}$

We consider scheme

$$\mathcal{S} = \{(s_1 = 0.0957, t_1 = 0.5, p_1 = 0.5), (s_2 = 0.5, t_2 = 0.9043, p_2 = 0.5)\}.$$

Denote the squares $S_1 = [(s_1, s_1), (0.5, 0.5)]$ and $S_2 = [(0.5, 0.5), (t_2, t_2)]$, and the corners of these squares by $A_1 = (0.0957, 0.0957)$, $A_2 = (0.5, 0.5)$, $A_3 = (0.9043, 0.9043)$, $B_1 = (0.5, 0.0957)$, $B_2 = (0.9043, 0.5)$, $C_1 = (0.0957, 0.5)$ and $C_2 = (0.5, 0.9043)$. (The constant 0.0957 and others have been determined through some computational experiment.) Fig. 4 illustrates this scheme.

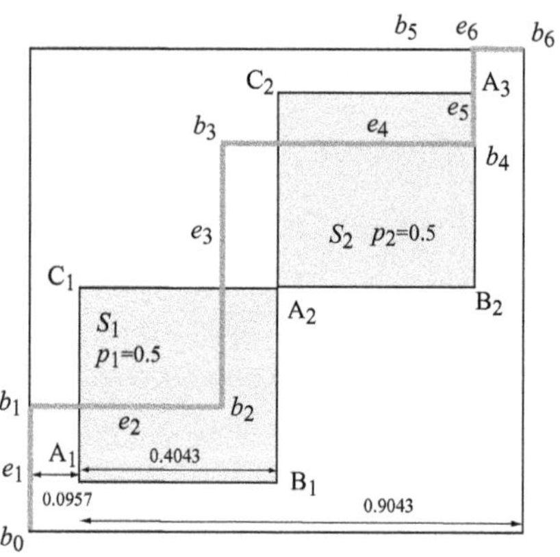

Fig. 4. Illustration of scheme $\mathcal{S} = \{(0.0957, 0.5, 0.5), (0.5, 0.9043, 0.5)\}$, where a grey line indicates an example of a monotone path P.

We first consider the case of (2). In (2), where $c_{uv} = 3$, $c_{vu} = 5$ and $p_{\mathcal{S}}(R_{up}) = p_{\mathcal{S}}(R_{dwn}) = 0.5$ holds, we have by Lemma 6, $E_{\mathcal{S}}[cross(u, v; \pi_\theta)] = p_{\mathcal{S}}(R_{dwn})c_{uv} + p_{\mathcal{S}}(R_{up})c_{vu} = 4 < 1.4664c_{uv}$.

Now consider nodes u and v in the general case. It is not difficult to see that an $\mathcal{S}$-maximal monotone path P consists of axis-parallel line segments, and that the resulting region $R_{up}(P)$ contains at most one convex corner in each subscheme S_i $(i = 1, 2)$. For simplicity, we consider a single subscheme S_i. It should be noted that $p_{\mathcal{S}}(R_{up})$ (or the contribution from S_i to $p_{\mathcal{S}}(R_{up})$) is given by $a(S_i \cap R_{up}(P))/a(S_i)$. As shown in Fig. 5(a), if a monotone path P does not satisfy these properties, then we can modify the path P into another monotone path P' such that $a(S_i \cap R_{up}(P')) = a(S_i \cap R_{up}(P))$ and $a(R_{up}(P')) \leq a(R_{up}(P))$. For such an axis-parallel piecewise linear monotone path P, we denote the sequence of the corner points by

$$b_0 = (0, 0), b_1, \ldots, b_k = (1, 1),$$

and the sequence of the edges by

$$e_1 = b_0b_1, e_2 = b_1b_2, \ldots, e_k = b_{k-1}b_k$$

(see Fig. 4). Let e be an edge on a path P, where e may be a partial segment of some edge e_i. Without loss of generality we further assume that an $\mathcal{S}$-maximal monotone path P is chosen so that the number of edges of squares in subschemes or of the entire unit square that are overlapped by the edges in P is maximized among all $\mathcal{S}$-maximal monotone paths.

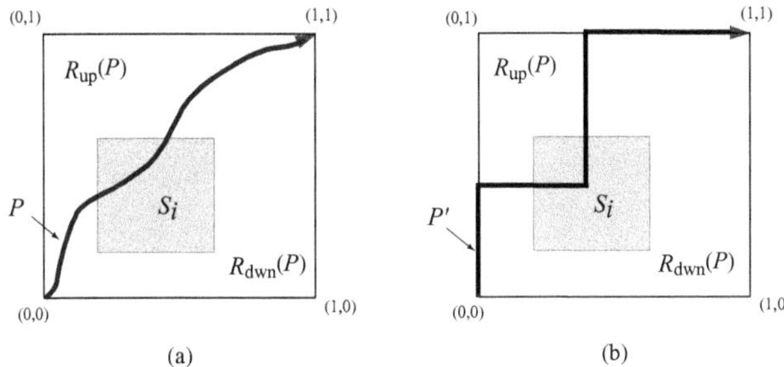

Fig. 5. A monotone path P that passes through a square S_i.

We define the *gain* of edge e at a subscheme $S_i = (s_i, t_i, p_i) \in \mathcal{S}$ as follows. Consider how much amount of $p_{\mathcal{S}}(R_{up})$ changes if we move the line segment L in its orthogonal direction by an infinitely small amount ϵ. The change in $p_{\mathcal{S}}(R_{up})$ is

$$\frac{\epsilon \cdot \ell(e \cap S_i) \cdot p_i}{(t_i - s_i)^2},$$

where $\ell(e \cap S_i)$ means the length of the intersection of e and S_i. On the other hand, the change in $a(R_{up}(P))$ is $\epsilon \cdot \ell(e)$. The *gain* is defined by the ratio of these two, i.e.,

$$g(e) = \frac{\ell(e \cap S_i) \cdot p_i}{(t_i - s_i)^2 \cdot \ell(e)}.$$

A vertical line segment e on a path P is called *incrementable* (resp., *decrementable*) if

- There is a real $\delta > 0$ such that e has the same gain $g(e)$ (with respect to a subscheme S_i) after translating it rightward (resp., leftward) by any amount $\delta' \in [0, \delta]$ (i.e., e remains to be intersecting S_i),
- For the rectangle R formed between e and the translated edge e' and the current path P, there is a monotone path P' such that $R_{up}(P') = R_{up}(P) \cup R$ (resp., $R_{up}(P') = R_{up}(P) - R$).

Analogously, the incrementability (resp., decrementability) of a horizontal line segment e is defined by replacing "rightward" with " downward" (resp., "leftward" with "upward").

An edge e_i between two corners in a path P is called a *free edge* if it does not overlap with any edge of square S_i in a subscheme or of the entire unit square S. For example, in Fig. 4, e_2, e_3 and e_4 are free edges, and e_5 is decrementable, but not incrementable.

By definition, we observe the following.

Lemma 8. *For an $\mathcal{S}$-maximal monotone path P, let e and e' be respectively an incrementable edge and a decrementable edge such that* $(0.5, 0.5)$ *is not an internal point in any of e and e'. Then if e and e' are not adjacent, then $g(e) < g(e')$. If e and e' are adjacent, then $g(e) = g(e')$.*

Proof. Otherwise we would have another monotone path P' such that $\beta(\mathcal{S}, P') > \beta(\mathcal{S}, P)$ or such that $\beta(\mathcal{S}, P') = \beta(\mathcal{S}, P)$ and P' overlaps with more edges of the squares than P does. □

In particular, there is no pair of non-adjacent free edges in an $\mathcal{S}$-maximal monotone path P.

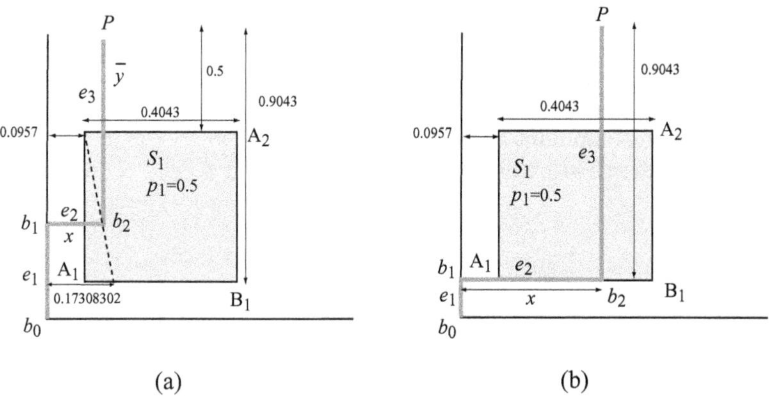

Fig. 6. Illustration for Case-(A,1), where (a) indicates the case where a corner point of P is inside S_1, and (b) indicates the case where edge A_1B_1 of S_1 is overlapped by edge e_2 of P.

In the sequel, P is assumed to be an $\mathcal{S}$-maximal monotone path, and for simplicity $R_{up}(P)$ is written by R_{up}. To prove that $\beta(\mathcal{S}, P) \leq 0.4664$ holds for our scheme $\mathcal{S}$, we distinguish the following cases.

Case-A: point $(0.5, 0.5)$ is not on the boundary of R_{up} or inside R_{up}; $\beta(\mathcal{S}, P)$ is given by $p_{\mathcal{S}}(R_{up})(\frac{1}{a(R_{up})} - 2)$.

Case-B: point $(0.5, 0.5)$ is on the boundary of R_{up} or inside R_{up}; $\beta(\mathcal{S}, P)$ is given by $p_{\mathcal{S}}(R_{up})(\frac{1.5}{a(R_{up})} - 2.5)$.

Case-1: R_{up} contains an internal point from exactly one of S_1 and S_2; Without loss of generality R_{up} contains no internal point in S_2.

Case-2: R_{up} contains an internal point from each of S_1 and S_2.

In the following, we only treat Case-(A,1) due to the space limitation.

In this case, R_{up} has no convex corner in S_2, and exactly one convex corner b_3 in S_1 (see Fig. 6(a)). Consider edges $e_2 = b_1b_2$ and $e_3 = b_2b_3$ in P. By $(0.5, 0.5) \in R_{up}$, e_3 does not overlaps with B_1A_2, and thereby e_3 is a free edge. Let $x = \ell(e_2)$ and $\bar{y} = \ell(e_3)$.

First consider the case where e_2 does not overlaps with A_1B_1, i.e., e_3 is a free edge. Then $g(e_2) = \frac{0.5}{(0.4043)^2} \times \frac{x-0.0957}{x}$, and $g(e_3) = \frac{0.5}{(0.4043)^2} \times \frac{\bar{y}-0.5}{\bar{y}}$. Since P is $\mathcal{S}$-monotone, it must hold $g(e_2) = g(e_3)$ for two free edges. Thus we have $\bar{y} = \frac{0.5}{0.0957}x$. Hence by $\bar{y} \leq 0.9043$, we have $x < 0.9043 \times \frac{0.0957}{0.5} = 0.17308302$. By $\bar{y} = \frac{0.5}{0.0957}x$, we have $\bar{y} - 0.5 = \frac{0.5}{0.0957}x - 0.5 = \frac{0.5}{0.0957}(x - 0.0957)$. We have $a(R_{up}) = x\bar{y}$ and $p_{\mathcal{S}}(R_{up}) = 0.5 \times \frac{(x-0.0957)(\bar{y}-0.5)}{(0.4043)^2} = \frac{0.5 \times 0.5}{(0.4043)^2 \times 0.0957}(x-0.0957)^2$. Then $\beta(\mathcal{S}, P) = p_{\mathcal{S}}(R_{up})(\frac{1}{a(R_{up})} - 2) = \frac{0.5 \times 0.5}{(0.4043)^2 \times 0.0957}(x - 0.0957)^2(\frac{1}{\frac{0.5}{0.0957}x^2} - 2)$. By Lemma 4 with $a = 1, b = 0.0957$ and $c = \frac{0.5}{0.0957}$, this takes the maximum at $x = \min\{0.17308302, (\frac{0.0957}{2 \times \frac{0.5}{0.0957}})^{\frac{1}{3}}\}$ (the latter is at least 0.209).

Cases-(A,2), (B,1) and (B,2) can be treated similarly. This establishes $\beta(\mathcal{S}) < 0.4664$ and hence Theorems 2 and 1.

References

1. G. Di Battista, P. Eades, R. Tamassia and I. G. Tollis, Graph Drawing: Algorithms for Visualization of Graphs, Prentice Hall (1999).
2. V. Dujmović and S. Whitesides, An efficient fixed parameter tractable algorithm for 1-sided crossing minimization, Graph Drawing 2002, Lecture Notes in Computer Science, 2528, Springer-Verlag, Berlin (2002) 118–129.
3. P. Eades and N. C. Wormald, Edge crossing in drawing bipartite graphs, Algorithmica, 11 (1994) 379–403.
4. M. R. Garey and D. S. Johnson, Crossing number is NP-complete, SIAM J. Algebraic Discrete Methods, 4 (1983) 312–316.
5. F. Harary and A. J. Schwenk, Trees with Hamiltonian square, Mathematiks, 18 (1971) 138–140.
6. F. Harary and A. J. Schwenk, A new crossing number for bipartite graphs, Utilitas Math., 1 (1972) 203–209.
7. M. Jünger and P. Mutzel, 2-layer straight line crossing minimization: performance of exact and heuristic algorithms, J. Graph Algorithms and Applications, 1 (1997) 1–25.
8. X. Y- Li and M. F. Sallmann, New bounds on the barycenter heuristic for bipartite drawing, Information Processing Letters, 82 (2002) 293–298.
9. E. Mäkinen, Experiments on drawing 2-level hierarchical graphs, International Journal of Computer Mathematics, 36 (1990) 175–181.

10. X. Muñoz, W. Unger and I. Vrt'o, One sided crossing minimization is NP-hard for sparse graphs, Graph Drawing 2001, Lecture Notes in Computer Science, 2265, Springer-Verlag, Berlin (2002) 115–123.
11. C. Sechen, VLSI Placement and Global Routing Using Simulated Annealing, Kluwer, Dordrecht (1988).
12. K. Sugiyama, S. Tagawa and M. Toda, Methods for visual understanding of hierarchical system structures, IEEE Transactions on Systems, Man, and Cybernetics, 11 (1981) 109–125.
13. A. Yamaguchi and A. Sugimoto, An approximation algorithm for the two-layered graph drawing problem, In: 5th Annual Intl. Conf. on Computing and Combinatorics, Lecture Notes in Computer Science, 1627, Springer-Verlag, Berlin (1999) 81–91.

Straight-Line Drawings of 2-Outerplanar Graphs on Two Curves
(Extended Abstract) *

Emilio Di Giacomo and Walter Didimo

Università di Perugia
({digiacomo,didimo}@diei.unipg.it).

Abstract. We study how to draw a 2-outerplane graph on two concentric circles, so that the internal (resp. external) vertices lie on the internal (resp. external) circle.

1 Introduction

A 1-*outerplanar embedded graph* (or simpler an *outerplane graph*) is an embedded planar graph where all vertices are on the external face. A k-*outerplane graph* G ($k > 1$) is an embedded planar graph such that the embedded graph obtained by removing all vertices of the external face is a $(k-1)$-outerplane graph. In other words, iteratively removing k times from G the vertices of the external face, the graph becomes empty. Each embedded planar graph is a k-outerplane graph for some constant k. The properties of k outerplane graphs have been studied in several papers (see, e.g., [1,2,4]).

In this paper we concentrate on the problem of drawing 2-outerplane graphs. A natural way to conceive a drawing of a 2-outerplane graph consists of mapping all its vertices on two concentric circles in such a way that all the external vertices lie on the external circle, and all the internal vertices lie on the internal circle. Again, it would be desirable to have straight-line edges with no crossings. We call such a drawing a *2-curve drawing* . The computation of a 2-curve drawing is a problem related to the automatic layout of graphs where vertices are constrained to be on a given set of lines or curves in the plane, and edges bend as little as possible. The main results of this paper are listed below:

- We show that every 2-outerplane graph whose internal vertices induce a biconnected subgraph admits a 2-curve drawing , which can be computed in linear time. Our result is in the middle of two opposite results known in the literature. Namely, in [5] it is shown that every planar graph can be drawn without crossings by mapping its vertices on a circular arc and by drawing every edge as a polyline with at most one bend. However, the class

* Research partially supported by "Progetto ALINWEB: Algoritmica per Internet e per il Web", MIUR Programmi di Ricerca Scientifica di Rilevante Interesse Nazionale.

G. Liotta (Ed.): GD 2003, LNCS 2912, pp. 419–424, 2004.

of graphs that have a planar straight-line drawing where vertices lie on a circular arc coincides with the class of outerplanar graphs, and this class is further reduced if the vertices must lie on two parallel straight lines [3]. Finally, note that if a plane graph admits a planar straight-line drawing where vertices lie on two circular arcs, it is necessarily a k-outerplanar graph with $k \in \{1, 2\}$. Therefore, k-outerplanar graphs with $k \in \{1, 2\}$ are the only graphs that one can hope to draw on two circular arcs with straight-line edges and without crossings.
- We extend our drawing technique to 2-outerplane graphs whose internal vertices induce a simply connected graph. In this case, the computed drawings may have a number of bends that is linear in the number of edges connecting external vertices to cut-vertices of the graph induced by the internal vertices. The problem of deciding whether a general 2-outerplane graph admits a 2-curve drawing remains open.

2 Preliminaries

Let $G = (V, E)$ be a 2-outerplane graph. $V_{ext} \subseteq V$ denotes the set of the vertices on the external face of G, and $V_{int} = V - V_{ext}$ is the set of its internal vertices. Also, $E_{int} = \{(u, v) \in E | u, v \in V_{int}\}$, $E_{ext} = \{(u, v) \in E | u, v \in V_{ext}\}$, and $E_{mix} = \{(u, v) \in E | u \in V_{int}, v \in V_{ext}\}$. Clearly, $E = E_{int} \cup E_{ext} \cup E_{mix}$. $G(V_{int})$ and $G(V_{ext})$ denote the subgraphs of G induced by the vertices of V_{int} and V_{ext}, respectively. Figure 1(a) shows an example of a 2-outerplane graph.

Given any two distinct concentric circles and a horizontal straight line l through the center, denote by Λ_{int} and Λ_{ext} the two semicircles that lie on the upper half-plane defined by l, where Λ_{int} is the semicircle with smaller radius. A *2-curve drawing* of G on Λ_{int}, Λ_{ext} is a planar drawing of G such that: (i) The vertices of V_{int} and V_{ext} are mapped to points of Λ_{int} and Λ_{ext}, respectively; (ii) Each edge of E is drawn as a straight-line segment.

A *biconnected-core 2-outerplane* graph is a 2-outerplane graph G such that $G(V_{int})$ is biconnected. For the sake of brevity, a biconnected-core 2-outerplane graph is called a *BC2O-graph* from now on.

Without loss of generality, in the following sections we always assume that G is a *BC2O-graph* graph where all internal faces are triangles. We call such a graph *quasi-triangulated*. Indeed, if G is not quasi-triangulated we can recursively decompose an internal non-triangular face of G avoiding that a vertex of V_{int} becomes an internal vertex of $G(V_{int})$, and avoiding multiple edges. In this way we obtain a 2-outerplane graph including G and preserving the 2-outerplanar embedding of G. More specifically, each internal face having only vertices of V_{int} or only vertices of V_{ext} can be recursively triangulated with a classical approach. Also, for each internal face f that has both a vertex $u \in V_{int}$ and a vertex $v \in V_{ext}$, such that u and v are not adjacent in G, we can split f with a new edge (u, v) (which will be a new edge of E_{mix}). The obtained graph is still a 2-outerplane graph with the same sets V_{int}, V_{ext} as in G. There exists a simple linear time procedure that can be used to recursively decompose all faces that have both internal and external vertices. Due to space limitation, we omit the detailed description of such a procedure.

3 The Drawing Algorithm for BC2O-Graphs

Let $G = (V, E)$ be a quasi-triangulated $BC2O$-graph, and let v be a vertex of $G(V_{int})$. Since $G(V_{int})$ is biconnected, v is encountered only once when traversing the external boundary of $G(V_{int})$. Let u and w be the two vertices that immediately precede and succeed v while walking counterclockwise the external boundary of $G(V_{int})$. Denote by $e_{in}(v)$ the edge (u, v) and by $e_{out}(v)$ the edge (v, w). Also, denote by $E(v)$ the circular list of edges incident on v in clockwise order. The following properties hold. Proofs are omitted for reasons of space.

Property 1. Let v be a vertex of $G(V_{int})$. There is no edge $e \in E_{mix} \cap E(v)$ between $e_{in}(v)$ and $e_{out}(v)$ in $E(v)$.

Property 2. Let v be a vertex of V_{ext}, and let $e_i = (v, v_i)$ $(i = 1, \ldots, k)$ be the edges in $E_{mix} \cap E(v)$ in the order they appear in $E(v)$. Vertices $v_1, v_2, \ldots, v_k$ are encountered in this order when traversing counterclockwise the external boundary of $G(V_{int})$.

Property 3. There exists a vertex $v \in V_{int}$ such that there are two consecutive edges of E_{mix} in $E(v)$.

Based on the above properties, a $BC2O$-graph G can be described in terms of a general structure (see also Figure 1(b)) that is useful to prove the correctness of the drawing algorithm. Let v be a vertex of V_{int} with two incident edges $e_w = (v, w')$ and $e_u = (v, u')$ of E_{mix}, that are consecutive in $E(v)$. This vertex exists by Property 3. If w' is adjacent to some vertex of V_{int} distinct from v, let $w = w'$, otherwise let w be the first vertex encountered in counterclockwise order in the adjacency list of v starting from w' such that it is adjacent to some vertex of V_{int} distinct from v. Analogously, if u' is adjacent to some vertex of V_{int} distinct from u, let $u = u'$, otherwise let u be the first vertex encountered in clockwise order in the adjacency list of v starting from u' such that it is adjacent to some vertex of V_{int} distinct from v. Note that u and w always exist and they are distinct since the graph is simple. Let h be the external boundary of $G(V_{int})$. Denote by $u_1, u_2, \ldots, u_r$ the vertices of V_{int} that are adjacent to u, in counterclockwise order walking on h. Also, denote by $w_1, w_2, \ldots, w_s$ the vertices of V_{int} that are adjacent to w, in counterclockwise order walking on h. In the general structure of G we distinguish among three subgraphs, G_1, G_2, and G_3, whose edges partition the set of edges of G (see Figure 1). The three subgraphs are outerplane and thus easy to embed. They are defined as follows: (i) G_1 is the subgraph induced by the following vertices: v, u, w, and all the vertices of V_{ext} that are between u and w while walking on the external boundary of G counterclockwise; (ii) G_2 coincides with $G(V_{int})$; (iii) G_3 is the subgraph whose edges are those of G that are not in G_1 and G_2, and whose vertices are exactly the end-vertices of these edges.

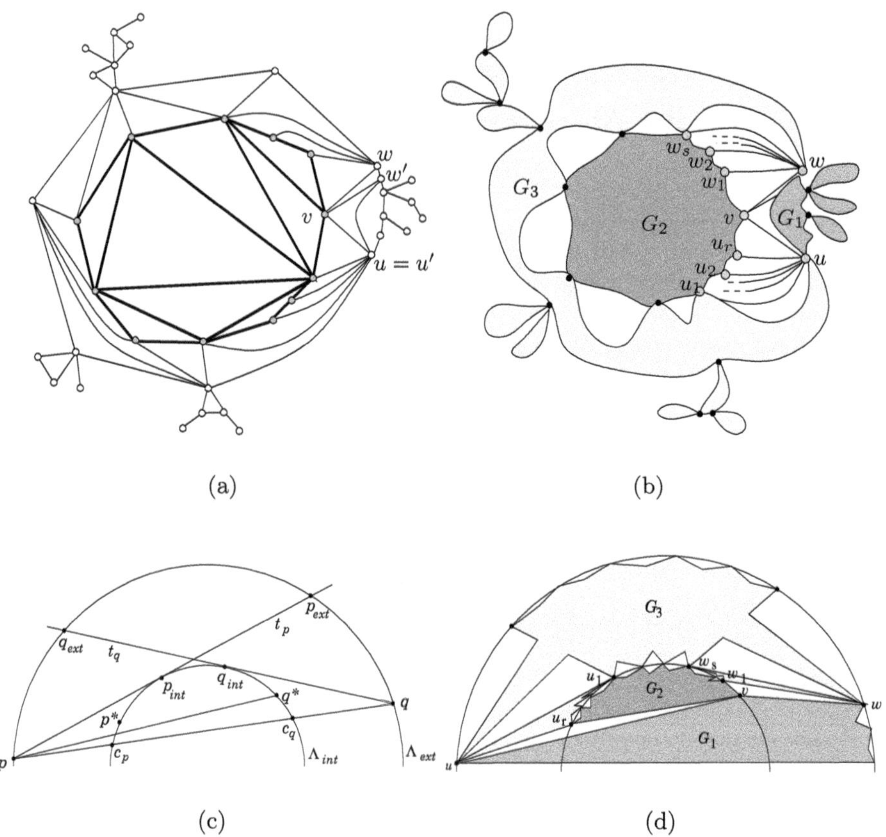

Fig. 1. (a) A 2-outerplane graph G where $G(V_{int})$ is biconnected, that is, a $BC2O$-graph. The edges of $G(V_{int})$ are thick. (b) Structure of G. (c) Notation used in the description of the drawing algorithm. (d) Structure of a drawing computed by the drawing algorithm.

We call 2-CURVE-DRAWER the algorithm that computes a 2-curve drawing of a $BC2O$-graph. The algorithm works as follows (see also Figure 1(c)). First it chooses two distinct points, p and q, of Λ_{ext} such that: (i) the x- and y-coordinate of p is less than the x- and the y-coordinate of q, respectively; (ii) segment $\overline{pq}$ is a chord of Λ_{ext} that has two intersection points c_p, c_q with Λ_{int}, where c_p is the first point encountered while walking on $\overline{pq}$ from p to q; (iii) there are two lines t_p and t_q passing for p and q, respectively, that are tangent to Λ_{int}, and intersecting in a point lying in the portion of the annulus delimited by Λ_{int} and Λ_{ext}. Denote by $p_{ext} \neq p$ and p_{int} the points where t_p intersects Λ_{ext} and Λ_{int}, respectively. Similarly, let $q_{ext} \neq q$ and q_{int} be the points where t_q intersects

Λ_{ext} and Λ_{int}, respectively. Also denote by q^* any point of Λ_{int} between q_{int} and c_q, and by p^* any point of Λ_{int} between p_{int} and the point $\overline{pq^*} \cap \Lambda_{int}$.

Then the algorithm proceeds as follows (see also Figure 1(d)): It maps all vertices of V_{int} to points of Λ_{int}, according to the clockwise order they appear on the external boundary of $G(V_{int})$, in such a way that: (i) u_r and u_1 are mapped to p^* and p_{int}, respectively; (ii) w_s is mapped to q_{int}; (iii) v is mapped to q^*.

Also, it maps all vertices of V_{ext} to points of Λ_{ext}, according to the clockwise order they appear on the external boundary of G, in such a way that: (i) u and w are mapped to p and q, respectively; (ii) all vertices from u to w are mapped to points between q_{ext} and p_{ext} (iii) all vertices from w to u are mapped to points below q.

Lemma 1. *Algorithm* 2-CURVE-DRAWER*(G) computes a 2-curve drawing of G.*

Proof. Denote by Γ_1, Γ_2, and Γ_3 the drawings of G_1, G_2, and G_3 in the drawing Γ of G. We first prove that these drawings lie in three distinct connected regions R_1, R_2, and R_3 of the plane whose interiors do not intersect (see Figure 1(d)). This guarantees that there is no crossing between edges of Γ_1, Γ_2, and Γ_3. Then, we prove that each Γ_i $(i = 1, 2, 3)$ is planar, thus proving that Γ is planar. Given two points a and b on a circular arc, we denote by $\widehat{ab}$ the sub-arc from a to b. All vertices of G_1 distinct from u, v, and w are mapped to points that are below q on Λ_{ext}. Let z be the lowest of these points. The boundary of R_1 consists of the segments $\overline{zp}$, $\overline{pq^*}$, $\overline{q^*q}$, and the arc $\widehat{qz}$. Since q^* is always above c_q, R_1 is a convex region. Hence, since all vertices of G_1 lie on the boundary of R_1, all edges of G_1 are inside R_1. The vertices of G_2 lie on the boundary of R_2, which is the region bounded by segment $\overline{p^*q^*}$ and arc $\widehat{q^*p^*}$. Since R_2 is convex, the edges of G_2 are inside R_2. The vertices of G_3 lie on the boundary of region R_3, which consists of segment $\overline{pp^*}$, arc $\widehat{p^*q^*}$ on Λ_{int}, segment $\overline{q^*q}$, and arc $\widehat{qp}$ on Λ_{ext}. Although R_3 is not convex, the edges of G_3 are still inside R_3. Indeed, all the vertices of G_3 that are on Λ_{ext}, except u and w, lie on arc $\widehat{q_{ext}p_{ext}}$, and hence edges among these vertices are inside R_3. Also, there is no edge of G_3 among vertices on Λ_{int}. The vertices connected to u are u_i $(i = 1, \dots, r)$ plus, possibly, some vertices on Λ_{ext} lying on arc $\widehat{q_{ext}p_{ext}}$. On the other hand, by construction, u is mapped on p, p_{ext} is the intersection point between t_p and Λ_{ext}, u_r is mapped to p^*, and u_1 is mapped to p_{int} (which is the intersection point between t_p and Λ_{int}). Therefore, all vertices connected to u are visible from u within R_3, that is, each segment from u to any of its adjacent vertices is inside R_3. The same reasoning can be applied to prove that all the edges incident on w are inside R_3. Finally, the edges of E_{mix} that belong to G_3 are represented by segments that have one end-point on arc $\widehat{q_{ext}p_{ext}}$ of Λ_{ext} and the other on arc $\widehat{p_{int}q_{int}}$ of Λ_{int}. Hence they are inside R_3. The planarity of each Γ_i $(i = 1, 2, 3)$ is guaranteed by the fact that: (i) we maintain the planar embedding of G_i, (ii) all the vertices of G_i are on the boundary of the region R_i, and (iii) any two adjacent vertices of G_i can be connected with a segment inside R_i, as proved above.

Theorem 1. *Let $G = (V, E)$ be a biconnected-core 2-outerplane graph. There exists an $O(|V|)$-time algorithm that computes a 2-curve drawing of G.*

4 Extending the Algorithm

Let $G = (V, E)$ be a 2-outerplane graph whose internal vertices induces a simply connected graph. A *1-bend-2-curve drawing* of G on Λ_{int}, Λ_{ext} is a planar drawing of G such that: (i) The vertices of V_{int} and V_{ext} are mapped to points of Λ_{int} and Λ_{ext}, respectively; (ii) Each edge of E is drawn as a polyline with at most one bend.

Algorithm 2-CURVE-DRAWER can be extended in order to compute a *1-bend-2-curve drawing* of G with a "small" number of bends. Namely, if $G(V_{int})$ is not biconnected, we apply on G a preprocessing step in order to make $G(V_{int})$ biconnected and still outerplane. We sketch an iterative procedure that can be used to do that. Let v be a cut-vertex of $G(V_{int})$ and let u and w be two vertices adjacent to v and such that: (a) u and w belong to distinct biconnected components of $G(V_{int})$, (b) u and w are consecutive in the counterclockwise list of vertices adjacent to v. Add a dummy edge (u, w), splitting the external face of $G(V_{int})$. The new dummy edge may cross a certain number of edges of E_{mix} that are incident on v and that are between u and v in the counterclockwise list of edges incident on v. No other kind of edges are crossed. Replace each crossing with a dummy vertex that will be considered as a new vertex of V_{int}. Note that, dummy vertices are not cut-vertices of the new $G(V_{int})$.

It can be proven that the new $G(V_{int})$ is biconnected and still outerplanar and that each edge of E_{mix} in the original graph is split at most by one dummy vertex. Also, it can be proven that the procedure above takes linear time. To the new embedded graph we apply the algorithm presented in the previous section, so to compute a planar drawing on two circles. At the end, dummy vertices are removed. The removal of each dummy vertex gives rise to one bend on an edge of E_{mix}, so that the resulting drawing is a 1-bend-2-curve drawing of G.

Corollary 1. *Let $G = (V, E)$ be a 2-outerplane graph whose internal vertices induces a simply connected graph. There exists an $O(n)$-time algorithm that computes a 1-bend-2-curve drawing of G with $O(E^{(c)}_{mix})$ bends, where $E^{(c)}_{mix}$ are those edges of E_{mix} incident on cut-vertices of $G(V_{int})$.*

References

1. B. S. Baker. Approximation algorithms for NP-complete problems on planar graphs. *JACM*, 41(1):153–180, 1994.
2. H. L. Bodlander. Some classes of graphs with bounded tree width. *Bulletin of the European Association for Theoretical Computer Science*, pages 116–126, 1988.
3. S. Cornelsen, T. Schank, and D. Wagner. Drawing graphs on two and three lines. In *Graph Drawing (Proc. GD '02)*, volume 2528 of *LNCS*, pages 31–41, 2002.
4. C. M. D. Bienstock. On the complexity of embedding planar graphs to minimize certain distance measures. *Algorithmica*, 5(1):93–109, 1990.
5. E. Di Giacomo, W. Didimo, G. Liotta, and S. K. Wismath. Drawing planar graphs on a curve. In *Proc. WG '03*, LNCS, to appear.

An Energy Model for Visual Graph Clustering

Andreas Noack

Institute of Computer Science
Brandenburg Technical University at Cottbus
PO Box 10 13 44, 03013 Cottbus, Germany
an@informatik.tu-cottbus.de

Abstract. We introduce an energy model whose minimum energy drawings reveal the clusters of the drawn graph. Here a cluster is a set of nodes with many internal edges and few edges to nodes outside the set. The drawings of the best-known force and energy models do not clearly show clusters for graphs whose diameter is small relative to the number of nodes. We formally characterize the minimum energy drawings of our energy model. This characterization shows in what sense the drawings separate clusters, and how the distance of separated clusters to the other nodes can be interpreted.

1 Introduction

Force-directed and energy-based methods are popular for creating straight-line drawings of undirected graphs. They are comparatively easy to implement, adaptable to different drawing criteria, and give satisfactory results for many graphs ([4, Chap. 10], [7]). Since the introduction of multi-scale algorithms they are even quite efficient [28,15,17,26,29].

Energy-based methods generally have two parts: an energy model, and an algorithm that searches a state with minimum total energy. In force-directed methods, the model is a force system, and an algorithm searches for an equilibrium state where the total force on each node is zero. Because force is the negative gradient of energy, this corresponds to searching a local minimum of energy. We will take the perspective of energy minimization in this paper.

Finding clusters, i.e. subsets of nodes with many internal edges and few edges to outside nodes, in graphs is an important problem in VLSI design [2], parallel computing [25], software engineering [23], and graph drawing [9]. The most popular force and energy models do not clearly isolate clusters, especially in graphs with small diameter. The first main result of this paper is an energy model whose minimum energy drawings reveal the clusters of such graphs.

The purpose of energy models is to create drawings from which a human viewer can infer properties of the drawn graph. To make the inference more efficient and its results more valid, the drawings are required to have certain properties, like small and uniform edge lengths, well-distributed nodes, or well-separated clusters. Empirical studies were performed to evaluate to what degree force and energy models fulfill such criteria (e.g. [6]), but few analytical results exist. A formal characterization of the minimum energy drawings of our energy model is the second main result of this paper.

G. Liotta (Ed.): GD 2003, LNCS 2912, pp. 425–436, 2004.

The new energy model and its formal analysis are presented in Sect. 2. Section 3 shows example drawings and compares them with drawings of the well-known Fruchterman-Reingold force model [14]. Section 4 compares our energy model with existing force and energy models, and proposes a generalization that includes among others our energy model and the Fruchterman-Reingold model. It discusses applications, the importance of interpretability, and the benefits of theoretical analyses of energy models.

1.1 Basic Definitions

A *graph* $G = (V, E)$ consists of a finite set V of *nodes* and a finite set E of *edges* with $E \subseteq V^{(2)}$, where $V^{(2)}$ is the set of all subsets of V which have exactly two elements. We only consider graphs with at least two nodes. Because layouts can be computed separately for different components of a graph, we restrict ourselves to connected graphs, i.e. graphs where every pair of nodes is connected by a path.

For two nodes $u, v \in V$, the length of the shortest path connecting u and v is the *graph-theoretic distance* of u and v. The *diameter* of the graph G is the greatest graph-theoretic distance between any two nodes. A *cut* of G is a pair (V_1, V_2) of nonempty, disjoint sets of nodes with $V_1 \cup V_2 = V$.

For $V_1, V_2 \subseteq V$ and $F \subseteq V^{(2)}$ let $F[V_1] = \{\{u, v\} \in F \mid u, v \in V_1\}$ and $F[V_1, V_2] = \{\{u, v\} \in F \mid u \in V_1, v \in V_2\}$. Then $E[V_1]$ is the set of edges *in* V_1, and $E[V_1, V_2]$ is the set of edges *between* V_1 and V_2.

A *d-dimensional drawing* of the graph G is a vector $p = (p_v)_{v \in V}$ of node positions $p_v \in \mathbb{R}^d$. For a drawing p and two nodes $u, v \in V$ the length of the difference vector $p_v - p_u$ is called the *(Euclidean) distance* of u and v in p and denoted by $||p_v - p_u||$.

For a subset F of $V^{(2)}$ and a drawing p, the arithmetic mean $\mathrm{arithmean}(F, p)$ of the distances of F is defined as

$$\mathrm{arithmean}(F, p) = \frac{1}{|F|} \sum\nolimits_{\{u,v\} \in F} ||p_v - p_u||,$$

the geometric mean $\mathrm{geomean}(F, p)$ of the distances of F is defined as

$$\mathrm{geomean}(F, p) = \sqrt[|F|]{\prod\nolimits_{\{u,v\} \in F} ||p_v - p_u||},$$

and the harmonic mean $\mathrm{harmmean}(F, p)$ of the distances of F is defined as

$$\mathrm{harmmean}(F, p) = \frac{|F|}{\sum_{\{u,v\} \in F} \frac{1}{||p_v - p_u||}}.$$

2 LinLog: A Clustering Energy Model

In this section, we first define a new energy model, called LinLog, and the cut ratio as measure of the coupling of two disjoint sets of nodes. Using this measure we can formalize and prove that in minimum energy drawings of the LinLog model, clusters are clearly separated from the remaining graph, and the distance of each cluster to the remaining graph is interpretable with respect to properties of the graph (more precisely, the distance is approximately inversely proportional to the coupling).

2.1 The LinLog Energy Model

The *LinLog energy* $U_{LinLog}(p)$ of a drawing p is defined as

$$U_{LinLog}(p) = \sum_{\{u,v\}\in E} ||p_u - p_v|| - \sum_{\{u,v\}\in V^{(2)}} \ln ||p_u - p_v||$$

The first term of the difference can be interpreted as attraction between adjacent nodes, and the second term can be interpreted as repulsion between any two (different) nodes. To avoid infinite energies we assume that different nodes have different positions. This is no serious restriction because we are interested in good drawings, i.e. drawings with low energy.

2.2 The Cut Ratio as a Measure of Coupling

Many different definitions of the term *cluster of a graph* have been proposed. Informally, we denote by cluster a set of nodes with many internal edges (high cohesion) and few edges to nodes outside the set (low coupling). Similar definitions are used in VLSI design [2], parallel computing [25], and software engineering [23].

To make more precise what we mean by a cluster, we use a formally defined measure of coupling: the cut ratio. For a cut (V_1, V_2), the *cut ratio* is the fraction of possible edges between V_1 and V_2 which actually occur:

$$\text{cutratio}(V_1, V_2) = \frac{|E[V_1, V_2]|}{|V_1| \cdot |V_2|}$$

The problem of finding a cut with minimum cut ratio is known as the sparsest cut problem. Section 4 shortly discusses work related to this problem.

The cut ratio is normalized with respect to the number $|V_1| \cdot |V_2|$ of possible edges between V_1 and V_2, which makes its interpretation independent of the size of V_1 and V_2. For all cuts (V_1, V_2) of a random graph the expected value of the cut ratio is equal, while the expected number of edges crossing the cut strongly depends on the size of V_1 and V_2.

2.3 LinLog Separates Clusters

In this subsection we show that in minimum energy drawings of the LinLog model, clusters are clearly separated from the remaining nodes, and the nodes of each cluster are close to each other. More precisely: If a set of nodes V_1 is loosely coupled to the remaining nodes $V \setminus V_1$, then the distance of V_1 to the remaining nodes is large. If V_1 is strongly coupled to $V \setminus V_1$, then V_1 is close to the nodes in $V \setminus V_1$, in particular to its adjacent nodes. This is achieved by *minimizing the ratio of the average edge length to the average distance of (all pairs of different) nodes*.

Let us illustrate this with two examples. First, consider a drawing where V_1 is close to the remaining nodes, although it is only loosely coupled to them. Loose coupling means that there are few edges and many non-edges between V_1 and the remaining nodes. Thus increasing the distance of V_1 to the remaining nodes will increase the average distance of nodes much more than the average edge length. In the opposite case, V_1 is strongly coupled to the remaining nodes, but is drawn at a great distance from them. Then decreasing the distance, especially to the nodes to which V_1 is adjacent, will make relatively many edges and few non-edges shorter. Thus the average edge length will decrease more than the average distance of nodes.

The following theorem states that *drawings with minimal LinLog energy have a minimal ratio of the arithmetic mean of the edge lengths to the geometric mean of the node distances.* So we formalize the average node distance as the geometric mean of the node distances. The reason is our intuition that increasing the distance of two non-adjacent nodes from 10 to 11 is not as good as increasing the distance from 1 to 2. Rather, an increase from 1 to 2 (by a factor of 2) is about as good as an increase from 10 to 20 (also by a factor of 2). This also captures that different nodes should not have the same position. Thus we preferred the geometric mean instead of the arithmetic mean, but we do not claim that this is the only possible choice.

Theorem 1. *Let $G = (V, E)$ be a connected graph, and let p^0 be a drawing of G with minimum LinLog energy. Then p^0 is a drawing of G that minimizes $\frac{\text{arithmean}(E,p)}{\text{geomean}(V^{(2)},p)}$.*

Proof: The basic idea is to fix the average edge length temporarily. This does not restrict generality, but only the scaling factor, and thus can be removed at the end of the proof. It allows to transform the minimization of energy into a minimization of the inverse geometric mean of the node distances.

Let the drawing p^0 be a solution of the minimization problem:

$$\text{Minimize } U_{LinLog}(p).$$

Let $c = \sum_{\{u,v\}\in E} ||p_u^0 - p_v^0||$. Note that $c \geq 0$. Then p^0 is also a solution of

$$\text{Minimize } U_{LinLog}(p) \text{ subject to } \sum\nolimits_{\{u,v\}\in E} ||p_u - p_v|| = c.$$

This is equivalent to

$$\text{Minimize } \; -\sum\nolimits_{\{u,v\}\in V^{(2)}} \ln ||p_u - p_v|| \text{ subject to } \sum\nolimits_{\{u,v\}\in E} ||p_u - p_v|| = c.$$

Because $f(x) = \sqrt[|V^{(2)}|]{\exp(x)}$ is a monotonic increasing function, p^0 is a solution of

$$\text{Min. } \sqrt[|V^{(2)}|]{\exp\left(-\sum\nolimits_{\{u,v\}\in V^{(2)}} \ln ||p_u - p_v||\right)} \text{ subj. to } \sum\nolimits_{\{u,v\}\in E} ||p_u - p_v|| = c.$$

This is equivalent to

$$\text{Minimize } \frac{1}{\text{geomean}(V^{(2)}, p)} \text{ subject to } \text{arithmean}(E, p) = \frac{c}{|E|}.$$

($|E| > 0$ because we only consider connected graphs with at least two nodes.) Because $\frac{c}{|E|}$ is nonnegative, p^0 is also a solution of

$$\text{Minimize } \frac{\text{arithmean}(E, p)}{\text{geomean}(V^{(2)}, p)} \text{ subject to } \text{arithmean}(E, p) = \frac{c}{|E|}.$$

For every drawing q^0 of G that minimizes $\frac{\text{arithmean}(E,p)}{\text{geomean}(V^{(2)},p)}$, we can construct a drawing $q^1 = \frac{c}{|E|\,\text{arithmean}(E,q^0)} q^0$ with $\frac{\text{arithmean}(E,q^1)}{\text{geomean}(V^{(2)},q^1)} = \frac{\text{arithmean}(E,q^0)}{\text{geomean}(V^{(2)},q^0)}$ and $\text{arithmean}(E,q^1) = \frac{c}{|E|}$. (Because $\text{geomean}(V^{(2)},q^0) > 0$, no two different nodes have the same position in q^0, so $\text{arithmean}(E,q^0) > 0$.) So p^0 is also a solution of

$$\text{Minimize } \frac{\text{arithmean}(E,p)}{\text{geomean}(V^{(2)},p)}.$$

□

This characterization of drawings with minimum LinLog energy is precise up to scaling. To show this, we use a lemma which states that the overall edge length in all minimum LinLog energy drawings is $\frac{(|V|-1)\cdot|V|}{2}$. So the overall edge length is independent from the number of edges and the structure of the graph. This lemma has a practical value in itself, because it helps to choose the scaling factor: For a drawing with an average edge length of 1, scale the minimum LinLog energy drawing by a factor of $\frac{2|E|}{(|V|-1)\cdot|V|}$.

Lemma 1. *Let $G = (V, E)$ be a connected graph, and let p be a drawing of G with minimum LinLog energy. Then*

$$\sum_{\{u,v\}\in E} ||p_u - p_v|| = \left|V^{(2)}\right|$$

Proof: The basic idea is to express the LinLog energy as a function of the scaling factor, where the scaling is applied to a minimum LinLog energy drawing. Then this function has a minimum at the minimum energy drawing, i.e. at a scaling factor of 1. The details of the proof are given in [24]. □

Theorem 2. *Let $G = (V, E)$ be a connected graph. Then p^0 is a drawing of G with minimum LinLog energy if and only if p^0 is a drawing that minimizes $\frac{\text{arithmean}(E,p)}{\text{geomean}(V^{(2)},p)}$ and satisfies $\sum_{\{u,v\}\in E} ||p^0_u - p^0_v|| = \left|V^{(2)}\right|$.*

Proof: ⇒: Follows from Theorem 1 and Lemma 1.
⇐: Similar to the proof of Theorem 1. The details are given in [24]. □

2.4 LinLog Creates Interpretable Distances

In this subsection we show that in minimum energy drawings of the LinLog model, the distance of each cluster to the remaining nodes is interpretable. More precisely, the distance approximates the inverse coupling of the cluster to remaining nodes.

The following theorem states that *wherever one cuts a one-dimensional minimum LinLog energy drawing into two nonempty parts, the harmonic mean of the distances between the nodes in the two parts equals the inverse cut ratio.* So we formalized the distance of two sets of nodes as the harmonic mean of the distances of their members. The harmonic mean corresponds better to our intuition than the arithmetic or geometric mean because it weights small distances higher than large distances. If the distances within the two sets of nodes are much smaller than the distances between the two sets, then the harmonic, the geometric and the arithmetic mean are roughly equal.

Theorem 3. *Let $G = (V, E)$ be a connected graph, and let p be a one-dimensional drawing of G with minimal LinLog energy. Let (V_1, V_2) be a cut of G such that the nodes in V_1 have smaller positions than the nodes in V_2 (i.e. $\forall v_1 \in V_1 \, \forall v_2 \in V_2 : p_{v_1} < p_{v_2}$). Then*

$$\text{harmmean}(V^{(2)}[V_1, V_2], p) = \frac{1}{\text{cutratio}(V_1, V_2)}.$$

Proof: The basic idea is to express the LinLog energy as a function of the relative positions of the two sets of nodes, and to exploit that the minimum energy drawing p is a minimum of this function.

The LinLog energy of the drawing p is:

$$\sum_{\{u,v\} \in E} |p_u - p_v| - \sum_{\{u,v\} \in V^{(2)}} \ln |p_u - p_v|$$

(To simplify the notation, we consider the positions as scalars here.) If some $d \in \mathbb{R}$ is added to the coordinates of all nodes in V_1, such that the condition that the nodes of V_1 have smaller coordinates then the nodes of V_2 is still satisfied (i.e. $d < \min\{p_v \mid v \in V_2\} - \max\{p_v \mid v \in V_1\}$), the LinLog energy is

$$\begin{aligned} U(d) = & \sum_{\{u,v\} \in E[V_1] \cup E[V_2]} |p_u - p_v| - \sum_{\{u,v\} \in V_1^{(2)} \cup V_2^{(2)}} \ln |p_u - p_v| \\ + & \sum_{\{u,v\} \in E[V_1, V_2]} (|p_u - p_v| + d) - \sum_{\{u,v\} \in V^{(2)}[V_1, V_2]} \ln(|p_u - p_v| + d) \end{aligned}$$

Because p is a drawing with minimal LinLog energy, this function has a global minimum at $d = 0$, so $U'(0) = 0$.

$$0 = U'(0) = \big|E[V_1, V_2]\big| - \sum_{\{u,v\} \in V^{(2)}[V_1, V_2]} \frac{1}{|p_u - p_v|}$$

Inserting the harmonic mean $\text{harmmean}(V^{(2)}[V_1, V_2], p)$ of the distances between V_1 and V_2 we get

$$0 = \big|E[V_1, V_2]\big| - \frac{|V_1| \cdot |V_2|}{\text{harmmean}(V^{(2)}[V_1, V_2], p)}$$

$$\text{harmmean}(V^{(2)}[V_1, V_2], p) = \frac{|V_1| \cdot |V_2|}{|E[V_1, V_2]|} = \frac{1}{\text{cutratio}(V_1, V_2)} \qquad \square$$

The generalization of the theorem to higher-dimensional drawings is: *For each hyperplane that cuts a minimum LinLog energy drawing into two nonempty parts, the harmonic mean of the distances between the corresponding sets of nodes equals their inverse cut ratio.* But it is not true. The reason is, loosely stated, that in two or more dimensions the directions of the edges between the two sets of nodes can be different from the directions of the non-edges. So when one set is moved, the average change of the edge lengths is different from the average change of the non-edge lengths. However, this difference is small in most practical cases. In particular, the one-dimensional case is a good approximation when the distance of a cluster to the remaining nodes is large.

3 Examples

Figures 1, 2 and 3 show drawings of pseudo-random graphs with clusters for the LinLog energy model and the well-known Fruchterman-Reingold force model ([14], further discussed in Sect. 4). In all figures, the LinLog model (left drawings) reveals the clusters more clearly than the Fruchterman-Reingold model (right drawings).

Figure 1 shows a pseudo-random graph with eight clusters of 100 nodes each. The probability of an edge $\{u, v\}$ is 0.16 if u and v belong to the same cluster and 0.02 otherwise. (All edges are chosen independently.) Overall, the graph has 6341 intra-cluster edges and 5625 inter-cluster edges. The eight clusters are clearly separated in the drawing for the LinLog model, but the borders of the clusters look fuzzy and some nodes do not seem to belong to any cluster. This is to be expected of a random graph: Some nodes have a small degree, and hence drift to the border of the drawing. Other nodes are equally connected to two clusters, and are drawn between these clusters.

Figure 2 shows a graph with four clusters, each having four subclusters of 50 nodes. The subclusters are cliques (edge probability 1), the edge probability between nodes of different subclusters of the same cluster is 0.32, and the edge probability between nodes of different clusters is 0.16. The drawing for the LinLog model clearly shows the hierarchical structure. With 19600 intra-subcluster edges and 57346 inter-subcluster edges, separating the (sub)clusters seems more difficult than in the first graph. Nevertheless, Fig. 2 looks more orderly than Fig. 1, because every node is guaranteed to be adjacent to all other nodes of its subcluster.

The graph of Fig. 3 is a pseudo-random graph with one central cluster of 200 nodes and three "satellite" clusters of 100 nodes each, called cluster A, B and C. The probability of an edge $\{u, v\}$ is

- 0.12 if u and v belong to the same cluster,
- 0.04 if u belongs to the central cluster and v belongs to cluster A,
- 0.02 if u belongs to the central cluster and v belongs to cluster B,
- 0.01 if u belongs to the central cluster and v belongs to cluster C, and
- 0 otherwise.

The first thought might be that the distance from the central cluster to cluster A should be half the distance from the central cluster to cluster B, which again should be half the distance from the central cluster to cluster C. But the actual central-to-B distance in the drawing for the LinLog energy model is greater than twice the central-to-A distance. This is because for cluster B, the central cluster and cluster A effectively form one big cluster of 300 nodes, yielding an effective edge probability of only about 0.013. On the other hand, cluster B has little influence on the central-to-A distance because it is relatively far away from both. So the LinLog model did not only separate different clusters, it also produced interpretable distances between the clusters.

4 Discussion

Related Work: Force and Energy Models. In Sect. 2.4 we have shown that in drawings with minimum LinLog energy, the distance of a cluster to the remaining graph is approximately inversely proportional to the coupling. For many of the best-known force

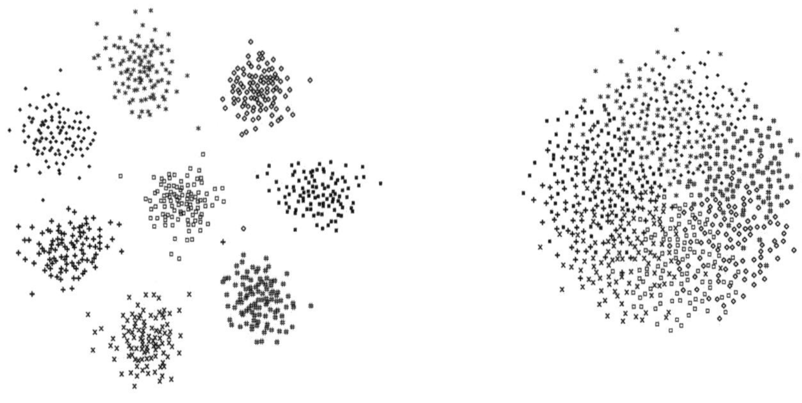

Fig. 1. Pseudo-random graph with intra-cluster edge probability 0.16 and inter-cluster edge probability 0.02; left: LinLog model, right: Fruchterman-Reingold model

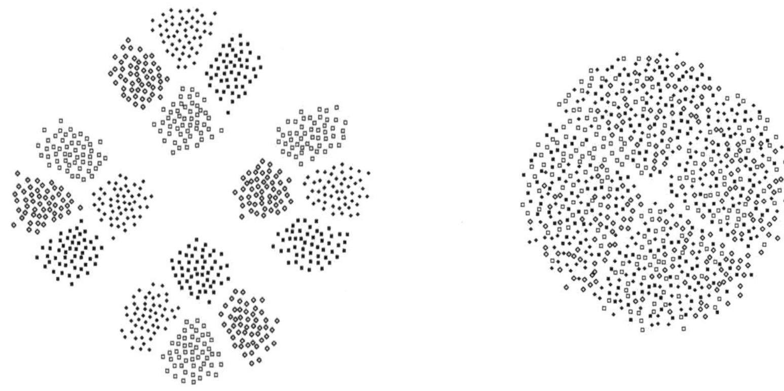

Fig. 2. Pseudo-random graph with hierarchical clusters; left: LinLog model, right: Fruchterman-Reingold model

Fig. 3. Pseudo-random "satellite" graph; left: LinLog model, right: Fruchterman-Reingold model

and energy models there is only a weak dependency of distance and coupling, and thus these energy models do not isolate clusters well. (See [24] for a more detailed discussion of these models.)

The models of Eades [12], Fruchterman and Reingold [14], and Davidson and Harel [10] strongly enforce uniform edge lengths. However, separating clusters requires some long edges. Section 3 contains example drawings of the Fruchterman-Reingold model. In multidimensional scaling [20] and the model of Kamada and Kawai [18], the distances between nodes in the drawing are determined by their graph-theoretic distances. But the graph-theoretic distances and the coupling are only weakly related. The energy model of Hall ([16], recently used e.g. in [8,19]) places some non-adjacent nodes and loosely coupled subgraphs very close or at the same position.

All mentioned energy models easily generalize to graphs with weighted edges. Given appropriate edge weights, the models create drawings which reveal clusters. But this means putting clusters in (in the form of edge weights) to get clusters out. Other approaches apply force-directed methods to draw graphs with a *given* hierarchical cluster structure [30,13]. But finding clusters or appropriate edge weights is difficult, because most variants of the graph clustering problem are $\mathcal{NP}$-hard.

In the LinLog energy model, the attraction and repulsion between the nodes reveal the clusters. It does not require knowledge about clusters as input, but provides it as output. Classical force and energy models and the LinLog model complement each other, as discussed in the third subsection.

Related Work: Clustering by Minimizing Distance Ratios. For a given graph, the cut with the minimum cut ratio is called the sparsest cut. Computing sparsest cuts is $\mathcal{NP}$-hard. In [22] and [3] approximation algorithms are given that compute sparse cuts from minima of $\frac{\mathrm{arithmean}(E,p)}{\mathrm{arithmean}(V^{(2)},p)}$ (with respect to appropriate metrics). So it is not surprising that minimizing the similar ratio $\frac{\mathrm{arithmean}(E,p)}{\mathrm{geomean}(V^{(2)},p)}$ reveals clusters. The proof that minima of the LinLog energy are also minima of the latter ratio (Theorem 2) is a main result of this paper, because it enables the application of algorithms from force-directed graph drawing to minimize this ratio. Lacking space for a detailed discussion, we can only note that the results concerning sparsest cut approximations from [22] and [3] cannot be directly transferred to minima of the LinLog model, and that the minima used in these approximations are generally less interpretable to humans than LinLog minima.

Generalization: The r-PolyLog Energy Models. The class of energy models r-PolyLog contains two energy models that were already discussed: The 1-PolyLog model is the LinLog model introduced in Sect. 2, and the 3-PolyLog model is equivalent to the Fruchterman-Reingold model [14] discussed in the first subsection. (The Fruchterman-Reingold model is usually expressed as force model, but we give the energy version for uniformity.) For all $r \in \mathbb{R}$ with $r \geq 1$, the *r-PolyLog energy* $U_{r\text{-}PolyLog}(p)$ of a drawing p is defined as

$$U_{r\text{-}PolyLog}(p) = \sum_{\{u,v\}\in E} \frac{1}{r}||p_u - p_v||^r - \sum_{\{u,v\}\in V^{(2)}} \ln ||p_u - p_v||$$

The class contains energy models which isolate clusters ($r = 1$), energy models which enforce uniform edge lengths ($r \to \infty$), and many compromises between both extremes ($1 < r < \infty$). Thus it contains suitable energy models for many kinds of undirected graphs from the small worlds described in the next subsection to the graphs addressed by classical force-directed graph drawing.

Application: Drawing Clustered Small-World Graphs. According to Albert and Barabási [1], three concepts are prominent in contemporary thinking about complex networks: small worlds, clustering, and degree distributions. In small-world graphs, the nodes have small graph-theoretic distances, relative to the number of nodes. For clustered small-world graphs with hundreds or thousands of nodes, drawings with uniform edge lengths or Euclidean distances proportional to graph-theoretic distances do not reveal the structure of the graph. That is why most well-known force and energy models are not suitable for drawing such graphs.

The LinLog model can reveal the clusters in small worlds. We applied it for drawing call graphs and similar models of object-oriented programs [21]. Further examples of clustered small-world graphs include models of computer networks, the World Wide Web, and social networks [27,1].

Interpretability of Drawings. Drawings of graphs are useful because we can infer properties of the graph from properties of the drawing. For valid inferences, we need precise statements which properties of the drawing correspond to which properties of the graph. For minimum energy drawings of the LinLog model, Theorem 1 states how the Euclidean distance of nodes corresponds to the adjacency of nodes, and Theorem 3 states how the Euclidean distance of sets of nodes corresponds to their coupling.

Empirical studies have shown that human viewers indeed attribute semantics to the placement of nodes, even if they are not told that such a correspondence exists. In a study of Dengler and Cowan, the most important semantic attribution is that observers view graph drawings hierarchically, that they separate them into interconnected subgraphs [11]. Blythe et al. concluded from their study in the context of social network analysis, that the Euclidean distance of nodes has a significant effect on the viewers' assignment of nodes to groups [5].

The Role of Theory in the Development and Evaluation of Energy Models. To date, force and energy models have been evaluated mainly empirically. Complementing this empirical evaluation with theoretical evaluation (as done in Sects. 2.3 and 2.4) is desirable for two reasons. First, while empirical studies can only examine a limited number of graphs, theory can prove results for large or infinite classes of graphs. Second, theory can explain why an approach does or does not work for a given class of graphs.

Unlike the strategy of presentation in this paper, the development process of the LinLog model was not first guessing and then evaluating it. Rather, we gradually developed the energy model from the informal requirement of having isolated clusters with interpretable distances. In this process, theoretical considerations were used for the stepwise reduction of the space of candidate energy models. (See [24] for details.)

5 Conclusion

The LinLog energy model was introduced. It was shown that in its minimum energy drawings, clusters are clearly separated from the remaining nodes and their distance to the remaining nodes is approximately the inverse of the coupling. The LinLog model complements well-known force and energy models like the Fruchterman-Reingold model which do not separate the clusters of graphs with small diameter well. The class of energy models r-PolyLog was proposed that includes suitable energy models for many graphs from clustered graphs with small diameter to the graphs addressed by classical force-directed graph drawing.

In the future, we will apply and extend our theoretical tools to develop special-purpose energy models for interpretable visualizations of software systems. Another direction of future work is the theoretical analysis of existing force and energy models. Initial results can be found in [24].

An important problem that was not addressed in this paper is the development of algorithms that reliably and efficiently find good energy minima. We have shown that we can draw valid inferences from minimum energy drawings, but interpreting drawings with high energy (e.g. most random drawings) in the same way can give invalid results. Different kinds of graphs not only require different energy models, but may also require different minimization algorithms. For example, some multi-scale algorithms are not expected to find good energy minima for dense graphs or graphs with small diameter [15, 17,29].

References

1. R. Albert and A.-L. Barabási. Statistical mechanics of complex networks. *Reviews of Modern Physics*, 74(1):47–97, 2002.
2. C. J. Alpert and A. B. Kahng. Recent directions in netlist partitioning: A survey. *Integration, the VLSI Journal*, 19(1-2):1–81, 1995.
3. Y. Aumann and Y. Rabani. An $O(\log k)$ approximate min-cut max-flow theorem and approximation algorithm. *SIAM Journal on Computing*, 27(1):291–301, 1998.
4. G. D. Battista, P. Eades, R. Tamassia, and I. G. Tollis. *Graph Drawing: Algorithms for the Visualization of Graphs*. Prentice Hall, 1999.
5. J. Blythe, C. McGrath, and D. Krackhardt. The effect of graph layout on inference from social network data. In F.-J. Brandenburg, editor, *Proc. Symposium on Graph Drawing (GD 1995)*, LNCS 1027, pages 40–51, 1996. Springer-Verlag.
6. F.-J. Brandenburg, M. Himsolt, and C. Rohrer. An experimental comparison of force-directed and randomized graph drawing algorithms. In F.-J. Brandenburg, editor, *Proc. Symposium on Graph Drawing (GD 1995)*, LNCS 1027, pages 76–87, 1996. Springer-Verlag.
7. U. Brandes. Drawing on physical analogies. In M. Kaufmann and D. Wagner, editors, *Drawing Graphs: Methods and Models*, LNCS 2025, pages 71–86. Springer-Verlag, 2001.
8. U. Brandes and S. Cornelsen. Visual ranking of link structures. In F. Dehne, J.-R. Sack, and R. Tamassia, editors, *Proc. 7th International Workshop on Algorithms and Data Structures (WADS 2001)*, LNCS 2125, pages 222–233, 2001. Springer-Verlag.
9. R. Brockenauer and S. Cornelsen. Drawing clusters and hierarchies. In M. Kaufmann and D. Wagner, editors, *Drawing Graphs: Methods and Models*, LNCS 2025. Springer-Verlag, 2001.

10. R. Davidson and D. Harel. Drawing graphs nicely using simulated annealing. *ACM Transactions on Graphics*, 15(4):301–331, 1996.
11. E. Dengler and W. Cowan. Human perception of laid-out graphs. In S. H. Whitesides, editor, *Proc. 6th International Symposium on Graph Drawing (GD 1998)*, LNCS 1547, pages 441–443, 1998. Springer-Verlag.
12. P. Eades. A heuristic for graph drawing. *Congressus Numerantium*, 42:149–160, 1984.
13. P. Eades and M. L. Huang. Navigating clustered graphs using force-directed methods. *Journal of Graph Algorithms and Applications*, 4(3):157–181, 2000.
14. T. M. J. Fruchterman and E. M. Reingold. Graph drawing by force-directed placement. *Software – Practice and Experience*, 21(11):1129–1164, 1991.
15. P. Gajer, M. T. Goodrich, and S. G. Kobourov. A multi-dimensional approach to force-directed layouts of large graphs. In J. Marks, editor, *Proc. 8th International Symposium on Graph Drawing (GD 2000)*, LNCS 1984, pages 211–221, 2001. Springer-Verlag.
16. K. M. Hall. An r-dimensional quadratic placement algorithm. *Management Science*, 17(3):219–229, 1970.
17. D. Harel and Y. Koren. A fast multi-scale method for drawing large graphs. In J. Marks, editor, *Proc. 8th International Symposium on Graph Drawing (GD 2000)*, LNCS 1984, pages 183–196, 2001. Springer-Verlag.
18. T. Kamada and S. Kawai. An algorithm for drawing general undirected graphs. *Information Processing Letters*, 31(1):7–15, 1989.
19. Y. Koren, L. Carmel, and D. Harel. Ace: A fast multiscale eigenvector computation for drawing huge graphs. In *Proc. IEEE Symposium on Information Visualization (InfoVis 2002)*, pages 137–144, 2002.
20. J. B. Kruskal and M. Wish. *Multidimensional Scaling*. Sage Publications, 1978.
21. C. Lewerentz and A. Noack. CrocoCosmos – 3D visualization of large object-oriented programs. In *Graph Drawing Software*, pages 279–297. Springer-Verlag, 2003.
22. N. Linial, E. London, and Y. Rabinovich. The geometry of graphs and some of its algorithmic applications. *Combinatorica*, 15(2):215–245, 1995.
23. S. Mancoridis, B. S. Mitchell, C. Rorres, Y. Chen, and E. R. Gansner. Using automatic clustering to produce high-level system organizations of source code. In *Proc. 6th IEEE International Workshop on Program Understanding (IWPC 1998)*, pages 45–52, 1998.
24. A. Noack. Energy models for drawing clustered small-world graphs. Technical Report 07/03, Institute of Computer Science, Brandenburg University of Technology at Cottbus, 2003.
25. A. Pothen. Graph partitioning algorithms with applications to scientific computing. In D. E. Keyes, A. Sameh, and V. Venkatakrishnan, editors, *Parallel Numerical Algorithms*, pages 323–368. Kluwer, 1997.
26. A. J. Quigley and P. Eades. FADE: Graph drawing, clustering, and visual abstraction. In J. Marks, editor, *Proc. 8th International Symposium on Graph Drawing (GD 2000)*, LNCS 1984, pages 197–210, 2001. Springer-Verlag.
27. S. H. Strogatz. Exploring complex networks. *Nature*, 410:268–276, 2001.
28. Daniel Tunkelang. *A Numerical Optimization Approach to General Graph Drawing*. PhD thesis, School of Computer Science, Carnegie Mellon University, 1999.
29. C. Walshaw. A multilevel algorithm for force-directed graph drawing. In J. Marks, editor, *Proc. 8th International Symposium on Graph Drawing (GD 2000)*, LNCS 1984, pages 171–182, 2001. Springer-Verlag.
30. X. Wang and I. Miyamoto. Generating customized layouts. In F.-J. Brandenburg, editor, *Proc. Symposium on Graph Drawing (GD 1995)*, LNCS 1027, pages 504–515, 1996. Springer-Verlag.

Simultaneous Graph Drawing: Layout Algorithms and Visualization Schemes*

Cesim Erten, Stephen G. Kobourov, Vu Le, and Armand Navabi

Department of Computer Science
University of Arizona
{cesim,kobourov,vle,navabia}@cs.arizona.edu

Abstract. In this paper we consider the problem of drawing and displaying a series of related graphs, i.e., graphs that share all, or parts of the same vertex set. We designed and implemented three different algorithms for simultaneous graph drawing and three different visualization schemes. The algorithms are based on a modification of the force-directed algorithm that allows us to take into account vertex weights and edge weights in order to achieve mental map preservation while obtaining individually readable drawings. The implementation is in Java and the system can be downloaded at `http://simg.cs.arizona.edu/`.

1 Introduction

Consider the problem of drawing a series of graphs that share all, or parts of the same vertex set. The graphs may represent different relations between the same set of objects. For example, in social networks, graphs are often used to represent different relations between the same set of entities. Alternatively, the graphs may be the result of a single relation that changes through time. For example, in software visualization, the inheritance graph in a Java program changes as the program is being developed. Consider the graphs in Fig. 1. There are two simultaneously displayed graphs that represent two snapshots of a file system structure rooted at the directory `graphs/`. The drawing conveys well both underlying structures and it is easy to identify the changes between the two snapshots.

In this paper, we attempt to address the following problem: Given a series of graphs that share all, or parts of the same vertex set, what is a natural way to layout and display them? The layout and display of the graphs are different aspects of the problem, but also closely related, as a particular layout algorithm is likely to be matched best with a specific visualization technique. As stated above, however, the problem is too general and it is unlikely that one particular layout algorithm will be best for all possible scenarios. Consider the case where we only have a pair of graphs in the series, and the case where we have hundreds of related graphs. The "best" way to layout and display the two series is likely going to be different. Similarly, if the graphs in the sequence are very closely

* This work is partially supported by the NSF under grant ACR-0222920.

G. Liotta (Ed.): GD 2003, LNCS 2912, pp. 437–449, 2004.

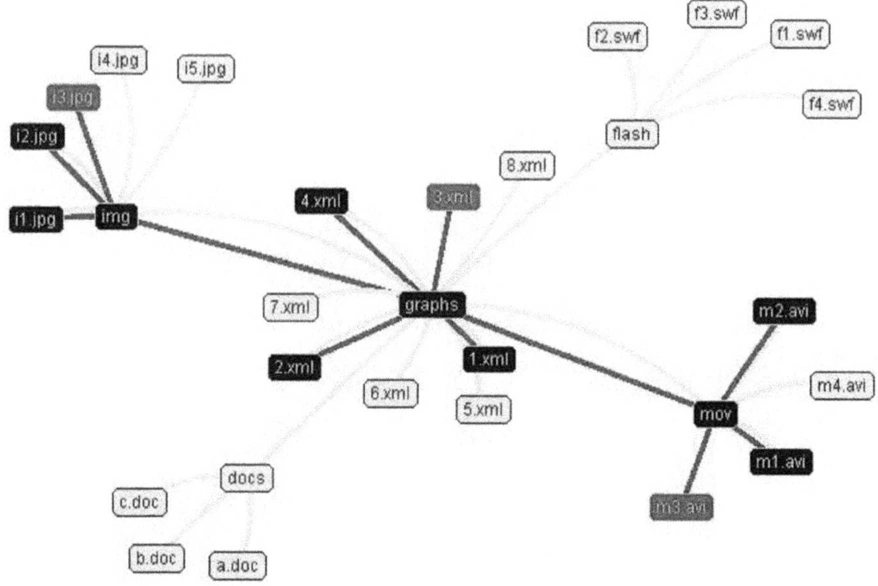

Fig. 1. Two snapshots of the file structure rooted at directory **graphs/**. Red vertices and edges belong to earlier snapshot. Dark blue vertices belong to both snapshots. Light blue vertices and edges belong to the later snapshot. The edges of later snapshot are curved.

related or not related at all, different layout and display techniques may be more appropriate. With this in mind, we consider several different algorithms and visualization models.

For the layout of the graphs, there are two important criteria to consider: the *readability* of the individual layouts and the *mental map preservation* in the series of drawings. The readability of individual drawings depends on aesthetic criteria such as display of symmetries, uniform edge lengths, and minimal number of crossings. Preservation of the mental map can be achieved by ensuring that vertices that appear in consecutive graphs in the series, remain in the same positions. These two criteria are often contradictory. If we individually layout each graph, without regard to other graphs in the series, we may optimize readability at the expense of mental map preservation. Conversely, if we fix the vertex positions in all graphs, we are optimizing the mental map preservation but the individual layouts may be far from readable.

For the visualization of the graphs there are numerous different possibilities. We could draw each graph in the series in its own 2D plane, in order of appearance, or we could show one graph at a time, and morph to the next one. If there are only a small number of graphs in the sequence, we could display all of them simultaneously, using different edge styles.

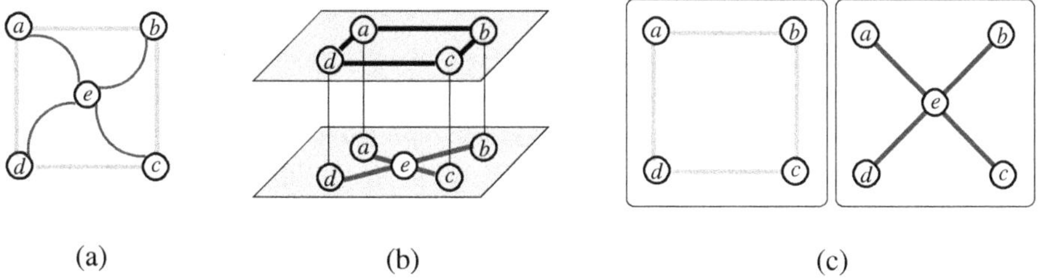

Fig. 2. Layout and visualization: (a) aggregate; (b) merged; (c) split.

We designed and implemented three layout algorithms and three visualization schemes; see Fig. 2. We summarize the layout algorithms below:

1. In the first layout algorithm we create an *aggregate* graph from the given sequence of graphs. The aggregate graph is node-weighted and edge-weighted and the node (edge) weight corresponds to the number of times a particular node (edge) appears in the sequence. A modified force-directed approach is used to layout the aggregate graph, taking into account the weights of the nodes and the edges.
2. In the second layout algorithm, we create a *merged* graph. The merged graph consists of all the graphs in the sequence, together with additional edges connecting the same vertices in all graphs. A modified force-directed layout is used to layout the merged graph by restricting each graph to its own 2D plane.
3. The third layout algorithm is designed for a pair of related graphs G_1 and G_2 but can be generalized to larger series of graphs. We use intelligent (rather than random) placement of the vertices, based on graph distances, to independently obtain initial drawings D_1 and D_2 for the two graphs. Next the placement of the vertices from $D_1(D_2)$ is used to "seed" an iteration of the force-directed layout for $G_2(G_1)$ and the process is repeated until the two layouts converge.

The three visualization schemes closely correspond to the algorithms above. However, different combinations of layout algorithms and visualization schemes can also be used. We summarize the layout models below:

1. In the *aggregate view model* we use the aggregate graph to show all the graphs in one combined drawing, using different edge(node) styles, to differentiate between the different graphs.
2. In the *merged view model* we create a 3D drawing, in which each graph is displayed in its own 2D plane, and the planes are arranged on top of each other in the order that the graphs appear in the sequence.
3. In the *split view model* each graph is displayed in its own drawing window.

2 Previous Work

Classical force-directed methods [9,15] for graph drawing use a random initial embedding of the graph and treat the graph as a system of interacting physical objects. Force-directed layout algorithms typically employ an energy function that characterizes the state of the system. The minimization of suitably chosen energy functions tends to produce aesthetically pleasing graph drawings. Several variations of force-directed methods for edge-weighted graphs have been proposed. In [7,12] edge-weighted graphs are drawn so that the length of edges is proportional to their weights. Similarly, layouts for vertex-weighted graphs have also been considered in the context of focus-vertices that apply repulsive force proportional to their weight, so that the neighborhoods of such vertices will not be too cluttered [14].

In dynamic graph drawing the goal is to maintain a nice layout of a graph that is modified via operations such as insert/delete edge and insert/delete vertex. Techniques based on static layouts have been used [3,13,17]. North [16] studies the incremental graph drawing problem in the DynaDAG system. Brandes and Wagner adapt the force-directed model to dynamic graphs using a Bayesian framework [2]. Diehl and Görg [6] consider graphs in a sequence to create smoother transitions. Brandes and Corman [1] present a system for visualizing network evolution in which each modification is shown in a separate layer of 3D representation with vertices common to two layers represented as columns connecting the layers. Thus, mental map preservation is achieved by precomputing good locations for the vertices and fixing the position throughout the layers.

Simultaneous planar graph embedding is a related problem that asks whether there exist locations for the vertices of two different planar graphs such that each of the graphs can be drawn with straight lines and no crossings. Recent theoretical results [4,8] imply that simultaneous embeddings exist only for special cases and relaxations of the problem (such as the one we address in this paper) should be considered. Along these lines, Collberg *et al* [5] describe a graph-based system for visualization of software evolution, which uses a modification of our algorithm for visualization of large graphs [10], while preserving the mental map by fixing the locations of all common vertices in the evolving graph.

3 Modified Force-Directed Method

We first review the basic force-directed graph layout algorithm and then describe the modifications for node-weighted and edge-weighted graphs. The modified force-directed algorithm is used in all three layout algorithms.

A standard force-directed layout algorithm begins with an initial random placement of the vertices. Then it iteratively computes the effect of repulsive and attractive forces on vertices and updates the temperature. The temperature controls the scale of each iteration. At the beginning the temperature is high and vertices move significant distances and with time, the temperature is decreased.

The attractive and repulsive forces are respectively defined as, $f_r(d) = -\kappa^2/d$ and $f_a(d) = d^2/\kappa$, where d is the distance between two vertices. The repulsive forces are calculated for each pair of vertices whereas the attractive forces are calculated for pairs of vertices connected by an edge. The ideal distance between vertices, κ, is defined as $\kappa = C\sqrt{A_{frame}/n}$, where A_{frame} is the area of the frame, C is a constant determining how the vertices fill the frame, and n is the total number of vertices.

Given a series of graphs $G_1, G_2, \ldots, G_k$, we create one node-weighted and edge-weighted *aggregate graph* $G_A = (V_A, E_A)$. A node $v \in V_A$ has weight w if it appears in w of the graphs in the series. Similarly, an edge $(u, v) \in E_A$ has weight proportional to the number of times edge (u, v) appears in the series. We use the node and edge weights to modify the standard force-directed algorithm as follows. If vertex v has large weight (it appears in many graphs) then it should to be placed close to the center in the final layout. If an edge (u, v) has large weight then the vertices u and v should be placed very close to each other in the final layout. This is a simple heuristic, but it ensures that:

- persistent vertices remain close to the center of the layout, while fleeting vertices appear and disappear on the periphery;
- vertices that are adjacent in many of the graphs in the series are placed close together.

In order to handle the vertex weights we place a dummy vertex in the center of the frame and ensure that it attracts all the other vertices in proportion to their weights. We formulate this new central attraction force as, $f_{ca}(d) = d^2 \times w/\kappa$, where w is the weight of the vertex and d is its distance from the center. To handle edge weights we scale the attractive forces by their edge weights and the new formulation of the attractive forces becomes, $f_a(d) = d^2 \times w_e/\kappa$, where w_e is the weight of the edge e.

4 Layout Algorithms and Visualization Schemes

Depending on mainly two factors, the number of graphs to be embedded simultaneously and how similar the individual graphs are, different layout methods and visualization techniques arise. If there are not too many graphs to be embedded and the graphs share a reasonably large common substructure, then a layout method that embeds common vertices of each individual graph at exactly the same locations and the common edges in a similar manner is preferable. In terms of visualization, it might be more advantageous to view the graphs on the same plane. However, if there are many graphs to be embedded, or if the individual graphs do not share many common substructures, then more flexible embeddings might be more visually appealing. In such cases, we do not insist on exactly the same locations for shared vertices of different graphs but rather try to locate them in close proximity, so that the mental map of the viewer is somewhat preserved. Not insisting on the exact same location for same vertices, allows for more freedom to draw each graph with higher readability. In terms of

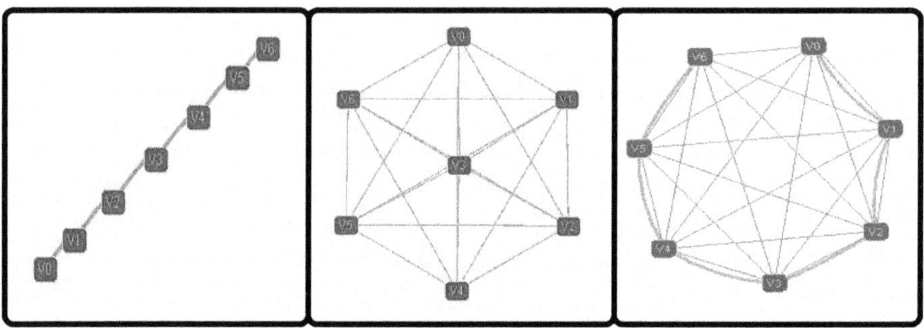

Fig. 3. Left: Individual layout of P_7 drawn with curved edges. Middle: Individual layout of K_7 drawn with straight-line segments. Right: Simultaneous embedding of P_7 and K_7 obtained from the aggregate layout method.

visualization, having each graph laid out on a separate 2D plane or morphing between consecutive 3D drawings seems most suitable.

Based on these observations we describe three different layout methods: *aggregate graph layout*, *merged graph layout* and *independent iterations layout*. After describing each layout method, we present a matching visualization scheme that seems most appropriate for it: *aggregate view*, *merged view* and *split view*. While the three visualization schemes closely correspond to their matching algorithms, different combinations of layout and visualization algorithms can also be used.

In the aggregate graph layout method we begin by creating the node-weighted and edge weighted graph $G_A = (V_A, E_A)$ from the graph sequence, $G_1, G_2, \ldots, G_k$, as described in the previous section. We then apply the modified force-directed layout algorithm to obtain a drawing for G_A. From this drawing we extract the drawings of each individual graph in the series. Thus, vertices and edges that are present more than once in the series are in the same position in all graphs that they appear in. This approach guarantees mental map preservation, possibly at the expense of good readability. Yet, since the vertex/edge weights are taken into account in the layout of the aggregate graph, the final layout will be close to an individual layout of a graph proportional to the importance of that one graph. In other words, if a graph G_i has many vertices/edges that exist in most of the graphs, then G_i is an important graph and the resulting layout will be similar to that of an independent layout of G_i.

4.1 Aggregate Graph Layout

Fig. 3 shows the simultaneous layout of K_7, the complete graph on seven vertices and P_7, the path with seven vertices. The edges that belong to the path are drawn using curved and thick edges. Note that although an individual layout of K_7 would place one of the vertices in the middle, the simultaneous embedding

with the aggregate layout method pushes that vertex out because of the presence of the path. A summary of the aggregate graph layout algorithm is in Fig. 4.
Aggregate View: The matching visualization scheme for the aggregate graph layout is the aggregate view. In this scheme we only display a vertex once, even though it may be in multiple graphs and we display all edges from all the graphs in the sequence; see Fig. 3. The graphs can be displayed in 2D or 3D and we employ different edge colors and edge styles to differentiate between the different graphs. Displaying all graphs using a single vertex set allows the viewer to see multiple graphs at the same time and view the difference in relationships more easily. Different edge colors and edge styles are used to distinguish between the relationships from each graph. For example in Fig. 3 the edges of one graph are drawn with green straight line segments, whereas the other graph is drawn with thicker curved edges in a different tone of green.

4.2 Merged Graph Layout

In contrast to the aggregate method, the merged graph layout method does not guarantee perfect mental map preservation. The algorithm begins with the creation of a merged graph $G_M = (V_M, E_M)$ from the given sequence of graphs $G_1, G_2, \dots, G_k$. The merged graph is obtained by taking $G_1, G_2, \dots G_k$ and interconnecting all corresponding vertices with a special class of edges, E_{new}. Thus, if a vertex v appears m times in the sequences, there will be m copies of it in G_M.

The positions of corresponding vertices in each layout depend on how we assign weights to the edges in E_{new}. The larger the weight of edges in E_{new} the closer the corresponding vertices in each separate layout will be. An important property of this layout method is the proximity of corresponding vertices in the final layout. Let $u_1, u_2, \dots u_j$ be all the vertices corresponding to u in the merged graph and $v_1, v_2, \dots v_m$ be the ones corresponding to v. If $j > m$, then the u vertices get placed closer to each other than the v vertices do in the final layout. Once the merged graph has been created and the weights assigned, the modified force-directed method is applied.

In our implementation we allow the user to interactively assign a weight for the edges in E_{new}, so that the user has an overall control on the relative

Aggregate Graph Layout

1 Construct $G_A = (V_A, E_A)$:
$V_A = V_1 \cup V_2 \cup \dots \cup V_k$, $E_A = E_1 \cup E_2 \cup \dots \cup E_k$
2 Assign weights to each $u \in V_A$ and $(u, v) \in E_A$:
$w(u) =$ number of appearances of u in $V_1, V_2, \dots, V_k$
$w(u, v) =$ number of appearances of (u, v) in $E_1, E_2, \dots, E_k$
3 Use the modified force-directed layout algorithm on G_A
4 Extract the layout of each G_i from the layout of G_A

Fig. 4. Aggregate Graph Layout. $G_1, G_2, \dots, G_k$ are the input graphs.

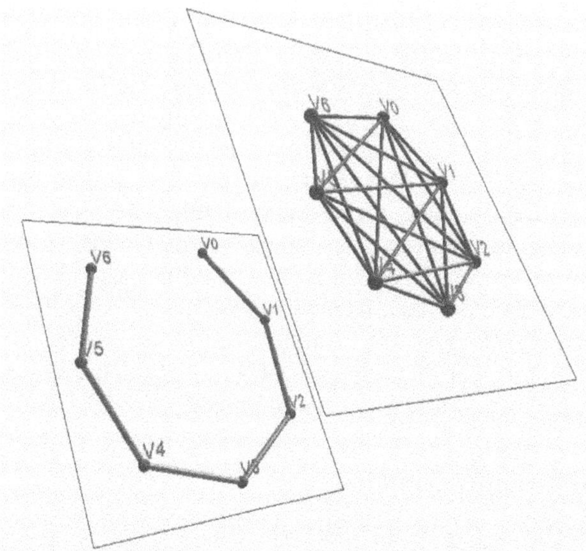

Fig. 5. Simultaneous embedding of K_7 and P_7 using the merged graph layout method. The visualization is done in 3D using a separate plane for each graph.

distances of corresponding vertices in different layers. Thus, in effect, the user has overall control over the extent of mental map preservation. Fig. 5 illustrates the simultaneous embedding resulting from the merged graph layout of K_7 and P_7. Note that although the locations of the corresponding vertices might not be the same, the mental map is still preserved since the relative locations of the corresponding vertices remain the same. A summary of the merged graph layout algorithm is in Fig. 6.

Merged Graph Layout

Rename the vertices in $V_1, V_2, \ldots, V_k$ so that each vertex is unique
Construct E_{new} by connecting corresponding vertices in $V_1, V_2, \ldots, V_k$
Construct $G_M = (V_M, E_M)$:
$V_M = V_1 \cup V_2 \cup \ldots \cup V_k$, $E_M = E_1 \cup E_2 \cup \ldots \cup E_k \cup E_{new}$
Assign weights for the edges in E_{new}
Apply the modified force-directed layout algorithm on G_M

Fig. 6. Merged Graph Layout. $G_1, G_2, \ldots, G_k$ are the input graphs.

Merged View: The matching visualization scheme for the merged graph layout is the merged view. In this scheme each of the graphs is drawn on its separate 2D plane, and the planes are layered in 3D in the order of appearance; see Fig. 5.

At the same time, all the graph layouts are shown on the same screen and since corresponding vertices from any two planes have the same approximate positions on their planes, this model provides a clear mental mapping between the two relationships represented by each graph.

This view model also allows the user to move and rotate the planes in 3D. This feature is useful in case the user wants to see a particular graph in more detail, in which case it is sufficient to rotate the view around a particular axis. In addition, to enhance the user's 3D view, the vertices are drawn as spheres and the edges as cylindrical pipes.

4.3 Independent Iterations Layout

The two approaches defined above construct a global graph and extract individual layout of each graph from this global layout. Our final layout method is quite different and we describe it here for only two graphs.

The algorithm begins by creating independent layouts for the two graphs $G_1 = (V_1, E_1)$ and $G_2 = (V_2, E_2)$. The layouts are obtained using intelligent (rather than random) placement of the vertices, based on the graph distance, as described in [11]. At this stage, we have the best drawings for each graph when they are drawn independently. As a result we obtain two different point-sets, P_1 and P_2 specifying the locations of the vertices in G_1 and G_2, respectively.

In the next step G_1 "borrows" the point-set P_2 of G_2 and treats it as an initial placement for the standard force-directed algorithm. Similarly, G_2 uses the point-set P_1 of G_1 and uses it as an initial placement for the standard force-directed algorithm. After applying force-directed iterations to both graphs (again independently) we arrive at two new point-sets P_1' and P_2'. We repeat the process of point-set swapping and force-directed calculations until the resulting point-sets converge to a given threshold minimum desirable distance between them or until the number of iterations exceed a fixed constant.

Given a mapping between two point-sets, the distance between them can be measured as the sum of Euclidean distances between each pair of corresponding points in the point-sets. This simple metric is not well-suited to our problem as the following example shows: Assume layout l_2 is just a 90° rotation of layout l_1. Even though the topology of the layouts is the same, calculating the distance between l_1 and l_2 as the sum of Euclidean distances between points would be misleadingly high. To overcome this problem, we first align the two layouts as best as possible using rigid 3D motion. In particular, we apply an affine linear transformation on l_1 so that the layout of l_1 after the transformation, is as close as possible to l_2. The transformation consists of translation, rotation, scaling, shearing and given a point $p = (x, y)$ on the plane it can be defined as:

$$f(p) = \begin{pmatrix} c_{x1} & c_{y1} \\ c_{x2} & c_{y2} \end{pmatrix} \begin{pmatrix} x \\ y \end{pmatrix} + \begin{pmatrix} c_{x3} \\ c_{y3} \end{pmatrix}$$

We would like to find the function $f(p)$, (i.e. all the constants c_{x1} etc.) that minimizes the distance between the transformed layout of l_1 and l_2, which is

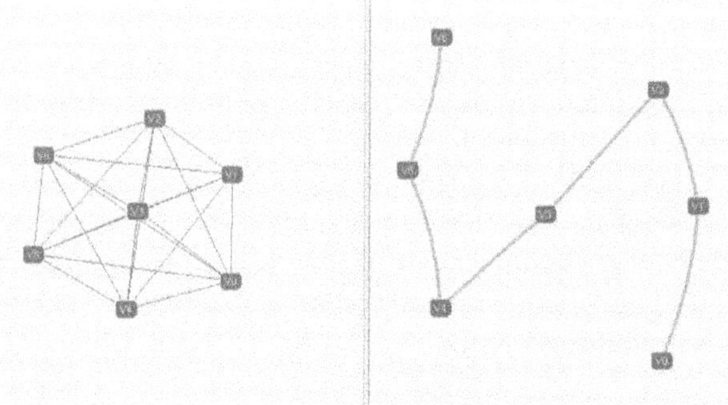

Fig. 7. Simultaneous embedding of K_7 and P_7 using independent iterations layout method and split view model for visualization.

Independent Iterations Layout

1 Using independent intelligent placement obtain layouts l_1 and l_2 for G_1 and G_2
2 Apply a linear transformation on l_1 to align it to l_2
3 Let $mindist = dist(l_1, l_2)$ and $bestl_1 = l_1$, $bestl_2 = l_2$
4 Repeat until ($mindist < threshold$) or ($iterationcount > maxiterationcount$):
 4.1 Apply layout algorithm on G_1 to get $l_{1'}$ using l_2 for initial placement
 4.2 Apply layout algorithm on G_2 to get $l_{2'}$ using l_1 for initial placement
 4.3 Apply a linear transformation to align l_1 to l_2
 4.4 If $dist(l_{1'}, l_{2'}) < mindist$
  $mindist = dist(l_{1'}, l_{2'})$
  $bestl_1 = l_{1'}$, $bestl_2 = l_{2'}$
 4.5 $l_1 = l_{1'}$, $l_2 = l_{2'}$, $iterationcount$ ++

Fig. 8. Independent Iterations Layout. G_1 and G_2 are the input graphs

then equivalent to minimizing $\sum_{p \in l_1} dist(f(p), p')$, where p' is the point in l_2 corresponding to p and $dist(f(p), p')$ is the Euclidean distance between $f(p)$ and p'. Then the minimization can easily be achieved by taking the derivative with respect to c_{xi} and solving for the resulting linear equations. Fig. 7 shows the simultaneous embedding of K_7 and P_7 resulting from independent iterations layout using the split view, described below. Note that the resulting layout for each graph is not the same as an individual layout for that graph. Instead, the independent iterations layout is a compromise between the two individual layouts. A summary of the layout algorithm is in Fig. 8.

The algorithm is well defined for two graphs but can be extended to handle more graphs. The point-set swapping can be extended to swapping the point-sets of neighboring graphs in the sequence and the distance measure between a pair of layouts can be extended to measure distances between multiple point-sets. Currently our implementation works for pairs of graphs only.

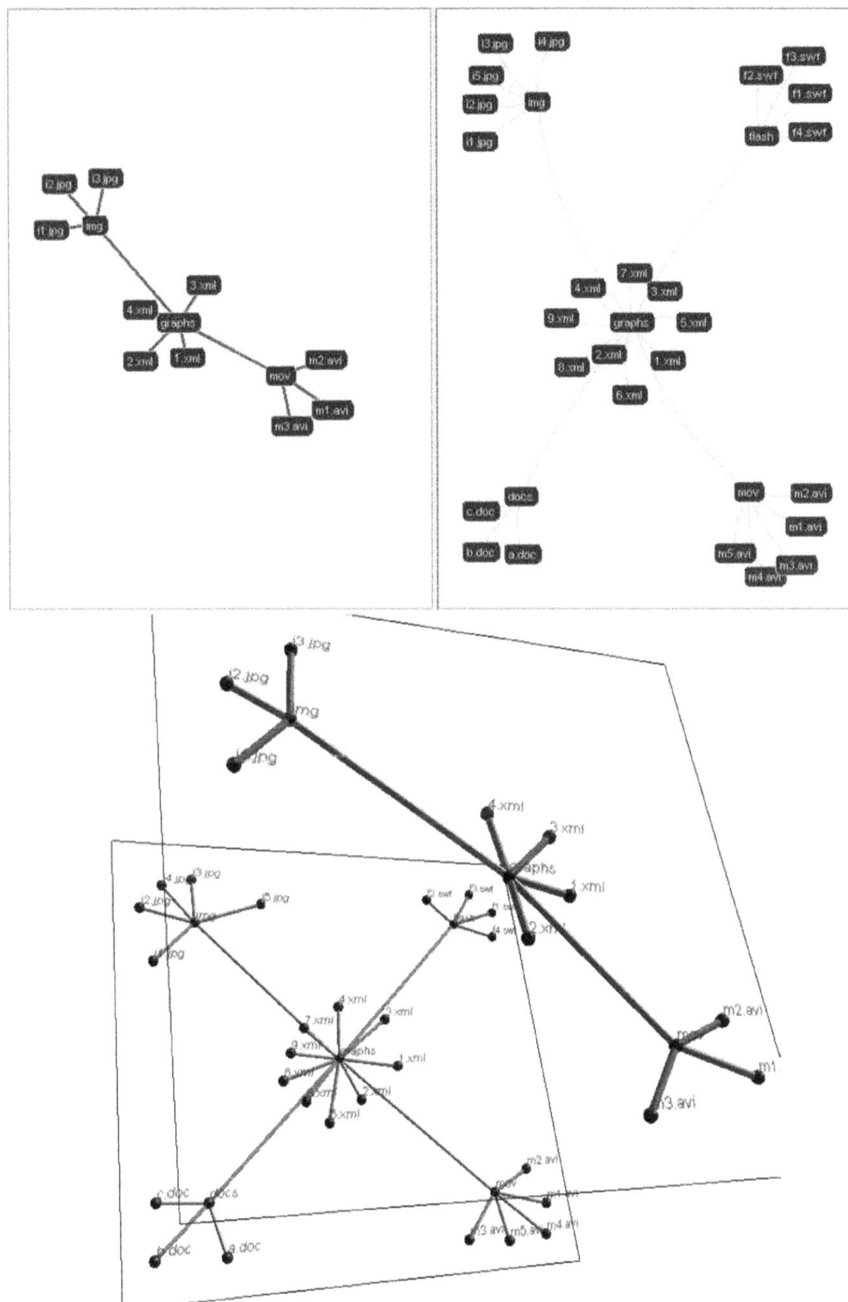

Fig. 9. A pair of graphs representing file system snapshots. The images on top show a split view and the images on the bottom show a merged view.

Split View: The two graphs are drawn separately in their own windows in 2-dimensions and both windows are on the same screen; see Fig. 7. The view model can be generalized to handle many graphs, in which case the screen would be split into many individual panes. Still, as the number of graphs to be visualized increases, the user's ability to read the relations between them greatly decreases in this case which makes the model more suitable for visualization of small number of graphs.

5 Implementation

We have implemented our layout methods and visualization schemes using Java and the system can be downloaded at `http://simg.cs.arizona.edu/`. In addition to the three layout methods and three visualization schemes, the system provides various capabilities such as graph editing, building some common classes of graphs (complete graphs, trees, paths), building random graphs, etc. Graphs in GML format can be loaded and stored. All images in the paper (except that in Fig. 2) are from our system. In Fig. 9 we show more layouts obtained with our system.

References

1. U. Brandes and S. R. Corman. Visual unrolling of network evolution and the analysis of dynamic discourse. In *IEEE Symposium on Information Visualization (INFOVIS '02)*, pages 145–151, 2002.
2. U. Brandes and D. Wagner. A bayesian paradigm for dynamic graph layout. In *Proceedings of the 5th Symposium on Graph Drawing (GD)*, volume 1353 of *LNCS*, pages 236–247, 1998.
3. J. Branke. Dynamic graph drawing. In M. Kaufmann and D. Wagner, editors, *Drawing Graphs: Methods and Models*, number 2025 in LNCS, chapter 9, pages 228–246. Springer-Verlag, Berlin, Germany, 2001.
4. P. Brass, E. Cenek, C. A. Duncan, A. Efrat, C. Erten, D. Ismailescu, S. G. Kobourov, A. Lubiw, and J. S. B. Mitchell. On simultaneous graph embedding. In *Proceedings of 8th Workshop on Algorithms and Data Structures (WADS)*, pages 243–255, 2003.
5. C. Collberg, S. G. Kobourov, J. Nagra, J. Pitts, and K. Wampler. A system for graph-based visualization of the evolution of software. In *ACM Symposium on Software Visualization*, pages 77–86, 2003.
6. S. Diehl and C. Görg. Graphs, they are changing. In *Proceedings of the 10th Symposium on Graph Drawing (GD)*, pages 23–30, 2002.
7. P. Eades and C. F. X. M. Neto. Vertex splitting and tension-free layout. In *Graph Drawing*, volume 1027 of *Lecture Notes in Computer Science*, pages 202–211. Springer, 1996.
8. C. Erten and S. G. Kobourov. Simultaneous embedding of a planar graph and its dual on the grid. In *13th Intl. Symp. on Algorithms and Computation (ISAAC)*, pages 575–587, 2002.
9. T. Fruchterman and E. Reingold. Graph drawing by force-directed placement. *Softw. - Pract. Exp.*, 21(11):1129–1164, 1991.

10. P. Gajer, M. T. Goodrich, and S. G. Kobourov. A multi-dimensional approach to force-directed layouts. In *Proceedings of the 8th Symposium on Graph Drawing (GD)*, pages 211–221, 2000.
11. P. Gajer and S. G. Kobourov. GRIP: Graph dRawing with Intelligent Placement. *Journal of Graph Algorithms and Applications*, 6(3):203–224, 2002.
12. D. Harel and Y. Koren. Drawing graphs with non-uniform vertices. In *Proceedings of Working Conference on Advanced Visual Interfaces (AVI'02)*, pages 157–166, 2002.
13. Herman, G. Melançon, and M. S. Marshall. Graph visualization and navigation in information visualization: A survey. *IEEE Transactions on Visualization and Computer Graphics*, 6(1):24–43, 2000.
14. M. L. Huang, P. Eades, and J. Wang. On-line animated visualization of huge graphs using a modified spring algorithm. *Journal of Visual Languages and Computing*, 9:623–645, 1998.
15. T. Kamada and S. Kawai. Automatic display of network structures for human understanding. Technical Report 88-007, Dept. of Inf. Science, University of Tokyo, 1988.
16. S. C. North. Incremental layout in DynaDAG. In *Proceedings of the 4th Symposium on Graph Drawing (GD)*, pages 409–418, 1996.
17. K.-P. Yee, D. Fisher, R. Dhamija, and M. Hearst. Animated exploration of dynamic graphs with radial layout. In *IEEE Symposium on Information Visualization (INFOVIS '01)*, pages 43–50, 2001.

Axis-by-Axis Stress Minimization

Yehuda Koren and David Harel

Dept. of Computer Science and Applied Mathematics
The Weizmann Institute of Science, Rehovot, Israel
{yehuda,dharel}@wisdom.weizmann.ac.il

Abstract. Graph drawing algorithms based on minimizing the so-called stress energy strive to place nodes in accordance with target distances. They were first introduced to the graph drawing field by Kamada and Kawai [11], and they had previously been used to visualize general kinds of data by multidimensional scaling. In this paper we suggest a novel algorithm for the minimization of the Stress energy. Unlike prior stress-minimization algorithms, our algorithm is suitable for a one-dimensional layout, where one axis of the drawing is already given and an additional axis needs to be computed. This 1-D drawing capability of the algorithm is a consequence of replacing the traditional node-by-node optimization with a more global axis-by-axis optimization. Moreover, our algorithm can be used for multidimensional graph drawing, where it has time and space complexity advantages compared with other stress minimization algorithms.

1 Introduction

A graph is a structure $G(V{=}\{1,\ldots,n\},E)$ representing a binary relation E over a set of nodes V. In [9] we addressed the problem of drawing a graph in one dimension. The major application is when one of the drawing coordinates is already given, and we want to compute an additional coordinate for each node. Traditional force-directed algorithms turned out to be unsuitable in this 1-D case, because of their local optimization methods, which become stuck in bad local minima. In this paper, we show how the familiar method of Kamada and Kawai [11], which minimizes an energy function known as *the stress energy*,[1] can be used for 1-D graph drawing. We introduce a novel method for axis-by-axis optimization of the stress energy, which is very suitable for 1-D graph drawing as it utilizes the special nature of this problem. Interestingly, using our technique it seems that minimizing the stress in only one dimension is definitely the easier case for stress minimization. Moreover, we harness the benefits of the new 1-D-based algorithm for computing a multidimensional drawing of the graph, by calculating each axis of the drawing separately. In this multidimensional case, our algorithm also has an advantage over the classical stress minimization suggested by Kamada and Kawai.

2 One-Dimensional Stress Optimization

Stress energy is a traditional measure of drawing quality, based on the heuristic that a nice drawing relates to good isometry. Accordingly, it calls for placing the nodes so

[1] Indeed, the stress energy was originally used for multidimensional scaling, where its name originated; see, e.g., [3,12].

G. Liotta (Ed.): GD 2003, LNCS 2912, pp. 450–459, 2004.

that the resulting pairwise Euclidean distances will approach the corresponding target (graph-theoretical) distances. Given a k-D layout $x = (x_1, \ldots, x_n)$, where the place of node i is $x_i \in \mathbb{R}^k$, the concrete form of the energy is:

$$E(x) \stackrel{\text{def}}{=} \sum_{i<j} k_{ij} \left(|x_i - x_j| - d_{ij}\right)^2 . \tag{1}$$

Here, the target distance d_{ij} is typically the graph-theoretical distance between nodes i and j. The normalization constant k_{ij} equals $d_{ij}^{-\alpha}$, where $0 \leqslant \alpha \leqslant 2$. Kamada and Kawai [11] picked $\alpha = 2$, whereas Cohen [3] also considered $\alpha = 0$ and $\alpha = 1$. Moreover, Cohen suggested setting d_{ij} to the linear-network distance.

For the rest of this paper, we explore axis-by-axis stress minimization. Hence, we assume x to be an 1-D layout, i.e., $x = (x_1, \ldots, x_n) \in \mathbb{R}^n$. Given two 1-D layouts, $x, \tilde{x} \in \mathbb{R}^n$, we define the following family of auxiliary functions:

$$\delta_{ij}^{\tilde{x}}(x) = \begin{cases} x_i - x_j & \tilde{x}_i \geqslant \tilde{x}_j \\ x_j - x_i & \tilde{x}_i < \tilde{x}_j \end{cases} \qquad 1 \leqslant i < j \leqslant n. \tag{2}$$

Next, we define the following energy function of the layout x:

$$E^{\tilde{x}}(x) \stackrel{\text{def}}{=} \sum_{i<j} k_{ij} \left(\delta_{ij}^{\tilde{x}}(x) - d_{ij}\right)^2 . \tag{3}$$

It is important to understand the relations between $E^{\tilde{x}}(x)$ and the stress energy, $E(x)$.

Lemma 1. *For every $x, \tilde{x} \in \mathbb{R}^n$, $E(x) \leqslant E^{\tilde{x}}(x)$.*

Proof. Let us pick some pair $i < j$, and analyze the corresponding terms of $E(x)$ and $E^{\tilde{x}}(x)$. Observe that $|\delta_{ij}^{\tilde{x}}(x)| = |x_i - x_j|$. In addition, it is always the case that $k_{ij}, d_{ij} \geqslant 0$. Therefore, $k_{ij}(\delta_{ij}^{\tilde{x}}(x) - d_{ij})^2 \geqslant k_{ij}(|x_i - x_j| - d_{ij})^2$. The lemma follows. □

Lemma 2. *For every $x \in \mathbb{R}^n$, $E(x) = E^x(x)$.*

Proof. Simply observe that $\delta_{ij}^x(x) = |x_i - x_j|$. □

By the last two Lemmas we conclude:

Corollary 1. *For every $x, \tilde{x} \in \mathbb{R}^n$, $E^x(x) \leqslant E^{\tilde{x}}(x)$.*

The usefulness of the energy $E^{\tilde{x}}(x)$ stems from the fact that it can be minimized optimally. To realize this we need some additional notations. First, we define a related $n \times n$ Laplacian matrix $\mathcal{L}$, where

$$\mathcal{L}_{ij} = \begin{cases} -k_{ij} & i \neq j \\ \sum_{j \neq i} k_{ij} & i = j \end{cases} \qquad i, j = 1, \ldots, n.$$

Note that the Laplacian depends only on the graph, regardless of its layout $\tilde{x}$. We also use the vector $b^{\tilde{x}} \in \mathbb{R}^n$, where:

$$b_i^{\tilde{x}} = \sum_{j \neq i:\ \tilde{x}_j \leqslant \tilde{x}_i} k_{ij} d_{ij} - \sum_{j \neq i:\ \tilde{x}_j > \tilde{x}_i} k_{ij} d_{ij} \qquad i = 1, \ldots, n.$$

Now, using some elementary algebra, it can be shown that:

$$E^{\tilde{x}}(x) = x^T \mathcal{L} x - 2x^T b^{\tilde{x}} + C\,,$$

where C is a constant that is independent of x. Since the Laplacian is known to be positive semi-definite, we conclude by differentiation:

Lemma 3. *The minimizer of $E^{\tilde{x}}(x)$ is the solution of the system of equations:*

$$\mathcal{L} x = b^{\tilde{x}}\,.$$

All these observations suggest the following process for minimizing the stress energy:

Function 1-D_stress_minimization $(G(V,E),\ x \in \mathbb{R}^n)$
  Compute the Laplacian $\mathcal{L}$
  do
    $\tilde{x} \leftarrow x$
    Compute $b^{\tilde{x}}$
    Compute x for which $\mathcal{L} x = b^{\tilde{x}}$
  while $(x \neq \tilde{x})$

Our main result is that each iteration of this process (except for the last one) must decrease the stress energy: $E(x) < E(\tilde{x})$. This stems from the above lemmas: By Lemma 2, $E(\tilde{x}) = E^{\tilde{x}}(\tilde{x})$, and by Lemma 3, $E^{\tilde{x}}(x) < E^{\tilde{x}}(\tilde{x})$ (since $\tilde{x} \neq x$, the energy must decrease). Now, by Corollary 1, $E^{x}(x) \leqslant E^{\tilde{x}}(x)$. Taken together, we obtain:

$$E(x) = E^{x}(x) \leqslant E^{\tilde{x}}(x) < E^{\tilde{x}}(\tilde{x}) = E(\tilde{x})\,.$$

Of course the energy is bounded below by zero, so the process must converge. Unlike node-by-node local optimization methods, the new optimization process does not suffer from working in a single dimension. Later, we introduce some encouraging experimental results. However, note that like other optimization methods, we can only guarantee convergence to a local minimum.

2.1 Working with a Sparse Laplacian

A significant drawback of stress optimization is that its space complexity is $\Theta(n^2)$, since we have to store all pairwise distances. This, of course, slows down running time, and even more importantly, prevents us from dealing with large graphs containing more than around 10,000 nodes. However, it appears that most graphs that can be drawn nicely possess much redundancy in the pairwise distances. We utilize this redundancy by neglecting most pairwise distances. Specifically, we define the set S of the "interesting" node pairs, so the stress energy is defined as:

$$\sum_{\{i,j\} \in S} k_{ij} \left(|x_i - x_j| - d_{ij}\right)^2\,.$$

Accordingly, we can use a sparse Laplacian, whose nonzero entries correspond to the pairs in $\mathcal{S}$:

$$\mathcal{L}_{ij} = \begin{cases} -k_{ij} & \{i,j\} \in \mathcal{S} \\ \sum_{j:\{i,j\}\in\mathcal{S}} k_{ij} & i = j \\ 0 & \text{otherwise} \end{cases} \qquad i,j = 1,\dots,n.$$

Similarly, the vector $b^{\tilde{x}}$ is defined as:

$$b_i^{\tilde{x}} = \sum_{\substack{j: \\ \{i,j\}\in\mathcal{S},\tilde{x}_j \leqslant \tilde{x}_i}} k_{ij}d_{ij} - \sum_{\substack{j: \\ \{i,j\}\in\mathcal{S},\tilde{x}_j > \tilde{x}_i}} k_{ij}d_{ij} \qquad i = 1,\dots,n.$$

How can we choose the pairs in $\mathcal{S}$? In fact, such sparsification of the pairwise distances was already used in the multiscale stress-minimization methods of [4,5,6], where *in each scale* only the relations between nodes and their close neighborhood are considered. Here, we would like to offer another, very efficient sparsification scheme, which is especially appropriate for the new optimization method. An important feature of our sparsification is that it guarantees linear space requirements for bounded degree graphs. Moreover, using more sophisticated techniques that will be described in a forthcoming paper [10], we can maintain the linear space requirements for any kind of graph.

We construct the pairs-set, $\mathcal{S}$, to consist of two kinds of node-pairs: (1) "pivot-based" pairs and (2) pairs of close nodes. This construction is accomplished in the following way. First, we construct a set $\mathcal{P}$ that contains m *pivot nodes* uniformly scattered over the graph. Such pivots can be chosen randomly. A better choice that we use in practice is described in [7]. We then initialize $\mathcal{S}$ with all the pairs $\{\{i,j\} \mid i \in P, j \in V, i \neq j\}$. The distances between these pairs preserve the overall structure of the graph. In [10] we describe a more advanced technique where these pivot-based pairs are enough. However, here, we should add to $\mathcal{S}$ all pairs that are close in the graph: $\{\{i,j\} \mid d_{ij} \leqslant B\}$. Typical values for B are 3 to 6. This way we account for the local aesthetics of the drawing, which are not captured well by the pivots.

Computation of the distances from the pivots is performed in time $O(m \cdot |E|)$, and the storage requirements are $O(m \cdot |V|)$. We cannot state here a satisfactory recipe for finding an adequate value of m, and of course increasing m never degrades the quality of the drawing. However, in our experiments, working with $30 \leqslant m \leqslant 100$ seems to provide good results. Consequently, our optimization process becomes efficient in terms of time and space, without affecting its drawing quality very much. Interestingly, such a sparsification approach completely fails with the node-by-node local stress optimization suggested by Kamada and Kawai, for reasons presently unknown. Anyway, this is an advantage of our new optimization method over the more traditional one.

Smoothing the pivots. One minor problem may arise when working with pivots. Since the pivots are related to many more pairwise distances than the rest of the nodes, for some graphs they appear to "break out" of the drawing. For example, consider the 2-D layout of the 64×16 torus shown in Fig. 1(a). (Details about the computation of 2-D layouts are given in the following sections.) Here, we have set $B = 5$ and used 50 pivots, which are designated in the layout as large red squares. As the reader can see, the pivots are not very well-placed.

There is an easy solution to this problem. At a postprocessing stage, we fix the places of all the non-pivots, and then apply our algorithm only to the pivots, taking into account just the short distances (smaller than B). Consequently, in each iteration we have to solve a system of only m equations. Furthermore, typically the pivots are not placed close together, so we can solve independently one equation for each of them, i.e., for each pivot node, i, we iteratively perform:

$$x_i \leftarrow \frac{\sum_{j\neq i:\ x_j \leqslant x_i, d_{ij} \leqslant B} k_{ij}(x_j + d_{ij}) + \sum_{j:\ x_j > x_i, d_{ij} \leqslant B} k_{ij}(x_j - d_{ij})}{\sum_{j\neq i:\ d_{ij} \leqslant B} k_{ij}}$$

until x_i no longer changes.

All this local smoothing takes negligible time. Figure 1(b) shows again the layout of the 64×16 torus, but now after using the pivot-smoothing.

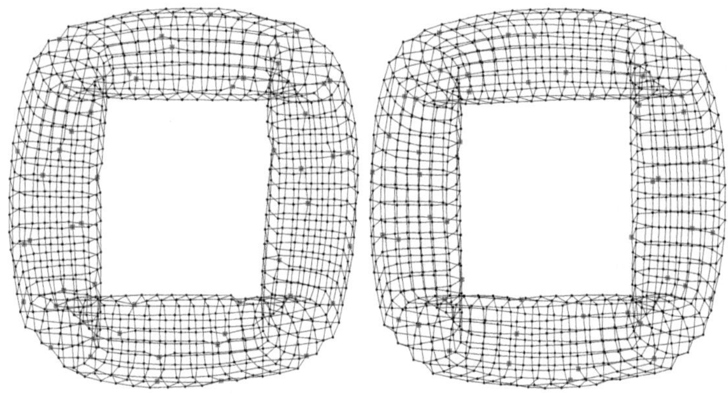

Fig. 1. Drawing a 64×16 torus; pivots are the red squares. (a) Pivots seem to slightly break out of the drawing. (b) Pivot-smoothing is performed, so that pivots are better placed.

3 Working in the Presence of Predefined Coordinates

The main application of 1-D graph drawing is where one axis of the drawing is given in advance. Therefore, we should account for the already computed coordinates. Note that a careless computation that ignores the precomputed coordinates can be very problematic. Such a computation might yield new coordinates that are very similar to the given ones, resulting in a drawing whose intrinsic dimensionality would really be 1, meaning that one axis would be wasted.

Fortunately, we have found a very effective way of taking care of the predefined axis, which we term in this section axis z. The stress minimization strategy strives to achieve the target distances in the drawing. Some fraction of the target distances is already achieved in the predefined axis, z. Hence, when computing a new axis we would like the achieve the *residual target distances* instead of the original ones. The new residual target distances are defined as:

$$d^z_{ij} = \begin{cases} \sqrt{d^2_{ij} - (z_i - z_j)^2} & d_{ij} > |z_i - z_j| \\ 0 & \text{otherwise} \end{cases} \qquad i, j = 1, \ldots, n\,. \tag{4}$$

Note that there is no reason to alter the normalization weights k_{ij}.

However, there is still one problem remaining: The target distances set the scale of the layout. Therefore, it is possible that the predefined axis, z, has an entirely different scale (e.g., it may be very small), so our computation of residual target distances makes no sense. To overcome this problem, we re-scale z in order to bring it to the scale of the target distances (alternatively, we could re-scale the target distances in order to bring them to the scale of z). More specifically, we want to compute a constant $c > 0$ that minimizes the stress energy of the scaled $c \cdot z$. To find the exact value of c, we differentiate:

$$\frac{\partial}{\partial c} E(c \cdot z) = 0$$

The solution is:

$$c = \frac{\sum_{\{i,j\} \in \mathcal{S}} k_{ij} d_{ij} |z_i - z_j|}{\sum_{\{i,j\} \in \mathcal{S}} k_{ij} (z_i - z_j)^2} . \tag{5}$$

To summarize, when an axis z is given in advance, we replace it with $c \cdot z$ to bring it into the scale of the target distances. The constant c is computed by (5). Then, we take into account the distances already captured by z, by computing the residual target distances, d_{ij}^z, as in (4). The additional axis x is computed in order to minimize the stress function $\sum_{\{i,j\} \in \mathcal{S}} k_{ij} \left(|x_i - x_j| - d_{ij}^z\right)^2$.

4 Experimental Results

4.1 Application for Drawing Digraphs

The digraph drawing algorithm in [1] computes the y-axis by minimizing the hierarchy energy in order to convey the overall directionality of the graph. The x-axis was computed by Eigen-projection or by minimum linear arrangement, and shows additional properties of the graph. The new 1-D stress minimization algorithm can serve in computing the x-axis. Our experiments show that in most cases using stress minimization improves the quality of the layouts by being able to accurately show sharp changes in the graph structure. Examples are shown in [2].

4.2 Two-Dimensional Layouts

Our 1-D stress minimization algorithm can be used to compute 2-D layouts. We first compute the x-coordinates, and then compute the y-coordinates given the x-coordinates, as explained in Sec. 3. Several results are shown in Fig. 2.

Comparing the drawings in Fig. 2 to the drawings of the same graphs in Figs. 1, and 3(a,b) reveals that the results given in Fig. 2 are sub-optimal. Since the x-coordinates were computed independently of the y-coordinates, the algorithm strives to minimize the stress energy as much as possible using only the x-axis. This kind of a "greedy" approach stretches the layout along the x-axis. However, when we minimize the stress in 2-D from the beginning (using a method we describe in Sec. 5 or by employing the Kamada and Kawai algorithm [11]), the overall result is better. We provide quantitative results by comparing the values of the stress energy in the following table. For each of the three

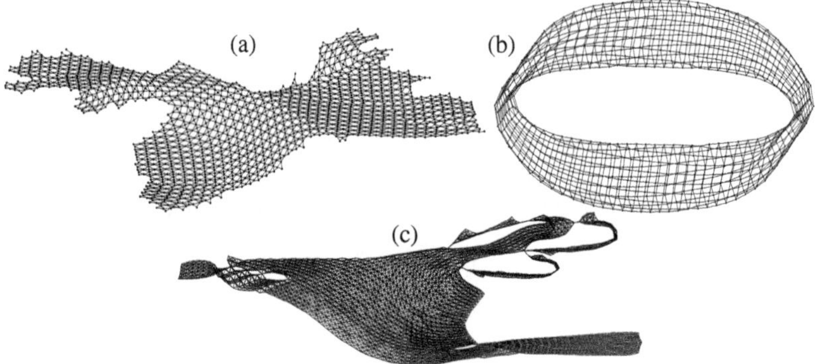

Fig. 2. 2-D layouts by 1-D stress minimization of: (a) Plsk1919, (b) 64 × 16 torus, (c) Shuttle.

graphs we measured its stress energy twice: once only along the x-axis and another time in the 2-D drawing. The results show that when computing the x-axis independently, the stress along it is smaller, but the overall 2-D stress is larger compared with a method that computes the two axes together. We have concluded that the algorithm as described so far is more suitable for achieving 1-D layouts.

Graph name	Separate computation of axes		Coupled computation of axes	
	x-axis stress	2-D stress	x-axis stress	2-D stress
Plsk1919	198,343.09	66,367.90	347,724.76	20,141.41
Torus 64 × 16	107,855.51	43,581.09	144,607.29	26,029.12
Shuttle	529,776.08	203,817.77	1,035,382.73	48,909.45

4.3 Speed of Computation for Selected Graphs

Smart Initialization

We recommend using our algorithm with a smart initialization. This way the probability of avoiding poor local minima is improved. Also, running time is significantly decreased as the number of iterations is reduced. Moreover, for a random initialization, we should enlarge the set of target distances $\mathcal{S}$ (by increasing the parameters m or B), which further increases running time. Such a smart initialization could result from faster methods like those of [7,8]. However, better results are obtained by initializing the algorithm by an approximation of itself that addresses a simplified problem. For such an approximation one can use the multiscale strategy and initialize the layout with a layout of a related, but much smaller graph. However, we have done it differently. We initialized the algorithm by first running a few iterations of the method described in the forthcoming paper [10], which accelerates the stress minimization by constraining the layout to lie in a unique small subspace. This significantly reduced the number of iterations, and allowed us to run the sparse version of the algorithm with parameters m=50 and B=3, yielding good results for all the given graphs. Note that sometimes increasing B improves the quality of the layouts.

Table 1 shows the number of iterations for several graphs. For each graph we have computed a 2-D layout, so the results refer to the computation of two axes. The overall running time, which includes the smart initialization, was measured on a Pentium IV

2GHz PC with 256MB RAM. As shown, for a medium-sized graph (up to $\sim 10,000$ nodes), running times are comparable to those of the multi-scale stress minimization algorithms [4,5,6]. However, due to the better space complexity, our new method might deal with graphs of $\sim 100,000$ nodes.

Table 1. Number of iterations and running time (in seconds) for various graphs

Graph name	|V|	|E|	x-axis iterations	y-axis iterations	Time (sec.)
Torus 64×16	1024	2048	11	8	0.33
4970	4970	7400	19	22	3.66
Shuttle (Data)†	2851	15093	64	61	5.28
Crack†	10240	30380	41	27	15.95
Fidap006§	1651	23914	49	23	2.42
Nos3§	960	7442	20	12	0.58
Nos5§	468	2352	18	20	0.27
Nos6§	675	1290	10	4	0.14
Nos7§	729	1944	15	29	0.36
Plat362§	362	2712	9	5	0.13
Plat1919§	1919	15240	17	31	1.73
Plsk1919§	1919	4831	8	19	1.16
Sierpinski (depth 10)	88575	177147	40	22	165.64
Grid 317×317	100489	200344	17	5	189.66

† From Walshaw's collection: `www.gre.ac.uk/~c.walshaw/partition`

§ From the Matrix Market collection: `math.nist.gov/MatrixMarket`

5 Alternating Two-Dimensional Stress Minimization

As we have seen, directly applying our algorithm to compute a 2-D layout produces imperfect results, in which the graph is often "stretched" along the first computed axis. However, we can overcome this artifact by coupling the computation of the two axes, thereby neutralizing any preference for one of them. Here is the precise algorithm:

Function 2-D_stress_minimization $(G(V,E),\ x, y \in \mathbb{R}^n)$
% Coordinates of node i are (x_i, y_i)
  Compute the Laplacian $\mathcal{L}$
  do
    $\tilde{x} \leftarrow x$
    Compute $b^{\tilde{x}}$ using the residual distances d_{ij}^{y}
    Compute x for which $\mathcal{L}x = b^{\tilde{x}}$
    $\tilde{y} \leftarrow y$
    Compute $b^{\tilde{y}}$ using the residual distances d_{ij}^{x}
    Compute y for which $\mathcal{L}y = b^{\tilde{y}}$
  while ($x \neq \tilde{x}$ or $y \neq \tilde{y}$)

This technique can be directly extended to higher dimensions. This way, we can use our algorithm to compute multidimensional layouts. The results are comparable to the Kamada-Kawai algorithm, but there is the advantage of the fast, pivot-based algorithm, whose memory requirements are lower than other implementations of the Kamada-Kawai algorithm. In [10] we describe another advantage of this technique, which enables an additional substantial reduction of time and space complexity.

The actual running times of the algorithm are slightly larger than the results given in Table 1 (mainly because of the added time for recomputing the residual distances at each iteration). Results are shown in Fig. 3 (and in Fig. 1).

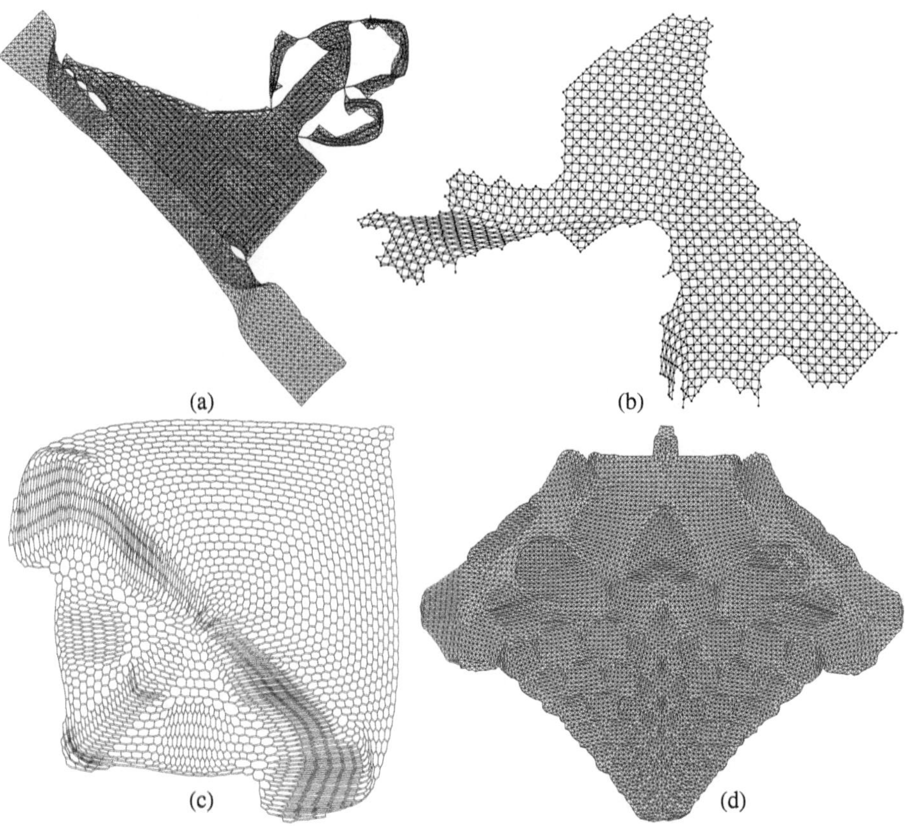

Fig. 3. Drawing by alternating stress minimization of: (a) Shuttle, (b) Plsk1919, (c) 4970, (d) Crack.

6 Discussion

Strategies based on minimizing the stress energy are known for their ability to produce nice graph layouts. Common optimization processes are based on node-by-node local optimization methods. Specifically, Kamada and Kawai [11], use a localized 2-dimensional Newton-Raphson process. However, working with a full Newton-Raphson process is impractical, since it requires a recomputation of the Hessian matrix in each step. Moreover, the Hessian might be singular or ill-conditioned, which may cause additional numerical difficulties. We have devised an algorithm for minimizing the stress energy needed for computing 1-D layouts. Technically, our algorithm introduces a rather global optimization process that replaces the usual local node-by-node optimization. Interestingly, it can be proved that this process is equivalent to performing a full n-dimensional Newton-Raphson optimization, unlike the local 2-dimensional process used by Kamada and Kawai. Hence, while the common wisdom is that stress-minimization is very difficult for 1-D layouts as optimization processes tend to become stuck in poor local minima, we have shown that the 1-D case is indeed the easier case allowing the use of a global optimization process. This motivates us to try to find ways to incorporate the powerful 1-D optimization process for computing multidimensional layouts. Consequently, we devised an algorithm for axis-by-axis computation of multidimensional layouts that has an advantage over other stress minimization approaches in terms of time and space complexity.

References

1. L. Carmel, D. Harel and Y. Koren, "Drawing Directed Graphs Using One-Dimensional Optimization", *Proc. Graph Drawing 2002*, LNCS 2528, pp. 193–206, Springer-Verlag, 2002.
2. L. Carmel, D. Harel and Y. Koren, "Combining Hierarchy and Energy for Drawing Directed Graphs", *IEEE Transactions on Visualization and Computer Graphics*, IEEE, in press.
3. J. D. Cohen, "Drawing Graphs to Convey Proximity: an Incremental Arrangement Method", *ACM Transactions on Computer-Human Interaction* **4** (1997), 197–229 .
4. P. Gajer, M. T. Goodrich and S. G. Kobourov, "A Multi-dimensional Approach to Force-Directed Layouts of Large Graphs", *Proc. Graph Drawing 2000*, LNCS 1984, pp. 211–221, Springer-Verlag, 2000.
5. R. Hadany and D. Harel, "A Multi-Scale Method for Drawing Graphs Nicely", *Discrete Applied Mathematics* **113** (2001), 3–21.
6. D. Harel and Y. Koren, "A Fast Multi-Scale Method for Drawing Large Graphs", *Journal of Graph Algorithms and Applications* **6** (2002), 179–202.
7. D. Harel and Y. Koren, "Graph Drawing by High-Dimensional Embedding", *Proc. Graph Drawing 2002*, LNCS 2528, pp. 207–219, Springer-Verlag, 2002.
8. Y. Koren, L. Carmel and D. Harel, "ACE: A Fast Multiscale Eigenvectors Computation for Drawing Huge Graphs", *Proc. IEEE Information Visualization (InfoVis'02)*, IEEE, pp. 137–144, 2002.
9. Y. Koren and D. Harel, "One-Dimensional Graph Drawing: Part I — Drawing Graphs by Axis Separation", Technical report MCS03-08, Faculty of Math. and Computer Science, The Weizmann Institute of Science, 2003.
10. Y. Koren, "Graph Drawing by Subspace Optimization", to be published.
11. T. Kamada and S. Kawai, "An Algorithm for Drawing General Undirected Graphs", *Information Processing Letters* **31** (1989), 7–15.
12. J. W. Sammon, "A Nonlinear Mapping for Data Structure Analysis", *IEEE Trans. on Computers* **18** (1969), 401–409.

Drawing Graphs with Nonuniform Nodes Using Potential Fields

Jen-Hui Chuang[1], Chun-Cheng Lin[2], and Hsu-Chun Yen[2]*

[1] Dept. of Computer and Information Science,
National Chiao-Tung University, Taiwan
jchuang@cis.nctu.edu.tw

[2] Dept. of Electrical Engineering, National Taiwan University, Taiwan
sanlin@cobra.ee.ntu.edu.tw, yen@cc.ee.ntu.edu.tw

Abstract. A *potential field* approach, coupled with force-directed methods, is proposed in this paper for drawing graphs with nonuniform nodes in 2-D and 3-D. In our framework, nonuniform nodes are uniformly or nonuniformly charged, while edges are modelled by springs. Using certain techniques developed in the field of potential-based *path planning*, we are able to find analytically tractable procedures for computing the repulsive force and torque of a node in the potential field induced by the remaining nodes. Our experimental results suggest this new approach to be promising, as drawings of good quality for a variety of graphs in 2-D and 3-D can be produced efficiently.

1 Introduction

As graphs are known to be one of the most important abstract models in various scientific and engineering areas, *graph drawing* (or *information visualization* in a broader sense) has naturally emerged as a fast growing research topic in computer science. In visualizing graphs associated with real-world entities, it is common to annotate each node with some labels (such as name, attribute, for instance), and a good way to display such information is to fill the information inside the drawing of the node. As a result, it is of importance and interest to be able to cope with vertices of different sizes and shapes in graph drawing.

Most of the conventional graph drawing algorithms only deal with graphs with zero-sized vertices. A naive extension of conventional drawing algorithms by plugging in objects of nonzero size for point vertices in the drawing is unlikely to produce a high quality drawing. In view of this, it becomes apparent that a new line of research has to be carried out in hope of narrowing the gap between the conventional algorithms and graphs consisting of nodes of various sizes and shapes that are frequently encountered in the real world. Along this line of research, a recent article [1] extends the conventionally force directed method [2] by developing two new algorithms (called *the elliptic spring method* and *the modified spring method*) to cope with graphs with nonuniform nodes

* Corresponding author

G. Liotta (Ed.): GD 2003, LNCS 2912, pp. 460–465, 2004.

(including circles, ellipses and rectangles). The underlying aesthetic criteria include preventing nonuniform nodes from being overlapped and edges from being crossed over with one another. Some encouraging results have been reported in [1].

In this paper, we present a new approach, combining the force-directed strategy as well as the theory of *potential fields*, for drawing graphs with nonuniform nodes. In our setting, each edge is modelled by a spring (same as that in conventional force-directed methods) and the boundary of every nonuniform node is uniformly or nonuniformly charged. We consider both 2-dimensional and 3-dimensional drawings of nonuniform nodes, each of which is represented by a polygon (in 2-D) or polyhedron (in 3-D). Other shapes of objects can be approximated by their enclosing polygons (or polyhedrons). For a given nonuniform node, the forces on the node come from two sources: one resulting from being connected by springs, and the other is caused by being present in the potential field induced by the remaining nodes. In comparison with the work of [1], our method deals with graphs in not only 2-D but 3-D, and makes every nonuniform node rotate with an angle of inclination to display a high degree of symmetry.

The rest of this paper is organized as follows. In Section 2, the theory behind potential fields in graph drawing is developed in depth. Section 3 illustrates a number of experimental results. Finally, a conclusion is given is Section 4.

2 Force-Directed Method Using Potential Fields

To draw graphs with nonuniform nodes, our force-directed method is based upon the idea of replacing each edge by a spring and assuming that the border of every nonuniform node is uniformly or nonuniformly charged. In this setting, the repulsive forces among nonuniform nodes avoid collision, and the spring forces pull nonuniform nodes closer. A nice drawing would be generated when the corresponding model reaches an equilibrium between the repulsive forces due to the charged nonuniform nodes and attractive forces due to springs.

In what follows, we mainly focus on the repulsive forces due to charged nonuniform nodes; the force formula of springs is based upon Hooke's Law [2] which uses the following logarithmic function as the force formula:

$$f_a(d) = C1 \log(d/C2) \tag{1}$$

where d is the spring length, and $C1$ and $C2$ are constants. In addition to repulsive forces, the repulsive torques need also be considered since every nonuniform node has some degree of inclination.

In what follows, we elaborate on the foundations of the theory behind the use of potential fields in graph drawing. The reader is referred to [3,4] for more about potential fields as well as some of the detailed derivations of formulas involved in our subsequent discussion.

Consider two polygons (representing two nonuniform nodes) A and B in 2-D connected by two edges $\overline{a_1b_2}$ and $\overline{a_3b_2}$ as shown in Figure 1-(a). A has vertices $a_1, ..., a_6$, and B has b_1, b_2, and b_3 along their boundaries. Each polygon

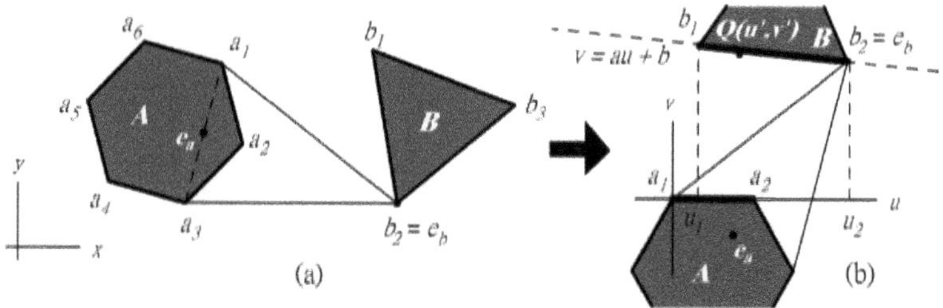

Fig. 1. Coordinate transformation of a 2-D graph with two nonuniform nodes A and B where e_a and e_b are the reference points. (a) The original coordinate system (xy-plane). (b) The new coordinate system (uv -plane) after the transformation.

is associated with a *reference point* exerted by repulsive and attractive forces, and shifts and rotations are carried out with respect to this reference point. For instance, in Figure 1 e_a and e_b are the reference points of A and B, respectively. For any two points of unit charge and of distance r apart, the *Newtonian potential* is defined as

$$V = \frac{1}{r} \tag{2}$$

and the Newtonian potential of a point x with respect to a line segment $\overline{a_1a_2}$ is

$$V = \int_{a_1}^{a_2} \frac{1}{r} dx \tag{3}$$

More generally, if the charge density increases (decreases) quadratically while x is moving along the line segment from a_1 to a_2, the charge density function can be expressed as $\alpha x^2 + \beta x + \gamma$ with some constants α, β, and γ; furthermore, the total potential at a point due to the charged line segment is equal to

$$V = \int_{a_1}^{a_2} \frac{\alpha x^2 + \beta x + \gamma}{r} dx \tag{4}$$

which can be evaluated analytically (4). Note that by letting $\alpha = 0$ (resp., $\alpha = \beta = 0$ and $\gamma = 1$), the above equation characterizes the linear (resp., Newtonian) potential field.

Consider two line segments $\overline{a_1a_2}$ and $\overline{b_1b_2}$ on the new coordinate system (uv-plane) after coordinate transformation as shown in Figure 1-(b). $Q(u\prime, v\prime)$ is a point on $\overline{b_1b_2}$, and $v = au + b$ (where a and b are constants) is the line equation representing $\overline{b_1b_2}$. d is the length of $\overline{a_1a_2}$, and u_1 and u_2 are the projection points of the end points of $\overline{b_1b_2}$. After a sequence of computations of the potential function of Eq. 4, such as negative gradient, integral, and coordinate transformation, the repulsive force on $\overline{b_1b_2}$ due to $\overline{a_1a_2}$, i.e. the repulsive force between two line segments, can be formulated as[1]

[1] For simplicity, only the u-component is considered for the rest of this paper.

$$F_u = \sqrt{1+a^2} \int_{u_2}^{u_1} \int_0^d \frac{u' - u}{r^3} \rho(u)\rho(u') du du'. \tag{5}$$

And the repulsive torque on $\overline{b_1 b_2}$ due to $\overline{a_1 a_2}$, i.e. the repulsive torque between two line segments, with respect to the reference point e_b can be formulated as

$$\begin{aligned}\tau_{e_b} = \sqrt{1+a^2}(& \int_{u_2}^{u_1} \int_0^d (u' - u_0) \frac{au' + b}{r^3} \rho(u)\rho(u') du du' \\ - & \int_{u_2}^{u_1} \int_0^d (au' + b - v_0) \frac{u' - u}{r^3} \rho(u)\rho(u') du du')\end{aligned} \tag{6}$$

Assume that the charge density $\rho(u)$ is equal to 1, u, or u^2 for a border line $\overline{a_1 a_2}$ of A, and $\rho(s) = 1$, s, or s^2 for a border line $\overline{b_1 b_2}$ of B, nine different combinations of charge distributions need to be considered in evaluating the repulsive force and torque between the two line segments. For example, form (5), the repulsive force along the u-axis for these nine combinations can be obtained from

$$\begin{aligned} F_u^{ij} &= F_u^{ij}(u_2) - F_u^{ij}(u_1) \\ &= (1+a^2)^{\frac{j+1}{2}} \int_{u_1}^{u_2} \int_0^d \frac{u' - u}{r^3} u^i du (u')^j du' \end{aligned} \tag{7}$$

where i (=0, 1, or 2) is equal to the order of the charge density of the border line $\overline{a_1 a_2}$, and j (=0, 1, or 2) is equal to that of the border line $\overline{b_1 b_2}$. It can be shown that analytic expressions exist for all these integral equations. Similar results can be obtained for F_v and τ_{e_b}.

In addition, the torques due to the spring forces should be considered. In Figure 1, suppose the attractive force on the point b_2 due to the spring edge $\overline{a_1 b_2}$ is equal to $f_a(d)$, where d is the length of $\overline{a_1 b_2}$ from (1), and B rotates with respect to the reference point e_b. The torque with respect to point e_b due to the attractive force $f_a(d)$ from spring $\overline{a_1 b_2}$, on point b_2, is equal to

$$T_{e_b} = f_a(d) \times (b_2 - e_b) \tag{8}$$

which is perpendicular to the plane containing A and B (i.e., the x-y plane).

Note that it is inappropriate to choose the center of shape as the reference point of a polygon because the forces and torques due to springs are computed by the spring lengths and each spring is not connected to the center of shape. Instead, the mean of the coordinates of the vertices connected by springs is selected as the reference point. Take Figure 1 for instance. A has vertices a_1 and a_3 connected by springs, and the midpoint of a_1 and a_3 is the reference point of A. B only has a single vertex b_2 connected by a spring, and hence b_2 is the reference point of B.

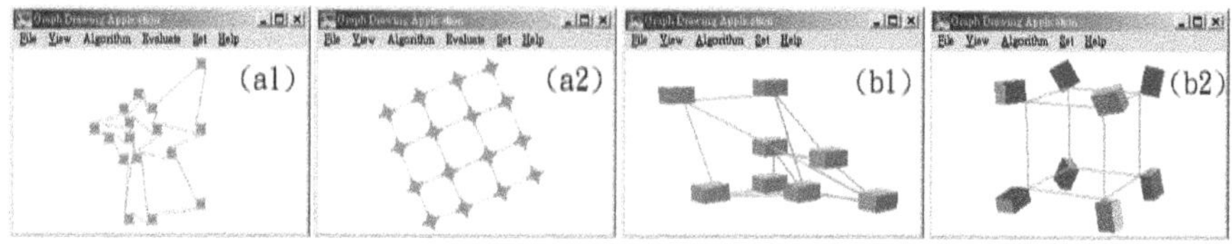

Fig. 2. (a) Mesh structure in 2D. (b) Cube structure in 3D. ($a1$, $b1$: initial drawings; $a2, b2$: final drawings)

In view of the above derivations, the force on a line segment due to a polygon can be derived. For example, in Figure 1, the total repulsive force on polygon B due to polygon A is equal to

$$F_u^r = \sum_{n \in \mathbf{B}(B)} F_u^n = \sum_{n \in \mathbf{B}(B)} \sum_{m \in \mathbf{B}(A)} F_u^{nm} \tag{9}$$

where n is a border line of B, m is a border line of A, $\mathbf{B}(A)$ is the set of border lines of A, and F_u^{mn} is the repulsive force on line segment n ($\overline{b_1 b_2}$) due to m along the u-axis from (7). On the other hand, the total attractive force on polygon B due to all of the springs is equal to

$$F_u^a = \sum_{p \in \mathbf{P}(B)} F_u^p = \sum_{p \in \mathbf{P}(B)} \sum_{e \in \mathbf{E}(G)} f_{au}(d(e)) \tag{10}$$

where p is a vertex of B, e is an edge of the graph, $\mathbf{E}(G)$ is the set of edges of the graph G, $d(e)$ is the length of spring edge e, $\mathbf{P}(B)$ is the set of vertices of B, and f_{au} is the spring force defined in (1) along the u-axis. In summary, the total force applied to polygon A due to the remaining polygons and springs is equal to the sum of the repulsive force (Eq. 9) and the attractive force (Eq. 10).

$$\mathbf{F}_u = \sum_{\mathbf{V}-\{B\}} (F_u^r + F_u^a) \tag{11}$$

Similar results can be obtained for the total force $\mathbf{F}_v$ along the v-axis and the total torque τ_{e_b} with respect to the reference point e_b.

Due to space limitations, the details of extending the potential field approach to 3D are omitted here.

3 Experimental Results

Based on the theory of potential fields detailed in the previous section, we develop a prototype system for drawing graphs with nonuniform nodes in 2D and 3D.

Figure 2 shows some drawing results in 2D and 3D. Figure 3 illustrates a strategy to handle the drawing of a clustered graph [5]. (See [6] for a related result.)

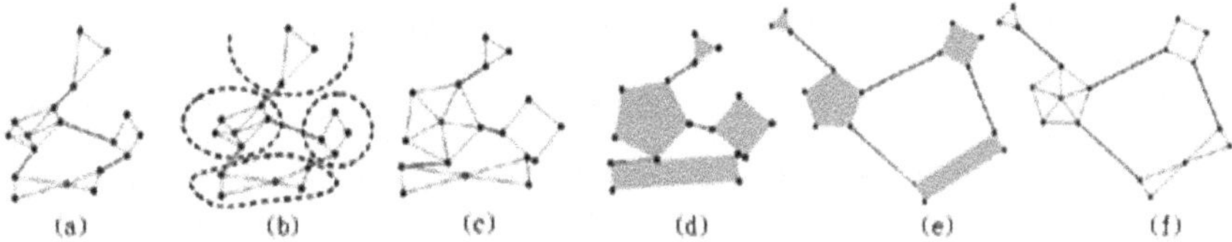

Fig. 3. Drawing a clustered graph. (a) A clustered graph. (b) Cluster partition. (c) Drawing individual clusters using the conventional force-directed method. (d) Convex hulls of clusters. (e) Drawing while treating clusters as nonuniform nodes. (f) Final drawing by plugging in details of clusters.

4 Conclusion

A potential-based approach, coupled with a force-directed method, has been proposed and implemented for drawing graphs with nodes of different sizes and shapes. A unique feature of our approach is that in the process of reaching equilibrium, the degree of inclination of a nonuniform can be adjusted while moving from one position to another. By doing so, the final drawing has a tendency to display a high degree of symmetry. An equally important aspect of our approach is that the formulas derived in our potential field model are analytically tractable, making our algorithm computationally efficient. Our experimental results look promising. It would also be of interest to compare our approach with existing algorithms designed specially for drawing graphs with nonuniform nodes, in spite of the scarcity of such algorithms.

References

1. D. Harel and Y. Koren. Drawing graphs with non-uniform vertices. *Proc. of Working Conference on Advanced Visual Interfaces (AVI'02)*, pp. 157–166, ACM Press, 2002.
2. P. Eades. A heuristic for graph drawing. *Congress Numerantium*, 42, pp. 149–160, 1984.
3. J.-H Chuang and N. Ahuja. An analytically tractable potential field model of free space and its application in obstacle avoidance. *IEEE Transactions on System, Man, and Cybernetics - Part B: Cybernetics*, Vol. 28, No. 5, pp. 729 –736, Oct, 1998.
4. J.-H Chuang. Potential-based modeling of three-dimensional workspace for obstacle avoidance. *IEEE Trans. Robotics and Automation*, Vol. 14, No. 5, pp. 778–785, Oct. 1998.
5. P. Eades and M. L. Huang. Navigating clustered graphs using force-directed methods. *Journal of Graph Algorithms and Applications*, Vol. 4, No. 3, pp. 157–181, 2000.
6. M. Huang, P. Eades and J. Wang. On-line animated visualization of huge graphs using a modified spring algorithm. *Journal of Visual Languages and Computing*, pp. 623–345, 1998.

Drawing Area-Proportional Venn and Euler Diagrams

Stirling Chow and Frank Ruskey

Department of Computer Science, University of Victoria, Canada
{schow,fruskey}@cs.uvic.ca

Abstract. We consider the problem of drawing Venn diagrams for which each region's area is proportional to some weight (e.g., population or percentage) assigned to that region. These area-proportional Venn diagrams have an enhanced ability over traditional Venn diagrams to visually convey information about data sets with interacting characteristics. We develop algorithms for drawing area-proportional Venn diagrams for any population distribution over two characteristics using circles and over three characteristics using rectangles and near-rectangular polygons; modifications of these algorithms are then presented for drawing the more general Euler diagrams. We present results concerning which population distributions can be drawn using specific shapes. A program to aid further investigation of area-proportional Venn diagrams is also described.

1 Introduction

In 1880, John Venn introduced a notation for representing logical propositions using intersecting Jordan curves [11]. These Venn diagrams were specialized instances of a more general notation for representing set relations which Leonard Euler developed in the 18th century [5].

Many people encounter the familiar 2-set and 3-set Venn diagrams shown in Fig. 1 at some point in their mathematical education; they are often used to introduce students to set theory. Informally, an n-set Venn diagram is comprised of n Jordan curves The curves divide the plane into 2^n connected sets of points called *regions* which uniquely represent all possible intersections of the (open) interiors and exteriors of the curves. For compactness, the intersection symbols ($\cap$) in Fig. 1 will be omitted from future diagrams.

The utility of Venn diagrams comes by mapping the curve interiors to sets in the problem domain. For example, if curves A, B, and C in Fig. 1(b) represent teenagers, smokers, and drinkers, respectively, then region $AB\overline{C}$ would represent teenage smokers who do not drink. If a region is mapped to an empty problem domain set, it is shaded as shown by regions $\overline{AB}C$ and $\overline{A}BC$ in Fig. 2(a).

Euler diagrams are similar to Venn diagrams except they omit shaded regions all-together. Figure 2(b) is an Euler diagram derived from Fig. 2(a) by removing regions $\overline{AB}C$ and $\overline{A}BC$. Euler diagrams are useful because they reduce some of

G. Liotta (Ed.): GD 2003, LNCS 2912, pp. 466–477, 2004.

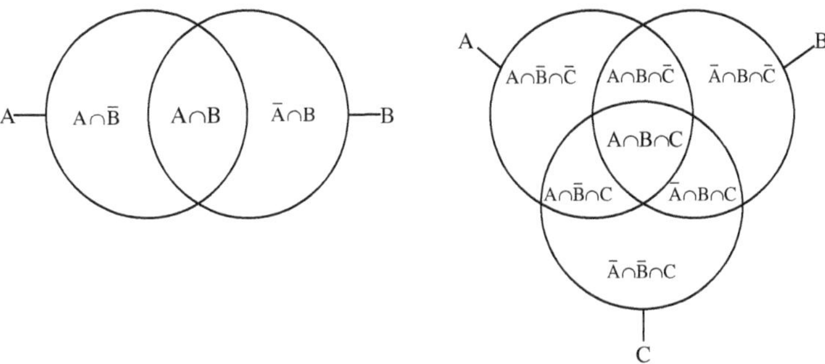

Fig. 1. Common 2-set and 3-set Venn diagrams

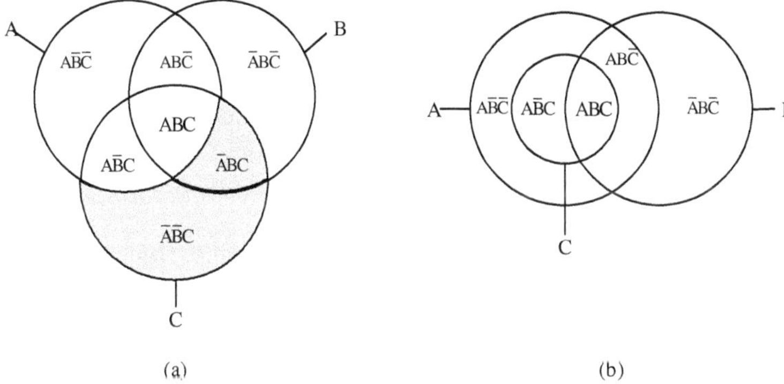

Fig. 2. A Venn diagram (a) with empty regions and the associated Euler diagram (b)

the clutter associated with Venn diagrams and make certain relationships more evident [8].

In addition to being useful as teaching aids, Venn and Euler diagrams have been used to solve problems in many domains including diagrammatic logic reasoning, finite set counting, and constraint modelling [7]. Irrespective of particular problems, Venn and Euler diagrams can also be used to effectively visualize data sets.

This paper focuses on techniques for drawing Venn and Euler diagrams with an improved ability to convey information about a data set. The work contained herein was inspired after one of the authors published a survey of Venn diagrams [10] and received numerous requests by readers for software to produce diagrams such as the one Fig. 3(a), but with the added constraint that each region's area should be representative of its associated data. Figure 3(b) is an example of such an area-proportional Venn diagram.

Although there are algorithms to produce Venn and Euler diagrams for any number of sets [11,4,6], the authors are unaware of any work on drawing algorithms that take area into account. As such, the forthcoming sections will formalize the concept of area-proportional Venn and Euler diagrams, provide

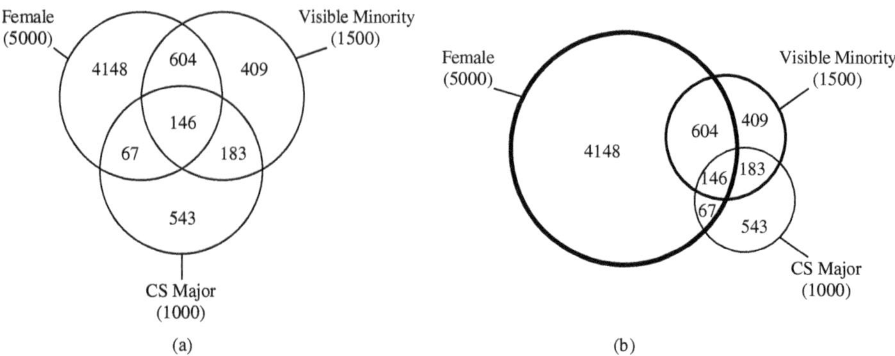

Fig. 3. A Venn diagram (a) representing a weighted data set and its corresponding area-proportional version (b)

algorithms for drawing 2-set and 3-set Venn and Euler diagrams, and discuss some of the aesthetic qualities desirable in area-proportional diagrams. It is the authors' hope that this paper will provide a solid foundation for future work in the area.

This work is related to graph drawing in the sense that a Venn or Euler diagram can be viewed as a graph embedded in the plane. In the graph drawing literature some attention has been paid to the areas of regions, mainly the aesthetic criteria that the area not get too small in relation to the area of other regions [1].

2 Formalisms

Before we proceed with describing the drawing algorithms, we must formally define area-proportional Venn and Euler diagrams. We begin by defining a division of the plane $\mathbb{R}^2$ into regions by a collection of Jordan curves.

Definition 1. *Let $C = \{C_1, C_2, \ldots, C_n\}$ be a collection of Jordan curves in $\mathbb{R}^2$ and let $int(C_i)$ and $ext(C_i)$ denote the points of $\mathbb{R}^2 - C_i$ that are interior and exterior, respectively, to C_i as defined by the Jordan Curve Theorem [9].*

The regions of C, denoted $R(C)$, are defined as follows:

$$R(C) = \{X_1 \cap X_2 \cap \cdots \cap X_n \mid X_i \in \{int(C_i), ext(C_i)\}\} . \qquad (1)$$

Note that $|R(C)| = 2^n$. Each region, $X_1 \cap X_2 \cap \cdots \cap X_n$, is labelled $L_1 L_2 \cdots L_n$ where $L_i = C_i$ if $X_i = int(C_i)$ and $L_i = \overline{C_i}$ if $X_i = ext(C_i)$. For example, region $int(A) \cap int(B) \cap ext(C)$ would be referred to as region $AB\overline{C}$.

Region $\overline{C_1}\,\overline{C_2} \ldots \overline{C_n}$ represents the part of the universe exterior to all curves and is designated U'.

Venn diagrams are a special instance of a collection of curves whose regions satisfy a non-empty and connected condition.

Definition 2. *A collection of curves, C, is a* Venn diagram *if for all $r \in R(C)$, r is a non-empty set of connected points.*

For example, Fig. 4 shows two collections of curves that are not Venn diagrams; (a) fails because region $\overline{A}B\overline{C}$ is empty and (b) fails because region $A\overline{B}$ is disconnected.

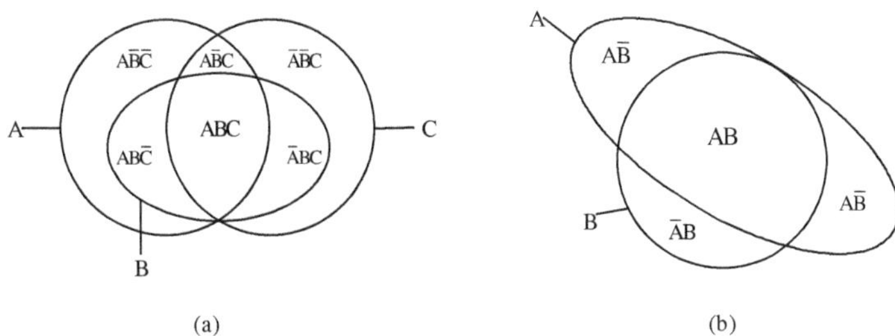

Fig. 4. Two collections of curves that are not Venn diagrams; however, (a) is an Euler diagram

By relaxing the non-empty constraint on Venn diagrams, we yield a definition for Euler diagrams. Note that because certain regions can be empty, an Euler diagram may have non-intersecting curves.

Definition 3. *A collection of curves, C, is an* Euler diagram *if for all $r \in R(C)$, the region r is connected (an empty region is vacuously connected).*

For example, returning to Fig. 4, (a) is an Euler diagram, but (b) remains neither a Venn nor an Euler diagram.

The aforementioned definitions of Venn and Euler diagrams are the broadest possible. A number of subclasses of Venn and Euler diagrams have been studied (interested readers should consult [10]), and some researchers define Venn and Euler diagrams at the exclusion of certain subclasses. Although we consider the most general cases of Venn and Euler diagrams, we will discuss the implications of certain characteristics on the diagrams' visual effectiveness. The two characteristics that we will consider are simplicity and finiteness.

Definition 4. *A Venn or Euler diagram is* simple *if no more than two curves intersect at a common point; otherwise, it is termed* non-simple.

For example, the Venn diagram in Fig. 1(b) is simple, but the Euler diagram in Fig. 4(a) is non-simple because of the common intersection of curves A, B, and C at the bottom of the diagram.

Definition 5. *A Venn or Euler diagram is* finite *if its curves intersect at finitely many points; otherwise, it is termed* infinite.

All the Venn and Euler diagrams discussed up to this point have been finite. In Section 4 we will encounter infinite Venn diagrams.

Now that we have formally defined Venn and Euler diagrams, we can proceed to formalize the notion of area-proportionality. Area-proportionality only has meaning once problem domain values have been mapped to the regions of a Venn or Euler diagram. The problem domain values may be related to the cardinality of the sets represented by each region as shown in Fig. 3 or they may be other values such as the probabilities shown in Fig. 5. Regardless of their meaning, the values can be treated simply as arbitrary weights assigned to the regions of a Venn or Euler diagram. The idea behind area-proportionality is that a region's area (that is, the smallest area of the plane that encloses the region's points), should be directly proportional to the weight assigned to the region. For example, if region A has a weight twice that of region B then region A should have an area twice that of region B.

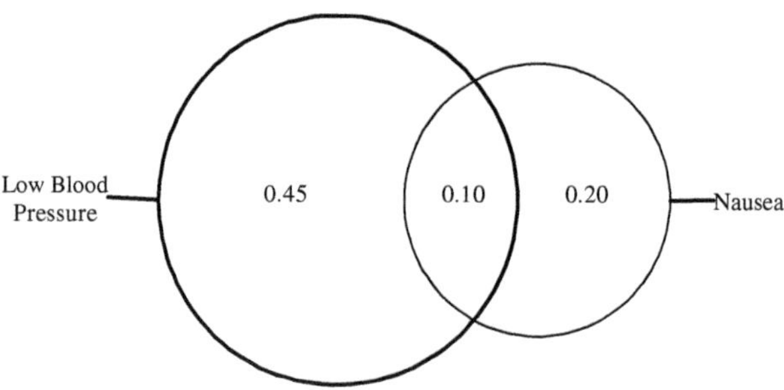

Fig. 5. A area-proportional Venn diagram showing the probabilities of certain complications for a trial drug

Definition 6. *Given a Venn or Euler diagram, C, and a non-negative function $\omega : R(C) \to \mathbb{R}$, a diagram C is an* area-proportional *with respect to ω if*

$$\frac{Area(r)}{\sum_{r' \in R(C)-U'} Area(r')} = \frac{\omega(r)}{\sum_{r' \in R(C)-U'} \omega(r')} \quad \textit{for all } r \in R(C) - U'.$$

Note that since the universe is represented by the unbounded plane, it does not factor into determining the proportions of the diagram. Also, by necessity, a Venn diagram can only be area-proportional to an ω with positive values.

3 2-Set Area-Proportional Venn and Euler Diagrams

Figure 5 shows a 2-set area-proportional Venn diagram with curves that are circles; to compute such a diagram, $G = \{A, B\}$, for a weight function ω, we first calculate the radii of the circles so that their areas equal the sum of their constituent regions' weights:

$$r_1 = \sqrt{\frac{\omega(A\overline{B}) + \omega(AB)}{\pi}}, \quad r_2 = \sqrt{\frac{\omega(\overline{A}B) + \omega(AB)}{\pi}}. \tag{2}$$

We next choose a canonical orientation for the diagram by centering A in the cartesian plane $\mathbb{R}^2$ at $(0, 0)$ and centering B at $(d, 0)$ where d is chosen so that the area of overlap between A and B is $\omega(AB)$ (see Fig. 6). Note that since πr_1^2 and πr_2^2 are at least $\omega(AB)$, there is always a d that yields the necessary overlap.

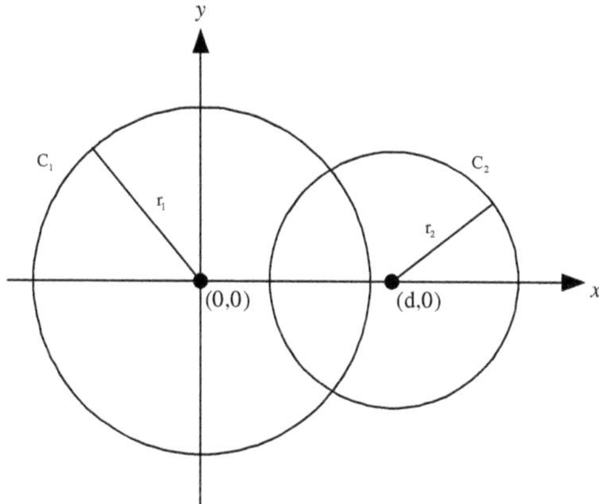

Fig. 6. Canonical form of two-circle diagram

In order to compute d, we first need a formula, σ, for the area of overlap between the two circles; this can be derived by applying Euclidean geometry and trigonometry to yield the following result:

$$\sigma(d) = \frac{1}{2}r_1^2\left(\alpha - \sin(\alpha)\right) + \frac{1}{2}r_2^2\left(\beta - \sin(\beta)\right) \tag{3}$$

where

$$\alpha = 2\arccos\left(\frac{d^2 + r_1^2 - r_2^2}{2r_1 d}\right) \quad \beta = 2\arccos\left(\frac{d^2 + r_2^2 - r_1^2}{2r_2 d}\right)$$

are in radians.

Because σ is not invertible, in order to solve $\sigma(d) = \omega(AB)$ we need to apply numerical methods. Note that σ is minimal at $d = r_1 + r_2$, maximal at $d = r_1 - r_2$, and monotonically non-decreasing within this range regardless of the relative sizes of A and B (although if $r_1 < r_2$, σ achieves a maximum at $d = r_2 - r_1$). Because of σ's monotonicity, we can apply a simple bisection algorithm [2] in the range $[r_1 - r_2, r_1 + r_2]$ to efficiently determine d.

This algorithm produces Euler diagrams without need for any changes. If regions $A\overline{B}$ and $\overline{A}B$ both have weight 0, the algorithm will produce two equal-sized circles centered at $(0, 0)$; if only one of these regions has weight 0, the algorithm will produce a diagram where the circles intersect at a single point and one circle lies within the other. Lastly, if region AB has weight 0, the algorithm will produce a diagram where the closed interiors of the circles intersect a single point.

The values of r_1, r_2, and d fully specify a $1:1$ scale diagram with respect to ω. In order to produce a diagram for an output device such as a printer, a standard scaling transformation can be applied. This algorithm provides a constructive proof for the following theorem.

Theorem 1. *A 2-set area-proportional Venn or Euler diagram, G, whose curves are circles exists with respect to any non-negative weight function $\omega : R(G) \to \mathbb{R}$.*

4 3-Set Area-Proportional Venn and Euler Diagrams

It is generally not possible to extend the results of Thm. 1 to three circles; this can be seen by considering Fig. 7. Given a weight function ω, the amount of overlap between circles A and B in Fig. 7(b) is proportional to $\omega(AB\overline{C}) + \omega(ABC)$; similarly, for B and C it is proportional to $\omega(ABC) + \omega(\overline{A}BC)$. Using the algorithm from Sec. 3, we can compute the distance between the centers of A and B so that the necessary overlap is achieved; if all the regions are non-empty (i.e., a Venn diagram), then there is only one solution. A similar computation is performed for B and C. At this point, we don't know the relative orientation of A and C as shown in Fig. 7(a). To determine the overlap of A and C while maintaining the overlap that has already been established between A and B, we revolve A in a fixed-distance orbit about B until region ABC has the proper area as shown in Fig. 7(b); this can be done using polar coordinates and another bisection. At the end of this process, regions $\overline{A}B\overline{C}$, $AB\overline{C}$, ABC, and $\overline{A}BC$ have the right area, but regions $A\overline{BC}$, $A\overline{B}C$, and $\overline{AB}C$ may be incorrect. There are no more degrees of freedom in positioning the circles, so in general, there are weight distributions that cannot be represented using three circles.

In order to increase the number of degrees of freedom, more complex curves are necessary. An alternative to circles are ellipses, but the area calculations become unwieldy, and it is unclear how to adjust the ellipse parameters to achieve the desired areas. Because of their amenability to area calculations and discrete manipulation, we will consider how to use axis-aligned rectangles to produce orthogonal drawings [1] of Venn and Euler diagrams. Following is an example to describe an algorithm for producing a 3-set finite Venn diagram, $G = \{A, B, C\}$, given the weight function ω defined in Tbl. 1.

The idea is to build the diagram region-by-region using rectangles whenever possible. We use W_r and H_r to refer to the width and height, respectively, of region r; if r is not rectangular, then W_r and H_r refer to the longest width and

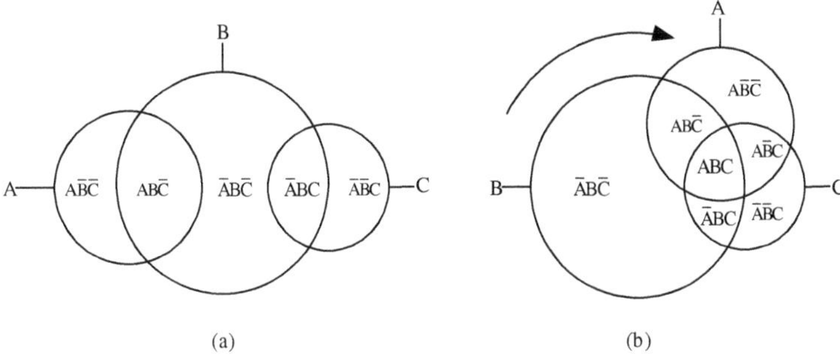

Fig. 7. A possible algorithm for 3-set area-proportional Venn diagrams

Table 1. Sample weight function

r	$\omega(r)$
ABC	2
$AB\overline{C}$	6
$A\overline{B}C$	6
$\overline{A}BC$	1
$A\overline{BC}$	3.5
$\overline{AB}C$	7
$\overline{A}B\overline{C}$	9

height. The first step is to choose a value for W_{ABC} and place the resulting rectangle on the plane. In this example we choose $W_{ABC} = 2$ and place a 2×1 rectangle on the plane.

Next, we place a $W_{ABC} \times \frac{\omega(A\overline{B}C)}{W_{ABC}}$ rectangle above ABC for $A\overline{B}C$, and similarly a $\frac{\omega(\overline{A}BC)}{H_{ABC}} \times H_{ABC}$ rectangle to the right of ABC for $\overline{A}BC$.

For $AB\overline{C}$ we need to decide how much of it to place below ABC and how much of it to place to the left of ABC. We let d be the vertical distance of $AB\overline{C}$ below ABC, and we can choose any value for this as long as $W_{ABC} \cdot d < \omega(AB\overline{C})$. Once d is chosen, $AB\overline{C}$ will be the union of a $\frac{\omega(AB\overline{C}) - W_{ABC} \cdot d}{H_{ABC} + d} \times (H_{ABC} + d)$ rectangle and a $W_{ABC} \times d$ rectangle. In this example we choose $d = 1$ and the resulting diagram is shown in Fig. 8(a).

Now that the core of the diagram has been established, the remaining regions are filled in as best as possible to create rectangular curves. Referring to Fig. 8(b), if $\omega(\overline{AB}C) > W_{\overline{A}BC} \cdot H_{A\overline{B}C}$, then $\overline{AB}C$ can fill the corner between $\overline{A}BC$ and $A\overline{B}C$ before overflowing to the right of $\overline{A}BC$. Similarly, if $\omega(\overline{A}B\overline{C}) > d \cdot W_{\overline{A}BC}$, then $\overline{A}B\overline{C}$ can fill the corner between $AB\overline{C}$ and $\overline{A}BC$ before overflowing below and to the left of $AB\overline{C}$.

In the case of $A\overline{BC}$, $\omega(A\overline{BC}) \leq W_{AB\overline{C}} \cdot H_{A\overline{B}C}$ so the corner between $AB\overline{C}$ and $A\overline{B}C$ cannot be filled to create a rectangular curve A. As a result, we

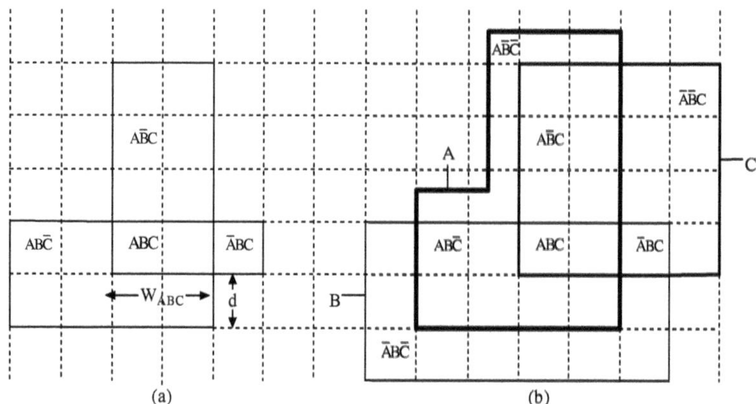

Fig. 8. Example of 3-set infinite area-proportional Venn diagram algorithm

evenly distribute $A\overline{B}\overline{C}$ around the top of $AB\overline{C}$ and top left corner of $A\overline{B}C$. An analogous rule would hold for $\overline{A}\overline{B}C$ and $\overline{A}B\overline{C}$ where they not sufficiently large to fill there associated corners.

There are a few things to note about this algorithm. First, if a curve cannot be made rectangular, then its shape is that of a rectangle with a corner cut out (see curve A in Fig. 8(b)); we call these shapes near-rectangular. Second, the choice to overflow $\overline{A}\overline{B}C$ to the right of $\overline{A}BC$ is arbitrary. $\overline{A}\overline{B}C$ could overflow on top of $A\overline{B}C$ instead which would have caused the subsequent regions to "rotate" counterclockwise 90 degrees, and although this results in a different diagram, the criteria concerning whether or not the resultant curves are rectangular does not change. Third, once W_{ABC} and d are chosen, the rest of the diagram is determined (according to the algorithm). Fourth, the choice of d can effect whether or not the resulting curves are rectangular. Larger values of d decrease the width of $AB\overline{C}$ to the left of ABC which might allow $A\overline{B}\overline{C}$ to fill its corner and create a rectangular curve A; however, there is a trade-off since larger values of d create a larger corner for $\overline{A}B\overline{C}$ to fill and may result in curve B becoming near-rectangular.

The last point is important because it suggests that the choice of d is critical to achieving a diagram comprised of rectangles. The following theorem is a direct result of this observation.

Theorem 2. *A simple 3-set area-proportional Venn diagram, $G = \{A, B, C\}$, whose curves are axis-aligned rectangles exists with respect to a positive weight function $\omega : R(G) \to \mathbb{R}$ if, and only if, there exists a permutation, π, of $\{A, B, C\}$ such that $\Omega(r) = \omega(\pi(r))$ (that is, Ω permutes the regions' labels) and the following conditions hold:*

$$max\left(0, \frac{\Omega(AB\overline{C}) \cdot \Omega(A\overline{B}C) - \Omega(A\overline{B}\overline{C}) \cdot \Omega(ABC)}{\Omega(A\overline{B}\overline{C}) + \Omega(A\overline{B}C)}\right)$$

$$< min\left(\Omega(AB\overline{C}), \frac{\Omega(\overline{A}B\overline{C})\cdot\Omega(ABC)}{\Omega(\overline{A}BC)}\right)$$

and

$$\frac{\Omega(\overline{A}BC)\cdot\Omega(A\overline{B}C)}{\Omega(ABC)} < \Omega(\overline{A}\overline{B}C) .$$

Proof. *For the sake of space, we provide only a sketch of the proof. For sufficiency, we establish the inequalities that must be satisfied in order for all curves to be rectangular; that is, we require the corners of Fig. 8(a) to be small enough to permit filling by the subsequent regions. By simultaneously solving these inequalities, we derive the conditions of the theorem. The addition of Ω is to allow for all possible orderings of the curves around the diagram.*

For necessity, we consider a three rectangle Venn diagram and remove one of the curves. The resulting diagram has two intersecting curves whose shared region is itself a rectangle. In order to be a Venn diagram, the third curve must bisect each region exactly once which results in a core that is isomorphic to Fig. 8(a) under rotation and reflection. The result then follows.

Even if a rectangular Venn diagram cannot be produced, the resulting curves will at least be near-rectangular. In addition, the algorithm can also be applied without change for Euler diagrams. As long as $\omega(ABC) > 0$, the algorithm functions correctly; however, it is not optimal in the sense that some infinite Euler diagrams are produced that could be finite. The case when $\omega(ABC) = 0$, remains an open problem.

5 Conclusion

In order to experiment with various drawing algorithms and to better understand some of the user interface issues concerning area-proportional diagrams, we have developed a Java application called *DrawVenn* that implements the 2-set and 3-set Venn and Euler algorithms described in this paper. As shown in Fig. 9, a DrawVenn user can dynamically adjust the weight function and view the updated diagram. There is also a control for adjusting algorithm-specific parameters such as the values of W_{ABC} and d in the 3-set algorithm. Within DrawVenn, the curves are filled with user-selectable transparent colors that are alpha-blended to provide an intuitive cue as to which curves are intersecting in a given region. DrawVenn can also output color bitmaps as well as scalable vector formats. In the case of vector formats, DrawVenn produces monochrome drawings and varies the thickness of the curves to indicate their relative sizes.

We are also interested in understanding which aesthetic characteristics of an area-proportional diagram affect its ability to convey information. A person viewing an area-proportional Venn or Euler diagram needs to be able to determine which curves enclose a region (i.e., what the region represents) and how the size of a region compares to other regions.

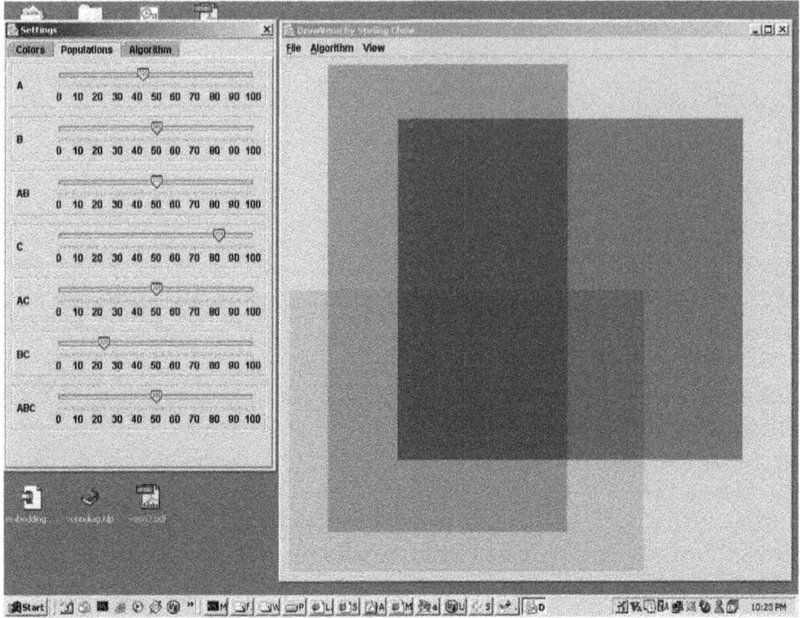

Fig. 9. Screenshot of DrawVenn showing sliders for dynamic adjustment of populations

In order to determine a region's enclosing curves, the curves must be distinguishable and easy to trace (i.e., follow with the eye). On the surface, it would seem that the curves of an infinite Venn or Euler diagram are less distinguishable than the curves of a finite one. Similarly, it would seem that the curves of a non-simple Venn or Euler diagram are more difficult to trace than the curves of a simple diagram since there are more options for where a curve exits a vertex. According to the gestalt "good continuation" principle [3], smooth curves should also be more traceable than ones whose direction changes sharply, thus providing some motivation for developing a 3-set algorithm that uses ellipses rather than rectangles.

In order to compare regions, their shapes should be similar. For example, it is easier to compare the relative size of a circle to another circle than it is to a triangle. The orthogonal diagrams hold an advantage in this area since their regions are largely rectangular; however, the areas of rectangles can be difficult to compare if their aspect ratios are vastly different. The measure of a diagram's regional uniformity may be a good indicator of its effectiveness.

Area-proportional Venn and Euler diagrams have an advantage over traditional diagrams because they leverage both an individual's perceptual and cognitive abilities. We hope that the algorithms described in this paper provide a first step in the direction of understanding how best to utilize area-proportional diagrams, and we hope others will share our interest in this area.

References

[1] G.D. Battista, P. Eades, R. Tomassia, and I.G. Tollis. *Graph Drawing: Algorithms for the Visualization of Graphs.* Prentice-Hall, 1999.
[2] Richard L. Burden and J. Douglas Faires. *Numerical Analysis: 4th Edition.* PWS Publishing Co., 1988.
[3] Stuart K. Card, Jock D. Mackinlay, and Ben Shneiderman. *Readings in Information Visualization: Using Vision to Think.* Morgan Kaufmann Publishers, 1999.
[4] Anthony W.F. Edwards. Venn diagrams for many sets. *New Scientist*, 7:51–56, January 1989.
[5] Leonard Euler. *Lettres a Une Princesse d'Allemagne*, volume 2. 1761. Letters no. 102–108.
[6] Jean Flower and John Howse. Generating Euler diagrams. In *Proceedings of Diagrams 2002*, pages 61–75. Springer-Verlag, April 2002.
[7] Joseph (Yossi) Gil, Stuart Kent, and John Howse. Formalizing spider diagrams. In *Proceedings of the IEEE Symposium on Visual Languages*, pages 130–137. IEEE Computer Society Press, September 1999.
[8] Joseph (Yossi) Gil, Stuart Kent, John Howse, and John Taylor. Projections in Venn-Euler diagrams. In *Proceedings of the IEEE Symposium on Visual Languages*, pages 119–126. IEEE Computer Society Press, September 2000.
[9] Konrad Knopp. *Theory of Functions, Parts I and II, Two Volumes Bound as One*, volume 1. Dover, 1996.
[10] Frank Ruskey. A survey of Venn diagrams. *Electronic Journal of Combinatorics*, 4, 1997 (update 2001). DS#5.
[11] John Venn. On the diagrammatic and mechanical representation of propositions and reasonings. *The London, Edinburgh, and Dublin Philosophical Magazine and Journal of Science*, 9:1–18, 1880.

Optimal Pants Decompositions and Shortest Homotopic Cycles on an Orientable Surface*

Éric Colin de Verdière[1] and Francis Lazarus[2]

[1] Laboratoire d'informatique de l'École normale supérieure, UMR 8548 (CNRS), Paris, France. Eric.Colin.de.Verdiere@ens.fr
[2] Laboratoire IRCOM-SIC, UMR 6615 (CNRS), Poitiers, France. Francis.Lazarus@sic.univ-poitiers.fr

Abstract. A *pants decomposition* of a compact orientable surface $\mathcal{M}$ is a set of disjoint simple cycles which cuts $\mathcal{M}$ into *pairs of pants*, i.e., spheres with three boundaries. Assuming $\mathcal{M}$ is a polyhedral surface, with weighted vertex-edge graph G, we consider combinatorial pants decompositions: the cycles are closed walks in G that may overlap but do not cross.

We give an algorithm which, given a pants decomposition, computes a homotopic pants decomposition in which each cycle is a shortest cycle in its homotopy class. In particular, the resulting decomposition is *optimal* (as short as possible among all homotopic pants decompositions), and any optimal pants decomposition is made of shortest homotopic cycles. Our algorithm is polynomial in the complexity of the input and in the longest-to-shortest edge ratio of G. The same algorithm can be applied, given a simple cycle C, to compute a shortest cycle homotopic to C which is itself simple.

1 Introduction

Let $\mathcal{M}$ be a connected, compact, orientable surface. A *pants decomposition* of $\mathcal{M}$ is a set of disjoint simple cycles in $\mathcal{M}$ which cuts $\mathcal{M}$ into *pairs of pants*, i.e., spheres with three boundaries. See Fig. 1A.

We shall consider pants decompositions on a polyhedral surface (a surface obtained by assembling simple polygons). In our combinatorial setting, the cycles are closed walks on the weighted vertex-edge graph G of $\mathcal{M}$; the cycles may share edges and vertices of G, provided that they can be spread apart with a thin space so that they become simple and disjoint.

We describe a conceptually simple, iterative scheme which takes a given pants decomposition and outputs a shorter homotopically equivalent pants decomposition. We prove that, at the end of the process, each cycle is a shortest cycle in its homotopy class. In particular, the resulting decomposition is *optimal* in

* A preliminary short version of this paper appeared in the Abstracts of the 19th European Workshop on Computational Geometry, 2003.

G. Liotta (Ed.): GD 2003, LNCS 2912, pp. 478–490, 2004.

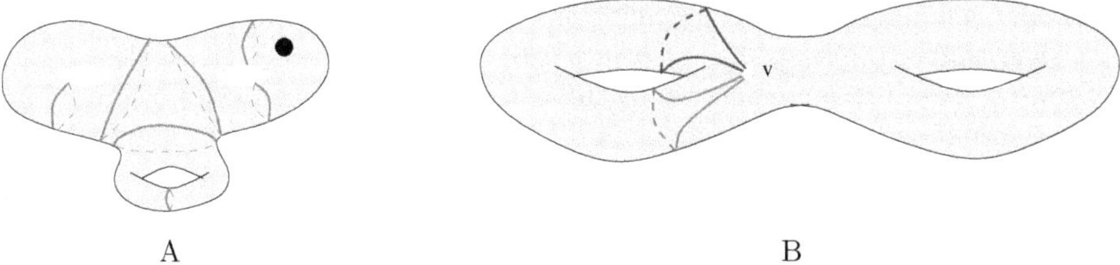

Fig. 1. A: A pants decomposition of a genus three surface with one boundary. B: The two cycles on this double-torus are freely homotopic, though non-homotopic when considered as loops with endpoint v.

the sense that it is as short as possible among all homotopic pants decompositions, and any optimal pants decomposition is made of shortest homotopic cycles. Furthermore, this scheme can be implemented, leading to an algorithm which is polynomial in the complexity of the surface and of the input pants decomposition, and in the longest-to-shortest edge ratio of G.

Let γ be a non-contractible simple cycle on $\mathcal{M}$, and let Γ be the set of all shortest cycles homotopic to γ. We can compute an element of Γ which is simple: the idea is to extend γ into a pants decomposition of $\mathcal{M}$; after optimization, this pants decomposition contains such a cycle. Even the existence of a simple cycle in Γ is non-obvious, and the fact that this optimization problem has polynomial complexity was previously unknown.

The problem of shortening a pants decomposition of a combinatorial surface was raised in the conclusion of [4]; to our knowledge, we present the first algorithm for this purpose. Concerning the optimization of a single cycle, our result extends [6] to more general surfaces in the case of simple cycles. The present work is also a natural extension of our former paper [2] where we treat the case of optimal simple loops in a given class of homotopy with fixed basepoint.

This paper is organized as follows. In Sect. 2, we review elementary topological notions, and present the framework and our main theorem. Its proof is given in the next three sections. Finally, we discuss the computational issues and give the complexity of our algorithm.

2 Framework and Result

2.1 Homotopy and Pants Decompositions

We begin with some useful definitions.

Let $\mathcal{M}$ be a connected, compact, orientable surface, possibly with boundary. A *path* is a continuous mapping $p : [0,1] \to \mathcal{M}$; its *endpoints* are $p(0)$ and $p(1)$. A *closed path*, or *loop*, is a path whose endpoints coincide. A *cycle* is a continuous mapping $\gamma : \mathbb{R} \to \mathcal{M}$, such that $\gamma(x) = \gamma(x+1)$ for all $x \in \mathbb{R}$. A path is *simple* if it is one-to-one; a cycle is *simple* if its restriction to $[0,1)$ is one-to-one.

Two paths p and q, both with endpoints a and b, are *homotopic* if there is a continuous family of paths with endpoints a and b which joins p and q. More formally, a *homotopy* between p and q is a continuous mapping $h : [0,1] \times [0,1] \to \mathcal{M}$ such that $h(0,.) = p$, $h(1,.) = q$, $h(.,0) = a$, and $h(.,1) = b$. A closed path is *contractible* if it is homotopic to the constant path. Two cycles γ and γ' are *homotopic* if there is a continuous family of cycles joining γ to γ'. Equivalently, if p and p' denote the restrictions of γ and γ' to $[0,1]$, there exists a path β joining $p(0)$ to $p'(0)$ such that the loop $\beta^{-1}.p.\beta.p'^{-1}$ is contractible ("." denotes paths concatenation).

Homotopy of cycles (also called free homotopy) and homotopy of loops (also called homotopy with basepoint) are two different equivalence relations (Fig. 1B).

A *pants decomposition* of $\mathcal{M}$ is an ordered set of simple, pairwise disjoint cycles which split $\mathcal{M}$ into pairs of pants (see [5]). Every compact orientable surface, except the sphere, disk, cylinder, and torus, admits a pants decomposition, obtained for example by cutting the surface iteratively along an essential cycle (a simple cycle which does not bound a disk nor a cylinder). Although pants decompositions do not exist for the torus and the cylinder, this paper applies to these surfaces as well (with minor changes) if we allow a pants decomposition to decompose the surface into pairs of pants *and/or cylinders*. If $\mathcal{M}$ has genus g and b boundary cycles, a pants decomposition is made of $3g + b - 3$ cycles.

We can augment any pants decomposition, s, of $\mathcal{M}$ to form a *doubled pants decomposition*. Just add to s a copy of each of its cycles and a copy of each of the boundaries of $\mathcal{M}$, slightly translated, in the same homotopy class, such that s is still a set of pairwise disjoint simple cycles. A doubled pants decomposition $s = (s_1, \ldots, s_N)$ is thus made of $N = 6g + 3b - 6$ cycles. A cycle of s and its translated copy, or a boundary of $\mathcal{M}$ and its translated copy, are called *twins*. For a cycle s_j in s, the closure of the component of $\mathcal{M} \setminus \{s \setminus s_j\}$ that contains s_j is a pair of pants.

2.2 Length of Cycles

$\mathcal{M}$ is assumed to be a polyhedral surface, whose edges have positive weights. Let G be the (weighted) vertex-edge graph of $\mathcal{M}$, and G^* be its dual graph embedded into $\mathcal{M}$.[1] We are interested in sets of piecewise linear (PL) curves (paths and cycles) drawn on $\mathcal{M}$ which are *regular* with respect to G^*. More precisely, the set of (self-)intersection points between the curves (resp. between the curves and G^*) is finite and, at such points, exactly two curve parts (resp. exactly one curve part and one edge of G^*) meet and actually cross. *Regularity is always assumed throughout this paper, although omitted in all statements.* If a curve c crosses the edges $e_1^*, \ldots, e_k^*$ of G^*, its *length* $|c|$ is defined to be the sum of the weights of $e_1, \ldots, e_k$, counting multiplicities.

Any set of simple, pairwise disjoint, PL cycles on $\mathcal{M}$ can be retracted onto closed walks on G without changing their homotopy classes, see Fig. 2AB; and

[1] This means that there is a vertex of G^* in each face of G and an edge of G^* crossing each interior edge of G. Furthermore, for each edge of G on the boundary of $\mathcal{M}$, we put a vertex of G^* on this edge and link it with an edge of G^* to the vertex of G^* in the incident face of G.

the length of a cycle is the length of the corresponding closed walk. The resulting walks can fail to be simple or disjoint as they can travel several times through a same vertex or edge of G; however, it is always possible to perturb them to get disjoint, simple cycles. This observation motivates the definition of length of a cycle: in this paper, we are interested in combinatorial sets of cycles in G. From an algorithmic point of view, it will be sufficient to work with cycles stored as closed walks on G, with the additional information, if several edges of these walks go along a same edge e of G, of their ordering, from left to right, along e.

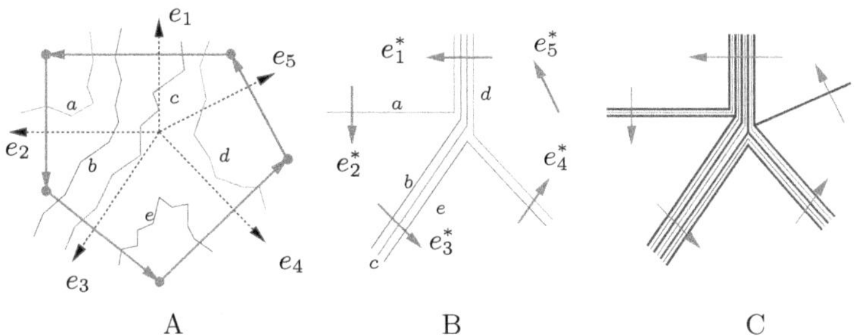

Fig. 2. A, B: A retraction of the set of simple, pairwise disjoint cycles (a, b, c, d, e) onto G, in the neighborhood of a vertex whose incident edges are $e_1, \ldots, e_5$. C: The construction of the graph $G(\mathcal{P}_j)$, represented in bold lines, for the computation of an Elementary Step (described in Sect. 6).

2.3 Our Result

Definition 1. *Let s be a doubled pants decomposition of $\mathcal{M}$. An* Elementary Step *$f_j(s)$ consists in replacing the jth cycle s_j by a shortest simple homotopic cycle in the pair of pants of $\mathcal{M} \setminus (s \setminus s_j)$ containing s_j. A* Main Step *$f(s)$ is the application of $f = f_N \circ f_{N-1} \circ \ldots \circ f_2 \circ f_1$ to s. These operations transform a doubled pants decomposition into another one, keeping the homotopy class of the decomposition.*

Here is our main theorem:

Theorem 2. *Let s^0 be a doubled pants decomposition of $\mathcal{M}$, and let $s^{n+1} = f(s^n)$. For some $m \in \mathbb{N}$, s^m and s^{m+1} have the same length and, in this situation, s^m is a doubled pants decomposition homotopic to s^0 made of simple cycles which are individually as short as possible among all cycles in their (free) homotopy class. In particular, s^m is an optimal doubled pants decomposition of $\mathcal{M}$, and contains an optimal pants decomposition.*

Since any non-contractible simple cycle can be extended to a doubled pants decomposition of $\mathcal{M}$, we obtain:

Corollary 3. *Let γ be a non-contractible simple cycle in $\mathcal{M}$. There exists a* simple *cycle γ' homotopic to γ which is as short as possible among all cycles homotopic to γ.*

3 Crossing Words

In this section, we will introduce the main ingredient of this paper: the crossing word between a set of disjoint, simple paths or cycles, and a given path or cycle.

3.1 Universal Cover and Lifts

We refer the reader to any textbook in algebraic topology (e.g. [7]) for the details.

The *universal cover* of $\mathcal{M}$ is a simply connected surface, $\tilde{\mathcal{M}}$, (i.e., each closed path is contractible) together with a continuous *projection* π from $\tilde{\mathcal{M}}$ onto $\mathcal{M}$ satisfying: each point x of $\mathcal{M}$ has an open, arcwise connected neighborhood U so that $\pi^{-1}(U)$ is a union of disjoint open sets $(U_i)_{i \in I}$ and $\pi|_{U_i} : U_i \to U$ is a homeomorphism. It is known that the universal cover of a surface is unique up to isomorphism. A *translation* τ of $\tilde{\mathcal{M}}$ is a projection-preserving homeomorphism: $\pi \circ \tau = \pi$. A *lift* of a path p is a path $\tilde{p}$ on $\tilde{\mathcal{M}}$ such that $\pi \circ \tilde{p} = p$. Analogously, a *lift* of a cycle γ is a continuous mapping $\tilde{\gamma} : \mathbb{R} \to \tilde{\mathcal{M}}$ such that $\pi \circ \tilde{\gamma} = \gamma$. The main properties of $\tilde{\mathcal{M}}$ used in this paper are:

- the *lift property*: let p be a path in $\mathcal{M}$ with source point y; let $x \in \tilde{\mathcal{M}}$ be such that $\pi(x) = y$. Then there is a unique path $\tilde{p}$ in $\tilde{\mathcal{M}}$, starting at x, such that $\pi \circ \tilde{p} = p$;
- the *homotopy property*: two paths p_1 and p_2 with the same endpoints are homotopic in $\mathcal{M}$ if and only if they have two lifts $\tilde{p}_1$ and $\tilde{p}_2$ with the same endpoints in $\tilde{\mathcal{M}}$;
- the *intersection property*: a path p in $\mathcal{M}$ self-intersects if and only if either a lift of p self-intersects, or two lifts of p intersect.

If $\tilde{\gamma}$ is a lift of a cycle γ, and $k \in \mathbb{Z}$, note that $\tilde{\gamma}'(.) := \tilde{\gamma}(k + .)$ and $\tilde{\gamma}$ are identical as points sets. We therefore define the *geometric lifts* of a cycle γ to be all lifts of γ, identifying two lifts $\tilde{\gamma}$ and $\tilde{\gamma}'$ whenever there exists $k \in \mathbb{Z}$ such that $\tilde{\gamma}'(.) = \tilde{\gamma}(k + .)$. If γ is simple, the geometric lifts of γ correspond precisely to the connected components of $\pi^{-1}(\gamma)$. By definition, for paths, the sets of lifts and of geometric lifts coincide.

A *lifted set* $\mathcal{C}$ is a set of simple, pairwise disjoint curves on $\mathcal{M}$ which are:

- either non-contractible cycles in the interior of $\mathcal{M}$,
- or paths in $\mathcal{M}$, whose intersections with the boundary of $\mathcal{M}$ are precisely their endpoints,

together with the data, for each curve c in $\mathcal{C}$, of an enumeration $(c^\alpha)^{\alpha \in \mathbb{N}}$ of its geometric lifts in $\tilde{\mathcal{M}}$.

In the rest of this paper, $\mathcal{C}$ is a lifted set. The proof of the following lemma is a consequence of the Jordan–Schönflies theorem (see [1, p. 417]).

Lemma 4. *Each geometric lift of a curve in $\mathcal{C}$ separates $\tilde{\mathcal{M}}$ in two connected components.*

3.2 Crossing Words for Paths

We consider words on the alphabet made of letters of the form c^α and $\bar{c}^\alpha$, where $c \in \mathcal{C}$ and $\alpha \in \mathbb{N}$. Let w be a word. If w contains a subword $c^\alpha \bar{c}^\alpha$ or $\bar{c}^\alpha c^\alpha$, let w' be the word resulting from removing this subword from w; we say that w' is deduced from w by an *elementary c-reduction*. An *elementary reduction* is an elementary c-reduction for some c. A word w is *(c-)irreducible* if it can be applied no elementary (c-)reduction, it *(c-)reduces* to w' if w' can be obtained from w by successive elementary (c-)reductions.

Let $\tilde{p}$ be a path in $\tilde{\mathcal{M}}$. Walk along $\tilde{p}$ and, at each crossing encountered with a geometric lift c^α in $\mathcal{C}$, write down the symbol c^α or $\bar{c}^\alpha$, according to the orientation of the crossing (with respect to a fixed orientation of $\tilde{\mathcal{M}}$ – recall that $\tilde{p}$ and the elements of $\mathcal{C}$ are oriented). The word we obtain is called the *crossing word* of $\tilde{p}$ with $\mathcal{C}$, and denoted by $\mathcal{C}/\tilde{p}$.

In all this paper, the following situation will often occur: $\tilde{p}$ is a lift of a path p, and an elementary reduction is possible on $\mathcal{C}/\tilde{p}$. This reduction corresponds to two intersection points, a and a', of a lift $\tilde{c}$ of $\mathcal{C}$ with $\tilde{p}$. The subpaths associated with this possible elementary reduction are the parts $\tilde{c}_1$ and $\tilde{p}_1$ of $\tilde{c}$ and $\tilde{p}$ which are between a and a'. We will often remove these two crossings, by replacing $\tilde{p}_1$ by a path with the same endpoints, going along $\tilde{c}$, and obtaining by projection onto $\mathcal{M}$ a new path p' which crosses $\mathcal{C}$ twice less than p. Obviously, $\mathcal{C}/\tilde{p}'$ is deduced from $\mathcal{C}/\tilde{p}$ by proceeding to the elementary reduction; and the new path p' is homotopic to p. See Fig. 3.

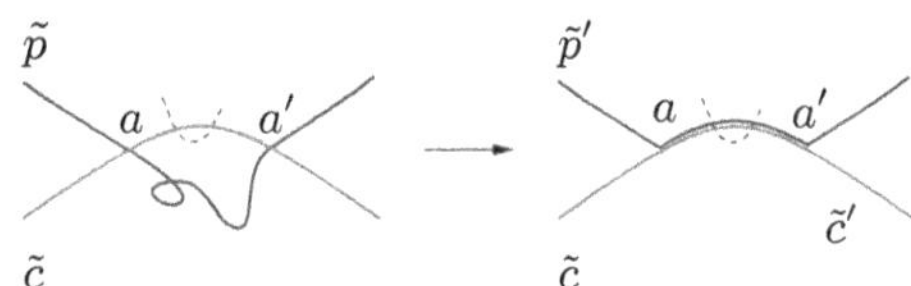

Fig. 3. The fundamental operation of uncrossing the parts of two curves c and p corresponding to an elementary reduction on $\mathcal{C}/\tilde{p}$. $\tilde{p}_1$ and $\tilde{c}_1$ are the parts of $\tilde{p}$ and $\tilde{c}$ between a and a'; $\tilde{p}_1$ is not necessarily simple, and $\tilde{c}_1$ can cross other pieces of $\tilde{p}$.

Lemma 5. *Any word w reduces to exactly one irreducible word.*

Of interest are the words which are *parenthesized*, i.e., those reducing to the empty word ε. The proof of the following lemma relies on Lemma 4.

Lemma 6. *Let $p : [0,1] \to \mathcal{M}$ be a contractible closed path, and let $\tilde{p}$ be a lift of p. Then, $\mathcal{C}/\tilde{p}$ is parenthesized.*

3.3 Crossing Words Sets for Cycles

The goal of this section is to define the analogue of the crossing word between $\mathcal{C}$ and a geometric lift $\tilde{\gamma}$ of a non-contractible cycle γ. A *lifted period* of $\tilde{\gamma}$ is a path which is the restriction of (an element of) $\tilde{\gamma}$ to $[a, a+1]$ for some $a \in \mathbb{R}$ (reparameterized over $[0,1]$). The *crossing words set* of $\tilde{\gamma}$ with $\mathcal{C}$, denoted by $[\mathcal{C}/\tilde{\gamma}]$, is the set of crossing words $\mathcal{C}/\tilde{p}$, over all lifted periods $\tilde{p}$ of $\tilde{\gamma}$ whose endpoints are not on lifts of $\mathcal{C}$. Our first task will be to show that the crossing words set $[\mathcal{C}/\tilde{\gamma}]$ is entirely determined once we know one of its elements.

We note that $\tilde{\gamma}$ induces a translation $\tau_{\tilde{\gamma}}$ in $\tilde{\mathcal{M}}$, as follows. Let $v \in \tilde{\mathcal{M}}$. Let $\tilde{p}$ be a lifted period of $\tilde{\gamma}$; consider a path β^0 joining $\tilde{p}(0)$ to v and call β^1 the lift of $\pi(\beta^0)$ starting at $\tilde{p}(1)$. The target v' of β^1 satisfies $\pi(v) = \pi(v')$; intuitively, $\tilde{\gamma}$ translates v to v'. It is readily seen that v' does not depend on the choice of β^0 and $\tilde{p}$. We therefore define $\tau_{\tilde{\gamma}}(v) := v'$. In particular, $\tau_{\tilde{\gamma}}$ sends a geometric lift of a curve $c \in \mathcal{C}$ to another geometric lift of c.

Define a permutation $\phi_{\tilde{\gamma}}$ over the set of words by: $\phi_{\tilde{\gamma}}(c^\alpha.w) = w.\tau_{\tilde{\gamma}}(c^\alpha)$, $\phi_{\tilde{\gamma}}(\bar{c}^\alpha.w) = w.\overline{\tau_{\tilde{\gamma}}(c^\alpha)}$, and $\phi_{\tilde{\gamma}}(\varepsilon) = \varepsilon$ (w is any word, and "." denotes concatenation).

Lemma 7. *For any word w in $[\mathcal{C}/\tilde{\gamma}]$, we have: $[\mathcal{C}/\tilde{\gamma}] = \{\phi^n_{\tilde{\gamma}}(w), n \in \mathbb{Z}\}$.*

If w is a word, we define $[w]_{\tilde{\gamma}}$ to be the set $\{\phi^n_{\tilde{\gamma}}(w), n \in \mathbb{Z}\}$. The sets of words having this form are called the *$\tilde{\gamma}$-words sets.* Note that $\phi_{\tilde{\gamma}}$ does not affect the length of a word, so that the *length* of a $\tilde{\gamma}$-words set is well-defined. Let W be a $\tilde{\gamma}$-words set. If there exists $w \in W$ containing a subword $c^\alpha \bar{c}^\alpha$ or $\bar{c}^\alpha c^\alpha$, denote by w' the word resulting from removing this subword from w; we say that W (which equals $[w]_{\tilde{\gamma}}$) *elementarily c-reduces* to $[w']_{\tilde{\gamma}}$.

Lemma and Definition 8. *Any $\tilde{\gamma}$-words set W c-reduces (resp. reduces) to exactly one c-irreducible (resp. irreducible) $\tilde{\gamma}$-words set. We define $g^{\tilde{\gamma}}_c(W)$ (resp. $g^{\tilde{\gamma}}(W)$) to be this $\tilde{\gamma}$-words set.*

When an elementary c-reduction is possible on $[\mathcal{C}/\tilde{\gamma}]$, exactly the same phenomenon occurs as in Fig. 3 (with $\tilde{\gamma}$ instead of $\tilde{p}$), and we may also proceed to the reduction by modifying γ, removing the two crossings.

Proposition 9. *Let γ be a cycle homotopic in $\mathcal{M}$ to some cycle γ' disjoint from $\mathcal{C}$. Let $\tilde{\gamma}$ be a geometric lift of γ. Then $g^{\tilde{\gamma}}([\mathcal{C}/\tilde{\gamma}]) = [\varepsilon]_{\tilde{\gamma}}$.*

Proof. Let p and p' be the restrictions of γ and γ' to $[0,1]$. There exists a path β joining $p(0)$ to $p'(0)$ such that the path $q := \beta^{-1}.p.\beta.p'^{-1}$ is contractible in $\mathcal{M}$. Let $\tilde{q}$ be a lift of q, concatenation of the inverse of β^0, $\tilde{p}$, β^1, and the inverse of $\tilde{p}'$ (respectively lifts of β, p, β, and p'). We choose $\tilde{q}$ so that $\tilde{p}$ is a lifted period of $\tilde{\gamma}$.

Since p' is disjoint from $\mathcal{C}$, $w := \mathcal{C}/\tilde{q}$ is the concatenation of $\mathcal{C}/(\beta^0)^{-1}$, $\mathcal{C}/\tilde{p}$, and $\mathcal{C}/\beta^1$. Furthermore, $\tau_{\tilde{\gamma}}(\beta^0)$ is equal to β^1; hence, if the kth symbol of $\mathcal{C}/\beta^0$ is equal to c^α (resp. $\bar{c}^\alpha$), then the kth symbol of $\mathcal{C}/\beta^1$ is equal to $\tau_{\tilde{\gamma}}(c^\alpha)$ (resp. $\overline{\tau_{\tilde{\gamma}}(c^\alpha)}$). It follows that $[w]_{\tilde{\gamma}}$ reduces to $[\mathcal{C}/\tilde{\gamma}]$. Now, by Lemma 6, w is parenthesized, so that $[w]_{\tilde{\gamma}}$ also reduces to $[\varepsilon]_{\tilde{\gamma}}$. Lemma 8 concludes. □

So far, we have introduced the notations $\mathcal{C}/\tilde{p}$ and $[\mathcal{C}/\tilde{\gamma}]$, where $\tilde{p}$ and $\tilde{\gamma}$ are lifts a path p and a cycle γ of $\mathcal{M}$, respectively. In the rest of this paper, when no risk of confusion arises, we will also use the notations $\mathcal{C}/p$ or $[\mathcal{C}/\gamma]$, meaning that we consider in fact $\mathcal{C}/\tilde{p}$ or $[\mathcal{C}/\tilde{\gamma}]$, where $\tilde{p}$ and $\tilde{\gamma}$ are *any* fixed geometric lifts of p or γ. Furthermore, the notation $[p/\gamma]$ will mean that we consider the crossing words set of γ with the geometric lifts of $\mathcal{C}$, this lifted set being made of only one path p in $\mathcal{M}$ with an arbitrary enumeration of its geometric lifts.

4 Curves on Pairs of Pants

In this section, we use crossing words to prove some basic facts regarding curves on pairs of pants.

Proposition 10. *Let K be a cylinder or a pair of pants, and γ be a cycle homotopic to a boundary of K. There exists a* simple *cycle homotopic to and not longer than γ.*

Proof. We will only give a proof when K is a pair of pants; the proof of the case where K is a cylinder is simpler. Let p be a shortest path between the two boundaries of K which are not homotopic to γ; let C be the cylinder obtained when cutting K along p; and let p' be a shortest path in C between its two boundaries.

$[p/\gamma]$ reduces to $[\varepsilon]_{\tilde{\gamma}}$ by Proposition 9; if it is not empty, let $\tilde{\gamma}_1$ and $\tilde{p}_1$ be the subpaths of lifts of γ and p corresponding to an elementary reduction. Since p is a shortest path, $|\tilde{p}_1| \leq |\tilde{\gamma}_1|$, and we can, like in Fig. 3, proceed to the elementary reduction by changing γ to another cycle, which is homotopic to and not longer than γ, and has two crossings less than γ with p. By induction, we obtain that there exists a cycle γ', homotopic to and not longer than γ, which does not cross p. Using similar techniques, we prove that there exists a cycle γ'', homotopic to and not longer than γ, which does not cross p and crosses p' only once, say at some point a.

Cutting C along p', we obtain two copies a' and a'' of a, and γ is transformed into a path between a' and a''. Hence, a shortest path between a' and a'' leads to a cycle in C which is simple, not longer than γ, and homotopic to γ in K. □

Proposition 11. *Let K be a cylinder or a pair of pants, and γ be one boundary of K. Assume γ is a shortest cycle among the simple cycles homotopic to γ. Let q be a path in K whose endpoints are on γ and which is homotopic to a path whose range (set of values) is included in the range of γ. Then the shortest path on γ homotopic to q is not longer than q.*

The proof is omitted and relies on similar ideas as the proof of Proposition 10.

Proposition 12. *Let s be a (doubled) pants decomposition of $\mathcal{M}$. Assume that a cycle γ is inside one component K of $\mathcal{M} \setminus s$ (a cylinder or a pair of pants), and homotopic in $\mathcal{M}$ to a cycle s_k. Then γ is homotopic,* in K, *to one boundary of K.*

Lemma 13. *Any cycle inside K which is contractible (in $\mathcal{M}$) is also contractible in K.*

Proof (of Proposition 12). Again, we assume that K is a pair of pants. Let S be a lifted set whose curves are the cycles in s, with an arbitrary enumeration of the geometric lifts. Let γ' be a simple cycle, disjoint from all cycles of s, and homotopic in $\mathcal{M}\backslash(s\backslash s_k)$ to s_k. Let p and p' be the restrictions of γ and γ' to $[0,1]$. There exists a path β joining $p(0)$ to $p'(0)$ such that the path $q := \beta^{-1}.p.\beta.p'^{-1}$ is contractible in $\mathcal{M}$. Without loss of generality, assume that S/β is irreducible. If this crossing word is empty, then q is contractible in K by Lemma 13, hence γ and γ' are homotopic in K; so are γ and s_k, and the proof is complete. Assume this crossing word is non-empty.

Since p and p' do not cross s, S/q is the concatenation of $S/(\beta^0)^{-1}$ and S/β^1, where β^0 is a lift of β, and $\beta^1 = \tau_{\tilde{\gamma}}(\beta^0)$. Because S/β is irreducible and S/q can be elementarily reduced, the first lifts of S crossed by β^0 and β^1 must be the same, say s_j^α. Let β' be the beginning of β before its first crossing with s; we get that $\beta'^{-1}.p.\beta'$ is homotopic to a power of s_j in $\mathcal{M}$, hence also in K by Lemma 13. By [3, Theorem 4.2], the nth power of s_j is homotopic to no simple cycle if $|n| \geq 2$. Hence γ is homotopic, in K, to s_j or its reverse. □

5 Proof of Theorem 2

Let b denote the set of boundary cycles of $\mathcal{M}$. As for a lifted set, we will assume that the geometric lifts of the cycles in b are enumerated. Consider now a doubled pants decomposition s of $\mathcal{M}$. We define a lifted set S whose curves in $\mathcal{M}$ are the curves in s.

Note that a cycle in s and its twin (in s or in b) are homotopic disjoint cycles, hence bound a cylinder by [3, Lemma 2.4]; the lifts of this cylinder in $\tilde{\mathcal{M}}$ are disjoint infinite strips which contain no geometric lift of s or b in their interior. Let $k \in [1,N]$, and let $s_{k'}$ or $b_{k'}$ be the twin of s_k (depending on whether it is an element of s or b). We can choose the enumeration of the geometric lifts of s_k so that, for each α, s_k^α and $s_{k'}^\alpha$ (or $b_{k'}^\alpha$) bound a strip which contains no lift of s and b in its interior.

Fix $j \in [1,N]$; let $r = f_j(s)$. We consider the lifted set R whose curves are the cycles in r; the enumeration of the geometric lifts of r is as follows. If $k \neq j$, then $r_k^\alpha = s_k^\alpha$ for any $k \in \mathbb{N}$. For the enumeration of the geometric lifts of r_j, we note that r_j and its twin, $r_{j'}$ (or $b_{j'}$), bound a cylinder in $\mathcal{M} \setminus r$; as above, we choose r_j^α so that r_j^α and $r_{j'}^\alpha = s_{j'}^\alpha$ (or $b_{j'}^\alpha$) bound an infinite strip containing no lift of r or b in its interior.

Finally, fix $i \in [1,N]$; let t_i be a shortest cycle among all cycles homotopic to s_i, and $\tilde{t}_i$ be a geometric lift of t_i. Henceforth, the words on the lifted sets R and S will be written differently as above, by omitting the "r" and the "s" (for example, we shall write ${}^3_1\,\overline{{}^7_5}\,\overline{{}^4_2}$ instead of $s_1^3\,\overline{s_5^7}\,\overline{s_2^4}$). This allows to say, for example, that $[R/\tilde{t}_i] = [S/\tilde{t}_i]$ if t_i does not cross r_j nor s_j. Let $\mathcal{P}_j$ be the pair of pants bounded by $s \setminus s_j$ in which s_j and r_j are.

Proposition 14. $g_j^{\tilde{t}_i}([R/\tilde{t}_i]) = g_j^{\tilde{t}_i}([S/\tilde{t}_i])$.

We note S_j the lifted set made of the cycle s_j, with the enumeration of its geometric lifts induced by S; the same holds for R_j.

Lemma 15. *Let p be a path in $\mathcal{P}_j$ whose endpoints are on the boundary of $\mathcal{P}_j$, and $\tilde{p}$ be a lift of p. Then $S_j/\tilde{p}$ and $R_j/\tilde{p}$ reduce to the same irreducible word.*

Proof (of Proposition 14). Assume first that t_i is contained in $\mathcal{P}_j$. By Proposition 9, we have $g^{\tilde{t}_i}([R/\tilde{t}_i]) = [\varepsilon]_{\tilde{t}_i} = g^{\tilde{t}_i}([S/\tilde{t}_i])$. But this also equals $g_j^{\tilde{t}_i}([R/\tilde{t}_i])$ and $g_j^{\tilde{t}_i}([S/\tilde{t}_i])$, and this concludes the proof. If t_i is not entirely contained in $\mathcal{P}_j$, then let t'_i be a maximal subpath of t_i which is inside $\mathcal{P}_j$, and $\tilde{t}'_i$ be a lift of t'_i; it is sufficient to prove that $R_j/\tilde{t}'_i$ and $S_j/\tilde{t}'_i$ reduce to the same irreducible word; but this follows from Lemma 15. □

Proposition 16. *There exists a cycle t'_i, homotopic to and not longer than t_i, and a geometric lift $\tilde{t}'_i$ of t'_i, such that $\tau_{\tilde{t}_i} = \tau_{\tilde{t}'_i}$ and $[R/\tilde{t}'_i] = g_j^{\tilde{t}_i}([S/\tilde{t}_i])$.*

Proof. By Proposition 14, $[R/\tilde{t}_i]$ j-reduces to $g_j^{\tilde{t}_i}([S/\tilde{t}_i])$. If $[R/\tilde{t}_i]$ is j-irreducible, there is nothing to show. Otherwise, an elementary j-reduction is possible on $[R/\tilde{t}_i]$. We can apply Proposition 11 to the subpath of t_i corresponding to this j-reduction, and apply the uncrossing operation to t_i. We obtain a geometric lift $\tilde{t}'_i$ of a cycle t'_i which is homotopic to and not longer than t_i. Clearly, $\tau_{\tilde{t}_i} = \tau_{\tilde{t}'_i}$ (which implies that $g^{\tilde{t}_i} = g^{\tilde{t}'_i}$). Furthermore, $[R/\tilde{t}'_i]$ results from $[R/\tilde{t}_i]$ by this elementary j-reduction. By induction, we obtain the desired t'_i. □

Proposition 17. *Assume t_i is disjoint from s, and that t_i and s_k are homotopic in the cylinder or pair of pants of $\mathcal{M} \setminus s$ containing t_i. Then, there exists a cycle t'_i, homotopic to and not longer than t_i, which is disjoint from r, and which is homotopic to r_k in the cylinder or pair of pants of $\mathcal{M} \setminus r$ containing t'_i.*

The (omitted) proof relies on Propositions 10 and 11. We now conclude the proof of our main theorem.

Proof (of Theorem 2). We consider a lifted set S^0 whose curves are s^0, the enumeration of the geometric lifts being chosen as described at the beginning of this section. By induction on $n \in \mathbb{N}$, we construct a lifted set S^n whose set of curves is s^n, with the enumeration of the lifts being chosen also as in the beginning of the section.

Let $\tilde{t}_i^0$ be a lift of a shortest cycle t_i^0 homotopic to s_i^0. By Proposition 9, $[S^0/\tilde{t}_i^0]$ reduces to $[\varepsilon]_{\tilde{t}_i^0}$. By Proposition 16, we can construct a sequence $(\tilde{t}_i^n)$ of lifts of shortest homotopic cycles such that the length of $[S^n/\tilde{t}_i^n]$ strictly decreases until it becomes empty at some stage n. By Proposition 12, t_i^n and a cycle s_k^n are homotopic in the cylinder or pair of pants of s^n containing t_i^n. By $k-1$ applications of Proposition 17, and then using Proposition 10, $|s_k^{n+1}| = |t_i^n|$. The cycle s_k^{n+1} is either s_i^{n+1} or its twin; in the latter case, since s_i^{n+1} and s_k^{n+1} bound a cylinder, $|s_i^{n+2}| = |s_k^{n+2}| = |t_i^n|$. From this discussion, it follows that the length of $(s_i^n)_{n\in\mathbb{N}}$ becomes stationary. It remains to prove that all lengths remain unchanged once s^n and s^{n+1} have the same lengths. The proof of this fact uses the same tools as above, and is omitted from this version.[2] □

6 Computational Issues

We describe here the combinatorial framework (which is similar to the one described in [2]) used to perform the optimization process.

6.1 Edge-Ordered Set of Cycles

We (temporarily) view G, the vertex-edge graph of $\mathcal{M}$, as a directed graph: each edge of G is replaced by two opposite directed edges. An *edge-ordered set of cycles* (EOSC for short) S on the graph G is a set of closed walks (without basepoint) in G, with the data, for each oriented edge e of G, of an order $\preceq_e$ over all edges of the walks in S corresponding to e or $-e$ (edge e with opposite orientation). These orders should be consistent in the following sense: $a \preceq_e b$ if and only if $b \preceq_{-e} a$. Intuitively, $a \preceq_e b$ if and only if a is on the left of b on e.

Let v be a vertex of G, and $e_1, \ldots, e_n$ be the clockwise-ordered list of oriented edges of G whose source is v. We define a cyclic order $\preceq_v$ over the edges of the walks in S meeting at v, by enumerating its elements in this order: first, the edges of the walks in S on e_1 or $-e_1$, in $\preceq_{e_1}$-order; then the edges of the walks in S on e_2 or $-e_2$, in $\preceq_{e_2}$-order; and so on. We say that two subpaths of length two, a_1, a_2 and b_1, b_2, of walks in S *cross* if the targets of a_1 and b_1 are the same vertex, v, and if, in the cyclic order $\preceq_v$, a_1 and a_2 separate b_1 and b_2. The EOSC S is *simple* if no crossing occurs in S.

Clearly, a set of disjoint simple cycles s can be retracted onto G to get a simple EOSC $S = \rho(s)$ (see Fig. 2AB), this retraction preserving the lengths (with the appropriate definitions) and homotopy classes of the cycles. The converse is also true: the closed walks of a simple EOSC S can be expanded along the edges to get a set of disjoint, simple cycles $s \in \rho^{-1}(S)$.

6.2 Computation of Shortest Paths

By the proof of Proposition 10, we can proceed to an Elementary Step f_j as follows: find the pair of pants $\mathcal{P}_j$ of $s \setminus s_j$ that contains s_j; find a shortest path p between the two boundaries of $\mathcal{P}_j$ which are not homotopic to s_j, and a shortest path p' between the two boundaries of $C := \mathcal{P}_j \setminus p$. Cutting C along p' yields a topological disk D, where points of p' on C correspond to pairs of points on D. The solution is found by considering the shortest paths between all such pairs of points and taking the shortest of these shortest paths.

[2] Note that this fact is unnecessary if we modify the algorithm as follows: if $f_i(s)_i$ has the same length as s_i, we choose $f_i(s)_i$ to be equal to s_i.

From an algorithmic point of view, let $S = \rho(s)$ be a simple EOSC. Define $G(\mathcal{P}_j)$ to be the weighted graph whose vertices are the components of $\mathcal{P}_j \setminus G^*$ and whose edges join two vertices separated by (a piece of) an edge e^* of G^*; such an edge has the same weight as e. This graph represents the vertex-edge graph of the surface $\mathcal{M}$ after cutting along the cycles of S. Its construction is easy and skipped in this abstract (see Fig. 2C). We can perform the computation of a shortest homotopic cycle in $\mathcal{P}_j$ by translating the operations of the previous paragraph to this combinatorial framework.

6.3 Complexity Analysis

Let R be a simple EOSC. For each vertex v of $\mathcal{M}$, consider the parenthesized word formed by the pairs of consecutive edges of R meeting at v. The *multiplicity* of vertex v (w.r.t. R) is the maximal number of nested parentheses in (any cyclic permutation of) this expression. Thus any 2-path a_1, a_2 crosses R a number of times which is at most the multiplicity of R at the target of a_1.

Let n be the complexity of $\mathcal{M}$ (total number of vertices, edges, and faces), g its genus and b its number of boundaries. Let α be the longest-to-shortest edge ratio of $\mathcal{M}$. Let s be a doubled pants decomposition of $\mathcal{M}$ composed of $O(g+b)$ cycles, and $S = \rho(s)$. Let μ be the maximum, over $j \in [1, N]$ and the vertices v of $\mathcal{M}$, of the multiplicity of S_j at v.

Bounding the number of Elementary Steps in the algorithm reduces to bounding the maximum, over i, of the minimal number of crossings between s_i and a shortest homotopic loop t_i. Doing so, we obtain (using Dijsktra's algorithm for the computation of shortest paths):

Theorem 18. *This algorithm computes an optimal pants decomposition homotopic to s in $O((g+b)^2\alpha^3\mu^4n^3\log(\alpha\mu n))$ time.*

Finally, assuming $\mathcal{M}$ is triangulated, and given an EOSC made of a single simple cycle γ, with multiplicity μ, we can compute a doubled pants decomposition containing γ which has multiplicity $O(\mu)$ (details omitted). This implies that computing a shortest cycle homotopic to γ is possible in time $O((g+b)^2\alpha^3\mu^4n^3\log(\alpha\mu n))$.

References

[1] P. Buser. *Geometry and spectra of compact Riemann surfaces*, volume 106 of *Progress in Mathematics*. Birhäuser, 1992.

[2] É. Colin de Verdière and F. Lazarus. Optimal system of loops on an orientable surface. In *IEEE Symp. Found. Comput. Sci.*, pages 627–636, 2002.

[3] D. Epstein. Curves on 2-manifolds and isotopies. *Acta Mathematica*, 115:83–107, 1966.

[4] J. Erickson and S. Har-Peled. Optimally cutting a surface into a disk. In *Proc. 18th Annu. ACM Symp. Comput. Geom.*, pages 244–253, 2002.

[5] A. Hatcher. Pants decompositions of surfaces. http://www.math.cornell.edu/˜hatcher/Papers/pantsdecomp.pdf, 2000.

[6] J. Hershberger and J. Snoeyink. Computing minimum length paths of a given homotopy class. *Computational Geometry: Theory and Applications*, 4:63–98, 1994.
[7] W. S. Massey. *Algebraic Topology: An Introduction*, volume 56 of *Graduate Texts in Mathematics*. Springer-Verlag, 1977.

Degree Navigator™: The Journey of a Visualization Software

Guy-Vincent Jourdan[1], Ivan Rival (1947–2002)[1*], and Nejib Zaguia[2]

[1] Decision Academic Graphics
University of Ottawa
gvj@DagSoft.com
http://www.IvanRival.com
[2] The School of Information Technology and Engineering,
University of Ottawa
zaguia@site.uottawa.ca

Abstract. In this paper, we follow the evolution of a tool that has its roots in academic research on visualization and graph drawing. From its ancestor *Order Explorer (1994)* and initial version of 1995, to its latest version of 2003, we analyze the significant steps of a software that has gone from being a relatively simple, very specialized research tool to being a successful, widely deployed, complex commercial software used for academic degree audits. As the tool has expanded over the years, the initial scope has been completely outgrown and finding back the original idea in the current version is not completely obvious. Never the less, Degree Navigator™ is a wonderful example of a research project that has gone a long way.

1 Order Explorer

At the root of the problem, in 1994, was the question of upward drawing of ordered sets, and more precisely the automatic, computer generated drawing of these sets. Although several methods were available in the literature (see e.g. [1], [3], [4], [5] for surveys of the question), none of the solutions were particularly satisfying. In fact, it was not even obvious to have a solution that was at least correct, if not aesthetic or practical. For example, most practical solutions, especially for large ordered sets would not prevent an edge between two vertices to go right through a third vertex, making it impossible for the viewer to know whether the third vertex was part of the relationships or not. Also, orders with large number of covering relations (i.e. with lots of edges to represent) were particularly difficult to render.

A simple and elegant solution was introduced with Order Explorer ([2]), an order manipulation tool that presented a rendering interface of a novel sort: Order Explorer simply omits the edges! With only vertices to draw, the problem became significantly easier. Of course, merely removing the edge doesn't quite resolve the problem. The

* Ivan Rival passed away in January of 2002. He was a central actor and partner in the design, the development and the success of the software described here.

G. Liotta (Ed.): GD 2003, LNCS 2912, pp. 491–493, 2004.

user still needs to access to the relationships information one way or the other. The way Order Explorer is providing this information is interactivity. By placing the mouse over a vertex, Order Explorer shows the adjacent vertices. The user has then to "explore" the order vertex after vertex to get a global view, if such a view was needed. Simple "mouse" clicks display comparabilities, adjacencies, upper and lower covers, etc. Order Explorer's solution is particularly suitable when the main emphasis is to display local information about specific vertices.

2 The Introduction of Degree Navigator

A typical department in a North American University will routinely offer over fifty courses, sometimes more than one hundred. These courses have a relationship among themselves, the *prerequisites.* If a course is a prerequisite for another one, then the first course must be completed before one can enroll in the second one. In its simplest form, the prerequisite relationship is an order.

The first goal of Degree Navigator ([6]) was to display the map of the courses of a department, along with the prerequisite information: the *prerequisite map.* The concept developed for Order Explorer was particularly suitable in the context: an algorithm to display quickly and clearly a relatively large order was needed. An easy access to the relationship information was necessary, but focusing on local information rather than on the full picture at once. In addition, the notion of groups of vertices (corresponding in that case to courses of the same year) was introduced: the vertices of the same group were drawn inside a patch a color, which was presented as an *island.*

The second step was to provide a visual representation of the *requirements* of a degree: the *degree map.* The requirements create a much more complex relationship between the vertices of our graph, and an upward drawing is not an option anymore. The type of information that need to be displayed include data such as "*take all the courses in a list*", "*take x courses/credits from a list*", "*take x courses/credits from a list, with at least y courses/credits from a sublist*", "*take either x courses/credits from this list, or y courses/credits from this other list*" and more.

In that visualization, we still use the metaphor of *islands* to group together the courses/vertices that belong to the same requirement. Because some of the requirements are broken down into sub-requirements, we generalize the concept and use *sub-islands*, or islands inside an island. Finally, in that map we may have to display a fairly large number of vertices inside one island and so we introduce the notion of *towers,* which is a stack of vertices of the same type grouped together on the same island.

The visual effect of the degree map resemble the visual chosen for the prerequisite map, and, as in Order Explorer, we make heavy usage of user interaction, mouse and click type of approach to extract the data out of a particular point.

3 The Evolution of Degree Navigator

The next step was to provide a web version of the product. A Java version of the drawing interface was introduced in 1998. The same user interactivity idea is used.

Note that the prerequisite relationship is usually more complex than the standard ordering relation. Never the less, similar idea is used, and multiple colors are used to distinguish the different prerequisites values.

It is worth noting that Degree Navigator generates all the pictures, including the course location and the island shapes in real time and doesn't store any rendering information in its database.

The current version of the software is a long way away for its primitive parent, but the original ideas are still present and effective. With hundreds of thousands of end users across North America and beyond, Degree Navigator is an encouraging example of a commercially successful product issued from the graph drawing research.

References

1. G. Di Battista, P. Eades, R. Tamassia, and I. G. Tollis, "Algorithms for Drawing Graphs: an Annotated Bibliography," *Computational Geometry: Theory and Applications*, vol. 4, no. 5. pp. 235–282 (1994).
2. G.-V. Jourdan, I. Rival, N. Zaguia, "Order Explorer, A System to See and Do in Four Dimensions", *International Conference on Ordinal and Symbolic Data Analysis '95*, Paris, France, June 1995.
3. I. Rival, "Reading, drawing, and order", *Algebras and orders* (Montreal, PQ, 1991), 359–404, NATO Adv. Sci. Inst. Ser. C Math. Phys. Sci., 389, Kluwer Acad. Publ., Dordrecht, 1993.
4. I. Rival, "The diagram", *Graphs and order* (Banff, Alta., 1984), 103–133, NATO Adv. Sci. Inst. Ser. C Math. Phys. Sci., 147, Reidel, Dordrecht, 1985.
5. I. Rival, Graphical Data Structures for Ordered Sets, in *Algorithms and Order*, ed. I. Rival, pp. 3–31, Kluwer Academic Publishers, 1989.
6. http://www.DagSoft.com/DegreeNavigator

HexGraph: Applying Graph Drawing Algorithms to the Game of Hex

Colin Murray, Carsten Friedrich, and Peter Eades

School of Information Technologies, The University of Sydney, Australia
{cmurray,carsten,peter}@it.usyd.edu.au

Hex [1] is a two player board game which is traditionally played on a rhombic hexagonal pattern (See Figure 1). Players are assigned a colour and make moves by putting a token of their colour onto an empty field on the board. The first player to connect the two borders of the board in his colour by a path of his tokens on the board wins the game. Alternatively, Hex is played on an undirected, tricoloured (Red, Blue, Unclaimed) graph G [2]. The fields are represented by nodes and adjacent fields on the board are connected by an edge. The four borders of the board are represented by one node of equivalent colour each (See Figure 1).

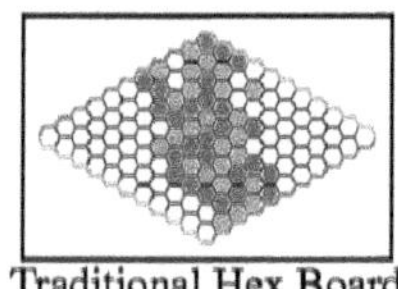

Traditional Hex Board

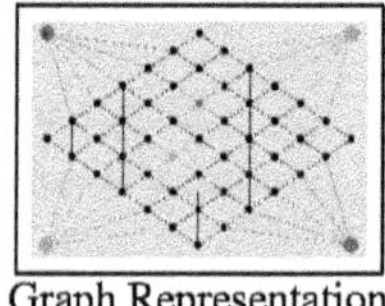

Graph Representation

Fig. 1. Traditional (Source [2]) & Graph representations.

Making a move triggers the following transformation f on the graph:

1. Edges to nodes held by the opponent are obsolete and are removed from the graph.
2. Edges which connect nodes of the same colour can be contracted.
3. Isolated nodes, that is, nodes which do not have any two non-intersecting paths to the two border nodes are obsolete (they cannot be part of a shortest winning path) and can thus be removed from the graph.

The graph continually simplifies during the course of the game. The challenge is to make the drawing simpler as well, and thus making it easier to play a winning strategy. Note that the transformation f preserves planarity. However, simple minded contraction of edges introduces edge-crossings. The HexGraph paradigm uses planar and embedding-preserving graph drawing algorithms, in combination with the graph transformation f above. We have experimented with a number of approaches. The best results are achieved by combining the Tutte [3,4] and the force directed PrEd [5] algorithms. The edge-crossings are removed by the Tutte algorithm and then the force directed PrEd algorithm

G. Liotta (Ed.): GD 2003, LNCS 2912, pp. 494–495, 2004.

improves symmetry and uniformity of edge lengths without introducing any edge crossings. These graph drawing features help to ensure that the player understands the state of the game and should improve the player's ability to play a winning strategy.

We also use graph animation [6] to show the transition from the initial graph drawing to the final graph drawing to help the user maintain the mental map [7] of the game.

In our experience, the HexGraph paradigm provides great potential to increase a player's ability to find good moves and strategies in a game of hex. Our implementation is available from `http://www.it.usyd.edu.au/~carsten/hex`.

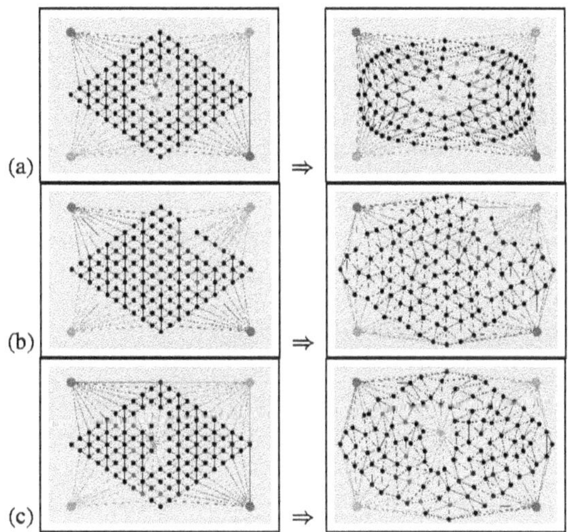

Fig. 2. (a) The Tutte Algorithm removes edges crossings from the graph. (b) Force directed PrEd makes edge crossings less confusing. (c) The edge-crossings are removed by the Tutte algorithm and then the force directed PrEd algorithm improves symmetry and uniformity of edge lengths

References

1. C. Browne. *Hex Strategy: Making the Right Connections.* A. K. Peters, Natick, Massachusetts, 2000.
2. Jack van Rijswijck. Search and evaluation in Hex. Tech report, University of Alberta.
3. W. T. Tutte. Convex representations of graphs. *Proc. Lond. Math. Soc.*, 10:304–320, 1960.
4. W. T. Tutte. How to draw a graph. *Proc. London Math Soc.*, 13:743–768, 1963.
5. F. Bertault. A force-directed algorithm that preserves edge-crossing properties. *Information Processing Letters*, 74(1–2):7–13, 2000.
6. Carsten Friedrich. *Animation in Relational Information Visualization.* Phd thesis, University of Sydney, 2002.
7. P. Gould and R. White. *Mental Maps.* Allen and Unwin, Winchester, 1986.

GLuskap: Visualization and Manipulation of Graph Drawings in 3-Dimensions*

Breanne Dyck, Jill Joevenazzo, Elspeth Nickle, Jon Wilsdon, and Stephen Wismath

Department of Mathematics and Computer Science, University of Lethbridge, Alberta, Canada T1K 3M4.

Abstract. GLuskap is a software tool for displaying graphs in 3 dimensions, interactively editing the resulting drawing and finally creating a high quality ray traced image. The software is written in C++ and uses OpenGL and SDL. The interactive editing and navigation features may also be used in stereoscopic mode providing an even more realistic environment. The drawing can be exported either to a standard graph format or to source code directly suppliable to the POV-Ray ray tracing package which then renders the drawing into any standard image format. This software should be of value to all researchers in the 3D graph drawing community and is freely available at: http://www.cs.uleth.ca/~vpak/demo.

1 Overview

Drawing graphs in 3 dimensions (3D) is a recent active research area. Whereas 2-dimensional graph drawing has been supported by a variety of powerful software tools for some time, there are only a few 3D graph drawing packages and most are tailored for specific applications. GLuskap[1] is a software package for interactively accomplishing the creative and exploratory graph drawing and modeling commonly required by practioners in this area, and providing a general utility for the manipulation and visualization of 3-dimensional drawings of graphs.

GLuskap serves as an easy-to-use graphical user interface that produces 3-dimensional graph drawings in OpenGL. See figure 1. The user is presented with numerous options for exploring 3D graph drawings, as well as having the ability to freely modify the representations. Of particular importance is the real-time interactive vertex and edge manipulation, which allows the user to modify the graph drawing, and immediately see the results in 3D. Import and export options accommodate various popular file formats, including GML and GraphML. A distinguishing feature of GLuskap is its ability to export a 3D graph drawing to the POV-Ray software package for subsequent rendering into a standard image format.

* Research supported by the Natural Sciences and Engineering Research Council of Canada.

[1] Gluskap is the Algonquin name for the *Creator force*

G. Liotta (Ed.): GD 2003, LNCS 2912, pp. 496–497, 2004.

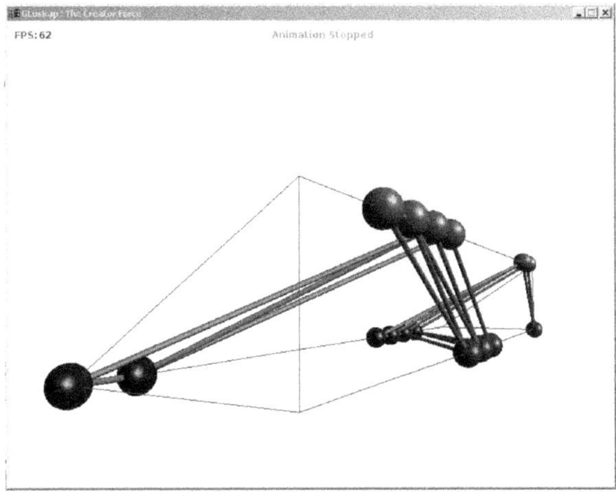

Fig. 1. Screenshot of GLuskap

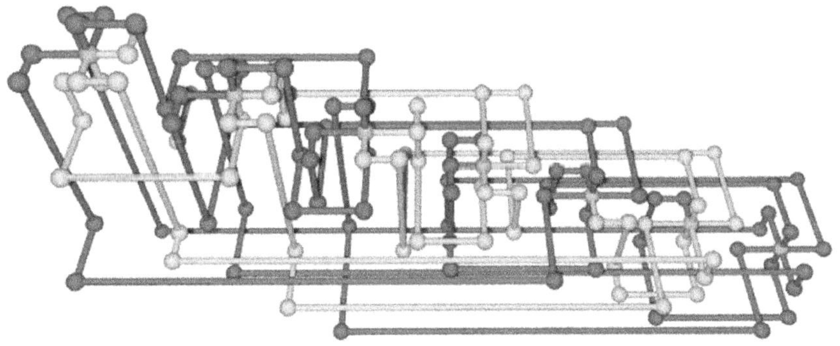

Fig. 2. A Ray-traced image

Figure 2 shows a rendered drawing produced through this process. Many users may wish to bypass the graphical editing/navigating process and concentrate solely on this post-processing step of creating POV-Ray source code. Nevertheless, the navigation aspects of the software are unique, as they provide interactive, 3-dimensional exploration of a graph drawing. Coupled with support for stereoscopic viewing, these features make the visualization aspects of GLuskap especially powerful. GLuskap is available for download as Windows executables and as C++ source code from `http://www.cs.uleth.ca/~vpak/demo`. See the web site for further screenshots and more detailed information.

Web-Linkage Viewer: Drawing Links in the Web Based on a Site-Oriented Framework

Yasuhito Asano and Takao Nishizeki

Graduate School of Information Sciences, Tohoku University,
Aoba-yama, Sendai, Japan

1 Introduction

In recent years, link-based information retrieval methods from the Web are developed, such as HITS and Trawling. Since these methods utilize characteristic graph structures of links in the Web, analyzing the links by drawing them understandably will play an important role in the link-based information retrieval. Moreover, Asano et al. [2], [3] have shown that a framework using a site as a unit of information is more natural and useful for link-based information retrieval than the existing framework using a page as a unit. Therefore, we should distinguish links inside a site (called *local-links*) between links between sites (called *global-links*) and analyze their own graph structures. However, existing drawing tools, such as Gravis [1] and H3Viewer [5], do not distinguish local-links between global-links, and therefore they cannot draw graph structures of these links understandably. In this paper, we propose a new drawing tool, named **Web-linkage Viewer**, in order to draw graph structures in the Web according to the site-oriented framework.

2 Web-Linkage Viewer

When a site is regarded as a unit of information as humans naturally consider so, every page must belong to some site, and then the following two kinds of links can be defined. (1) Let A and B be two distinct sites. If there is a link from a page v in A to a page w in B, we say there is a *global-link* from A to B. (2) A link from a page v to a page w with v and w in A is called a *local-link* inside A.

In order to show relationships between a site and other sites and relationships between pages and sites understandably, we draw global-links and local-links distinctively in the same space and avoid overlaps of the local-links in some site with the local-links in another site as much as possible. Concretely, we draw sites as cubes and global-links as arrows on a spherical surface, and for each site, draws the pages inside the site as cubes and the local-links as lines in a cone emanating from a cube representing the site. The sites are arranged by using the spring model and the pages in each site are arranged from the apex of cone to the base according to the order determined by the breadth first search of the local-links in the site. See [2] for the details.

G. Liotta (Ed.): GD 2003, LNCS 2912, pp. 498–499, 2004.

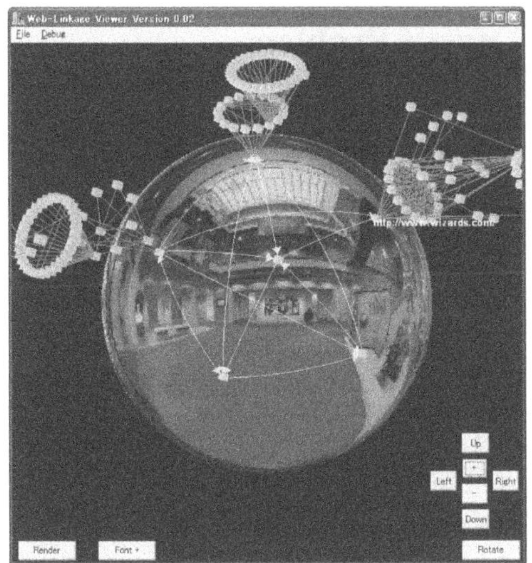

Fig. 1. Screenshot of Web-linkage Viewer

By drawing the global-links and the local-links distinctively in the same space, we can understand a relationship between a site and pages inside it.

As a result, we conclude that our system provides us with a more understandable drawing of structures of links in the Web. In particular, by using our system we can understand that a large part of local-links inside each site forms a tree structure frequently (i.e. other kinds of links, such as links connecting siblings, are fewer), while global-links have another kind of structure.

We implemented our software on Microsoft Windows XP by using Borland C++ Builder 5.0 and Microsoft DirectX 9.0a. It can zoom in and out, rotate the drawing by mouse, and display URLs of the nodes and open a page with Internet Explorer by click. We have compared our drawing with drawings of Gravis and H3Viewer and shown that our drawing is more understandable than them [2]. We have developed also a tool named **Neighbor Community Finder** for finding Web communities related to given URLs [4]. As a future work, we are trying to combine this tool with our viewer, in order to provide this tool with a GUI and to visualize graph structures in neighbor sites and pages of found communities.

Acknowledgments. We thank Prof. Hiroshi Imai and Ms. Sonoko Moriyama for their comments and help for computational experiments of comparison of our drawing with other tools.

References

1. Gravis. `http://www-pr.informatik.uni-tuebingen.de/`.
2. Y. Asano. *A New Framework for Link-based Information Retrieval from the Web.* PhD thesis, The University of Tokyo, December 2002.
3. Y. Asano, H. Imai, M. Toyoda, and M. Kitsuregawa. Applying the site information to the information retrieval from the Web. In *Proceedings of the 3rd International Conference on Web Information Systems Engineering*, pages 83–92, 2002.
4. Y. Asano, H. Imai, M. Toyoda, and M. Kitsuregawa. Finding neighbor communities in the Web using an inter-site graph. In *Proceedings of the 14th International Conference on Database and Expert Systems Applications*, page to appear, 2003.
5. T. Munzer. Drawing large graphs with H3Viewer and site manager. In *6th International Symposium, GD98, LNCS1547*, pages 384–393, 1998.

The Puzzle Layout Problem

Kozo Sugiyama[1], Seok-Hee Hong[2], and Atsuhiko Maeda[3]

[1] School of Knowledge Science, Japan Advanced Institute of Science and Technology, Asahidai 1-1, Tatsunokuchi, Nomi, Ishikawa, 923-1292, Japan
sugi@jaist.ac.jp
[2] School of Information Technologies, University of Sydney, NSW 2006, Australia
shhong@it.usyd.edu.au
[3] NTT Network Innovation Labs, NTT Corporation, Hikarinooka 1-1, Yokosuka, Kanagawa, 239-0847, Japan
maeda.atsuhiko@lab.ntt.co.jp

Abstract. This poster describes an implementation of the puzzle generators and puzzle layouts introduced in [1].

1 Introduction

We present a new application of graph drawing, puzzle layout. We analyze the operations of the puzzles and derive two abstract models which can be modeled as puzzle graphs. Based on these models, we implement two puzzle generators and produce various layouts of the puzzles using various graph drawing algorithms. Using these puzzle generators, we can parametrically change the levels of difficulty of the puzzles and create new puzzles. Further by applying various graph drawing algorithms, we can create new user interfaces for a puzzle. For details, see [1].

The puzzle layout should be very beautiful and attractive to the users. Further, the layout should operate as a puzzle interactively by the users. We applied various standard graph drawing methods to produce puzzle layouts such as spring embedder algorithm, orthogonal drawing and visibility representations. This poster describes an implementation of the puzzle generators and puzzle layouts presented in [1].

2 The Puzzle Generator and the Layouts of Puzzles

Figure 1 shows a user interface of the puzzle generator and the layout. Using the interface, puzzles can be defined in a text-based dialog box. We use a spring algorithm with manual change to produce the layout.

To operate the layout as a puzzle, the user can simply use drag and drop to move an element or to rotate a cycle. Then the cycle which contains the element and the other cycles which are included in the operation rotate together.

G. Liotta (Ed.): GD 2003, LNCS 2912, pp. 500–501, 2004.

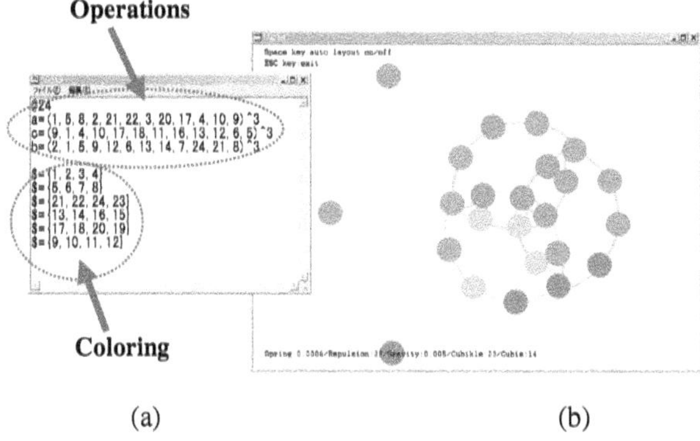

Fig. 1. User interface of a permutation puzzle generator and its layout.

Figure 2 shows three different layouts of the 2^3 Rubik's Cube and the Pyraminx.

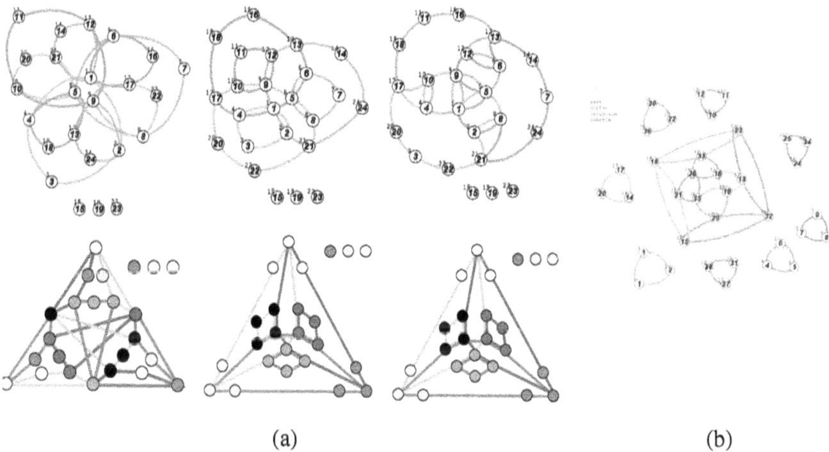

Fig. 2. Layouts of (a) 2^3 Rubik's Cube and (b) Pyraminx.

3 Symmetric Layout Algorithm for Puzzles

We observed that symmetry is the far most important aesthetic criteria in the puzzle layout problem. However, the spring algorithm sometimes fails to achieve a good symmetric layout. Our current work is to implement the symmetric puzzle layout algorithm presented in [1].

References

1. K. Sugiyama, S. Hong and A. Maeda, The Puzzle Conversion and Layout Problem, KS-RR-2003-002, School of Knowledge Science, JAIST, 2003.

Visual Data Mining with ILOG Discovery

Thomas Baudel, Bruno Haible, and Georg Sander

ILOG SA, 9 rue de Verdun - BP 85, 94253 Gentilly Cedex, FRANCE
{baudel|haible|sander}@ilog.fr
http://www2.ilog.com/preview/Discovery

1 Introduction

Data mining deals with the discovery of useful and previously unknown knowledge from large data sets [3]. Traditional data mining tools use a combination of machine learning, statistical analysis, modeling techniques and database technology to find patterns, exceptions and subtle relationships in data. Typical applications include market segmentation, customer profiling, fraud detection, credit risk analysis, and business data development.

ILOG Discovery is different to traditional data mining tools: It is a interactive visual information exploration tool. From data sets gathered from your information systems (such as spreadsheets, statistical or usage databases, usage logs or accounting data), ILOG Discovery provides simple yet powerful ways to view this data in many different ways. It allows you to quickly change the display parameters, so that you can interactively search for views of your data that highlight trends, spot general patterns, or pinpoint exceptions. ILOG Discovery can easily handle large amounts of data and produce a wide variety of visualizations with very few manipulations, which makes it a completely novel exploration tool.

2 Highlights

ILOG Discovery reads a set of records, each record containing many fields. It displays the records while mapping the fields to different visual parameters such as color, width, position, etc. The mapping methods include:

- 2-D Charts: traditional x-y charts that allow to quickly exchange and adapt the x and y axes and visual parameters (color, size, chart layout).
- Parallel histograms (similar to table lens [4]) that allow to sort, filter and compare different fields of the records.
- Clustered Treemaps [2,5] where the nested clustering is shown by an inclusion drawing standard while size, color, position and structure of each cluster can be used to express certain aspects of the data.
- and many combinations of these techniques.

While ILOG Discovery's technical origin is more close to 2-D charting, it uses some graph layout techniques to manage clustered treemaps. A treemap

G. Liotta (Ed.): GD 2003, LNCS 2912, pp. 502–503, 2004.

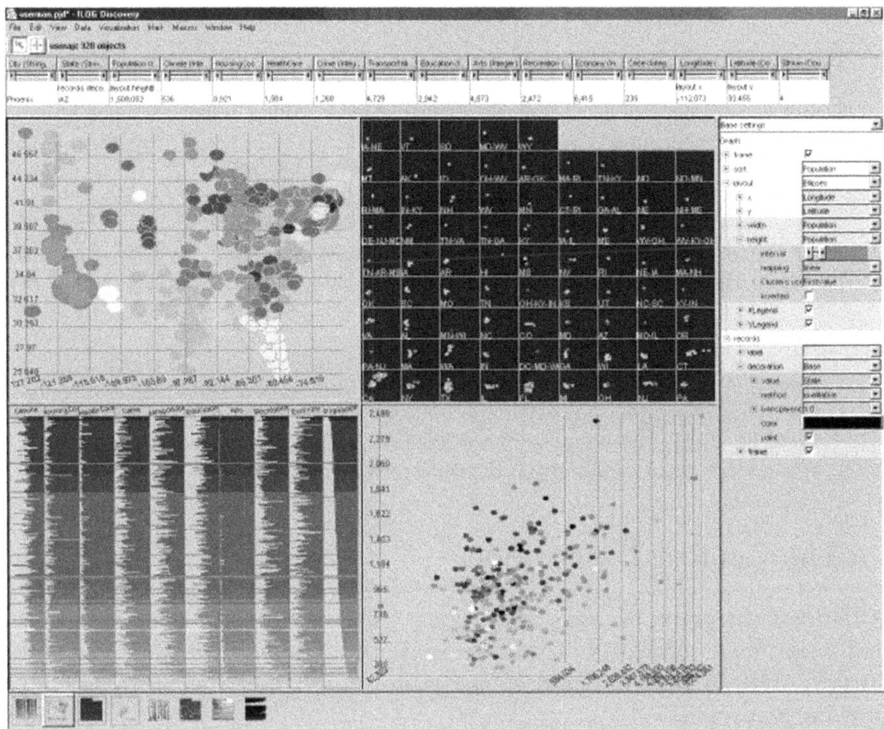

Fig. 1. ILOG Discovery

consists of clusters that are shown as rectangular areas. The cluster nesting is an inclusion tree that must be laid out with respect to space constraints, e.g. each cluster should approximately cover a square, and the area for the entire drawing is given, similar to [1]. Also, ILOG Discovery uses a simple still powerful labeling technology to annotate the diagrams.

ILOG Discovery is implemented in pure Java. A stand alone preview version can be downloaded for free at `http://www2.ilog.com/preview/Discovery`.

References

1. P. Eades, Lin T., and Lin X. Two tree drawing conventions. *Int. Journal of Computational Geometry and Applications*, 3(2):133–153, 1993.
2. J.-D. Fekete and C. Plaisant. Interactive information visualization of a million items. In *Proc. IEEE Symp. InfoVis 2002*, pages 117–124. IEEE Press, 2002.
3. D.J. Hand, H. Mannila, and P. Smyth. *Principles of Data Mining.* MIT Press, 2001.
4. R. Rao and S. K. Card. The table lens: Merging graphical and symbolic representations in an interactive focus+context visualization for tabular information. In *Proc. ACM SIGCHI'94*, pages 318–322. ACM Press, 1994.
5. B. Shneiderman. Tree visualization with tree-maps: 2-d space-filling approach. *ACM Transactions on Graphics*, 11(1):92–99, 1992.

Graph Drawing Contest Report

Franz J. Brandenburg[1], Ulrik Brandes[2], Peter Eades[3], and Joe Marks[4]

[1] Universität Passau 94030 Passau, Germany
brandenb@informatik.uni-passau.de
[2] Universität Konstanz, 78457 Konstanz, Germany
[3] Basser Department of Computer Science, University of Sydney, Sydney, Australia
[4] Mitsubishi Research Laboratories (MERL), 201 Broadway, Cambridge, MA, USA

Abstract. This report describes the Tenth Annual Graph Drawing Contest, held in conjunction with the 2003 Graph Drawing Symposium in Perugia, Italy. The purpose of the contest is to monitor and challenge the current state of the graph-drawing technology.

1 The Contest Description

In the past the graph drawing contest had "challenge graphs" and the primary goal were "nice" drawings of these graphs. In addition there was an artistic category. Details can be found in the contest reports of the previous Graph Drawing Symposia.

The concept for this year's contest has been revised with a focus on research directions and the hope for more entries. The contest theme was "drawing graphs within graphs". This emphasizes the largely open question of how to visualize distinguished graph structures that are contained in larger graphs in a distinct way, such as: Describe algorithms that make explicitly or implicity given subgraphs visible and display them, e.g., a (or all) $K_{3,3}$ and K_5 in a non-planar graph. Embed a distinguished subgraph in a given partial drawing of a graph in a suitable way. Make all occurrences of a small graph visible and show how they distribute over the graph, or how they relate to each other. Draw a graph using templates of certain subgraphs. Given a nested graph, place the contained graphs in a layout of their containers or vice versa. Make isomorphic subgraphs recognizable.

Two real world data sets were provided, that could be used for the illustration of the approaches. The first network shows the transcriptional regulation of Escherichia Coli. This network contains many occurrences of so-called motifs, which are described in [1]. Motifs are distinguished subgraphs which should be used in a graph drawing algorithm to clarify the structure of the network. The second network is a social network that represents organizations involved in drug policy making. There are confirmed and unconfirmed lines of communication between the organizations, and the inherent structure of the organization should be displayed. Text descriptions for the 2003 contest were made available via the World Wide Web and announced with the Graph Drawing Symposium. The data on the graphs was provided in several graph formats.

G. Liotta (Ed.): GD 2003, LNCS 2912, pp. 504–508, 2004.

2 The Decisions

The seven entries address the contest theme in different ways, using hierachical approaches and reduction techniques, 3D, and nice curves. The entries are of a high scientific level and describe research work at an early stage. They contained interesting new ideas with promising concepts for graph drawing. The authors spent much time and effort for the preparation of their entries, and students were strongly involved in the preparation of most entries.

The contest jury, formed by the authors of this report, evaluated the entries and made a pre-decision based on the originality and closeness towards the contest theme. Due to conflicting interests, the final decision on the first prize winners was made by an independent jury with Patrick Healy, Joe Marks, and Sue Whitesides.

2.1 The Winners

There are two first prize winning entries: "Drawing graphs within graphs: A contribution to the graph drawing contest 2003" by Daniel Gmach, Paul Holleis, and Thomas Zimmermann, students at the University of Passau, Germany, and "Graph pattern analysis with PatternGraffiti" by Christian Kuklas, Dirk Koschützki, and Falk Schreiber from the Bioinformatics Center, Gatersleben, Germany. Both teams are experienced graph drawing contest participants, since the Passau students were among the winners in 2002, and F. Schreiber was a contest winner in 1995.

Gmach, Holleis, and Zimmermann directly approach the contest theme and describe how to draw graphs within graphs. Distinguished subgraphs are mapped into supernodes, such that a summary graph remains from the given input graph. The subgraphs and the summary graph may be layouts separately, using even different drawing algorithms. Overlapping subgraphs are treated separately, taking their union or their symmetric difference as new subgraphs, or making copies of the overlapping parts. The edges between the supernodes are computed from exclusive connections between the subgraphs behind the supernodes. This process is illustrated in Fig. 1. The approach has been applied to the largest connected component of the transcriptional network of Escherichia Coli with then reduces from 328 vertices to a graph with just 16 supernodes, each representing a special motif (SIM). The summary graph clearly displays the connections between the SIMs, see Fig. 2. There is also a prototype implementation in Graffiti, a graph drawing tool from the University of Passau. A more detailed description of the approach is supposed to appear in a forthcoming paper.

Motivated by the GD2003 contest, Kuklas, Koschützki and Schreiber have developed PatternGraffiti as is a tool for the analysis and display of pattern. A pattern is a distinguished labeled subgraph. It is extended by regular expressions for the edge and node labels. A matching is a subgraph isomorphism of the pattern and a match of the labels. Matchings are computed by an adaptation of Ullmann's algorithm [4]. Different pattern are distinguished by a colouring, and all occurrences of a pattern are drawn in the same way using an adapted

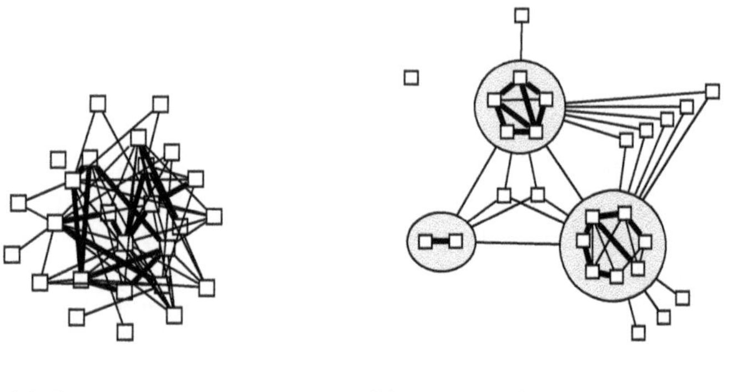

(a) Original graph (b) Layout of the summary graph

Fig. 1. Social network and connection sets

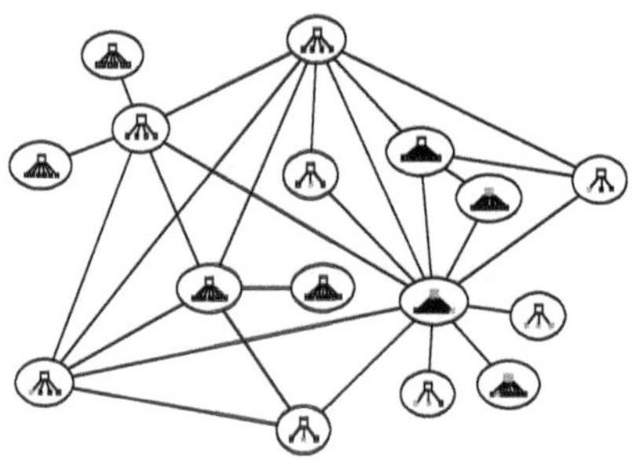

Fig. 2. SIMs in transcriptional network

spring embedder algorithm. PatternGraffiti is a plug-in to Graffiti. It enables to the user to specify a set of patterns, which are then searched in the input graph and visualized by the algorithm. A screenshot with a part of the transcriptional network and a motif (the original is coloured) is shown in Fig. 3. The approach has also been applied to the data from the social network. A more detailed description of the approach is supposed to appear in a forthcoming paper.

2.2 The Runners-Up

The following entries are the second prize winners:
Breanne Byck, Jill Joevenazzo, Elspeth Nickle, Jon Wilsdon, and Steve Wismath from the University of Lethbridge, Canada, address the social network and use a 3D approach to highlighten the inherent structure within the given network. Green and red edge colourings are used to emphasize the confirmed and unconfirmed relationships. Their central image in a stereogram, a 2D drawing which appears to be 3D. See also their poster in this volume.

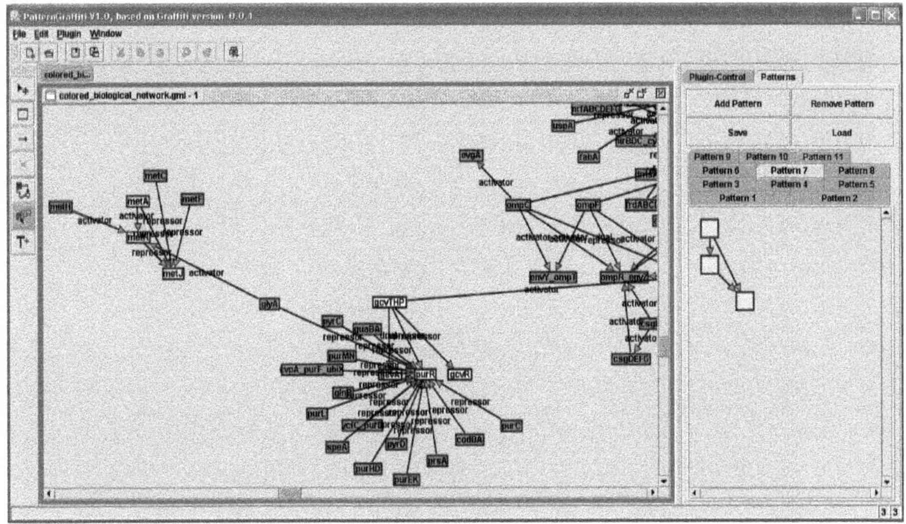

Fig. 3. Screenshot from PatternGraffiti

David Eppstein, Michael T. Goodrich, and Jeremy Y. Meng, University of California, Irvine, CA, applied their recent approach on confluent drawings [3] to reduce the visual complexity of drawings. The outcome are "dancing bicliques", showing the famous Tower of Pisa in a new way, see Fig. 4. Given a layered graph, the edges between two adjacent layers are covered by bicliques i.e., complete bipartite graphs, which are then displayed by confluent drawings. Bicliques are drawn in colours, since the planarity constraint from [3] cannot be preserved.

Ken Frank, Michigan State University, East Lansing, MI, uses his KliqueFinder clustering algorithm to produce crystalized sociograms, and has applied it to express and represent friendships among the French financial elite.

Marco Gärtler from the University of Karlsruhe, Germany, has introduced an advanced reduction technique to decompose a network into smaller pieces. Besides the motifs from [1] he has found other remarkable patterns in the regulatory network which are collapsed for the contraction of the network. The remainder is drawn in a DAG layout style. This technique is suitable for the interactive exploration of other structural data.

Nestor Sosa, Andreas Kremling, Esther Ratsch, and Isabel Rojas from the European Media Laboratory, Heidelberg, Germany, and the Max Planck Institute for Dynamic Complex Technical Systems in Magdeburg describe the development of an application for the visualization of the transcriptional networks. The main concept towards a visualization is a simplification of the network using the given motifs and other substructures. The reduction is enormous, so that a three layer network remains from the transcriptional network of Escherichia Coli, which can be drawn right away.

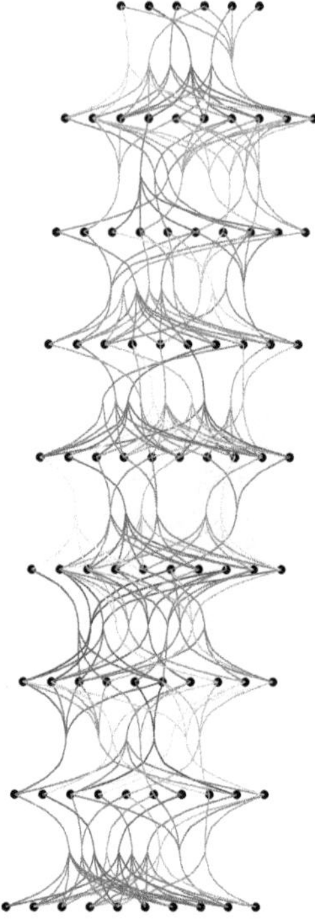

Fig. 4. Tower of Pisa

Acknowledgement. We wish to thank Uri Alon, The Weizman Institute, Tel Aviv for providing the data on the transcriptional network and Patrick Kenis from Tilburg University for providing the data on the social network.

The 2003 Graph Drawing Contest has been sponsered by AT&T Research Laboratories, Murray Hill, MERL Mitsubishi Electric Research Laboratories, Cambridge, and Tom Sawyer Software, Oakland.

References

1. S.S. Shen-Orr, R. Milo, S. Mangan, and U. Alon, Network motifs in the transcriptional regulation network of escherichia coli, *Nature Genetics*, 31 (2002) 64–68.
2. R. Milo, S. Shen-Orr, S. Itzkovitz, N. Kashtan, D. Chklovskii, and U. Alon, *Network motifs: Simple building blocks of complex networks, Science*, 298 (2002) 824–827.
3. M. Dickerson, D. Eppstein, M. T. Goodrich, and J.Y. Meng, Confluent drawings, visualizing nonplanar diagrams in a planar way, *these proceedings.*
4. J. R. Ullmann, An algorithm for subgraph isomorphism, *J. Assoc. Comput. Mach. 23*, (1976), 31–42.

Engineering and Visualizing Algorithms*

Camil Demetrescu[1], Irene Finocchi[2], and Giuseppe F. Italiano[2]

[1] Dipartimento di Informatica e Sistemistica,
Università di Roma "La Sapienza",
Via Salaria 113, 00198 Roma, Italy. demetres@dis.uniroma1.it,
http://www.dis.uniroma1.it/~demetres/

[2] Dipartimento di Informatica, Sistemi e Produzione,
Università di Roma "Tor Vergata",
Via del Politecnico 1, 00133 Roma, Italy.
{finocchi,italiano}@disp.uniroma2.it,
http://www.info.uniroma2.it/~{finocchi,italiano}/

Abstract. We discuss some relevant issues in Algorithm Engineering, focussing on the interplay between theory and practice, and showing how it can integrate and reinforce the traditional theoretical approaches to the design and analysis of algorithms and data structures, while devising methodologies and tools for developing and engineering efficient algorithmic codes.

1 Introduction

The whole process of designing, analyzing, implementing, tuning, debugging and experimentally evaluating algorithms can be referred to as *Algorithm Engineering*. Algorithm Engineering views algorithmics also as an engineering discipline rather than a purely mathematical discipline. Implementing algorithms and engineering algorithmic codes is a key step for the transfer of algorithmic technology, which often requires a high-level of expertise, to different and broader communities, and for its effective deployment in industry and real applications. Moreover, experiments often raise new conjectures and theoretical questions, opening unexplored research directions that may lead to further theoretical improvements and eventually to more practical algorithms. Theoretical breakthroughs that answer fundamental algorithmic questions are a crucial step towards the solution of a problem. However, they often lead to algorithms that are far from efficient implementations, raising the question of whether more practical solutions exist. We believe that Algorithm Engineering should take into account the *whole* process, from the early design stage to the realization of efficient implementations.

* Work partially supported by the IST Programmes of the EU under contract numbers IST-1999-14186 (ALCOM-FT) and IST-2001-33555 (COSIN), and by the Italian Ministry of University and Scientific Research (Project "ALINWEB: Algorithmics for Internet and the Web").

G. Liotta (Ed.): GD 2003, LNCS 2912, pp. 509–513, 2004.

An important aspect of algorithm engineering, usually referred to as *Experimental Algorithmics*, is related to performing empirical studies for comparing actual relative performance of algorithms so as to study their amenability for use in specific applications. This may lead to the discovery of algorithm separators, i.e., families of problem instances for which the performances of solving algorithms are clearly different, and to identifying and collecting problem instances from the real world. Other important results of empirical investigations include assessing heuristics for hard problems, characterizing the asymptotic behavior of complex algorithms, discovering the speed-up achieved by parallel algorithms and studying the effects of the memory hierarchy and of communication on real machines, thus helping in predicting performance and finding bottlenecks in real systems. Another goal of algorithm engineering is to define standard methodologies and realistic computational models. For instance, there is more and more interest in defining models for the memory hierarchy and for Web algorithmics. Issues related to Web algorithmics are received a lot of attention, as the Internet is nowadays a primary motivation for several problems: security infrastructure, Web caching, Internet searching and information retrieval are just a few of the hot topics. Devising realistic models for the Internet and Web graphs is thus essential for testing the algorithmic solutions proposed in this settings.

2 Algorithm Engineering and Visualization

Algorithms must be implemented and tested in order to have a practical impact. Experiments can help measure practical indicators, such as implementation constant factors, real-life bottlenecks, locality of references, cache effects and communication complexity, that may be extremely difficult to predict theoretically. We remark here that locality and cache effects are particularly important for graph algorithms: graphs are typically de-localized and pointer-based data structures. Many clear examples of the value of the experimentation for graph algorithms are addressed in the literature: among them, the implementation issues of the push-relabel algorithm for the maximum flow problem by Goldberg and Tarjan [9] stand out.

Unfortunately, as in any empirical science, it may be sometimes difficult to draw general conclusions about algorithms from experiments. Towards this aim, some researchers have proposed accurate and comprehensive guidelines on different aspects of the empirical evaluation of algorithms maturated from their own experience in the field (see, for example [1, 14, 15, 18]). The interested reader may find in [17] an annotated bibliography of experimental algorithmics sources addressing methodology, tools and techniques.

Among the tools useful in algorithm engineering, we cite visualization systems, which exploit interactive graphics to enhance the development, presentation, and understanding of computer programs [21]. Thanks to the capability of conveying a large amount of information in a compact form that is easily perceivable by a human observer, visualization systems can help developers gain insight about algorithms, test implementation weaknesses, and tune suitable heuristics for improving the practical performances of algorithmic codes. Some examples of this kind of usage are described in [11].

Systems for algorithm animation have matured significantly since the rise of modern computer graphic interfaces and dozens of algorithm animation systems have been developed in the last two decades [2–8, 10, 12, 16, 19, 20, 22]. For a comprehensive survey we refer the interested reader to [13, 21] and to the references therein. In the following we limit to discuss the features of algorithm visualization systems that appear to be most appealing for their deployment in engineering algorithms.

From the viewpoint of the algorithm developer, it is desirable to rely on systems that offer visualizations at a *high level of abstraction*. Namely, one would be more interested in visualizing the behavior of a complex data structure, such as a graph, than in obtaining a particular value of a given pointer.

Fast prototyping of visualizations is another fundamental issue: algorithm designers should be allowed to create visualization from the source code at hand with little effort and without heavy modifications. At this aim, *reusability* of visualization code could be of substantial help in speeding up the time required to produce a running animation.

One of the most important aspects of algorithm engineering is the development of *libraries*. It is thus quite natural to try to interface visualization tools to algorithmic software libraries: libraries should offer default visualizations of algorithms and data structures that can be refined and customized by developers for specific purposes.

Software visualization tools should be able to animate *not just "toy programs"*, but significantly complex algorithmic codes, and to test their behavior on large data sets. Unfortunately, even those systems well suited for large information spaces often lack advanced navigation techniques and methods to alleviate the screen bottleneck. Finding a solution to this kind of limitations is nowadays a challenge.

Advanced debuggers take little advantage of sophisticated graphical displays, even in commercial software development environments. Nevertheless, software visualization tools may be very beneficial in addressing problems such as finding memory leaks, understanding anomalous program behavior, and studying performance. In particular, environments that provide interpreted execution may more easily integrate advanced facilities in support to *debugging and performance monitoring*, and many recent systems attempt at exploring this research direction.

There is a general consensus that algorithm visualization systems can strongly benefit from the potentialities offered by the *World Wide Web*. Indeed, the use of the Web for easy communication, education, and distance learning can be naturally considered a valid support for improving the cooperation between students and instructors, and between algorithm engineers.

3 Concluding Remarks

Algorithm engineering refers to the process of designing, analyzing, implementing, tuning, debugging, and experimentally evaluating algorithms. It refines and reinforces the traditional theoretical approach with experimental studies, fitting the general models and techniques used by theoreticians to actual existing machines and leading to robust and efficient implementations of algorithms and data structures that can be used by non-experts. Trends in engineering algorithms include the following. First of all, programs should be considered as first class citizens: going from efficient algorithms to programs is not a trivial task and the development of algorithmic libraries may be helpful for this. Furthermore, in may applications seconds matter, and thus experimentation can add new insights to theoretical analyses. Moreover, current computing machines are far away from RAMs: if algorithms are to have a practical utility, issues such as effects of the memory hierarchy (cache, external memory), implications of communication complexity, numerical precision must be considered. Machine architecture, compiler optimization, operating system, programming language are just a few of the technical issues that may substantially affect the execution performance. Algorithmic software libraries, tools for algorithm visualization, program checkers, generators of test sets for experimenting with algorithms may be of great help throughout this process.

References

1. R. Anderson. The role of experiment in the theory of algorithms. In *Proceedings of the 5th DIMACS Challenge Workshop*, 1996. Available over the Internet at the URL: `http://www.cs.amherst.edu/~dsj/methday.html`.
2. J.E. Baker, I. Cruz, G. Liotta, and R. Tamassia. A New Model for Algorithm Animation over the WWW. ACM Computing Surveys, 27(4):568–572, 1996.
3. J.E. Baker, I. Cruz, G. Liotta, and R. Tamassia. Animating Geometric Algorithms over the Web. In *Proceedings of the 12th Annual ACM Symposium on Computational Geometry*, pages C3–C4, 1996.
4. J.E. Baker, I. Cruz, G. Liotta, and R. Tamassia. The Mocha Algorithm Animation System. In *Proceedings of the 1996 ACM Workshop on Advanced Visual Interfaces*, pages 248–250, 1996.
5. R.S. Baker, M. Boilen, M.T. Goodrich, R. Tamassia, and B. Stibel. Testers and visualizers for teaching data structures. *SIGCSEB: SIGCSE Bulletin (ACM Special Interest Group on Computer Science Education)*, 31, 1999.
6. M.H. Brown. *Algorithm Animation.* MIT Press, Cambridge, MA, 1988.
7. M.H. Brown. Zeus: a System for Algorithm Animation and Multi-View Editing. In *Proceedings of the 7-th IEEE Workshop on Visual Languages*, pages 4–9, 1991.
8. G. Cattaneo, G.F. Italiano, and U. Ferraro-Petrillo. CATAI: Concurrent Algorithms and Data Types Animation over the Internet. *Journal of Visual Languages and Computing*, 13(4):391–419, 2002. System Home Page: `http://isis.dia.unisa.it/catai/`.
9. B.V. Cherkassky and A.V. Goldberg. On implementing the push-relabel method for the maximum flow problem. *Algorithmica*, 19:390–410, 1997.

10. P. Crescenzi, C. Demetrescu, I. Finocchi, and R. Petreschi. Reversible Execution and Visualization of Programs with Leonardo. *Journal of Visual Languages and Computing*, 11(2), 2000. System home page: `http://www.dis.uniroma1.it/~demetres/Leonardo/`.
11. C. Demetrescu, I. Finocchi, G.F. Italiano, and S. Naeher. Visualization in algorithm engineering: Tools and techniques. In *Dagstuhl Seminar on Experimental Algorithmics 00371*. Springer Verlag, 2001.
12. C. Demetrescu, I. Finocchi, and G. Liotta. Visualizing Algorithms over the Web with the Publication-driven Approach. In *Proc. of the 4-th Workshop on Algorithm Engineering (WAE'00)*, LNCS 1982, pages 147–158, 2000.
13. S. Diehl. *Software Visualization*. LNCS 2269. Springer Verlag, 2001.
14. A.V. Goldberg. Selecting problems for algorithm evaluation. In *Proc. 3-rd Workshop on Algorithm Engineering (WAE 99), LNCS 1668*, pages 1–11, 1999.
15. D. Johnson. A theoretician's guide to the experimental analysis of algorithms. In *Proceedings of the 5th DIMACS Challenge Workshop*, 1996. Available over the Internet at the URL: `http://www.cs.amherst.edu/~dsj/methday.html`.
16. A. Malony and D. Reed. Visualizing Parallel Computer System Performance. In M. Simmons, R. Koskela, and I. Bucher, editors, *Instrumentation for Future Parallel Computing Systems*, pages 59–90. ACM Press, 1999.
17. C. McGeoch. A bibliography of algorithm experimentation. In *Proceedings of the 5th DIMACS Challenge Workshop*, 1996. Available over the Internet at the URL: `http://www.cs.amherst.edu/~dsj/methday.html`.
18. B.M.E. Moret. Towards a discipline of experimental algorithmics. In *Proceedings of the 5th DIMACS Challenge Workshop*, 1996. Available over the Internet at the URL: `http://www.cs.amherst.edu/~dsj/methday.html`.
19. G.C. Roman, K.C. Cox, C.D. Wilcox, and J.Y Plun. PAVANE: a System for Declarative Visualization of Concurrent Computations. *Journal of Visual Languages and Computing*, 3:161–193, 1992.
20. J.T. Stasko. Animating Algorithms with X-TANGO. *SIGACT News*, 23(2):67–71, 1992.
21. J.T. Stasko, J. Domingue, M.H. Brown, and B.A. Price. *Software Visualization: Programming as a Multimedia Experience*. MIT Press, Cambridge, MA, 1997.
22. A. Tal and D. Dobkin. Visualization of Geometric Algorithms. *IEEE Transactions on Visualization and Computer Graphics*, 1(2):194–204, 1995.

Report on the Invited Lecture by Pat Hanrahan, Titled "On Being in the Right Space"

The invited lecturer Pat Hanrahan is a professor at the Computer Science Department of Stanford University, Stanford CA 94305-9025, USA.

The talk given by Prof. Hanrahan was about the connection between semantic constraints and aesthetics in graph drawing and information visualization. Examples were given of different application contexts where, in order to convey the right semantic information in a diagram, it is essential to respect geometric constraints concerning the shape of the edges and /or the relative position of the vertices. These examples include drawing maps (such as the map of the metro), visualizing (portions of) the Internet, and representing chemical maps.

G. Liotta (Ed.): GD 2003, LNCS 2912, p. 514, 2004.

Selected Open Problems in Graph Drawing

Franz Brandenburg, David Eppstein, Michael T. Goodrich,
Stephen Kobourov, Giuseppe Liotta, and Petra Mutzel

[1] Univ. of Passau, Germany.
brandenburg@informatik.uni-passau.de
[2] Univ. of California-Irvine, USA.
eppstein@ics.uci.edu
[3] Univ. of California-Irvine, USA.
goodrich@acm.org
[4] Univ. of Arizona, USA.
kobourov@cs.arizona.edu
[5] Univ. of Perugia, Italy.
liotta@diei.unipg.it
[6] Vienna Univ. of Technology, Austria.
mutzel@ads.tuwien.ac.at

Abstract. In this manuscript, we present several challenging and interesting open problems in graph drawing. The goal of the listing in this paper is to stimulate future research in graph drawing.

1 Introduction

This paper is devoted to exploring some interesting and challenging open problems in graph drawing. We have specifically chosen topics motivated from research themes of current interest in graph drawing with an emphasis on computational geometry, including proximity drawings, 2D and 3D straight-line drawings, 2D and 3D orthogonal representations, crossing minimization, and graph drawing checkers. Our choice of open problems and themes should not be seen as exhaustive, however, as graph drawing is a rich area of research with many open problems and research themes. For example, there are interesting open problems related to the routing of multiple curves in the plane, dynamic graph drawing, drawing of hypergraphs and Venn diagrams, applied graph drawing, and label placement in graph drawings, which we do not specifically address in this paper.

In any survey of open problems, there is naturally going to be a certain emphasis that reflects the perspective of the authors, and this paper is no exception. Nevertheless, we have striven to include open problems motivated by topics we believe are of general interest. That is, we feel that the solution to any of the open problems we include in this paper would be of wide interest in the graph drawing community.

G. Liotta (Ed.): GD 2003, LNCS 2912, pp. 515–539, 2004.

2 Proximity Drawability Problems

Recently, much attention has been devoted to the study of the combinatorial properties of different types of proximity graphs. Proximity graphs, originally defined in the context of computational geometry and pattern recognition, are typically used to describe the shape of a set of points. In the survey by Toussaint [84], such graphs are classified by using the notion of *proximity* between points: In a proximity graph, two points are connected by an edge if and only if they are deemed *close* by some proximity measure. Examples of proximity graphs include well-known geometric graphs such as minimum spanning trees, Gabriel graphs, minimum weight triangulations, rectangle of influence graphs, visibility graphs, and Delaunay diagrams [116].

Proximity graphs can be regarded as straight-line drawings that satisfy some additional geometric constraints. Thus the problem of analyzing the combinatorial properties of a given type of proximity graph naturally raises the question of the characterization of those graphs admitting the given type of drawing. This, in turn, leads to the investigation of the design of efficient algorithms for computing such a drawing when one exists. These questions are far from being resolved in general, and only partial answers have appeared in the literature [46, 48,84,95,97,102,103,105].

2.1 Realizability of Minimum Spanning Trees in 3D Space

Among the most challenging questions in this family of graph drawing problems it has to be mentioned one related with drawing trees as minimum spanning trees in 3D. A *minimum spanning tree* of a set P of points is a connected, straight-line drawing that has P as vertex set and minimizes the total edge length. A tree T can be drawn as a minimum spanning tree if there exists a set P of points such that the minimum spanning tree of P is isomorphic to T.

The problem of testing whether a tree can be drawn as a Euclidean minimum spanning tree in the plane is essentially solved. Monma and Suri [110] proved that each tree with maximum vertex degree at most five can be drawn as a minimum spanning tree of some set of vertices by providing a linear time (real RAM) algorithm. In the same paper it is shown that no tree having at least one vertex with degree greater than six can be drawn as a minimum spanning tree. As for trees having maximum degree equal to six, Eades and Whitesides [57] showed that it is NP-hard [65] to decide whether such trees can be drawn as minimum spanning trees. The 3-dimensional counterpart of the problem is however not yet solved. In [101] it is shown that no trees having at least one vertex with degree greater than twelve can be drawn as a Euclidean minimum spanning tree in 3D while all trees with vertex degree at most nine are drawable.

Problem 1: *Let T be a tree with maximum vertex degree at most twelve. Is there a polynomial time algorithm to decide whether T can be drawn as a Euclidean minimum spanning tree in 3D-space and, if so, compute such a drawing?*

2.2 Drawing Minimum Weight Triangulations

A *minimum weight drawing* of a planar triangulated graph G is a straight-line drawing Γ of G with the additional property that Γ is a minimum weight triangulation of the points representing the vertices. If a graph admits a minimum weight drawing it is called *minimum weight drawable*, else it is called *minimum weight forbidden.*

Little is known about the problem of constructing a minimum weight drawing of a planar triangulation. Moreover, it is still not known whether computing a minimum weight triangulation of a set of points in the plane is NP-hard. Several papers have been published on this last problem, either providing partial solutions, or giving efficient approximation heuristics. A limited list of references includes the work by Meijer and Rappaport [118], Lingas [100], Keil [89], Dickerson et al. [43], Kirkpatrick [90], Aichholzer et al. [2], Cheng and Xu [22], Dickerson and Montague [44], and Levcopoulos and Krznaric [98,99].

In [95] Lenhart and Liotta show that all maximal outerplanar triangulations are minimum weight drawable and a linear time (real RAM) drawing algorithm for computing a minimum weight drawing of these graphs was given. In [97] Lenhart and Liotta examine the *endoskeleton*—or *skeleton*, for short—of a triangulation: that is, the subgraph induced by the internal vertices of the triangulation. They construct skeletons that cannot appear in any minimum weight drawable triangulation; skeletons that do appear in minimum weight drawable triangulations; and skeletons that guarantee minimum weight drawability.

Besides the natural (and ambitious) goal of characterizing those planar graphs that are minimum weight drawable, we recall here two open problems whose solution would represent an interesting contribution to the investigation of the combinatorial structure of minimum weight triangulations.

In [97] it is shown that any forest can be the skeleton of a minimum weight drawable triangulation. On the other hand, Wang, Chin, and Yang [133] also focus on the minimum weight drawability of triangulations with acyclic skeletons and show examples of triangulations of this type that do not admit a minimum weight drawing. These two results naturally lead to the following problem.

Problem 2: *Characterize the class of triangulations with acyclic skeleton that admit a minimum weight drawing.*

Lenhart and Liotta [97] also show that there exist an infinite class of minimum weight drawable triangulations that cannot be realized as Delaunay triangulations (that is, for any triangulation T of the class it does not exist a set P of points such that the Delaunay triangulation of P is isomorphic to T). It is worth remarking that the study of the geometric differences between the minimum-weight and Delaunay triangulations of a given set of points in order to compute good approximations of the former has a long tradition (see, e.g., [90,98,107]), yet little is known about the combinatorial difference between Delaunay triangulations and minimum-weight triangulations.

Problem 3: *Further investigate the combinatorial relationship between minimum weight and Delaunay drawable triangulations. Are there any Delaunay drawable and minimum weight forbidden triangulations?*

2.3 Gabriel Drawability

Let P be a set of n distinct points in the plane. The Gabriel graph of P is a geometric graph whose vertices are the points of P such that there exists an edge (u, v) if and only if the disk whose antipodal points are u and v does not contain any points of P except u and v (the disk is assumed to be a closed set). Gabriel graphs have been first studied in the context of pattern recognition and computational morphology, where one is given a set of points on the plane and is asked to display the underlying *shape* of the set by constructing a graph whose vertices are the points and whose edges are segments connecting pairs of points. For references on this topics see, e.g., the survey by Toussaint [84].

More recently, the problem of computing straight-line drawings that have the additional property of being Gabriel graphs of their vertices has been considered. In [13], the problem of characterizing Gabriel-drawable trees has been addressed and an algorithm to compute Gabriel drawings of trees in the plane is given. Lubiw and Sleumer [105] proved that maximal outerplanar graphs admit both relative neighborhood drawings and Gabriel drawings. This result has been extended in [96] to all biconnected outerplanar graphs. Several problems about Gabriel drawings remain open. Among the most fascinating and challenging, we mention the following.

Problem 4: *Characterize the family of Gabriel drawable triangulations, that is, the family of those triangulations that admit a straight-line drawing where the angles of each triangular face are less than $\pi/2$.*

3 Straight-Line Drawability in 2D and 3D

A classical area of investigation deals with crossing-free straight-line drawings of planar graphs in two and three dimensions. Given a graph G, the vertices in a drawing of G are constrained to be located at integer grid points and the optimization goal is that of computing drawings whose area/volume is small. The area/volume of a drawing Γ is measured as the number of grid points contained in a *bounding box* of Γ, i.e., the smallest axis-aligned box enclosing Γ.

3.1 Compact Straight-Line Drawings in 2D Space

A rich body of literature has been published on computing straight-line drawings of graphs, such that the vertices are the intersection points of an integer 2D grid and the overall area of the drawing is kept small. Typically, papers that deal with this subject focus on lower bounds on the area required by straight-line drawings of specific classes of graphs and on the design of algorithms that possibly match these lower bounds. Schnyder [122,123] and de Fraysseix, Pach, and Pollack [29] independently showed that every n-vertex triangulated planar graph has a crossing-free straight-line $O(n) \times O(n)$ grid drawing, and that this is worst case optimal. This seminal contribution was followed by related work by Kant [86,87], Chrobak and Kant [25], Schnyder and Trotter [124], Felsner [62], and Chrobak, Goodrich, and Tamassia [23]. Results on classes of drawings of trees include the results by Garg, Goodrich and Tamassia [68], by Chan [17], and by Garg and Rusu [70]. Summarizing tables and more references can be found in the book by Di Battista, Eades, Tamassia, and Tollis [31].

The following problem is motivated by the observation that trees admit straight-line grid drawings with linear or almost-linear area, while triangulations may require a grid of quadratic size.

Problem 5: *Find nontrivial classes of graphs with n vertices richer than trees that admit straight-line planar drawings on an integer 2D grid of size $o(n^2)$.*

Some evidence that Problem 5 might be solvable for families of graphs other than trees was given in a recent paper by Biedl [9], where she shows that all outerplanar graphs admit an $O(n \log n)$ area drawing; however, the algorithm presented in [9] does not compute straight-line drawings. Very recently, a paper presented at GD 2003 by Garg and Rusu [71] proves that an outerplanar graph with n vertices and degree d has a straight-line grid drawing in area $O(dn^{1.693}$.

Additional problems in this area include the following well-known problem, which has been popularized by Xin He:

Problem 6: *Given an n-vertex plane triangular graph G, what is the smallest area straight-line planar drawing of G such that vertices are drawn at integer grid points?*

This problem was first posed by Rosenstiehl and Tarjan [120]. A well-known upper bound is $(n-1) \times (n-1)$ (many such algorithms achieve this), which was recently improved by Zhang and He [136] to $(n - \Delta - 1) \times (n - \Delta - 1)$, where $0 \leq \Delta \leq \lfloor (n-1)/2 \rfloor$ is a value derived from the cycle structure of G. A known lower bound for this problem is provided by the plane graph consisting of a sequence of $n/3$ nested triangles connected to each other by six edges. It is not hard to see that this graph requires a drawing area of at least $2n/3 \times 2n/3$ [29, 106]. Note, that this is not really a lower bound for planar graphs, only for plane graphs (i.e., the combinatorial embedding and the outer face is fixed). Since the nested triangle graph is 3-connected, it has only one combinatorial embedding (up to its mirror image). However, it is not clear what happens if one is free to pick the outer face.

Problem 7: *What is the smallest area grid drawing for the nested triangles graph, when the outer face is not given?*

Many planar graph drawing algorithms incrementally construct a drawing based on inserting vertices according to a particular ordering, such as an *st*-numbering or canonical ordering of the graph's vertices [87]. Such drawings place vertices in a *numbering-upward* fashion, so that listing the vertices according to the ordering lists the vertices by non-decreasing y-coordinates.

Problem 8: *Is there a polynomial algorithm for numbering and embedding the vertices of a given planar graph G so as to minimize the area of a numbering-upward planar drawing of G (taken over all possible numberings of G)?*

3.2 Extensibility and Universality for Planar Graphs

Consider the following problem: given a planar graph in which some vertices have already been placed in the plane (i.e., a *partial embedding*), place the remaining vertices to form a straight line embedding of the graph.

Problem 9: *Is there a polynomial time algorithm for extending a partial straight-line embedding to a planar straight-line drawing?*

This problem also comes up in mesh generation, where the already-placed vertices can be assumed to form a simple polygon and the graph can be assumed to have all interior faces triangles; do these assumptions simplify the problem?

A related problem is the following.

Problem 10: *(Drawing with fixed vertex positions) Suppose you are given as input an unlabeled planar graph with n vertices, and a set of n points in the plane. You wish to assign vertices to points to create a planar straight line drawing. What is the computational complexity of this problem?*

Concerning the problem above, Bose [12] conjectured it to be NP-complete and solved some special cases in polynomial time. Kaufmann and Wiese [88] show NP-completeness of a related problem in which some bends are allowed.

In general, a set S of points is called a *universal set* for a family of a graphs $\mathcal{F}$ if every graph in $\mathcal{F}$ can be drawn crossing-free using the points from S for vertex locations.

Problem 11: *Does there exist a small universal point set for planar graphs? That is, given n, is there a set of $O(n)$ points in the plane, so that every planar graph of n vertices can be drawn with straight-lines and no crossings on this point set? If not, what is the smallest $f(n)$ such that there exists a set of $f(n)$ points such that every n-vertex graph can be drawn using these points?*

There are several algorithms that show $f(n) \leq n^2$. Chrobak and Karloff [26] show that $f(n) \geq 1.098n$, for sufficiently large n. Likewise, the following bi-universal set question is also of interest.

Problem 12: *Given two planar graphs each on n vertices, can one always find a set of n points, so that each of the two graphs can be embedded with straight-lines and no crossings on this set?*

3.3 Universal Sets in 3D

While the problem of computing small-sized crossing-free straight-line drawings in the plane has a long tradition, its 3D counterpart has become the subject of much attention only in recent years. Chrobak, Goodrich, and Tamassia [23] gave an algorithm for constructing 3D convex drawings of triconnected planar graphs with $O(n)$ volume and non-integer coordinates. Cohen, Eades, Lin and Ruskey [27] showed that every graph admits a straight-line crossing-free 3D drawing on an integer grid of $O(n^3)$ volume, and proved that this is asymptotically optimum. Calamoneri and Sterbini [15] showed that all 2-, 3-, and 4-colorable graphs can be drawn in a 3D grid of $O(n^2)$ volume with $O(n)$ aspect ratio and proved a lower bound of $\Omega(n^{1.5})$ on the volume of such graphs. For r-colorable graphs, Pach, Thiele and Tóth [113] showed a bound of $\theta(n^2)$ on the volume. Garg, Tamassia, and Vocca [74] showed that all 4-colorable graphs (and hence all planar graphs) can be drawn in $O(n^{1.5})$ volume and with $O(1)$ aspect ratio but using a grid model where the coordinates of the vertices may not be integer.

In a recent paper, Felsner, Liotta, and Wismath [63] approached the drawing problem with the following point of view: Instead of "squeezing" a drawing onto a small portion of a grid of unbounded dimensions, it is assumed that a grid of *specified* dimensions (involving a function of n) is given and we consider what the graphs are whose drawings fit that restricted grid. For example, it is well-known that there are families of graphs that require $\Omega(n^2)$ area to be drawn in the plane, the canonical example being a sequence of $n/3$ nested triangles (see [29, 24,123]), as mentioned above. Such graphs can be drawn on the surface of a three dimensional triangular prism of linear volume and using integer coordinates. Thus a natural question is whether there exist specific restrictions of the 3D integer grid of linear volume that can support straight-line crossing-free drawings of meaningful families of graphs. The following open problem is mentioned in the paper by Felsner, Liotta, and Wismath [63].

Problem 13: *Does there exist a 3D universal grid set of linear volume that supports all planar graphs?*

Partial results on Problem 13 are as follows. In [63] it is shown that all outerplanar graphs can be drawn in a restricted integer 3D grid of linear volume, called *prism*. A prism is a subset of the integer 3D grid consisting of three parallel lines through the points $(0,1,0)$, $(0,0,0)$, and $(0,0,1)$. In the same paper it is also shown that a prism does not support all planar graphs and that even adding a fourth parallel line does not suffice in general. Dujmović, Morin, and Wood [50] present $O(n \log^2 n)$ volume drawings of graphs with bounded tree-width and $O(n)$ volume for graphs with bounded path-width. Wood [135] shows that also graphs with bounded queue number have 3D straight-line grid drawings of $O(n)$ volume. A recent result by Dujmović and Wood [51,56] shows that linear volume can also be achieved for graphs with bounded tree-width; they show 3D straight-line grid drawings of volume $c \times n$ for these graphs, where c is a constant whose value exponentially depends on the tree-width. The value of constant c is reduced in the work by Di Giacomo, Liotta, and Meijer [39], although the constant still depends exponentially on the tree-width. For the special case of 2-trees (every series-parallel graph can be augmented to be a 2-tree), Dujmović and Wood [51,56] show $36 \times 37 \times 37\lceil \frac{n}{18} \rceil$ volume. This bound is reduced to $30 \times 31 \times 31\lceil \frac{n}{15} \rceil$ by Di Giacomo, Liotta, and Meijer [39]. Different families of graphs with constant queue number are studied by Di Giacomo and Meijer [42] who present algorithms that compute linear volume 3D drawings of for these graphs where the multiplicative constants in the volume bounds significantly improve those obtained by directly applying the results in [135]. Di Giacomo [38] and Di Giacomo, Liotta, and Wismath [41] show $4 \times n$ and $32 \times n$ volume for two subclasses of series-parallel graphs. A result related to Problem 13 is the elegant work presented at GD 2003 by Dujmović and Wood [54] who describe different families of graphs, including planar graphs, for which a sub-quadratic (but still super-linear) volume upper bound can be found. A list of very recent papers related to Problem 13 also includes [52,55,53].

4 Orthogonal Representations

An *orthogonal drawing* of a graph maps its vertices to points on an integer grid and its edges to a sequence of alternating axis-parallel segments. The study of the orthogonal drawing convention has a very long tradition in the graph drawing literature, because of the several applications of this convention in a variety of fields. For an introduction on orthogonal drawings and their several applications, see [31]. It is immediate to see that in order for an orthogonal drawing to exist, the degree of the vertices must not exceed four in the plane and six in three dimensions.

4.1 Orthogonal Representations in 2D Space

The *topology-shape-metrics approach* [31] for constructing a planar orthogonal drawing of a planar graph in $2D$ consists of three main steps, called planarization, orthogonalization, and compaction. The planarization step determines an embedding, i.e., the face cycles, for the graph in the plane. The orthogonalization step determines an *orthogonal representation* of the input graph, i.e., a labeling for each edge (u, v) of the graph that defines the shape of (u, v) in the final drawing. For example, (u, v) could be labeled $RLLLR$, which would say "starting from u first turn right, then turn left three times, then right." Finally, the compaction step computes the drawing, giving coordinates to vertices and bends while preserving the shape of the edges determined in the orthogonalization step. For each step of the approach, different optimization problems (for example minimizing the number of bends, minimizing the area, minimizing the maximum edge length) have been studied, and papers providing optimal algorithms and effective heuristics have been presented (see, e.g., [6,111,92,91]).

A *bend-minimum* planar orthogonal drawing of a plane graph G is one which has the minimum number of bends along the edges among all possible planar orthogonal drawings of G. The problem of computing bend-minimum planar orthogonal drawings is among the most famous in the graph drawing literature and has been studied both in the *fixed embedding setting* and in the *variable embedding setting*.

In the fixed embedding setting, the input is a planar graph G together with a planar embedding (i.e., a circular ordering of the edges incident on each vertex of G and an outer face); the output is a planar orthogonal drawing of G that has the minimum number of bends among all drawings which maintain the given embedding of G. A well-known seminal paper by Tamassia [128] shows that such a drawing of G can be computed in $O(n^2 \log n)$ time by mapping the problem to computing a flow of minimum cost on a suitable network. The time complexity bound is further improved by Garg and Tamassia [72] who present an $O(n^{1.75} \log n)$ time algorithm.

Problem 14: *Let G be a 4-planar graph (i.e., a planar graph whose maximum vertex degree is at most four) with a given planar embedding Φ. Is there an algorithm that computes a bend-minimum orthogonal representation of G preserving Φ and has time complexity $o(n^{1.75} \log n)$?*

Notice that the choice of a planar embedding of G can deeply affect the resulting number of bends in the orthogonal representation. For example, the algorithm in [128] can give rise to orthogonal representations of the same graph that, depending on the choice of its planar embedding, differ in the number of bends by even a linear factor [36]. Thus, there is natural interest in algorithms that work in a *variable embedding setting*, i.e., algorithms that are allowed to change the planar embedding of the input graph in order to minimize the bends.

The variable embedding version of the bend-minimization problem is unfortunately harder than the fixed embedding one. Namely, Garg and Tamassia [73] have proved that computing a bend-minimum planar orthogonal drawing in the variable embedding setting is NP-hard. Di Battista, Liotta, and Vargiu [36] proved that the problem can be solved in $O(n^3)$ time if the input graph is a series-parallel graph and that it can be solved in $O(n^5 \log n)$ time if the input is a 3-planar graph (i.e., a graph that has vertex degree at most three). Garg and Liotta [69] further reduced this time complexity, but at the expenses of a few extra bends in the computed representation. Namely, they present an $O(n^2)$-time algorithm that receives as input a 3-planar graph G with n vertices and computes an orthogonal representation of G with at most 3 bends more than the minimum number of bends.

The natural counterpart of Problem 14 in the variable embedding setting is therefore the following.

Problem 15: *Let G be a planar graph whose vertices have degree at most three. Is there an algorithm to compute a planar bend-minimum orthogonal drawing of G in $o(n^5 \log n)$ time?*

Very recently, a paper presented at GD 2003 by Rahman, Egi, and Nishizeki [117] partially answers Problem 15 by showing a linear-time algorithm for subdivisions of planar triconnected cubic graphs.

Because of the NP-hardness of finding the best planar embedding concerning the number of bends, recent papers have focussed on identifying *good* embeddings. Pizzonia and Tamassia [115] have suggested to consider planar embeddings with the minimum depth. The *depth of an embedding* measures the topological nesting of the biconnected components of G in the embedding. They have also given a polynomial time algorithm for computing the minimum depth embedding in the case when the embeddings of the biconnected components are fixed. Recently, Gutwenger and Mutzel [80] have suggested a linear time algorithm which computes the minimum depth embedding over the class of all planar embeddings without any restriction. They also suggest to look for embeddings with maximum outer face, and provide a linear time algorithm. Moreover, they give a linear time algorithm for the combined problem, namely, to search the embedding with maximum outer face among the set of all minimum depth embeddings. However, so far, no experiments or proofs exist that show that embeddings satisfying these criteria (almost) always lead to good drawings. The research in this direction has just started.

Problem 16: *Identify (provable) criteria for planar embeddings that lead to good drawings, e.g., that lead to drawings with a relatively small number of bends and/or small area. Moreover, come up with corresponding polynomial time algorithms.*

4.2 Compact Orthogonal Layouts

As in other drawing problems, there is considerable interest in the exact area bounds for various orthogonal drawing problems.

Problem 17: *What are exact bounds for the area of binary tree layouts in each of the following scenarios:*

1. *straight-line drawings on the grid*
2. *rectangular drawings on the grid (H-layouts)*
3. *straight-line upward drawings on the grid*
4. *rectangular upward drawings on the grid (T-layouts)*

Problem 18: *Does every binary tree have a drawing with straight-line edges in $O(n)$ area such that all nodes are placed on grid points?*

In the rectangular case, the straight-line edges are either horizontal or vertical. In the upward case, the edges are y-monotone or go in directions west, east, and north (south is excluded). Known results are $O(n \log \log n)$ area for straight-line drawings, $O(n \log \log n)$ for straight-line upward drawings [68,125], $O(n \log \log n)$ for rectangular drawings [18], and $O(n \log n)$ for upward rectangular (from hv drawings).

Problem 19: *What are exact bounds for the area of upward (strictly upward) layouts of binary trees. Specifically, what is the complexity of determining, given a binary tree T and a bound K, whether there exists a rectangular upward (strictly upward, hv) layout with area at most K? Are these problems NP-hard?*

Known results: NP-hardness is known for "H"-tree layouts (all 4 directions) [7] and improved to binary trees by Gregori [76], for upward drawings of ternary trees by Edler [58], and for rectilinear layouts by Garg and Tamassia [73].

4.3 Orthogonal Representations in 3D Space

An essential prerequisite of the topology-shape-metrics approach to drawing graphs in 2D is a characterization of 2D orthogonal representation, i.e., a characterization of those graphs with edges labeled by orthogonal directions that can be drawn without crossings, while respecting the desired shapes for the edges. This problem has been studied in several papers, including [128,130], and has also been generalized to non-orthogonal polygons and graphs in [131,37,67]. However, while the literature on 3D orthogonal drawings is quite rich, the extension of the topology-shape-metrics approach to 3D has so far remained an elusive target. A major difficulty is that in 3D, there is no counterpart to the 2D characterization of orthogonal representations.

A *direction label* is a label in the set $\{E, W, N, S, U, D\}$, where each label specifies a direction *East*, *West*, *North*, *South*, *Up*, or *Down*, respectively. Let G be a graph such that each edge (u, v) of G is associated with two opposite orientations called *darts* of (u, v). A *3D shape graph* is a labeling σ of the darts of G such that:

- Each dart is associated with exactly one direction label.
- For each edge e of G the two opposite darts of e have labels which specify opposite directions.
- Each vertex does not have two entering darts with the same label.

Shape graph γ is said to be *a 3D orthogonal representation* if there exists an orthogonal drawing Γ of G so that Γ is *simple* (i.e., no two edges of Γ share any points except common endpoints) and satisfies the direction constraints on its edges as specified by σ.

Problem 20: *Characterize those shape graphs that are 3D orthogonal representations.*

Only very preliminary results toward finding such characterization have so far been discovered. Di Battista, Liotta, Lubiw, and Whitesides [35] solve the problem for paths (with the additional constraint that one endpoint must be drawn at the origin and the other in a given octant) and for shape cycles [34]. As for structurally more complex graphs, Di Giacomo, Liotta, and Patrignani [40] show that the characterization for shape cycles does not extend to general graphs, even for apparently simple structures such as theta graphs (a *theta graph* is a simple graph consisting of three cycles). In the same paper, the authors present a sufficient condition under which a shape theta graph is a 3D orthogonal representation. A consequence of the work in [40] is the following question, whose answer may be an important intermediate step toward the ambitious goal of solving Problem 20.

Problem 21: *Characterize those shape theta graphs that are 3D orthogonal representations.*

5 Crossing Minimization

The *crossing number* $cr(G)$ of a graph G is the minimum number of edge crossings in any drawing of G. It represents a fundamental measure of nonplanarity of graphs and has been studied for more than 40 years. Nevertheless, so far only few infinite classes of graphs exist for which the crossing number is known. We do not even know if the long open standing conjectures for the complete graph K_n and for the complete bipartite graph $K_{m,n}$ are true. So far, the correctness has been verified only for K_n with $n \leq 10$ and $K_{m,n}$ with $m \leq 6$ as well as a few more special cases (see, e.g., [119]).

Problem 22: *Determine the crossing number for the complete graph K_n and for the complete bipartite graph $K_{m,n}$.*

The crossing number problem and many of its variants are NP-hard. So far, no algorithms exist which are able to compute the crossing number of even small graphs in practice (like, e.g., branch-and-bound algorithms). The main problem with exact, enumeration-based, methods is the lack of tight lower bounds. Most lower bounds only depend on the number of vertices and edges of the graph, but not on its particular structural properties. Research results in the latter direction have been provided by Leighton [93], Pach, Shahrokhi, and Szegedy [112], Sýkora and Vrťo [127], and Dijdjev and Vrťo [45] who have shown bounds depending on the bisection width and the cut width, respectively. Both problems are NP-hard, too, but they are easier to deal with.

Problem 23: *Find tight lower bounds for the crossing number based on structural properties of graphs.*

So far, not even efficient approximation algorithms have been found. For graphs with bounded degrees, Even, Guha, and Schieber [60] have suggested an approximation algorithm in which the sum of the numbers of vertices and crossings is $O(\log^3 |V|)$ times the minimum sum thus improving the results of $O(\log^4 |V|)$ by Bhatt and Leighton [8] and Leighton and Rao [94]. A promising approach could be to search for efficient fixed-parameter algorithms. Grohe [77] has given an exact quadratic time algorithm if the number of crossings is consider to be a fixed (small) parameter. Unfortunately, the constant in the worst case time analysis is too big to be useful in practice.

Problem 24: *Provide approximation algorithms or efficient fixed-parameter algorithms for the crossing number problem.*

Since the crossing number is so hard to address, researchers started to study variants of the crossing number that might be easier to attack. Pach and Toth [114] have introduced the *pairwise crossing number* $cr_p(G)$, which is the minimum number of pairs of edges that cross at least once, and the *odd crossing number* $cr_o(G)$, which is the minimum number of pairs of edges that cross an odd number of times. They have shown that

$$cr_o(G) \leq cr_p(G) \leq cr(G) \leq 2(cr_o(G))^2.$$

So far, it is not clear if equality holds.

Problem 25: *Prove or disprove:* $\mathrm{cr}_o(G) = \mathrm{cr}_p(G) = \mathrm{cr}(G)$.

Better known is the *rectilinear crossing number*, which is the minimum number of crossings in a straight-line drawing. It has been shown in Bienstock and Dean [10] that there are graphs with crossing number 4, whose rectilinear crossing numbers are arbitrarily large. Although, it seems easier to study the rectilinear crossing number, also here, the value of the complete graph K_n is not known for $n > 12$ [3].

Problem 26: *What is the rectilinear crossing number of* K_n*?*

Closely related to the rectilinear crossing number of complete graphs are Sylvester's *Four Point Problem* and the problem of finding the minimum number of convex quadrilaterals determined by n points in general position in the plane [104,121].

Aronov et al. [4] showed an $\Omega(\sqrt{n})$ bound for the size of crossing families (i.e., each edge crosses each other), and asked whether a linear sized crossing family always exists.

Problem 27: *In a straight-line drawing of the complete graph* K_n*, how large must a set of mutually crossing edges be?*

A successful approach for computing a small crossing number in practice is the planarization approach (see, e.g., [32,79]). Here, in a first step, the minimum number of edges is deleted from G in order to obtain a planar graph P. In a second step, the edges are reinserted into P while trying to keep the number of crossings small.

Finding the largest cardinality planar subgraph of a graph is NP-hard, and even MAXSNP-hard, so one cannot expect to approximate it arbitrarily well, but there may be room for reducing the best known approximation ratio of 2.25, which has been achieved by an algorithm by Calinescu et al. [16].

Problem 28: *Find an efficient approximation algorithm for the maximum cardinality planar subgraph problem with factor below 2.25.*

Also the edge reinsertion step is a NP-hard optimization problem. The standard algorithm reinserts the edges iteratively into a fixed combinatorial embedding of the planar subgraph P. This can be done easily via simple shortest-path computations in the extended dual graph. However, the quality of the resulting drawing highly depends on the chosen embedding for P. Gutwenger, Mutzel and Weiskircher [81] have suggested a linear time algorithm for inserting one edge into a planar graph P so that the number of crossings in $P \cup \{e\}$ over the set of all possible planar embeddings of P is minimized. A generalization of this approach for fixed k would be of interest theoretically and practically.

Problem 29: *Find a polynomial time algorithm for inserting a fixed number of edges* $e_1, \ldots, e_k$ *into a planar graph such that the number of crossings in* $P \cup \{e_1, \ldots, e_k\}$ *over the set of all possible planar embeddings of P is minimized.*

6 Graph Drawing Checkers

The intrinsic structural complexity of the implementation of geometric algorithms makes the problem of formally proving the correctness of the code unfeasible in most of the cases. This has been motivating the research on *checkers*. A checker is an algorithm that receives as input a geometric structure and a predicate stating a property that should hold for the structure. The task of the checker is to verify whether the structure satisfies or not the given property. Here, the expectation is that it is often easier to evaluate the quality of the output than the correctness of the software that produces it. Different papers (see, e.g., [30,109]) have agreed on the basic features that a "good" checker should have:

Correctness: The checker should be correct beyond any reasonable doubt. Otherwise, one would incur in the problem of checking the checker.

Simplicity: The implementation should be straightforward.

Efficiency: The expectation is to have a checker that is more efficient than the algorithm that produces the geometric structure.

Robustness: The checker should be able to handle degenerate configurations of the input and should not be affected by errors in the flow of control due to round-off approximations.

Checking is especially relevant in the graph drawing context. Indeed, graph drawing algorithms are among the most sophisticated of the entire computational geometry field, and their goal is to construct complex geometric structures with specific properties. Also, because of their immediate impact on application areas, graph drawing algorithms are usually implemented right after they have been devised. Of course, the checking problem becomes crucial when the drawing algorithm deals with very large data sets, when a simple complete visual inspection of the drawing is difficult or unfeasible.

Devising graph drawing checkers involves answering only apparently innocent questions like: "is this drawing planar?" or "is this drawing upward?" or "are the faces convex polygons?".

The problem of checking the planarity of a subdivision has been independently studied by Mehlhorn et al. [109] and by Devillers et al. [30]. In these papers linear time algorithms are given to check the planarity of a subdivision composed by convex faces. The inputs are the subdivision plus its topological embedding in terms of the ordered adjacency lists of the edges. Unfortunately, extending the above techniques to checking the planarity of a subdivision whose faces are not constrained to be convex, relies on the usage of algorithms for testing the simplicity of a polygon. The only general linear time algorithm known for this problem is the well-known and fairly complex algorithm by Chazelle [20]. Hence, devising a checker based on such algorithm would not satisfy the simplicity requirement. The algorithm in [20] tests the simplicity of a polygon by means of an intermediate triangulation step. Alternative algorithms that can triangulate in linear time special classes of polygons have been devised. See e.g. [59,66, 134]. Other almost optimal algorithms can be found in [21,83,129].

Problem 30: *Let Γ be a straight-line drawing of a connected graph G with its combinatorial embedding (i.e., the circular ordering of the edges incident on each vertex of Γ). Devise a simple, robust, and efficient checker for testing the planarity of Γ.*

An example of checker of the type described by Problem 30 can be found in a paper by Di Battista and Liotta [33] who check the *upward planarity* of straight-line oriented drawings whose faces may not be convex polygons. They introduce *regular* upward planar embeddings and show that such embeddings coincide with those having a "unique" including planar st-digraph. The concept of regularity is exploited to investigate the relationships between topology and geometry of upward planar drawings. In particular, it is shown that an upward drawing of a regular planar upward embedding satisfies strong constraints on the left-to-right ordering of the edges. Based upon the above results and under the assumption of regularity, a simple and robust linear time checker is presented that tests whether a given drawing Γ is upward planar without using any polygon triangulation routine.

Besides planarity, several other checking problems remain open in graph drawing. Indeed, all graph drawing algorithms guarantee certain geometric properties for the drawings they produce. Such properties are usually called "graphic standards" or "drawing conventions". Some of them appear to be easy to check, while others like checking proximity drawings seem to be more challenging. For example, consider the following problem.

Problem 31: *Let Γ be a straight-line drawing of a tree (or even a simple chain). Is there a robust and simple algorithm to check in $o(n \log n)$ time whether Γ is a Euclidean minimum spanning tree of the set of its vertices?*

7 Visualizing Graph Properties

In this section we consider open problems that are related to the visualization of various graph properties or additional information associated with a graph.

7.1 Layered Graphs

Some graphs are labeled so as to partition the set of vertices into layers. Drawing algorithms that deal with this additional information usually seek to assign the vertices of each layer to a shared y-coordinate that is greater than that of the previous layer. Algorithms for drawing layered graphs usually are based on an approach that is commonly referred to as the Sugiyama algorithm [126].

A crucial problem is the *layered crossing minimization problem*, in which for each layer L permutations of the vertices in L are searched with the goal of obtaining the minimum nmber of crossings when the layers are parallel horizontal lines, all vertices are drawn on the line corresponding to their layers with x-coordinate compatible with their position in the layer permutation and all edges are drawn as straight line segments between consecutive layers. The layered crossing minimization problem is NP-hard, even when restricted to two layers. Exact approaches have been suggested by Healy and Kuusik [82] and Jünger et al. [85], but so far they are not yet efficient enough to be used in practical graph layout systems. So far, no polynomial time approximation algorithms exist. Relaxing the layering condition and only requiring that all the directed edges point downwards, then the algorithm by Dujmovic et al. [49] can be used to determine whether a given graph on n vertices can be drawn on h layers with at most k crossings, where h and k are fixed parameters. However, due to the high constants, this algorithm is rather of theoretical nature.

Problem 32: *Design a practically efficient approximation algorithm (or fixed-parameter algorithm or exact algorithm) for layered crossing minimization.*

In practice, the layered crossing minimization problem is solved via a series of two-layer problems. In this context, the task of efficiently counting the crossings in a bilayered graph $G = (V, E)$ is essential. The best known algorithm for counting the crossings for a set of straight line segments in the plane is due to Chazelle [19] and runs in time $O(|E|^{1.695})$. For the *bilayer cross counting problem,* a popular alternative in graph drawing is a sweep-line algorithm running in time $O(|C| + |E|)$, where $|C|$ denotes the number of crossings. A breakthrough has been achieved by Waddle and Malhotra [132] who have provided the first algorithm running in time $O(|E| \log |V|)$. The idea is based on a sweep-line procedure and involves complicated case distinctions. Recently, Barth, Jünger and Mutzel [5] have observed that the bilayer cross counting problem can be reduced to counting the inversions of a certain sequence, which directly lead to a set of simple $O(|E| \log |V|)$ algorithms (e.g., a variant of merge sort). Using the accumulator tree, they can improve the running time to $O(|E| \log |V_s|)$, where V_s is the smaller set of the two layers. In this context, the following problems arise:

Problem 33: *Is it possible to count the number of crossings in a bilayered graph faster than in $\Theta(|E| \log |V|)$ time? Is it possible to count the inversions in a sequence of length n faster than $\Theta(n \log n)$?*

Another issue regarding layered drawings is the representational complexity of describing a layered drawing.

Problem 34: *If a multi-layer drawing of a graph is made in which all the vertices have integer coordinates, how big do those integers need to be? (For planar graphs, it is known that all coordinates can be $O(n)$.)*

7.2 Clustered Graph Drawing

A clustered graph $C = (G, T)$ consists of an undirected graph G and a rooted tree T in which the leaves of T correspond to the vertices of $G = (V, E)$. Each vertex c in T corresponds to a subset of vertices of the graph called a *cluster*. c-planarity is a natural extension of graph planarity for clustered graphs, and plays an important role in automatic graph drawing. However, the complexity status of c-planarity testing is unknown. Feng, Cohen, and Eades [64] and Dahlhaus [28] have shown that c-planarity can be tested in polynomial time for graphs in which all the cluster induced subgraphs are connected. Recently, Gutwenger et al. [78] have further extended the class of graphs.

Problem 35: *Can c-planarity for clustered graphs be tested in polynomial time?*

7.3 Geometric Thickness

The *geometric thickness* of a graph is the minimum number of subgraphs into which the graph can be partitioned, in such a way that all subgraphs have planar straight line drawings with the same vertex positions in each drawing. Useful references for the problems that follow are [47,108].

Problem 36: *The complete graphs K_n should require a number of layers of the form $c\,n$, for some constant c. What is c? Similarly, what is the asymptotic number of layers needed for $K_{n,n}$?*

For arbitrary graphs, computing the thickness is known to be NP-complete.

Problem 37: *Is it equally hard to compute the geometric thickness? Can geometric thickness be approximated efficiently?*

Problem 38: *Which minor-closed graph families have bounded geometric thickness? Do bounded-degree graphs have bounded geometric thickness?*

The solution of the first part of Problem 38 has been announced by Blankenship and Oporowski [11]. Using results of Heath and a structure theorem of Robertson and Seymour, they have shown an even stronger result, namely, that for every minor-closed class of graphs, other than the class of all graphs, there is an integer K such that all members of the class have book thickness at most K.

7.4 Alternative Graph Representations

There are many additional research themes in graph drawing that consider alternative representations than the one that assigns vertices to points and edges to curves joining those points. The following problems fall into the category of *visibility graph recognition* problems.

Problem 39: *Is there a polynomial algorithm that takes as input a graph and outputs a representation in the form of a closed simple polygon, with the graph's vertices placed at polygon corners, such that an edge is present in the graph if and only if the line segment between the vertices is contained in the polygon?*

An interesting class of problems looks for small representations of dense graphs. For example, given a simple polygon P, the *visibility graph* G for P is the graph having vertex set the same as P and such that there is an edge (v, w) in G if and only if the line segment joining v and w never crosses the boundary of P. The following problem[1] is classic [1,61,75].

Problem 40: *Is there a polynomial algorithm for determining, given a graph G and Hamiltonian cycle C in G, if there exists a polygon P with C as its boundary and having G as its visibility graph?*

The following problem falls into the class of *incidence graph recognition* problems.

Problem 41: *Which graphs can be represented as incidence graphs of line segments in the plane? Does every planar graph have such a representation*[2]*?*

This problem is related to *Conway's famous thrackle conjecture.* A *thrackle* is a planar drawing of a graph of n vertices by edges which are smooth curves between the vertices, with the condition that any two edges intersect at exactly one point (including adjacent edges). Conway conjectured that the number of edges in a thrackle cannot exceed n. The best proven upper bound is $1.5(n-1)$ (see [14]).

Problem 42: *Prove or disprove Conway's thrackle conjecture.*

Acknowledgements. We wish to thank Emilio Di Giacomo for useful discussions and comments while preparing this collection of open problems. We also want to thank the anonymous reviewers for their valuable comments and suggestions.

---

[1] http://cs.smith.edu/~orourke/TOPP/P30.html#Problem.17

[2] http://www.cs.unt.edu/~cbms/CBMS

References

1. J. Abello and K. Kumar. Visibility graphs and oriented matroids. In R. Tamassia and I. G. Tollis, editors, *Graph Drawing (Proc. GD '94)*, volume 894 of *Lecture Notes Comput. Sci.*, pages 147–158. Springer-Verlag, 1995.
2. O. Aichholzer, F. Aurenhammer, S.-W. Chen, N. Katoh, M. Taschwer, G. Rote, and Y.-F. Xu. Triangulations intersect nicely. *Discrete Comput. Geom.*, 16:339–359, 1996.
3. O. Aichholzer, F. Aurenhammer, and H. Krasser. On the crossing number of complete graphs. In *Proceedings of the 18th ACM Symposium on Computational Geometry*, pages 19–24. ACM Press, 2002.
4. B. Aronov, P. Erdős, W. Goddard, D. J. Kleitman, M. Klugerman, J. Pach, and L. J. Schulman. Crossing families. *Combinatorica*, 14:127–134, 1994.
5. W. Barth, M. Jünger, and P. Mutzel. Simple and efficient cross counting. In M. Goodrich and S. Kobourov, editors, *Graph Drawing (Proc. GD'02)*, volume 2528 of *Lecture Notes Comput. Sci.*, pages 130–141. Springer-Verlag, 2002.
6. P. Bertolazzi, G. Di Battista, and W. Didimo. Computing orthogonal drawings with the minimum number of bends. In *WADS '97*, volume 1272 of *Lecture Notes in Comput. Sci.*, pages 331–344, 1998.
7. S. Bhatt and S. Cosmadakis. The complexity of minimizing wire lengths in VLSI layouts. *Inform. Process. Lett.*, 25:263–267, 1987.
8. S. N. Bhatt and F. T. Leighton. A framework for solving VLSI graph layout problems. *J. Comput. Syst. Sci.*, 28:300–343, 1984.
9. T. Biedl. Drawing outer-planar graphs in $O(n \log n)$ area. In M. Goodrich, editor, *Graph Drawing (Proc. GD'02)*, volume 2528 of *Lecture Notes Comput. Sci.*, pages 54–65. Springer-Verlag, 2002.
10. D. Bienstock and N. Dean. Bounds for rectilinear crossing numbers. *J. Graph Theory*, 17:333, 1993.
11. R. Blankenship and B. Oporowski. Book embeddings of graphs and minor-closed classes. In *Thirty-Second Southeastern Internat. Conf. Combinatorics, Graph Theory and Computing*, page 30, Baton Rouge, Dept. of Mathematics, Louisiana State University, 2001. http://www.math.lsu.edu/~conf_se/program.pdf.
12. P. Bose. On embedding an outer-planar graph in a point set. *Comput. Geom.*, 23(3):303–312, 2002.
13. P. Bose, W. Lenhart, and G. Liotta. Characterizing proximity trees. *Algorithmica*, 16:83–110, 1996.
14. G. Cairns and Y. Nikolayevsky. Bounds for generalized thrackles. *Discrete Comput. Geom.*, 23(2):191–206, 2000.
15. T. Calamoneri and A. Sterbini. 3D straight-line grid drawing of 4-colorable graphs. *Inform. Process. Lett.*, 63(2):97–102, 1997.
16. G. Calinescu, C. G. Fernandes, U. Finkler, and H. Karloff. A better approximation algorithm for finding planar subgraphs. In *Proc. 7th ACM-SIAM Symp. on Discrete Algorithms (SODA)*, pages 16–25, 1996.
17. T. Chan. A near-linear area bound for drawing binary trees. *Algorithmica*, 34:1–13, 2001.
18. T. M. Chan, M. T. Goodrich, S. R. Kosaraju, and R. Tamassia. Optimizing area and aspect ratio in straight-line orthogonal tree drawings. In S. North, editor, *Graph Drawing (Proc. GD '96)*, volume 1190 of *Lecture Notes Comput. Sci.*, pages 63–75. Springer-Verlag, 1997.

19. B. Chazelle. Reporting and counting segment intersections. *J. Comput. Syst. Sci.*, 32:156–182, 1986.
20. B. Chazelle. Triangulating a simple polygon in linear time. *Discrete Comput. Geom.*, 6:485–524, 1991.
21. B. Chazelle and J. Incerpi. Triangulation and shape-complexity. *ACM Trans. Graph.*, 3(2):135–152, 1984.
22. S.-W. Cheng and Y.-F. Xu. On β-skeleton as a subgraph of the minimum weight triangulation. *Theoretical Computer Science*, 262(1-2):459–471, 2001.
23. M. Chrobak, M. T. Goodrich, and R. Tamassia. Convex drawings of graphs in two and three dimensions. In *Proc. 12th Annu. ACM Sympos. Comput. Geom.*, pages 319–328, 1996.
24. M. Chrobak and S. ichi Nakano. Minimum-width grid drawings of plane graphs. *Comput. Geom. Theory Appl.*, 11:29–54, 1998.
25. M. Chrobak and G. Kant. Convex grid drawings of 3-connected planar graphs. *Internat. J. Comput. Geom. Appl.*, 7(3):211–223, 1997.
26. M. Chrobak and H. Karloff. A lower bound on the size of universal sets for planar graphs. *SIGACT News*, 20, 1989.
27. R. F. Cohen, P. Eades, T. Lin, and F. Ruskey. Three-dimensional graph drawing. *Algorithmica*, 17:199–208, 1997.
28. E. Dahlhaus. Linear time algorithm to recognize clustered planar graphs and its parallelization. In C. Lucchesi, editor, *3rd Latin American Symposium on Theoretical Informatics (LATIN '98)*, volume 1380 of *Lecture Notes Comput. Sci.*, pages 239–248. Springer-Verlag, 1998.
29. H. de Fraysseix, J. Pach, and R. Pollack. How to draw a planar graph on a grid. *Combinatorica*, 10(1):41–51, 1990.
30. O. Devillers, G. Liotta, F. P. Preparata, and R. Tamassia. Checking the convexity of polytopes and the planarity of subdivisions. *Comput. Geom. Theory Appl.*, 11:187–208, 1998.
31. G. Di Battista, P. Eades, R. Tamassia, and I. G. Tollis. *Graph Drawing.* Prentice Hall, Upper Saddle River, NJ, 1999.
32. G. Di Battista, A. Garg, G. Liotta, R. Tamassia, E. Tassinari, and F. Vargiu. An experimental comparison of four graph drawing algorithms. *Comput. Geom. Theory Appl.*, 7:303–325, 1997.
33. G. Di Battista and G. Liotta. Upward planarity checking: "Faces are more than polygons". In S. Whitesides, editor, *Graph Drawing (Proc. GD '98)*, Lecture Notes Comput. Sci., pages 72–86. Springer-Verlag, 1998.
34. G. Di Battista, G. Liotta, A. Lubiw, and S. Whitesides. Orthogonal drawings of cycles in 3D space. In J. Marks, editor, *Graph Drawing (Proc. GD '00)*, Lecture Notes Comput. Sci. Springer-Verlag, 2000.
35. G. Di Battista, G. Liotta, A. Lubiw, and S. Whitesides. Embedding problems for paths with direction constrained edges. *Theoretical Computer Science*, 289(2):897–917, 2002.
36. G. Di Battista, G. Liotta, and F. Vargiu. Spirality and optimal orhogonal drawings. *SIAM J. Comput.*, 27(6):1764–1811, 1998.
37. G. Di Battista and L. Vismara. Angles of planar triangular graphs. *SIAM J. Discrete Math.*, 9(3):349–359, 1996.
38. E. Di Giacomo. Drawing series-parallel graphs on restricted integer 3D grids. In *Graph Drawing (Proc. GD 03)*, Lecture Notes in Comput. Sci., 2003. These proceedings.
39. E. Di Giacomo, G. Liotta, and H. Meijer. 3D straight-line drawings of k-trees. Technical Report TR-2003-473, Queen's University, School of Computing, 2003.

40. E. Di Giacomo, G. Liotta, and M. Patrignani. Orthogonal 3D shapes of theta graphs. In M. Goodrich, editor, *Graph Drawing (Proc. GD'02)*, volume 2528 of *Lecture Notes Comput. Sci.*, pages 142–149. Springer-Verlag, 2002.
41. E. Di Giacomo, G. Liotta, and S. Wismath. Drawing series-parallel graphs on a box. In *14th Canadian Conference On Computational Geometry (CCCG '02)*, 2002.
42. E. Di Giacomo and H. Meijer. Track drawings of graphs with constant queue number. These proceedings, 2003.
43. M. Dickerson, J. Keil, and M. Montague. A large subgraph of the minimum weight triangulation. *Discrete and Computational Geometry*, 18:289–304, 1997.
44. M. T. Dickerson and M. H. Montague. A (usually?) connected subgraph of the minimum weight triangulation. In *Proc. 12th Annu. ACM Sympos. Comput. Geom.*, pages 204–213, 1996.
45. H. Dijdjev and I. Vrťo. An improved lower bound for crossing numbers. In M. Jünger, S. Leipert, and P. Mutzel, editors, *Graph Drawing (Proc. GD '01)*, Lecture Notes in Comput. Sci. Springer-Verlag, 2001. to appear.
46. M. B. Dillencourt. Realizability of Delaunay triangulations. *Inform. Process. Lett.*, 33(6):283–287, Feb. 1990.
47. M. B. Dillencourt, D. Eppstein, and D. S. Hirschberg. Geometric thickness of complete graphs. *J. Graph Algorithms & Applications*, 4(3):5–17, 2000.
48. M. B. Dillencourt and W. D. Smith. Graph-theoretical conditions for inscribability and Delaunay realizability. In *Proc. 6th Canad. Conf. Comput. Geom.*, pages 287–292, 1994.
49. V. Dujmović, M. Fellows, M. Hallett, M. Kitching, G. Liotta, C. McCartin, N. Nishimura, P. Radge, F. Rosamond, M. Suderman, S. Whitesides, and D. Wood. On the parameterized complexity of layered graph drawing. In F. M. auf der Heide, editor, *Algorithms, 9th European Symposium (ESA '01)*, volume 2161 of *Lecture Notes in Comput. Sci.*, pages 488–499. Springer-Verlag, 2001.
50. V. Dujmović, P. Morin, and D. Wood. Pathwidth and three-dimensional straight line grid drawings of graphs. In M. T. Goodrich and S. Koburov, editors, *Graph Drawing (Proc. GD '02)*, volume 2528 of *Lecture Notes Comput. Sci.*, pages 42–53. Springer-Verlag, 2002.
51. V. Dujmović and D. R. Wood. Tree-partitions of k-trees with applications in graph layout. Technical Report 02-03, Dept. Comput. Sci., Carleton Univ., Ottawa, Canada, 2002.
52. V. Dujmović and D. R. Wood. On linear layouts of graphs. Technical Report 03-05, Dept. Comput. Sci., Carleton Univ., Ottawa, Canada, 2003.
53. V. Dujmović and D. R. Wood. Stacks, queues and tracks: Layouts of graphs subdivisions. Technical Report 03-07, Dept. Comput. Sci., Carleton Univ., Ottawa, Canada, 2003.
54. V. Dujmović and D. R. Wood. Three-dimensional grid drawings with subquadratic volume. In *Graph Drawing (Proc. GD 03)*, Lecture Notes in Comput. Sci., 2003. These proceedings.
55. V. Dujmović and D. R. Wood. Track layouts of graphs. Technical Report 03-06, Dept. Comput. Sci., Carleton Univ., Ottawa, Canada, 2003.
56. V. Dujmović and D. R. Wood. Tree-partitions of k-trees with application in graph layout. In *Workshop on Graph Theoretic Concepts in Computer Science (Proc. WG '03)*, Lecture Notes Comput. Sci. Springer-Verlag, to appear.
57. P. Eades and S. Whitesides. The realization problem for Euclidean minimum spanning trees is NP-hard. *Algorithmica*, 16:60–82, 1996. (special issue on Graph Drawing, edited by G. Di Battista and R. Tamassia).

58. B. Edler. Effiziente Algorithmen für flächenminimale Layouts von Bäumen. Diplomarbeit, Universität Passau, 1999.
59. H. ElGindy, D. Avis, and G. T. Toussaint. Applications of a two-dimensional hidden-line algorithm to other geometric problems. *Computing*, 31:191–202, 1983.
60. G. Even, S. Guha, and B. Schieber. Improved approximations of crossings in graph drawing and VLSI layout area. In *Proceedings of the 32nd ACM Symposium on Theory of Computing (STOC'00)*, pages 296–305. ACM Press, 2000.
61. H. Everett. Visibility graph recognition. Report 231/90, Dept. Comput. Sci., Univ. Toronto, Toronto, ON, 1990. Ph.D. Thesis.
62. S. Felsner. Convex drawings of planar graphs and the order dimension of 3-polytopes. *Order*, 18(1):19–37, 2001.
63. S. Felsner, G.Liotta, and S. Wismath. Straight line drawings on restricted integer grids in two and three dimensions. In P. Mutzel, M. Jünger, and S. Leipert, editors, *Graph Drawing (Proc. GD '01)*, volume 2265 of *Lecture Notes Comput. Sci.*, pages 328–342. Springer-Verlag, 2001.
64. Q.-W. Feng, R. F. Cohen, and P. Eades. Planarity for clustered graphs. In P. G. Spirakis, editor, *Algorithms—ESA '95, Third Annual European Symposium*, volume 979 of *Lecture Notes in Computer Science*, pages 213–226, Corfu, Greece, 25–27 Sept. 1995. Springer.
65. M. R. Garey and D. S. Johnson. *Computers and Intractability: A Guide to the Theory of NP-Completeness*. W. H. Freeman, New York, NY, 1979.
66. M. R. Garey, D. S. Johnson, F. P. Preparata, and R. E. Tarjan. Triangulating a simple polygon. *Inform. Process. Lett.*, 7:175–179, 1978.
67. A. Garg. New results on drawing angle graphs. *Comput. Geom. Theory Appl.*, 9(1–2):43–82, 1998.
68. A. Garg, M. T. Goodrich, and R. Tamassia. Planar upward tree drawings with optimal area. *Internat. J. Comput. Geom. Appl.*, 6:333–356, 1996.
69. A. Garg and G. Liotta. Almost bend-optimal planar orthogonal drawings of biconnected degree-3 planar graphs in quadratic time. In J. Kratochvil, editor, *Graph Drawing (Proc. GD '99)*, Lecture Notes Comput. Sci. Springer-Verlag, 1999.
70. A. Garg and A. Rusu. Straight-line drawings of binary trees with linear area and good aspect ratio. In M. Goodrich, editor, *Graph Drawing (Proc. GD'02)*, volume 2528 of *Lecture Notes Comput. Sci.*, pages 320–331. Springer-Verlag, 2002.
71. A. Garg and A. Rusu. Area-efficient drawings of outerplanar graphs. In *Graph Drawing (Proc. GD 03)*, Lecture Notes in Comput. Sci., 2003. These proceedings.
72. A. Garg and R. Tamassia. A new minimum cost flow algorithm with applications to graph drawing. In S. C. North, editor, *Graph Drawing (Proc. GD '96)*, Lecture Notes Comput. Sci. Springer-Verlag, 1997.
73. A. Garg and R. Tamassia. On the computational complexity of upward and rectilinear planarity testing. *SIAM J. Comput.*, 31(2):601–625, 2001.
74. A. Garg, R. Tamassia, and P. Vocca. Drawing with colors. In *Proc. 4th Annu. European Sympos. Algorithms*, volume 1136 of *Lecture Notes Comput. Sci.*, pages 12–26. Springer-Verlag, 1996.
75. S. K. Ghosh. On recognizing and characterizing visibility graphs of simple polygons. *Discrete Comput. Geom.*, 17:143–162, 1997.
76. A. Gregori. Unit length embedding of binary trees on a square grid. *Inform. Process. Lett.*, 31:167–172, 1989.
77. M. Grohe. Computing crossing numbers in quadratic time. In *Proceedings of the 32nd ACM Symposium on Theory of Computing (STOC'00)*, pages 231–236. ACM Press, 2000.

78. C. Gutwenger, M. Jünger, S. Leipert, P. Mutzel, M. Percan, and R. Weiskircher. Advances in c-planarity testing of clustered graphs. In M. Goodrich and S. Kobourov, editors, *Graph Drawing (Proc. GD'02)*, volume 2528 of *Lecture Notes Comput. Sci.*, pages 220–325. Springer-Verlag, 2002.
79. C. Gutwenger and P. Mutzel. An experimental study of crossing minimization heuristics. In *Graph Drawing (Proc. GD 03)*, Lecture Notes in Comput. Sci., 2003. These proceedings.
80. C. Gutwenger and P. Mutzel. Graph embedding with minimum depth and maximum external face. In *Graph Drawing (Proc. GD 03)*, Lecture Notes in Comput. Sci., 2003. These proceedings.
81. C. Gutwenger, P. Mutzel, and R. Weiskircher. Inserting an edge into a planar graph. In *Proc. Ninth Annual ACM-SIAM Symp. Discrete Algorithms (SODA '2001)*, pages 246–255, Washington, DC, 2001. ACM Press.
82. P. Healy and A. Kuusik. The vertex-exchange graph: a new concept for multi-level crossing minimization. In J. Kratochvíl, editor, *Graph Drawing (Proc. GD '99)*, volume 1731 of *Lecture Notes in Comput. Sci.*, pages 205–216. Springer-Verlag, 1999.
83. S. Hertel and K. Mehlhorn. Fast triangulation of the plane with respect to simple polygons. *Inform. Control*, 64:52–76, 1985.
84. J. W. Jaromczyk and G. T. Toussaint. Relative neighborhood graphs and their relatives. *Proc. IEEE*, 80(9):1502–1517, Sept. 1992.
85. M. Jünger, E. Lee, P. Mutzel, and T. Odenthal. A polyhedral approach to the multi-layer crossing number problem. In G. Di Battista, editor, *Graph Drawing (Proc. GD '97)*, volume 1353 of *Lecture Notes in Comput. Sci.*, pages 13–24. Springer-Verlag, 1997.
86. G. Kant. A new method for planar graph drawings on a grid. In *Proc. 33rd Annu. IEEE Sympos. Found. Comput. Sci.*, pages 101–110, 1992.
87. G. Kant. Drawing planar graphs using the canonical ordering. *Algorithmica*, 16:4–32, 1996. (special issue on Graph Drawing, edited by G. Di Battista and R. Tamassia).
88. M. Kaufmann and R. Wiese. Embedding vertices at points: Few bends suffice for planar graphs. *Journal of Graph Algorithms and Applications*, 6(1):115–129, 2002.
89. M. Keil. Computing a subgraph of the minimum weight triangulation. *Comput. Geom. Theory Appl.*, 4:13–26, 1994.
90. D. G. Kirkpatrick. A note on Delaunay and optimal triangulations. *Inform. Process. Lett.*, 10(3):127–128, 1980.
91. G. W. Klau, K. Klein, and P. Mutzel. An experimental comparison of orthogonal compaction algorithms. In *Graph Drawing (Proc. GD 2000)*, Lecture Notes Comput. Sci. Springer Verlag, 2001.
92. G. W. Klau and P. Mutzel. Optimal compaction of orthogonal grid drawings. In G. Cornuejols, R. E. Burkard, and G. J. Woeginger, editors, *Integer Programming and Combinatorial Optimization*, volume 1610 of *Lecture Notes Comput. Sci.*, pages 304–319. Springer-Verlag, 1999.
93. F. T. Leighton. New lower bound techniques for VLSI. In *Proc. 22nd Annu. IEEE Sympos. Found. Comput. Sci.*, pages 1–12, 1981.
94. F. T. Leighton and S. Rao. Multicommodity max-flow min-cut theorems and their use in designing approximation algorithms. *Journal of the ACM*, 46(6):787–832, 1999.
95. W. Lenhart and G. Liotta. Drawing outerplanar minimum weight triangulations. *Inform. Process. Lett.*, 57(5):253–260, 1996.

96. W. Lenhart and G. Liotta. Proximity drawings of outerplanar graphs. In *Graph Drawing (Proc. GD '96)*, volume 1190 of *Lecture Notes Comput. Sci.* Springer-Verlag, 1997.
97. W. Lenhart and G. Liotta. The drawability problem for minimum weight triangulations. *Theoret. Comp. Sci.*, 270:261–286, 2002.
98. C. Levcopoulos and D. Krznaric. Tight lower bounds for minimum weight triangulation heuristic. *Inform. Process. Lett.*, 57:129–135, 1996.
99. C. Levcopoulos and D. Krznaric. A linear-time approximation schema for minimum weight triangulation of convex polygons. *Algorithmica*, 21:285–311, 1998.
100. A. Lingas. A new heuristic for minimum weight triangulation. *SIAM J. Algebraic Discrete Methods*, 8(4):646–658, 1987.
101. G. Liotta and G. Di Battista. Computing proximity drawings of trees in the 3-dimensional space. In *Proc. 4th Workshop Algorithms Data Struct.*, volume 955 of *Lecture Notes Comput. Sci.*, pages 239–250. Springer-Verlag, 1995.
102. G. Liotta, A. Lubiw, H. Meijer, and S. H. Whitesides. The rectangle of influence drawability problem. *Comput. Geom. Theory Appl.*, 10(1):1–22, 1998.
103. G. Liotta and H. Meijer. Voronoi drawings of trees. *Comput. Geom. Theory Appl.*, 2003. to appear.
104. L. Lovasz, K. Vesztergombi, U. Wagner, and E. Welzl. Convex quadrilaterals and k-sets. *Contemporary Mathematics*, 2003.
105. A. Lubiw and N. Sleumer. Maximal outerplanar graphs are relative neighborhood graphs. In *Proc. 5th Canad. Conf. Comput. Geom.*, pages 198–203, 1993.
106. S. Malitz and A. Papakostas. On the angular resolution of planar graphs. In *Proc. 24th Annu. ACM Sympos. Theory Comput.*, pages 527–538, 1992.
107. G. K. Manacher and A. L. Zobrist. Neither the greedy nor the Delaunay triangulation of a planar point set approximates the optimal triangulation. *Inform. Process. Lett.*, 9:31–34, 1979.
108. A. Mansfield. Determining the thickness of a graph is NP-hard. *Math. Proc. Cambridge Philos. Soc.*, 93(9):9–23, 1983.
109. K. Mehlhorn, T. S. S. Näher, S. Schirra, M. Seel, R. Seidel, , and C. Uhrig. Checking geometric programs or verification of geometric structures. *Comput. Geom. Theory Appl.*, 12:85–113, 1999.
110. C. Monma and S. Suri. Transitions in geometric minimum spanning trees. *Discrete Comput. Geom.*, 8:265–293, 1992.
111. P. Mutzel and R. Weiskircher. Computing optimal embeddings for planar graphs. In D.-Z. Du, P. Eades, V. Estivill-Castro, X. Lin, and A. Sharma, editors, *Computing and Combinatorics, Proc. Sixth Annual Internat. Conf. (COCOON '2000)*, volume 1858 of *Lecture Notes in Comput. Sci.*, pages 95–104. Springer-Verlag, 2000.
112. J. Pach, F. Shahrokhi, and M. Szegedy. Applications of the crossing number. *Algorithmica*, 16:111–117, 1996. (special issue on Graph Drawing, edited by G. Di Battista and R. Tamassia).
113. J. Pach, T. Thiele, and G. Tóth. Three-dimensional grid drawings of graphs. In G. Di Battista, editor, *Graph Drawing (Proc. GD '97)*, volume 1353 of *Lecture Notes Comput. Sci.*, pages 47–51. Springer-Verlag, 1997.
114. J. Pach and G. Tóth. Graphs drawn with few crossings per edge. *Combinatorica*, 17:427–439, 1997.
115. M. Pizzonia and R. Tamassia. Minimum depth graph embedding. In *Proc. European Symposium on Algorithms (ESA 2000)*, Lecture Notes in Computer Science. Springer-Verlag, 2000.

116. F. P. Preparata and M. I. Shamos. *Computational Geometry: An Introduction.* Springer-Verlag, 3rd edition, Oct. 1990.
117. S. Rahman, N. Egi, and T. Nishizeki. No-bend orthogonal drawings of subdivisions of planar triconnected cubic graphs. In *Graph Drawing (Proc. GD 03)*, Lecture Notes in Comput. Sci., 2003. These proceedings.
118. D. Rappaport and H. Meijer. Computing the minimum weight triangulation of a set of linearly ordered points. *Information Processing Letters*, 42:35–38, 1992.
119. R. Richter and C. Thomassen. Relations between crossing numbers of complete and complete bipartite graphs, February 1997.
120. P. Rosenstiehl and R. E. Tarjan. Rectilinear planar layouts and bipolar orientations of planar graphs. *Discrete Comput. Geom.*, 1(4):343–353, 1986.
121. E. R. Scheinerman and H. S. Wilf. The rectilinear crossing number of a complete graph and Sylvester's "four point problem" of geometric probability. *Amer. Math. Monthly*, 101(10):939–943, 1994.
122. W. Schnyder. Planar graphs and poset dimension. *Order*, 5:323–343, 1989.
123. W. Schnyder. Embedding planar graphs on the grid. In *Proc. 1st ACM-SIAM Sympos. Discrete Algorithms*, pages 138–148, 1990.
124. W. Schnyder and W. T. Trotter. Convex embeddings of 3-connected plane graphs. *Abstracts of the AMS*, 13(5):502, 1992.
125. C.-S. Shin, S. K. Kim, and K.-Y. Chwa. Area-efficient algorithms for straight-line tree drawings. *Comput. Geom. Theory Appl.*, 15:175–202, 2000.
126. K. Sugiyama, S. Tagawa, and M. Toda. Methods for visual understanding of hierarchical systems. *IEEE Trans. Syst. Man Cybern.*, SMC-11(2):109–125, 1981.
127. O. Sýkora and I. Vrťo. On VLSI layouts of the star graph and related networks. *The VLSI Journal*, 17:83–93, 1994.
128. R. Tamassia. On embedding a graph in the grid with the minimum number of bends. *SIAM J. Comput.*, 16(3):421–444, 1987.
129. G. Toussaint. Efficient triangulation of simple polygons. *Visual Comput.*, 7:280–295, 1991.
130. G. Vijayan and A. Wigderson. Rectilinear graphs and their embeddings. *SIAM J. Comput.*, 14:355–372, 1985.
131. V. Vijayan. Geometry of planar graphs with angles. In *Proc. 2nd Annu. ACM Sympos. Comput. Geom.*, pages 116–124, 1986.
132. V. Waddle and A. Malhotra. An E log E line crossing algorithm for levelled graphs. In J. Kratochvil, editor, *Graph Drawing (Proc. GD '99)*, Lecture Notes Comput. Sci., pages 59–70. Springer-Verlag, 1999.
133. C. A. Wang, F. Y. Chin, and B. Yang. Triangulations without minimum weight drawing. In *Algorithms and Complexity (Proc. CIAC 2000)*, volume 1767 of *Lecture Notes Comput. Sci.*, pages 163–173. Springer-Verlag, 2000.
134. T. C. Woo and S. Y. Shin. A linear time algorithm for triangulating a point-visible polygon. *ACM Trans. Graph.*, 4(1):60–70, 1985.
135. D. R. Wood. Queue layouts, tree-width, and three-dimensional graph drawing. In *22nd Foundations of Software Technology and Theoretical Comput. Sci. (FSTTCS '02)*, volume 2556 of *Lecture Notes in Comput. Sci.*, pages 348–359, 2002.
136. H. Zhang and X. He. Compact visibility representations and straight-line grid embedding of plane graphs. In *Proc. 8th International Workshop on Algorithms and Data Structures (WADS '03)*, volume 2748 of *LNCS*, pages 493–504. Springer, 2003.

Author Index

GPSR Compliance
The European Union's (EU) General Product Safety Regulation (GPSR) is a set of rules that requires consumer products to be safe and our obligations to ensure this.

If you have any concerns about our products, you can contact us on

ProductSafety@springernature.com

In case Publisher is established outside the EU, the EU authorized representative is:

Springer Nature Customer Service Center GmbH
Europaplatz 3
69115 Heidelberg, Germany

www.ingramcontent.com/pod-product-compliance
Ingram Content Group UK Ltd.
Pitfield, Milton Keynes, MK11 3LW, UK
UKHW021901190726
13853UKWH00003B/1369

9783662183991